Nanotechnology in the Life Sciences

Series Editor
Ram Prasad
Department of Botany
Mahatma Gandhi Central University
Motihari, Bihar, India

Nano and biotechnology are two of the 21st century's most promising technologies. Nanotechnology is demarcated as the design, development, and application of materials and devices whose least functional make up is on a nanometer scale (1 to 100 nm). Meanwhile, biotechnology deals with metabolic and other physiological developments of biological subjects including microorganisms. These microbial processes have opened up new opportunities to explore novel applications, for example, the biosynthesis of metal nanomaterials, with the implication that these two technologies (i.e., thus nanobiotechnology) can play a vital role in developing and executing many valuable tools in the study of life. Nanotechnology is very diverse, ranging from extensions of conventional device physics to completely new approaches based upon molecular self-assembly, from developing new materials with dimensions on the nanoscale, to investigating whether we can directly control matters on/in the atomic scale level. This idea entails its application to diverse fields of science such as plant biology, organic chemistry, agriculture, the food industry, and more.

Nanobiotechnology offers a wide range of uses in medicine, agriculture, and the environment. Many diseases that do not have cures today may be cured by nanotechnology in the future. Use of nanotechnology in medical therapeutics needs adequate evaluation of its risk and safety factors. Scientists who are against the use of nanotechnology also agree that advancement in nanotechnology should continue because this field promises great benefits, but testing should be carried out to ensure its safety in people. It is possible that nanomedicine in the future will play a crucial role in the treatment of human and plant diseases, and also in the enhancement of normal human physiology and plant systems, respectively. If everything proceeds as expected, nanobiotechnology will, one day, become an inevitable part of our everyday life and will help save many lives.

More information about this series at https://link.springer.com/bookseries/15921

Jin-Chul Kim • Madhusudhan Alle
Azamal Husen
Editors

Smart Nanomaterials in Biomedical Applications

Springer

Editors
Jin-Chul Kim
Department of Biomedical Science &
Institute of Bioscience and Biotechnology
Kangwon National University
Chuncheon, Republic of Korea

Azamal Husen
Wolaita Sodo University
Wolaita, Ethiopia

Madhusudhan Alle
Institute of Forest Science
Kangwon National University
Chuncheon, Republic of Korea

Department of Biomedical Science &
Institute of Bioscience and Biotechnology
Kangwon National University
Chuncheon, Republic of Korea

ISSN 2523-8027 ISSN 2523-8035 (electronic)
Nanotechnology in the Life Sciences
ISBN 978-3-030-84261-1 ISBN 978-3-030-84262-8 (eBook)
https://doi.org/10.1007/978-3-030-84262-8

This Springer imprint is published by the registered company Springer Nature Switzerland AG
The registered company address is: Gewerbestrasse 11, 6330 Cham, Switzerland

Preface

At the beginning of year 2020, the wrath of the COVID-19 pandemic urged the scientific community to develop more advanced drug delivery approaches for biomedical applications, endowing conventional drugs with additional therapeutic benefits and minimum side effects. Although significant advancements have been done in the field of drug delivery, there is a need to focus towards strategizing novel and improved drug delivery systems that should be convenient and cost-effective to the patients, and simultaneously they should also provide financial benefits to pharmaceutical companies. Controlled drug delivery technology offers ample opportunities and scope for improvising the therapeutic efficacy of drugs *via* optimizing the drug release rate and time. For this endeavor, smart nanomaterials have served as remarkable candidates in the biomedical applications, owing to their groundbreaking properties and design. The development of such nanomaterials requires a broad knowledge related to their physio-chemical properties, molecular structure, mechanisms by which the nanomaterials interact with the cells, and methods by which drugs are released at the site of action. This knowledge must also be allied with the knowledge of signaling crosstalk mechanisms that are modulated by the nanomaterial-drugs composite. It can be anticipated that these emerging drug delivery technologies can facilitate the world to successfully encounter such pandemic outbursts in the future in a cost-effective and time-effective manner. This book consists of 19 chapters dealing with the advanced strategies to synthesize active smart nanoparticles for biomedical application.

Since the last few decades, there have been numerous developments in nanomaterials-based cancer therapy. Metallic nanoparticles have been applied for the treatment of cancer due to their diverse physical and chemical properties. Gold nanoparticles are among such metallic nanoparticles that are used for various therapeutic purposes, such as in cancer therapy. Understanding the mechanism of action of gold nanoparticles can enhance their therapeutic versatility. A brief introduction on the biomedical applications of gold nanoparticles for cancer therapy is discussed in Chaps. 1 and 2. Since decades, natural compounds and their synthetic derivatives are well-known for their immense therapeutic benefits. Phytochemicals, such as flavonoids, are known to prevent, or even cure, various forms of cancers *via* acting through multi-targeted approach, such as modulation of genes and cell signaling pathways responsible for initiation, progression, and metastasis of cancer. Chapter 3 accumulates the recent developments of commonly used flavonoids as chemotherapeutic and/or chemopreventive agents.

The carbon-based smart nanomaterials have their unique chemical and physical properties that enable them to gain special attention for various biomedical applications, including drug delivery, tissue engineering, biosensing, diagnosis, imaging, and cancer therapy. Carbon dots are novel type of carbon-based nanomaterials that have fascinated scientists for years, because of their varied biological and physicochemical attributes. Metal ion doping of carbon dots has enhanced their intrinsic properties and has received a great interest in the field of bioimaging and cancer therapy, which is discussed in Chap. 4. The therapeutic aspects of other carbon-based smart nanomaterials, such as graphene and carbon nanotubes, for cancer treatment are discussed in Chaps. 5 and 6.

Hydrogel-based drug delivery systems have already shown enormous therapeutic benefits in clinics. Hydrogels are impregnated with drugs before their application, followed by sustained release of drugs at the site of application. The recent developments addressing the use of hydrogels for drug delivery is discussed in Chap. 7.

Recent advances in the vast use of polymeric nanomaterials in biomedical applications have suggested the use of natural polymers in replacement of synthetic polymers due to their low toxicity, excellent biological properties, biodegradability, and biocompatibility. As witnessed, cellulose has emerged as one of the widely used biopolymers in the field of drug delivery. It is important to understand the properties of cellulose so at to explore the possible functional modifications of nanocellulose to enhance their drug loading and therapeutic potential. In Chap. 8, an overview about the advances of nanocellulose-based materials in the biomedical field is discussed.

One of the extensively explored applications is microneedle-based prolonged drug delivery through skin. Microneedles array patches have been investigated to deliver various therapeutic molecules and vaccines, including influenza vaccine, chemotherapeutics, antibiotics, and vitamins, among others, which is discussed in Chap. 9.

Over the years, the site-specific controlled drug release has shown significant potential as it helped in boosting therapeutic efficacy and minimized undesired adverse effects. The "stimuli-responsive" or smart nanomaterials have been designed to target compromised cells *via* exploiting the differences in the intrinsic properties between pathologically challenged and normal cells. Such polymeric materials release the guest molecule or loading drug at the required site under a given environmental stimulus. Chapters 10 and 11 focus on the biomedical application of pH-responsive, ion-sensitive, and thermos-responsive materials.

Various drugs and therapeutically active molecules are not soluble in aqueous phase, leading to increase their dose regime causing possible toxic side effects upon oral administration. To increase the loading efficiency of hydrophobic drugs, polymeric micelles have occupied a colossal position owing to the presence of block copolymers in their composition. The drug delivery aspect of polymeric micelles is described in Chap. 12. Dendrimers are also unimolecular micelles with engineerable structure. Dendrimers are branched polymers that differ in physical properties than those of linear polymers. Thus, opening new opportunities for featured modifications enabling controlled drug delivery, as discussed in Chap. 13.

Lipid-based nanoparticles have been developed as promising colloidal carriers for these kinds of drugs or biologically active therapeutic molecules. The chemical modification of lipid-based nanoparticles can allow the nanoparticle-drug composite to increase the water solubility of drug and also avoid detection by immune cells. The application of lipid-based nanoparticles, such as solid lipid nanoparticles, in the improvement of oral bioavailability of drugs is summarized in Chap. 14. The applications of other lipid-based nanoparticles, such as liposomes in drug delivery, are illustrated in Chap. 15. Similar to liposomes, niosomes are also vesicular structures. However, they are composed of non-ionic surfactants and are highly stable than liposomes, increasing the bioavailability of the loaded drug and reducing its clearance from body fluids. The niosomes have gained a huge interest as drug carriers as they can improve the therapeutic efficacy of drug by increasing their circulation time in the body fluid. The concept and biomedical application of niosomes are discussed in Chap. 16.

In the past few years, non-invasive drug delivery carriers have gained considerable attention for transdermal drug delivery. Ethosomes are ethanol-based specifically designed vesicular carriers that can efficiently deliver therapeutic molecules into deep skin layers and across the skin. However, to ensure the safety of ethosomes, it is utmost important to understand their chemical composition and behavior. Chapter 17 deals with the transdermal drug delivery application of ethosomes. Drug delivery can be enhanced using natural polysaccharide-based devices, as discussed in Chap. 18. Nanoparticles can also be used as nanosensors to detect the toxic chemicals and microorganisms with high sensitivity. These novel applications of nanosensors are discussed in Chap. 19.

This book provides a comprehensive summary of smart novel drug delivery strategies for various biomedical applications. This book covers the fundamentals, applications, and future possibilities to modify nanomaterials for developing more advanced nanomaterials. The chapters in this book deal with the advanced technologies and approaches that can benefit the readers in developing smart and intelligent nanomaterials for future biomedical applications. We hope that this book will help advance drug delivery and promote human health.

We express our sincere thanks to our family and parents for their overall support. We also deeply appreciate the distinguished authors who have shared their knowledge and contributed chapters for this book. We feel indebted to Dr. Eric Stannard at Springer and all his associates for their sustained cooperation. We shall be happy receiving comments and criticism, if any, from subject experts and general readers of this book.

Chuncheon, Republic of Korea — Jin-Chul Kim
Chuncheon, Republic of Korea — Madhusudhan Alle
Wolaita, Ethiopia — Azamal Husen
June 2021

Contents

About the Editors

Jin-Chul Kim has been working as a professor in the Department of Biomedical Science and Institute of Bioscience and Biotechnology, Kangwon National University, South Korea, since 2003. Earlier he worked as a senior researcher at LG Household & Healthcare, Republic of Korea (1999–2003). He completed his postdoctoral research in the Department of Industrial and Physical Pharmacy (IPPH), Purdue University, USA (1997–1998). He received his master's and doctorate degree from the Department of Chemical Engineering, Korea Advanced Institute of Science and Technology (KAIST), Republic of Korea (1991–1997) and his bachelor's degree from the Department of Chemical Engineering, Yonsei University, Republic of Korea (1987–1991). Professor Kim has more than 25 years of experience in teaching and research in both academia and industry. He has published more than 300 scientific papers in SCI/SCIE-grade journals as a lead author/corresponding author. He has many awards to his credit, to name a few, Outstanding Teacher Award (Kangwon National University, 2014), Teaching Award (Kangwon National University, 2008), Outstanding Researcher Award (LG household & healthcare, 2001), Patent Award (LG Household & Healthcare, 2001), Patent Award (LG Chem., 2000), and Award in Championship Transport Phenomenon Competition for University Student (Korean Chemical Engineering Society, 1990). Professor Kim was the leader of a team named "Education and research team for cultivation of graduate students talented with development of drug carriers for biopharmaceuticals delivery," under BK21 PLUS (2016–2020). He is leading a research part "Development of biomedical materials based on nanocellulose" of the Priority Research Centers Program of the National Research Foundation of Korea (NRF) funded by the Ministry of Education (2018-2027). Professor Kim is the leader of a team named "Education and research team for cultivation of graduate students globally and convergently talented with development of drug carriers for biopharmaceuticals delivery," under BK21 FOUR (2020–2027). His current research interests include drug delivery systems, colloid and interfaces science, and polymer science.

Madhusudhan Alle is currently working as a research professor in the Department of Biomedical Science and Institute of Bioscience and Biotechnology, Kangwon National University, South Korea (since 2018). Earlier, he worked as an associate professor in the Department of Chemistry, Gondar University, Gondar, Ethiopia (2014–2018). During this period, he also served as a coordinator of MSc (chemistry) programs and as the head of the Chemistry Department. Four students submitted their master's thesis under his supervision. He received his doctorate from the Department of Chemistry, Osmania University, India (2013), where he worked on research projects sponsored by University Grants Commission, India. He has more than 18 years of experience in teaching and research in both academia and industry. He has published more than 50 scientific papers in SCI/SCIE-grade journals as a first and corresponding author and also published four book chapters (Springer and Elsevier). He is also listed as a potential reviewer for many reputed international journals. He is also a member of the Ethiopian Chemical Society and the Indian Chemical Society. He has presented his work in several national and international conferences in India, Ethiopia, and South Korea. He has conducted two research projects sponsored by the University Grants Commission (UGC), New Delhi, India. His current research interests include biosciences, polymeric nano-drug delivery systems, cancer biology, biomaterials for sustainability, and catalysis. Extraction and chemical modification of nanocellulose from lignocellulosic biomass, green synthesis of metal and metal oxide nanoparticles on nanocellulose support and their novel applications in biomedicine, biosensors, and catalysis.

Azamal Husen (BSc from Shri Murli Manohar Town Post Graduate College, Ballia, UP, MSc from Hamdard University, New Delhi, and PhD from Forest Research Institute, Dehra Dun, India) is a Foreign Delegate at Wolaita Sodo University, Wolaita, Ethiopia. He has served the University of Gondar, Ethiopia, as a Full Professor of Biology, and also worked as the coordinator of the MSc programme and as Head of the Department of Biology. He was a visiting faculty at the Forest Research Institute and the Doon College of Agriculture and Forest at Dehra Dun, India. He has more than 20 years' experience of teaching, research, and administration. Dr. Husen specializes in biogenic nanomaterial fabrication and their application; plant responses to nanomaterials; plant production and adaptation to harsh environments at the physiological, biochemical, and molecular levels; herbal medicine; and clonal propagation and improvement of tree species. He has conducted several research projects sponsored by various funding agencies, including the World Bank, the Indian Council of Agriculture Research (ICAR), the Indian Council of Forest Research Education (ICFRE), and the Japan Bank for International Cooperation (JBIC). He has published over 150 research papers, review articles, and book chapters, edited books of international repute, presented papers in several conferences, and produced over a dozen manuals and monographs. Dr. Husen received four fellowships from India and a recognition award from the University of Gondar, Ethiopia, for excellent teaching, research, and community service. An active organizer of seminars/conferences and an efficient evaluator of research projects and book proposals, Dr. Husen has been on the editorial board and the panel of reviewers of several reputed journals published by Elsevier, Frontiers Media SA, Taylor & Francis, Springer Nature, RSC, Oxford University Press, Sciendo, The Royal Society, CSIRO, PLOS, and John Wiley & Sons. He is on the advisory board of Cambridge Scholars Publishing, UK. He is a Fellow of the Plantae group of the American Society of Plant Biologists, and a member of the International Society of Root Research, Asian Council of Science Editors, ISDS, and INPST. Dr. Azamal Husen is editor-in-chief of the *American Journal of Plant Physiology*. He is also working as a series editor of *Exploring Medicinal Plants*, published by Taylor & Francis Group, USA; and '*Plant Biology, Sustainability, and Climate Change*' published by Elsevier Inc., MA 02139, USA.

Chapter 1
Current Trends in Engineered Gold Nanoparticles for Cancer Therapy

Madhusudhan Alle, Rajkumar Bandi, Garima Sharma, Ramakrishna Dadigala, Azamal Husen, and Jin-Chul Kim

1.1 Introduction

Cancer continues to be one of the most common causes of death for people all over the world (Wen et al. 2020). Since the last two decades, many advances in conventional chemotherapy, radiation therapy, and surgery have been approved in the clinics (Bosset et al. 2006). Although scientists have made enormous efforts to enhance these traditional regimens, these selective methods still possess several constraints such as low efficacy, serious side effects, and high-risk cancer recurrence (Kim and Tomé 2006). Thus, it is necessary to develop approaches with fewer adverse impacts and enhance therapeutic performance. Currently, the development of nanomedicine has simplified our ability to develop various new nanomaterials (NMs) for simultaneous diagnosis and treatment (Jorfi and Foster 2015).

A series of new NMs have been widely used for cancer therapy, i.e., quantum dots, polymers, liposomes, dendrimers, nanotubes, inorganic nanoparticles, and so on (Yao

M. Alle (✉)
Institute of Forest Science, Kangwon National University, Chuncheon, Republic of Korea

Department of Biomedical Science & Institute of Bioscience and Biotechnology, Kangwon National University, Chuncheon, Republic of Korea

R. Bandi · R. Dadigala
Institute of Forest Science, Kangwon National University, Chuncheon, Republic of Korea

G. Sharma · J.-C. Kim
Department of Biomedical Science & Institute of Bioscience and Biotechnology, Kangwon National University, Chuncheon, Republic of Korea

A. Husen
Wolaita Sodo University, Wolaita, Ethiopia

J.-C. Kim et al. (eds.), *Smart Nanomaterials in Biomedical Applications*, Nanotechnology in the Life Sciences,
https://doi.org/10.1007/978-3-030-84262-8_1

et al. 2016). Of these, nanoparticles with inorganic core have been thoroughly explored for several applications, including neurological disease management, bioimaging, cancer diagnosis, and treatment (Husen and Siddiqi 2014; Madhusudhan et al. 2014; Husen 2017, 2019, 2020; Siddiqi and Husen 2017; Siddiqi et al. 2018; Yang et al. 2019; Husen and Iqbal 2019; Alle et al. 2021; Bachheti et al. 2021). For example, iron oxide nanoparticles can be used as magnetic resonance imaging (MRI) contrast agents and magnetic hyperthermia agent owing to their remarkable magnetic properties (Lee and Hyeon 2012; Siddiqi et al. 2016). Due to the excellent tunable fluorescent emission, photostability, carbon dots (CDs) have shown great potential in labelling the target (Liu et al. 2018a). Besides, gold nanoparticles (AuNPs) have been extensively used for biosensing, computed tomography (CT) imaging, and photothermal therapy (PTT) because of their high X-ray absorption efficiency, size, and shape-dependent optical properties (Liu et al. 2018b).

AuNPs have the ability to be used in biomedical applications, according to new research. AuNPs have been shown to be easy to synthesize with various shapes and sizes and to have several advantages, including unique tunable optical and electrical properties, a high X-ray absorption coefficient, the ability to be functionalized by a variety of biomolecules (e.g., drugs, genes, and targeting ligands), nontoxicity, biocompatibility, and a distinct surface effect. In view of all these unique features, AuNPs can be considered the more sensitive material for various biomedical applications.

AuNPs are well known for their photothermal therapy (PTT), which has emerged as an important therapeutic section for cancer treatment due to its noninvasive nature, low cost, large surface to volume ratio, simple synthesis, and negligible side effects or high biocompatibility (Fratoddi et al. 2015). Upon irradiation with near-infrared region (NIR), the surface electrons of AuNPs are excited and resonate, converting NIR light into heat within picoseconds, and finally increased the temperature of the cancer cell to about 41–47 °C due to SPR resonant peaks, which causes irreversible cell damage, such as denaturation of protein and cell membrane destruction (Yuan et al. 2014; Kim et al. 2021a, b). AuNPs used in PTT have various advantages, including ease of administration to the local tumor area while minimizing nonspecific dissemination (Alle et al. 2020). They can be made to penetrate biological tissues with NIR laser light and modulated to create a multifaceted cancer PTT and drug delivery system.

Another therapeutic advantage of AuNPs is the treatment of cancer via photodynamic therapy (PDT). Whenever the surgery is not possible, PDT has been considered a promising early-stage tumor and helps prolonged survival of cancer patients (García Calavia et al. 2018). The PDT is cost-effective and is a repeatable process that supports the long-term management of the tumor (Agostinis et al. 2011). In PDT, the hydrophobic light-sensitive molecules, known as photosensitizers (PS), generate singlet oxygen species ($^1O_2$) or reactive oxygen species (ROS) that kills cancer cells (Li et al. 2013). AuNPs are known to enhance the photosensitizing properties of PSs, possibly due to localized surface plasmon resonance (LSPR) of the metal. Thus, AuNPs can increase the PDT efficiency of the PSs (Hone et al. 2002). Studies showed that AuNPs could induce a synergistic PDT and PTT effect against cancer cells.

Cancer cells become more radiosensitive in the presence of AuNPs. The radiosensitivity is dependent on the type, size, concentration, and degree of internal localization of AuNPs (Mesbahi 2010). It also depends on the type of cancer cells involved and the type of energy source for the radio waves. In addition, the surface of AuNPs can be conjugated with a variety of moieties like peptides, folate, ligands, antibodies, etc. (Huang et al. 2010c). Hence, AuNPs can be used as biocompatible carriers for delivering a variety of drugs and genes for cancer therapy (Mieszawska et al. 2013). The high drug delivery efficiency of AuNPs will ensure the better accumulation of the payload at the target. Moreover, by activating the PTT effect of AuNPs, the drug release can also be regulated. The functionalization of AuNPs by thiolated polymers has also shown to regulate intracellular release (Amina and Guo 2020).

Advance researches on cancer therapy PTT, PDT, radiotherapy, and glucose starvation have increased in the last few years (Fig. 1.1). The main aim of the book chapter is to elucidate the usage of AuNPs in cancer therapy. Thus, in this book chapter, we focused on the cancer therapy strategies rather than synthesis and surface modification of various shapes of AuNPs (nanospheres, nanorods, nanoflowers, nanostars) used for cancer therapy (Table 1.1). These strategies will be mainly categorized by different mechanisms, including PTT, PDT, and synergistic administration of PTT along with chemotherapy and starvation therapy.

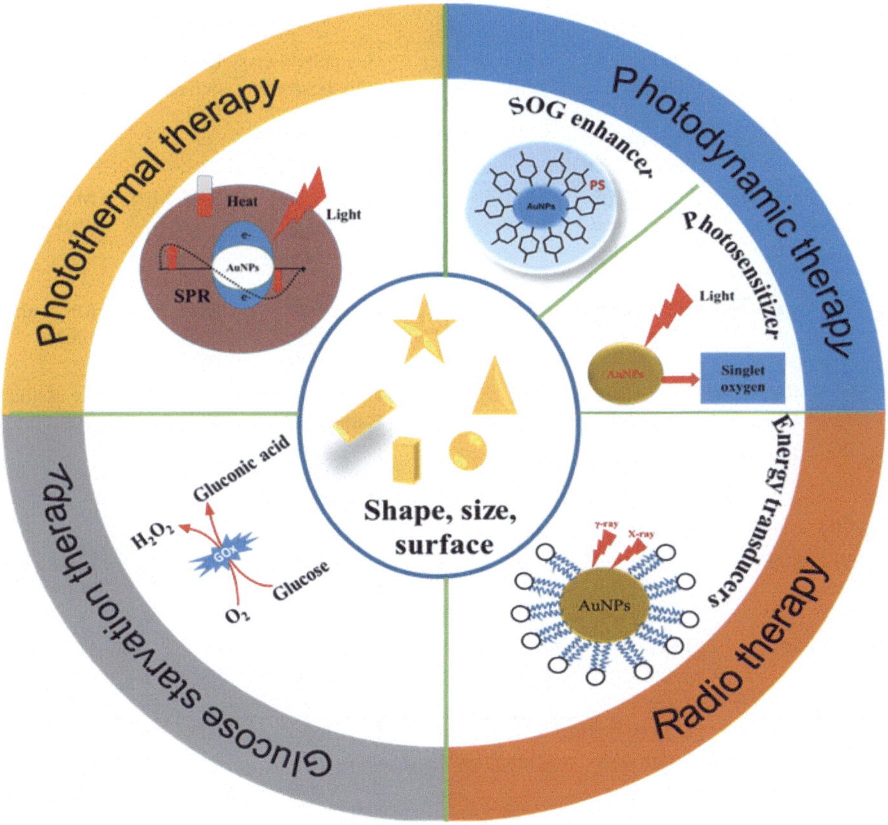

Fig. 1.1 Schematic representation of different AuNPs in various types of cancer therapies

Table 1.1 Properties and biomedical application of gold nanoparticles

Size (nm)	SPR peak (nm)	Surface molecules	Target	Cell line	Biomedical applications	Reference
AuNSs						
65	820	PEG	–	SKBR3 cells/ CB17-prkdcscid/ Jmice bearing canine TVT	PTT	Hirsch et al. (2003)
130	800	Anti-HER-2 antibody-conjugated PEG	HER-2	SKBR3 cells	PTT	Loo et al. (2005)
110	820	PEG, anti-HER-2 nanobodies	HER-2	KB cells	PTT	Sheikholeslami et al. (2011)
160	800	Cystine-conjugated A54 peptides	BEL-7404	BEL-7404 cells	PTT	Liu et al. (2010)
240	810	DOX	–	SMM-7721 cells	Drug delivery, PTT	Chu and Wu (2011)
37	800	PEG c(TNYL-RAW) peptide-conjugated SH-PEG/DOX	EphB4	Hey cells/Swiss mice bearing Hey tumors	Drug delivery, PTT	You et al. (2012)
90	800	PEG,C225 antibodies/ SPIO NPs	EGFR	A431 cells/ nude mice bearing A431 tumors	PTT, MRI	Melancon et al. (2011)
279	400 ~ 900	PEG/DOX, SPIO NPs	–	HeLa cells	Drug delivery, PTT, MRI	Ma et al. (2013)
370	400 ~ 900	PEG/PFOB, SPIO NPs	–	HT-1080 cells/ Balb/c nude mouse implanted with HT-1080	PTT, MRI, US imaging	Ke et al. (2014)
AuNRs						
12/50 (width/ length)	800	PSS, anti-EGFR antibodies	EGFR	HOC 313, HSC-3 cells	PTT	Huang et (al. 2006)
12/50	800	PEG	–	Nu/nu mice implanted with HSC-3 tumors	PTT	Dickerson et al. (2008)
23/45	708	CTAB, CD11b antibodies	Leukocyte integrin Mac-1	Raw 264,7 murine macrophage cells	PTT	Pissuwan et al. (2007a, b)

(continued)

Table 1.1 (continued)

Size (nm)	SPR peak (nm)	Surface molecules	Target	Cell line	Biomedical applications	Reference
20/40	692	CTAB, anti-*Toxoplasma gondii* antibodies	*T. gondii*	*T. gondii* tachyzoite	PTT	Pissuwan et al. (2007a, b)
10/40	800	CTAB, chitosan-conjugated, pluronic-based nanocarriers	–	SCC7 cells/ athymic nude mice bearing SSC7 tumors	PTT	Choi et al. (2011)
14/53	760	Hyaluronic acid-conjugated nanographene oxide/DOX	Hyaluronan receptor, cluster determinant 44	Huh-7 cells	Drug delivery, PTT	Xu et al. (2013)
13/47	810	PEG	–	Nude mice bearing MDA-MB-435 tumors	PTT, CT imaging	von Maltzahn et al. (2009)
12/38	735	Poly-l-lysine, silica shells	–	Atherosclerotic human right coronary artery sections	US imaging, PA imaging, PTT	Yeager et al. (2014)
17/62	783	Polydopamine, anti-EGFR antibodies	EGFR	OSCC15, MDA-MB-231, MCF7 cells	OCT, PT	Black et al. (2013)
AuNSTs						
60	635	Anti-HER-2 antibodies	HER-2	SKOV3 cells	PTT	Van de Broek et al. (2011)
35	810	PEG	–	SKBR3 cells	PTT	Yuan et al. (2012b)
50	800	PEG,TAT peptides	–	BT549 cells	PTT	Yuan et al. (2012a)
58	780	Cyclic RGD peptides, DOX	~ Integrin αv β_3	MDA-MB-231 cells	Drug delivery, PT	Chen et al. (2013)

CT Computed tomography, *CTAB* Cetyltrimethylammonium bromide, *DOX* Doxorubicin, *EGFR* EGF receptor, *AuNC* Gold nanocage, *AuNP* Gold nanopopcorn, *AuNR* Gold nanorod, *AuNS* Gold nanoshell, *AuNST* Gold nanostar, *AuV* Gold vesicle, *HB* Hypocrellin B, *NP* Nanoparticle, *OCT* Optical coherence tomography, *PA* Photoacoustic, *PFOB* Perfluorooctyl bromide, *PNIPAM* Poly(N-isopropylacrylamide), *PSMA* Prostate-specific membrane antigen, *PSS* Poly(sodium-p-styrenesulfonate), *PT-OCT* Photothermal optical coherence tomography, *PTT* Photothermal therapy, *SAN* "Smart" gold nanoparticle, *SPIO* Superparamagnetic iron oxide, *SPR* Surface plasmon resonance, *SWCNT* Single-wall carbon nanotube, *TCaP* Calcium phosphate, *TVT* Transmissible venereal tumor, *US* Ultrasound

1.2 Key Properties of AuNPs Important for Cancer Therapy

1.2.1 Importance of NIR in Biological Applications

In the cancer cells, AuNPs show significant light absorption at the wavelength of the NIR region, ranging between 650 and 1350 nm (Vijayaraghavan et al. 2014). However, healthy tissues do not absorb light at this spectral region (Riley and Day 2017). Furthermore, light scattering is reduced at longer wavelengths. As a result of the limited scattering and NIR absorption, deep tissue penetration is possible (Dang et al. 2019). Depending on the tissue type, the penetration depth will range from 1 to 10 cm (Zhang et al. 2016a).

1.2.2 The Influence of SPR on AuNP-Based Cancer Application

The oscillating electric field of light causes the conduction band electrons on the surface of AuNPs to polarize, and the polarized electrons undergo mutual coherent oscillations with respect to the positive ions in the metallic lattice; these oscillations are known as surface plasmon oscillations (Fig. 1.2). Since it has the same frequency as the incident light, it is also known as surface plasmon resonance (SPR) (Huang et al. 2010b).

The frequency of SPR is strongly dependent on the size and shape of the nanoparticles. There is only one SPR frequency for spherical AuNPs, i.e., about 520 nm in the visible region, responsible for the spherical AuNPs' intense red color. Two SPR frequencies, defined as the longitudinal and transverse bands, occur for gold nanorods (AuNRs) (Fig. 1.2). Similar to the SPR frequency of the spherical AuNPs, the transverse band is a weak absorption band in the visible field for electron oscillations, which occur in the transverse direction (Fig. 1.2). The longitudinal band is aligned with the longitudinal path of electron oscillations and has a heavy absorption band in the vis-NIR region (Madhusudhan et al. 2019).

As the excited surface plasmon electrons relax, they may either emit light with the same energy as the incident light (SPR dispersion) or transfer the absorbed energy to the metallic lattice in the form of thermal energy. The hot lattice is then cooled by phonon-phonon interactions, and the heat is transferred to the surrounding medium (Sperling et al. 2008). This is the basis for all applications of plasmonic nanoparticle-based PPT. Because both scattering and absorption components are included in the SPR band of NPs, tuning the NP shapes and sizes will significantly change their dispersion and absorption properties (Hu et al. 2006). It is known that with an increase in the aspect ratio of the AuNRs, the longitudinal band undergoes a redshift, and the absorption portion of the longitudinal band decreases while the scattering component increases (Lee and El-Sayed 2005). As a result, for NIR imaging applications, larger AuNRs with high scattering efficiency are preferred, whereas smaller AuNRs with high absorption efficiency are preferred for photothermal applications.

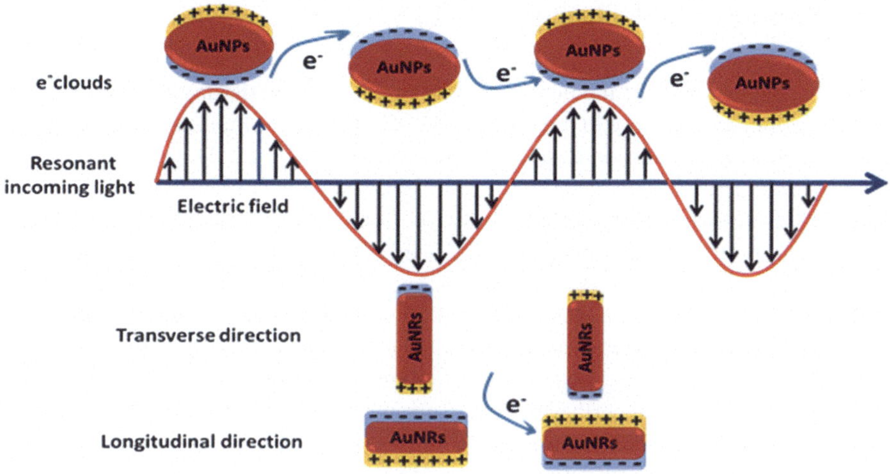

Fig. 1.2 Surface plasmon resonances of spherical gold nanoparticles (AuNPs) and gold nanorods (AuNRs)

1.2.3 Comparison of the NIR-Absorbing AuNPs

For cancer cell sensing and therapies, some commonly used AuNPs in NIR absorption are gold nanoshells (AuNSs), gold nanocages (AuCNs), and AuNRs. Since this wavelength range of light can safely and deeply penetrate normal tissue to reach AuNPs embedded inside tumors, the AuNPs used in PTT may be engineered based on absorbing light within the first (650–850 nm) or second (950–1350 nm) NIR windows. The efficacy of AuNPs in cancer treatment is determined by using light scattering and absorption within one of these two regions of light. The common AuNPs have been designed to maximally absorb within 800 nm, which can safely penetrate 2–3 cm of tissue. The discrete dipole approximation (DDA) method is used to compare these properties after adjusting their structure to tune their SPR band to 800 nm. To obtain the SPR peak at 800 nm, AuNRs' width and length should have an aspect ratio of 3.6 and 20 nm, respectively. AuNSs' shell thickness and silica core diameter are 3.2 nm and 50 nm, respectively. AuNCs' inner edge length and wall thickness are 50 nm and 6 nm, respectively. As per DDA results, the absorption and scattering section of AuNRs and AuNCs are similar, and they are also more significant than that of AuNRs (Li and Chen 2015). The higher absorption section of AuNRs and AuNCs is considered ideal candidates for the PTT agents (Hu et al. 2006). For instance, per gram of gold producing heat of AuNRs is six times faster higher than AuNCs (Von Maltzahn et al. 2009). However, structural modification of the Au nanostructures during synthesis provides a shift of the absorption band to the desirable therapeutic window (650–950 nm).

1.3 Photothermal Therapy (PTT)

Hyperthermia is a type of medical treatment in which body tissues are exposed to a slightly higher temperature in the range of 41–47 °C than normal conditions to damage and destroy the tumor tissues (D'Acunto et al. 2021). This therapy might increase therapeutic efficacy and reduce the undesired side effects. The excision of cancer cells at the temperature of 41–47 °C is possibly due to irreversible DNA damage and protein distortion in the cellular membrane due to their poor blood supply. To deliver heat to tumors, various sources are used, including hot water, ultrasound, microwave, and radiofrequency radiation, and have been studied (Zhu et al. 2019). Although these methods effectively heat the cancer cells, unfortunately, they lack the specificity for the cancer cells, which results in systemic toxicity. This results in whole-body hyperthermia exposure which can cause cardiovascular side effect and gastrointestinal symptoms (Robins et al. 1985). In this regard, another alternative therapy needs to solve these problems.

PTT is non-invasive hyperthermia related emerging cancer treatment method. This method aims to destroy cancer cells while preventing damage to the healthy cells (Ma et al. 2016). In PTT, the absorbed light turns into heat or thermal energy, and tumor cells induce necrosis or apoptosis due to higher temperature. Heat can be selectively oriented to cancer cells by directed and directionally controlling incident radiation and by region-specific administration of photosensitizers, resulting in localized heat transfer to the surrounding area, making PTT more effective than other hyperthermia. The use of laser wavelength and the tissue penetration of light are the major concerns for the efficacy of PTT. The NIR light is believed to have a maximum ability to penetrate tissues because the majority constituents of biotissues, including water, skin, hemoglobin, and other pigments, have shown negligible absorption and scattering of light in this region (Zhang et al. 2016c). Hyperthermia is possible in the presence of a photo-absorbing agent and NIR light; however, PTT itself causes significantly less damage to normal cells. Recently, PTT has been the focus as compared to chemotherapy, radiotherapy, and surgery for tumor treatment because of its low invasive risk and high specificity.

The natural chromophores in the tissue or externally added dye molecules such as indocyanine green, naphthalocyanines, and porphyrins coordinated with transition metals have been explored as the photo-absorbing agents. Normal chromophores, on the other hand, have a very low absorption rate. The best photothermal agents were chosen based on their large absorption cross sections and high light-to-heat conversion efficiency. This reduces the amount of laser energy needed to kill diseased cells locally, making the therapy process less invasive. Nevertheless, these PSs have specific limitations such as vulnerability to photobleaching, low absorption cross section, and poor specificity for cancer cells. As a result, incorporating nanomaterials into PTT could be advantageous (Huang et al. 2008). PTT or their combination therapy has been tested in animal models of the lung and bone, including lymph metastasis from various cancers, and has shown to be successful.

1.3.1 AuNPs for PTT

To date, various NMs are used in PTT. All AuNPs have been considered the most important inorganic NMs for cancer treatment by PTT due to their plasmon resonance tunability, simple surface functionalization, and high photothermal conversion efficiency. Additionally, AuNPs have higher photostability, and they do not suffer from photobleaching (Hwang et al. 2014). Other advantage to use AuNPs in PTT is that the optical properties of AuNPs can be tuned by controlling their structure dimension so that they adsorb NIR light (650–1350 nm), which is suitable for PTT because it can safely penetrate deeply through normal tissue to reach AuNPs embedded within tumor. The absorbed NIR light by AuNPs is then converted as heat which is effectively used for PTT. Besides, such plasmonic AuNPs have not been found with a severe side effect, thermal instability, and preclinical intolerance; hence, they can quickly enter into the cancer cells due to strong enhanced permeability and retention (EPR) effect (Her et al. 2017). Importantly, tumor blood vessels contain a high number of pores; thus, injected AuNPs may leak out, resulting in aggregation of AuNPs around the tumor interstitial fluid (Baeza et al. 2016). Due to this phenomenon, AuNPs can stay long within the tumor cells, which stimulates high radiation absorption and provides effective treatment without serious side effects (Abadeer and Murphy 2016).

Since the early twentieth centuries, various type of AuNPs, i.e., Au nanospheres, AuNRs, AuNCs, AuNSs, and Au nanocage, have been used in PTT owing to their strongly enhanced absorption in the NIR regions on account of their SPR oscillations (Table 1.2). Out of these first three have a great deal of scientific attention in PTT, because of their easy synthesis process and functionalization and adjustable optical properties. The tumor treatment efficiency mainly depends on the light scattering and absorption properties of the AuNPs, which can be compared by the DDA method after tuning the SPR peak at 800 nm with adjusting structure (Yasun et al. 2013). The gold nanoparticles' SPR peak is 800 nm, and if AuNRs may have width

Table 1.2 Shows the characteristics of AuNPs. A description of the properties of different AuNPs

Particles	Sketch	Size	λ_{spr} (nm)	C_{abs}	η_{abs}	$Cabs/C_{ext}$
Au sphere		r = 20 nm	528	2827	2.3	0.94
Si core/Au shell		r_1 = 20 nm r_2 = 75 nm	800	45,769	2.6	0.53
Au_2S core/Au shell		r_1 = 21 nm r_2 = 25 nm	~780	2750	1.4	˃0.9
Au rod[a]		r_1 = 13 nm r_2 = 49 nm	797	5674	13.8	0.93
Au cage[b]		r_1 = 36.7 nm r_2 = 3.3 nm	800	7260	5.4	0.90

Note: The terms in the table are SPR wavelength (λ_{spr}), absorption cross section (C_{abs}), absorption efficiency ($\eta_{abs = Cabs/A}$), and the ratio of absorption over extinction (C_{abs}/C_{ext})
Adopted from Stabile et al. (2020)
[a]L_1 and L_2 are the width and length of the rod. Effective radius (sphere with the same volume r_{eff} = 11.43 nm) is used to calculate the geometric cross section
[b]L is the outer length of the cubic and t is the thickness of the cage. The geometric cross section used is L^2

and length aspect ratio of 30.6 and 20 nm, AuNSs may have gold shell thickness, and silica core diameter is 3.2 nm and 50 nm, and AuNCs may have an inner edge length and wall thickness of 50 nm and 6 nm, respectively. According to DDA results, the absorption and scattering cross sections of AuNRs and AuNCs are similar, and they are considerably larger than those of AuNSs (Hu et al. 2006). AuNRs and AuNCs have extinction cross-sectional coefficients that are more than twice as high as AuNSs. AuNRs and AuNCs are ideal candidates for PTT because of their broad absorption cross section (Hu et al. 2006). Heating per gram of gold for AuNRs, for example, is at least six times faster than for AuNSs (von Maltzahn et al. 2009).

AuNPs in the PTT treatment process involve the collision of laser radiation into AuNPs and the electron excitation of NPs; then, the electrons achieve new equilibrium Fermi electron distribution during energy redistribution (Qin and Bischof 2012). Hot electrons send their energy to the lattice, and heat is transmitted to the atmosphere in picoseconds by phonon-phonon dispersion, and AuNPs cool rapidly (Huang et al. 2008). Thus, fast local heat is produced due to excitation; the temperature of hot electrons rises to thousands of degrees Kelvin, and the lattice temperature exceeds 10 °C when AuNPs are irradiated by NIR (Mocan et al. 2013). This makes AuNPs ideal for the treatment of cancer using PTT. In particular, a low-power laser can be used in AuNP-mediated PTT in which damage to normal tissue surrounding it and consequent side effects decrease in normal cells while cancer cell death increases (Rau et al. 2016). Due to temperature variations, various physical and biological responses occur based on the laser intensity, including phase modification of the lipid membrane, accompanied by nanoscale membrane melting, denaturation of the proteins surrounding AuNPs, rapid melting of DNA, the release of surface AuNP molecules, and water evaporation around the nanoparticles (Qin and Bischof 2012). The combination of the special properties of AuNPs with PTT that leads to cancer cell death is seen in Fig. 1.3.

The photothermal conversion efficacy of AuNPs in PTT is dependent on many factors, among which the structural dimension is one of the key factors. In general, the AuNPs' size reduces the ability to convert light into heat increase; thus, PTT's efficiency is enhanced (Du et al. 2017). Other studies reported that in gold silica nanoshells, the thickness of gold in nanoshells increases with the absorption of light; however, the therapeutic heat effect was reduced (Park et al. 2015). Laser flux distribution is the second aspect that affects cell death based on the duration and strength of laser influence. Lasers in pulsed mode (and especially with high pulse energy) are associated with powerful heating effects that can have many implications. Compared to continuous-wave lasers (CW), the pulsed laser with a brief pulse length provides a larger amount of energy to tissue. Pulsed lasers often induce a phase shift in the water due to heat transfer to water, and hence bubbles are formed in the biological system (Rau et al. 2016). Bubbles have been reported to be substantially created by AuNPs irradiated by pulsed lasers in the cytoplasm. Therefore, the gain is the use of pulsed lasers in the presence of AuNPs, so that the pulse laser intensity can be decreased while having the same effects (Qin and Bischof 2012).

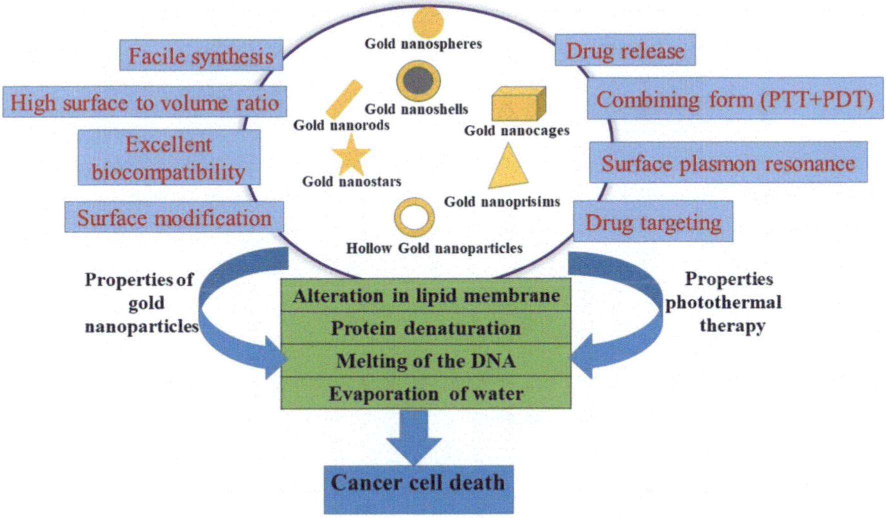

Fig. 1.3 Illustration of various forms of gold nanoparticles used in PTT

It has been seen that the laser power density directly influences cellular disruption, where the cell death rate increases with an increase in the laser power density (Zhang et al. 2016a). In addition to the high PPT effect, high-energy laser irradiation also causes nonspecific damage to cells (Park et al. 2015). These results show the significance of the suitable selecting power to reduce the harmful effect in the healthy tissues near cancer. The laser power in the wavelength range 808–980 nm, which has about 330–350 mW/cm^2 with a radiation time of 10 to 100 s, has shown skin tolerance threshold (Vankayala et al. 2014b). The threshold level of cell damage occurs when optimized laser power is used in PTT. In addition, the laser irradiation time increases when the temperature is enhanced; this simultaneously increased cell destruction.

The efficacy of AuNP-mediated PTT depends on the amount of AuNPs accumulated within the tumor cells (Wang et al. 2016). In addition to this, size, shape, surface modification and incubation time may also affect the efficacy of PTT (Liu et al. 2015b). For example, the incubation time of AuNPs with cancer cells increases, the amount of AuNPs' internalization into cancer cells increases, and hence, the PTT efficiency is also increased (Yang et al. 2015). Various parameters used in AuNP-mediated PTT and also their results are shown in Table 1.3.

1.3.2 Surface Modification of AuNPs for PPT

AuNPs used in PTT mainly depend on two parts, i.e., core and peripheral part. The core part is responsible for converting the absorbed light into heat, which is utilized for hyperthermia treatment. The peripheral part acts in stabilizing AuNPs to the physiological environments and is also used as loading of targeting agents. It can

Table 1.3 Shows some in vitro research on the use of lasers and nanoparticles for cancer PTT

Nanoparticles	Cell line	Radiation time (min)	Laser power density (W/cm^2)	Wavelength of laser (nm)	Emission mode of laser	Outcome	NIR peak absorption of AuNPs (nm)	References
Anti-EGFR-AuNRs	1321 N1 U373-MG	40, 75	2.5	808	Continuous wave	Cell death increased after laser irradiation with conjugated GNRs compared to unconjugated GNRs	808	Fernández-Cabada et al. (2016)
PEG-Pro G-AuNPs	SK-BR-3	3	3, 5	817	–	Due to localized hyperthermia, disruptions of cell membrane resulted that provided a mechanism for killing cancer cells	883 ± 26	Sun et al. (2013)
Anti-EGFR-AuNRs	HaCaT HOCHSC 3	4	5, 10, 15, 20, 25	800	Continuous wave	Threshold energy to kill the cancer cells is found to be 10 W/cm^2, which is HSC and HOC are injured while the HaCat is not affected	~ 800	Huang et al. (2006)
PEG-AuNRs	A-549 BC-37 Hela HepG2 HL-7702	8	3–11	808	–	Compared to the normal liver cell, NIR irradiations in the presence of PEG-GNs induced much more damage to the cancer cells	~ 800	Zhan et al. (2012)
PEG-AuNRs	PC3-PSMA	15	20	800	Continuous wave	Extracellular hyperthermia significantly induced cancer cell death	800	Huang et al. (2010a)

(continued)

AuNCs AuNPs-shell AuNRs-shell	HeLa	40	0.054	940	Continuous wave	Gold nanoshells can serve as excellent multifunctional theranostic agents (fluorescence imaging + PDT + PTT) upon single-photon NIR light excitation under ultralow laser doses	–	Vankayala et al. (2014b)
S-AuNSC M-AuNSC L-AuNSC	B16-F10	–	6	800	Pulsed wave	Antitumor ability of these three GNSCs depends on their morphology and size	800	Wang et al. (2016)

Reprint with permission from Norouzi et al. (2018). Copyright (2018) Springer

Anti-EGFR-AuNRs anti-epidermal growth factor receptor gold nanorods, *PEG-Pro G-AuNPs* PEGylated protein-G gold nanoparticles, *PEG-AuNRs* PEGylated gold nanorods, *12P-PEG-AuNRs* the peptide 12P PEGylated gold nanorods, *AuNCs* gold nanocages, *AuNPs-shell* gold nanoparticle in shell, *AuNRs-shell* gold nanorod in shell, *S-AuNSC* small gold nanoshell capsule, *M-AuNSC* medium gold nanoshell capsule, *L-AuNSC* large gold nanoshell capsule enhance cancer cell identification and subsequently increase their uptake by tumor cells (Du et al. 2017). Even though AuNPs can enter tumor cells without targeting agent on their surface, they cannot accurately determine the location to kill tumor cells (Liu et al. 2010). If non-targeted AuNPs enter the normal cells, unwanted side effects occur in the normal cells. Therefore, the effective treatment of cancer is not possible using PTT without the differential cellular uptake of the AuNPs between normal and tumor cells. Thus, the attachment of AuNPs to effective targeting ligand is the best strategy for the treatment of cancer with minimal side effect and reducing laser power which consequently enhances the safety of AuNP-mediated PPT (Park et al. 2015).

Here, we focus on how AuNPs can be surface modified with thiol (-SH) and amine (-NH_2) groups in order to actively target the cancer sites (Mendoza-Nava et al. 2013). The tagging ligands may be proteins, antibodies (Abs), aptamers, peptides, folic acid, and small molecules, which improve the stability and facilitate the identification and attachment of specific receptors present on the tumors (Zhang et al. 2016a). The tagged molecules are biocompatible and sustained their biological activity after binding to the surface of the AuNPs (Liu et al. 2015b). Due to the targeting functionalization, the selective delivery of AuNPs to the tumors offers localized heating at the tumors sites. In this way, the unwanted side effects associated with conventional tumor treatment are averted. Once AuNPs accumulate into tumors, NIR irradiation can trigger photothermal heating. For instance, tagged AuNPs selectively bonded to the CD-44 cells, and after NIR light irradiation, it has shown a significant therapeutic effect compared to untagged AuNPs (Zhang et al. 2016a). These results indicated that after functionalizing AuNPs with appropriate targeting molecules, the therapeutic efficacy of the AuNPs in PTT had been increased five times compared to without tagged one (Du et al. 2017). Other recent studies have also shown that PTT's efficiency is increased after attaching targeting molecules on AuNPs' surface.

The recognition and attachment of the AuNPs to the tumor-targeting molecule is responsible for their particular PTT activity. For this purpose, the various surface modifying strategies for AuNPs have been adopted. One of the strategies is coating AuNPs with polyethylene glycol (PEG). The PEGylation is used to increase the hydrophilicity, passive targeting ability, and blood retention time of AuNPs (Ma et al. 2016). The citrate and cetyltrimethylammonium bromide (CTAB) covered AuNPs, AuNRs, and AuNSs and have inadequate stability when they are dispersed in buffer solution due to the aggregation effect of the salt ion. By covering AuNPs with PEG, the biocompatibility is greatly enhanced and also prevents the aggregation of AuNPs. The PEGylation occurs via thiol functionalization between PEG and AuNPs. In addition, folic acid (folate) can also be used as an active tumor-targeting agent due to the folate receptor on the tumor cells. Folic acid is immunogenic, a water-soluble vitamin. Foliated AuNPs are widely used to detect various types of tumor cells (ovarian, kidney, lung, and thyroid cancerous tissues). Foliation may occur through either esterification or ammoniation between folic acid and capped AuNPs (Khoshgard et al. 2014). For example, Wei and coworkers showed that foliated AuNRs can be used for the specific targeting of the surface of the KB tumor cells. The tagged AuNPs preferentially enter into tumor sites through the EPR effect (Hirsch et al. 2003).

The aptamer-based targeting is more active, specific, and efficient. Aptamers are oligonucleotides of DNA or RNA, binding to specific proteins of cell membrane with a three-dimensional structure (Sun et al. 2013). It can be tagged to the surface of the AuNPs by the thiol functional group. Aptamer used as a targeting agent has many advantages such as simple synthesis, smaller size, easy surface modification procedure, and storage. Recently, aptamer is replaced by other targeting agents such as antibodies due to its advantages (Azhdarzadeh et al. 2016). These aptamers can also detect lower-density tumor cells with a higher sensitivity and higher affinity for binding to the AuNPs' surface. Due to the higher affinity, the concentration of AuNPs in tumor cells increases, and PTT's efficacy is also increased simultaneously (Du et al. 2017).

The Abs-conjugated AuNPs provide an effective and selective tumor targeting method. The Abs is selected to target a specific tumor marker. For example, AuNPs with anti-VEGF Abs absorbed on the surface provided to be more efficient for the induction of apoptosis in β-chronic lymphocytic leukemia cells as compared to AuNPs without Abs. AuNPs immobilized Abs should retain their optimal activities, which is vital to the active targeted imaging and therapy where the conjugated AuNPs can target specific cells (Huang et al. 2008). The conjugation of the Abs depends upon several factors, including particle shape, protein structure, and conjugation strategies. The commonly accepted strategies are physical absorption (hydrophobic or electrostatic interaction) and chemical binding between the functional group on the Abs and the metal surface of NPs (Jazayeri et al. 2016). Accumulation of the Ab-conjugated AuNPs in tumor cells is selective to PPT. In addition to specific binding to the tumor cells, Abs also damages the tumor cells by blocking the signal pathway.

Polymeric peptide amino acid sequences also have good targeting properties. Therefore, it may be possible to achieve innovations in targeted PTT applications. Compared with a monoclonal antibody, the peptides are applied as targeting agents in PTT because they have some advantages: less immunogenicity, highly biocompatibility, and better pharmacokinetic properties (Heidari et al. 2015). The peptide-tagged AuNPs can have higher internalization into cells as compared to Abs-tagged AuNPs. AuNRs conjugated with peptides significantly minimize cancer cell survival in PTT with minimal damage to surrounding cells, according to a report (Sun et al. 2013).

1.3.3 Cellular Uptake of AuNPs for PTT

In addition to developments in the production of suitable AuNPs, it is important to examine their destination for treatment as a deciding factor in the efficiency of light to heat conversion (Espinosa et al. 2016). AuNPs have a therapeutic effect based on their location within or outside the cells (Zhang et al. 2016b). AuNPs may either stay on the cell surface or may cause biological responses via various cellular absorption pathways. The biggest barrier to entering AuNPs into the cell is the cell membrane (Qin and Bischof 2012). A non-invasive and permanent approach for the death of tumor cells is causing damage to the cell membrane by PTT. It has also been shown that AuNPs attached to the cell membrane require lower laser power than the internalized AuNPs in the cells for photo-thermolysis induction (Sun et al. 2013).

On the other hand, tumor cells are weakened merely because of the presence of AuNPs in the membrane of the cells, through which AuNPs are unable to reach the cells through endocytosis; whereas the necrotic phase happens through their introduction into the cells during PTT (Rau et al. 2016). Another gain of incorporating AuNPs into PTT cancer cells is the elimination of side effects, as it eliminates the need for elevated doses of AuNPs. The investigation of intracellular events will also help to better understand the PTT process mediated by AuNP (Kodiha et al. 2015).

The endocytosis of NPs is primarily achieved by the phagocyte system, an energetic operation, and by non-phagocytic processes (pinocytosis), including multiple mechanisms like endocytosis mediated by caveolae (Stern et al. 2012). Therefore, in various cells, the function of the NPs' entry into cells may be different. Often, because of their high metabolic activity, tumor cells have a greater potential for endocytosis relative to normal cells (Aşık et al. 2016). The internalization of NPs into cells plays a significant role in causing permanent damage and cytotoxicity. The internalization ability of AuNPs into a cell depends on many factors, including the size morphology, surface charge, and concentration of AuNPs, length of exposure of AuNPs to tumor cells, as well as the type of cell and type of ligand (Wang et al. 2011). A significant factor affecting their cellular uptake, intracellular dissemination, and cytotoxicity is the scale of the AuNPs. Cellular responses to PTT are also dependent on the size of AuNPs; a study of different sizes of spherical AuNPs has shown that the size of 50 nm has the greatest efficiency in accessing the cell (Aşık et al. 2016). AuNPs of different sizes show various degrees of contact with the cells' membrane, and hence a different cellular uptake occurs. Owing to electrostatic interactions with the cell membrane, the positive charge on the surface of the AuNPs increases cellular absorption (Du et al. 2017). The addition of ligands can increase the positive charge on the AuNP surface to the AuNP surface (Sun et al. 2013). So the internalization of the AuNPs into the cells increases.

Moreover, PEGylation also increases cellular absorption to improve the stability of AuNPs. Functionalized AuNPs can also bind to receptors on the surface of the membrane and normally reach the cells through receptor-mediated endocytosis (Lee et al. 2015). Cellular uptake therefore relies on both the incubation time of the AuNP cells and the type of cell. The longer the AuNPs are incubated, the probability of more AuNPs entering the cell increases (Zhang et al. 2016b). These properties also influence the intracellular position of AuNPs (Alfranca et al. 2016). The use of high-intensity lasers will improve the membrane cells' permeability so that more AuNPs can reach the cells.

For AuNP-mediated PTT using NIR light, cellular absorption of AuNPs is essential, and targeting particular intracellular organelles may also enhance PTT efficacy.

The mechanical damage and consequent effectiveness of PTT will increase if AuNPs accumulate in the subcellular components (Kodiha et al. 2015). In cell apoptosis activation, attacking lysosomes and mitochondria as two intracellular organelles plays a crucial role (Zhang et al. 2016b). Functionalized AuNPs can bind to the cells' surface receptors, as stated, and the endocytosis process mediated by the receptor occurs (Wang and Thanou 2010). The endosome that has been formed will ultimately be paired with lysosomes (Lee et al. 2015). There are acidic pH and separate hydrolytic enzymes in the lysosome, a stable condition for AuNPs. In general experience, in the presence or absence of AuNPs, equilibrium in normal cells' lysosome membrane avoids lysosomal damage. Although the lysosomal membrane of tumor cells is relatively vulnerable, hydrolytic materials are released and can be transferred to other organelles, such as mitochondria, due to their degradation by AuNPs (Zhang et al. 2016b). The development of hydrolytic enzymes increases through the aggregation of AuNPs in lysosomes, contributing to an increase in lysosomal activity and ultimately leading to a breach of the lysosomal membrane (Mkandawire et al. 2015).

As the main site of energy production, mitochondria are an effective focus for the treatment of tumors (Kodiha et al. 2015). Tumor cells have displayed substantial metabolic activity in contrast to normal cells, resulting in mitochondrial damage (Mkandawire et al. 2015). NIR excites AuNP electrons, and, as a result, the phonophonic process interfaces with the electron transport chain of mitochondria. The mitochondria have also undergone oxygen deficiency (Mocan et al. 2013). This decrease in the amount of oxygen and the other effects of AuNPs on mitochondria, including the activation of cytochrome c enzymes and active oxygen species, triggers cell apoptosis (Zhang et al. 2016b). In other words, through local accumulation in mitochondria, AuNPs can cause heat oxidation during PTT. Cavities greater than 2 nm in the mitochondrial internal membrane that inhibit the cellular uptake of the AuNPs to the mitochondrial matrix are the remaining problems in mitochondrial therapy (Mkandawire et al. 2015). The research investigated the apoptosis process in tumor cells in the presence of AuNPs and found that only a small number of AuNPs were involved in mitochondrial disruption, unlike lysosomes. This non-selective accumulation revealed that AuNPs could not be directly involved in mitochondrial damage; thus, lysosomal protease could be the source of the damage (Zhang et al. 2016b). It has been documented that AuNPs migrate toward lysosomes with certain particular ligands, and other ligands move AuNPs toward the mitochondria; in other words, AuNP surface ligands play a significant role in the movement of NPs within cells (Mkandawire et al. 2015).

Since the cell nucleus is such an important organelle for survival and proliferation in comparison to other organelles within the cell, considering AuNPs' impact on the nucleus is critical (Kodiha et al. 2014), and the nucleus can be chosen as a suitable target for AuNP-mediated PTT. AuNPs are present in other organelles such as endosomes and lysosomes before being internalized into cells, as previously mentioned. As a consequence, in order to enter the nucleus, they must escape these organelles via the cytoplasm (Yang et al. 2014). If AuNPs are to be delivered to the nucleus, the size of the nuclear pore complex (NPC) transport channels must also be considered. Since the number of NPCs differs depending on a cell's activity level and the nuclear transport machinery is often altered in different cancer cells, there is no specific procedure for NP internalization into the nucleus in various tumoral cells (Kodiha et al. 2015). AuNPs can alter nuclear morphology, reorganize nuclear laminae, and ultimately inhibit nuclear functions, reducing the subnuclear compartments that produce ribosomes in particular. PTT has been shown to improve some of these effects, resulting in an increase in cancer cell proliferation (Kodiha et al. 2014).

1.4 Photodynamic Therapy (PDT)

Both PTT and PDT are examples of light-triggered non-invasive tumor treatments against several types of cancers and non-oncology diseases (Vijayaraghavan et al. 2014). The US Food and Drug Administration approved the PDT as a promising medically recognized treatment method with a selective cytotoxic activity to malignant cells (Konan et al. 2002). In this method, non-toxic PS are irradiated with a specific wavelength of light that leads to the destruction of the target cell via oxidative damage (Murphy et al. 2010). In PDT light source, PS and oxygen are the key components that act simultaneously; none of this is individually toxic to the target cells. However, when the PS are excited with a certain wavelength of light and generate ROS, they cause significant toxicity to the targeting cells through which cell apoptosis and necrosis occur (Lucky et al. 2015). PS are initially transferred to the target cells to be treated, subsequently photoactivated by illumination of light at specific wavelength. The activated PS energy is transited to oxygen molecules, which releases reactive oxygen species (ROS) or singlet oxygen ($^1O_2$), which initiates a series of biochemical events that cause cytotoxicity of neoplastic cells and the regression of tumor (Wang and Lu 2018).

1.4.1 Mechanism of the PS in PDT

Figure 1.4 depicts the exact process of PDT using PS in a simplified Jablonski diagram. A PS receives a photon of light and transitions to the excited singlet state ($^1PS^*$). The excited PS can either decay to the ground state, releasing fluorescence for clinical imaging and photodetection, or undergo intersystem crossing (ISC), in which the spin of its exciting electron inverts to form a relatively long-lived triplet state ($^3PS^*$) (Castano et al. 2004). The triplet-excited PS can then interact directly with a substrate, such as a cell membrane or a molecule, and transfer a proton or an electron to form a radical anion or cation, which then interacts with oxygen to form oxygenated products such as superoxide anion radicals, hydroxyl radicals, and hydrogen peroxides (type I reaction). Alternatively, the energy of the excited PS can be transferred directly to molecular oxygen (a triplet in the ground state) to form $^1O_2$

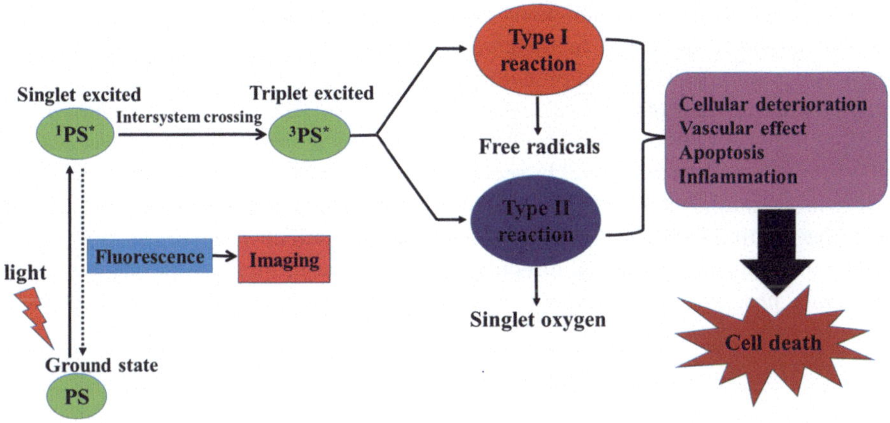

Fig. 1.4 Schematic illustration of a typical photodynamic reaction

(type II reaction). The transition of oxygen from the triplet ground state to the excited singlet state requires 22 kcal mol^{-1}, which corresponds to a wavelength of 1274 nm (Yamaguchi et al. 1981). As a result, only a small amount of energy is needed to generate $^1O_2$ (MACDONALD and DOUGHERTY 2001). The cell-killing and therapeutic effects of PDT are due to the byproducts produced as a result of the type I and type II reactions. It's worth noting that both type I and type II reactions can occur at the same time, and the ratio between them is determined by the PS type, as well as the molecular oxygen and substrate concentrations present. However, most studies show type II reactions, implying that $^1O_2$ plays a significant role in PDT (Ding et al. 2011).

By transferring an electron from the triplet PS to the oxygen molecule, the type I reaction produces a superoxide anion. While superoxide does not cause oxidative damage in and of itself, it can react with itself to produce H_2O_2 and oxygen. The superoxide dismutase enzyme will catalyze this reaction, which is now known as "dismutation." In this case, superoxide serves as a reducing agent rather than an oxidizing agent. This is because superoxide donates an electron to reduce metal ions (such as Fe^{3+}) that serve as catalysts in the conversion of H_2O_2 to hydroxyl radicals (OH·). The reduced metal (Fe^{2+}) or MNPs (AuNPs) act as a catalyst to break the oxygen-oxygen bond of H_2O_2 to produce a hydroxyl radical ($OH^•$) and a hydroxide ion (OH^-) (Alle et al. 2021). The highly reactive oxygen species of $_1O^2$ and peroxynitrite ($OONO^{-)}$ may form when superoxide reacts with the $OH^•$ and nitric oxide (NO^-). This oxidizing agent directly reacts with tumor tissue and irreversible damage to tumor cells.

The classical approach for the destruction of tumor cells is when PS utilize light energy to produce cytotoxic oxygen-based species, such as $_1O^2$, free radical, and peroxides, leading to tumor cell death. The generated cytotoxic oxygen-based species react with DNA, protein, or other cellular compartments, which results in the irreversible damage of the cell structure and function (Zhou et al. 2016). Besides this, PDT itself destroys malignant microvessels, resulting in the scarcity of the oxygen and nutrition supply to the tumor tissues. Unfortunately, most available PSs, such as hematoporphyrin and phthalocyanine, absorb light energy only over wavelengths shorter than 600 nm, a severe drawback for PDT. In addition, they have another drawback: poor solubility in water, nonspecific targeting, and low chemical stability. As a result, they quickly accumulate under physiological conditions, lowering the quantum yields of ROS output dramatically (Dolmans et al. 2003). Even with the modified PS for increased solubility, their selective aggregation at target cells is inadequate for clinical effectiveness. As a result, much work has gone into designing new PSs or chemically modifying existing ones with high absorption cross sections in the NIR region.

1.4.2 AuNPs for PDT

Engineered AuNPs have been used successfully as a PS in PDT treatment. As AuNPs are photoexcited with a specific wavelength of light, the exciting light energy may be passed to neighboring molecular oxygen (O_2) or organic PS, generating $^1O_2$ (Lucky et al. 2015). In terms of their active roles in PDT, AuNPs could act as catalysts for the production of free radicals from dissolved oxygen, as well as ROS modulators and light sources for photosensitizer activation. It is known that the PS activity of AuNPs can occur in two ways, such as AuNPs encouraging the photosensitization of PS to produce $^1O_2$ and AuNPs themselves sensitizing $^1O_2$ formation (Bai et al. 2020). Due to NIR lights' absorption by the LSPR AuNPs, it is passed to a PS or oxygen molecule, which addresses typical disadvantages of organic PS, such as less photo-stability, fast enzyme oxidation, and less effective conversion of light energy (Vijayaraghavan et al. 2014). Furthermore, AuNP-mediated PS delivery systems have a high surface-to-volume ratio, resulting in high PS loading capacity and reduced transition into normal tissues. This is accomplished either by adding ligands such as tumor-specific antibodies or proteins or by an EPR effect, preventing unwanted nonspecific distribution. AuNPs can be used as photosensitive enhancers, passing absorbed light energy to organic PS and increasing $^1O_2$ sensitization (Jang et al. 2011). In another scenario, AuNPs themselves act as a PS; here, AuNPs absorb light energy directly transferred to oxygen molecules in the local environment to generate $^1O_2$. The two reactions can occur simultaneously and rely on the tissue's oxygenation, killing the cancer cells (Vankayala et al. 2011).

Vankayala and colleagues were the first to show that $^1O_2$ can be formed by direct sensitization by AuNPs under visible light exposure in 2011 (Vankayala et al. 2011); they were also the first to show that this sensitization can be implemented under NIR (915 nm) light excitation and can exert dramatic PDT effects on the destruction of solid tumors in mice and that $^1O_2$ yield is related to its shape (Vankayala et al. 2014a). Recently, various research groups are exploring that the efficacy of $_1O^2$ generation is increased by placing various PSs (phthalocyanines, toluidine blue O, indocyanine green, etc.) on AuNPs' surface (Gamaleia et al. 2010). For instance,

AuNP-capped phthalocyanine may produce cytotoxic $^1O_2$ (Hone et al. 2002). In another study, PS functionalized with AuNPs and TOCAB (phase transfer reagent), phthalocyanine was used during preparation. The combination of PS/AuNPs/phase transfer reagent has shown higher production of $^1O_2$ compared with single PS. They suggested that AuNPs efficiently deliver the PS in PDT and enhance the PS's cytotoxicity (Wieder et al. 2006). In another study, PEG-capped AuNPs conjugated with silicon and phthalocyanine-4 (Pc-4) were used for the hydrophobic drug delivery to its site of action of PDT due to its water-soluble and biocompatibility properties. The in vitro and in vivo drug released study indicates that the Pc-4 is efficiently delivered and passively accumulated at the malignant site. PEG-capped AuNPs have shown two times higher efficiency than conventional PDT delivery of the drug in vivo methods. The in vivo findings showed that there were no obvious side effects and that Pc-4 was distributed across the mouse body, including the lungs and kidneys. Hence, the hybrid PSs had shown enhanced generation of the $^1O_2$ and PDT efficacy against tumors (Yao et al. 2016).

The PDT activity of the AuNPs may depend on size; targeting ability and surface charge are the key factors. In general, AuNPs enter the cell membrane through diffusion and alter the endocytotic process of NPs (Dhanalekshmi et al. 2020). As the size range between 40 and 50 nm, the internalization of AuNPs is increased into the cells (Asharani et al. 2011). The production of the $^1O_2$ or ROS with AuNPs is size-dependent; AuNPs with bigger size have a greater tendency to $^1O_2$ PS, which is produced as a result of a stronger scattering EM field around the particles as compared to smaller particles. For example, it has been reported that AuNPs' mean size of 40 nm could produce $_1O^2$ upon either pulse or CW laser irradiation. Although these reports demonstrated that less than 50 nm size AuNPs are potentially used as PS in the future, the wavelength used in these experiments is 514 and 532 nm. In another study, it has been demonstrated that generation of ROS depends on the size of the AuNPs in protoporphyrin IX (PpIX) (Khaing Oo et al. 2012). Because of the stronger scattering EM field around the particles, larger AuNPs have a greater ability to increase PS ROS generation than smaller AuNPs. Surface charge is another crucial factor influencing cellular uptake AuNPs that allows either specific or nonspecific interactions within the cellular lipid bilayer. The positively charged AuNPs enter the cell membrane through electrostatic interaction (Verma and Stellacci 2010). The negatively charged AuNPs inertly reach the cytosol by targeting the lipophilic domains. In addition, these conjugated AuNPs enable intercellular penetration owing to their tendency to bind with thiols, disulfides, and amines.

The PDT efficiency depends on the wavelength of the light used to activate the PS. Generally, the light's wavelength up to 600–700 nm has higher activation of the PS than the light above 800 nm wavelength. The wavelength of light up to 800 nm is sufficient to activate the PS and produce $^1O_2$ due to the triplet state of the PS, which is above the energy level of the $^1O_2$ (Allison and Sibata 2010). Previously, discharge lamps, metal halogen lamps, and short-arc xenon lamps were used to enable the PS, but these lamps have a shallow penetration depth and are only useful for superficial tumors (Nitzan et al. 2004). However, in the last 20 years, X-ray and self-luminescence have been used as a light source because these lamps have a greater ability to penetrate deep tissues and treat tumors.

1.4.3 Surface Modification of AuNPs for PDT

The primary aim of PDT is to cause efficient destruction of malignant cells while minimizing PS side effects on healthy cells. Surface modification of AuNPs to bind active targeting molecules such as ligands, aptamers, peptides, and antibodies, which improve PDT efficiency, can significantly increase the PS's selective therapeutic effect. For example, Russell et al. first reported that anti-HER2 monoclonal antibody attached to PEGylated AuNP-phthalocyanine (C11Pc) conjugates was targeted to breast cancer cells. When tumor-associated antigens are present on the cytoplasmic membrane of malignant cells, in vitro cellular experiments showed that the targeting capability of the four-component NP (antibody-C11Pc-PEGAuNPs) conjugates increases the efficacy of PDT cell death (Stuchinskaya et al. 2011). As a result, in another study, they used jacalin to target cancer cells. They observed that the two biofunctionalized C11Pc-PEG AuNPs (jacalin and anti-HER2 antibody) had equal targeted PDT efficacies in both tests (Obaid et al. 2015). To target tumor cells, Savarimuthu et al. used folic acid attached to protoporphyrin (PpIX)-conjugated AuNPs. This study's in vitro results revealed that foliated PpIX-NP complex is more phototoxic than the free PpIX and PpIX-NP complex (Savarimuthu et al. 2015).

Meyers et al. devised a new targeted PDT approach on the basis of epidermal growth factor peptide-targeted AuNPs (EGFpep-AuNPs). After increasing localization in early endosomes, in vitro studies of PDT showed that EGFpep-AuNP-Pc4 kills tumor cells twofold better than pure Pc4. In vivo studies show that EGFpep-AuNP-Pc4 causes a threefold increase in fluorescence Pc4 accumulation in the subcutaneous relative to untagged AuNPs. The EGFpep-AuNP-Pc4 fluorescence test in vivo revealed that after intravenous injection, EGFpep-AuNP-Pc4 could reduce the initial uptake by the reticuloendothelial system (RES) and increase the amount of AuNP circulation in the blood, which impacts biodistribution AuNPs (Meyers et al. 2015).

Cheng et al. compared in vitro drug release study with HeLa cancer cells through covalent and non-covalent attached silicon phthalocyanine 4 (SiPc4) with PEGylated AuNPs. It was found that non-covalently attached SiPc4 with AuNPs showed efficient PDT activity compared to covalent thiol-bonded AuNPs due to the slow intracellular release of the drug (Cheng et al. 2010). Later they investigated the drug delivery and pharmacokinetic mechanisms after 7 days of intravenous administration of non-covalently bound PEG-SiPc4-AuNP conjugates. Non-covalent attachment of PS to AuNP resulted in efficient release and penetration of the PEGSiPc4-AuNP conjugate deep into tumor in vivo experiments. Even though AuNPs have a relatively long retention period in the body, especially in the liver and spleen, it is discovered that renal clearance and the hepatobiliary system excrete the drug and AuNPs quickly from the body (Cheng et al. 2011).

1.4.4 Combined PTT/PDT Therapy

Due to the combination of different light-triggered cell killing mechanisms that may resolve the challenges of single model therapies, combined PTT/PDT is often an excellent therapeutic outcome for all cancer treatments. PDT uses light irradiation to excite the PS, resulting in the production of reactive oxygen species (ROS) or $^1O_2$, which causes cellular toxicity, while AuNPs can convert light to heat, resulting in a synergistic effect with PDT treatment. Currently, several research groups are working on combining gold-mediated PTT and PDT for cancer treatment. Examples of this approach include AuNR-PS complex (AuNR-A1PcS4) for the combination of PDT/PTT therapy. The AuNR-A1PcS4 complex is excited by 810 and 610 nm lasers. In vitro results revealed that AuNR-A1PcS4-treated cells showed four times higher intracellular uptake and enhanced therapeutic efficiency compared to pure A1PcS4-treated cells. After intravenous injection of AuNR-A1PcS4, both PTT and PDT effect can be induced by irradiation of 670 and 810 nm lasers separately. Compared to PDT alone, the combination of PDT/PTT enhanced the tumor's inhibition growth from 79 to 95% (Jang et al. 2011). Kuo et al. used AuNRs simultaneously as imaging and killing A-549 tumor cells. Here, AuNRs are used not only as PDT and PTT therapy but also as optical contrast agents. In their work, AuNRs conjugated indocyanine green (ICG) as PS treated with A549 cancer cells followed by IR laser light irradiated at 808 nm to implement PDT and hyperthermia. The results showed that the combined PDT and PTT abolish tumor cells more effectively than PDT or PTT alone and improved photodestruction efficiency (Kuo et al. 2010). The same group demonstrated that both AuNPs and AuNRs conjugated with ICG simultaneously sever dual mediated PDT and PTT to destroy tumor cells (Kuo et al. 2012).

Many studies also use AuNSs to achieve combined PTT/PDT performance. For example, Kah et al. used anti-EGFR-conjugated AuNSs for combined PTT and PDT. The AuNSs' absorption spectrum reveals a strong extinction band (wavelength > 580 nm) and hypericin (peak absorbance at 595 nm) excited by a 100 × 50 cm light with a wide band illumination above 585 nm. In vitro findings revealed that combining PTT and PDT treatment was a more successful treatment technique than PDT or PTT treatment alone (Kah et al. 2008). In another study, chlorin e6-conjugated AuNSs are used to operate simultaneous PDT/PTT upon single laser irradiation. For this purpose, AuNSs' LSPR peak was tuned to match the absorbance of chlorin e6. The experimental results showed that the difference in photostability between PS and AuNSs is different, and thus the ratio between PDT and PTT can be modulated by adjusting irradiation time. The combined PDT/PTT has shown a synergetic effect, significantly enhanced anticancer efficacy, and inhibited breast cancer growth in animal models (Wang et al. 2013).

Recently, many research groups worked in PTT's synergistic influence on PDT with various lights, various PS, various Au nanostructures, and various cell lines. Gao et al. developed a new nanocomplex by coating AuNCs with lipid-loaded PS and hypocrellin B (HB). The nanocomplex has tunable heavy scattering and absorption in the NIR transparent window (790 nm), allowing it to be used as a possible contrast agent and photothermal converter for cancer diagnosis and treatment. They showed that when the nanocomplex is internalized by tumor cells, the PTT effect significantly increases PDT under two-photon illumination (Gao et al. 2012). Wang et al. developed an aptamer switch probe (ASP)-chlorin e6 (Ce6)-AuNR complex for combined PDT/PTT targeting cancer cells. As the composite comes in contact with target tumor cells, the ASP changes conformation to drive Ce6 away from the Au surface, producing $^1O_2$ for PDT under light irradiation. Furthermore, because of their high absorption efficiencies, AuNRs can kill tumor cells using the PTT process. Wang et al. created a Ce6-pHLIPss-AuNR conjugate with pH-driven targeting capacity for synergistic PDT/PTT. AuNRs are used as a PTT agent, as well as a carrier and quencher of Ce6 for PDT therapy, in this design, preventing nonspecific activation and self-destruction of PSs (Wang et al. 2014a).

1.4.4.1 In Vivo Studies for Combined PDT/PTT Therapy

For in vivo PDT studies, Khlebtsov et al. developed a novel nanocomposite with a gold-silver nanocage core and a mesoporous SiO_2 shell-modified Yb-2,4-dimethoxyhematoporphyrin (Yb-Hp) as a PS. Under 630 nm excitation, the fabricated nanocomposite provided $^1O_2$ and created heat when irradiated with laser light at the plasmon resonance wavelength (750–800 nm). Multifunctional nanocomposites tend to be a smart theranostic platform for synchronized IR-luminescence diagnostic and PDT due to Yb-Hp and plasmonic PDT due to Au-Ag nanocages in this design (Khlebtsov et al. 2011). Wang et al. conducted another in vivo analysis. AuNRs conjugated with rose bengal (RB) were used in the PDT/PTT of oral cancer in vitro and in vivo. When irradiated with 532 nm green light, the fabricated RB-AuNRs provided $^1O_2$ and produced high heat upon irradiation with 810 nm NIR light. Tumor cell death is caused by these two distinct mechanisms (Wang et al. 2014a).

Lin et al. created a multifunctional theranostic platform for successful tumor imaging and therapy, in which a monolayer of assembled AuNPs was used as vesicles to be filled with Ce6 as PS, as part of his ongoing study. The Au vesicles have high absorption in the near-infrared range (650–800 nm), allowing neighboring AuNPs in the vesicular membranes to plasmonically couple. This helps the 671 nm laser to excite both Au vesicles and Ce6 to produce heat and $^1O_2$ for tumor cell killing. The therapeutic efficiency of GV-Ce6 was increased in vitro and in vivo as a result of the synergetic effect of PTT/PDT treatment with single wavelength CW laser irradiation (Lin et al. 2013). In vitro and in vivo, Wang et al. used gold nanostars-PEG-Ce6 for combined PDT/PTT therapy with a single CW laser (Wang et al. 2013). Terentyuk et al. investigated in vivo tumor treatment using the AuNRs/SiO_2-HP (hematoporphyrin) complex for the first time in order to increase the efficiency of tumor treatment by combining PDT/PTT treatment of massive (about 3 cm^3) solid tumors in a xenografted tumor rat model (Terentyuk et al. 2014). As compared to alone PDT treatment, combined PDT/PTT treatment showed a large-area tumor necrosis and led to tumor volume decreasing dramatically.

Vankayala et al. showed that AuNSs could mediate the bimodal PDT/PTT effect in vivo at ultralow doses of NIR light (28–150 mW/cm^2). Following that, much work was done to boost the PDT's in vitro and in vivo efficacy (Vankayala et al. 2014b). Vijayaraghavan et al. were the first to show that dual-modal multibranched gold nanoechinus-mediated PDT/PTT would fully kill cancer in the NIR window II (1000–1350 nm) (Wang et al. 2013). Yu et al. recently published a HAuNS-pHLIP-Ce6 antitumor platform in which hollow gold nanospheres (HAuNs) were prepared to move the LSPR peak and then the pH (low) insertion peptide (pHLIP) and chlorin e6 (Ce6) were adhered to HAuNs through absorption. Because of pHLIP's pH-directed capacity, the antitumor complex will enter the tumor site. Following irradiation, the HAuNs were heated, resulting in PTT, while Ce6 was released from the gold sphere, generating ROS. PTT/PDT is performed synchronously by HAuNs in this study (Terentyuk et al. 2014).

Despite the PTT effect, increased electromagnetic (EM) field at the surface of AuNPs can be achieved with SPR resonant excitation (Schwartzberg et al. 2004). Currently, some reports claim that AuNPs can increase $^1O_2$ species, especially ROS, by the excited PS, which improves EM and PDT treatment efficacy (Geng et al. 2014). When Oo and his colleagues used AuNPs as carriers to transport PpIX as a PS, they discovered this (Khaing Oo et al. 2008). Similar experimental findings were observed when AuNPs were used to deliver 5-aminolevulinic acid (Xu et al. 2012). The size of AuNPs that increased the output of ROS from PpIX was demonstrated by Wang et al. Since bigger-size AuNPs have a large scattering EM field around particles, they developed more ROS from PS than smaller-size AuNPs (Khaing Oo et al. 2012). As a result, using AuNPs as a carrier will boost cellular uptake and increase the output of ROS from PS, thereby increasing the PDT's treatment efficacy (Zhang et al. 2015).

It was also discovered that when AuNPs were irradiated at LSPR, they could generate $^1O_2$, in addition to increasing the ROS output of PS (Vankayala et al. 2011). Pasparakis demonstrated the development of $^1O_2$ by irradiating bare AuNPs with a CW or pulsed laser. The development of $^1O_2$ is mediated by two separate mechanisms. One is a plasmon-activated pathway, which is activated when plasmons and hot electrons interact with molecular oxygen. The second is the indirect photothermal pathway, which occurs when AuNPs are exposed to intense heat from a pulsed laser, causing fragmentation and increased thermionic electron emission (Pasparakis 2013). Vankayal's team also investigated photosensitization and $^1O_2$ production by Au and Ag nanoparticles and discovered that it was highly morphology-dependent. They demonstrated that instead of Au (111) and Au (100), $^1O_2$ could be photosensitized and produced by irradiating Au nanostructures with an Au (110) surface. AuNPs made specifically for PDT could serve as a photosensitizer (Vankayala et al. 2013).

AuNPs as a PS have a 4–6 order higher extinction coefficient than traditional organic dyes and are much more resistant to enzymatic and photochemical degradation (Jain et al. 2006). It means that a much smaller volume of AuNPs as a PS must be taken up by cancer cells in order to reach the PDT treatment threshold concentration. Furthermore, Xu's team discovered that AuNRs have a high $^1O_2$ efficacy under two-photon excitation, implying that AuNPs can be used as the PS in two-photon photodynamic therapy (TP PDT). As a result of the PTT effect, AuNPs can be used not only as a drug delivery vector but also as a promising dual-functional photome (Zhao et al. 2012).

1.5 Radiotherapy

The ultimate aim of radiotherapy is to destroy cancer cells by using high-energy rays such as X-rays, γ-rays, electron beams, and protons to prevent tumor development. X-ray radiotherapy, like PTT and PDT, only affects clear irradiated areas and penetrates much deeper than NIR rays used in PTT and PDT treatment.

It has been observed that ionizing radiation does not differentiate between cancerous and normal cells, thus causing damage to normal cells along with cancerous ones during radiation therapy. To overcome this, tumor-specific nanoparticles have been used in radiation therapy to cause more toxicity in tumors and fewer in normal tissues (Dorsey et al. 2013). For this purpose, various forms of radiosensitizers have been designed. In fact, AuNPs have been shown to act as radiosensitizers for X-ray radiotherapy due to their high density, high coefficient of energy absorption, and low toxicity (Kobayashi et al. 2010). Hainfeld et al. showed the increased therapeutic effect of X-ray radiotherapy when treated along with an intravenously injected high dose of 1.9 nm Au clusters (2.7 g of Au/kg of mice body weight) into tumor-bearing mice (Hainfeld et al. 2004). The experimental findings show that AuNPs and X-ray irradiation-treated mice contribute to a 1-year survival of 86% compared to 20% with either X-rays or AuNPs alone. Moreover, the small AuNPs could be quickly cleared via the kidney after therapy, reducing the possibility of Au-related side effects.

1.5.1 Surface Modification of AuNPs for Radiotherapy

The attachment of a cell targeting ligand to the surfaces of AuNP-based radiosensitizers would significantly boost cellular uptake and thus treatment efficiency. Xing et al. functionalized the surface of 10.8 nm AuNPs with thioglucose (Glu) to improve absorption by a breast cancer cell line (MCF-7). They used a benign breast cancer cell line (MCF-10A) as a control in their study. The absorption of functional Glu-AuNPs by tumor cells is much greater than that of pure AuNPs, as shown by transmission electron microscopy (TEM) imaging. Furthermore, radiotherapy findings showed that MCF-7 cells were killed faster than MCF-10A cells in the presence of Glu-conjugated AuNPs. This result indicates that the Glu-AuNPs only entered malignant cancer cells and improved their radiation sensitivity, rather than benign cells, which may be useful for cancer treatment (Kong et al. 2008). Geng et al. found similarly targeted radio enhancement in ovarian cancer and cervical cancer (Geng et al. 2011). Despite the fact that glioma cells and brain tumors are kept out of circulation by the blood-brain barrier, PEGylated AuNPs may be used to attack them and radiosensitize them, causing DNA damage enhancement, cancer cell eradication, and enhanced survival (Dorsey et al. 2013). Thus, modified AuNPs that target molecules like glucose, folate, cisplatin, peptides, anti-EGFR, and thioglucose in conjunction with imaging and other therapies are becoming more popular (Khoshgard et al. 2014; Geng et al. 2014).

Furthermore, Polf et al. discovered that proton beam radiotherapy increased the killing of AuNP-laden prostate cancer cells by 15–20 percent relative to cancer cells alone (Polf et al. 2011). Khoshgard et al. subjected HeLa tumor cells to various low energy levels of orthovoltage X-rays and megavoltage gamma-rays to examine cell killing efficacy in the presence of folate-modified AuNPs and non-functionalized AuNPs (Co-60). When using the same dose enhancement factor, there was a substantial difference between experimental groups with and without folate modification from their study. Researchers were able to obtain the highest-dose enhancement factor for both AuNPs using the 180 kVp X-ray laser. Since several cancer cells express the folate receptor, folate-modified AuNPs may be used to improve the killing of cancerous cells using orthovoltage X-ray energies in superficial radiotherapy techniques (Khoshgard et al. 2014).

As previously mentioned, several groups have studied AuNP-aided radiosensitization (AuNP-RS), but most of the work has been based on experimental phenomena rather than sensitization mechanisms, so the mechanism is still unknown. It was previously thought that AuNP-RS could improve photoelectric photon absorption by high-Z materials at kilovoltage photon energies. However, in clinical practice, megavoltage energies are used, and this physical process did not predict the therapeutic effect (Safaee et al. 2015). It will be necessary to understand the effect of AuNPs' concentration, scale, distance from the target, and surface coating on AuNP-aided radiosensitization in order to incorporate this new technology into the clinic and maximize the effect. Normally, the arguments were predicted using Monte Carlo simulations. Geng et al. discovered that AuNP-mediated radiotherapy would improve ovarian cancer therapeutic performance, with levels of ROS output enhancement observed when X-ray radiation was combined with AuNPs (Geng et al. 2011). Some researchers went into great detail about the physics and possible biological pathways involved in AuNP-mediated radiotherapy (Butterworth et al. 2012).

1.6 Glucose-Starvation Cancer Therapy

Cancer-starvation therapy, especially glucose-starvation therapy, is now widely regarded as an effective treatment method by blocking nutrient supply to suppressing cancer cell growth while causing minimal side effects. Glucose is one of the most essential nutrients for both tumor and normal cells (Wang et al. 2014b). Due to the Warburg effect, tumor cells consume ten times more glucose than normal cells, making them more vulnerable to glucose deficiency. As a result, depleting intracellular glucose through a glucose-metabolic reaction can turn a healthy tumor cell into a starving cell, resulting in cancer cell death (Wang et al. 2020). This is an alternative method of preventing tumor proliferation. Glucose oxidase (GOx), an oxidoreductase enzyme, uses molecular oxygen to oxidize glucose to gluconic acid, leading to H_2O_2-mediated ROS formation, causing cancer cell death (Fu et al. 2018). In addition, the GOx enzyme also enhances hypoxia for hypoxia-activated chemotherapy, amplifies acidity for enhanced meta-based chemodynamic therapy, and increases the content of H_2O_2 for oxidation therapy. Thus, the transfer of GOx enzyme within cancer cells may expect to kill tumor cells through multiple approaches.

Furthermore, the overexpressed superoxide dismutase produces a large amount of H_2O_2 through a catalytic process involving superoxide ions released by mitochondria (López-Lázaro 2007). Using a nanocatalytic process, the synergistically derived H_2O_2 from glucose oxidation and intra-tumor process is converted into more toxic ROS under mild acid conditions (Lin et al. 2018). However, natural GOx has a number of drawbacks, including a high cost of preparation and purification, as well as poor operational reliability, which will limit its functional biomedical application in complex and harsh physiological environments (Kotov 2010). As a result, it is critical to look into alternative nanozymes that can improve stability while also lowering costs. Artificial enzymes, also known as nanozymes, are nanomaterials used in catalytic processes that have many advantages over natural enzymes, including facial fabrication, low cost, and robust stability under extreme conditions (Lin et al. 2014). Currently, carbon-based NPs, metal NPs, and metal oxide NPs are used as enzyme-mimic catalytic processes in a variety of fields, including biosensors, cancer diagnosis and treatment, immunoassays, and biomolecular detection, among others (Song et al. 2010; Fan et al. 2011; Liu et al. 2015a).

AuNPs are a common form of nanozyme with GOx activity among all metal NPs and are frequently used to catalyze the oxidation of glucose. Rossi and his colleagues recently demonstrated that, in the presence of dissolved oxygen, AuNPs could catalyze the oxidation of glucose to glucono delta-lactone (GDL) and H_2O_2. This reaction is quite similar to the one catalyzed by GOx, implying that AuNPs could function as a GOx mimic (Comotti et al. 2004). Because of their unusual GOx mimic operation, AuNPs have been studied for a variety of applications since then. Liebana et al. discovered that an Au-silica nanohybrid could be used as a nanozyme replacement with a GOx-like response. It was also demonstrated that dendritic mesoporous silica (DMSN) to construct the composite nanoplatforms (DMSN-Au-Fe_3O_4) with multi-enzyme mimicking catalytic activities could be used to treat tumors (Gao et al. 2019a, b). Here in situ grown AuNPs within the large mesopores of DMSN NPs as a GOx-mimic nanozyme specifically catalyze β-D-glucose oxidation into gluconic acid and H_2O_2 under aerobic conditions, while the produced H_2O_2 can subsequently be catalyzed by the co-loaded Fe_3O_4-based Fenton nanocatalysts to produce high-toxic hydroxyl radicals, which substantially induce tumor cell death (Fig. 1.5).

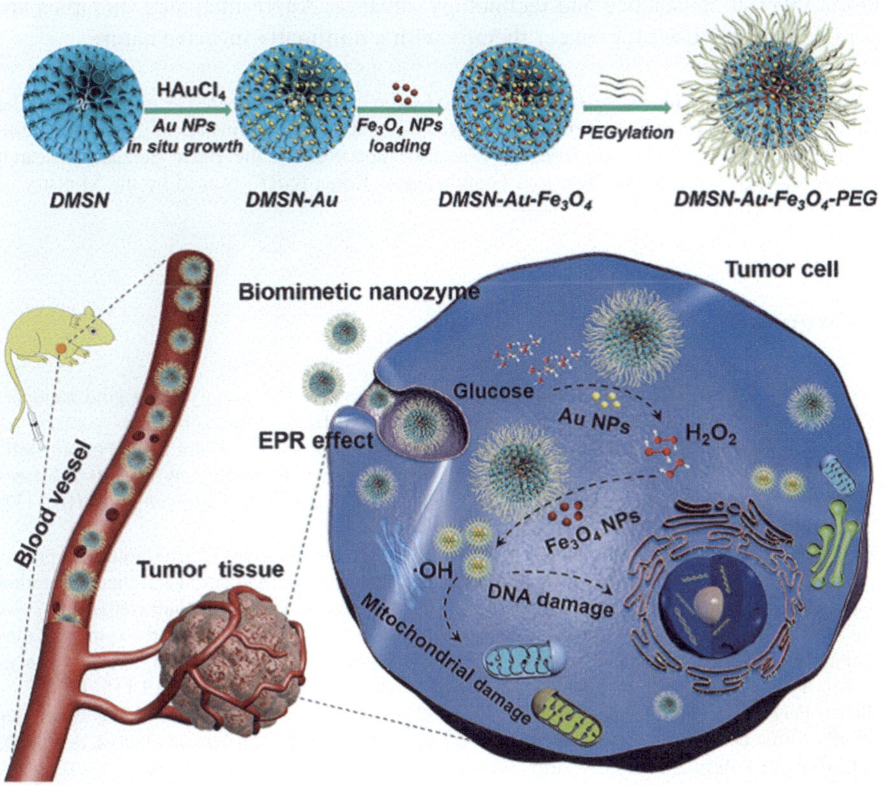

Fig. 1.5 Schematic illustration of a biomimetic inorganic nanomedicine-triggered amplification catalytic reaction, which is used to create a "toxic drug-free" nanocatalytic tumor therapy. (Adopted from Gao et al. 2019a, b)

1.7 Conclusion

Latest systematic studies show that gold nanoparticle (AuNP)-mediated cancer therapies have gained much attention in recent years due to their high selectivity and low side effects. In addition, AuNPs have shown many groundbreaking properties such a light-to-heat converter, local field enhancement, and radiation sensitizer for cancer treatment to effective damage cancerous cell lines both in vitro and in vivo. AuNPs are small in size and can penetrate widely throughout the body due to the EPR effect, preferentially accumulating at tumor sites. They can bind various proteins and drugs, and they can be used to actively target cancer cells with overexpressed cell surface receptors. Thanks to the fast-growing technology, different types of AuNPs are applied to various types of cancer therapies such as photothermal therapy, photodynamic therapy, radiotherapy, and glucose-starvation therapy. Furthermore, significant progress has been made in developing AuNP-mediated multifunctional nanoparticle systems, suggesting that a combination therapy strategy for improving cancer therapeutic efficiency may be introduced in the near future. Overall, as science and technology advance, AuNP-mediated therapies are more likely to be used for cancer therapy with a minimally invasive nature.

Acknowledgments This research was supported by Basic Science Research Program through the National Research Foundation of Korea (NRF) funded by the Ministry of Education (No. 2018R1A6A1A03025582). This research was also supported by the Basic Science Research Program through the National Research Foundation of Korea (NRF) funded by the Ministry of Education (NRF-2018R1D1A1B07043439).

References

Abadeer NS, Murphy CJ (2016) Recent progress in cancer thermal therapy using gold nanoparticles. J Phys Chem C 120:4691–4716. https://doi.org/10.1021/acs.jpcc.5b11232

Agostinis P, Berg K, Cengel KA, Foster TH, Girotti AW, Gollnick SO, Hahn SM, Hamblin MR, Juzeniene A, Kessel D, Korbelik M, Moan J, Mroz P, Nowis D, Piette J, Wilson BC, Golab J (2011) Photodynamic therapy of cancer: an update. CA Cancer J Clin. https://doi.org/10.3322/caac.20114

Alfranca G, Artiga Á, Stepien G, Moros M, Mitchell SG, de la Fuente JM (2016) Gold nanoprism-nanorod face off: comparing the heating efficiency, cellular internalization and thermoablation capacity. Nanomedicine (London) 11:2903–2916. https://doi.org/10.2217/nnm-2016-0257

Alle M, Reddy GB, Kim TH, Park SH, Lee S-H, Kim J-C (2020) Doxorubicin-carboxymethyl xanthan gum capped gold nanoparticles: microwave synthesis, characterization, and anti-cancer activity. Carbohydr Polym 229:115511. https://doi.org/10.1016/j.carbpol.2019.115511

Alle M, Park SC, Bandi R, Lee S-H, Kim J-C (2021) Rapid in-situ growth of gold nanoparticles on cationic cellulose nanofibrils: recyclable nanozyme for the colorimetric glucose detection. Carbohydr Polym 253:117239. https://doi.org/10.1016/j.carbpol.2020.117239

Allison RR, Sibata CH (2010) Oncologic photodynamic therapy photosensitizers: a clinical review. Photodiagn Photodyn Ther 7:61–75. https://doi.org/10.1016/j.pdpdt.2010.02.001

Amina SJ, Guo B (2020) A review on the synthesis and functionalization of gold nanoparticles as a drug delivery vehicle. Int J Nanomedicine 15:9823–9857. https://doi.org/10.2147/IJN.S279094

Asharani PV, Lianwu Y, Gong Z, Valiyaveettil S (2011) Comparison of the toxicity of silver, gold and platinum nanoparticles in developing zebrafish embryos. Nanotoxicology 5:43–54. https://doi.org/10.3109/17435390.2010.489207

Aşık E, Akpınar Y, Güray NT, İşcan M, Demircigil GÇ, Volkan M (2016) Cellular uptake, genotoxicity and cytotoxicity of cobalt ferrite magnetic nanoparticles in human breast cells. Toxicol Res (Camb) 5:1649–1662. https://doi.org/10.1039/C6TX00211K

Azhdarzadeh M, Atyabi F, Saei AA, Varnamkhasti BS, Omidi Y, Fateh M, Ghavami M, Shanehsazzadeh S, Dinarvand R (2016) Theranostic MUC-1 aptamer targeted gold coated superparamagnetic iron oxide nanoparticles for magnetic resonance imaging and photothermal therapy of colon cancer. Colloids Surf B: Biointerfaces 143:224–232. https://doi.org/10.1016/j.colsurfb.2016.02.058

Bachheti RK, Abate L, Deepti, Bachheti A, Madhusudhan A, Husen H (2021) Algae-, fungi-, and yeast-mediated biological synthesis of nanoparticles and their various biomedical applications. In: Kharisov B, Kharissova O (eds) Handbook of greener synthesis of nanomaterials and compounds, volume 1: fundamental principles and methods. Elsevier Inc, Cambridge, MA, pp 701–734. https://doi.org/10.1016/B978-0-12-821938-6.00022-0

Baeza A, Manzano M, Colilla M, Vallet-Regí M (2016) Recent advances in mesoporous silica nanoparticles for antitumor therapy: our contribution. Biomater Sci 4:803–813. https://doi.org/10.1039/C6BM00039H

Bai X, Wang Y, Song Z, Feng Y, Chen Y, Zhang D, Feng L (2020) The basic properties of gold nanoparticles and their applications in tumor diagnosis and treatment. Int J Mol Sci 21:2480. https://doi.org/10.3390/ijms21072480

Black KCL, Yi J, Rivera JG, Zelasko-Leon DC, Messersmith PB (2013) Polydopamine-enabled surface functionalization of gold nanorods for cancer cell-targeted imaging and photothermal therapy. Nanomedicine 8:17–28. https://doi.org/10.2217/nnm.12.82

Bosset J-F, Collette L, Calais G, Mineur L, Maingon P, Radosevic-Jelic L, Daban A, Bardet E, Beny A, Ollier J-C (2006) Chemotherapy with preoperative radiotherapy in rectal cancer. N Engl J Med. https://doi.org/10.1056/nejmoa060829

Butterworth KT, McMahon SJ, Currell FJ, Prise KM (2012) Physical basis and biological mechanisms of gold nanoparticle radiosensitization. Nanoscale 4:4830–4838. https://doi.org/10.1039/c2nr31227a

Castano AP, Demidova TN, Hamblin MR (2004) Mechanisms in photodynamic therapy: part one – photosensitizers, photochemistry and cellular localization. Photodiagn Photodyn Ther 1:279–293. https://doi.org/10.1016/S1572-1000(05)00007-4

Chen H, Zhang X, Dai S, Ma Y, Cui S, Achilefus S, Gu Y (2013) Multifunctional gold nanostar conjugates for tumor imaging and combined photothermal and chemo-therapy. Theranostics 3:633–649. https://doi.org/10.7150/thno.6630

Cheng Y, Samia AC, Li J, Kenney ME, Resnick A, Burda C (2010) Delivery and efficacy of a cancer drug as a function of the bond to the gold nanoparticle surface. Langmuir 26:2248–2255. https://doi.org/10.1021/la902390d

Cheng Y, Meyers JD, Broome A-M, Kenney ME, Basilion JP, Burda C (2011) Deep penetration of a PDT drug into tumors by noncovalent drug-gold nanoparticle conjugates. J Am Chem Soc 133:2583–2591. https://doi.org/10.1021/ja108846h

Choi W II, Kim J-Y, Kang C, Byeon CC, Kim YH, Tae G (2011) Tumor regression in vivo by photothermal therapy based on gold-nanorod-loaded, functional nanocarriers. ACS Nano 5:1995–2003. https://doi.org/10.1021/nn103047r

Chu M, Wu (2011) A gold nanoshell with a silica inner shell synthesized using liposome templates for doxorubicin loading and near-infrared photothermal therapy. Int J Nanomedicine 6:807. https://doi.org/10.2147/IJN.S16701

Comotti M, Della Pina C, Matarrese R, Rossi M (2004) The catalytic activity of? Naked? Gold particles. Angew Chem Int Ed 43:5812–5815. https://doi.org/10.1002/anie.200460446

D'Acunto M, Cioni P, Gabellieri E, Presciuttini G (2021) Exploiting gold nanoparticles for diagnosis and cancer treatments. Nanotechnology 32:192001. https://doi.org/10.1088/1361-6528/abe1ed

Dang X, Bardhan NM, Qi J, Gu L, Eze NA, Lin CW, Kataria S, Hammond PT, Belcher AM (2019) Deep-tissue optical imaging of near cellular-sized features. Sci Rep. https://doi.org/10.1038/s41598-019-39502-w

Dhanalekshmi KI, Sangeetha K, Magesan P, Johnson J, Zhang X, Jayamoorthy K (2020) Photodynamic cancer therapy: role of Ag- and Au-based hybrid nano-photosensitizers. J Biomol Struct Dyn:1–8. https://doi.org/10.1080/07391102.2020.1858965

Dickerson EB, Dreaden EC, Huang X, El-Sayed IH, Chu H, Pushpanketh S, McDonald JF, El-Sayed MA (2008) Gold nanorod assisted near-infrared plasmonic photothermal therapy (PPTT) of squamous cell carcinoma in mice. Cancer Lett 269:57–66. https://doi.org/10.1016/j.canlet.2008.04.026

Ding H, Yu H, Dong Y, Tian R, Huang G, Boothman DA, Sumer BD, Gao J (2011) Photoactivation switch from type II to type I reactions by electron-rich micelles for improved photodynamic therapy of cancer cells under hypoxia. J Control Release 156:276–280. https://doi.org/10.1016/j.jconrel.2011.08.019

Dolmans DEJGJ, Fukumura D, Jain RK (2003) Photodynamic therapy for cancer. Nat Rev Cancer 3:380–387. https://doi.org/10.1038/nrc1071

Dorsey JF, Sun L, Joh DY, Witztum A, Zaki A Al, Kao GD, Alonso-Basanta M, Avery S, Tsourkas A, Hahn SM (2013) Gold nanoparticles in radiation research: potential applications for imag-

ing and radiosensitization. Transl Cancer Res 2:280–291. https://doi.org/10.3978/j.issn.2218-676X.2013.08.09
Du B, Ma C, Ding G, Han X, Li D, Wang E, Wang J (2017) Cooperative strategies for enhancing performance of Photothermal Therapy (PTT) agent: optimizing its photothermal conversion and cell internalization ability. Small 13:1603275. https://doi.org/10.1002/smll.201603275
Espinosa A, Silva AKA, Sánchez-Iglesias A, Grzelczak M, Péchoux C, Desboeufs K, Liz-Marzán LM, Wilhelm C (2016) Cancer cell internalization of gold nanostars impacts their photothermal efficiency in vitro and in vivo: toward a plasmonic thermal fingerprint in tumoral environment. Adv Healthc Mater 5:1040–1048. https://doi.org/10.1002/adhm.201501035
Fan J, Yin J-J, Ning B, Wu X, Hu Y, Ferrari M, Anderson GJ, Wei J, Zhao Y, Nie G (2011) Direct evidence for catalase and peroxidase activities of ferritin-platinum nanoparticles. Biomaterials 32:1611–1618. https://doi.org/10.1016/j.biomaterials.2010.11.004
Fernández-Cabada T, de Pablo CS-L, Pisarchyk L, Serrano-Olmedo JJ, Ramos-Gómez M (2016) Optical hyperthermia using anti-epidermal growth factor receptor-conjugated gold nanorods to induce cell death in glioblastoma cell lines. J Nanosci Nanotechnol 16:7689–7695. https://doi.org/10.1166/jnn.2016.12570
Fratoddi I, Venditti I, Cametti C, Russo MV (2015) How toxic are gold nanoparticles? The state-of-the-art. Nano Res. https://doi.org/10.1007/s12274-014-0697-3
Fu L-H, Qi C, Lin J, Huang P (2018) Catalytic chemistry of glucose oxidase in cancer diagnosis and treatment. Chem Soc Rev 47:6454–6472. https://doi.org/10.1039/c7cs00891k
Gamaleia N, Shishko E, Dolinsky GA, Shcherbakov AB, Usatenko AV, Kholin VV (2010) U photodynamic activity of hematoporphyrin conjugates with gold nanoparticles: experiments in vitro. Exp Oncol 32(1):44–47
Gao L, Fei J, Zhao J, Li H, Cui Y, Li J (2012) Hypocrellin-loaded gold nanocages with high two-photon efficiency for photothermal/photodynamic cancer therapy in vitro. ACS Nano 6:8030–8040. https://doi.org/10.1021/nn302634m
Gao S, Lin H, Zhang H, Yao H, Chen Y, Shi J (2019a) Nanocatalytic tumor therapy by biomimetic dual inorganic nanozyme-catalyzed cascade reaction. Adv Sci 6:1801733. https://doi.org/10.1002/advs.201801733
García Calavia P, Bruce G, Pérez-García L, Russell DA (2018) Photosensitiser-gold nanoparticle conjugates for photodynamic therapy of cancer. Photochem Photobiol Sci 17(11)
Geng F, Song K, Xing JZ, Yuan C, Yan S, Yang Q, Chen J, Kong B (2011) Thio-glucose bound gold nanoparticles enhance radio-cytotoxic targeting of ovarian cancer. Nanotechnology 22:285101. https://doi.org/10.1088/0957-4484/22/28/285101
Geng F, Xing JZ, Chen J, Yang R, Hao Y, Song K, Kong B (2014) Pegylated glucose gold nanoparticles for improved in-vivo bio-distribution and enhanced radiotherapy on cervical cancer. J Biomed Nanotechnol 10:1205–1216. https://doi.org/10.1166/jbn.2014.1855
Hainfeld JF, Slatkin DN, Smilowitz HM (2004) The use of gold nanoparticles to enhance radiotherapy in mice. Phys Med Biol 49:N309–N315. https://doi.org/10.1088/0031-9155/49/18/n03
Heidari Z, Salouti M, Sariri R (2015) Breast cancer photothermal therapy based on gold nanorods targeted by covalently-coupled bombesin peptide. Nanotechnology 26:195101. https://doi.org/10.1088/0957-4484/26/19/195101
Her S, Jaffray DA, Allen C (2017) Gold nanoparticles for applications in cancer radiotherapy: mechanisms and recent advancements. Adv Drug Deliv Rev 109:84–101. https://doi.org/10.1016/j.addr.2015.12.012
Hirsch LR, Stafford RJ, Bankson JA, Sershen SR, Rivera B, Price RE, Hazle JD, Halas NJ, West JL (2003) Nanoshell-mediated near-infrared thermal therapy of tumors under magnetic resonance guidance. Proc Natl Acad Sci 100:13549–13554. https://doi.org/10.1073/pnas.2232479100
Hone DC, Walker PI, Evans-Gowing R, FitzGerald S, Beeby A, Chambrier I, Cook MJ, Russell DA (2002) Generation of cytotoxic singlet oxygen via phthalocyanine-stabilized gold nanoparticles: a potential delivery vehicle for photodynamic therapy. Langmuir 18:2985–2987. https://doi.org/10.1021/la0256230

Hu M, Chen J, Li Z-Y, Au L, Hartland GV, Li X, Marquez M, Xia Y (2006) Gold nanostructures: engineering their plasmonic properties for biomedical applications. Chem Soc Rev 35:1084–1094. https://doi.org/10.1039/b517615h

Huang X, El-Sayed IH, Qian W, El-Sayed MA (2006) Cancer cell imaging and photothermal therapy in the near-infrared region by using gold nanorods. J Am Chem Soc 128:2115–2120. https://doi.org/10.1021/ja057254a

Huang X, Jain PK, El-Sayed IH, El-Sayed MA (2008) Plasmonic photothermal therapy (PPTT) using gold nanoparticles. Lasers Med Sci 23:217–228. https://doi.org/10.1007/s10103-007-0470-x

Huang H-C, Rege K, Heys JJ (2010a) Spatiotemporal temperature distribution and cancer cell death in response to extracellular hyperthermia induced by gold nanorods. ACS Nano 4:2892–2900. https://doi.org/10.1021/nn901884d

Huang X, El-Sayed IH, El-Sayed MA (2010b) Applications of gold nanorods for cancer imaging and photothermal therapy. Methods Mol Biol. https://doi.org/10.1007/978-1-60761-609-2_23

Huang X, Kang B, Qian W, Mackey MA, Chen PC, Oyelere AK, El-Sayed IH, El-Sayed MA (2010c) Comparative study of photothermolysis of cancer cells with nuclear-targeted or cytoplasm-targeted gold nanospheres: continuous wave or pulsed lasers. J Biomed Opt. https://doi.org/10.1117/1.3486538

Husen A (2017) Gold nanoparticles from plant system: synthesis, characterization and their application. In: Ghorbanpourn M, Manika K, Varma A (eds) Nanoscience and plant–soil systems, vol 48. Springer, Cham, pp 455–479. https://doi.org/10.1007/978-3-319-46835-8_17

Husen A (2019) Medicinal plant-product based fabrication nanoparticles (Au and Ag) and their anticancer effect. In: Kintzios SE, Barberaki M, Flampouri (eds) Plants that fight cancer – second edition. Taylor & Francis/CRC Press, pp 133–147

Husen A (2020) Introduction and techniques in nanomaterials formulation. In: Husen A, Jawaid M (eds) Nanomaterials for agriculture and forestry applications. Elsevier Inc, Cambridge, MA, pp 1–14. https://doi.org/10.1016/B978-0-12-817852-2.00001-9

Husen A, Iqbal M (2019) N*anomaterials* and plant potential: an overview. In: Husen A, Iqbal M (eds) N*anomaterials* and plant potential. Springer, Cham, pp 3–29. https://doi.org/10.1007/978-3-030-05569-1_1

Husen A, Siddiqi KS (2014) Phytosynthesis of nanoparticles: concept, controversy and application. Nanoscale Res Lett 9:229. https://doi.org/10.1186/1556-276X-9-229

Hwang S, Nam J, Jung S, Song J, Doh H, Kim S (2014) Gold nanoparticle-mediated photothermal therapy: current status and future perspective. Nanomedicine (London) 9:2003–2022. https://doi.org/10.2217/nnm.14.147

Jain PK, Lee KS, El-Sayed IH, El-Sayed MA (2006) Calculated absorption and scattering properties of gold nanoparticles of different size, shape, and composition: applications in biological imaging and biomedicine. J Phys Chem B 110:7238–7248. https://doi.org/10.1021/jp057170o

Jang B, Park J-Y, Tung C-H, Kim I-H, Choi Y (2011) Gold nanorod-photosensitizer complex for near-infrared fluorescence imaging and photodynamic/photothermal therapy in vivo. ACS Nano 5:1086–1094. https://doi.org/10.1021/nn102722z

Jazayeri MH, Amani H, Pourfatollah AA, Pazoki-Toroudi H, Sedighimoghaddam B (2016) Various methods of gold nanoparticles (GNPs) conjugation to antibodies. Sens Bio-Sensing Res 9:17–22. https://doi.org/10.1016/j.sbsr.2016.04.002

Jorfi M, Foster EJ (2015) Recent advances in nanocellulose for biomedical applications. J Appl Polym Sci 132:1–19. https://doi.org/10.1002/app.41719

Kah JCY, Wan RCY, Wong KY, Mhaisalkar S, Sheppard CJR, Olivo M (2008) Combinatorial treatment of photothermal therapy using gold nanoshells with conventional photodynamic therapy to improve treatment efficacy: an in vitro study. Lasers Surg Med 40:584–589. https://doi.org/10.1002/lsm.20674

Ke H, Wang J, Tong S, Jin Y, Wang S, Qu E, Bao G, Dai Z (2014) Gold nanoshelled liquid perfluorocarbon magnetic nanocapsules: a nanotheranostic platform for bimodal ultrasound/magnetic resonance imaging guided photothermal tumor ablation. Theranostics 4:12–23. https://doi.org/10.7150/thno.7275

Khaing Oo MK, Yang X, Du H, Wang H (2008) 5-aminolevulinic acid-conjugated gold nanoparticles for photodynamic therapy of cancer. Nanomedicine 3:777–786. https://doi.org/10.2217/17435889.3.6.777

Khaing Oo MK, Yang Y, Hu Y, Gomez M, Du H, Wang H (2012) Gold nanoparticle-enhanced and size-dependent generation of reactive oxygen species from protoporphyrin IX. ACS Nano 6:1939–1947. https://doi.org/10.1021/nn300327c

Khlebtsov B, Panfilova E, Khanadeev V, Bibikova O, Terentyuk G, Ivanov A, Rumyantseva V, Shilov I, Ryabova A, Loshchenov V, Khlebtsov NG (2011) Nanocomposites containing silica-coated gold-silver nanocages and Yb-2, 4-dimethoxyhematoporphyrin: multifunctional capability of IR-luminescence detection, photosensitization, and photothermolysis. ACS Nano 5:7077–7089. https://doi.org/10.1021/nn2017974

Khoshgard K, Hashemi B, Arbabi A, Rasaee MJ, Soleimani M (2014) Radiosensitization effect of folate-conjugated gold nanoparticles on HeLa cancer cells under orthovoltage superficial radiotherapy techniques. Phys Med Biol 59:2249–2263. https://doi.org/10.1088/0031-9155/59/9/2249

Kim Y, Tomé WA (2006) Risk-adaptive optimization: selective boosting of high-risk tumor subvolumes. Int J Radiat Oncol Biol Phys. https://doi.org/10.1016/j.ijrobp.2006.08.032

Kim TH, Alle M, Park SC, Zhao F, Long W, Samala S, Kim J-C (2021a) Self-assembly prepared using an ion pair of poly(ethylene imine) and (phenylthio) acetic acid as a drug carrier for oxidation, temperature, and NIR-responsive release. Chem Eng J 415:128954. https://doi.org/10.1016/j.cej.2021.128954

Kobayashi K, Usami N, Porcel E, Lacombe S, Le Sech C (2010) Enhancement of radiation effect by heavy elements. Mutat Res 704:123–131. https://doi.org/10.1016/j.mrrev.2010.01.002

Kodiha M, Hutter E, Boridy S, Juhas M, Maysinger D, Stochaj U (2014) Gold nanoparticles induce nuclear damage in breast cancer cells, which is further amplified by hyperthermia. Cell Mol Life Sci 71:4259–4273. https://doi.org/10.1007/s00018-014-1622-3

Kodiha M, Wang YM, Hutter E, Maysinger D, Stochaj U (2015) Off to the organelles – killing cancer cells with targeted gold nanoparticles. Theranostics 5:357–370. https://doi.org/10.7150/thno.10657

Konan YN, Gurny R, Allémann E (2002) State of the art in the delivery of photosensitizers for photodynamic therapy. J Photochem Photobiol B Biol 66:89–106. https://doi.org/10.1016/S1011-1344(01)00267-6

Kong T, Zeng J, Wang X, Yang X, Yang J, McQuarrie S, McEwan A, Roa W, Chen J, Xing JZ (2008) Enhancement of radiation cytotoxicity in breast-cancer cells by localized attachment of gold nanoparticles. Small 4:1537–1543. https://doi.org/10.1002/smll.200700794

Kotov NA (2010) Chemistry. Inorganic nanoparticles as protein mimics. Science 330:188–189. https://doi.org/10.1126/science.1190094

Kuo W-S, Chang C-N, Chang Y-T, Yang M-H, Chien Y-H, Chen S-J, Yeh C-S (2010) Gold nanorods in photodynamic therapy, as hyperthermia agents, and in near-infrared optical imaging. Angew Chem Int Ed Eng 49:2711–2715. https://doi.org/10.1002/anie.200906927

Kuo W-S, Chang Y-T, Cho K-C, Chiu K-C, Lien C-H, Yeh C-S, Chen S-J (2012) Gold nanomaterials conjugated with indocyanine green for dual-modality photodynamic and photothermal therapy. Biomaterials 33:3270–3278. https://doi.org/10.1016/j.biomaterials.2012.01.035

Lee KS, El-Sayed MA (2005) Dependence of the enhanced optical scattering efficiency relative to that of absorption for gold metal nanorods on aspect ratio, size, end-cap shape, and medium refractive index. J Phys Chem B. https://doi.org/10.1021/jp054385p

Lee N, Hyeon T (2012) Designed synthesis of uniformly sized iron oxide nanoparticles for efficient magnetic resonance imaging contrast agents. Chem Soc Rev. https://doi.org/10.1039/c1cs15248c

Lee H, Dam DHM, Ha JW, Yue J, Odom TW (2015) Enhanced human epidermal growth factor receptor 2 degradation in breast cancer cells by lysosome-targeting gold nanoconstructs. ACS Nano 9:9859–9867. https://doi.org/10.1021/acsnano.5b05138

Li W, Chen X (2015) Gold nanoparticles for photoacoustic imaging. Nanomedicine 10:299–320. https://doi.org/10.2217/nnm.14.169

Li L, Nurunnabi M, Nafiujjaman M, Lee YK, Huh KM (2013) GSH-mediated photoactivity of pheophorbide a-conjugated heparin/gold nanoparticle for photodynamic therapy. J Control Release. https://doi.org/10.1016/j.jconrel.2013.07.002

Lin J, Wang S, Huang P, Wang Z, Chen S, Niu G, Li W, He J, Cui D, Lu G, Chen X, Nie Z (2013) Photosensitizer-loaded gold vesicles with strong plasmonic coupling effect for imaging-guided photothermal/photodynamic therapy. ACS Nano 7:5320–5329. https://doi.org/10.1021/nn4011686

Lin Y, Ren J, Qu X (2014) Catalytically active nanomaterials: a promising candidate for artificial enzymes. Acc Chem Res 47:1097–1105. https://doi.org/10.1021/ar400250z

Lin L-S, Song J, Song L, Ke K, Liu Y, Zhou Z, Shen Z, Li J, Yang Z, Tang W, Niu G, Yang H-H, Chen X (2018) Simultaneous Fenton-like ion delivery and glutathione depletion by MnO 2-based nanoagent to enhance chemodynamic therapy. Angew Chem 130:4996–5000. https://doi.org/10.1002/ange.201712027

Liu S-Y, Liang Z-S, Gao F, Luo S-F, Lu G-Q (2010) In vitro photothermal study of gold nanoshells functionalized with small targeting peptides to liver cancer cells. J Mater Sci Mater Med 21:665–674. https://doi.org/10.1007/s10856-009-3895-x

Liu B, Sun Z, Huang P-JJ, Liu J (2015a) Hydrogen peroxide displacing DNA from nanoceria: mechanism and detection of glucose in serum. J Am Chem Soc 137:1290–1295. https://doi.org/10.1021/ja511444e

Liu J, Liang Y, Liu T, Li D, Yang X (2015b) Anti-EGFR-conjugated hollow gold nanospheres enhance radiocytotoxic targeting of cervical cancer at megavoltage radiation energies. Nanoscale Res Lett 10:218. https://doi.org/10.1186/s11671-015-0923-2

Liu H, Li Z, Sun Y, Geng X, Hu Y, Meng H, Ge J, Qu L (2018a) Synthesis of luminescent carbon dots with ultrahigh quantum yield and inherent folate receptor-positive cancer cell targetability. Sci Rep. https://doi.org/10.1038/s41598-018-19373-3

Liu Y, Crawford BM, Vo-Dinh T (2018b) Gold nanoparticles-mediated photothermal therapy and immunotherapy. Immunotherapy 10:1175À1188

Loo C, Lowery A, Halas N, West J, Drezek R (2005) Immunotargeted nanoshells for integrated cancer imaging and therapy. Nano Lett 5:709–711. https://doi.org/10.1021/nl050127s

López-Lázaro M (2007) Dual role of hydrogen peroxide in cancer: possible relevance to cancer chemoprevention and therapy. Cancer Lett 252:1–8. https://doi.org/10.1016/j.canlet.2006.10.029

Lucky SS, Soo KC, Zhang Y (2015) Nanoparticles in photodynamic therapy. Chem Rev 115:1990–2042. https://doi.org/10.1021/cr5004198

Ma Y, Liang X, Tong S, Bao G, Ren Q, Dai Z (2013) Gold nanoshell nanomicelles for potential magnetic resonance imaging, light-triggered drug release, and photothermal therapy. Adv Funct Mater 23:815–822. https://doi.org/10.1002/adfm.201201663

Ma N, Jiang Y-W, Zhang X, Wu H, Myers JN, Liu P, Jin H, Gu N, He N, Wu F-G, Chen Z (2016) Enhanced radiosensitization of gold nanospikes via hyperthermia in combined cancer radiation and photothermal therapy. ACS Appl Mater Interfaces 8:28480–28494. https://doi.org/10.1021/acsami.6b10132

Macdonald IJ, Dougherty TJ (2001) Basic principles of photodynamic therapy. J Porphyrins Phthalocyanines 05:105–129. https://doi.org/10.1002/jpp.328

Madhusudhan A, Reddy G, Venkatesham M, Veerabhadram G, Kumar D, Natarajan S, Yang M-Y, Hu A, Singh S (2014) Efficient pH dependent drug delivery to target cancer cells by gold nanoparticles capped with carboxymethyl chitosan. Int J Mol Sci 15:8216–8234. https://doi.org/10.3390/ijms15058216

Madhusudhan A, Reddy GB, Krishana IM (2019) Green synthesis of gold nanoparticles by using natural gums. In: Nanomaterials and plant potential. Springer, Cham, pp 111–134

Melancon MP, Lu W, Zhong M, Zhou M, Liang G, Elliott AM, Hazle JD, Myers JN, Li C, Jason Stafford R (2011) Targeted multifunctional gold-based nanoshells for magnetic resonance-guided laser ablation of head and neck cancer. Biomaterials 32:7600–7608. https://doi.org/10.1016/j.biomaterials.2011.06.039

Mendoza-Nava H, Ferro-Flores G, Ocampo-García B, Serment-Guerrero J, Santos-Cuevas C, Jiménez-Mancilla N, Luna-Gutiérrez M, Camacho-López MA (2013) Laser heating of gold nanospheres functionalized with octreotide: in vitro effect on HeLa cell viability. Photomed Laser Surg 31:17–22. https://doi.org/10.1089/pho.2012.3320

Mesbahi A (2010) A review on gold nanoparticles radiosensitization effect in radiation therapy of cancer. Reports Pract. Oncol. Radiother

Meyers JD, Cheng Y, Broome A-M, Agnes RS, Schluchter MD, Margevicius S, Wang X, Kenney ME, Burda C, Basilion JP (2015) Peptide-targeted gold nanoparticles for photodynamic therapy of brain cancer. Part Part Syst Charact 32:448–457. https://doi.org/10.1002/ppsc.201400119

Mieszawska AJ, Mulder WJM, Fayad ZA, Cormode DP (2013) Multifunctional gold nanoparticles for diagnosis and therapy of disease. Mol Pharm 10(3):831–847

Mkandawire MM, Lakatos M, Springer A, Clemens A, Appelhans D, Krause-Buchholz U, Pompe W, Rödel G, Mkandawire M (2015) Induction of apoptosis in human cancer cells by targeting mitochondria with gold nanoparticles. Nanoscale 7:10634–10640. https://doi.org/10.1039/c5nr01483b

Mocan L, Ilie I, Tabaran FA, Dana B, Zaharie F, Zdrehus C, Puia C, Mocan T, Muntean V, Teodora P, Ofelia M, Marcel T, Iancu C (2013) Surface plasmon resonance-induced photoactivation of gold nanoparticles as mitochondria-targeted therapeutic agents for pancreatic cancer. Expert Opin Ther Targets 17:1383–1393. https://doi.org/10.1517/14728222.2013.855200

Murphy CJ, Thompson LB, Alkilany AM, Sisco PN, Boulos SP, Sivapalan ST, Yang JA, Chernak DJ, Huang J (2010) The many faces of gold nanorods. J Phys Chem Lett 1:2867–2875. https://doi.org/10.1021/jz100992x

Nitzan Y, Salmon-Divon M, Shporen E, Malik Z (2004) ALA induced photodynamic effects on Gram positive and negative bacteria. Photochem Photobiol Sci 3:430. https://doi.org/10.1039/b315633h

Norouzi H, Khoshgard K, Akbarzadeh F (2018) In vitro outlook of gold nanoparticles in photothermal therapy: a literature review. Lasers Med Sci 33:917–926. https://doi.org/10.1007/s10103-018-2467-z

Obaid G, Chambrier I, Cook MJ, Russell DA (2015) Cancer targeting with biomolecules: a comparative study of photodynamic therapy efficacy using antibody or lectin conjugated phthalocyanine-PEG gold nanoparticles. Photochem Photobiol Sci 14:737–747. https://doi.org/10.1039/c4pp00312h

Park S-E, Lee J, Lee T, Bae S-B, Kang B, Huh Y-M, Lee S-W, Haam S (2015) Comparative hyperthermia effects of silica-gold nanoshells with different surface coverage of gold clusters on epithelial tumor cells. Int J Nanomedicine 10:261–271. https://doi.org/10.2147/IJN.S88309

Pasparakis G (2013) Light-induced generation of singlet oxygen by naked gold nanoparticles and its implications to cancer cell phototherapy. Small 9:4130–4134. https://doi.org/10.1002/smll.201301365

Pissuwan D, Valenzuela SM, Miller CM, Cortie MB (2007a) A golden bullet? Selective targeting of toxoplasma gondii Tachyzoites using antibody-functionalized gold nanorods. Nano Lett 7(12):3808–3812. https://doi.org/10.1021/NL072377+

Pissuwan D, Valenzuela SM, Killingsworth MC, Xu X, Cortie MB (2007b) Targeted destruction of murine macrophage cells with bioconjugated gold nanorods. J Nanopart Res 9:1109–1124. https://doi.org/10.1007/s11051-007-9212-z

Polf JC, Bronk LF, Driessen WHP, Arap W, Pasqualini R, Gillin M (2011) Enhanced relative biological effectiveness of proton radiotherapy in tumor cells with internalized gold nanoparticles. Appl Phys Lett 98:193702. https://doi.org/10.1063/1.3589914

Qin Z, Bischof JC (2012) Thermophysical and biological responses of gold nanoparticle laser heating. Chem Soc Rev 41:1191–1217. https://doi.org/10.1039/c1cs15184c

Rau L-R, Huang W-Y, Liaw J-W, Tsai S-W (2016) Photothermal effects of laser-activated surface plasmonic gold nanoparticles on the apoptosis and osteogenesis of osteoblast-like cells. Int J Nanomedicine 11:3461–3473. https://doi.org/10.2147/IJN.S108152

Riley RS, Day ES (2017) Gold nanoparticle-mediated photothermal therapy: applications and opportunities for multimodal cancer treatment. Wiley Interdiscip. Rev. Nanomedicine Nanobiotechnology

Robins HI, Dennis WH, Neville AJ, Shecterle LM, Martin PA, Grossman J, Davis TE, Neville SR, Gillis WK, Rusy BF (1985) A nontoxic system for 41.8°C whole-body hyperthermia: results of a phase I study using a radiant heat device. Cancer Res 45:3937–3944

Safaee SMR, Janipour M, Karami MA (2015) Modeling and analysis of optical properties of a gold nanoring based on electric and magnetic dipoles. Appl Opt 54:8313. https://doi.org/10.1364/AO.54.008313

Savarimuthu WP, Gananathan P, Rao AP, Manickam E, Singaravelu G (2015) Protoporphyrin IX-gold nanoparticle conjugates for targeted photodynamic therapy – an in-vitro study. J Nanosci Nanotechnol 15:5577–5584. https://doi.org/10.1166/jnn.2015.10302

Schwartzberg AM, Grant CD, Wolcott A, Talley CE, Huser TR, Bogomolni R, Zhang JZ (2004) Unique gold nanoparticle aggregates as a highly active surface-enhanced Raman scattering substrate. J Phys Chem B 108:19191–19197. https://doi.org/10.1021/jp048430p

Sheikholeslami F, Fekrazad R, Rasaee AS, Hakimiha F, Kalhori (2011) Treatment of oral squamous cell carcinoma using anti-HER2 immunonanoshells. Int J Nanomedicine 6:2749. https://doi.org/10.2147/ijn.s24548

Siddiqi KS, Husen A (2017) Recent advances in plant-mediated engineered gold nanoparticles and their application in biological system. J Trace Elem Med Biol 40:10–23. https://doi.org/10.1016/j.jtemb.2016.11.012

Siddiqi KS, Rahman A, Tajuddin HA (2016) Biogenic fabrication of iron/iron oxide nanoparticles and their application. Nanoscale Res Lett 11:498. https://doi.org/10.1186/s11671-016-1714-0

Siddiqi KS, Husen A, Sohrab SS, Yassin MO (2018) Recent status of nanomaterial fabrication and their potential applications in neurological disease management. Nanoscale Res Lett 13:231. https://doi.org/10.1186/s11671-018-2638-7

Song Y, Qu K, Zhao C, Ren J, Qu X (2010) Graphene oxide: intrinsic peroxidase catalytic activity and its application to glucose detection. Adv Mater 22:2206–2210. https://doi.org/10.1002/adma.200903783

Sperling RA, Gil PR, Zhang F, Zanella M, Parak WJ (2008) Biological applications of gold nanoparticles. Chem Soc Rev. https://doi.org/10.1039/b712170a

Stabile J, Najafali D, Cheema Y, Inglut CT, Liang BJ, Vaja SJ, Sorrin A, Huang H-C (2020) Engineering gold nanoparticles for photothermal therapy, surgery, and imaging. In: Nanoparticles for biomedical applications. Elsevier, pp 175–193

Stern ST, Adiseshaiah PP, Crist RM (2012) Autophagy and lysosomal dysfunction as emerging mechanisms of nanomaterial toxicity. Part Fibre Toxicol 9:20. https://doi.org/10.1186/1743-8977-9-20

Stuchinskaya T, Moreno M, Cook MJ, Edwards DR, Russell DA (2011) Targeted photodynamic therapy of breast cancer cells using antibody–phthalocyanine–gold nanoparticle conjugates. Photochem Photobiol Sci 10:822. https://doi.org/10.1039/c1pp05014a

Sun X, Zhang G, Keynton RS, O'Toole MG, Patel D, Gobin AM (2013) Enhanced drug delivery via hyperthermal membrane disruption using targeted gold nanoparticles with PEGylated Protein-G as a cofactor. Nanomedicine 9:1214–1222. https://doi.org/10.1016/j.nano.2013.04.002

Terentyuk G, Panfilova E, Khanadeev V, Chumakov D, Genina E, Bashkatov A, Tuchin V, Bucharskaya A, Maslyakova G, Khlebtsov N, Khlebtsov B (2014) Gold nanorods with a hematoporphyrin-loaded silica shell for dual-modality photodynamic and photothermal treatment of tumors in vivo. Nano Res 7:325–337. https://doi.org/10.1007/s12274-013-0398-3

Van de Broek B, Devoogdt N, D'Hollander A, Gijs H-L, Jans K, Lagae L, Muyldermans S, Maes G, Borghs G (2011) Specific cell targeting with nanobody conjugated branched gold nanoparticles for photothermal therapy. ACS Nano 5:4319–4328. https://doi.org/10.1021/nn1023363

Vankayala R, Sagadevan A, Vijayaraghavan P, Kuo C-L, Hwang KC (2011) Metal nanoparticles sensitize the formation of singlet oxygen. Angew Chem Int Ed 50:10640–10644. https://doi.org/10.1002/anie.201105236

Vankayala R, Kuo C-L, Sagadevan A, Chen P-H, Chiang C-S, Hwang KC (2013) Morphology dependent photosensitization and formation of singlet oxygen (1Δg) by gold and silver nanoparticles and its application in cancer treatment. J Mater Chem B 1:4379. https://doi.org/10.1039/c3tb20806k

Vankayala R, Huang Y-K, Kalluru P, Chiang C-S, Hwang KC (2014a) First demonstration of gold nanorods-mediated photodynamic therapeutic destruction of tumors via near infra-red light activation. Small 10:1612–1622. https://doi.org/10.1002/smll.201302719

Vankayala R, Lin C-C, Kalluru P, Chiang C-S, Hwang KC (2014b) Gold nanoshells-mediated bimodal photodynamic and photothermal cancer treatment using ultra-low doses of near infra-red light. Biomaterials 35:5527–5538. https://doi.org/10.1016/j.biomaterials.2014.03.065

Verma A, Stellacci F (2010) Effect of surface properties on nanoparticle-cell interactions. Small 6:12–21. https://doi.org/10.1002/smll.200901158

Vijayaraghavan P, Liu C-H, Vankayala R, Chiang C-S, Hwang KC (2014) Designing multi-branched gold nanoechinus for NIR light activated dual modal photodynamic and photothermal therapy in the second biological window. Adv Mater 26:6689–6695. https://doi.org/10.1002/adma.201400703

von Maltzahn G, Park J-H, Agrawal A, Bandaru NK, Das SK, Sailor MJ, Bhatia SN (2009) Computationally guided photothermal tumor therapy using long-circulating gold nanorod antennas. Cancer Res 69:3892–3900. https://doi.org/10.1158/0008-5472.CAN-08-4242

Wang S, Lu G (2018) Applications of gold nanoparticles in cancer imaging and treatment. In: Noble and precious metals – properties, nanoscale effects and applications. InTech

Wang M, Thanou M (2010) Targeting nanoparticles to cancer. Pharmacol Res 62:90–99. https://doi.org/10.1016/j.phrs.2010.03.005

Wang L, Liu Y, Li W, Jiang X, Ji Y, Wu X, Xu L, Qiu Y, Zhao K, Wei T, Li Y, Zhao Y, Chen C (2011) Selective targeting of gold nanorods at the mitochondria of cancer cells: implications for cancer therapy. Nano Lett 11:772–780. https://doi.org/10.1021/nl103992v

Wang S, Huang P, Nie L, Xing R, Liu D, Wang Z, Lin J, Chen S, Niu G, Lu G, Chen X (2013) Single continuous wave laser induced photodynamic/plasmonic photothermal therapy using photosensitizer-functionalized gold nanostars. Adv Mater 25:3055–3061. https://doi.org/10.1002/adma.201204623

Wang N, Zhao Z, Lv Y, Fan H, Bai H, Meng H, Long Y, Fu T, Zhang X, Tan W (2014a) Gold nanorod-photosensitizer conjugate with extracellular pH-driven tumor targeting ability for photothermal/photodynamic therapy. Nano Res 7:1291–1301. https://doi.org/10.1007/s12274-014-0493-0

Wang Y-XJ, Zhu X-M, Liang Q, Cheng CHK, Wang W, Leung KC-F (2014b) In vivo chemoembolization and magnetic resonance imaging of liver tumors by using iron oxide nanoshell/doxorubicin/poly(vinyl alcohol) hybrid composites. Angew Chem 126:4912–4915. https://doi.org/10.1002/ange.201402144

Wang H, Zhao R, Li Y, Liu H, Li F, Zhao Y, Nie G (2016) Aspect ratios of gold nanoshell capsules mediated melanoma ablation by synergistic photothermal therapy and chemotherapy. Nanomedicine 12:439–448. https://doi.org/10.1016/j.nano.2015.11.013

Wang C, Yang J, Dong C, Shi S (2020) Glucose oxidase-related cancer therapies. Adv Ther 3:2000110. https://doi.org/10.1002/adtp.202000110

Wen H, Tamarov K, Happonen E, Lehto V, Xu W (2020) Inorganic nanomaterials for photothermal-based cancer theranostics. Adv Ther 4(2):2000207. https://doi.org/10.1002/adtp.202000207`

Wieder ME, Hone DC, Cook MJ, Handsley MM, Gavrilovic J, Russell DA (2006) Intracellular photodynamic therapy with photosensitizer-nanoparticle conjugates: cancer therapy using a 'Trojan horse.'. Photochem Photobiol Sci 5:727–734. https://doi.org/10.1039/B602830F

Xu H, Liu C, Mei J, Yao C, Wang S, Wang J, Li Z, Zhang Z (2012) Effects of light irradiation upon photodynamic therapy based on 5-aminolevulinic acid-gold nanoparticle conjugates in K562 cells via singlet oxygen generation. Int J Nanomedicine 7:5029–5038. https://doi.org/10.2147/IJN.S33261

Xu C, Yang D, Mei L, Li Q, Zhu H, Wang T (2013) Targeting chemophotothermal therapy of hepatoma by gold nanorods/graphene oxide core/shell nanocomposites. ACS Appl Mater Interfaces 5:12911–12920. https://doi.org/10.1021/am404714w

Yamaguchi K, Yabushita S, Fueno T, Houk KN (1981) Mechanism of photooxygenation reactions. Computational evidence against the diradical mechanism of singlet oxygen ene reactions. J Am Chem Soc 103:5043–5046. https://doi.org/10.1021/ja00407a013

Yang C, Uertz J, Yohan D, Nanoscale BC (2014) undefined Peptide modified gold nanoparticles for improved cellular uptake, nuclear transport, and intracellular retention. pubs.rsc.org

Yang L, Tseng Y-T, Suo G, Chen L, Yu J, Chiu W-J, Huang C-C, Lin C-H (2015) Photothermal therapeutic response of cancer cells to aptamer-gold nanoparticle-hybridized graphene oxide under NIR illumination. ACS Appl Mater Interfaces 7:5097–5106. https://doi.org/10.1021/am508117e

Yang G, Phua SZF, Bindra AK, Zhao Y (2019) Degradability and clearance of inorganic nanoparticles for biomedical applications. Adv Mater 31(10):1805730

Yao C, Zhang L, Wang J, He Y, Xin J, Wang S, Xu H, Zhang Z (2016) Gold nanoparticle mediated phototherapy for cancer. J Nanomater 2016:1–29. https://doi.org/10.1155/2016/5497136

Yasun E, Kang H, Erdal H, Cansiz S, Ocsoy I, Huang Y-F, Tan W (2013) Cancer cell sensing and therapy using affinity tag-conjugated gold nanorods. Interface Focus 3:20130006. https://doi.org/10.1098/rsfs.2013.0006

Yeager D, Chen Y-S, Litovsky S, Emelianov S (2014) Intravascular photoacoustics for image-guidance and temperature monitoring during plasmonic photothermal therapy of atherosclerotic plaques: a feasibility study. Theranostics 4:36–46. https://doi.org/10.7150/thno.7143

You J, Zhang R, Xiong C, Zhong M, Melancon M, Gupta S, Nick AM, Sood AK, Li C (2012) Effective photothermal chemotherapy using doxorubicin-loaded gold nanospheres that target EphB4 receptors in tumors. Cancer Res 72:4777–4786. https://doi.org/10.1158/0008-5472.CAN-12-1003

Yuan H, Fales AM, Vo-Dinh T (2012a) TAT peptide-functionalized gold nanostars: enhanced intracellular delivery and efficient NIR photothermal therapy using ultralow irradiance. J Am Chem Soc 134:11358–11361. https://doi.org/10.1021/ja304180y

Yuan H, Khoury CG, Wilson CM, Grant GA, Bennett AJ, Vo-Dinh T (2012b) In vivo particle tracking and photothermal ablation using plasmon-resonant gold nanostars. Nanomedicine 8:1355–1363. https://doi.org/10.1016/j.nano.2012.02.005

Yuan F, Chen H, Xu J, Zhang Y, Wu Y, Wang L (2014) Aptamer-based luminescence energy transfer from near-infrared-to-near- infrared upconverting nanoparticles to gold nanorods and its application for the detection of thrombin. Chem – A Eur J. https://doi.org/10.1002/chem.201304556

Zhan T, Li P, Bi S, Dong B, Song H, Ren H, Wang L (2012) 12P-conjugated PEG-modified gold nanorods combined with near-infrared laser for tumor targeting and photothermal therapy. J Nanosci Nanotechnol 12:7198–7205. https://doi.org/10.1166/jnn.2012.6502

Zhang Z, Wang S, Xu H, Wang B, Yao C (2015) Role of 5-aminolevulinic acid-conjugated gold nanoparticles for photodynamic therapy of cancer. J Biomed Opt 20:51043. https://doi.org/10.1117/1.JBO.20.5.051043

Zhang A, Guo W, Qi Y, Wang J, Ma X, Yu D (2016a) Synergistic effects of gold nanocages in hyperthermia and radiotherapy treatment. Nanoscale Res Lett 11:279. https://doi.org/10.1186/s11671-016-1501-y

Zhang F, Zhu X, Gong J, Sun Y, Chen D, Wang J, Wang Y, Guo M, Li W (2016b) Lysosome-mitochondria-mediated apoptosis specifically evoked in cancer cells induced by gold nanorods. Nanomedicine (London) 11:1993–2006. https://doi.org/10.2217/nnm-2016-0139

Zhang H, Salo D, Kim DM, Komarov S, Tai Y-C, Berezin MY (2016c) Penetration depth of photons in biological tissues from hyperspectral imaging in shortwave infrared in transmission and reflection geometries. J Biomed Opt 21:126006. https://doi.org/10.1117/1.JBO.21.12.126006

Zhao T, Shen X, Li L, Guan Z, Gao N, Yuan P, Yao SQ, Xu Q-H, Xu GQ (2012) Gold nanorods as dual photo-sensitizing and imaging agents for two-photon photodynamic therapy. Nanoscale 4:7712–7719. https://doi.org/10.1039/c2nr32196c

Zhou Z, Song J, Nie L, Chen X (2016) Reactive oxygen species generating systems meeting challenges of photodynamic cancer therapy. Chem Soc Rev 45:6597–6626. https://doi.org/10.1039/c6cs00271d

Zhu L, Altman MB, Laszlo A, Straube W, Zoberi I, Hallahan DE, Chen H (2019) Ultrasound hyperthermia technology for radiosensitization. Ultrasound Med Biol 45:1025–1043. https://doi.org/10.1016/j.ultrasmedbio.2018.12.007

Further Reading

Gao S, Lin H, Zhang H, Yao H, Chen Y, Shi J (2019b) Nanocatalytic tumor therapy by biomimetic dual inorganic nanozyme-catalyzed cascade reaction. Adv Sci 6:1801733. https://doi.org/10.1002/advs.201801733

Kim TH, Alle M, Park SC, Zhao F, Long W, Samala S, Kim J-C (2021b) Self-assembly prepared using an ion pair of poly(ethylene imine) and (phenylthio) acetic acid as a drug carrier for oxidation, temperature, and NIR-responsive release. Chem Eng J 415:128954. https://doi.org/10.1016/j.cej.2021.128954

Chapter 2
Biomedical Applications of Gold Nanoparticles

Mallikarjun Vasam, Raja Abhilash Punagoti, and Rita Mourya

Abbreviations

Ab	Antibodies
AuNPs	Gold nanoparticles
EGFR	Epidermal growth factor receptor
EMS	Electron microscope
MTX	Methotrexate
NP	Nanoparticles
PEG	Polyethylene glycol

2.1 Introduction

Nanotechnology is a field of science used nowadays to increase bioavailability and therapeutic action for a longer period and site-specific action. Nanoparticles are those which are in the size range of 1–10 nm (Khan et al. 2014). Owing to their nanosize and large surface area, a plethora of applications in biomedical field can be found. A unique group of nanoparticles in recent years are applied for various fields. Recently, the advances in the synthesis of nanoparticles enabled the treatment of

M. Vasam (✉)
Chaitanya (Deemed to be University)-Pharmacy, Warangal, Telangana, India

R. A. Punagoti
Department of Pharmaceutical Chemistry, College of Medicine and Health Sciences, University of Gondar, Gondar, Ethiopia

R. Mourya
Department of Pharmaceutical Chemistry, School of Pharmacy, SAM Global University, Bhopal, India

J.-C. Kim et al. (eds.), *Smart Nanomaterials in Biomedical Applications*, Nanotechnology in the Life Sciences,
https://doi.org/10.1007/978-3-030-84262-8_2

cancer and other applications in the drug delivery field (Pal et al. 2013; Khan et al. 2014). Among the various kinds of nanomaterials, gold nanoparticles (AuNPs) are playing a promising role in various categories of drug delivery due to their unique physicochemical properties (Alle et al. 2021).

Gold is one of the first metals to be found, with an important application history that dates back at least a thousand years. The earliest evidence of colloidal gold as a therapeutic agent can be found in literature written by Chinese, Arabian and Indian scientists around fifth and fourth centuries BC. Colloidal gold was studied and used in alchemist laboratories in Europe throughout the Middle Ages. The therapeutic properties of gold quintessence—*quinta essentia auri*, which were obtained by reducing gold chloride with vegetable extracts in alcohols or oils—were described by Paracelsus and were reported to be useful in the treatment of mental illnesses and syphilis (Dykman and Khlebtsov 2011). Despite its long history, the "revolution in immunochemistry" associated with the use of AuNPs in biological studies did not begin until 1971. Faulk and Taylor extensively reviewed on gold nanoparticles and concluded where conjugation of antibody with colloidal gold can visualize through direct electron microscopy (Faulk and Taylor 1971).

One of the unique properties of AuNP is that it shows surface plasmon resonance (SPR) effect which occurs due to interaction between electromagnetic waves and condition of electrons in metals. Due to its low or negligible toxicity, the SPR peaks are useful in bio-diagnosis and bio-imaging field. Moreover, AuNPs can be used in chemotherapy and diagnosis of cancer due to scattering of nanoparticles in the tumour cells. Furthermore, AuNPs have multiple surfaces for attachment of antibodies, oligonucleotides, drug and protein enabling them to qualify as versatile drug carrier. These are synthesized using an inorganic, organic or combination of both, but inorganic platforms are mostly used for diagnosis and therapy due to their high drug loading capacity, easy modification and stability (Alle et al. 2020).

Extensive research has been conducted on AuNPs and concluded that they are effective in gene therapy, biosensing and photothermal therapy. Gold nanocages can be used as drug carrier with a temperature-sensitive polymer. Several lower and higher plants have been also used to produce AuNPs; and they have shown several applications (Chung et al. 2016 Husen and Siddiqi 2014; Siddiqi and Husen 2016a, b; Siddiqi et al. 2018a; Husen 2020; Husen and Iqbal 2019; Husen et al. 2019).

AuNPs can serve as drug carriers to exert desired therapeutic action on targeted cells and tissues due to their excellent binding properties, as it can hold biological and organic molecules with relatively low toxicity and high sensitivity (Siddiqi and Husen 2017). By modifying the upper surface of AuNPs, it can attach/combine with biomolecules or drugs. The attached drug release takes place by two techniques: external stimuli and internal stimuli operations using a biologically sustained/catalyst process. There were many findings that supported this phenomenon, for instance, in one of the recent studies. In the reduction of dilute gold chloride solution, phosphorus is used as catalyst, where they obtained solutions of AuNPs which are unstable and have a colour purple red to blue (Faraday et al. 1857). The scientist

Faraday showed a reduction of gold can be done by heating gold chloride and studied other metals like palladium, silver, lead, tin, iron and mercury and their sedimentation and aggregation. Zsigmondy began an investigation with colour to glass ruby by sedimentation method called a nuclear method with fruitful results (Zsigmondy 1909). In a similar fashion, using formaldehyde alkaline solution of chloride gold was reduced. Zsigmondy invented ultra-microscope for envisioning of colloidal aurum (Au) particles to study the diffusion behaviour of nanoparticles. When NaCl and the red-gold solution were combined, it produces a blue colour. This change of colour can be defined as the gold number (Zsigmondy 1909) of mg of the hydrophilic colloid per 10 cm^3 of gold solution, and these studies are useful in characterizing different properties.

Moreover, Svedberg and Tiselius (1929)utilized hydrochloroauric acid for aurum (Au) particles by electrochemical technique that illustrated the change in colour for end product from red to blue colour dispersion. Apart from the binding property, the movement of particles described as Brownian motion is a diffusion process which is worth enough to describe molecular dispersion (Einstein 1956). Smoluchowskis' theories (for diffusion and coagulation) stated collision and coalescence issues in colloidal solutions. If diffusing particle's movement region was outside the sphere then the movement of the particles was unaffected, if it moves region outside the sphere and stick to each other (or) another particle by moving inside the sphere, this explains the phase behaviour and stability of colloids. Thus, all these research findings inferred that AuNPs are efficient in terms of drug formulation and are an alternative to conventional drug delivery without compromising the safety.

2.2 Classification of Nanoparticles

In its natural state, gold is yellow in colour and inert, while AuNPs are red in colour and have antioxidant properties. The classification of nanoparticles was done according to their shapes: spherical, triangle, nanocluster, nanosphere, nanorods, nanoshells, nanostar, etc., as shown in Fig. 2.1.

2.2.1 Gold (Au) Nanospheres

Au colloids are also known as "Au nanospheres". These are made by adding a reducing agent, such as citrate, to an aqueous $HAuCl_4$ solution, resulting in monodispersed Au nanospheres with sizes ranging from 2 to 100 nm (Turkevich et al. 1951). If the amount of citrate used is reduced, a large number of nanospheres can be generated, and the size of the nanospheres can be regulated by changing the citrate-to-Au ratio.

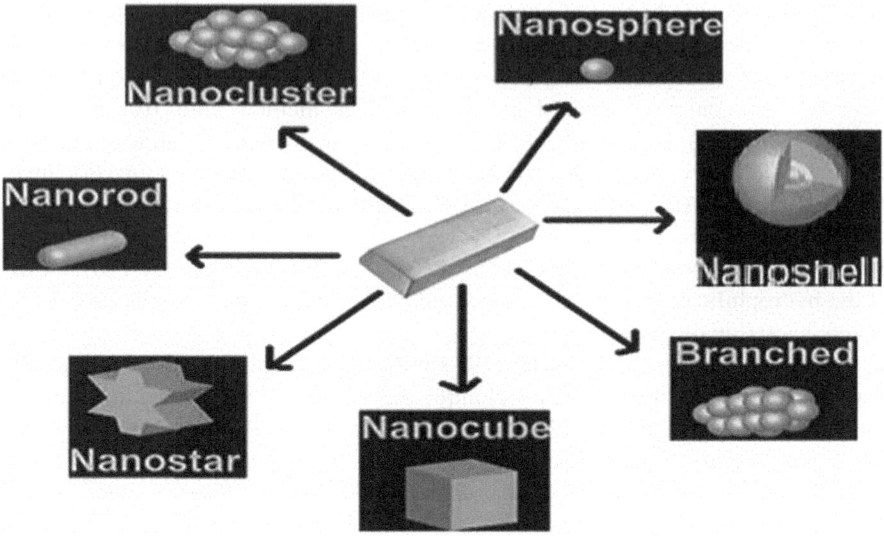

Fig. 2.1 Various shapes of gold nanoparticles. (Adopted from Khan et al. 2014)

2.2.2 Gold (Au) Nanorods

Au nanorods were produced using the template method, based on the electrochemical deposition of Au within the pores of nanoporous polycarbonate or alumina template membranes. The length of the nanorods depends on the Au deposition in the pores of the membrane.

2.2.3 Gold (Au) Nanoshells

Nanoshell is one of the spherical nanoparticles that consist of dielectric core, which is covered with gold shell (Loo et al. 2004). These nanoshells are comprised of weakly interaction particles called plasmon, derived from quantum plasma oscillations, in which the electrons simultaneously oscillate around all ions. The shell thickness could be controlled by the amount of deposited Au on the core surface (Minakshi et al. 2011).

2.2.4 Gold (Au) Nanocages

Gold nanocages were synthesized by the galvanic replacement reaction of truncated gold nanocubes and aqueous $HAuCl_4$ (Chen et al. 2006). These are primarily used for the loading of drugs and genes into their cavities.

2.3 Synthesis Methods of AuNPs

A variety of techniques available for the synthesis of AuNPs can be broadly divided into two methods, namely, synthetic and biological methods.

2.3.1 Synthetic Methods

Usually, the synthesis of AuNPs by the chemical reduction method involves two steps, namely, reduction followed by stabilization with the help of suitable agents. Many theories or methods are proposed by various researchers for the fabrication of AuNPs.

2.3.1.1 Turkevich Method

This method is also known as conventional chemical synthesis proposed by Turkevich in 1951. This method involves the boiling of $HAuCl_4$ solution, and then trisodium citrate dihydrate is then added rapidly under vigorous stirring. After a few minutes, the colour of the solution changes from light yellow to wine red, which indicates the formation of AuNPs measuring about 20 nm in diameter. In this technique, citrate ions play a double role, as both stabilizing and reducing agents (Turkevich et al. 1951). The schematic route for synthesis of AuNPs is shown in Fig. 2.2. The reducing and stabilizing agent ratio influences the size of the nanoparticles produced (Roya et al. 2016). For an instance, taking high concentrations of citrates leads to stabilization of AuNPs of smaller sizes and low concentration leads to large particles by aggregation of small particles (Kimling et al. 2006). In another study) have concluded that the addition of reagents in the reverse order (i.e. addition of $HAuCl_4$ into a boiling sodium citrate solution) leads to the formation of AuNPs with small size and a narrow size distribution.

2.3.1.2 Brust Method

Thermally stable and air-stable AuNPs are produced by Brust method, where $HAuCl_4$ is transformed from aqueous phase to toluene phase with the addition of phase-transfer agent like tetraoctylammonium bromide (TOAB) and reduced by $NaBH_4$, in the presence of dodecanethiol. Colour change of orange to deep brown was observed upon addition of reducing agent, indicating the formation of AuNPs (Brust et al. 1994). The step-by-step process is illustrated in Fig. 2.3. This method yields AuNPs from 1.5 to 5 nm by controlling the conditions such as gold-to-thiol ratio, rate of reduction and temperature at which reaction takes place.

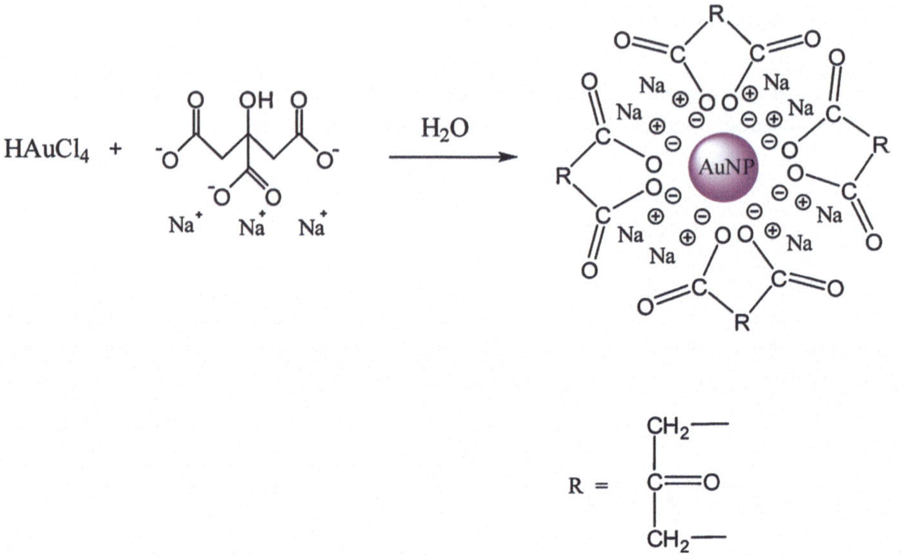

Fig. 2.2 Production of AuNPs by Turkevich method. (Adopted from Zhao et al. 2013)

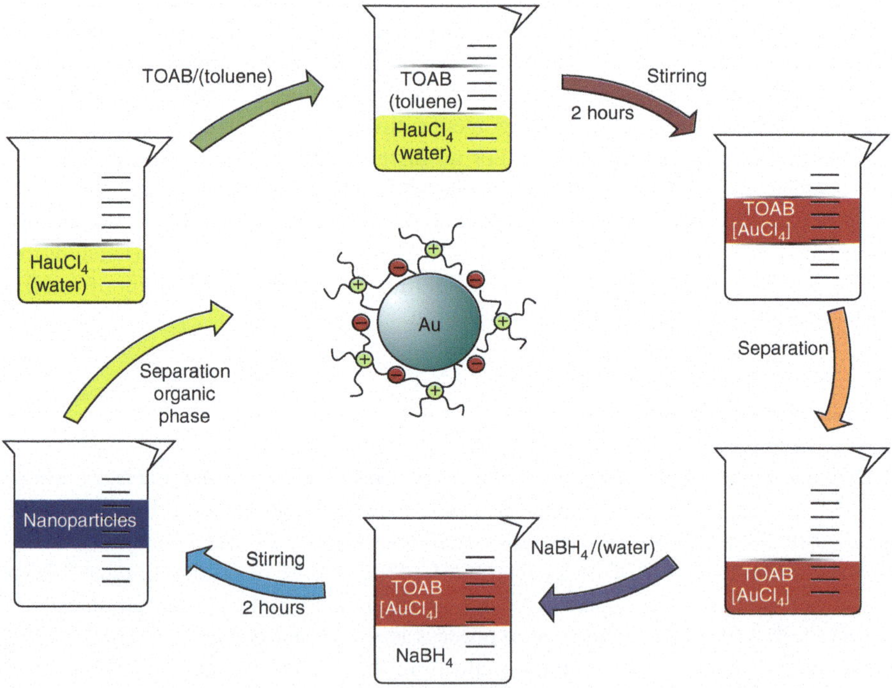

Fig. 2.3 Process involved in the synthesis of AuNPs using Brust method. (Adopted from Calandra et al. 2010)

2.3.1.3 Synthesis of Colloidal Gold Using Polymers

AuNP synthesis accomplished in the presence of poly-(methylphenylphosphazene) (PMPP). Firstly, $HAuCl_4$ was dissolved in water and a phase-transfer reagent (tetra-octylammonium bromide) is added to make the solution organic, then PMPP is added to this organic solution and stirred for 1 h, followed by addition of $NaBH_4$ and stirred for 24 h, which results in the formation of dark purple polymer-Au nanoparticles (Walker et al. 2001).

2.3.2 *UV -Induced Photochemical Synthesis Method*

The colloidal gold particles with different sizes were prepared in aqueous solution system containing PEG, acetone and $HAuCl_4$ with different concentrations. To the solution of $HAuCl_4$, if UV radiations of different wavelengths are induced, it enhances polymerization which in turn affects particle size and the dimensions of Au particles. Different sized gold particles with improved monodispersity and uniform shape were obtained by varying the concentration of surfactant/polymer (Sau et al. 2001). The schematic representation of this process is shown in Fig. 2.4.

2.3.2.1 Ultrasound-Assisted Synthesis of Colloidal Gold

A conventional ultrasonic bath is used for this method. It was filled with 4.5 L of water and maintained at temperature of 80 °C. After reaching the temperature, the two conical flasks containing aqueous $HAuCl_4$ solutions were immersed in an ultrasonic bath for 10 min, and then sodium citrate was added to the two aqueous $HAuCl_4$ solutions and sonicated to produce AuNPs. The procedure was represented pictorially as shown in Fig. 2.5. The extent of sonication determines the size of AuNPs,

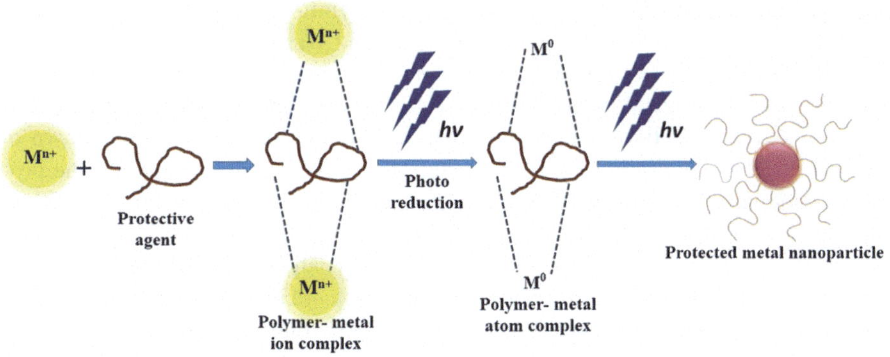

Fig. 2.4 UV irradiation to the polymers. (Adopted from Kim et al. 2002)

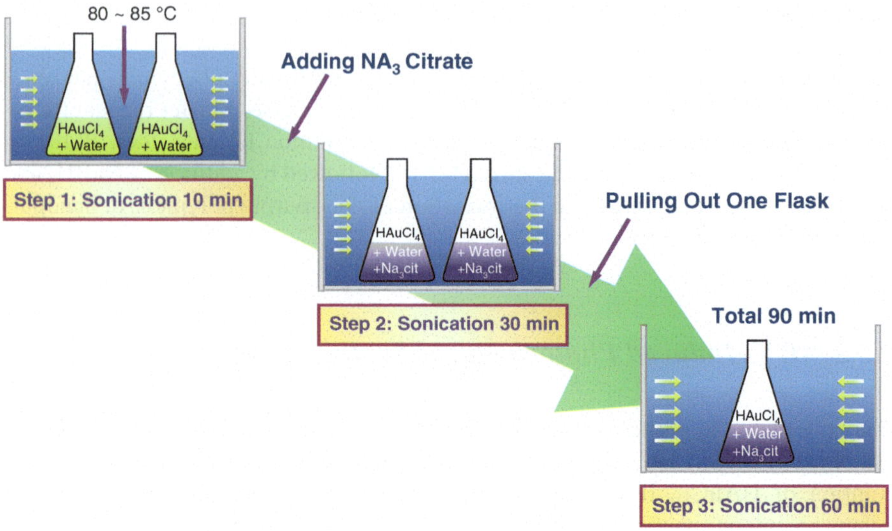

Fig. 2.5 Various steps involved in the ultrasonic-mediated synthesis. (Adopted from Lee et al. 2012)

and moreover, the stability of AuNPs was increased with various stabilizers like citrate and disulphide (Grzelczak and Liz-Marzán 2014).

2.3.3 Biosynthetic Methods of AuNPs

The methods described above lead to toxic by-products that may prove to have environmental consequences during large-scale production. In response to these concerns, new strategies to generate AuNPs without toxic chemicals are being actively developed with principles of green chemistry, like rapidly biodegradable reagents, limiting waste products, pressure and synthesis at ambient temperature and low toxicity of chemical products (Anastas and Warner 1998). In recent years, biological synthesis has been proposed as an alternative method for the synthesis of nanoparticles. Biological synthesis involves using environmentally friendly materials such as bacteria, fungus and plant extract. Plant extracts have been shown to perform the biogenic reduction of the metal, thus capping and stabilizing the produced nanoparticles (Husen 2019a; Bachheti et al. 2019, 2021; Joshi et al. 2019; Siddiqi et al. 2018b; Siddiqi and Husen 2020; Alle et al. 2021). Plant extracts generally contain sugars, green terpenoids, polyphenols, alkaloids, phenolic acids and proteins that act as excellent reducing agents, which are useful in the fabrication of AuNPs (Husen 2017, 2019b). Various studies show that AuNPs can be obtained successfully from the different parts of the plants, viz., leaf, flower, fruit, bark, seed and gums (Madhusudhan et al. 2019; Bachheti et al. 2020a, b, c; Painuli et al. 2020; Beshah et al. 2020; Sharma et al. 2020).

The general procedure is quite simple, where the plant parts or plant extracts are added to the bulk metal solution and due to the ability of plant extracts to act both as reducing and stabilizing agents assist in the formation of AuNPs, which are later purified and then characterized (Anastas and Warner 1998). These eco-friendly synthesis methods have gained more importance in the recent years because of its simplicity, single-step process and low toxicity (Elavazhagan and Arunachalam 2011). In an interesting study, two important medicinal plants *Cucurbita pepo* and *Malva crispa* have been utilized in the synthesis of AuNPs and possessed antibacterial activity against food spoilage pathogens. Rai et al. in 2007 proposed a novel approach, where Au core-Ag shell nanoparticles in triangle shape were obtained by the reduction of gold ions with the help of lemongrass extract as reducing agent, followed by reduction of the surface-bound Ag + ions by ascorbic acid, and these were successfully evaluated for its anticancer activity.

When the aqueous seed extract of *Abelmoschus esculentus* was used to synthesize AuNPs, its antifungal activities were improved to a greater extent and were effective for a broad range of spectrum (Sujitha and Kannan 2013). An Indian medicinal plant *Dalbergia sissoo* (Roy et al. 2011) and its plant bark extract were utilized for the reduction of gold ions and showed promising effect as antioxidants. The colloidal AuNPs prepared with fruit and fruit pulp is having been reported to play a significant role in preventing diabetes and cancer (Daisy and Saipriya 2012). A plethora of other plants used for the preparation of AuNPs are listed in Table 2.1.

2.3.4 *Microbial Synthesis of AuNPs*

Many microbes have the capability to assist in the production of inorganic nanostructures and metal nanoparticles with properties similar to nanoparticles derived from chemical synthesis, with control over the size, shape and composition of the

Table 2.1 A list of some of the different plants which have been used as whole or a part of it for synthesis of AuNPs

Plant	Part of the plant used	Size of AuNPs	References
Sapindus mukorossi	Fruit	9–19 nm	Reddy et al. (2013)
Prunus domestica	Fruit	14–26 nm	Dauthal and Mukhopadhyay (2012)
Magnolia kobus	Leaf	5–300 nm	Song et al. (2009)
Diospyros kaki	Leaf	5–300 nm	Song et al. (2009)
Abelmoschus esculentus	Seed	45–75 nm	Jayaseelan et al. (2013)

Table 2.2 Compilation of different microorganisms used for synthesis of AuNPs of different sizes and shapes

Microorganism	Genus	Size of AuNPs(nm)	References
Pseudomonas fluorescens	Bacterium	50–70	Radhika et al. (2012)
Shewanella algae	Bacterium	10–20	Konishi et al. (2007)
Marinobacter pelagius	Bacterium	10	Girilal et al. (2013)
Stoechospermum marginatum	Algae	18.7–93.7	Arockiya et al. (2012)
Candida albicans	Fungi	20–80	Chauhan et al. (2011)

particles. Many researchers adopted this alternative approach to chemical synthesis procedures for the preparation of AuNPs.

Cultures like fungi, bacteria and yeast were grown in test tubes containing 10 mL nutrient medium in shaker incubators at 28 °C for 24–48 h. After the specific period of incubation time, the biomass was separated from the medium by centrifugation with 7500 rpm for 10 min. The obtained biomass was washed three times in sterile distilled water. Then the biomass was resuspended in 10 mL distilled water and the pH adjusted to between 5 and 6 with 0.2 M NaOH. $HAuCl_4$ was added to give an overall Au concentration of 250 mg/L. The mixture was left for a further 24–72 h in a shaker incubator at 28 °C. The accumulation and reduction of gold were followed by visual observation of the biomass turning purple, an indication of the formation of AuNPs.

This method is eco-friendly where hazardous chemicals and toxic substances/derivatives are not used and antimicrobial properties can also increase using this method. For instance, antimicrobial properties of *Klebsiella pneumoniae* were increased with conjugation of AuNPs, which shows the synergetic effect of antimicrobial property. Synthesis of AuNPs from different microorganisms is illustrated in Table 2.2.

2.4 Applications of AuNPs

Gold nanoparticles are gaining importance progressively in medicine because of their potential uses in a broad variety of areas that encompass nanotechnology, biomedicine and electronics. More recently, the unique properties of AuNPs have been researched and utilized in high technology applications such as organic photovoltaics, sensor probes, therapeutic agents, drug delivery in biological and medical applications, electronic conductors and catalysis. In this chapter, we review vast and vital applications of AuNPs in medicine, and summary of its applications is pictorially represented in Fig. 2.6.

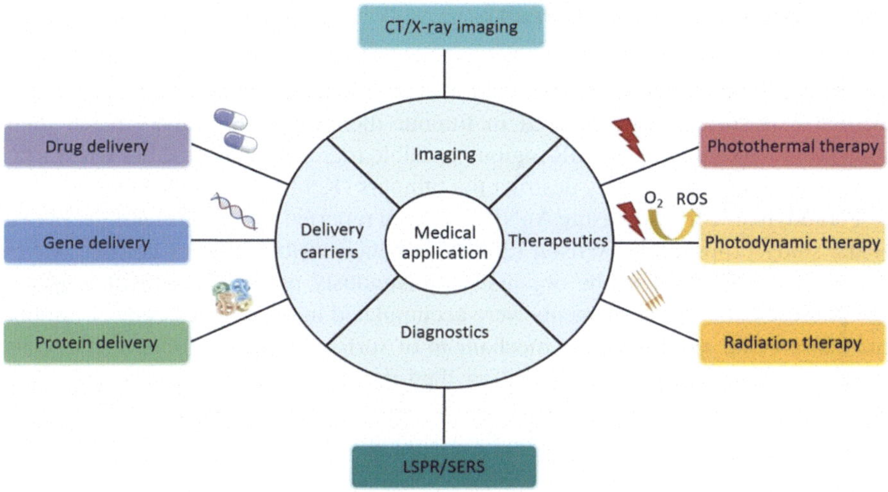

Fig. 2.6 Medical applications of AuNP. (Adopted from Baek et al. 2016)

2.4.1 *In Vitro Assays*

The advent of AuNPs opened a plethora of opportunities for in vitro assay of proteins and for the detection of cancer cells that were studied by various scientists. Atomic force microscopy (Han et al. 2000; Jin et al. 2007), scanometric assay (Son and Lee 2007) chronocoulometry, gel electrophoresis (Qin and Yung 2007), amplified voltametric detection and Raman spectroscopy methods were used to detect protein or polynucleotide present on oligo-nano particle-capped AuNPs (Han et al. 2000; Jin et al. 2007). Moreover, cancer cells were also successfully detected by protein assay, immunoassay and capillary electrophoresis (Medley et al. 2008).

2.4.2 *Gold Nanoparticles as Diagnostic Agents*

Many researchers assured AuNPs as potential and safe alternative to traditional diagnostic tools. Drygin et al. (2009) and Naja et al. (2008) used AuNPs in the identification of chemical and biological infectious agents. These unique biomacromolecules were also used in various analytic methods of clinical diagnostics by solution particle immunoassay (SPIA) in which solution changes from red into purple colour through absorption spectrum detected by spectrophotometric method (Khlebtsov et al. 2004). This method was subsequently used for the direct detection of cancer cells (Medley et al. 2008) and *Leptospira* cells in urine and for determining markers of Alzheimer's disease.

2.4.3 AuNPs in Phototherapy

An interesting and important application of AuNPs is photothermal therapy. This thermal exposure therapy is used in tumour therapy where AuNPs reach their absorption maximum in near-IR region or visible region and become hot when irradiated at cells and eventually destruct the tumours (Kennedy et al. 2011).

Photodynamic therapy using AuNPs has been reported by various authors. In one of the study proposed by Wilson (2010), in which visible light of definite wavelength was introduced in the organism intravenously using light-sensitive agents like photosensitizers, these agents were accumulated in the targeted cells or tumour and cause death to cells with a mechanism of surface plasma resonance (i.e. cells present in tumour lead to malnutrition then damage to micro vessels and death occurs). This method was used in dermal and oncological diseases (Baek et al. 2016). The only drawback of this method is photosensitizers remain in patient's tissue and tissue becomes sensitive to light.

2.4.4 AuNPs as Cell and Phantom Imaging Agents

Durr et al. (2007) utilized the versatile properties of AuNPs and optical image of cells. In their observation, by utilizing gold nanorods with fluorescent dyes, 3D fluorescent imaging of cancer cells with a depth of 75 μm deep was developed by two-photon luminescence. By imaging of AuNPs, its associated drawbacks such as photobleaching and blinking can be prevented because of the formation of fluorophores with low toxicity (Yao et al. 2005). In vivo targeted cancer imaging using nanoparticles has been achieved, and even some studies exhibited tumour targeting efficacy that is sufficient for potential molecular therapy applications in the clinical setting. Both preclinical and clinical magnetic resonance (MR) imaging can be done by paramagnetic nanoparticles (Thorek et al. 2006). In MR angiography, nanoparticles were used as blood-pool agents that are capable for contrast of blood vessels seen by in vivo imaging (Su et al. 2007).

Gold nanospheres and protease-sensitive quenched probes are used for visual monitoring of both proteases and protease inhibitors in vitro and in vivo (Lee et al. 2008). Some other methods revealed that antibody-conjugated gold nanorods were reported to give a Raman spectrum that is greatly enhanced, sharpened and polarized. In the other study, reported by Keren et al. (2008), enlargement of image and stabilization were done by in vivo Raman imaging with surface-enhanced Raman scattering (SERS) NP, where it consists of gold core, a Raman active molecular and silica coating. Many researchers, including Bremer et al. (2001), and Jaffer et al. (2007), assured that AuNPs stabilized by thiolated PEG enhance the resolution and optical properties of the image.

Singh et al. (2020) concluded in their study that, when conjugated to tumour-targeting ligands, the conjugated SERS nanoparticles were able to target tumour markers such as epidermal growth factor receptor (EGFR) on human cancer cells and in xenograft tumour models.

2.4.5 AuNPs as Therapeutic Agents

The AuNPs are not only used in diagnostic purpose and cell photolysis experiments but also used as therapeutic agent. The transport of therapeutic agents to the cells by AuNPs is a critical process in biomedical treatment. Several research groups have used functionalized AuNPs to investigate the interactions with cell membrane to improve delivery efficiency. In the recent past, plenty of research has been conducted on AuNPs as therapeutic agents. For instance, Abraham and Himmel (1997) used colloidal gold particle rheumatoid arthritis, and the study concluded that anti-angiogenic activity was increased by binding AuNPs with vascular EGF, therefore decreasing inflammation and macrophage infiltration.

Additionally, in cancer therapy (Amrita et al. 2011; Renat et al. 2006; Limei et al. 2006), nanospheres/nanorods/nanoshells/nanocages were irradiated with focused laser pulses of suitable wavelength to kill the cancer cells. Combination of paclitaxel, a chemotherapeutic drug, and AuNPs has showed elevated anti-proliferation activity on human hepatoma cells. Moreover, AuNPs with PEG chains of optimized size with circulation half-life at least a few hours may be most efficacious for cancer therapy (Liu et al. 2007).

Methotrexate, an inhibitor of dihydrofolate, was a chemotherapeutic agent for treating a variety of cancer types. Huennekens (1994), in their study, conjugate of methotrexate with AuNPs was prepared and the cytotoxic effect was examined in vitro and in vivo. Administration of the conjugate prevents tumour growth in a mouse, whereas free methotrexate equal dose as conjugate had no antitumour effect. Modulated drug delivery of methylene blue, insulin and lysozyme was achieved by irradiation of drug-loaded nanoshell- hydrogel combination and release of drug depends on therapeutic weight of drug molecule (Chen 2007). In high ionic strength media, the nanoparticle size, PEG chain length and monolayer composition provide the stability of AuNPs. The PEG chains present on AuNP and half-life circulation are the most important in cancer therapy (Madhusudhan et al. 2014).

2.4.6 AuNPs as Drug Carriers

The targeted delivery of drugs is one of the most promising and actively developing commands in therapeutic use of AuNPs including, but not limited to, antibiotics and antitumour agents in targeted delivery (Duncan et al. 2010; Pissuwan et al. 2011). The antibacterial agents and antibiotics such as vancomycin and ciprofloxacin can be delivered in the form of AuNPs acting as carriers by forming a stable complex, which shows action against strains like *E. coli* and *E. faecalis* (Gu et al. 2003).

Rosemary et al. (2006) showed increased antibacterial activity of ciprofloxacin gold nanoshells towards *E. coli*. Anti-leukaemia drug 5-fluorouracil, when conjugated with colloidal gold, illustrated a noticeable antibacterial and antifungal effect against *P. aeruginosa, E. coli and S. aureus* (Selvaraj and Alagar 2007). This effect was observed due to complete suppression of function of damaged or overexpressed gene. The ill-functioning of gene can be adjusted by delivering genetic material by AuNPs to the cytoplasm and cell nucleus.

2.4.7 AuNPs in Drug Delivery

The advent of various synthetic approaches for the preparation of AuNPs opened a myriad of opportunities for novel drug delivery and pharmaceutical applications. Due to the versatility and affordability of AuNPs, their use is dramatically increased in the recent past. AuNPs as drug delivery have been already evaluated in various routes of drug administration systems and yielded promising results.

2.5 Conclusions

In the field of bio-nanomedicine, gold nanoparticles (AuNPs) have made an impact in terms of applications in medicine and development of new methods for the functionalization of AuNP with different classes of molecules providing stabilization in vivo, and directed interaction with biological targets is of significance. Both unconventional and conventional synthesis techniques are discussed for AuNPs, and biological-based synthesis is promising, because of its eco-friendliness, tenability and free from toxicity when compared with synthetic techniques.

The tremendous potential of nanotechnology-guided AuNPs represents an attractive research field for current science community, including the biomedical research. Depending on the versatility and physicochemical features, AuNPs possess an enormous potential for copious biomedical-related applications such as controlled drug delivery, biomedical imaging, bio-detection platforms and gene therapy. By combing the antibiotic with AuNPs, the antibacterial efficiency was improved, and it became a solution to combat several dreadful diseases caused by various microorganisms. In addition to its nontoxicity, use of AuNPs in humans, as an antibacterial agent, warrants an undisputable knowledge about the mechanisms of antibiotic action of AuNPs, so as to avoid any risk it may cause to normal tissue/organ.

The unique properties of AuNPs recommend its broad applications such as in photodynamic therapy, drug and gene delivery, photothermal therapy and imaging. Optical properties of AuNPs with SPR bands are used as carrier in cancer therapy. The cancer imaging can be done by conjugating the cancer-seeking peptide with AuNPs. AuNPs are without any doubt, due to its versatile attributes, potential tools for next-generation biomedical applications.

References

Abraham GE, Himmel PB (1997) Management of rheumatoid arthritis: rationale for the use of colloidal metallic Gold. J Nutr Med 7:295–305

Alle M, Reddy GB, Kim TH, Park SH, Lee S-H, Kim J-C (2020) Doxorubicin-carboxymethyl xanthan gum capped gold nanoparticles: microwave synthesis, characterization, and anti-cancer activity. Carbohydr Polym 229:115511. https://doi.org/10.1016/j.carbpol.2019.115511

Alle M, Park SC, Bandi R, Lee S-H, Kim J-C (2021) Rapid in-situ growth of gold nanoparticles on cationic cellulose nanofibrils: recyclable nanozyme for the colorimetric glucose detection. Carbohydr Polym 253:117239. https://doi.org/10.1016/j.carbpol.2020.117239

Amrita M, Suraj KT, Rizwan W, Song HJ, Inho H, You-B Y, Young-Soon K, Hyung S, Soon Y (2011) Microbial synthesis of gold nanoparticles using the fungus Penicillium brevicompactum and their cytotoxic effects against mouse mayo blast cancer C2C12 cells. Appl Microbiol Biotechnol 92:617–630

Anastas PT, Warner JC (1998) Green chemistry: theory and practice. Oxford University Press, New York, p 30

Arockiya F, Aarthi R, Parthiban C, Ganeshkumar V, Anantharamn P (2012) Biosynthesis of antibacterial gold nanoparticles using brown alga, Stoechospermum marginatum. Spetrochim Acta A Mol Biomol Spectrosc 99:166–173

Bachheti RK, Gupta V, Husen A, Joshi A, Konwarh R (2019) Green synthesis of iron oxide nanoparticles: cutting edge technology and multifaceted applications. In: Husen A, Iqbal M (eds) Nanomaterials and plant potential. Springer, Cham, pp 239–259. https://doi.org/10.1007/978-3-030-05569-1_9

Bachheti RK, Fikadu A, Bachheti A, Husen A (2020a) Biogenic fabrication of nanomaterials from flower-based chemical compounds, characterization and their various applications: a review. Saudi J Biol Sci 27:2551–2562. https://doi.org/10.1016/j.sjbs.2020.05.012

Bachheti RK, Godebo Y, Joshi A, Yassin MO, Husen A (2020b) Root-based fabrication of metal and or metal-oxide nanomaterials and their various applications. In: Husen A, Jawaid M (eds) Nanomaterials for agriculture and forestry applications. Elsevier Inc, Cambridge, MA, pp 135–166. https://doi.org/10.1016/B978-0-12-817852-2.00006-8

Bachheti RK, Sharma A, Bachheti A, Husen A, Shanka GM, Pandey DP (2020c) Nanomaterials from various forest tree species and their biomedical applications. In: Husen A, Jawaid M (eds) Nanomaterials for agriculture and forestry applications. Elsevier Inc, Cambridge, MA, pp 81–106. https://doi.org/10.1016/B978-0-12-817852-2.00004-4

Bachheti RK, Abate L, Deepti, Bachheti A, Madhusudhan A, Husen H (2021) Algae-, fungi-, and yeast-mediated biological synthesis of nanoparticles and their various biomedical applications. In: Kharisov B, Kharissova O (eds) Handbook of greener synthesis of nanomaterials and compounds, Volume 1: fundamental principles and methods. Elsevier Inc, Cambridge, MA, pp 701–734. https://doi.org/10.1016/B978-0-12-821938-6.00022-0

Baek SM, Singh RK, Kim TH, Seo JW, Shin US, Chrzanowski W (2016) Triple hit with drug carriers: pH- and temperature-responsive theranostics for multimodal chemo- and photothermal-therapy and diagnostic applications. ACS Appl Mater Interfaces 8:8967–8979

Beshah F, Hunde Y, Getachew M, Bachheti RK, Husen A, Bachheti A (2020) Ethnopharmacological, phytochemistry and other potential applications of *Dodonaea* genus: a comprehensive review. Curr Res Biotechnol 2:103–119. https://doi.org/10.1016/j.crbiot.2020.09.002

Bremer C, Tung CH, Weissleder R (2001) In vivo molecular target assessment of matrix metalloproteinase inhibition. Nat Med 7:743–748

Brust M, Walker M, Bethell D, Schiffrin DJ, Whyman RJ (1994) Synthesis of thiol- derivatised gold nanoparticles in a two-phase liquid–liquid system. Chem Soc Chem Commun 7:801–802

Calandra P, Calogero G, Sinopoli A, Gucciardi PG (2010) Metal nanoparticles and carbon based nanostructures as advance materials for cathode applications in dye sensitized solar cells. Int J Photoenergy 1:1–15

Chauhan A, Zubair S, Tufail S, Sherwani A, Sajid M, Raman SC, Azam A, Owais M (2011) Fungus mediated biological synthesis of gold nanoparticles: potential in detection of liver cancer. Int J Nanomedicine 6:2305–2319

Chen YH (2007) Methotrexate conjugated to Au nanoparticles inhibits tumour growth in a syngeneic lung tumour model. Mol Pharm 4:713–722

Chen J, Joseph M, Mclellan AS, Yujie X, Li Z-y, Younan X (2006) Facile synthesis of Au-silver nanocages with controllable pores on the surface. J Am Chem Soc 128:14776–14777

Chung M, Park I, Seung-Hyun K, Thiruvengadam M, Rajakumar G (2016) Plant-mediated synthesis of silver nanoparticles their characteristic properties and therapeutic applications. Nanoscale Res Lett 11:1–14

Daisy P, Saipriya K (2012) Biochemical analysis of Cassia fistula aqueous extract and phyto chemically synthesized gold nanoparticles as hypoglycaemic treatment for diabetes mellitus. Int J Nanomedicine 7:1189

Dauthal P, Mukhopadhyay M (2012) Prunus domestica fruit extract- mediated synthesis of gold nanoparticles and its catalytic activity for 4-nitrophenol reduction. Ind Eng Chem Res 51:13014–13020

Drygin YF, Blintsov AN, Osipov AP (2009) High-sensitivity express Immuno chromatographic method for detection of plant infection by tobacco mosaic virus. Biochem Mosc 74:986–983

Duncan B, Kim C, Rotello VM (2010) Gold nanoparticle platforms as drug and biomacro molecule delivery systems. J Control Release 148:122–127

Durr NJ, Larson T, Smith DK (2007) Two-photon luminescence imaging of cancer cells using molecularly targeted gold nanorods. Nano Lett 7:941–945

Dykman LK, Khlebtsov NG (2011) Gold nanoparticles in biology and medicine: recent advances and prospects. Acta Nat 3:34–45

Einstein A (1956) Investigation on the theory of Brownian movement. Dover Publications, pp 17–19

Elavazhagan T, Arunachalam KD (2011) Memecylon edule leaf extract mediated green synthesis of silver and gold nanoparticles. Int J Nanomedicine 6:1265–1278

Faraday M et al (1857) The Bakerian lecture – experimental relations of gold (and other metals) to light. Philos Trans R Soc Lond 147:36

Faulk WP, Taylor GM (1971) An immune colloid method for the electron microscope. Immunochemistry 8:1081–1083

Girilal M, Fayaz M, Mohan P, Kalaichelvan PT (2013) Augmentation of PCR efficiency using highly thermostable gold nanoparticles synthesized from a thermophilic bacterium, Geobacillus stearothermophilus. Colloids Surf B: Biointerfaces 106:165–169

Grzelczak M, Liz-Marzán LM (2014) The relevance of light in the formation of colloidal metal nanoparticles. Chem Soc Rev 4:2089–2097

Gu H, Ho PL, Tong E, Wang L, Xu B (2003) Presenting vancomycin on nanoparticles to enhance antimicrobial activities. Nano Lett 3:1261–1263

Han S, Lin J, Zhou F (2000) Oligonucleotide-capped gold nanoparticles for improved atomic force microscopic imaging and enhanced selectivity in polynucleotide detection. Biochem Bio Phys Res Commun 279:265–269

Huennekens FM (1994) The methotrexate story: a paradigm for development of cancer chemotherapeutic agents. Adv Enzym Regul 34:397–419

Husen A (2017) Gold nanoparticles from plant system: synthesis, characterization and their application. In: Ghorbanpourn M, Manika K, Varma A (eds) Nanoscience and plant–soil systems, vol 48. Springer, Cham, pp 455–479. https://doi.org/10.1007/978-3-319-46835-8_17

Husen A (2019a) Natural product-based fabrication of zinc oxide nanoparticles and their application. In: Husen A, Iqbal M (eds) Nanomaterials and plant potential. Springer, Cham, pp 193–291. https://doi.org/10.1007/978-3-030-05569-1_7

Husen A (2019b) Medicinal plant-product based fabrication nanoparticles (Au and Ag) and their anticancer effect. In: Kintzios SE, Barberaki M, Flampouri (eds) Plants that fight cancer, 2nd edn. Taylor & Francis/CRC Press, pp 133–147

Husen A (2020) Introduction and techniques in nanomaterials formulation. In: Husen A, Jawaid M (eds) Nanomaterials for agriculture and forestry applications. Elsevier Inc, Cambridge, MA, pp 1–14. https://doi.org/10.1016/B978-0-12-817852-2.00001-9

Husen A, Iqbal M (2019) Nanomaterials and plant potential: an overview. In: Husen A, Iqbal M (eds) Nanomaterials and plant potential. Springer, Cham, pp 3–29. https://doi.org/10.1007/978-3-030-05569-1_1

Husen A, Siddiqi KS (2014) Phytosynthesis of nanoparticles: concept, controversy and application. Nanoscale Res Lett 9:229. https://doi.org/10.1186/1556-276X-9-229

Husen A, Rahman QI, Iqbal M, Yassin MO (2019) Plant-mediated fabrication of gold nanoparticles and their application in some cutting-edge areas. In: Husen A, Iqbal M (eds) Nanomaterials and plant potential. Springer, Cham, pp 71–110. https://doi.org/10.1007/978-3-030-05569-1_3

Jaffer FA, Kim DE, Quinti L (2007) Optical visualization of cathepsin K activity in atherosclerosis with a novel, protease-activatable fluorescence sensor. Circulation 15:2292–2298

Jayaseelan C, Ramkumar R, Rahuman A, Perumal P (2013) Green synthesis of gold nanoparticles using seed aqueous extract of Abelmoschus esculentus and its antifungal activity. Ind Crop Prod 45:423–429

Jin R, He X, Wang K (2007) Characterization of different sequences of DNA on si substrate by atomic force microscopy and gold nanoparticle labeling. J Nanosci Nanotechnol 7:418–423

Joshi A, Sharma A, Bachheti RK, Husen A, Mishra VK (2019) Plant-mediated synthesis of copper oxide nanoparticles and their biological applications. In: Husen A, Iqbal M (eds) Nanomaterials and plant potential. Springer, Cham, pp 221–237. https://doi.org/10.1007/978-3-030-05569-1_8

Kennedy LC, Bickford LR, Lewinski NA, Coughlin AJ, Hu Y, Day ES, West JL, Drezek RA (2011) A new era for cancer treatment: gold-nanoparticle-mediated thermal therapies. Small 7:169–183

Keren S, Zavaleta C, Cheng Z (2008) Noninvasive molecular imaging of small living subjects using Raman spectroscopy. Proc Natl Acad Sci 105:5844–5849

Khan AK, Rashid R, Murtaza G, Zahra A (2014) Gold nanoparticles: synthesis and applications in drug delivery. Trop J Pharm Res 13:1169–1177

Khlebtsov NG, Melnikov AG, Dykman LA, Bogatyrev VA (2004) In: Videen G, Yatskiv S, Mishchenko MI (eds) Photopolarimetry in remote sensing. Kluwer Academic Publication, Dordrecht, pp 265–308

Kim F, Song JH, Yang P (2002) Photochemical synthesis of gold nanorods. J Am Chem Soc 124:14316–14317

Kimling J, Maier M, Okenve B, Kotaidis V, Ballot H, Plech A (2006) Turkevich method for gold nanoparticle synthesis revisited. J Phys Chem B 110:15700–15707

Konishi Y, Sukiyama T, Tachimi T, Saitoh N, Nomura T, Nagamine S (2007) Microbial deposition of gold nanoparticles by the metal reducing bacterium Shewanella algae. Electrochim Acta 53:186–192

Lee S, Cha EJ, Park K (2008) A near-infrared-fluorescence-quenched gold-nanoparticle imaging probe for in vivo drug screening and protease activity determination. Angew Chem Int Ed Eng 47:2804–2807

Lee JH, Choi SUS, Lee SY (2012) Production of aqueous spherical gold nanoparticles using conventional ultrasonic bath. Nanoscale Res Lett 7:420

Limei A, Feng G, Bifeng P, Rong H, Daxiang C (2006) Fluoro immunoassay for antigen based on fluorescence quenching signal of gold nanoparticles. Anal Chem 78:1104–1106

Liu Y, Shipton MK, Ryan J (2007) Stability and cellular internalization of Au nanoparticles containing mixed peptide poly (ethylene glycol) monolayers. Anal Chem 79:2221–2229

Loo C, Alex L, Leon HBS, Min-Ho LMS (2004) Nano shell-enabled photonics- based imaging and therapy of cancer. Technol Cancer Res Treat 3:33–40

Madhusudhan A, Reddy G, Venkatesham M, Veerabhadram G, Kumar D, Natarajan S, Yang M-Y, Hu A, Singh S (2014) Efficient pH dependent drug delivery to target cancer cells by gold nanoparticles capped with carboxymethyl chitosan. Int J Mol Sci 15:8216–8234. https://doi.org/10.3390/ijms15058216

Madhusudhan A, Reddy GB, Krishana IM (2019) Green synthesis of gold nanoparticles by using natural gums. In: Nanomaterials and plant potential. Springer, Cham, pp 111–134
Medley CD, Smith JE, Tang Z, Wu Y, Bamrungsap S, Tan W (2008) Gold nanoparticle-based colorimetric assay for the direct detection of cancerous cells. Anal Chem 80:1067–1072
Minakshi D, Kyu HS, Seong SA, Dong KY (2011) Review on gold nanoparticles and their applications. Toxicol Environ Heal Sci 3:193–205
Naja G, Hrapovic S, Male K, Bouvrette P, Luong JHT (2008) Rapid detection of microorganisms with nanoparticles and electron microscopy. Microsc Res Tech 71:742–748
Painuli S, Semwal P, Bachheti A, Bachheti RK, Husen A (2020) Nanomaterials from non-wood forest products and their applications. In: Husen A, Jawaid M (eds) Nanomaterials for agriculture and forestry applications. Elsevier Inc, Cambridge, MA, pp 15–40. https://doi.org/10.1016/B978-0-12-817852-2.00002-0
Pal R, Panigrahi S, Bhattacharyya D, Chakraborti AS (2013) Characterization of citrate capped gold nanoparticles quercitin complex experimental and quantum chemical approach. J Mol Struct 1046:153–163
Pissuwan D, Niidome T, Cortie MB (2011) The forthcoming applications of gold nano particles in drug and gene delivery systems. J Control Release 149:65–71
Qin WJ, Yung LY (2007) Nanoparticle-based detection and quantification of DNA with single nucleotide polymorphism (SNP) discrimination selectivity. Nucleic Acids Res 35:111
Radhika, Rajashree SR, Suman TY (2012) Extracellular biosynthesis of gold nanoparticles using a gram negative bacterium Pseudomonas fluorescens. As Pac J Trop Dis 2:796–799
Reddy V, Torati RS, Kim CG (2013) Biosynthesis of gold nanoparticles assisted by sapindus mukorossi Gaertn. Fruit pericarp and their catalytic application for the reduction of p-nitroaniline. Ind Eng Chem Res 52:556–564
Renat RL, Charles J, Thomas FG, Vladimir PZ (2006) Laser-induced explosion of gold nanoparticles: potential role for nano photo thermolysis of cancer. Nano Med 1:473–480
Rosemary MJ, Maclaren I, Pradeep T (2006) Investigations of antibacterial properties of ciprofloxacin@SiO_2. Langmuir 22:10125–10129
Roy N, Laskar RA, Kumari D, Ghosh T, Begum NA (2011) A detailed study on the antioxidant activity of the stem bark of Dalbergia sissoo Roxb, an Indian medicinal plant. Food Chem 126:1115–1121
Roya H, Elham A, Morteza M, Abolfazl A (2016) Current methods for synthesis of gold nanoparticles. Artif Cells Nanomed Biotechnol 44:596–602
Sau TK, Pal A, Jana NR, Wang ZL, Pal T (2001) Size controlled synthesis of gold nanoparticles using photo chemically prepared seed particles. J Nanopart Res 3:257–261
Selvaraj V, Alagar M (2007) Analytical detection and biological assay of anti-leukemic drug 5-fluorouracil using gold nanoparticles as probe. Int J Pharm 337:275–281
Sharma A, Bachheti A, Sharma P, Bachheti RK, Husen A (2020) Phytochemistry, pharmacological activities, nanoparticle fabrication, commercial products and waste utilization of *Carica papaya* L.: a comprehensive review. Curr Res Biotechnol 2:145–160. https://doi.org/10.1016/j.crbiot.2020.11.001
Siddiqi KS, Husen A (2016a) Fabrication of metal and metal oxide nanoparticles by algae and their toxic effects. Nanoscale Res Lett 11:363. https://doi.org/10.1186/s11671-016-1580-9
Siddiqi KS, Husen A (2016b) Fabrication of metal nanoparticles from fungi and metal salts: scope and application. Nanoscale Res Lett 11:98. https://doi.org/10.1186/s11671-016-1311-2
Siddiqi KS, Husen A (2017) Recent advances in plant-mediated engineered gold nanoparticles and their application in biological system. J Trace Elem Med Biol 40:10–23. https://doi.org/10.1016/j.jtemb.2016.11.012
Siddiqi KS, Husen A (2020) Current status of plant metabolite-based fabrication of copper/copper oxide nanoparticles and their applications: a review. Biomat Res 24:11. https://doi.org/10.1186/s40824-020-00188-1

Siddiqi KS, Husen A, Sohrab SS, Yassin MO (2018a) Recent status of nanomaterial fabrication and their potential applications in neurological disease management. Nanoscale Res Lett 13:231. https://doi.org/10.1186/s11671-018-2638-7

Siddiqi KS, Husen A, Rao RAK (2018b) A review on biosynthesis of silver nanoparticles and their biocidal properties. J Nanobiotechnol 16:14. https://doi.org/10.1186/s12951-018-0334-5

Singh RK, Kurian AG, Patel KD, Mandakhbayar N, Knowles JC, Kim HW (2020) Label-free fluorescent mesoporous bio glass for drug delivery, optical triple-mode imaging, and photo thermal/photodynamic synergistic cancer therapy. ACS Appl Bio Mater 3:2218–2229

Son SJ, Lee SB (2007) A platform for ultrasensitive and selective multiplexed marker protein assay toward early-stage cancer diagnosis. Nanomedicine 2:79–82

Song JY, Jang HK, Kim BS (2009) Biological synthesis of gold nanoparticles using Magnolia kobus and Diospyros kaki leaf extracts. Process Biochem 44:1133–1138

Su CH, Sheu HS, Lin CY (2007) Nano shell magnetic resonance imaging contrast agents. J Am Chem Soc 129:2139–2146

Sujitha MV, Kannan S (2013) Green synthesis of gold nanoparticles using citrus fruits (Citrus limon, Citrus reticulata and Citrus sinensis) aqueous extract and its characterization, Spectrochim. Acta A Mol Biomol Spectrosc 102:15–23

Svedberg T, Tiselius A (1929) Colloid chemistry. The chemical catalogue company. Angew Chem 42:87–87

Thorek DL, Chen AK, Czupryna J (2006) Superparamagnetic iron oxide nanoparticle probes for molecular imaging. Ann Biomed Eng 34:23–38

Turkevich J, Stevenson PC, Hillier J (1951) The nucleation and growth processes in the synthesis of colloidal Au. Discuss Faraday Soc 11:55–75

Walker CH, St. John JV, Wisian-Neilson P (2001) Synthesis and size control of gold nanoparticles stabilized by poly (methylphenylphosphazene). J Am Chem Soc 123:3846–3847

Wilson BC (2010) In: Tuchin VV (ed) Handbook of photonics for biomedical science. CRC Press, Boca Raton, pp 649–686

Yao J, Larson DR, Vishwasrao HD (2005) Blinking and non-radiant dark fraction of water-soluble quantum dots in aqueous solution. Proc Natl Acad Sci 102:14284–14289

Zhao P, Li N, Didier A (2013) State of the art in gold nanoparticle synthesis. Coord Chem Rev 257(3):638–665

Zsigmondy R (1909) Colloids and ultra-microscope. J Am Chem Soc 31:951–952

Chapter 3
Nanoparticle-Mediated Delivery of Flavonoids for Cancer Therapy: Prevention and Treatment

Garima Sharma, Shalu Nim, Madhusudhan Alle, Azamal Husen, and Jin-Chul Kim

3.1 Introduction

Cancer is the world's leading cause of death and remains the greatest concern of the twenty-first century (Bray et al. 2018). Cancer is a progressive multistage, chronic illness, which renders the condition nearly incurable, contributing to second most deaths worldwide after coronary illness (Bray et al. 2018). Most of the cancer patients die in the early years after diagnosis and therapy of cancer, and about 20 percent are treated following surgery. With the progression of cancer illness, the chances of cancer patient's survival become difficult. Thus, the more serious the illness, the less chance of survival (Chen et al. 2015; Ganesh et al. 2008; Prat et al. 2014; Song et al. 2015b).

Late-stage cancer may be very challenging to manage in the therapeutics, and several vibrant factors such as extent in severity, care modality, comorbidity, and the genetic profile of an individual play an important role for a final result in anticancer

G. Sharma · J.-C. Kim (✉)
Department of Biomedical Science & Institute of Bioscience and Biotechnology, Kangwon National University, Chuncheon, Republic of Korea
e-mail: jinkim@kangwon.ac.kr

S. Nim
Shaheed Rajguru College of Applied Sciences for Women, Delhi University, Delhi 110096, India

M. Alle
Institute of Forest Science, Kangwon National University, Chuncheon, Republic of Korea

Department of Biomedical Science & Institute of Bioscience and Biotechnology, Kangwon National University, Chuncheon, Republic of Korea

A. Husen
Wolaita Sodo University, Wolaita, Ethiopia

J.-C. Kim et al. (eds.), *Smart Nanomaterials in Biomedical Applications*, Nanotechnology in the Life Sciences,
https://doi.org/10.1007/978-3-030-84262-8_3

therapy. Various genetic and epigenetic factors influence the tumor progression (Herbst et al. 2014). In addition, cancer growth and life expectancy of the cancer population are also influenced by risk factors, such as smoking, sedentary life, and western lifestyles (Jemal et al. 2011). Regardless of improvements in early detection approaches and available therapies, the incidences of cancer patients have risen in the recent years (Miller et al. 2016). The rising cancer incidences are however contributing to a major global decline.

Conventional cancer therapies include chemotherapy, radiation therapy, and immunotherapy where chemotherapy has raised issues regarding inadequate exposure to tumor tissue. A variety of adverse drug reactions (ADRs) are related to the broad and nonselective doses of the chemotherapeutic drugs (Jain 2001; Jang et al. 2003). Chemotherapy has proven to be a successful anticancer technique over the past six to seven decades. It is well-known that chemotherapeutic drugs induce cancer cell death via modulating various signaling pathways, such as genetic damage (i.e., DNA damage), cell cycle arrests, profound oxidative stress, and disruption of cytoskeleton (Annovazzi et al. 2017; Belayachi et al. 2017; Falzone et al. 2018; Goldstein and Kastan 2015; Pereira et al. 2017). The chemotherapeutic agent induces cell death via apoptosis or necrosis. Chemotherapy has a therapeutic function in early phases, but in late stages it may play a role as palliative or neoadjuvant manner (Alvarado-Miranda et al. 2009; Johnson et al. 2011; Robati et al. 2008). In addition to chemotherapy, radiation therapy is also given to cancer victims (Zheng et al. 2017). Radiation therapy is frequently paired with chemotherapy and surgery to optimize the improvement in cancer elimination. In addition to traditional chemotherapeutic medications, the targeted drugs and small molecule inhibitors have been proposed as a potential means of treating cancer. However, these drugs and inhibitors are costly and can cause serious side effects to the patients (Ozdelikara and Tan 2017; Vyas et al. 2010). Moreover, these drugs and inhibitors have low efficacy. Thus, there is demand for more innovative and reliable treatment methods for cancer.

Since decades, natural compounds and their synthetic derivatives are well-known for their various therapeutic benefits, including cancer. Natural compounds have been known for their immense biological and pharmacological properties. Although the mechanistic action of majority of natural products is not fully elucidated, they are used as antioxidants, anti-inflammatory, anti-mutagenic, and epigenetics modulators via modulating a variety of pathways associated with apoptosis, cell growth, angiogenesis, and metastasis (Kuntz et al. 2017; Rafieian-Kopaei and Movahedi 2017; Vaishnavi et al. 2012). Recent progresses in molecular biology, target selection, biomarker identifications, and genetic approaches have facilitated the studies to understand the anticancer properties of phytochemicals that are obtained from plants (Jin et al. 2016; Tao et al. 2015). It has been suggested that phytochemicals can be considered as alternative therapy to chemotherapy/radiotherapy or in combination with them (Hazra et al. 2011). The in vitro and in vivo studies showed the chemopreventive and chemotherapeutic efficacy of both dietary and non-dietary phytochemicals (Pande et al. 2017; Panjamurthy et al. 2009; Weng and Yen 2012), suggesting their potential as anticancer agent with reduced side effects to normal cells.

Phytochemicals, such as flavonoids, are considered to be beneficial in various forms of cancers (Chahar et al. 2011; Khan et al. 2019; Sak 2014; Suganuma et al. 2011; Ullah and Khan 2018). Phytochemicals exert their pharmacological actions via multi-targeted approach, such as modulation of genes and cell signaling pathways responsible for initiation, progression, and metastasis on cell cancer. Flavonoids are known to inhibit the activity of pro-carcinogens (Benavente-Garcia and Castillo 2008; Liu-Smith and Meyskens 2016). Some of the well-known phytochemicals possessing antitumor potential are shown in Fig. 3.1 (Bailon-Moscoso et al. 2017).

Because of their recorded antiproliferative, antiangiogenic, and antioxidant properties, several plant derivatives are under clinical trial. This book chapter accumulates the recent developments of commonly used flavonoids as chemotherapeutic and/or chemopreventive agents. In addition, here we focus on the methods to improve the efficacy of flavonoids for chemotherapeutic gain.

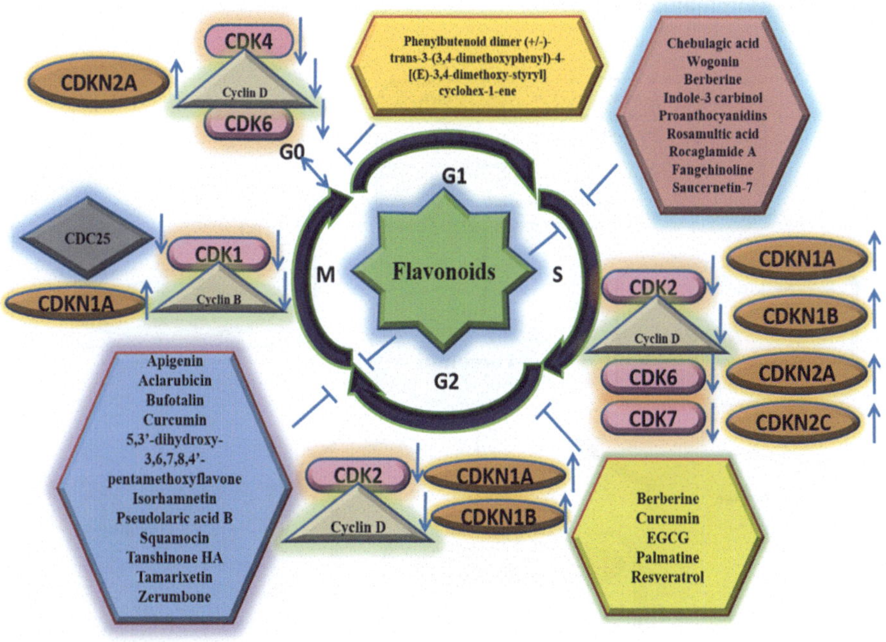

Fig. 3.1 Signaling pathways modulated by flavonoids

3.2 General Properties and Structures of Some Common Flavonoids

The structural components of flavonoid attributes to their anticancer/antioxidant properties. Here, we discuss some of the flavonoids and their structures.

3.2.1 Curcumin

Curcumin, (1E, 6E) -1,7-bis (4-hydroxy-3-methoxyphenyl)hepta-1,6-diene-3,5-dione); $C_{21}H_{20}O_6$], a phenolic metabolite or flavonoid, is isolated from *Curcuma longa* (turmeric) (Golonko et al. 2019; Salehi et al. 2020; Shabaninejad et al. 2020). As a traditional medicine, turmeric has been well identified for its role against a variety of illnesses, including inflammation, infection, diabetes, cancer, etc. (Selvam et al. 2019). The presence of O-methoxyphenol and methylene group in the structure of curcumin attributes to its antioxidant property (Priyadarsini proposed mechanisms that are responsible for the antioxidant potential of curcumin (Del Prado-Audelo et al. 2019). Curcumin is known to modulate various signaling pathways and cellular process, regulating the proliferation and growth of cancer cells (Fig. 3.2) (Marchiani et al. 2014).

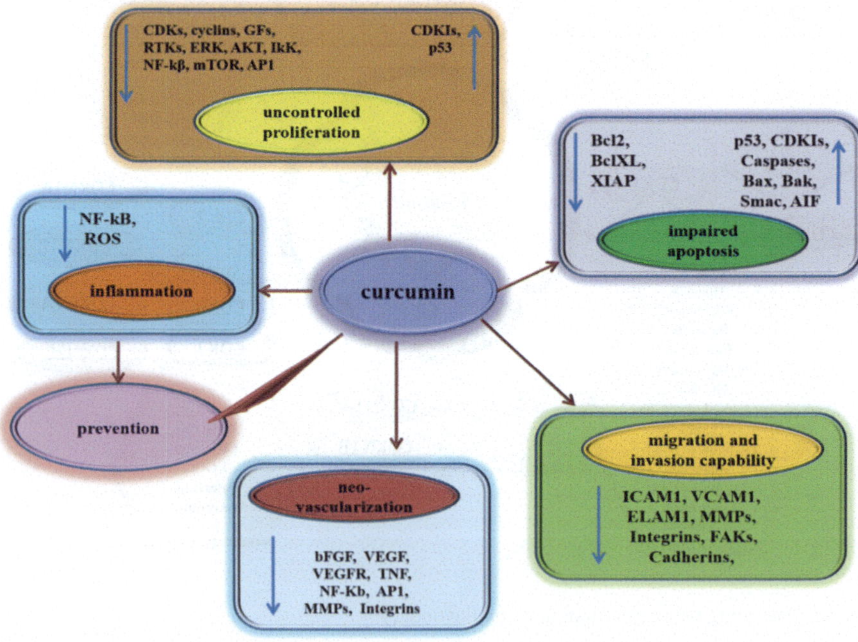

Fig. 3.2 Various cellular processes modulated by curcumin on cancer cells

3.2.2 *Quercetin*

Quercetin, [3,31,41,5,7-pentahydroxyflavone], a derivative of flavone (2-phenylchromen-4-one), contains five hydroxyl groups. The structure of quercetin O-glycosides showed the presence of O-glycosidic bonds at the C-3 carbon hydroxyl group. Quercetin possesses antioxidant property due to the presence of hydroxyl group and a catechol group in the A ring and B ring, respectively (Jana et al. 2004). These groups facilitate the transfer of hydrogen/electron, thus eliminating free reactive oxygen species (ROS) in the cells and, thereby, inhibiting enzymatic activities. The removal of ROS protects the cells against oxidative stress and reduces inflammation. In addition, quercetin inhibits protein peroxidation and protects against damage to cellular lipid membrane (Liu et al. 2010) and elevates glutathione level. The in vitro and in vivo studies showed that quercetin exerts growth inhibitory action on various types of cancer growth, i.e., L1210 and P-388 leukemia cell lines (Suolinna et al. 1975), breast cancer cells (Kim et al. 2013), colon cancer cells COLO 20DM (Hosokawa et al. 1990), pancreatic cancer cells (Angst et al. 2013), ovarian cancer cells OVCA 433 (Scambia et al. 1990), liver cancer cells HepG2 (Maurya and Vinayak 2015a), epidermoid cancer cells A431 (Huang et al. 1999), and gastric cancer cells (Yoshida et al. 1990). Moreover, quercetin also exerts antineoplastic effect against Walker carcinoma 256 (Edwards et al. 1979). It is plausible that the interaction between cancer growth signaling molecules and quercetin might lead to cancer growth inhibition (Fig. 3.3) (Nam et al. 2016). Therefore, it can be

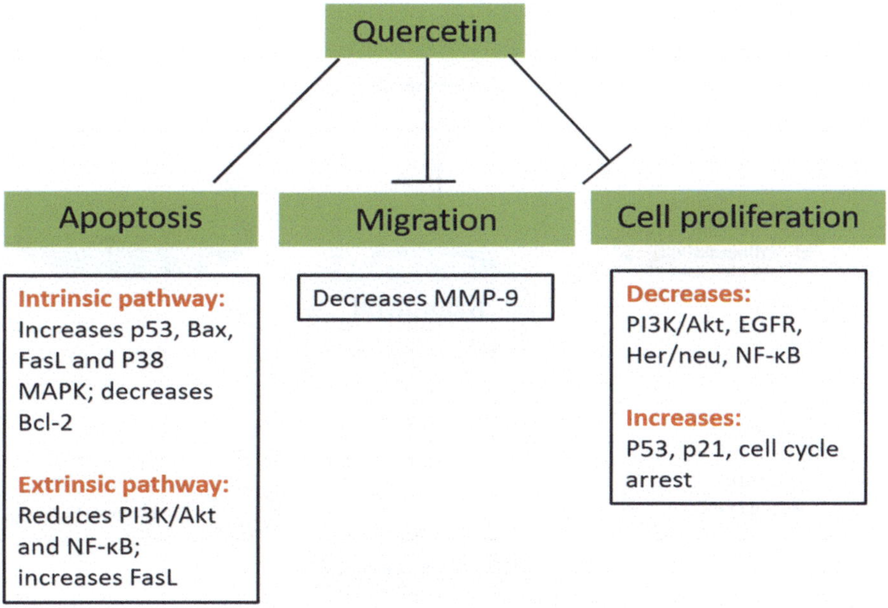

Fig. 3.3 Various cellular processes modulated by quercetin on cancer cells

considered that quercetin possesses free radical scavenging and antioxidant properties that might display the potential therapeutic potentials.

3.2.3 *Kaempferol*

Kaempferol, a tetrahydroxyflavone, contains four hydroxy groups at positions 3, 5, 7, and 4′ (Xiao et al. 2011). Kaempferol is present in a variety of plant parts and possesses therapeutic values, such as neuroprotective, cardioprotective, antidiabetic, anti-inflammatory, anticancer, antimicrobial, and antioxidant properties (Formica and Regelson 1995; Haddad et al. 2010; Kumar and Pandey 2013; Moon et al. 2001). Epidemiological studies reflected that kaempferol intake is associated with decreases in the occurrence of various cancer types (Bors et al. 1994; Prior 2003), suggesting the application of kaempferol in cancer therapy (Fig. 3.4) (Bors et al. 1994; Prior 2003; Ren et al. 2019).

3.2.4 *Epigallocatechin-3-Gallate (EGCG)*

EGCG is among the maximum studied chemopreventive flavonoid due to its effectiveness. EGCG is largely procured from green tea (Hayakawa et al. 2020). Besides chemoprevention, EGCG is also known for its activity against diabetes (Thielecke

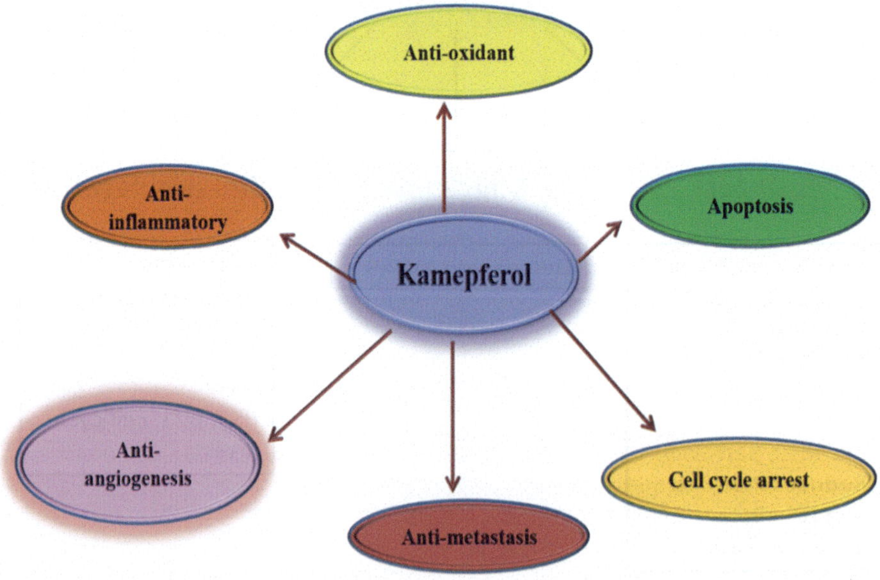

Fig. 3.4 Various cellular processes modulated by kaempferol on cancer cells

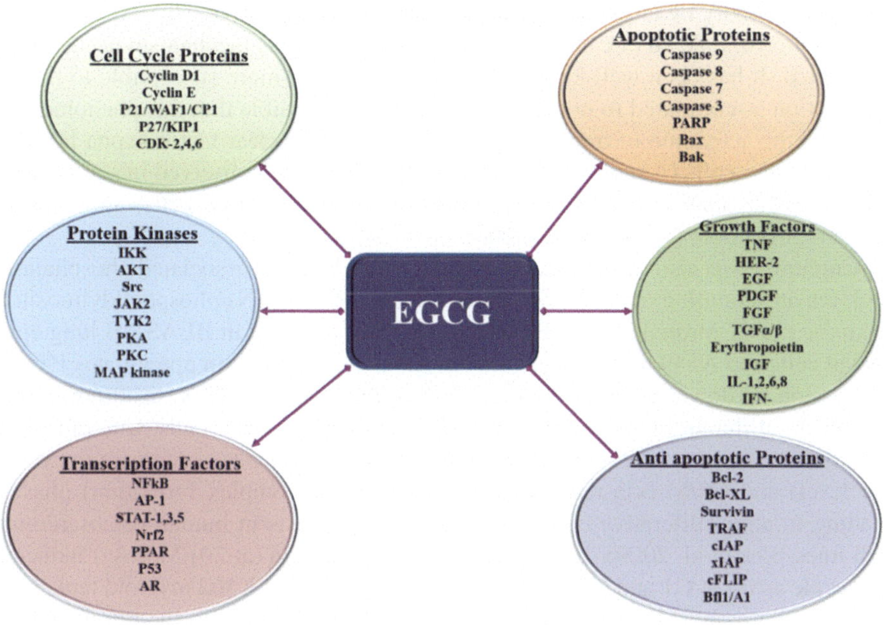

Fig. 3.5 Various cellular processes modulated by EGCG on cancer cells

and Boschmann 2009), cardiovascular (Tipoe et al. 2007), liver (Xiao et al. 2014), and neurodegenerative diseases (Hugel and Jackson 2012). The therapeutic benefits of EGCG are attributed by its structure, which is composed of three aromatic rings, i.e., A, B, and D, linked together by a pyran ring C. It has been known that the hydroxyl groups of the B and/or D rings are responsible for the single-electron transfer reactions and the transfer of hydrogen, resulting in the antioxidant activity of EGCG (Lambert and Elias 2010). In addition, the B and D rings are also involved in the inhibition of proteasome activity (Landis-Piwowar et al. 2005). The inhibition of heat-shock protein 90 is mediated by the A ring of EGCG (Khandelwal et al. 2013). The anticancer potential of EGCG has been attributed to its functionality to modulate various signaling pathways associated with the growth and proliferation of cancer cells (Fig. 3.5) (Singh et al. 2011).

3.3 Flavonoids as Anticancer Agent

3.3.1 Flavonoids Suppress Cancer Cell Growth via Regulating Cell Cycle

Genetic mutations in signaling proteins are the major cause for the formation of cancer cells. Therefore, modern therapies are aiming to restore the normal functioning of mutated genes for cancer interventions. Flavonoids offer its selective antitumor activity by targeting signaling mechanisms that are mostly related to cell proliferation and death of cancer cells, although the precise mechanism needs to be elucidated. It has been well-known that cancer cells are more susceptible to flavonoid action as compared to noncancerous cells. It is possible that this phenomenon is due to the activation of transcription factor, such as nuclear factor kappa B (NF-κB), in cancer cells (Aggarwal 2004). However, it has been observed in some cases that curcumin may induce growth inhibitory effect in normal cells, although at much higher doses (Watson et al. 2008).

Curcumin was also found to regulate tumor development associated phosphatase and tensin homolog deleted on chromosome 10 (PTEN)/phosphatidylinositol 3-kinase (PI3K)/protein kinase B (AKT) signaling pathway in BEAS-2B lung epithelial cells and A549 lung cancer cells. In A549 cells, curcumin upregulates PTEN expression by downregulating DJ1, a negative regulator of PTEN, resulting in the inhibition of the proliferation and growth of A549 lung cancer cells. Curcumin (at 27 μM dose) can cause sequential arrest in cell cycle in G1/S (via decrease in cyclin D1 level) and G2/M (via reduction in cdc2/cyclin B complex formation) phases leading to antiproliferative effect and inducing apoptosis in human osteosarcoma cell lines (Lee et al. 2009). In a different study, curcumin (at 20 μM dose) induced cell cycle arrest at G1 and G2/M phases in HL-60 cells and K562 myeloid leukemia cells, respectively (Martinez-Castillo et al. 2018). In addition to the regulation of cell cycle phases, curcumin is well-known to regulate various signaling pathways that are responsible for the growth and proliferation of cancer cells. It was found that curcumin arrests cell cycle in G2/M phase via regulating ataxia-telangiectasia mutated (ATM)/checkpoint protein 2 (CHK2)/tumor protein P53 signaling pathway in head and squamous cell carcinoma (HNSCC), both in vivo and in vitro. It is well-known that ATM acts as an indicator of DNA damage and its autophosphorylation at Ser-1981 reduces HNSCC proliferation (Hu et al. 2017). Various signaling pathways modulated by curcumin are shown in Fig 3.6 (Wang et al. 2019).

Quercetin, another flavonoid, regulates cell growth by inhibiting mammalian target of rapamycin (mTOR) activity. In addition, quercetin also regulates biosynthesis of proteins and autophagy. Quercetin downregulates P13K/Akt pathway in breast cancer cell HCC1937 (Gulati et al. 2006), SkBr3 cells (Jeong et al. 2008), liver cancer cell HepG2 (Maurya and Vinayak 2015a), and HL-60 leukemia (Yuan et al. 2012). In Dalton's lymphoma mice quercetin also inhibits glycolytic metabolism along with P13K-Akt-p53 pathway (Maurya and Vinayak 2015b). In addition, quercetin can also inhibit the growth of human epidermal growth factor receptor 2 (Her-2/neu) overexposing breast cancer cells *via* downregulating Her-2/neu gene (Jeong et al. 2008). Other signaling mechanisms observed to be modulated by quercetin are epidermal growth factor receptor (EGFR) signaling pathway in Sprague-Dawley male rats (Firdous et al. 2014) and Wnt/β-catenin signaling pathway via activation of Dickkopf (DKK), a Wnt antagonist, in 4T1 murine breast cancer cells (Kim et al. 2013), and in SW-480 colon cancer cells (Shan et al. 2009).

Quercetin regulates cell cycle via modulating various signaling molecules which can lead to the suppression of cancer cell growth. In addition, quercetin regulates various biochemical events that can arrest cell cycle events and inhibit cancer

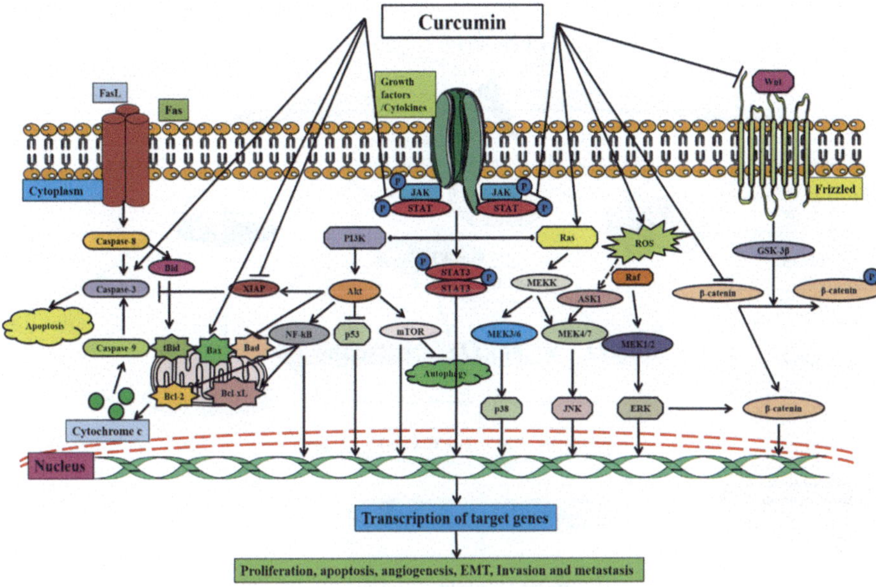

Fig. 3.6 Various signaling pathways regulated by curcumin for cancer growth inhibition

growth. It was observed that in HepG2 cells, quercetin arrests cell cycle in G1 phase and inhibits cell growth via regulating the expression of Cdk inhibitors, i.e., p21, p27, and p53 (Mu et al. 2007).Moreover, quercetin arrested cell cycle at G2/M in HeLa cells (Vidya Priyadarsini et al. 2010), A549 lung cancer cells (Yeh et al. 2011), SK0V ovarian cancer cells, U2OS osteosarcoma cells (Catanzaro et al. 2015), and TW206 and HSC-3 oral cancer cells (Huang et al. 2013a). Various signaling pathways modulated by quercetin are shown in Fig. 3.7 (Lai et al. 2013).

Kaempferol-induced cell cycle arrest at phase G2/M was also observed in SK-HEP-1 human hepatic cancer cells (Huang et al. 2013b), HT-29 human colon cancer cells (Cho and Park 2013), HeLa cervical cancer cells (Imran et al. 2019), MKN28 and SGC7901 gastric cancer cells (Song et al. 2015a), MDA-MB-231 breast cancer cells (Zhu and Xue 2019), and A2780/CP70 human ovarian cancer cells (Gao et al. 2018). Various signaling pathways modulated by kaempferol are shown in Fig. 3.8 (Rajendran et al. 2014).

EGCG was also found to induce cell cycle arrest and increases the percentage of cells in the G1 phase and reduces the percentage of cells in the S phase and G2 phase in the SKOV-3 human ovarian cancer cells (Rao and Pagidas 2010). EGCG was also observed to arrest cell cycle at G2/M phase, slightly decrease G0/G1 ratio of cells, and significantly decrease the ratio of cell population in S phase in MCF-7 cells. In addition, EGCG decreased mitochondrial membrane potential and promoted cell cycle arrest at G0/G1 phase in human HCC metastatic cell line HCCLM6 (Zhang et al. 2015). EGCG also promoted G1 phase arrest in mouse xenograft models of esophageal squamous cell carcinoma (Liu et al. 2015) and oral squamous

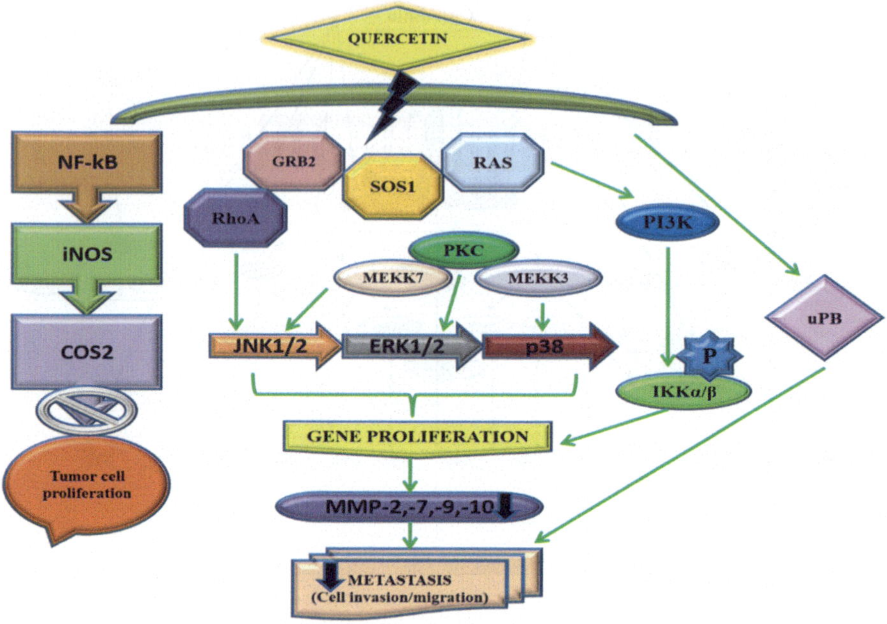

Fig. 3.7 Various signaling pathways regulated by quercetin for cancer growth inhibition

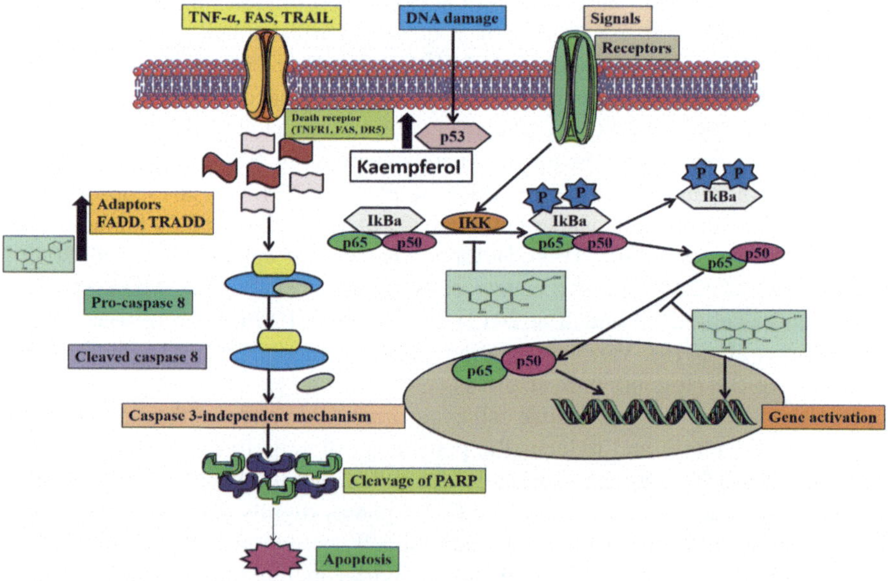

Fig. 3.8 Various signaling pathways regulated by kaempferol for cancer growth inhibition

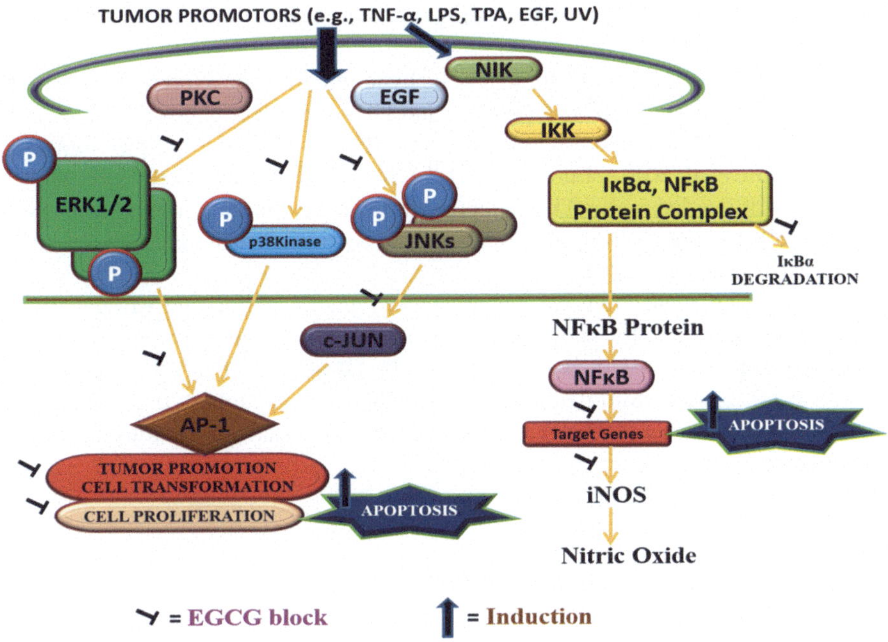

Fig. 3.9 Various signaling pathways regulated by EGCG for cancer growth inhibition

cancer cells (Yoshimura et al. 2019). Various signaling pathways modulated by EGCG are shown in Fig. 3.9 (Chen and Zhang 2007).

3.3.2 Flavonoids Inhibits Cell Proliferation via Angiogenesis

Angiogenesis is the formation of new blood vessels. In case of tumor, angiogenesis is responsible for the supply of nutrients and oxygen to the fast growing tumor cells (Jeong et al. 2020). Vascular endothelial growth factor (VEGF), angiopoietin-1 and angiopoietin-2, hypoxia-inducible factor-1α (HIF), and interleukin-2, interleukin-8, and interleukin-17 are important factors for angiogenesis (Lin et al. 2016; Norooznezhad and Norooznezhad 2017). Flavonoids can modulate these factors, thus regulating tumor growth as a result. It is noted that curcumin can regulate of VEGF expression, inhibiting its downstream signaling process (Saberi-Karimian et al. 2019). Antiangiogenesis ability of curcumin via targeting c-Met/PI3K/Akt/mTOR signaling pathway is reported in lung cancer cell lines (Jiao et al. 2016).

Quercetin can also possess antiangiogenic property, resulting in the inhibition of tumor growth. Quercetin targets VEGFR-2 and suppresses downstream signaling pathway via repressing AKT in prostate and breast cancers (Balakrishnan et al. 2016; Pratheeshkumar et al. 2012). Quercetin also increased the effect of anticancer drugs, limiting the angiogenesis of drug-resistant cells (Long et al. 2013; Oh et al. 2010).

Kaempferol can also inhibit angiogenesis by modulating the expression of VEGF through regulating both HIF-independent and HIF-dependent pathways (Luo et al. 2009) and extracellular signal-regulated kinase (ERK)-NF-κB-cMyc-p21 signaling pathways (Luo et al. 2012a; Luo et al. 2012b), in human ovarian cancer cells. Under hypoxic condition, kaempferol treatment inhibited the expression of HIF-1α protein in Huh7 cells via inactivation of p44/42 mitogen-activated protein kinase (MAPK) pathways (Mylonis et al. 2010).

Similarly, EGCG also regulates angiogenesis via downregulating VEGF expression in HT-29 human colon cancer cells (Jung et al. 2001), human pancreatic cancer cells (Shankar et al. 2013), breast cancer cells (Braicu et al. 2013), and human colorectal cancer cells (Shimizu et al. 2010). EGCG also modulates platelet-derived growth factor receptor (PDGFR) and EGFR, which are also contributing factors in cancer cell growth.

3.3.3 Flavonoids Induce Apoptosis

Apoptosis is well-known as programmed cell death regulating various critical physiological processes (Chong et al. 2020). Cross talk between various signaling molecules and pathways is responsible for both normal growth and apoptosis in cells. However, the dysregulation in pathways responsible for apoptosis might lead to the abnormal cell growth causing cancer progression (Lowe and Lin 2000). Apoptosis are of two types, mitochondria mediated and intrinsic or extrinsic (mediated by signals from other cells). It has been demonstrated that flavonoids regulate both extrinsic and intrinsic factors resulting in the cancer cell apoptosis. Indeed, caspases modulate both intrinsic and extrinsic apoptotic pathways. In the caspase-dependent apoptotic pathways, the cellular susceptibility of cancer cells against flavonoids or any chemotherapeutic drug is decided by the ratio between proapoptotic (Bax and Bak) and antiapoptotic (Bcl-2 and Bcl-xL) proteins.

3.3.3.1 EGCG

It has been reported that EGCG induces apoptosis via modulating the expression of the Bcl-2 family (downregulation) and Bax (upregulation) in nasopharyngeal cancer cells (Zhao et al. 2004), breast cancer cells (Roy et al. 2005), prostate cancer cells (Hastak et al. 2005), hepatic cancer cells (Nishikawa et al. 2006), bladder cancer cells (Qin et al. 2007), and ovarian cancer cells (Chen et al. 2013). Indeed, Bax is the downstream signaling molecule of p53 (Hastak et al. 2003). It is known that p53 gene plays a regulatory role in apoptosis, DNA repair, cell cycle, and checkpoints in the cells. Therefore, EGCG also induces cancer cell death via activation of p53 tumor suppressor protein. In addition, EGCG also induces proapoptotic and antitumorigenic nonsteroidal anti-inflammatory drug-activated gene-1 expression in head and neck cancer cells via inducing p53 (Kang et al. 2013). EGCG has been known to phosphorylate p53 at serine residue, resulting in the expression and activation of p53. Acetylation of p53 via EGCG is also reported that can increase p53 transcriptional activity by class 1 histone deacetylase inhibition (Thakur et al. 2012). Interestingly, p53 knockout activated alternative apoptosis pathway after EGCG administration in cervical cancer cells (Berindan-Neagoe et al. 2012).

In addition, EGCG upregulates a negative regulator of PI3K/Akt signaling, i.e., phosphatase and tensin homolog, in human pancreatic cancer cells resulting in the modulation of PI3K/Akt signaling pathway and apoptosis (Liu et al. 2013). EGCG also induces apoptosis via PI3K/Akt signaling pathway inhibition in bladder cancer cells (Qin et al. 2007). EGCG also induces caspase-independent apoptosis via cytochrome c-mediated translocation of endonuclease G (EndoG) and apoptosis-inducing factor (AIF) in the nucleus in laryngeal epidermoid cancer cells (Lee et al. 2010). It was observed that despite failure in caspase activation, cell viability is markedly decreased in laryngeal epidermoid cancer cells after EGCG treatment (Lee et al. 2010). It has been known that AIF and EndoG are involved in caspase-independent apoptotic pathway.

3.3.3.2 Curcumin

Curcumin can activate ATF6 transcription factor-C/EBP homologous protein 10 (CHOP) pathways in human papillary thyroid carcinoma (Zhang et al. 2018). CHOP is a proapoptotic protein that acts as endoplasmic reticulum (ER)-induced apoptosis indicator. Curcumin-activated CHOP can suppress the expression of Bcl-2 and can increase the ratio of Bax/Bcl-2 resulting in the induction of apoptotic pathway. Curcumin can also induce apoptosis by acting as proteasome inhibitor that causes increase in p53 in the cells, resulting in the caspase-3 and caspase-7 activation (Jana et al. 2004). Kuttikrishnan et al. (2019) showed that curcumin-mediated ROS generation can activate intrinsic apoptotic pathway in B-precursor lymphoblastic leukemia (ALL) (B-Pre-ALL) cell lines. Moreover, they showed that curcumin can potentiate the therapeutic effect of cisplatin, an anticancer drug. In addition, the inhibitors of apoptosis protein (IAPs) are also downregulated by curcumin in B-Pre-ALL cell lines. It has also been demonstrated that curcumin induces apoptosis via inhibition of PI3K/Akt signaling pathway (Kuttikrishnan et al. 2019).

Apoptosis is also mediated by ER stress, which is mainly related with the synthesis and formation of spatial structure of proteins. ER stressors play key role in adjusting the balance between the survival and apoptotic signaling (Sano and Reed 2013). Increase in ER stress can trigger mitochondrial-dependent or mitochondrial-independent apoptotic cell death (Wang et al. 2014b). Flavonoids have been shown to play a regulatory role in ER stress, although it may vary among different cell types (Wang et al. 2016), while it may enhance oxidative stress in cancer cells (Basile et al. 2013). It was observed that bis-hydroxy-curcumin (30 μM), a curcumin derivative, induces apoptosis in HCT116 human colon cancer cells via caspase-4-mediated ER stress induction (Basile et al. 2013). In addition, caspase-3-mediated apoptosis was observed in human osteosarcoma cells after curcumin treatment (Lee et al. 2009). Moreover, curcumin decreases Ca^{2+} concentration in ER via decreased Ca^{2+}-ATPase activity, resulting in the stress-induced efflux of Ca^{2+} from ER and accumulation of Ca^{2+} in mitochondrial matrix. In mitochondria, the accumulation of Ca^{2+} causes alterations in transmembrane mitochondrial potential, resulting in the generation of ROS followed by rupture of outer mitochondrial membrane and release of cytochrome C in the cytoplasm (Zhang et al. 2018).

3.3.3.3 Quercetin

Quercetin induces apoptosis by increasing the levels of proapoptotic protein, such as Bax, and decreasing the levels of antiapoptotic protein, such as Bcl-2, in MDA-MB-231 breast cancer cells (Chien et al. 2009). In HT-29 colon cancer cells, quercetin induced both p53-dependent apoptotic cell death (Kim et al. 2010) and p53-independent apoptotic pathway (Kim et al. 2014), via activating protein kinase (AMPK). In addition, in ER alpha-negative breast cancer cells also quercetin induced p53-dependent apoptotic cell death (Seo et al. 2011). Therefore, although quercetin showed induction of apoptosis in various cancer cells, further studies are needed to understand more mechanistic details.

3.3.3.4 Kaempferol

Kashafi and coworkers showed that kaempferol (12–100 μM) induces dose-dependent apoptotic and cell death in HeLa cervical cancer cells. In addition, kaempferol also reduced cell viability and downregulated PI3K/AKT and human telomerase reverse transcriptase (hTERT) pathways (Kashafi et al. 2017). Another study showed that orally administered kaempferol can induce dose-dependent apoptosis in cervical cancer cells and suppress its growth by disrupting mitochondrial membrane potential and elevating intracellular free calcium (Tu et al. 2016). Kaempferol can also induce apoptosis; decrease expressions of Bcl-2, p-TIMP2, p-AKT, and p-MMP2; and increase expressions of Fas, Bax, cleaved caspase-3, caspase-8, caspase-9, and cleaved PARP, in HCCC9810 and QBC939 lung metastasis model (Qin et al. 2016). Kaempferol was also observed to induce apoptosis via increasing expressions of p21 and Bax, decreasing expressions of cyclin E and D1 and cathepsin D, and reducing phosphorylation of AKT, IRS-1, ERK, and MEK1/2, resulting in the tumor growth inhibition in MCF-7 breast cancer xenograft mouse model (Jo et al. 2015; Kim et al. 2016). In addition, kaempferol also showed dose-dependent apoptotic activity in MIA PaCa-2, Panc-1, and SNU-213 human pancreatic cancer cell lines via inhibition of ERK1/2, EGFR-related Src, and AKT pathways (Lee and Kim 2016).

3.3.4 Flavonoids Affect Oxidative Stress

Flavonoids are known to act as scavenger of ROS, superoxide anion, hydroxyl radical, nitrogen dioxide, and reactive nitrogen species (RNS), which results in the prevention of oxidative damage. ROS/RNS are normally produced during the cellular metabolism and play a dual role, both as protective and as damaging agents, depending upon their concentration. ROS can induce transformation of normal cells into malignant cells. Therefore, inhibition of ROS production can protect and even inhibit this transformation (Lopez-Lazaro 2010; Salehi et al. 2018; Sharifi-Rad et al. 2018). The superoxide anions are converted into H_2O_2 by superoxide dismutase that can produce highly reactive hydroxyl radical after reacting with reduced metals (e.g., cuprous or ferrous ions) or produce peroxynitrite after reacting with nitric oxide. Both of these highly reactive species may cause damages in the biomolecules, such as lipids, proteins, or DNA (Wang et al. 2006). It has been observed that flavonoids can (i) scavenge hydroxyl radical, superoxide anion, and peroxynitrite (Heijnen et al. 2001; Klaunig and Kamendulis 2004; Wang et al. 2006), (ii) inhibit the expression of pro-oxidant enzymes (Ozyurek et al. 2009), (iii) induce the expression of antioxidant enzymes (Heijnen et al. 2001; Hong et al. 2009), and (iv) chelate ferrous or cuprous inhibiting the hydroxyl radical generation (Mira et al. 2002; Ren et al. 2008) . In addition, flavonoids can activate antioxidant enzymatic system by activating genes encoding glutathione peroxidase (GPx), superoxide dismutase (SOD), S-glutathione transferase (GST), etc. The imbalance between ROS/RNS generation and antioxidant enzymatic system may cause cell damage that might result in the onset of critical pathological conditions.

It has been observed that ROS activates Kelch-like ECH-associated protein 1 (KEAP1)-nuclear factor erythroid 2-related factor 2 (NRF2)-antioxidant response elements (AREs) antioxidant system, which initiates by conformational changes in KEAP1, which lead to the bond cleavage between KEAP1 and NRF2. This results in the release of NRF2, followed by the activation of antioxidant system. It was found that curcumin acts as an electrophile that binds to KEAP1 via sulfhydryl group (-SH) and modifies the cysteine group (Trujillo et al. 2014). Curcumin inhibits IKK/NF-κB pathway and inhibits serine-threonine kinase (Bharti et al. 2003). Curcumin also plays a role in regulating oxidative redox status in human leukemia (K562) cell line by binding and regulating several enzymes involved in ROS metabolism, such as glutathione-S-transferase phi 1 (GSTP1), glyoxalase I (GLO1), aldo-keto reductase family 1 member 1 (AKR1C1), NAD(P)H dehydrogenase [quinone] 1 (NQO1), and carbonyl reductase 1 (CBR1) (Cairns et al. 2011; Gorrini et al. 2013; Larasati et al. 2018; Panieri and Santoro 2016).

It has been observed that quercetin activates NRF2, leading to an increase in the expressions of GPx and GST in liver carcinoma cells HepG2 (Granado-Serrano et al. 2012). In addition, quercetin also inhibited lung tumors in mouse xenograft model via p53-mediated ROS pathway (Xingyu et al. 2016), suggesting the possibility of flavonoids to act in cancer chemoprevention and treatment via regulating oxidative stress pathways.

3.3.5 Inhibition of Metastasis

Metastasis is the stage of cancer when it spreads from one organ to another and is perhaps the most dangerous form of cancer. Therefore, inhibition of metastasis could be a possible target for cancer therapy. It has been found that matrix metalloproteinase's (MMPs) enzymes, such as MMP-9 and MMP-2, are involved in the degradation of extracellular matrix proteins in a variety of cells or tissues (Liabakk et al. 1996), suggesting that the suppression of MMPs can minimize the possibility of metastasis of cancer. It has been found that flavonoids inhibit the secretion of MMPs. For example, in MCF-7 breast cancer cells treated with 12-O-tetradecanoylphorbol-13-acetate (TPA), quercetin suppresses the activation and migration of MMP-9, resulting in the prevention of metastasis (Lin et al. 2008). Moreover, in JB6 P+ mouse epidermal cells, quercetin also inhibited tumor necrosis factor (TNF)-α-mediated upregulation of MMP-9 (Hwang et al. 2009). Along with MMP-9, decrease in the secretion of MMP-2 was also observed in MIA PaCa-2 pancreatic cancer cells (Lee et al. 2004) and A431 epidermoid cancer cells (Huang et al. 1999), post-quercetin treatment.

In addition to quercetin, other flavonoids also inhibited the metastasis of cancer. Curcumin also inhibited MMP-9 in prostate cancer (Hong et al. 2006), colon cancer (Su et al. 2006), human promyelocytic leukemia (Tan et al. 2006), breast cancer (Bachmeier et al. 2007), human laryngeal squamous carcinoma (Mitra et al. 2006), metastatic colorectal cancer (Kunnumakkara et al. 2009), U373 cells (Abdullah Thani et al. 2012), lung cancer (Chen et al. 2014; Lin et al. 2009), nasopharyngeal carcinoma (Xie et al. 2014), and liver metastasis from colorectal cancer (Shen et al. 2014). It was observed that curcumin inhibited lung metastasis in a breast cancer nude mouse model (Aggarwal et al. 2005). It was found that curcumin inhibits human colon cancer colo205 cells migration via inhibiting NF-κB and downregulation of cyclooxygenase-2 and MMP-2 expression (Su et al. 2006). Another study showed that curcumin attenuated Wnt/β-catenin pathway in meduloblastoma cell lines (He et al. 2014) and in non-small cell lung cancer (Lu et al. 2014). The migratory ability of highly metastatic mouse hepatoma Hca-F cells is also modulated by curcumin inactivating EGFR and Cav-1 signaling pathways (Wang et al. 2011). Curcumin also inhibited the metastatic potential and migration of human ovarian CaOV3 cancer cell line via attenuated EGF-induced aquaporin water channels in cells (Ji et al. 2008).

In addition, EGCG was also shown to inhibit metastasis and invasion via decreasing the expression of hepatocyte growth factor (HGF)-induced MMP- 9 and activating urokinase-type plasminogen activator (uPA) in hypopharyngeal carcinoma cells (Lim et al. 2008). EGCG was also observed to suppress HGF-Met signaling and decrease uPA activation and expression of MMP-2 and MMP-9 in oral cavity cancer cells and melanoma cells (Koh et al. 2011; Kwak et al. 2011). EGCG also reduces activation of small guanosine triphosphatase proteins, i.e., Rac and Rho. Inhibiting the activity of Rho inhibited the invasion in three-dimensional oral squamous cell carcinoma models (Hwang et al. 2013), while inhibition of Rac1 activity blocked cell migration and invasion via downregulating vasodilator-stimulated phosphoprotein expression in breast carcinoma cells (Zhang et al. 2009). In addition, EGCG prevents migration and inhibition in heregulin 1-induced breast carcinoma cells (Kushima et al. 2009) and thrombin-induced hepatocellular carcinoma cells (Kaufmann et al. 2009), respectively. Therefore, it can be suggested that EGCG can inhibit cancer cell metastasis by inhibiting the growth and invasion of cancer cells (Min and Kwon 2014).

3.3.6 Induction of Autophagy

In contrast to apoptosis, autophagy is involved in the lysosomal degradation of cellular organelles and macromolecules (Ma et al. 2020). Autophagy leads to degradation of cellular constituents by sequestering them into double membrane vesicles, i.e., autophagosomes, followed by fusion with lysosomes and degradation. Autophagy plays a dual role both as pro-death and pro-survival factor and is essential for cellular homeostasis (Liu et al. 2020; Unal et al. 2020). Although autophagy shows pro-survival effect on both cancer and normal cells, deregulation of autophagy leads to the activation of cell death signals. It has been observed that in chloroquine (an autophagy inhibitor)-treated mesothelioma cells, EGCG enhances LC3 II form and degrades p62, leading to autophagy in the cells (Satoh et al. 2013).This suggests that flavonoids can induce autophagy in cancer cells (Min and Kwon 2014). However, more mechanistic details on flavonoid-mediated autophagy in cancer cells need to be elucidated.

3.4 Enhancing the Bioavailability of Flavonoids in Cancer Therapeutics

Although flavonoids show various biological activities, they have poor bioavailability. Due to the poor bioavailability, the use of flavonoids as anticancer agent is limited. Various methods have been employed to increase the bioavailability of flavonoids, such as chemical modifications and conjugation with nanoparticle and/or antibodies for targeted delivery which are discussed below.

3.4.1 EGCG

Despite the various biological advantages of EGCG, it possesses several drawbacks. Firstly, EGCG lacks stability in both water and physiological fluids (Hong et al. 2002). Secondly, EGCG has low oral bioavailability, reducing the absorption of

EGCG by humans and laboratory animals (Lambert and Yang 2003; Warden et al. 2001). Thirdly, EGCG is rapidly metabolized and degraded by the enzymes present in the tissues. Fourthly, the target specificity of EGCG is low for cancer cells. Therefore, to overcome these vital issues, several attempts have been made to improve the bioavailability, stability, and target specificity of EGCG for cancer cells and improve its anticancer activities (Lambert and Yang 2003; Warden et al. 2001). As noted, the stability of EGCG can be improved by nanoliposomes (de Pace et al. 2013; Wang et al. 2014a; Zhang et al. 2013) and can be prevented from early degradation (Zhang et al. 2013). Moreover, encapsulation of EGCG in nanoparticles can reduce the effective dosage of EGCG. In addition, the attachment of targeting moieties on EGCG-encapsulated nanoparticle surface can provide the target specificity toward cancer cells, resulting in enhanced delivery of EGCG (Allen and Cullis 2004).

In a study, micellar nanocomplexes carrying and supplying Herceptin to breast cancer cells were made from EGCG derivatives. It was observed that these nanocomplexes decreased the viability of cancer cells in vitro and in vivo (Chung et al. 2014), suggesting the increased uptake and stability of EGCG in cancer cells. Moreover, EGCG-Ca/Al-NO_3-layered double hydroxide nanoparticles also showed increased apoptosis and reduced cell viability in human PC3 prostate cancer cells (Shafiei et al. 2015). Another study recorded that EGCG chitosan nanoparticles showed anticancer effects on Mel-928 human melanoma cells by apoptosis through increased level of Bax, decreased levels of caspase-3 and caspase-9 and Bcl-2, increased cleavage of poly(ADP ribose) polymerase (PARP), inhibition of cyclin D1 and D3, cell cycle arrest at G2/M phase, and increased expression in p21 and p27.

Anticancer effect of EGCG chitosan nanoparticles via suppression of tumor growth and proliferation, CDK4 and CDK6 inhibitions, and an increase in apoptosis was observed in the xenograft athymic mouse model of melanoma (Siddiqui et al. 2014). In combination with paclitaxel, EGCG core-shell PLGA-casein nanoparticles also showed anticancer activity by increasing apoptosis and decreasing NF-κB activation in human MDA-MB-231 and MCF-7 breast cancer cells (Narayanan et al. 2015). EGCG-containing green tea polyphenols were nanoformulated by graphene nanosheets as one of the major components and showed growth inhibitory effects on HT-29 and SW48 colon cancer cells (Abdolahad et al. 2013).

Nanoparticles may therefore increase the stability and bioavailability of EGCG, boost its sustained release, and deliver EGCG to cancer cells in a targeted manner, which could suggest a novel approach for cancer prevention and treatment (Davatgaran-Taghipour et al. 2017; Li et al. 2015). Some of the studies demonstrating enhanced uptake and anticancer ability of EGCG encapsulated in various nanoparticles are shown in Table 3.1.

Table 3.1 Nanoparticle-mediated delivery of EGCG

Cancer	Cancer type	Nanoparticle type/polymer	Size (nm)	Loading efficiency	Outcome	Reference
Prostate	Du145	Maltodextrin/gum arabic NPs	120 ± 28	85%	Decrease (10–20%) in cell proliferation at lower concentrations	Rocha et al. (2011)
	22Rv1	Chitosan NPs	150–200	10%	Increase in poly(ADP-ribose) polymerase cleavage, Bax Decrease in Bcl-2, caspases Growth inhibitory effect	Khan et al. (2014)
Colon	Caco-2	Chitosan-caseinophosphopeptide NPs	150 ± 4.3	24–53%	Elevation of EGCG bioavailability	Hu et al. (2012)
	Caco-2	Ovalbumin-dextran NPs	285	23.45%	Improved absorption of EGCG	Li and Gu (2014)
	Caco-2	Glutaraldehyde-ovalbumindextran conjugate NPs	339	30%	Improved absorption of EGCG	Li and Gu (2014)
	Caco-2	Chitosan- coated bovine serum albumin NPs	300	32.7%	Improved absorption of EGCG	Li et al. (2014)
Gastric	BGC823	Genipin-cross-linked caseinophosphopeptide chitosan NPs	236 ± 26.3	71%	Strong cytotoxicity, high biocompatibility, slow release of EGCG	Hu et al. (2014)
	MKN45, Luc MKN45	Fucose-chitosan/polyethylene glycol-chitosan/gelatin NPs	257.5 ± 5.1	53.7 ± 1.8%	Cell growth inhibition Induction in cell apoptosis Reduction in VEGF protein	Lin et al. (2015)
Breast	MCF-7	Chitosan-coated liposomes	85 ± 6.6	90%	Enhancement in the chemopreventive effect Decreased immunogenicity and side effects	de Pace et al. (2013)

(continued)

Table 3.1 (continued)

Cancer	Cancer type	Nanoparticle type/polymer	Size (nm)	Loading efficiency	Outcome	Reference
	MDA-MB-231, MCF-7	Poly(lactic-co-glycolic acid) and Casein Nanocarriers	190 ± 12	76.8 ± 9.1%	Induction in apoptosis Inhibition in activation of NF-κB, Inhibition of angiogenesis and tumor metastasis	Narayanan et al. (2015)
Melanoma	Mel-928	Chitosan NPs	150–200	10%	Induction of apoptosis Inhibition of cell proliferation Cell cycle arrest	Siddiqui et al. (2014)
Liver	HepG2	Genipin-cross-linked caseinophosphopeptidechitosan NPs	236 ± 26.3	71%	Increased cytotoxicity Slow release and biocompatible	Hu et al. (2014)

3.4.2 *Quercetin*

Studies have been performed to overcome the restrictions linked with the use of quercetin as anticancer agent. Quercetin loaded with phenylboronic acid (PBA)-conjugated zinc oxide nanoparticles (PBA-ZnO-Q) has been synthesized that enhanced oxidative stress and induced mitochondrial damage in human breast cancer cells (MCF-7), leading to apoptotic cell death in vitro (Sadhukhan et al. 2019). Additionally, PBA-ZnO-Q also showed reduction in tumor growth in tumor-bearing mice with reduced toxicity associated with liver, kidney, and spleen (Sadhukhan et al. 2019). In another study, poly(lactic acid) (PLA) nanoparticle-loading quercetin (PLA-Q) also showed sustained release kinetics and anticancer efficacy against human breast cancer cells (Pandey et al. 2015). Orally administration of PLGA-NPs coencapsulated with quercetin and tamoxifen (PLGA-NPs-Tmx-Q) efficiently normalized levels of MMP-2 and MMP-9, preventing tumor angiogenesis in DMBA-induced breast cancer model in female SD rats (Jain et al. 2013). In addition, no measurable hepatotoxicity was observed by PLGA-NPs-Tmx-Q nanoparticle (Jain et al. 2013). Similarly, MPEG-PLA encapsulated quercetin nanoparticle (MPEG-PLA-Q) efficiently induce apoptosis in triple negative breast cancer cell line (MDA-MB-231) and inhibited the tumor growth in xenograft model for breast cancer after post-tumoral injection (Sharma et al. 2015). Liposomal quercetin also increased the bioavailability and solubility of quercetin, in addition to enhanced cellular uptake in human MCF-7 breast cancer cells (Rezaei-Sadabady et al. 2016). The nanoribbons of quercetin are also suggested for their antiproliferatory effects on cancer cell (Han et al. 2012).

The PLGA nanoparticles, encapsulating etoposide and quercetin dihydrate separately, significantly increased the cytotoxicity in A549 (human lung adenocarcinoma epithelial cell line) (Pimple et al. 2012). A stable polyethylene glycol (PEG)-derivatized phosphatidylethanolamine (PE) (PEG-PE) nanomicellar formulation of quercetin significantly improved the anticancer activity of quercetin in A549 cancer cell line and murine xenograft model and no apparent toxicity to the intestinal epithelium (Tan et al. 2012). In JR8 human melanoma cells, aminopropyl functionalized mesoporous silica nanoparticles (NH2-MSN) were used as a topical carrier system for quercetin delivery, which showed antiproliferative effect on cancer cells (Sapino et al. 2015). The active targeting of PEG-PLGA-quercetin nanoparticles was also observed using folic acid as targeting agent in HeLa cells (El-Gogary et al. 2014).

Chitosan nanoparticles loaded with quercetin and 5-fluorouracil showed significant growth inhibitory effect on pancreatic cancer cells, both in 2D and 3D cultures (David et al. 2015). It was observed that 5-fluorouracil was localized in the periphery, while quercetin was located towardthe core of chitosan nanoparticles (David et al. 2015). It was also observed that quercetin-loaded liposomes showed necrosis without apoptosis in C6 glioma (Wang et al. 2012). Oral delivery of paclitaxel, an anticancer drug, was also improved by an amphiphilic carboxymethyl chitosan-quercetin (CQ) conjugate (Wang et al. 2014c). Table 3.2 shows some of the nanoparticle-based quercetin delivery against various types of cancer cells.

Table 3.2 Nanoparticle-mediated delivery of quercetin

Cancer	Cancer type	Nanoparticle type/ polymer	Size d. (nm)	Loading efficiency	Outcome	References
Breast	MDA-MB231	PLA	46 ± 6	62% ± 3%	~40% decrease in cell viability in 5 days	Jain et al. (2013)
	DMBA-induced breast cancer	PLGA (coencapsulated quercetin and tamoxifen)	185.3 ± 1.2	67.16% ± 1.24% (tamoxifen), 68.60% ± 1.58% (quercetin)	Increase in cell cytotoxicity Tumor was reduced to ~32.36% after 30 days	Jain et al. (2013)
	MDA-MB-231 and 4T1	MPEG-PLA	155.3 ± 3.2		~38% lower cell viability compared to control and reduced tumor size as compared to control	Sharma et al. (2015)
	MCF-7	Phosphatidylcholine	100.974 ± 0.3	40.7% ± 3.1%	~83% inhibition in cell proliferation at 48 h	Rezaei-Sadabady et al. (2016)
	MCF-7 and MDA-MB-231	Soy lecithin, glyceryl tridecanoate, glyceryl tripalmitate, vitamin E acetate, Kolliphor HS15	32	95%	~13.7% reduction in viability of MCF-7 cells and ~13.4% reduction in viability of MDA-MB-231 cells at 48 h	Sun et al. (2014)
	4T1	Nanoribbon fabricated by atmospheric pressure PVD	100–200	NA	~57% reduction in cell viability	Han et al. (2012)
Melanoma	JR8	Aminopropyl functionalized mesoporous silica nanoparticle	250 ± 50	NA	~50% inhibition of cell proliferation at 72 h and higher amount of quercetin was retained in the skin as compared to control at 24 h	Sapino et al. (2015)
(Sapino et al. 2015) Lung	A549	PLGA (combination treatment of quercetin and etopside)	153.4 ± 4.2 (etopside), 148.6 ± 1.6 (quercetin)	63.88% ± 1.5% (etopside), 41.36% ± 3.4% (quercetin)	Enhanced cytotoxic effect compared to free drug combination at 72 h	Pimple et al. (2012)

(continued)

	A549	DSPE-PEG2000 nanomicelle	15.4–18.5	>88.9%	Decreased cell viability at 72 h and ~1.5-fold higher tumor growth inhibition than free quercetin control group	Tan et al. (2012)
Cervical/ ovarian	HeLa/IGROV-1	PEG-PLGA and folic acid as targeting ligand	155.0 ± 1.2	97.8 ± 0.14	Folic acid enhances selective uptake of nanoparticles by folate receptor enriched cancer cells	El-Gogary et al. (2014)
Pancreatic	MiaPaCa2,	Chitosan	300	91%	Dose-dependent cell inhibition	David et al. (2015)
	MiaPaCa2	Chitosan (quercetin and 5-flourouracil dual drug loading)	400	95% (quercetin), 75% (flourouracil)	~70% decrease in cell viability	David et al. (2015)
Liver	HepG2	Chitosan-quercetin conjugate loaded with paclitaxel	185.8 ± 4.6	85.63% ± 1.26%	Dose-dependent cytotoxic effect with IC50 0.11 μg/mL and ~71.22% reduction in tumor size	Wang et al. (2012)
Glioma	C6	Glyceryl behenate, soy lecithin, and cholesterol	116.7	NA	Induced necrotic cell death	Wang et al. (2014c)

3.4.3 Curcumin

Although, curcumin possesses anticancer property on several kinds of cancer cells, such as lung cancer, breast cancer, pancreatic cancer, etc., it is hydrophobic and has less bioavailability. The dispersibility and chemical stability of curcumin can be done by encapsulating curcumin in solid lipid nanoparticles (SLNs), prolonging its antitumor activity and cellular uptake, and enhancing its bioavailability (Sun et al. 2013). It was observed that curcumin-loaded SLNs showed profound tumor inhibition on lung cancer (Wang et al. 2013). SLNs loaded with curcumin and layered double hydroxides loaded with 5-fluorouracil showed synergetic effect on SMMC-7721 liver cancer cell lines (Zhu et al. 2013). The combination of curcumin/5-fluorouracil loaded thiolated chitosan nanoparticles also showed enhanced anticancer effects on colon cancer cells in vitro and improved the bioavailability of the drugs in vivo (Anitha et al. 2014). The oral bioavailability of curcumin was increased using nanostructured lipid carriers, with increased cell growth inhibition in prostate cancer cells (Aditya et al. 2013).

The therapeutic potential of PLGA-curcumin nanoparticles against prostate cancer was also observed. It was observed that PLGA-curcumin nanoparticles were efficiently internalized in prostate cancer cells and release biologically active curcumin in cytosolic compartment of cells for effective therapeutic activity (Yallapu et al. 2014). Similarly, nanocurcumin synthesized using PLGA showed enhanced aqueous solubility and higher anticancer efficacy(Nair et al. 2012). Intravenously injectable aqueous formulation of curcumin encapsulated into monomethoxy poly(ethylene glycol)-poly(ε-caprolactone) micelles showed cytotoxicity C-26 colon carcinoma cells in vitro through inhibiting angiogenesis and directly killing cancer cells (Gou et al. 2011). It was suggested that nanocurcumin could limit the progression of the cancer and prevent metastasis by modulating vascular inflammation (Palange et al. 2014).

The antitumor study of a liposomal curcumin formulation showed reduced human pancreatic cancer growth against human MIA PaCa pancreatic cancer cells both in vitro and in xenograft studies, with no limiting side effects, suggesting that liposomal curcumin may be beneficial in patients with pancreatic cancer (Ranjan et al. 2013). Some of the studies related with nanoparticle-mediated enhanced delivery and antitumor efficacy of curcumin are shown in Table 3.3.

3.4.4 Kaempferol

Nanoparticles loaded with kaempferol have been studied for the enhanced anticancer activities (Luo et al. 2012a) and developed poly(ethylene oxide)-poly(propylene oxide)-poly(ethylene oxide) (PEO-PPO-PEO) and poly(DL-lactic acid-co-glycolic acid) (PLGA) nanoparticles loaded with kaempferol against ovarian cells. These nanoparticles significantly reduced cell viability of cancerous cells (Luo et al.

Table 3.3 Nanoparticle-mediated delivery of curcumin

Cancer	Cancer type/ cells	Nanoparticle type/ polymer	Size d. (nm)	Outcome	References
Breast	MDA-MB-231	Lipid-polymer nanoparticles	172	Decreases metastasis and inflammation	Palange et al. (2014)
	MCF-7	Solid lipid nanoparticles	153	Increases intracellular uptake, bioavailability, and anticancer activity	Sun et al. (2013)
Colon	HT-29	Thiolated chitosan nanoparticles	150	Increases sustained release, cellular uptake, apoptosis, anticancer activity, and bioavailability and decreases cancer cell viability	Anitha et al. (2014) and Anitha et al. (2019)
		MPEG-PCL micelles	27	Decreases tumor cell-induced angiogenesis and cancer cell growth	Gou et al. (2011)
Prostate	C4-2, DU-145, PC3	PLGA nanoparticles		Decreases proliferation of cancer cells, tumor growth, key oncogenic proteins, and oncogenic miR21 and increases apoptosis and miR-205	Yallapu et al. (2014)
	PC3	Nanostructured lipid carriers	108	Increases sustained release and anticancer activity and decreases cell viability	Aditya et al. (2013)
Liver	SMMC-7721	Solid lipid nanoparticles	100	Increases sustained release and apoptosis and decreases cell viability	Zhu et al. (2013)
	HeLa	PLGA nanoparticles	120-190	Increases aqueous solubility and sustained release, cellular uptake, and anticancer efficacy	Nair et al. (2012)
Lung	A549	Solid lipid nanoparticles	20-80	Increases the stability of curcumin, apoptosis and concentration in plasma, and tumor distribution of curcumin and decreases growth of lung cancer cells in vivo and tumor volume	Wang et al. (2013)
Pancreatic	MIA PaCa	Liposomes	100	Decreases cell viability, tumor growth, and angiogenesis	Ranjan et al. (2013)
Cervical	KB-V1 and KB-3-1	PLGA nanoparticles	130	Increase in targeting to cancer cells and cellular uptake and decrease in cell viability	Punfa et al. (2012)

2012a). Kaempferol was found to conjugate with small-size gold nanoclusters (AuNCs) that may target and damage the nuclei of the A549 lung cancer cells (Govindaraju et al. 2019). Intranasal administration of kaempferol chitosan nanoemulsions showed effective brain delivery and antitumor activity against glioma cells (Colombo et al. 2018).

3.5 Conclusion

The advancements in the nanoparticle-mediated uptake of phytochemicals indicate that nanoparticles loaded with phytochemicals might soon enter into the therapeutic regime as a sole or as an adjuvant therapy against various types of cancer. Therefore, we can conclude that the anticancer property of flavonoids can be enhanced using nanoparticle-mediated delivery approach. This is due to the increased bioavailability, stability, and target specific delivery that enhances the uptake of flavonoids in the cells. However, this kind of nanoparticle-based delivery of flavonoids still remains to be clinically validated.

Acknowledgments This research was supported by Basic Science Research Program through the National Research Foundation of Korea (NRF) funded by the Ministry of Education (No. 2018R1A6A1A03025582) and supported by the Technology Innovation Program (20009663) funded by the Ministry of Trade, Industry and Energy (MOTIE, Korea).

Conflicts of Interest The authors declare no conflicts of interests.

References

Abdolahad M, Janmaleki M, Mohajerzadeh S, Akhavan O, Abbasi S (2013) Polyphenols attached graphene nanosheets for high efficiency NIR mediated photodestruction of cancer cells. Mater Sci Eng C Mater Biol Appl 33:1498–1505. https://doi.org/10.1016/j.msec.2012.12.052

Abdullah Thani NA et al (2012) Induction of apoptosis and reduction of MMP gene expression in the U373 cell line by polyphenolics in Aronia melanocarpa and by curcumin. Oncol Rep 28:1435–1442. https://doi.org/10.3892/or.2012.1941

Aditya NP, Shim M, Lee I, Lee Y, Im MH, Ko S (2013) Curcumin and genistein coloaded nanostructured lipid carriers: in vitro digestion and antiprostate cancer activity. J Agric Food Chem 61:1878–1883. https://doi.org/10.1021/jf305143k

Aggarwal BB (2004) Nuclear factor-kappaB: the enemy within. Cancer Cell 6:203–208. https://doi.org/10.1016/j.ccr.2004.09.003

Aggarwal BB, Shishodia S, Takada Y, Banerjee S, Newman RA, Bueso-Ramos CE, Price JE (2005) Curcumin suppresses the paclitaxel-induced nuclear factor-kappaB pathway in breast cancer cells and inhibits lung metastasis of human breast cancer in nude mice. Clin Cancer Res 11:7490–7498. https://doi.org/10.1158/1078-0432.CCR-05-1192

Allen TM, Cullis PR (2004) Drug delivery systems: entering the mainstream. Science 303:1818–1822. https://doi.org/10.1126/science.1095833

Alvarado-Miranda A et al (2009) Concurrent chemo-radiotherapy following neoadjuvant chemotherapy in locally advanced breast cancer. Radiat Oncol 4:24. https://doi.org/10.1186/1748-717X-4-24
Angst E et al (2013) The flavonoid quercetin inhibits pancreatic cancer growth in vitro and in vivo. Pancreas 42:223–229. https://doi.org/10.1097/MPA.0b013e318264ccae
Anitha A, Deepa N, Chennazhi KP, Lakshmanan VK, Jayakumar R (2014) Combinatorial anticancer effects of curcumin and 5-fluorouracil loaded thiolated chitosan nanoparticles towards colon cancer treatment. Biochim Biophys Acta 1840:2730–2743. https://doi.org/10.1016/j.bbagen.2014.06.004
Anitha A, Deepa N, Chennazhi KP, Lakshmanan VK, Jayakumar R (2019) Corrigendum to "Combinatorial anticancer effects of curcumin and 5-fluorouracil loaded thiolated chitosan nanoparticles towards colon cancer treatment" [Biochimica et Biophysica Acta 1840 (2014) 2730-2743]. Biochim Biophys Acta Gen Subj 1863:992. https://doi.org/10.1016/j.bbagen.2019.03.008
Annovazzi L, Mellai M, Schiffer D (2017) Chemotherapeutic drugs: DNA damage and repair in glioblastoma. Cancers 9. https://doi.org/10.3390/cancers9060057
Bachmeier B et al (2007) The chemopreventive polyphenol Curcumin prevents hematogenous breast cancer metastases in immunodeficient mice. Cell Physiol Biochem 19:137–152. https://doi.org/10.1159/000099202
Bailon-Moscoso N, Cevallos-Solorzano G, Romero-Benavides JC, Orellana MI (2017) Natural compounds as modulators of cell cycle arrest: application for anticancer chemotherapies. Curr Genomics 18:106–131. https://doi.org/10.2174/1389202917666160808125645
Balakrishnan S et al (2016) Gold nanoparticle-conjugated quercetin inhibits epithelial-mesenchymal transition, angiogenesis and invasiveness via EGFR/VEGFR-2-mediated pathway in breast cancer. Cell Prolif 49:678–697. https://doi.org/10.1111/cpr.12296
Basile V et al (2013) bis-Dehydroxy-Curcumin triggers mitochondrial-associated cell death in human colon cancer cells through ER-stress induced autophagy. PloS one 8:e53664. https://doi.org/10.1371/journal.pone.0053664
Belayachi L, Aceves-Luquero C, Merghoub N, de Mattos SF, Amzazi S, Villalonga P, Bakri Y (2017) Induction of cell cycle arrest and apoptosis by ormenis eriolepis a morrocan endemic plant in various human cancer cell lines. Afr J Trad Complement Altern Med AJTCAM 14:356–373. https://doi.org/10.21010/ajtcam.v14i2.37
Benavente-Garcia O, Castillo J (2008) Update on uses and properties of citrus flavonoids: new findings in anticancer, cardiovascular, and anti-inflammatory activity. J Agric Food Chem 56:6185–6205. https://doi.org/10.1021/jf8006568
Berindan-Neagoe I, Braicu C, Irimie A (2012) Combining the chemotherapeutic effects of epigallocatechin 3-gallate with siRNA-mediated p53 knock-down results in synergic pro-apoptotic effects. Int J Nanomed 7:6035–6047. https://doi.org/10.2147/IJN.S36523
Bharti AC, Donato N, Singh S, Aggarwal BB (2003) Curcumin (diferuloylmethane) down-regulates the constitutive activation of nuclear factor-kappa B and IkappaBalpha kinase in human multiple myeloma cells, leading to suppression of proliferation and induction of apoptosis. Blood 101:1053–1062. https://doi.org/10.1182/blood-2002-05-1320
Bors W, Michel C, Saran M (1994) Flavonoid antioxidants: rate constants for reactions with oxygen radicals. Methods Enzymol 234:420–429. https://doi.org/10.1016/0076-6879(94)34112-5
Braicu C, Gherman CD, Irimie A, Berindan-Neagoe I (2013) Epigallocatechin-3-Gallate (EGCG) inhibits cell proliferation and migratory behaviour of triple negative breast cancer cells. J Nanosci Nanotechnol 13:632–637. https://doi.org/10.1166/jnn.2013.6882
Bray F, Ferlay J, Soerjomataram I, Siegel RL, Torre LA, Jemal A (2018) Global cancer statistics 2018: GLOBOCAN estimates of incidence and mortality worldwide for 36 cancers in 185 countries. CA Cancer J Clin 68:394–424. https://doi.org/10.3322/caac.21492
Cairns RA, Harris IS, Mak TW (2011) Regulation of cancer cell metabolism. Nat Rev Cancer 11:85–95. https://doi.org/10.1038/nrc2981

Catanzaro D, Ragazzi E, Vianello C, Caparrotta L, Montopoli M (2015) Effect of quercetin on cell cycle and cyclin expression in ovarian carcinoma and osteosarcoma cell lines. Nat Prod Commun 10:1365–1368

Chahar MK, Sharma N, Dobhal MP, Joshi YC (2011) Flavonoids: a versatile source of anticancer drugs. Pharmacog Rev 5:1–12. https://doi.org/10.4103/0973-7847.79093

Chen L, Zhang HY (2007) Cancer preventive mechanisms of the green tea polyphenol (-)-epigallocatechin-3-gallate. Molecules 12:946–957. https://doi.org/10.3390/12050946

Chen H, Landen CN, Li Y, Alvarez RD, Tollefsbol TO (2013) Epigallocatechin gallate and sulforaphane combination treatment induce apoptosis in paclitaxel-resistant ovarian cancer cells through hTERT and Bcl-2 down-regulation. Exp Cell Res 319:697–706. https://doi.org/10.1016/j.yexcr.2012.12.026

Chen QY et al (2014) Curcumin inhibits lung cancer cell migration and invasion through Rac1-dependent signaling pathway. J Nutr Biochem 25:177–185. https://doi.org/10.1016/j.jnutbio.2013.10.004

Chen PC, Lee JC, Wang JD (2015) Estimation of life-year loss and lifetime costs for different stages of colon adenocarcinoma in Taiwan. PloS one 10:e0133755. https://doi.org/10.1371/journal.pone.0133755

Chien SY et al (2009) Quercetin-induced apoptosis acts through mitochondrial- and caspase-3-dependent pathways in human breast cancer MDA-MB-231 cells. Hum Exp Toxicol 28:493–503. https://doi.org/10.1177/0960327109107002

Cho HJ, Park JH (2013) Kaempferol induces cell cycle arrest in HT-29 human colon cancer cells. J Cancer Prev 18:257–263. https://doi.org/10.15430/jcp.2013.18.3.257

Chong SJF, Marchi S, Petroni G, Kroemer G, Galluzzi L, Pervaiz S (2020) Noncanonical cell fate regulation by Bcl-2 proteins. Trends Cell Biol 30:537–555. https://doi.org/10.1016/j.tcb.2020.03.004

Chung JE et al (2014) Self-assembled micellar nanocomplexes comprising green tea catechin derivatives and protein drugs for cancer therapy. Nat Nanotechnol 9:907–912. https://doi.org/10.1038/nnano.2014.208

Colombo M, Tueiro F, de Fraga Dias A, Teixeira HF, AMO B, Koester LS (2018) Kaempferol-loaded mucoadhesive nanoemulsion for intranasal administration reduces glioma growth in vitro. Int J Pharm 543:214–223. https://doi.org/10.1016/j.ijpharm.2018.03.055

Davatgaran-Taghipour Y, Masoomzadeh S, Farzaei MH, Bahramsoltani R, Karimi-Soureh Z, Rahimi R, Abdollahi M (2017) Polyphenol nanoformulations for cancer therapy: experimental evidence and clinical perspective. Int J Nanomed 12:2689–2702. https://doi.org/10.2147/IJN.S131973

David KI, Jaidev LR, Sethuraman S, Krishnan UM (2015) Dual drug loaded chitosan nanoparticles-sugar--coated arsenal against pancreatic cancer. Colloids Surf B Biointerfaces 135:689–698. https://doi.org/10.1016/j.colsurfb.2015.08.038

de Pace RC et al (2013) Anticancer activities of (-)-epigallocatechin-3-gallate encapsulated nanoliposomes in MCF7 breast cancer cells. J Liposome Res 23:187–196. https://doi.org/10.3109/08982104.2013.788023

Del Prado-Audelo ML et al (2019) Formulations of curcumin nanoparticles for brain diseases. Biomolecules 9. https://doi.org/10.3390/biom9020056

Edwards JM, Raffauf RF, Le Quesne PW (1979) Antineoplastic activity and cytotoxicity of flavones, isoflavones, and flavanones. J Nat Prod 42:85–91. https://doi.org/10.1021/np50001a002

El-Gogary RI et al (2014) Polyethylene glycol conjugated polymeric nanocapsules for targeted delivery of quercetin to folate-expressing cancer cells in vitro and in vivo. ACS Nano 8:1384–1401. https://doi.org/10.1021/nn405155b

Falzone L, Salomone S, Libra M (2018) Evolution of cancer pharmacological treatments at the turn of the third millennium. Front pharmacol 9:1300. https://doi.org/10.3389/fphar.2018.01300

Firdous AB, Sharmila G, Balakrishnan S, RajaSingh P, Suganya S, Srinivasan N, Arunakaran J (2014) Quercetin, a natural dietary flavonoid, acts as a chemopreventive agent against prostate cancer in an in vivo model by inhibiting the EGFR signaling pathway. Food Funct 5:2632–2645. https://doi.org/10.1039/c4fo00255e

Formica JV, Regelson W (1995) Review of the biology of Quercetin and related bioflavonoids. Food Chem Toxicol 33:1061–1080. https://doi.org/10.1016/0278-6915(95)00077-1

Ganesh B, Talole SD, Dikshit R, Badwe RA, Dinshaw KA (2008) Estimation of survival rates of breast cancer patients--a hospital-based study from Mumbai. Asian Pac J Cancer Prev 9:53–57

Gao Y, Yin J, Rankin GO, Chen YC (2018) Kaempferol induces G2/M cell cycle arrest via checkpoint kinase 2 and promotes apoptosis via death receptors in human ovarian carcinoma A2780/CP70 cells. Molecules 23 https://doi.org/10.3390/molecules23051095

Goldstein M, Kastan MB (2015) The DNA damage response: implications for tumor responses to radiation and chemotherapy. Annu Rev Med 66:129–143. https://doi.org/10.1146/annurev-med-081313-121208

Golonko A, Lewandowska H, Swislocka R, Jasinska UT, Priebe W, Lewandowski W (2019) Curcumin as tyrosine kinase inhibitor in cancer treatment. Eur J Med Chem 181:111512. https://doi.org/10.1016/j.ejmech.2019.07.015

Gorrini C, Harris IS, Mak TW (2013) Modulation of oxidative stress as an anticancer strategy. Nat Rev Drug Discov 12:931–947. https://doi.org/10.1038/nrd4002

Gou M et al (2011) Curcumin-loaded biodegradable polymeric micelles for colon cancer therapy in vitro and in vivo. Nanoscale 3:1558–1567. https://doi.org/10.1039/c0nr00758g

Govindaraju S, Roshini A, Lee MH, Yun K (2019) Kaempferol conjugated gold nanoclusters enabled efficient for anticancer therapeutics to A549 lung cancer cells. Int J Nanomed 14:5147–5157. https://doi.org/10.2147/IJN.S209773

Granado-Serrano AB, Martin MA, Bravo L, Goya L, Ramos S (2012) Quercetin modulates Nrf2 and glutathione-related defenses in HepG2 cells: involvement of p38. Chem Biol Interact 195:154–164. https://doi.org/10.1016/j.cbi.2011.12.005

Gulati N, Laudet B, Zohrabian VM, Murali R, Jhanwar-Uniyal M (2006) The antiproliferative effect of Quercetin in cancer cells is mediated via inhibition of the PI3K-Akt/PKB pathway. Anticancer Res 26:1177–1181

Haddad AQ, Fleshner N, Nelson C, Saour B, Musquera M, Venkateswaran V, Klotz L (2010) Antiproliferative mechanisms of the flavonoids 2,2'-dihydroxychalcone and fisetin in human prostate cancer cells. Nutr Cancer 62:668–681. https://doi.org/10.1080/01635581003605524

Han Q et al (2012) Enhancement of biological activities of nanostructured hydrophobic drug species. Nanoscale 4:2078–2082. https://doi.org/10.1039/c2nr12013e

Hastak K, Gupta S, Ahmad N, Agarwal MK, Agarwal ML, Mukhtar H (2003) Role of p53 and NF-kappaB in epigallocatechin-3-gallate-induced apoptosis of LNCaP cells. Oncogene 22:4851–4859. https://doi.org/10.1038/sj.onc.1206708

Hastak K, Agarwal MK, Mukhtar H, Agarwal ML (2005) Ablation of either p21 or Bax prevents p53-dependent apoptosis induced by green tea polyphenol epigallocatechin-3-gallate. FASEB J 19:789–791. https://doi.org/10.1096/fj.04-2226fje

Hayakawa S, Ohishi T, Miyoshi N, Oishi Y, Nakamura Y, Isemura M (2020) Anti-cancer effects of green tea epigallocatchin-3-Gallate and Coffee Chlorogenic Acid. Molecules 25 https://doi.org/10.3390/molecules25194553

Hazra B, Ghosh S, Kumar A, Pandey BN (2011) The prospective role of plant products in radiotherapy of cancer: a current overview. Front Pharmacol 2:94. https://doi.org/10.3389/fphar.2011.00094

He M et al (2014) Curcumin suppresses cell proliferation through inhibition of the Wnt/beta-catenin signaling pathway in medulloblastoma. Oncol Rep 32:173–180. https://doi.org/10.3892/or.2014.3206

Heijnen CG, Haenen GR, van Acker FA, van der Vijgh WJ, Bast A (2001) Flavonoids as peroxynitrite scavengers: the role of the hydroxyl groups. Toxicol In Vitro 15:3–6. https://doi.org/10.1016/s0887-2333(00)00053-9

Herbst RS et al (2014) Predictive correlates of response to the anti-PD-L1 antibody MPDL3280A in cancer patients. Nature 515:563–567. https://doi.org/10.1038/nature14011

Hong J, Lu H, Meng X, Ryu JH, Hara Y, Yang CS (2002) Stability, cellular uptake, biotransformation, and efflux of tea polyphenol (-)-epigallocatechin-3-gallate in HT-29 human colon adenocarcinoma cells. Cancer Res 62:7241–7246

Hong JH, Ahn KS, Bae E, Jeon SS, Choi HY (2006) The effects of curcumin on the invasiveness of prostate cancer in vitro and in vivo. Prostate Cancer Prostatic Dis 9:147–152. https://doi.org/10.1038/sj.pcan.4500856

Hong JT, Yen JH, Wang L, Lo YH, Chen ZT, Wu MJ (2009) Regulation of heme oxygenase-1 expression and MAPK pathways in response to kaempferol and rhamnocitrin in PC12 cells. Toxicol Appl Pharmacol 237:59–68. https://doi.org/10.1016/j.taap.2009.02.014

Hosokawa N et al (1990) Flavonoids inhibit the expression of heat shock proteins. Cell Struct Funct 15:393–401. https://doi.org/10.1247/csf.15.393

Hu B, Ting Y, Zeng X, Huang Q (2012) Cellular uptake and cytotoxicity of chitosan-caseinophosphopeptides nanocomplexes loaded with epigallocatechin gallate. Carbohydr Polym 89:362–370. https://doi.org/10.1016/j.carbpol.2012.03.015

Hu B, Xie M, Zhang C, Zeng X (2014) Genipin-structured peptide-polysaccharide nanoparticles with significantly improved resistance to harsh gastrointestinal environments and their potential for oral delivery of polyphenols. J Agric Food Chem 62:12443–12452. https://doi.org/10.1021/jf5046766

Hu A et al (2017) Curcumin induces G2/M cell cycle arrest and apoptosis of head and neck squamous cell carcinoma in vitro and in vivo through ATM/Chk2/p53-dependent pathway. Oncotarget 8:50747–50760. https://doi.org/10.18632/oncotarget.17096

Huang YT et al (1999) Effects of luteolin and quercetin, inhibitors of tyrosine kinase, on cell growth and metastasis-associated properties in A431 cells overexpressing epidermal growth factor receptor. Br J Pharmacol 128:999–1010. https://doi.org/10.1038/sj.bjp.0702879

Huang CY, Chan CY, Chou IT, Lien CH, Hung HC, Lee MF (2013a) Quercetin induces growth arrest through activation of FOXO1 transcription factor in EGFR-overexpressing oral cancer cells. J Nutr Biochem 24:1596–1603. https://doi.org/10.1016/j.jnutbio.2013.01.010

Huang WW et al (2013b) Kaempferol induces autophagy through AMPK and AKT signaling molecules and causes G2/M arrest via downregulation of CDK1/cyclin B in SK-HEP-1 human hepatic cancer cells. Int J Oncol 42:2069–2077. https://doi.org/10.3892/ijo.2013.1909

Hugel HM, Jackson N (2012) Redox chemistry of green tea polyphenols: therapeutic benefits in neurodegenerative diseases. Mini Rev Med Chem 12:380–387. https://doi.org/10.2174/138955712800493906

Hwang MK, Song NR, Kang NJ, Lee KW, Lee HJ (2009) Activation of phosphatidylinositol 3-kinase is required for tumor necrosis factor-alpha-induced upregulation of matrix metalloproteinase-9: its direct inhibition by quercetin. Int J Biochem Cell Biol 41:1592–1600. https://doi.org/10.1016/j.biocel.2009.01.014

Hwang YS, Park KK, Chung WY (2013) Epigallocatechin-3 gallate inhibits cancer invasion by repressing functional invadopodia formation in oral squamous cell carcinoma. Eur J Pharmacol 715:286–295. https://doi.org/10.1016/j.ejphar.2013.05.008

Imran M et al (2019) Chemo-preventive and therapeutic effect of the dietary flavonoid kaempferol: a comprehensive review. Phytother Res 33:263–275. https://doi.org/10.1002/ptr.6227

Jain RK (2001) Delivery of molecular medicine to solid tumors: lessons from in vivo imaging of gene expression and function. J Control Release 74:7–25. https://doi.org/10.1016/s0168-3659(01)00306-6

Jain AK, Thanki K, Jain S (2013) Co-encapsulation of tamoxifen and quercetin in polymeric nanoparticles: implications on oral bioavailability, antitumor efficacy, and drug-induced toxicity. Mol Pharm 10:3459–3474. https://doi.org/10.1021/mp400311j

Jana NR, Dikshit P, Goswami A, Nukina N (2004) Inhibition of proteasomal function by curcumin induces apoptosis through mitochondrial pathway. J Biol Chem 279:11680–11685. https://doi.org/10.1074/jbc.M310369200

Jang SH, Wientjes MG, Lu D, Au JL (2003) Drug delivery and transport to solid tumors. Pharm Res 20:1337–1350. https://doi.org/10.1023/a:1025785505977

Jemal A, Bray F, Center MM, Ferlay J, Ward E, Forman D (2011) Global cancer statistics. CA Cancer J Clin 61:69–90. https://doi.org/10.3322/caac.20107

Jeong JH, An JY, Kwon YT, Li LY, Lee YJ (2008) Quercetin-induced ubiquitination and downregulation of Her-2/neu. J Cell Biochem 105:585–595. https://doi.org/10.1002/jcb.21859

Jeong JH, Ojha U, Lee YM (2020) Pathological angiogenesis and inflammation in tissues. Arch Pharm Res. https://doi.org/10.1007/s12272-020-01287-2

Ji C et al (2008) Curcumin attenuates EGF-induced AQP3 up-regulation and cell migration in human ovarian cancer cells. Cancer Chemother Pharmacol 62:857–865. https://doi.org/10.1007/s00280-007-0674-6

Jiao D, Wang J, Lu W, Tang X, Chen J, Mou H, Chen QY (2016) Curcumin inhibited HGF-induced EMT and angiogenesis through regulating c-Met dependent PI3K/Akt/mTOR signaling pathways in lung cancer. Mol Ther Oncolytics 3:16018. https://doi.org/10.1038/mto.2016.18

Jin L, Quan C, Hou X, Fan S (2016) Potential pharmacological resources: natural bioactive compounds from marine-derived fungi. Mar Drugs:14. https://doi.org/10.3390/md14040076

Jo E, Park SJ, Choi YS, Jeon WK, Kim BC (2015) Kaempferol suppresses transforming growth factor-beta1-induced epithelial-to-mesenchymal transition and migration of A549 lung cancer cells by inhibiting Akt1-mediated phosphorylation of smad3 at threonine-179. Neoplasia 17:525–537. https://doi.org/10.1016/j.neo.2015.06.004

Johnson N, Bryant A, Miles T, Hogberg T, Cornes P (2011) Adjuvant chemotherapy for endometrial cancer after hysterectomy. Cochrane Database Syst Rev CD003175 https://doi.org/10.1002/14651858.CD003175.pub2

Jung YD et al (2001) EGCG, a major component of green tea, inhibits tumour growth by inhibiting VEGF induction in human colon carcinoma cells. Br J Cancer 84:844–850. https://doi.org/10.1054/bjoc.2000.1691

Kang SU, Lee BS, Lee SH, Baek SJ, Shin YS, Kim CH (2013) Expression of NSAID-activated gene-1 by EGCG in head and neck cancer: involvement of ATM-dependent p53 expression. J Nutr Biochem 24:986–999. https://doi.org/10.1016/j.jnutbio.2012.07.003

Kashafi E, Moradzadeh M, Mohamadkhani A, Erfanian S (2017) Kaempferol increases apoptosis in human cervical cancer HeLa cells via PI3K/AKT and telomerase pathways. Biomed Pharmacother 89:573–577. https://doi.org/10.1016/j.biopha.2017.02.061

Kaufmann R, Henklein P, Henklein P, Settmacher U (2009) Green tea polyphenol epigallocatechin-3-gallate inhibits thrombin-induced hepatocellular carcinoma cell invasion and p42/p44-MAPKinase activation. Oncol Rep 21:1261–1267. https://doi.org/10.3892/or_00000349

Khan N et al (2014) Oral administration of naturally occurring chitosan-based nanoformulated green tea polyphenol EGCG effectively inhibits prostate cancer cell growth in a xenograft model. Carcinogenesis 35:415–423. https://doi.org/10.1093/carcin/bgt321

Khan H, Ullah H, Nabavi SM (2019) Mechanistic insights of hepatoprotective effects of curcumin: Therapeutic updates and future prospects. Food Chem Toxicol 124:182–191. https://doi.org/10.1016/j.fct.2018.12.002

Khandelwal A, Hall JA, Blagg BS (2013) Synthesis and structure-activity relationships of EGCG analogues, a recently identified Hsp90 inhibitor. J Org Chem 78:7859–7884. https://doi.org/10.1021/jo401027r

Kim HJ et al (2010) Apoptotic effect of quercetin on HT-29 colon cancer cells via the AMPK signaling pathway. J Agric Food Chem 58:8643–8650. https://doi.org/10.1021/jf101510z

Kim H et al (2013) Regulation of Wnt signaling activity for growth suppression induced by quercetin in 4T1 murine mammary cancer cells. Int J Oncol 43:1319–1325. https://doi.org/10.3892/ijo.2013.2036

Kim GT, Lee SH, Kim JI, Kim YM (2014) Quercetin regulates the sestrin 2-AMPK-p38 MAPK signaling pathway and induces apoptosis by increasing the generation of intracellular ROS in a p53-independent manner. Int J Mol Med 33:863–869. https://doi.org/10.3892/ijmm.2014.1658

Kim SH, Hwang KA, Choi KC (2016) Treatment with kaempferol suppresses breast cancer cell growth caused by estrogen and triclosan in cellular and xenograft breast cancer models. J Nutr Biochem 28:70–82. https://doi.org/10.1016/j.jnutbio.2015.09.027

Klaunig JE, Kamendulis LM (2004) The role of oxidative stress in carcinogenesis. Annu Rev Pharmacol Toxicol 44:239–267. https://doi.org/10.1146/annurev.pharmtox.44.101802.121851

Koh YW et al (2011) Green tea (-)-epigallocatechin-3-gallate inhibits HGF-induced progression in oral cavity cancer through suppression of HGF/c-Met. J Nutr Biochem 22:1074–1083. https://doi.org/10.1016/j.jnutbio.2010.09.005

Kumar S, Pandey AK (2013) Chemistry and biological activities of flavonoids: an overview. Sci World J 2013:162750. https://doi.org/10.1155/2013/162750

Kunnumakkara AB et al (2009) Curcumin sensitizes human colorectal cancer to capecitabine by modulation of cyclin D1, COX-2, MMP-9, VEGF and CXCR4 expression in an orthotopic mouse model. Int J Cancer 125:2187–2197. https://doi.org/10.1002/ijc.24593

Kuntz S, Kunz C, Rudloff S (2017) Inhibition of pancreatic cancer cell migration by plasma anthocyanins isolated from healthy volunteers receiving an anthocyanin-rich berry juice. Eur J Nutr 56:203–214. https://doi.org/10.1007/s00394-015-1070-3

Kushima Y et al (2009) Inhibitory effect of (-)-epigallocatechin and (-)-epigallocatechin gallate against heregulin beta1-induced migration/invasion of the MCF-7 breast carcinoma cell line. Biol Pharm Bull 32:899–904. https://doi.org/10.1248/bpb.32.899

Kuttikrishnan S et al (2019) Curcumin Induces Apoptotic Cell Death via Inhibition of PI3-Kinase/AKT Pathway in B-Precursor Acute Lymphoblastic Leukemia. Front Oncol 9:484. https://doi.org/10.3389/fonc.2019.00484

Kwak IH et al (2011) Epigallocatechin-3-gallate inhibits paracrine and autocrine hepatocyte growth factor/scatter factor-induced tumor cell migration and invasion. Exp Mol Med 43:111–120. https://doi.org/10.3858/emm.2011.43.2.013

Lai WW et al (2013) Quercetin inhibits migration and invasion of SAS human oral cancer cells through inhibition of NF-kappaB and matrix metalloproteinase-2/-9 signaling pathways. Anticancer Res 33:1941–1950

Lambert JD, Elias RJ (2010) The antioxidant and pro-oxidant activities of green tea polyphenols: a role in cancer prevention. Arch Biochem Biophys 501:65–72. https://doi.org/10.1016/j.abb.2010.06.013

Lambert JD, Yang CS (2003) Mechanisms of cancer prevention by tea constituents. J Nutr 133:3262S–3267S. https://doi.org/10.1093/jn/133.10.3262S

Landis-Piwowar KR, Kuhn DJ, Wan SB, Chen D, Chan TH, Dou QP (2005) Evaluation of proteasome-inhibitory and apoptosis-inducing potencies of novel (-)-EGCG analogs and their prodrugs. Int J Mol Med 15:735–742

Larasati YA, Yoneda-Kato N, Nakamae I, Yokoyama T, Meiyanto E, Kato JY (2018) Curcumin targets multiple enzymes involved in the ROS metabolic pathway to suppress tumor cell growth. Sci Rep 8:2039. https://doi.org/10.1038/s41598-018-20179-6

Lee J, Kim JH (2016) Kaempferol inhibits pancreatic cancer cell growth and migration through the blockade of EGFR-related pathway in vitro. PloS one 11:e0155264. https://doi.org/10.1371/journal.pone.0155264

Lee LT et al (2004) Transinactivation of the epidermal growth factor receptor tyrosine kinase and focal adhesion kinase phosphorylation by dietary flavonoids: effect on invasive potential of human carcinoma cells. Biochem Pharmacol 67:2103–2114. https://doi.org/10.1016/j.bcp.2004.02.023

Lee DS, Lee MK, Kim JH (2009) Curcumin induces cell cycle arrest and apoptosis in human osteosarcoma (HOS) cells. Anticancer Res 29:5039–5044

Lee JH et al (2010) EGCG induces apoptosis in human laryngeal epidermoid carcinoma Hep2 cells via mitochondria with the release of apoptosis-inducing factor and endonuclease G. Cancer Lett 290:68–75. https://doi.org/10.1016/j.canlet.2009.08.027

Li Z, Gu L (2014) Fabrication of self-assembled (-)-epigallocatechin gallate (EGCG) ovalbumin-dextran conjugate nanoparticles and their transport across monolayers of human intestinal epithelial Caco-2 cells. J Agric Food Chem 62:1301–1309. https://doi.org/10.1021/jf404621f

Li Z, Ha J, Zou T, Gu L (2014) Fabrication of coated bovine serum albumin (BSA)-epigallocatechin gallate (EGCG) nanoparticles and their transport across monolayers of human intestinal epithelial Caco-2 cells. Food Funct 5:1278–1285. https://doi.org/10.1039/c3fo60500k

Li C et al (2015) Biocompatible and biodegradable nanoparticles for enhancement of anti-cancer activities of phytochemicals. Chin J Nat Med 13:641–652. https://doi.org/10.1016/S1875-5364(15)30061-3

Liabakk NB, Talbot I, Smith RA, Wilkinson K, Balkwill F (1996) Matrix metalloprotease 2 (MMP-2) and matrix metalloprotease 9 (MMP-9) type IV collagenases in colorectal cancer. Cancer Res 56:190–196

Lim YC et al (2008) (-)-Epigallocatechin-3-gallate (EGCG) inhibits HGF-induced invasion and metastasis in hypopharyngeal carcinoma cells. Cancer Lett 271:140–152. https://doi.org/10.1016/j.canlet.2008.05.048

Lin CW, Hou WC, Shen SC, Juan SH, Ko CH, Wang LM, Chen YC (2008) Quercetin inhibition of tumor invasion via suppressing PKC delta/ERK/AP-1-dependent matrix metalloproteinase-9 activation in breast carcinoma cells. Carcinogenesis 29:1807–1815. https://doi.org/10.1093/carcin/bgn162

Lin SS et al (2009) Curcumin inhibits the migration and invasion of human A549 lung cancer cells through the inhibition of matrix metalloproteinase-2 and -9 and Vascular Endothelial Growth Factor (VEGF). Cancer Lett 285:127–133. https://doi.org/10.1016/j.canlet.2009.04.037

Lin YH, Chen ZR, Lai CH, Hsieh CH, Feng CL (2015) Active targeted nanoparticles for oral administration of gastric cancer therapy. Biomacromolecules 16:3021–3032. https://doi.org/10.1021/acs.biomac.5b00907

Lin Z, Zhang Q, Luo W (2016) Angiogenesis inhibitors as therapeutic agents in cancer: challenges and future directions. Eur J Pharmacol 793:76–81. https://doi.org/10.1016/j.ejphar.2016.10.039

Liu LP et al (2010) The protective effects of Polygonum multiflorum stilbeneglycoside preconditioning in an ischemia/reperfusion model of HUVECs. Acta Pharmacol Sin 31:405–412. https://doi.org/10.1038/aps.2010.7

Liu S, Wang XJ, Liu Y, Cui YF (2013) PI3K/AKT/mTOR signaling is involved in (-)-epigallocatechin-3-gallate-induced apoptosis of human pancreatic carcinoma cells. Am J Chin Med 41:629–642. https://doi.org/10.1142/S0192415X13500444

Liu L et al (2015) Molecular mechanism of epigallocatechin-3-gallate in human esophageal squamous cell carcinoma in vitro and in vivo. Oncol Rep 33:297–303. https://doi.org/10.3892/or.2014.3555

Liu H et al (2020) Curcumin inhibits proliferation and epithelial-mesenchymal transition in lens epithelial cells through multiple pathways. BioMed Res Int 2020:6061894. https://doi.org/10.1155/2020/6061894

Liu-Smith F, Meyskens FL (2016) Molecular mechanisms of flavonoids in melanin synthesis and the potential for the prevention and treatment of melanoma. Mol Nutr Food Res 60:1264–1274. https://doi.org/10.1002/mnfr.201500822

Long Q et al (2013) Induction of apoptosis and inhibition of angiogenesis by PEGylated liposomal quercetin in both cisplatin-sensitive and cisplatin-resistant ovarian cancers. J Biomed Nanotechnol 9:965–975. https://doi.org/10.1166/jbn.2013.1596

Lopez-Lazaro M (2010) A new view of carcinogenesis and an alternative approach to cancer therapy. Mol Med 16:144–153. https://doi.org/10.2119/molmed.2009.00162

Lowe SW, Lin AW (2000) Apoptosis in cancer. Carcinogenesis 21:485–495. https://doi.org/10.1093/carcin/21.3.485

Lu Y, Wei C, Xi Z (2014) Curcumin suppresses proliferation and invasion in non-small cell lung cancer by modulation of MTA1-mediated Wnt/beta-catenin pathway In Vitro. Cell Dev Biol Anim 50:840–850. https://doi.org/10.1007/s11626-014-9779-5

Luo H, Rankin GO, Liu L, Daddysman MK, Jiang BH, Chen YC (2009) Kaempferol inhibits angiogenesis and VEGF expression through both HIF dependent and independent pathways in human ovarian cancer cells. Nutr Cancer 61:554–563. https://doi.org/10.1080/01635580802666281

Luo H, Jiang B, Li B, Li Z, Jiang BH, Chen YC (2012a) Kaempferol nanoparticles achieve strong and selective inhibition of ovarian cancer cell viability. Int J Nanomed 7:3951–3959. https://doi.org/10.2147/IJN.S33670

Luo H, Rankin GO, Juliano N, Jiang BH, Chen YC (2012b) Kaempferol inhibits VEGF expression and in vitro angiogenesis through a novel ERK-NFkappaB-cMyc-p21 pathway. Food Chem 130:321–328. https://doi.org/10.1016/j.foodchem.2011.07.045

Ma Q et al (2020) Autophagy-dependent cell cycle arrest in esophageal cancer cells exposed to dihydroartemisinin. Chin Med 15:37. https://doi.org/10.1186/s13020-020-00318-w

Marchiani A, Rozzo C, Fadda A, Delogu G, Ruzza P (2014) Curcumin and curcumin-like molecules: from spice to drugs. Curr Med Chem 21:204–222. https://doi.org/10.2174/09298673210213120611581O

Martinez-Castillo M et al (2018) Curcumin differentially affects cell cycle and cell death in acute and chronic myeloid leukemia cells. Oncol Lett 15:6777–6783. https://doi.org/10.3892/ol.2018.8112

Maurya AK, Vinayak M (2015a) Anticarcinogenic action of quercetin by downregulation of phosphatidylinositol 3-kinase (PI3K) and protein kinase C (PKC) via induction of p53 in hepatocellular carcinoma (HepG2) cell line. Mol Biol Rep 42:1419–1429. https://doi.org/10.1007/s11033-015-3921-7

Maurya AK, Vinayak M (2015b) Quercetin regresses Dalton's lymphoma growth via suppression of PI3K/AKT signaling leading to upregulation of p53 and decrease in energy metabolism. Nutr Cancer 67:354–363. https://doi.org/10.1080/01635581.2015.990574

Miller KD et al (2016) Cancer treatment and survivorship statistics, 2016. CA Cancer J Clin 66:271–289. https://doi.org/10.3322/caac.21349

Min KJ, Kwon TK (2014) Anticancer effects and molecular mechanisms of epigallocatechin-3-gallate. Integr Med Res 3:16–24. https://doi.org/10.1016/j.imr.2013.12.001

Mira L, Fernandez MT, Santos M, Rocha R, Florencio MH, Jennings KR (2002) Interactions of flavonoids with iron and copper ions: a mechanism for their antioxidant activity. Free Radic Res 36:1199–1208. https://doi.org/10.1080/1071576021000016463

Mitra A, Chakrabarti J, Banerji A, Chatterjee A, Das BR (2006) Curcumin, a potential inhibitor of MMP-2 in human laryngeal squamous carcinoma cells HEp2. J Environ Pathol Toxicol Oncol 25:679–690. https://doi.org/10.1615/jenvironpatholtoxicoloncol.v25.i4.70

Moon JH, Tsushida T, Nakahara K, Terao J (2001) Identification of quercetin 3-O-beta-D-glucuronide as an antioxidative metabolite in rat plasma after oral administration of quercetin. Free Radic Biol Med 30:1274–1285. https://doi.org/10.1016/s0891-5849(01)00522-6

Mu C, Jia P, Yan Z, Liu X, Li X, Liu H (2007) Quercetin induces cell cycle G1 arrest through elevating Cdk inhibitors p21 and p27 in human hepatoma cell line (HepG2). Methods Find Exp Clin Pharmacol 29:179–183. https://doi.org/10.1358/mf.2007.29.3.1092095

Mylonis I, Lakka A, Tsakalof A, Simos G (2010) The dietary flavonoid kaempferol effectively inhibits HIF-1 activity and hepatoma cancer cell viability under hypoxic conditions. Biochem Biophys Res Commun 398:74–78. https://doi.org/10.1016/j.bbrc.2010.06.038

Nair KL, Thulasidasan AK, Deepa G, Anto RJ, Kumar GS (2012) Purely aqueous PLGA nanoparticulate formulations of curcumin exhibit enhanced anticancer activity with dependence on the combination of the carrier. Int J Pharm 425:44–52. https://doi.org/10.1016/j.ijpharm.2012.01.003

Nam JS, Sharma AR, Nguyen LT, Chakraborty C, Sharma G, Lee SS (2016) Application of bioactive quercetin in oncotherapy: from nutrition to nanomedicine. Molecules 21:E108. https://doi.org/10.3390/molecules21010108

Narayanan S, Mony U, Vijaykumar DK, Koyakutty M, Paul-Prasanth B, Menon D (2015) Sequential release of epigallocatechin gallate and paclitaxel from PLGA-casein core/shell nanoparticles sensitizes drug-resistant breast cancer cells. Nanomedicine 11:1399–1406. https://doi.org/10.1016/j.nano.2015.03.015

Nishikawa T et al (2006) A green tea polyphenol, epigalocatechin-3-gallate, induces apoptosis of human hepatocellular carcinoma, possibly through inhibition of Bcl-2 family proteins. J Hepatol 44:1074–1082. https://doi.org/10.1016/j.jhep.2005.11.045

Norooznezhad AH, Norooznezhad F (2017) Cannabinoids: Possible agents for treatment of psoriasis via suppression of angiogenesis and inflammation. Med Hypotheses 99:15–18. https://doi.org/10.1016/j.mehy.2016.12.003

Oh SJ et al (2010) Inhibition of angiogenesis by quercetin in tamoxifen-resistant breast cancer cells. Food Chem Toxicol 48:3227–3234. https://doi.org/10.1016/j.fct.2010.08.028

Ozdelikara A, Tan M (2017) The Effect of Reflexology on Chemotherapy-induced Nausea, Vomiting, and Fatigue in Breast Cancer Patients. Asia-Pac J Oncol Nurs 4:241–249. https://doi.org/10.4103/apjon.apjon_15_17

Ozyurek M, Bektasoglu B, Guclu K, Apak R (2009) Measurement of xanthine oxidase inhibition activity of phenolics and flavonoids with a modified cupric reducing antioxidant capacity (CUPRAC) method. Anal Chim Acta 636:42–50. https://doi.org/10.1016/j.aca.2009.01.037

Palange AL, Di Mascolo D, Carallo C, Gnasso A, Decuzzi P (2014) Lipid-polymer nanoparticles encapsulating curcumin for modulating the vascular deposition of breast cancer cells. Nanomedicine 10:991–1002. https://doi.org/10.1016/j.nano.2014.02.004

Pande AN, Biswas S, Reddy ND, Jayashree BS, Kumar N, Rao CM (2017) In vitro and in vivo anticancer studies of 2'-hydroxy chalcone derivatives exhibit apoptosis in colon cancer cells by HDAC inhibition and cell cycle arrest. EXCLI J 16:448–463. https://doi.org/10.17179/excli2016-643

Pandey SK, Patel DK, Thakur R, Mishra DP, Maiti P, Haldar C (2015) Anti-cancer evaluation of quercetin embedded PLA nanoparticles synthesized by emulsified nanoprecipitation. Int J Biol Macromol 75:521–529. https://doi.org/10.1016/j.ijbiomac.2015.02.011

Panieri E, Santoro MM (2016) ROS homeostasis and metabolism: a dangerous liason in cancer cells. Cell Death Dis 7:e2253. https://doi.org/10.1038/cddis.2016.105

Panjamurthy K, Manoharan S, Nirmal MR, Vellaichamy L (2009) Protective role of Withaferin-A on immunoexpression of p53 and bcl-2 in 7,12-dimethylbenz(a)anthracene-induced experimental oral carcinogenesis. Investig New Drugs 27:447–452. https://doi.org/10.1007/s10637-008-9199-z

Pereira LP, Silva P, Duarte M, Rodrigues L, Duarte CM, Albuquerque C, Serra AT (2017) Targeting colorectal cancer proliferation, stemness and metastatic potential using brassicaceae extracts enriched in isothiocyanates: a 3D cell model-based study. Nutrients 9. https://doi.org/10.3390/nu9040368

Pimple S, Manjappa AS, Ukawala M, Murthy RS (2012) PLGA nanoparticles loaded with etoposide and quercetin dihydrate individually: in vitro cell line study to ensure advantage of combination therapy. Cancer Nanotechnol 3:25–36. https://doi.org/10.1007/s12645-012-0027-y

Prat A et al (2014) Predicting response and survival in chemotherapy-treated triple-negative breast cancer. Br J Cancer 111:1532–1541. https://doi.org/10.1038/bjc.2014.444

Pratheeshkumar P et al (2012) Quercetin inhibits angiogenesis mediated human prostate tumor growth by targeting VEGFR- 2 regulated AKT/mTOR/P70S6K signaling pathways. PloS one 7:e47516. https://doi.org/10.1371/journal.pone.0047516

Prior RL (2003) Fruits and vegetables in the prevention of cellular oxidative damage. Am J Clin Nutr 78:570S–578S. https://doi.org/10.1093/ajcn/78.3.570S

Punfa W, Yodkeeree S, Pitchakarn P, Ampasavate C, Limtrakul P (2012) Enhancement of cellular uptake and cytotoxicity of curcumin-loaded PLGA nanoparticles by conjugation with anti-P-glycoprotein in drug resistance cancer cells. Acta Pharmacol Sin 33:823–831. https://doi.org/10.1038/aps.2012.34

Qin J et al (2007) A component of green tea, (-)-epigallocatechin-3-gallate, promotes apoptosis in T24 human bladder cancer cells via modulation of the PI3K/Akt pathway and Bcl-2 family proteins. Biochem Biophys Res Commun 354:852–857. https://doi.org/10.1016/j.bbrc.2007.01.003

Qin Y, Cui W, Yang X, Tong B (2016) Kaempferol inhibits the growth and metastasis of cholangiocarcinoma in vitro and in vivo. Acta Biochim Biophys Sin (Shanghai) 48:238–245. https://doi.org/10.1093/abbs/gmv133

Rafieian-Kopaei M, Movahedi M (2017) Breast cancer chemopreventive and chemotherapeutic effects of Camellia Sinensis (green tea): an updated review. Electronic Phys 9:3838–3844. https://doi.org/10.19082/3838

Rajendran P, Rengarajan T, Nandakumar N, Palaniswami R, Nishigaki Y, Nishigaki I (2014) Kaempferol, a potential cytostatic and cure for inflammatory disorders. Eur J Med Chem 86:103–112. https://doi.org/10.1016/j.ejmech.2014.08.011

Ranjan AP, Mukerjee A, Helson L, Gupta R, Vishwanatha JK (2013) Efficacy of liposomal curcumin in a human pancreatic tumor xenograft model: inhibition of tumor growth and angiogenesis. Anticancer Res 33:3603–3609

Rao SD, Pagidas K (2010) Epigallocatechin-3-gallate, a natural polyphenol, inhibits cell proliferation and induces apoptosis in human ovarian cancer cells. Anticancer Res 30:2519–2523

Ren J, Meng S, Lekka Ch E, Kaxiras E (2008) Complexation of flavonoids with iron: structure and optical signatures. J Phys Chem B 112:1845–1850. https://doi.org/10.1021/jp076881e

Ren J, Lu Y, Qian Y, Chen B, Wu T, Ji G (2019) Recent progress regarding kaempferol for the treatment of various diseases. Exp Ther Med 18:2759–2776. https://doi.org/10.3892/etm.2019.7886

Rezaei-Sadabady R, Eidi A, Zarghami N, Barzegar A (2016) Intracellular ROS protection efficiency and free radical-scavenging activity of quercetin and quercetin-encapsulated liposomes. Artif Cells Nanomed Biotechnol 44:128–134. https://doi.org/10.3109/21691401.2014.926456

Robati M, Holtz D, Dunton CJ (2008) A review of topotecan in combination chemotherapy for advanced cervical cancer. Therapeutics Clin Risk Manag 4:213–218. https://doi.org/10.2147/tcrm.s1771

Rocha S, Generalov R, Pereira Mdo C, Peres I, Juzenas P, Coelho MA (2011) Epigallocatechin gallate-loaded polysaccharide nanoparticles for prostate cancer chemoprevention. Nanomedicine (Lond) 6:79–87. https://doi.org/10.2217/nnm.10.101

Roy AM, Baliga MS, Katiyar SK (2005) Epigallocatechin-3-gallate induces apoptosis in estrogen receptor-negative human breast carcinoma cells via modulation in protein expression of p53 and Bax and caspase-3 activation. Mol Cancer Ther 4:81–90

Saberi-Karimian M, Katsiki N, Caraglia M, Boccellino M, Majeed M, Sahebkar A (2019) Vascular endothelial growth factor: an important molecular target of curcumin. Crit Rev Food Sci Nutr 59:299–312. https://doi.org/10.1080/10408398.2017.1366892

Sadhukhan P, Kundu M, Chatterjee S, Ghosh N, Manna P, Das J, Sil PC (2019) Targeted delivery of quercetin via pH-responsive zinc oxide nanoparticles for breast cancer therapy. Mater Sci Eng C Mater Biol Appl 100:129–140. https://doi.org/10.1016/j.msec.2019.02.096

Sak K (2014) Site-specific anticancer effects of dietary flavonoid quercetin. Nutr Cancer 66:177–193. https://doi.org/10.1080/01635581.2014.864418

Salehi B et al (2018) Antioxidants: positive or negative actors? Biomolecules 8. https://doi.org/10.3390/biom8040124

Salehi M et al (2020) Therapeutic potentials of curcumin in the treatment of non-small-cell lung carcinoma. Phytother Res. https://doi.org/10.1002/ptr.6704

Sano R, Reed JC (2013) ER stress-induced cell death mechanisms. Biochim Biophys Acta 1833:3460–3470. https://doi.org/10.1016/j.bbamcr.2013.06.028

Sapino S, Ugazio E, Gastaldi L, Miletto I, Berlier G, Zonari D, Oliaro-Bosso S (2015) Mesoporous silica as topical nanocarriers for quercetin: characterization and in vitro studies. Eur J Pharm Biopharm 89:116–125. https://doi.org/10.1016/j.ejpb.2014.11.022

Satoh M, Takemura Y, Hamada H, Sekido Y, Kubota S (2013) EGCG induces human mesothelioma cell death by inducing reactive oxygen species and autophagy. Cancer Cell Int 13:19. https://doi.org/10.1186/1475-2867-13-19

Scambia G et al (1990) Inhibitory effect of quercetin on OVCA 433 cells and presence of type II oestrogen binding sites in primary ovarian tumours and cultured cells. Br J Cancer 62:942–946. https://doi.org/10.1038/bjc.1990.414

Selvam C, Prabu SL, Jordan BC, Purushothaman Y, Umamaheswari A, Hosseini Zare MS, Thilagavathi R (2019) Molecular mechanisms of curcumin and its analogs in colon cancer prevention and treatment. Life Sci 239:117032. https://doi.org/10.1016/j.lfs.2019.117032

Seo HS, Ju JH, Jang K, Shin I (2011) Induction of apoptotic cell death by phytoestrogens by up-regulating the levels of phospho-p53 and p21 in normal and malignant estrogen receptor alpha-negative breast cells. Nutr Res 31:139–146. https://doi.org/10.1016/j.nutres.2011.01.011

Shabaninejad Z et al (2020) Therapeutic potentials of curcumin in the treatment of glioblastoma. Eur J Med Chem 188:112040. https://doi.org/10.1016/j.ejmech.2020.112040

Shafiei SS, Solati-Hashjin M, Samadikuchaksaraei A, Kalantarinejad R, Asadi-Eydivand M, Abu Osman NA (2015) Epigallocatechin gallate/layered double hydroxide nanohybrids: preparation, characterization, and in vitro anti-tumor study. PloS one 10:e0136530. https://doi.org/10.1371/journal.pone.0136530

Shan BE, Wang MX, Li RQ (2009) Quercetin inhibit human SW480 colon cancer growth in association with inhibition of cyclin D1 and survivin expression through Wnt/beta-catenin signaling pathway. Cancer Invest 27:604–612. https://doi.org/10.1080/07357900802337191

Shankar S, Marsh L, Srivastava RK (2013) EGCG inhibits growth of human pancreatic tumors orthotopically implanted in Balb C nude mice through modulation of FKHRL1/FOXO3a and neuropilin. Mol Cell Biochem 372:83–94. https://doi.org/10.1007/s11010-012-1448-y

Sharifi-Rad J et al (2018) In vitro and in vivo assessment of free radical scavenging and antioxidant activities of Veronica persica. Poir Cell Mol Biol (Noisy-le-grand) 64:57–64

Sharma G et al (2015) Methoxy poly(ethylene glycol)-poly(lactide) nanoparticles encapsulating quercetin act as an effective anticancer agent by inducing apoptosis in breast cancer. Pharm Res 32:723–735. https://doi.org/10.1007/s11095-014-1504-2

Shen F et al (2014) Synergism from the combination of ulinastatin and curcumin offers greater inhibition against colorectal cancer liver metastases via modulating matrix metalloproteinase-9 and E-cadherin expression. Onco Targets Ther 7:305–314. https://doi.org/10.2147/OTT.S57126

Shimizu M et al (2010) (-)-Epigallocatechin gallate inhibits growth and activation of the VEGF/VEGFR axis in human colorectal cancer cells. Chem Biol Interact 185:247–252. https://doi.org/10.1016/j.cbi.2010.03.036

Siddiqui IA et al (2014) Excellent anti-proliferative and pro-apoptotic effects of (-)-epigallocatechin-3-gallate encapsulated in chitosan nanoparticles on human melanoma cell growth both in vitro and in vivo. Nanomedicine 10:1619–1626. https://doi.org/10.1016/j.nano.2014.05.007

Singh BN, Shankar S, Srivastava RK (2011) Green tea catechin, epigallocatechin-3-gallate (EGCG): mechanisms, perspectives and clinical applications. Biochem Pharmacol 82:1807–1821. https://doi.org/10.1016/j.bcp.2011.07.093

Song H et al (2015a) Kaempferol inhibits gastric cancer tumor growth: an in vitro and in vivo study. Oncol Rep 33:868–874. https://doi.org/10.3892/or.2014.3662

Song X, Zhao Z, Barber B, Farr AM, Ivanov B, Novich M (2015b) Overall survival in patients with metastatic melanoma. Curr Med Res Opin 31:987–991. https://doi.org/10.1185/03007995.2015.1021904

Su CC, Chen GW, Lin JG, Wu LT, Chung JG (2006) Curcumin inhibits cell migration of human colon cancer colo 205 cells through the inhibition of nuclear factor kappa B /p65 and down-regulates cyclooxygenase-2 and matrix metalloproteinase-2 expressions. Anticancer Res 26:1281–1288

Suganuma M, Saha A, Fujiki H (2011) New cancer treatment strategy using combination of green tea catechins and anticancer drugs. Cancer Sci 102:317–323. https://doi.org/10.1111/j.1349-7006.2010.01805.x

Sun J, Bi C, Chan HM, Sun S, Zhang Q, Zheng Y (2013) Curcumin-loaded solid lipid nanoparticles have prolonged in vitro antitumour activity, cellular uptake and improved in vivo bioavailability. Colloids Surf B Biointerfaces 111:367–375. https://doi.org/10.1016/j.colsurfb.2013.06.032

Sun M, Nie S, Pan X, Zhang R, Fan Z, Wang S (2014) Quercetin-nanostructured lipid carriers: characteristics and anti-breast cancer activities in vitro. Colloids Surf B Biointerfaces 113:15–24. https://doi.org/10.1016/j.colsurfb.2013.08.032
Suolinna EM, Buchsbaum RN, Racker E (1975) The effect of flavonoids on aerobic glycolysis and growth of tumor cells. Cancer Res 35:1865–1872
Tan TW et al (2006) Curcumin-induced cell cycle arrest and apoptosis in human acute promyelocytic leukemia HL-60 cells via MMP changes and caspase-3 activation. Anticancer Res 26:4361–4371
Tan BJ, Liu Y, Chang KL, Lim BK, Chiu GN (2012) Perorally active nanomicellar formulation of quercetin in the treatment of lung cancer. Int J Nanomed 7:651–661. https://doi.org/10.2147/IJN.S26538
Tao W et al (2015) CancerHSP: anticancer herbs database of systems pharmacology. Sci Rep 5:11481. https://doi.org/10.1038/srep11481
Thakur VS, Gupta K, Gupta S (2012) Green tea polyphenols increase p53 transcriptional activity and acetylation by suppressing class I histone deacetylases. Int J Oncol 41:353–361. https://doi.org/10.3892/ijo.2012.1449
Thielecke F, Boschmann M (2009) The potential role of green tea catechins in the prevention of the metabolic syndrome – a review. Phytochemistry 70:11–24. https://doi.org/10.1016/j.phytochem.2008.11.011
Tipoe GL, Leung TM, Hung MW, Fung ML (2007) Green tea polyphenols as an anti-oxidant and anti-inflammatory agent for cardiovascular protection. Cardiovasc Hematol Disord Drug Targets 7:135–144. https://doi.org/10.2174/187152907780830905
Trujillo J, Granados-Castro LF, Zazueta C, Anderica-Romero AC, Chirino YI, Pedraza-Chaverri J (2014) Mitochondria as a target in the therapeutic properties of curcumin. Arch Pharm (Weinheim) 347:873–884. https://doi.org/10.1002/ardp.201400266
Tu LY, Bai HH, Cai JY, Deng SP (2016) The mechanism of kaempferol induced apoptosis and inhibited proliferation in human cervical cancer SiHa cell: From macro to nano. Scanning 38:644–653. https://doi.org/10.1002/sca.21312
Ullah H, Khan H (2018) Anti-Parkinson potential of silymarin: mechanistic insight and therapeutic standing. Front Pharmacol 9:422. https://doi.org/10.3389/fphar.2018.00422
Unal O, Akkoc Y, Kocak M, Nalbat E, Dogan-Ekici AI, Yagci Acar H, Gozuacik D (2020) Treatment of breast cancer with autophagy inhibitory microRNAs carried by AGO2-conjugated nanoparticles. J Nanobiotechnol 18:65. https://doi.org/10.1186/s12951-020-00615-4
Vaishnavi K et al (2012) Differential activities of the two closely related withanolides, Withaferin A and Withanone: bioinformatics and experimental evidences. PloS one 7:e44419. https://doi.org/10.1371/journal.pone.0044419
Vidya Priyadarsini R, Senthil Murugan R, Maitreyi S, Ramalingam K, Karunagaran D, Nagini S (2010) The flavonoid quercetin induces cell cycle arrest and mitochondria-mediated apoptosis in human cervical cancer (HeLa) cells through p53 induction and NF-kappaB inhibition. Eur J Pharmacol 649:84–91. https://doi.org/10.1016/j.ejphar.2010.09.020
Vyas P, Thakar AB, Baghel MS, Sisodia A, Deole Y (2010) Efficacy of Rasayana Avaleha as adjuvant to radiotherapy and chemotherapy in reducing adverse effects. Ayu 31:417–423. https://doi.org/10.4103/0974-8520.82029
Wang L, Tu YC, Lian TW, Hung JT, Yen JH, Wu MJ (2006) Distinctive antioxidant and antiinflammatory effects of flavonols. J Agric Food Chem 54:9798–9804. https://doi.org/10.1021/jf0620719
Wang S, Yu S, Shi W, Ge L, Yu X, Fan J, Zhang J (2011) Curcumin inhibits the migration and invasion of mouse hepatoma Hca-F cells through down-regulating caveolin-1 expression and epidermal growth factor receptor signaling. IUBMB Life 63:775–782. https://doi.org/10.1002/iub.507
Wang G et al (2012) Effects of quercetin nanoliposomes on C6 glioma cells through induction of type III programmed cell death. Int J Nanomed 7:271–280. https://doi.org/10.2147/IJN.S26935

Wang P, Zhang L, Peng H, Li Y, Xiong J, Xu Z (2013) The formulation and delivery of curcumin with solid lipid nanoparticles for the treatment of on non-small cell lung cancer both in vitro and in vivo. Mater Sci Eng C Mater Biol Appl 33:4802–4808. https://doi.org/10.1016/j.msec.2013.07.047

Wang S, Su R, Nie S, Sun M, Zhang J, Wu D, Moustaid-Moussa N (2014a) Application of nanotechnology in improving bioavailability and bioactivity of diet-derived phytochemicals. J Nutr Biochem 25:363–376. https://doi.org/10.1016/j.jnutbio.2013.10.002

Wang WA, Groenendyk J, Michalak M (2014b) Endoplasmic reticulum stress associated responses in cancer. Biochim Biophys Acta 1843:2143–2149. https://doi.org/10.1016/j.bbamcr.2014.01.012

Wang X, Chen Y, Dahmani FZ, Yin L, Zhou J, Yao J (2014c) Amphiphilic carboxymethyl chitosan-quercetin conjugate with P-gp inhibitory properties for oral delivery of paclitaxel. Biomaterials 35:7654–7665. https://doi.org/10.1016/j.biomaterials.2014.05.053

Wang L, Zhang B, Huang F, Liu B, Xie Y (2016) Curcumin inhibits lipolysis via suppression of ER stress in adipose tissue and prevents hepatic insulin resistance. J Lipid Res 57:1243–1255. https://doi.org/10.1194/jlr.M067397

Wang M et al (2019) Potential Mechanisms of Action of Curcumin for Cancer Prevention: Focus on Cellular Signaling Pathways and miRNAs. Int J Biol Sci 15:1200–1214. https://doi.org/10.7150/ijbs.33710

Warden BA, Smith LS, Beecher GR, Balentine DA, Clevidence BA (2001) Catechins are bioavailable in men and women drinking black tea throughout the day. J Nutr 131:1731–1737. https://doi.org/10.1093/jn/131.6.1731

Watson JL, Hill R, Lee PW, Giacomantonio CA, Hoskin DW (2008) Curcumin induces apoptosis in HCT-116 human colon cancer cells in a p21-independent manner. Exp Mol Pathol 84:230–233. https://doi.org/10.1016/j.yexmp.2008.02.002

Weng CJ, Yen GC (2012) Chemopreventive effects of dietary phytochemicals against cancer invasion and metastasis: phenolic acids, monophenol, polyphenol, and their derivatives. Cancer Treatment Rev 38:76–87. https://doi.org/10.1016/j.ctrv.2011.03.001

Xiao ZP, Peng ZY, Peng MJ, Yan WB, Ouyang YZ, Zhu HL (2011) Flavonoids health benefits and their molecular mechanism. Mini Rev Med Chem 11:169–177. https://doi.org/10.2174/138955711794519546

Xiao J et al (2014) Epigallocatechin gallate attenuates fibrosis, oxidative stress, and inflammation in non-alcoholic fatty liver disease rat model through TGF/SMAD, PI3 K/Akt/FoxO1, and NF-kappa B pathways. Eur J Nutr 53:187–199. https://doi.org/10.1007/s00394-013-0516-8

Xie YQ, Wu XB, Tang SQ (2014) Curcumin treatment alters ERK-1/2 signaling in vitro and inhibits nasopharyngeal carcinoma proliferation in mouse xenografts. Int J Clin Exp Med 7:108–114

Xingyu Z et al (2016) Quercetin suppresses lung cancer growth by targeting Aurora B kinase. Cancer Med 5:3156–3165. https://doi.org/10.1002/cam4.891

Yallapu MM et al (2014) Anti-cancer activity of curcumin loaded nanoparticles in prostate cancer. Biomaterials 35:8635–8648. https://doi.org/10.1016/j.biomaterials.2014.06.040

Yeh SL, Yeh CL, Chan ST, Chuang CH (2011) Plasma rich in quercetin metabolites induces G2/M arrest by upregulating PPAR-gamma expression in human A549 lung cancer cells. Planta Med 77:992–998. https://doi.org/10.1055/s-0030-1250735

Yoshida M et al (1990) The effect of quercetin on cell cycle progression and growth of human gastric cancer cells. FEBS Lett 260:10–13. https://doi.org/10.1016/0014-5793(90)80053-l

Yoshimura H et al (2019) The therapeutic potential of epigallocatechin3gallate against human oral squamous cell carcinoma through inhibition of cell proliferation and induction of apoptosis: In vitro and in vivo murine xenograft study. Mol Med Rep 20:1139–1148. https://doi.org/10.3892/mmr.2019.10331

Yuan Z, Long C, Junming T, Qihuan L, Youshun Z, Chan Z (2012) Quercetin-induced apoptosis of HL-60 cells by reducing PI3K/Akt. Mol Biol Rep 39:7785–7793. https://doi.org/10.1007/s11033-012-1621-0

Zhang Y, Han G, Fan B, Zhou Y, Zhou X, Wei L, Zhang J (2009) Green tea (-)-epigallocatechin-3-gallate down-regulates VASP expression and inhibits breast cancer cell migration and invasion by attenuating Rac1 activity. Eur J Pharmacol 606:172–179. https://doi.org/10.1016/j.ejphar.2008.12.033
Zhang J, Nie S, Wang S (2013) Nanoencapsulation enhances epigallocatechin-3-gallate stability and its antiatherogenic bioactivities in macrophages. J Agric Food Chem 61:9200–9209. https://doi.org/10.1021/jf4023004
Zhang Y, Duan W, Owusu L, Wu D, Xin Y (2015) Epigallocatechin-3-gallate induces the apoptosis of hepatocellular carcinoma LM6 cells but not non-cancerous liver cells. Int J Mol Med 35:117–124. https://doi.org/10.3892/ijmm.2014.1988
Zhang L, Cheng X, Xu S, Bao J, Yu H (2018) Curcumin induces endoplasmic reticulum stress-associated apoptosis in human papillary thyroid carcinoma BCPAP cells via disruption of intracellular calcium homeostasis. Medicine (Baltimore) 97:e11095. https://doi.org/10.1097/MD.0000000000011095
Zhao Y, Yang LF, Ye M, Gu HH, Cao Y (2004) Induction of apoptosis by epigallocatechin-3-gallate via mitochondrial signal transduction pathway. Prev Med 39:1172–1179. https://doi.org/10.1016/j.ypmed.2004.04.042
Zheng W, Zhao Y, Luo Q, Zhang Y, Wu K, Wang F (2017) Multi-targeted anticancer agents. Curr Top Med Chem 17:3084–3098. https://doi.org/10.2174/1568026617666170707124126
Zhu L, Xue L (2019) Kaempferol suppresses proliferation and induces cell cycle arrest, apoptosis, and DNA damage in breast cancer cells. Oncol Res 27:629–634. https://doi.org/10.3727/096504018X15228018559434
Zhu R, Wu X, Xiao Y, Gao B, Xie Q, Liu H, Wang S (2013) Synergetic effect of SLN-curcumin and LDH-5-Fu on SMMC-7721 liver cancer cell line. Cancer Biother Radiopharm 28:579–587. https://doi.org/10.1089/cbr.2012.1445

Chapter 4
Role of Metal-Doped Carbon Dots in Bioimaging and Cancer Therapy

Rajkumar Bandi, Srikrishna Tummala, Ramakrishna Dadigala, Madhusudhan Alle, and Seung-Hwan Lee

4.1 Introduction

Optical nanotechnology has grabbed the attention in various fields especially in biomedical diagnosis and therapy due to its high efficiency, nondestructive nature. Moreover, the optical imaging technology is not only for diagnosis but also applied for real-time monitoring of drug delivery and phototherapy (photodynamic and photothermal therapy) as well (Cai et al. 2019). Photosensitizers play an active role in photodynamic therapy (PDT) and photothermal therapy (PTT), and these photosensitizers react at certain wavelength of light producing active medium with reactive oxygen species to kill cancer cells. Furthermore, when light-irradiated photosensitizer which generated hyperthermia-based apoptosis or necrosis for treatment of cancer cells.

CDs were first discovered in 2004 by Scrivens and his group during the electrophoretic purification (Xu et al. 2004). The name carbon dots were firstly proposed by Sun et al. (2006). Carbon dots (CDs) are a fascinating class of fluorescent carbon nanomaterials with sizes below 10 nm, which particularly consist of sp^2-hybridized carbon atoms and the surface contains abundant functional groups (Baker and Baker 2010). CDs have gained stupendous attention due to their strong fluorescence,

R. Bandi (✉) · R. Dadigala · S.-H. Lee
Institute of Forest Science, Kangwon National University, Chuncheon, Republic of Korea

S. Tummala
Department of Chemistry, National Dong Hwa University, Hualien, Taiwan

M. Alle
Institute of Forest Science, Kangwon National University, Chuncheon, Republic of Korea

Department of Biomedical Science & Institute of Bioscience and Biotechnology, Kangwon National University, Chuncheon, Republic of Korea

J.-C. Kim et al. (eds.), *Smart Nanomaterials in Biomedical Applications*, Nanotechnology in the Life Sciences,
https://doi.org/10.1007/978-3-030-84262-8_4

nontoxicity, high photostability, chemical resistance, and hydrophilicity (Bandi et al. 2016, 2018b, a). These stunning properties were explored in various biomedical applications like bioimaging (Swathi et al. 2020), bio-labeling (Unnikrishnan et al. 2020), drug delivery (Lee et al. 2014), gene transfections (Dou et al. 2015), and photodynamic therapy (PDT) (Ge et al. 2014). Moreover CDs find wide applicability in catalysis (Bhagavanth Reddy et al. 2021; Bandi et al. 2021), forensic (Bandi et al. 2020b), sensors (Bandi et al. 2020a; Mohammed et al. 2021), and photocatalysis (Dadigala et al. 2017). In general, CDs can be obtained by methods such as laser ablation, electrochemical synthesis, hydrothermal/solvothermal treatment, and microwave irradiation (Tummala et al. 2021).

The optical and electronic characteristics and functionalities of CDs rely on three key factors: size, composition, and surface chemistry (Park et al. 2016). Heteroatom doping has been explored to enhance quantum yields and manipulate optical properties (emission peak position and excitation-dependent nature) of CDs (Xu et al. 2016). In comparison with nonmetallic heteroatoms such as N, P, and S, metal ions have more number of electrons to share and have unoccupied orbitals and possess large atomic radius (Lin et al. 2018). Particularly, metal ion-doped CDs having a great opportunity for charge transferring between metal ion and carbon core (graphene) are created to improve physicochemical properties of CDs (Tejwan et al. 2021).

4.2 Synthesis of Metal-Doped Carbon Dots

4.2.1 Hydrothermal Method

Hydrothermal method is the most efficient and extensively employed approach to synthesize metal-doped CDs. Hydrothermal approach has been considered environmentally friendly, inexpensive, and nontoxic method in order to synthesize the novel carbon materials from several raw constituents. Typically in this method, an aqueous solution of metal precursor and carbon source was sealed in a Teflon-lined stainless steel autoclave and treated at high temperature and pressure. Reports on the metal-doped CDs using hydrothermal approach are listed in Table 4.1.

Xu et al. prepared carbon quantum dot-stabilized gadolinium nanoprobe (Gd-CQDs) for dual modality imaging (Xu et al. 2014). Typical synthesis involves the hydrothermal reaction of citric acid (CA), ethylenediamine (EDA), and $GdCl_3$ at 200 °C for 4 h. Transmission electron microscope (TEM) images of Gd-CQDs revealed a flowerlike structure with average diameter of 47 nm. High-resolution TEM (HRTEM) analysis of this flowerlike structure disclosed the existence of small dots in the size of 2 to 8 nm. X-ray photoelectron spectroscopy (XPS) results revealed the presence of 33.70% Gd in Gd-CQDs. Both undoped CQDs and Gd-CQDs exhibited excitation-dependent emission in the range of 470 to 525 nm under the excitation of 340–460 nm. No apparent shift in the peak position was observed, but a significant decrease in the emission intensity was observed after Gd doping. Ren et al. prepared Gd^{3+}-doped CDs (Gd^{3+}-CQDs) as MR/FL bimodal probes by the hydrothermal treatment of gadopentetate monomeglumine (GdPM) as

Table 4.1 List of reports on hydrothermal synthesis of metal-doped carbon dots

S. no	Material	Precursors	Dopant	Q.Y (%)	λex/λem (nm)	References
1	Gd-CQDs	Citrate acid, ethanediamine, and $GdCl_3$	Gd	21	380/480	Xu et al. (2014)
2	Gd-doped CDs	Glycine and Gd-DTPA	Gd	13.4	370/475	Du et al. (2017)
3	Gd^{3+}-CQDs	Gadopentetate monomeglumine (GdPM)	Gd	7.1	365/445	Ren et al. (2016)
4	Gd-CDs	Citrate acid, Gd-nitrate, ethanediamine	Gd	43.6	350/450	Chen et al. (2016b)
5	C-CD and L-CD	Pcyclen, citric acid, $GdCl_3.6H_2O$ Ptaea, citric acid, DOTA	Gd	–	380/460 360/445	He et al. (2019)
6	Gd-CDs	Citric acid, $GdCl_3$, diethylenetriamine	Gd	69.86	448 nm	Yu et al. (2016)
7	Gd-CDs	Gd-DTPA, L-arginine	Gd	57.78 ± 2.14	–	Zheng et al. (2019)
8	Gd@GCNs	p-Phenylenediamine $Gd(NO_3)_3$	Gd	10.2 (λex 550 nm)	400/550 520/570 550/580	Chen et al. (2018)
9	Gd-CDs	Citric acid $GdCl_3$	Gd	4.06	430 nm	Liao et al. (2015)
10	Gd/Yb@ CDs	Na_2EDTA, L-arginine, $GdCl_3$, $YbCl_3$	Gd, and Yb	16.84	340/418	Zhao et al. (2018)
11	CQDs	Citric acid, $GdCl_3$, $YbCl_3$, ethylenediamine	Gd and Yb	66 ± 7	350/440	Bouzas-Ramos et al. (2019)
12	Cu,N-CDs	EDTA·2Na, $CuCl_2$	Cu	6	376/436	Guo et al. (2018)
13	HoBCDs	Citric acid, BPEI, Ho-DTPA	Ho	8.2	359/441	Fang et al. (2018)
14	Ce-doped CDs	Citric acid, $Ce(NO_3)_3$, glycine	Ce	–	36/440	Zhang et al. (2019)
15	Mn-CQDs	Waste green tea powder $MnCl_2$	Mn	12	360/430	Irmania et al. (2020)
16	Mg-EDA-CDs	Citric acid, $mg(OH)_2$ ethylenediamine	Mg	83	360/437	Li et al. (2014)
17	Mn-NCDs	Nigrosin, manganese acetate	Mn	10.57	Em 580 nm	Yang et al. (2018)
18	Ln-CQDs	Citrate acid, ethanediamine, $YbCl_3$, $NdCl_3$	Yb or Nd	52.32 (Yb) 49.06 (Nd)	360/450	Wu et al. (2016)
19	S, se-codoped CDs	Polythiophene, diphenyl diselenide	Se	18	460/731&820	Lan et al. (2017)

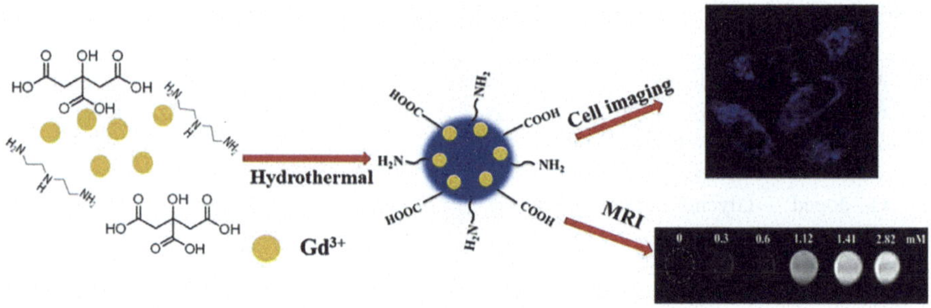

Fig. 4.1 Schematic illustration of one-step hydrothermal synthesis of Gd-CDs as a bimodal imaging probe. (Adopted from Yu et al. 2016)

the sole source material (Ren et al. 2016). The authors carefully studied the synthesis process by varying the reaction temperature and time. They observed that the solution pH increased after carbonization. The optimal conditions were 4 mg/mL GdPM aqueous solution, 220 °C, 2.0 MPa pressure, and 8 h of reaction time. Gd^{3+}-CQDs exhibited smaller size of ~1.6 nm and Gd^{3+} ion content of 16.7 wt%. Typical excitation-dependent emission is observed with optimal emission at 445 nm and quantum yield (QY) of 7.1%. As shown in Fig. 4.1, Yu et al. prepared Gd-CDs for bimodal imaging probe (Yu et al. 2016). The typical synthesis involved the hydrothermal reaction of CA, diethylenetriamine (DETA), and $GdCl_3$ at 180 °C for 3 h. The authors have optimized the synthetic conditions using fluorescence QY (FLQY) as a parameter and achieved a high QY of 69.86%. In TEM image, Gd-CDs appeared nearly spherical with sizes in the range of 2.6 to 6.5 nm. Dynamic light scattering revealed the average diameter as 5.5 nm. XPS analysis revealed the Gd content as 1.54 at %. Gd-CDs exhibited bright blue fluorescence centered at 448 nm. Interestingly the emission peak is same with undoped CDs and emission intensity is improved after doping.

Liao et al. produced Gd-CDs via hydrothermal treatment of CA and $GdCl_3$ at 200 °C for 8 h (Liao et al. 2015). Gd-CDs were purified by gel chromatography to remove toxic Gd^{3+} ions and other impurities. Eriochrome Black T indicator was used to probe the free Gd^{3+} ions in Gd-CDs. Compared to undoped CDs, the TEM image of Gd-CDs exhibited irregular particles with much larger size. XPS analysis revealed the doping ratio of approximately 0.5%. Gd-CDs exhibited excitation tunable emission with optimal emission at 446 nm and QY of 4.06%. Zhao et al. prepared double rare earth-doped CDs, i.e., Gd and Yb-doped CDs (Gd/Yb@CDs) (Zhao et al. 2018). The typical experiment involved the hydrothermal treatment of Na_2EDTA, $GdCl_3$, $YbCl_3$, and L-arginine at 200 °C for 10 h. TEM revealed the average particle size of Gd/Yb@CDs as 5.26 ± 0.93 nm, and elemental analysis revealed the contents of Gd and Yb as 7.40% and 21.38%. Excitation-dependent emission with maximum emission at 418 nm and QY of 16.84% was observed. Wu et al. reported the doping of CDs with NIR emissive lanthanide ions (Yb and Nd) and applied for bioimaging (Wu et al. 2016). The synthesis involves the dissolution of CA, EDA, and $LnCl_3$ in deionized water followed by its hydrothermal treatment at

200 °C for 4 h. Both Yb-CQDs and Nd-CQDs exhibited sizes in the range of 2–5 nm with average size of 3.3 nm and 3.4 nm, respectively. Both exhibited excitation-dependent emission with maximum emission peak located at 450 nm (λ_{ex} 360 nm). The FLQY of Yb-CQDs and Nd-CQDs was 52.32% and 49.06%, respectively, and it was lower than that of CQDs (65.53%). There is some overlap between the emission of Yb-CQDs and Nd-CQDs with the excitation of both Yb^{3+} (464 nm) and Nd^{3+} (432 nm) ions which leads to their NIR emission. Hence, Yb-CQDs and Nd-CQDs exhibited additional NIR emissions at 998 and 1068 nm, respectively.

Fang et al. prepared holmium(III)-doped CDs for fluorescence/magnetic dual-modal bioimaging (Fang et al. 2018). Hydrothermal treatment of CA, BPEI, and diethylenetriamine pentaacetic acid hydrate holmium (III) dihydrogen salt (Ho-DTPA) at 180 °C for 2 h produced Ho^{3+}-doped CDs (HoBCDs) with a mean diameter of 4.3 ± 1.0 nm. Interestingly, Ho doping has no significant influence on the optical properties. Both HoBCDs and undoped blue CDs (BCDs) exhibited excitation-dependent emission with optimal emission at 441 nm and at λ_{ex} 359 nm. There is no big difference in the FLQY of HoBCDs (8.2%) and BCDs (8.15%). Lan et al. reported the Se-doped CDs with NIR emission. The synthetic method involved the hydrothermal treatment of polythiophene (PT2) and diphenyl diselenide in an alkaline solution at 180 °C for 24 h (Lan et al. 2017). As obtained CDs have mean diameter of 20 nm and Se content of 1.13% from XPS analysis. The CDs presented excitation-independent emission nature and presented two peaks at 731 and 820 nm under the excitation of 430 nm. NIR emission is attributed to Se doping. The FLQY of Se-doped CDs is 0.2%, which is significantly less than undoped CDs (10%) and N,S-codoped CDs (17.5%). Du et al. prepared Gd-doped CDs by hydrothermal treatment of glycine and Gd-DTPA solution at 180 °C for 6 h (Du et al. 2017). As obtained Gd-CDs were quasi-spherical and exhibited narrow size distribution with average diameter of 18 nm. Gd-CDs exhibited excitation-dependent emission with maximum emission at 475 nm and QY of 13.4%. Chen et al. used Gd-functionalized CDs for FL/MR dual modality imaging of mesenchymal stem cells. Gd-CDs were obtained by the hydrothermal treatment of citric acid monohydrate, gadolinium nitrate hexahydrate, and EDA at 200 °C for 5 h (Chen et al. 2016b). Gd-CDs were spherical with diameter distribution of 2.3 to 4.5 nm and average diameter of 2.9 nm. The Gd^{3+} content in Gd-CDs was found to be 18.2% (w/w) using inductively coupled plasma-optical emission spectroscopy (ICP-OES). Study of optical properties revealed their excitation-independent nature with emission at 450 nm and QY of 43.6%. Zheng et al. prepared Gd-doped CDs (Gd-CDs) for enhanced MR imaging and cell fluorescence labeling (Zheng et al. 2019). Gd-CDs were produced by one-step hydrothermal treatment of Gd-DTPA and L-arginine at 200 °C for 1.5 h. As obtained Gd-CDs possess average diameter of 5.38 nm and 12% (wt%) Gd (ICP-MS). Gd-CDs exhibited typical excitation-dependent nature and FLQY as high as 57.78%. Zhang et al. reported the hydrothermal preparation of Ce-doped CQDs (Zhang et al. 2019). Typical synthesis involved the hydrothermal treatment of aqueous dispersion of CA, cerium nitrate, and glycine at 200 °C for 4 h. As obtained CQDs exhibited the average size of 2.5 nm and excitation tunable emission with optimal emission located at 440 nm (λ_{ex} 360 nm).

4.2.2 Solvothermal Method

Solvothermal approach is similar to hydrothermal; the only difference is that in solvothermal method, water is replaced with some other organic solvent. Fang et al. reported the preparation of red fluorescent Gd-doped CDs using solvothermal approach (Fang et al. 2020). They have taken CA, urea, and $GdCl_3$ in dimethylformamide (DMF) solvent and treated in an autoclave at 160 °C for 6 h. As obtained CDs have the average size of 2.0 ± 1.0 nm and possess 32.6% (w/w) Gd according to ICPMS analysis. Further they exhibited pH-dependent, excitation-independent emission with peak centered at 620 nm and FLQY of 2.3%. Gd doping caused a redshift in the emission peak position (580 nm for undoped) and an obvious decline in the FLQY. As illustrated in Fig. 4.2, Liu et al. prepared Cu- and N-doped CDs by solvothermal carbonization of folic acid and $CuCl_2$ in ethanol at 180 °C for 6 h (Liu et al. 2020). As obtained CDs have the average size of 3.57 nm and 1.35% of Cu according to XPS elemental content analysis. Interestingly they exhibited dual emission with peaks centered at 410 and 470 nm under the excitation of 340 nm, whereas the only N-doped CDs exhibited a single emission peak at 450 nm. The emission was excitation independent and pH dependent. In another work, Gao et al. prepared Si-doped CDs by one-pot solvothermal treatment of a silane molecule N-[3-(trimethoxysilyl)propyl]ethylenediamine (DAMO) in glycerol solvent at 260 °C for 12 h (Gao et al. 2018). These CDs possess an average size of 6.1 ± 1.4 nm, exhibited excitation-independent emission with optimal emission peak centered at 445 nm, and a high fluorescence QY of 45%. Han et al. described the preparation of highly fluorescent Mn-doped CDs (Mn-CDs) with QY as high as 90.79% and excitation-dependent blue emission (Han et al. 2016). These Mn-CDs were achieved by the solvothermal treatment of EDTA, triethylenetetramine, and $MnCl_2$ in ethylene glycol at 150 °C for 10 h and possess mean size of 4.6 + 0.9 nm. From XPS analysis the Mn content was found to be 1.17%. Interestingly, with increasing the amount of Mn, the authors observed an increase in the FL intensity with no change in the peak position. Sun et al. prepared gadolinium-rose Bengal coordination polymer nanodots (GRDs) by using solvothermal approach (Sun et al. 2020). They have taken Gd nitrate and rose Bengal in equimolar ratio in ethanol and treated in an autoclave at 180 °C for 12 h. The nanodots have the average hydrodynamic size of 7.7 ± 1.4 nm, and Gd doping ratio could be as high as 30% from ICP-MS analysis. Optimal λ_{ex} and λ_{em} of GRDs were 530/560 nm and exhibited 7.7-fold higher FLQY

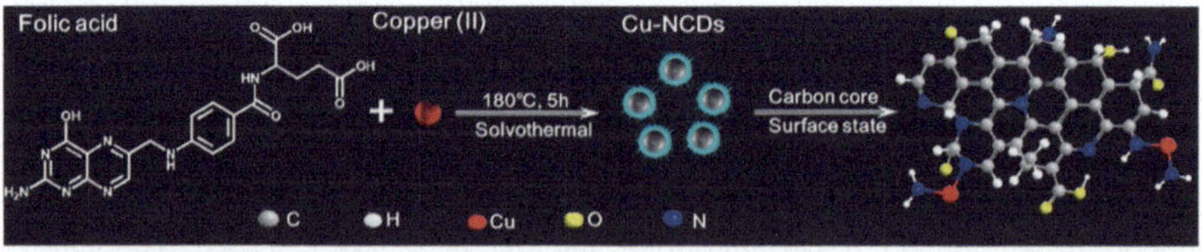

Fig. 4.2 Schematic diagram for the preparation of Cu-NCDs using solvothermal method. (Adopted from Liu et al. 2020)

than rose Bengal. In another study, Chen et al. prepared Gd-encapsulated CDs using solvothermal method (Chen et al. 2018). In the typical synthesis, p-phenylenediamine (pPD) and $Gd(NO_3)_3$ were taken in ethanol and treated at 180 °C for 12 h. As obtained Gd@GCNs exhibited average diameter of 5 nm in TEM. Energy dispersive X-ray analysis (EDX) shows the Gd content as 37.6 wt% and from ICP-MS it was found to be 30.8 wt%. Gd@GCNs exhibited red emission with multiple peaks centered at 550, 570, and 580 nm corresponding to the excitations of 400, 520, and 550 nm. The highest QY was 10.2% at the λ_{ex} of 550 nm.

4.2.3 Microwave Irradiation Method

Microwave-assisted methods have the advantages of rapid reaction times and eco-friendly nature. In this method, the precursors dissolved in either water or other solvent will be irradiated with microwaves which leads to the rapid and localized superheating and results in carbonization.

It has been reported S, N, and Gd tri-element-doped CDs (GdNS@CDs) use microwave method (Chiu et al. 2016). Sodium citrate, $GdCl_3$, and N-acetyl-L-cysteine were dissolved in distilled water by sonication and heated in a microwave oven at 220 °C for 10 min. The GdNS@CDs exhibited the average size of 4.3 ± 0.9 nm and excitation-dependent blue emission with QY of 12.8%. No shift in the emission peak position is observed with incorporation of Gd. However, the emission intensity is slightly quenched due to the disruption of radiative recombination of CDs by Gd^{3+} ions. Gong et al. reported the microwave-assisted polyol method for the synthesis of Gd-doped CDs (Gd-CDs) (Gong et al. 2014). Here, $GdCl_3$ obtained by dissolving Gd_2O_3 in HCl is used as Gd source, and it was mixed with sucrose solution and concentrated H_2SO_4 in diethylene glycol (DEG) and subjected to microwave irradiation in a domestic microwave oven operating at 700 W for 50 s. Gd-CDs had a mean diameter of 5 nm and the Gd content was found to be 18.2% (w/w) by ICP-OES. Gd-CDs exhibited excitation-independent bright green emission with peak at 521 nm and QY of 5.4%. Interestingly an unexpected weak blue emission peak was also observed indicating the formation of secondary CDs by the carbonization of DEG alone. Lin et al. reported the preparation of red emitting P^{3+}- and Mn^{2+}-doped CDs (PMn@CDs) using microwave-assisted pyrolysis method (Lin et al. 2019). In the typical synthesis, pPD, phosphoric acid, and Mn(OAc)n were dissolved in deionized water and heated to 220 °C for 10 min in a microwave reactor. Purification of PMn@CDs can be easily achieved by the addition of isopropanol as anti-solvent. PMn@CDs with the size of 6.47 nm exhibited optimal emission at 600 nm with corresponding excitation spectrum showing two peaks at 530 and 570 nm. FLQY of PMn@CDs was found to be 23% at optimal excitation of 530 nm using rhodamine 6G as reference. The authors further study the influence of dopants on the optical properties. The FLQY increased with increase in the P content, but decreased with increase in Mn content. Decrease in the QY with transition metal doping is likely due to the transfer of photoexcited electrons in the CDs to the d orbitals of the metal atoms resulting in nonradiative recombination pathways. Ji et al. reported the synthesis of Mn-doped CDs. CA, urea, and $MnCl_2$ were dissolved in distilled water and placed in a domestic microwave oven operating at 750 W and treated for 3.5 min (Ji et al. 2018). Mn-doped CDs exhibited narrow size distribution of 3–5 nm and excitation tunable emission. Changing λ_{ex} resulted in two prominent emissions (blue and green). The absolute QY of green emission measured using integrated sphere was about 10%. Rub Pakkath et al. reported a simple green method for the preparation of transition metal ion (Mn^{2+}, Fe^{2+}, Co^{2+}, and Ni^{2+})-doped CDs (TMCDs) (Rub Pakkath et al. 2018). In this method, the metal precursor is mixed with filtered lemon juice and EDA and subjected to microwave irradiation at 800 W for 6 min (Fig. 4.3). TMCDs possess a size of 3.2 ± 0.485 nm and exhibited excitation-dependent emission with optimal emission peak at 460 nm. In comparison to undoped C-dots (QY 48.31%), as synthesized Mn/CDs, Fe/CDs, Co/CDs, and Ni/CDs exhibited FLQY of 35.71%, 41.72%, 75.07%, and 50.84%, respectively. EDX analysis revealed the wt% of 0.16% Mn in Mn/CDs, 0.56% Fe in Fe/CDs, 0.38% Co in Co/CDs, and 0.13% Ni in Ni/CDs. Another green method for the preparation of transition metal ion (Gd^{3+}, Mn^{2+}, and Eu^{3+})-doped CDs was reported by Yao et al. Chitin from crab shell waste was dissolved in 1% acetic acid for 12 h at room temperature (Yao et al. 2017). Then the soluble fraction was mixed with transition metal chloride precursor and heated in a microwave reactor at 220 °C for 10 min. From TEM, the average sizes of Gd@CQDs, Mn@CQDs, and Eu@CQDs were 4.0 ± 0.7, 4.5 ± 1.0, and 4.2 ± 0.6 nm, respectively. All displayed excitation tunable fluorescence with optimal emission peak at 425 nm. However, the FLQY were 19.84%, 12.86%, and 14.97% for Gd-doped, Mn-doped, and Eu-doped, respectively.

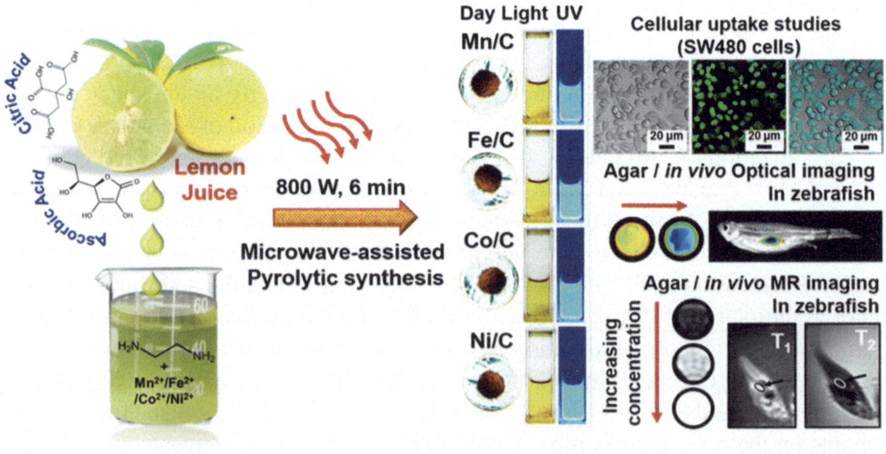

Fig. 4.3 Schematic illustration of microwave-assisted synthesis of TMCDs as potential nanoprobe for magneto-fluorescent bioimaging. (Adopted from Rub Pakkath et al. 2018)

4.2.4 Thermal Pyrolysis

Thermal carbonization or thermal pyrolysis involves the high temperature heating of precursors. Wang et al. reported the synthesis of Cu-doped CDs by the pyrolysis of Cu complex of poly(acrylic acid) at 400 °C for 1.5 h (Wang et al. 2019). Cu-CDs exhibited the mean size of 2.8 nm and 0.65% atomic percentage of copper from XPS. Cu-CDs displayed bright FL at 410 nm (λ_{ex} 280 nm) with excitation dependence nature. The absolute QY was 24.4% and is less than undoped CDs (32.78%). Pan et al. described the preparation of Gd-doped CDs (GCDs) by one-pot pyrolysis for high-performance multimodal bioimaging (Pan et al. 2017). In the typical synthesis, freshly prepared gadopentetic acid (Gd-DTPA) solution was added to an aqueous solution of CA and BPEI (Fig. 4.4). The homogeneous solution was heated at ~180 °C (by oil heating) until a pale-yellow gel was obtained and then 1.0 mL of cold water was added. Under continuous heating, this process was repeated for ten times until the gel color turned to orange. Then the product was dissolved in ultrapure water and purified. TEM analysis of as obtained GCDs revealed the diameter of 15.0 ± 5.0 nm and Gd content was 0.58% from XPS and 1 wt% by ICPMS. GCDs presented excitation-independent emission at 445 nm (λ_{ex} 354 nm) with an absolute QY of 40%.

In another study, Bourlinos et al. prepared Gd-doped quaternized CDs (Gd-QCDs) by mixing Gd-DTPA into Tris base and betaine hydrochloride followed by pyrolysis in air at 250 °C for 2 h (Bourlinos et al. 2012). Gd-QCDs with approximate size of ~3 nm exhibited excitation tunable emission with maximum emission at 445 nm. Gd-QDs contain 6.85% (w/w) of Gd based on elemental analysis. Chen et al. prepared Gd-encapsulated CDs with effective renal clearance for MRI (Chen et al. 2014). Gd@C-dots were obtained upon the calcination of Gd-DTPA solution at 300 °C for 2 h. The average particle size was 11 nm and contains 6.7 wt% of Gd (EDX). Gd@C-dots exhibited excitation-dependent emission with optimal emission peak at 425 nm (λ_{ex} 360 nm) and FLQY of 19.7%. The same author group came up with a size-controlled synthesis of Gd-encapsulated carbonaceous dots (Gd@C-dots) using mesoporous silica as nanoreactors (MSN) (Chen et al. 2016a). MSN with different pore sizes was first incubated in precursor solution (100 mM $Gd(NO_3)_3$

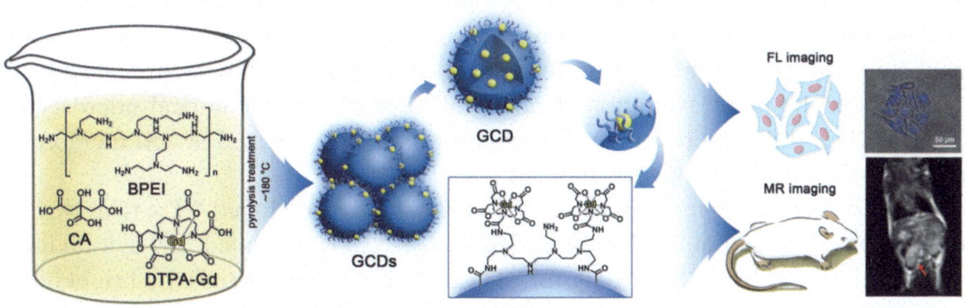

Fig. 4.4 Schematic illustration of the synthesis and dual modality imaging of GCDs. (Adopted from Pan et al. 2017)

and 10 mM Gd-DTPA) and subjected to calcination at 200 °C in air for 2 h. Then the silica framework was degraded by treating with 6 M NaOH. Gd@C-dots with mean diameters of 3.0 ± 0.5, 7.4 ± 1.2, and 9.6 ± 2.0 were successfully prepared using MSN with pore sizes of 3, 7, and 11 nm, respectively. All the three Gd@C--dots exhibited bright blue emission with excitation-dependent nature. The FLQY was decreased with increased particle size, and it was found to be 30.2%, 12.3%, and 1.6% for 3.0, 7.4, and 9.6 nm Gd@C-dots, respectively.

4.3 Metal-Doped Carbon Dots in Bioimaging

4.3.1 Fluorescence Imaging

Metal doping strategy is used to tune the optical properties such as shifting the emission peak position to longer wavelength and improving the FLQY. In these cases, the obtained MCDs were applied for only FL bioimaging. The use of Eu-doped CDs as FL cell imaging probe was reported by Chen et al. (Chen et al. 2016c). S180 sarcoma cells were employed to study the in vitro cellular uptake of Eu-CDs. Doping of Eu tuned the optical properties and resulted in red fluorescence. The cells incubated with Eu-CDs exhibited bright red emission, whereas the undoped CDs showed blue emission. Wu et al. prepared NIR emissive Yb^{3+} or Nd^{3+} ion-doped CDs (Wu et al. 2016). The obtained Yb-CQDs and Nd-CQDs were found less toxic and exhibited cell viabilities over 85% and 80% even at a concentration of 1 mg/mL. With this high biocompatibility, their cell imaging potential was evaluated by treating HeLa cells at a concentration of 0.2 mg/mL. The cells exhibited blue, green, and red emissions under 405 nm, 458 nm, and 514 nm excitations. However, the emission is mainly confined to cytoplasm leaving the nucleus dark. The use of Mn(II)-coordinated functional knots (MCDs) as FL imaging agents is reported by Wang et al. (Wang et al. 2015). HeLa cells incubated with MCDs were clearly observed with good morphologies indicating their good biocompatibility. The cytosol of cells exhibited bright blue, green, and red emissions under three laser excitations. Li et al. prepared Mg/N-doped CDs with high QY of 83% (Li et al. 2014). These CDs did not exhibit any cytotoxic effects, and the viabilities of L929 cell remained over 90% upon treating with 250 μg/mL CDs for 24 h. L929 cells incubated with 100 μg/mL CDs were brightly illuminated with blue, green, and red colors under the three excitation channels of laser scanning confocal microscopy (LSCM) (Fig. 4.5). Interestingly, the emission is mainly located at cell membrane and cytoplasm only, leaving the nucleus weakly stained.

Multicolor cellular imaging potential of Cu- and N-codoped CDs (Cu-NCDs) was evaluated by Liu et al. (Liu et al. 2020). Cu-NCDs exhibited low cytotoxicity with a high IC_{50} value of 812.96 μg/mL. The cells were illuminated with bright blue, cyan, and green colors under the excitations of 300, 360, and 440 nm, respectively. Cellular uptake and bioimaging potential of Ce-doped CQDs was studied by Zhang

Fig. 4.5 LSCM images of L929 cells labeled with CDs excited using (**a**) 405 nm, (**b**) 488 nm, and (**c**) 543 nm laser. (Adopted from Li et al. 2014)

et al. (Zhang et al. 2019). The authors first evaluated the biocompatibility using CCK-8 assay with MEF and NIH-3 T3 cells where no toxic effects were observed even at high dose of 800 μg/mL and for incubation of 24 h. Further in cell imaging experiments, the 3 T3 cells maintained their morphological features. The cells exhibited bright green emission under blue excitation and the emission was mainly located at cytoplasmic region and the nucleus shown weak emission. The authors assumed that the cell uptake of Ce-doped CQDs may occur via clathrin-mediated endocytosis and low permeability of the nucleus might be the reason for the weak emission.

4.3.2 Multimodal Imaging

Compared to simple FL imaging, multifunctional imaging is more efficient and attracting recent attention. With their inherent FL property, CDs can serve as FL imaging probe, and when CDs doped with magnetically active metal ions, they can serve as a magnetic resonance imaging probes. Furthermore, metal ion doping can also provide X-ray computed tomography (CT) imaging ability. As this doping process covers the metal ions inside the carbon lattice, it reduces the possible toxicity of metal ions. Literature reports on the usage of metal-doped CDs in bioimaging are listed in Table 4.2.

Han et al. used Mn-doped CDs for the imaging of ovarian cancer cells (Han et al. 2016). They first studied the in vitro toxicity of Mn-CDs on two cell lines and with the impressive results they studied the in vivo toxicity. There was no evidence of toxicity, and histological examinations also showed no negative effects on the major organs comprising the kidneys, liver, heart, and spleen. Anti-HE4 mAb as a targeting agent was covalently attached to Mn-CD surface to improve delivery efficiency and ovarian cancer cell specificity. In vitro and in vivo experiments revealed the good biocompatibility, admirable tumor targeting performance, and greater MR and FL imaging effects (Fig. 4.6). In another study, Ji et al. used Mn-doped CDs for MR/optical dual-mode imaging (Ji et al. 2018). In vivo MR imaging and ex vivo optical

Table 4.2 Metal-doped CDs in bioimaging

S. no.	Material	Precursor	Type of imaging	In vitro	In vivo	References
1	PMn@Cdots	1,4-phenylenediamine, $Mn(OAc)_2$	FL and MRI	B16F1 cancer cells and HeLa cells	*Danio rerio* AB line (zebrafish line)	Lin et al. (2019)
2	Gd-CQDs	Citric acid, ethanediamine, $GdCl_3$	FL and MRI	U87 cancer cells, HepG2 cell line	Zebrafish embryos	Xu et al. (2014)
3	N, Gd, and Yb CDs	Citric acid, ethanediamine, Gd_3^+ complex, $YbCl_3$	FL, MRI, and CT	NIH-3 T3, HeLa cells, CHO cells	Mice	Bouzas-Ramos et al. (2019)
4	Gd-doped CDs	Glycine, Gd-DTPA	FL and MRI	HepG2 cells	BALB/c mice MRI, tumor imaging	Du et al. (2017)
5	Europium (Eu)-doped carbon dots (CDs)	PEG-2000, glucose, and $Eu(NO_3)_3$	FL imaging	S180 sarcoma	–	Chen et al. (2016c)
6	Gd/Yb@CDs	Na_2EDTA, L-arginine, $GdCl_3$, $YbCl_3$	FL, MRI, and CT	HeLa cells	ICR female mice	Zhao et al. (2018)
7	Gd(III)/CQDs	Gadopentetate monomeglumine (GdPM)	FL and MRI	HeLa cells	–	Ren et al. (2014)
8	HoBCDs	Citric acid, BPEI and Ho-DTPA	FL and MRI	HeLa cells	–	Fang et al. (2018)
9	Ce-doped CDs	Citric acid, glycine, and Ce $(NO_3)_3.6H_2O$	FL	MEF and NIH-3 T3 cells	–	Zhang et al. (2019)
10	MnII -functionalized CDs	1-(2-pyridylazo)-2-naphthol, $MnCl_2$	FL	HeLa cells	–	Wang et al. (2015)
11	Gd-CDs	Citrate acid, ethanediamine, and $Gd(NO_3)_3.6H_2O$	FL and MRI	HeLa, A549, 293 T, and murine breast cancer cell line 4 T1	–	Chen et al. (2016b)
12	Gd-CDs	Pcyclen, citric acid, and $GdCl_3{\cdot}6H_2O$	FL and MRI	HeLa cells	BALB/c mice MRI	He et al. (2019)

(continued)

Table 4.2 (continued)

S. no.	Material	Precursor	Type of imaging	In vitro	In vivo	References
13	Gd-CDs	Citric acid, diethylenetriamine, and $GdCl_3$	FL and MRI	HeLa cells	–	Yu et al. (2016)
14	Gd@C-dots	Gd-DTPA	FL and MRI	U87MG cells.	U87MG tumor-bearing mice	Chen et al. (2014)
15	Gd@CQDs, Mn@CQDs, and Eu@ CQDs,	Crab shells, $GdCl_3$, $MnCl_2$, $EuCl_3$	FL and MRI	HeLa cells	–	Yao et al. (2017)
16	Mn-doped CDs	Citric acid, urea, and $MnCl_2$	FL and MRI	C6 cell	Mouse brain with tiny glioma	Ji et al. (2018)
17	Gd@C-dots	$Gd(NO_3)_3$, Gd-DTPA	FL and MRI	U87MG (human glioma) cells	U87MG tumor-bearing mice	Chen et al. (2016a)
18	Mg-EDA-CDs	Citric acid, ethylenediamine, $mg(OH)_2$	FL	L929 cells	–	Li et al. (2014)
19	Gd-CDs	Sucrose solution, H_2SO_4 and $GdCl_3$	FL and MRI	C6 glioma cells	–	Gong et al. (2014)
20	Ln-CQDs	Citric acid, ethanediamine, and $LnCl_3$	FL	HeLa cells	–	Wu et al. (2016)
21	Cu-NCDs	Folic acid and $CuCl_2$	FL	HepG2 cells.	–	Liu et al. (2020)
22	Gd-CDs	Citric acid, $GdCl_3$	FL and MRI	NCI-H446	Kunming mice	Liao et al. (2015)
23	GCDs	CA, BPEI, and Gd-DTPA	FL and MRI	HeLa cells	C57BL mice	Pan et al. (2017)
24	Gd-CDs	Gd-DTPA, L-arginine	FL and MRI	786-O human renal cancer cells, HK-2 tubular epithelial cells, and HO-8910 human ovarian cancer cells	Kunming mice	Zheng et al. (2019)

(continued)

Table 4.2 (continued)

S. no.	Material	Precursor	Type of imaging	In vitro	In vivo	References
25	FA-GdNS@ CQDs	Sodium citrate acid, N-acetyl-L-cysteine, Na_2S, and $GdCl_3$	FL and MRI	HeLa and HepG2	Zebrafish embryos	Chiu et al. (2016)
26	Mn-CDs	EDTA, triethylenetetramine, and $MnCl_2$	FL and MRI	HE4-positive HO-8910 cells and HE4-negative EA.hy926 cells	Kunming mice	Han et al. (2016)
27	TMCDs	Lemon juice, EDA, and $Mn(OAc)_2 \cdot 4H_2O$, $co(OAc)_2 \cdot 2H_2O$, $Ni(OAc)_2 .2H_2O$, $(NH4)_2 Fe(SO_4)_2 \cdot 6H_2O$	FL and MRI	SW480 cells	Wild-type zebrafish	Rub Pakkath et al. (2018)

Note: *FL* fluorescence, *MRI* magnetic resonance imaging, *CT* computed tomography

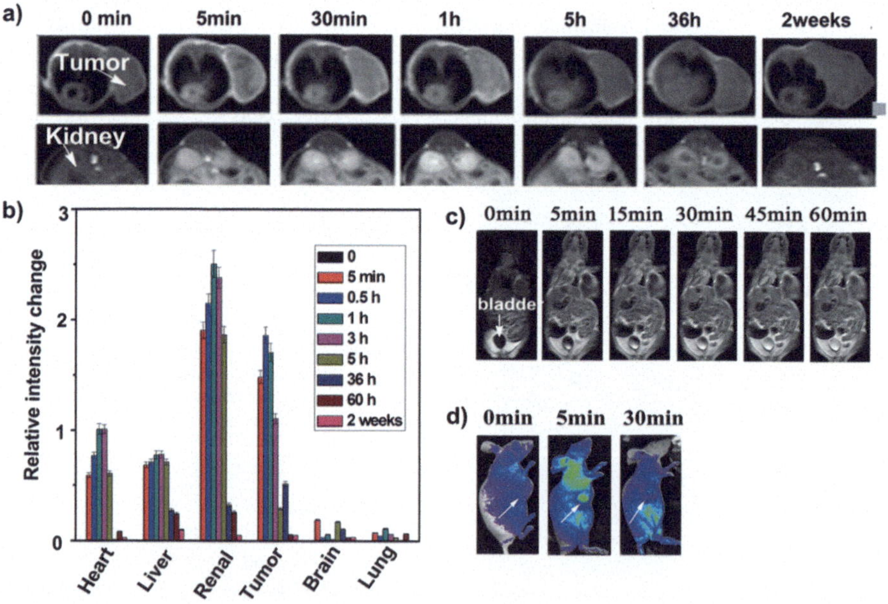

Fig. 4.6 (**a**) Axial T1-MRI of HO-8910 tumor-bearing nude mice before and after the administration of Mn-CDs recorded at different time intervals. (**b**) Biodistribution of Mn-CDs by MRI in tumor and major organs compared with muscle, including the lung, liver, kidney, heart, and brain. (**c**) Signal changes in bladder. (**d**) In vivo fluorescence images. (Adopted from Han et al. 2016)

imaging of a mouse brain with tiny glioma show that Mn-CDs can enhance MR T_1 contrast in the tiny brain glioma region, revealing the Mn-CDs have great promise as MR/optical dual-modal imaging nanoprobe for recognition and intraoperative position of tiny brain gliomas. Zhao et al. explored double rare earth-doped CDs for MR/CT/FL multimodal imaging (Zhao et al. 2018). Along with the excitation-dependent emission and superhigh photostability, the Gd/Yb@CDs also exhibited high longitudinal relaxivity ($r_1 = 6.65$ mM^{-1} s^{-1}) and admirable X-ray absorption efficiency (45.43 HU L g^{-1}). Superior biocompatibility of Gd/Yb@CDs was revealed by MTT assay, tissue section analysis, and body weight investigations. On the xenografted tumor model, MRI/CT imaging revealed that Gd/Yb@CDs could produce a clearly enhanced contrast effect. Additionally, biodistribution studies revealed that they were easily excreted from mice via the kidney elimination pathway. In a similar study by Bouzas-Ramos et al., CQDs codoped with N and lanthanides (i.e., Gd and Yb) were used as multimodal imaging probes (Bouzas-Ramos et al. 2019). These doped CQDs exhibited exceptional magnetic resonance (MR) and computed tomography (CT) characters along with bright fluorescence emission with FLQY of 66 ± 7%. Furthermore, no appreciable cytotoxicity was observed even after the exposure to three cell lines for 24 and 72 h. At first these doped CQDs were successfully used in in vitro FL as well as MR and CT cell imaging. Then their accumulation in the urinary organs after intravenous injection was monitored by in vivo MR and CT bimodal imaging. Both these imaging techniques showed a great loading of CQDs in the kidneys, ureters, and bladder.

For the first time, Bourlinos et al. reported the preparation of ultrafine-sized Gd(III)-doped CDs with dual fluorescence/MRI character (Bourlinos et al. 2012). The dots exhibited good water dispersibility, bright fluorescence, strong T_1-weighted MRI contrast comparable to commercial Gadovist, and low toxicity. With these characters, Gd-doped CDs could be used in biomedical research for multimodal imaging where the fluorescence and MRI modalities can be sued in combination to enhance image processing. Ren et al. reported the synthesis of a new kind of CQDs functionalized by Gd(III) chelates termed as Gd(III)/CQDs via a one-step pyrolysis of gadopentetate monomeglumine (Ren et al. 2014). With good longitudinal relaxation rate (5.5–6.4 mM $L^{-1}S^{-1}$) and FLQY (2.6–8.9%), low cytotoxicity, and small size, the Gd(III)/CQDs bestow a great potential for use as an MRI/FL contrast agent. Han Liao et al. used Gd-CDs for both cellular imaging and in vivo MR imaging agent (Liao et al. 2015). With great biocompatibility Gd-CDs were directly applied for in vivo MRI in mice. It was observed that these Gd-CDs were quickly excreted by kidney and accumulated in bladder of 30 min of intravenous injection. Strong contrast enhancement in the kidney and urinary bladder suggested that the Gd-CD agent could be easily excreted through renal filtration.

4.4 Metal-Doped Carbon Dots in Cancer Therapy

4.4.1 Photodynamic Therapy (PDT)

Wang et al. prepared Cu-doped CDs (Cu-CDs), with their strong emission and less toxicity. Cu-CDs were applied for the FL imaging of HeLa cells and SH-SY5Y multicellular spheroids (3DMCs) (Wang et al. 2019). Furthermore, Cu-CDs exhibited a high singlet oxygen ($^1O_2$) QY (36%), strong light-induced cytotoxicity, and efficient suppression of 3DMCs' growth. Under the same irradiation, Cu-CDs exhibited concentration-dependent cytotoxicity, as well as the irradiation time-dependent cytotoxicity. The authors feel that the Cu-CDs are encouraging for imaging-guided PDT for future cancer therapy. Sun et al. prepared Gd-rose Bengal coordination polymer nanodots (GRDs) (Sun et al. 2020). Compared to free rose Bengal, these GRDs presented a 7.7-fold improvement in FL and 1.9-fold rise in $^1O_2$ production ability. Furthermore, they demonstrated a twofold rise in r_1 relaxivity over Gd-DTPA and greater X-ray absorption than rose Bengal. These admirable features of GRDs are substantiated both in vitro and in vivo. The in vivo studies revealed that GRDs can amass in tumors via the EPR effect, as real time observed by MR/FL imaging. Compounding of PDT and radiation therapy (RT) suppresses tumor growth efficiently than monotherapies (PDT and RT). The GRDs presented greater antitumor efficiency devoid of systemic toxicity and long-term side effects (Fig. 4.7). In an interesting study, Jia et al. used Mn-doped CDs (Mn-CDs) to regulate tumor hypoxia and enhance PDT (Jia et al. 2018). Both NIR FL (λ_{em} 745 nm)

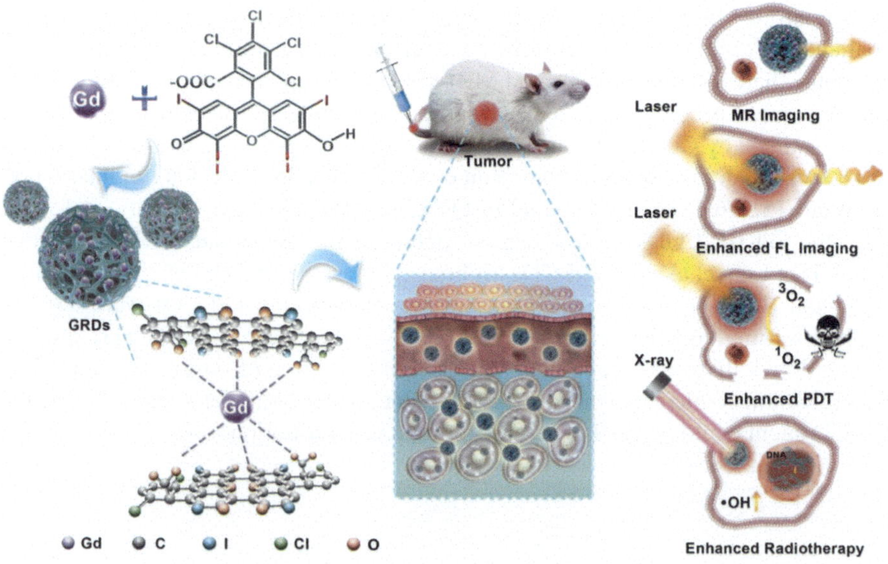

Fig. 4.7 Schematic illustration of preparation of GRDs and the in vivo fluorescence/MR imaging-guided PDT and radiotherapy for 4 T1 tumor cells using GRDs. (Adopted from Sun et al. 2020)

and T_1-weighted magnetic resonance (MR) (relaxivity of 6.97 mM^{-1} s^{-1}) imaging will benefit from Mn-CD assembly. More intriguingly, the Mn-CDs not only produce singlet oxygen (QY 40%) but also catalyze H_2O_2 to produce oxygen, demonstrating their ability to alleviate tumor hypoxia and improve tumor PDT. All these properties enable Mn-CD assembly to be used as an acidic H_2O_2-driven oxygenator to rise oxygen concentration in hypoxic solid tumors for simultaneous bimodal FL/MR imaging and enhanced PDT. Furthermore, the Mn-CD assembly was able to be excreted from the body, suggesting low biotoxicity. Irmania et al. conjugated FA and chlorin e6 (Ce6) to Mn-doped CDs (MnCQDs) to produce a magnetofluorescent PDT agent (MnCQDs@FA/Ce6) (Irmania et al. 2020). The specific targeting effect of the hybrid (MnCQDs@FA/Ce6) toward the folate receptor was evaluated by in vitro experiments on three cell lines. MnCQDs@FA/Ce6 exhibited good MRI signal with r_2/r_1 ratio of 5.77. PDT using 5 minutes of radiation (671 nm, 1 W cm^{-2}) killed >90% of the cells. The light-activated MnCQDs@FA/Ce6 can be used as a dual-modal FL/MRI probe as well as an effective PDT agent for remotely detecting and eradicating cancer cells. Chen et al. reported a nanoparticle-based photosensitizer with a high $^1O_2$ QY made of Gd-encapsulated graphene carbon nanoparticles (Gd@GCNs) (Chen et al. 2018). Gd@GCNs exhibited bright FL and high T_1 relaxivity (16.0 mM^{-1} s^{-1}, 7 T) making them an intrinsically dual-modal imaging agent. With their small size of 5 nm, Gd@GCNs can amass in tumors via EPR effect. Since the Gd is securely embedded inside an inert carbon shell and the dots are effectively excreted by renal clearance, no toxicity is expected from the unbound Gd@GCNs. Further, the ability of Gd@GCNs to facilitate image-guided PDT for cancer treatment has been demonstrated in rodent tumor models.

4.4.2 Photothermal Therapy (PTT)

As a new therapeutic strategy, photothermal therapy (PTT) gained tremendous attention in nanomaterial-mediated cancer therapy. PTT mainly involves the NIR light-absorbing material that converts the photoenergy into heat to kill cancer cells, resulting in irreversible cellular damage and tumor disruption. In comparison with chemotherapy, PTT has the advantages of less side effects, higher specificity, and minimal invasiveness. However, for efficient PTT the material must possess high NIR absorption and photothermal conversion along with minimal systemic toxicity. Currently, noble metal-based nanostructures and organic NIR chromophores are employed as PTT agents. However, the noble metal nanostructures suffer from potential metal toxicity and nondegradable nature with high cost, whereas the photobleaching and less hydrophilicity are the limitations of NIR dyes. In this context, with high tolerance to photobleaching, excellent biocompatibility CDs are preferable. Furthermore with the ultra-small size, CDs can be excreted from the body through renal clearance.

Minhuan et al. reported the hydrothermal preparation of S,Se-codoped CDs (Lan et al. 2017). Doping resulted in excitation-dependent NIR emission property with peaks at 731 and 820 nm. Importantly, the CDs presented a high photothermal conversion of ~58.2% that is almost near to Au nanostructures. Further, the CDs exhibited large two-photon absorption cross section of ~30,045 GM that permitted NIR emission and photothermal conversion via two-photon excitation mechanism. The cellular toxicity of CDs was studied on HeLa, A549, and KB cells under dark under irradiation. The viabilities of all three cell lines were decreased with increasing the irradiation time and CD concentration which indicate the photothermal efficiency of CDs. Simultaneous two-photon emission FL imaging and PTT of CDs were studied in vitro on HeLa cells using two-photon LSCM. Further, the in vivo studies were also performed on Balb/c nude mice with subcutaneous 4 T1 murine breast cancer cells by intravenous tail injection of CDs.

4.4.3 Other Therapies

Lin et al. used magnetofluorescent red-emitting P^{3+}- and Mn^{2+}-doped CDs (PMnCDs) in dual-modal imaging and free radical scavenging (Lin et al. 2019). To target the CD44 receptor-overexpressing cells, PMnCDs were conjugated with HA. The biocompatibility of PMnCDs/HA was assessed both in vitro (cell viability and blood compatibility) and in vivo (zebrafish embryos). B16F1 and HeLa cells treated using PMnCDs/HA were then analyzed by means of T_2-weighted images to check the in vitro MR capability PMnCDs/HA. In vitro studies on B16F1, HeLa, and HEL cells confirmed that the PMnCDs/HA scavenged excess intercellular ROS and thus provided cytoprotection against H_2O_2-induced oxidative stress. Furthermore, PMnCDs/HA exhibited a concentration-dependent cytoprotective potential against hydrogen peroxide-induced oxidative stress in B16F1, HeLa, and HEL cells.

Yao et al. prepared three types of magnetofluorescent CDs (Gd@CQDs, Mn@CQDs, and Eu@CQDs) and explored the dual modality imaging and drug delivery applications (Yao et al. 2017). Gd@CQDs had a high r_1 relaxivity of 4.78 mM^{-1} s^{-1} and a low r_2/r_1 ratio of 1.33, designating their ability as exceptional T_1 contrast agent. The r_2 relaxivity values of Mn@CQDs and Eu@CQDs were 140.7 and 28.32 mM^{-1} s^{-1}, respectively, indicating that they could be used as T_2 contrast agents. In vitro MR and FL imaging verified that additional attachment of Gd@CQDs with FA allowed the targeting of folate receptor-positive HeLa cells. Furthermore, a model drug doxorubicin (DOX) was loaded on FAGd@CQDs. Compared to free DOX, DOX-loaded nanocomposites exhibited substantial toxicity on HeLa cells. Toxicity was not detected for all three CQDs in in vivo (zebrafish embryos) and in vitro (cell viability) experiments, implying their potential as diagnostic probes or theranostic agents. Guo et al. prepared Cu,N-doped CDs (Cu,N-CDs) that upon NIR absorption generated reactive oxygen species and produced heat that enabled both PDT and PTT cancer therapy, respectively (Guo et al. 2018) (Fig. 4.8). Synergistic PDT/PTT growth inhibition of melanoma tumors was supported by in vitro and in vivo results. Furthermore, Cu,N-CDs can be applied for FL cell imaging and IR thermal imaging to illustrate the in vitro and in vivo therapeutic progression, respectively.

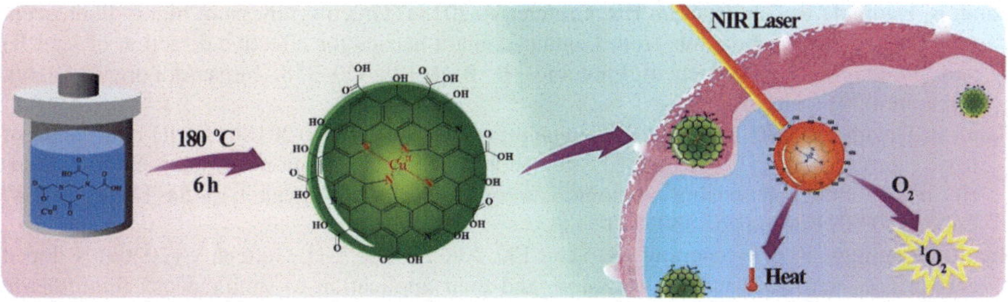

Fig. 4.8 **Schematic** illustration of *Cu,N*-CD preparation and their PTT/PDT of mouse melanoma (B16) cells. (Adopted from Guo et al. 2018)

4.5 Conclusions

CDs are a new class of fluorescent carbon nanomaterials that sparked a lot of interest among scientists because of their ease of synthesis, better aqueous solubility, higher biocompatibility, and excellent optical properties including excitation tunable emission, tolerance to photobleaching, and NIR absorption capacity. Moreover, these physicochemical properties can be strengthened further by metal or nonmetal doping. CDs not only possess high biocompatibility, but they can also be quickly excreted from the body, primarily by urine. In this book chapter, we have summarized the advancement of research in the field of metal-doped CDs. Various methods of metal-doped CD synthesis, effect of synthetic conditions and doping, and physicochemical properties are discussed first. Further the applications of metal-doped CDs in bioimaging and cancer therapy were also discussed in detail. While significant improvement in physicochemical properties and biomedical applications was achieved by metal doping, there are still few challenges that remained. For instance, even though metal ions were embedded in nontoxic carbon core, possible metal ion leaching and metal-induced toxicity cannot be ignored. Though metal-doped CDs exhibited promise in both PTT and PDT therapies, PDT is less explored. As a whole, with the promising future potential, metal doping into CDs opens new avenues in CD research as well as cancer therapy.

Acknowledgments This research was supported by Basic Science Research Program through the National Research Foundation of Korea (NRF) funded by the Ministry of Education (No. 2018R1A6A1A03025582).

References

Baker SN, Baker GA (2010) Luminescent carbon nanodots: emergent nanolights. Angew Chemie - Int Ed 49:6726–6744. https://doi.org/10.1002/anie.200906623

Bandi R, Gangapuram BR, Dadigala R, Eslavath R, Singh SS, Guttena V (2016) Facile and green synthesis of fluorescent carbon dots from onion waste and their potential applications as sensor and multicolour imaging agents. RSC Adv 6:28633–28639. https://doi.org/10.1039/c6ra01669c

Bandi R, Dadigala R, Gangapuram BR, Guttena V (2018a) Green synthesis of highly fluorescent nitrogen – doped carbon dots from Lantana camara berries for effective detection of lead(II) and bioimaging. J Photochem Photobiol B Biol 178:330–338. https://doi.org/10.1016/j.jphotobiol.2017.11.010

Bandi R, Devulapalli NP, Dadigala R, Gangapuram BR, Guttena V (2018b) Facile conversion of toxic cigarette butts to N,S-Codoped carbon dots and their application in fluorescent film, security ink, bioimaging, sensing and logic gate operation. ACS Omega 3:13454–13466. https://doi.org/10.1021/acsomega.8b01743

Bandi R, Dadigala R, Gangapuram BR, Sabir FK, Alle M, Lee S-H, Guttena V (2020a) N-doped carbon dots with pH-sensitive emission, and their application to simultaneous fluorometric determination of iron(III) and copper(II). Microchim Acta 187:30. https://doi.org/10.1007/s00604-019-4017-1

Bandi R, Gavash KH, Dadigala R, Reddy GB, Rao VJ, Guttena V (2020b) One step synthesis of hydrophobic carbon dots powder with solid state emission and application in rapid visualization of latent fingerprints. Opt Mater (Amst) 109:110349. https://doi.org/10.1016/j.optmat.2020.110349

Bandi R, Alle M, Park C, Han S, Kwon G, Kim N-H, Kim J, Lee S (2021) Cellulose nanofibrils/carbon dots composite nanopapers for the smartphone-based colorimetric detection of hydrogen peroxide and glucose. Sensors Actuators B Chem 330:129330. https://doi.org/10.1016/j.snb.2020.129330

Bhagavanth Reddy G, Dadigala R, Bandi R, Seku K, Koteswararao D, Mangatayaru KG, Shalan AE (2021) Microwave-assisted preparation of a silver nanoparticles/N-doped carbon dots nanocomposite and its application for catalytic reduction of rhodamine B, methyl red and 4-nitrophenol dyes. RSC Adv 11:5139–5148. https://doi.org/10.1039/D0RA10679H

Bourlinos AB, Bakandritsos A, Kouloumpis A, Gournis D, Krysmann M, Giannelis EP, Polakova K, Safarova K, Hola K, Zboril R (2012) Gd(III)-doped carbon dots as a dual fluorescent-MRI probe. J Mater Chem 22:23327–23330. https://doi.org/10.1039/c2jm35592b

Bouzas-Ramos D, Cigales Canga J, Mayo JC, Sainz RM, Ruiz Encinar J, Costa-Fernandez JM (2019) Carbon quantum dots Codoped with nitrogen and lanthanides for multimodal imaging. Adv Funct Mater 29:1903884. https://doi.org/10.1002/adfm.201903884

Cai Y, Wei Z, Song C, Tang C, Han W, Dong X (2019) Optical nano-agents in the second near-infrared window for biomedical applications. Chem Soc Rev 48:22–37. https://doi.org/10.1039/c8cs00494c

Chen H, Wang GD, Tang W, Todd T, Zhen Z, Tsang C, Hekmatyar K, Cowger T, Hubbard RB, Zhang W, Stickney J, Shen B, Xie J (2014) Gd-encapsulated carbonaceous dots with efficient renal clearance for magnetic resonance imaging. Adv Mater 26:6761–6766. https://doi.org/10.1002/adma.201402964

Chen H, Wang GD, Sun X, Todd T, Zhang F, Xie J, Shen B (2016a) Mesoporous silica as Nanoreactors to prepare Gd-encapsulated carbon dots of controllable sizes and magnetic properties. Adv Funct Mater 26:3973–3982. https://doi.org/10.1002/adfm.201504177

Chen H, Wang L, Fu H, Wang Z, Xie Y, Zhang Z, Tang Y (2016b) Gadolinium functionalized carbon dots for fluorescence/magnetic resonance dual-modality imaging of mesenchymal stem cells. J Mater Chem B 4:7472–7480. https://doi.org/10.1039/c6tb01422d

Chen Y, Xu J, Liu B, Li J, Fang X, Xiong L, Peng L, Han T, Tu M (2016c) Enhanced photoluminescence properties of carbon dots by doping with europium. J Nanosci Nanotechnol 16:3735–3738. https://doi.org/10.1166/jnn.2016.11800

Chen H, Qiu Y, Ding D, Lin H, Sun W, Wang GD, Huang W, Zhang W, Lee D, Liu G, Xie J, Chen X (2018) Gadolinium-encapsulated graphene carbon Nanotheranostics for imaging-guided photodynamic therapy. Adv Mater 30:1–9. https://doi.org/10.1002/adma.201802748

Chiu S-H, Gedda G, Girma WM, Chen J-K, Ling Y-C, Ghule AV, Ou K-L, Chang J-Y (2016) Rapid fabrication of carbon quantum dots as multifunctional nanovehicles for dual-modal targeted imaging and chemotherapy. Acta Biomater 46:151–164. https://doi.org/10.1016/j.actbio.2016.09.027

Dadigala R, Bandi RK, Gangapuram BR, Guttena V (2017) Carbon dots and ag nanoparticles decorated g-C3N4 nanosheets for enhanced organic pollutants degradation under sunlight irradiation. J Photochem Photobiol A Chem 342:42–52. https://doi.org/10.1016/j.jphotochem.2017.03.032

Dou Q, Fang X, Jiang S, Chee PL, Lee TC, Loh XJ (2015) Multi-functional fluorescent carbon dots with antibacterial and gene delivery properties. RSC Adv 5:46817–46822. https://doi.org/10.1039/c5ra07968c

Du F, Zhang M, Gong A, Tan Y, Miao J, Gong Y, Zou S, Zhang L, Zhang L, Wu C, Sun M, Ju H (2017) Engineered gadolinium-doped carbon dots for magnetic resonance imaging-guided radiotherapy of tumors. Biomaterials 121:109–120. https://doi.org/10.1016/j.biomaterials.2016.07.008

Fang Y, Jia J, Yang J, Zheng J, Yi C (2018) Facile preparation of holmium(III)-doped carbon nanodots for fluorescence/magnetic resonance dual-modal bioimaging. Chinese Chem Lett 29:1277–1280. https://doi.org/10.1016/j.cclet.2017.10.023

Fang Y, Zhou L, Zhao J, Zhang Y, Yang M, Yi C (2020) Facile synthesis of pH-responsive gadolinium(III)-doped carbon nanodots with red fluorescence and magnetic resonance properties for dual-readout logic gate operations. Carbon N Y 166:265–272. https://doi.org/10.1016/j.carbon.2020.05.060

Gao G, Jiang YW, Jia HR, Yang J, Wu FG (2018) On-off-on fluorescent nanosensor for Fe3+ detection and cancer/normal cell differentiation via silicon-doped carbon quantum dots. Carbon N Y 134:232–243. https://doi.org/10.1016/j.carbon.2018.02.063

Ge J, Lan M, Zhou B, Liu W, Guo L, Wang H, Jia Q, Niu G, Huang X, Zhou H, Meng X, Wang P, Lee CS, Zhang W, Han X (2014) A graphene quantum dot photodynamic therapy agent with high singlet oxygen generation. Nat Commun 5:4596. https://doi.org/10.1038/ncomms5596

Gong N, Wang H, Li S, Deng Y, Chen X, Ye L, Gu W (2014) Microwave-assisted polyol synthesis of gadolinium-doped green luminescent carbon dots as a bimodal nanoprobe. Langmuir 30:10933–10939. https://doi.org/10.1021/la502705g

Guo XL, Ding ZY, Deng SM, Wen CC, Shen XC, Jiang BP, Liang H (2018) A novel strategy of transition-metal doping to engineer absorption of carbon dots for near-infrared photothermal/photodynamic therapies. Carbon N Y 134:519–530. https://doi.org/10.1016/j.carbon.2018.04.001

Han C, Xu H, Wang R, Wang K, Dai Y, Liu Q, Guo M, Li J, Xu K (2016) Synthesis of a multifunctional manganese(II)-carbon dots hybrid and its application as an efficient magnetic-fluorescent imaging probe for ovarian cancer cell imaging. J Mater Chem B 4:5798–5802. https://doi.org/10.1039/c6tb01250g

He X, Luo Q, Zhang J, Chen P, Wang HJ, Luo K, Yu XQ (2019) Gadolinium-doped carbon dots as nano-theranostic agents for MR/FL diagnosis and gene delivery. Nanoscale 11:12973–12982. https://doi.org/10.1039/c9nr03988k

Irmania N, Dehvari K, Gedda G, Tseng PJ, Chang JY (2020) Manganese-doped green tea-derived carbon quantum dots as a targeted dual imaging and photodynamic therapy platform. J Biomed Mater Res - Part B Appl Biomater 108:1616–1625. https://doi.org/10.1002/jbm.b.34508

Ji Z, Ai P, Shao C, Wang T, Yan C, Ye L, Gu W (2018) Manganese-doped carbon dots for magnetic resonance/optical dual-modal imaging of tiny brain glioma. ACS Biomater Sci Eng 4:2089–2094. https://doi.org/10.1021/acsbiomaterials.7b01008

Jia Q, Ge J, Liu W, Zheng X, Chen S, Wen Y, Zhang H, Wang P (2018) A Magnetofluorescent carbon dot assembly as an acidic H2O2-driven Oxygenerator to regulate tumor hypoxia for simultaneous bimodal imaging and enhanced photodynamic therapy. Adv Mater 30:1–10. https://doi.org/10.1002/adma.201706090

Lan M, Zhao S, Zhang Z, Yan L, Guo L, Niu G, Zhang J, Zhao J, Zhang H, Wang P, Zhu G, Lee C-S, Zhang W (2017) Two-photon-excited near-infrared emissive carbon dots as multifunctional agents for fluorescence imaging and photothermal therapy. Nano Res 10:3113–3123. https://doi.org/10.1007/s12274-017-1528-0

Lee HU, Park SY, Park ES, Son B, Lee SC, Lee JW, Lee YC, Kang KS, Kim M II, Park HG, Choi S, Huh YS, Lee SY, Lee KB, Oh YK, Lee J (2014) Photoluminescent carbon nanotags from harmful cyanobacteria for drug delivery and imaging in cancer cells. Sci Rep 4:1–7. https://doi.org/10.1038/srep04665

Li F, Liu C, Yang J, Wang Z, Liu W, Tian F (2014) Mg/N double doping strategy to fabricate extremely high luminescent carbon dots for bioimaging. RSC Adv 4:3201–3205. https://doi.org/10.1039/c3ra43826k

Liao H, Wang Z, Chen S, Wu H, Ma X, Tan M (2015) One-pot synthesis of gadolinium(iii) doped carbon dots for fluorescence/magnetic resonance bimodal imaging. RSC Adv 5:66575–66581. https://doi.org/10.1039/C5RA09948J

Lin L, Luo Y, Tsai P, Wang J, Chen X (2018) Metal ions doped carbon quantum dots: synthesis, physicochemical properties, and their applications. TrAC - Trends Anal Chem 103:87–101. https://doi.org/10.1016/j.trac.2018.03.015

Lin JS, Tsai YW, Dehvari K, Huang CC, Chang JY (2019) A carbon dot based theranostic platform for dual-modal imaging and free radical scavenging. Nanoscale 11:20917–20931. https://doi.org/10.1039/c9nr05746c

Liu Y, Wu P, Wu X, Ma C, Luo S, Xu M, Li W, Liu S (2020) Nitrogen and copper (II) co-doped carbon dots for applications in ascorbic acid determination by non-oxidation reduction strategy and cellular imaging. Talanta 210:1–9. https://doi.org/10.1016/j.talanta.2019.120649

Mohammed A, Gugulothu Y, Bandi R, Dadigala R, Utkoor UK (2021) Ultraspeed synthesis of highly fluorescent N-doped carbon dots for the label-free detection of manganese (VI). J Chin Chem Soc 202000545. https://doi.org/10.1002/jccs.202000545

Pan Y, Yang J, Fang Y, Zheng J, Song R, Yi C (2017) One-pot synthesis of gadolinium-doped carbon quantum dots for high-performance multimodal bioimaging. J Mater Chem B 5:92–101. https://doi.org/10.1039/c6tb02115h

Park Y, Yoo J, Lim B, Kwon W, Rhee SW (2016) Improving the functionality of carbon nanodots: doping and surface functionalization. J Mater Chem A 4:11582–11603. https://doi.org/10.1039/c6ta04813g

Ren X, Liu L, Li Y, Dai Q, Zhang M, Jing X (2014) Facile preparation of gadolinium(iii) chelates functionalized carbon quantum dot-based contrast agent for magnetic resonance/fluorescence multimodal imaging. J Mater Chem B 2:5541–5549. https://doi.org/10.1039/c4tb00709c

Ren XY, Yuan XX, Wang YP, Liu CL, Qin Y, Guo LP, Liu LH (2016) Facile preparation of Gd3+ doped carbon quantum dots: photoluminescence materials with magnetic resonance response as magnetic resonance/fluorescence bimodal probes. Opt Mater (Amst) 57:56–62. https://doi.org/10.1016/j.optmat.2016.04.018

Rub Pakkath SA, Chetty SS, Selvarasu P, Vadivel Murugan A, Kumar Y, Periyasamy L, Santhakumar M, Sadras SR, Santhakumar K (2018) Transition metal ion (Mn2+, Fe2+, Co2+, and Ni2+)-doped carbon dots synthesized via microwave-assisted pyrolysis: a potential Nanoprobe for magneto-fluorescent dual-modality bioimaging. ACS Biomater Sci Eng 4:2581–2596. https://doi.org/10.1021/acsbiomaterials.7b00943

Sun YP, Zhou B, Lin Y, Wang W, Fernando KAS, Pathak P, Meziani MJ, Harruff BA, Wang X, Wang H, Luo PG, Yang H, Kose ME, Chen B, Veca LM, Xie SY (2006) Quantum-sized carbon dots for bright and colorful photoluminescence. J Am Chem Soc 128:7756–7757. https://doi.org/10.1021/ja062677d

Sun W, Luo L, Feng Y, Qiu Y, Shi C, Meng S, Chen X, Chen H (2020) Gadolinium–rose Bengal coordination polymer Nanodots for MR-/fluorescence-image-guided radiation and photodynamic therapy. Adv Mater 32:1–8. https://doi.org/10.1002/adma.202000377

Swathi R, Bhagavanth Reddy G, Rajkumar B, Yadagiri Swamy P (2020) Rapid fabrication of carbon dots from babul seed powder as green precursor: antioxidant activity and multicolor imaging. Mater Today Proc. https://doi.org/10.1016/j.matpr.2020.09.174

Tejwan N, Saini AK, Sharma A, Singh TA, Kumar N, Das J (2021) Metal-doped and hybrid carbon dots: a comprehensive review on their synthesis and biomedical applications. J Control Release 330:132–150. https://doi.org/10.1016/j.jconrel.2020.12.023

Tummala S, Lee C-H, Ho Y-P (2021) Boron, and nitrogen co-doped carbon dots as a multiplexing probe for sensing of p-nitrophenol, Fe (III), and temperature. Nanotechnology 32:265502. https://doi.org/10.1088/1361-6528/abeeb6

Unnikrishnan B, Wu RS, Wei SC, Huang CC, Chang HT (2020) Fluorescent carbon dots for selective labeling of subcellular organelles. ACS Omega 5:11248–11261. https://doi.org/10.1021/acsomega.9b04301

Wang Y, Zhang Y, Jia M, Meng H, Li H, Guan Y, Feng L (2015) Functionalization of carbonaceous Nanodots from MnII-coordinating functional knots. Chem - A Eur J 21:14843–14850. https://doi.org/10.1002/chem.201502463

Wang J, Xu M, Wang D, Li Z, Primo FL, Tedesco AC, Bi H (2019) Copper-doped carbon dots for optical bioimaging and photodynamic therapy. Inorg Chem 58:13394–13402. https://doi.org/10.1021/acs.inorgchem.9b02283

Wu F, Su H, Zhu X, Wang K, Zhang Z, Wong WK (2016) Near-infrared emissive lanthanide hybridized carbon quantum dots for bioimaging applications. J Mater Chem B 4:6366–6372. https://doi.org/10.1039/c6tb01646d

Xu X, Ray R, Gu Y, Ploehn HJ, Gearheart L, Raker K, Scrivens WA (2004) Electrophoretic analysis and purification of fluorescent single-walled carbon nanotube fragments. J Am Chem Soc 126:12736–12737. https://doi.org/10.1021/ja040082h

Xu Y, Jia XH, Yin XB, He XW, Zhang YK (2014) Carbon quantum dot stabilized gadolinium nanoprobe prepared via a one-pot hydrothermal approach for magnetic resonance and fluorescence dual-modality bioimaging. Anal Chem 86:12122–12129. https://doi.org/10.1021/ac503002c

Xu Q, Kuang T, Liu Y, Cai L, Peng X, Sreenivasan Sreeprasad T, Zhao P, Yu Z, Li N (2016) Heteroatom-doped carbon dots: synthesis, characterization, properties, photoluminescence mechanism and biological applications. J Mater Chem B 4:7204–7219. https://doi.org/10.1039/C6TB02131J

Yang C, Li Y, Yang Y, Tong R, He L, Long E, Liang L, Cai L (2018) Multidimensional theranostics for tumor fluorescence imaging, photoacoustic imaging and photothermal treatment based on manganese doped carbon dots. J Biomed Nanotechnol 14:1590–1600. https://doi.org/10.1166/jbn.2018.2565

Yao YY, Gedda G, Girma WM, Yen CL, Ling YC, Chang JY (2017) Magnetofluorescent carbon dots derived from crab Shell for targeted dual-modality bioimaging and drug delivery. ACS Appl Mater Interfaces 9:13887–13899. https://doi.org/10.1021/acsami.7b01599

Yu C, Xuan T, Chen Y, Zhao Z, Liu X, Lian G, Li H (2016) Gadolinium-doped carbon dots with high quantum yield as an effective fluorescence and magnetic resonance bimodal imaging probe. J Alloys Compd 688:611–619. https://doi.org/10.1016/j.jallcom.2016.07.226

Zhang M, Zhao L, Du F, Wu Y, Cai R, Xu L, Jin H, Zou S, Gong A, Du F (2019) Facile synthesis of cerium-doped carbon quantum dots as a highly efficient antioxidant for free radical scavenging. Nanotechnology:30. https://doi.org/10.1088/1361-6528/ab12ef

Zhao Y, Hao X, Lu W, Wang R, Shan X, Chen Q, Sun G, Liu J (2018) Facile preparation of double rare earth-doped carbon dots for mri/ct/fi multimodal imaging. ACS Appl Nano Mater 1:2544–2551. https://doi.org/10.1021/acsanm.8b00137

Zheng S, Yu N, Han C, Xie T, Dou B, Kong Y, Zuo F, Shi M, Xu K (2019) Preparation of gadolinium doped carbon dots for enhanced MR imaging and cell fluorescence labeling. Biochem Biophys Res Commun 511:207–213. https://doi.org/10.1016/j.bbrc.2019.01.098

Further Reading

Tejwan N, Saini AK, Sharma A, Singh TA, Kumar N, Das J (2021) Metal-doped and hybrid carbon dots: a comprehensive review on their synthesis and biomedical applications. J Control Release 330:132–150

Lin L, Luo Y, Tsai P, Wang J, Chen X (2018) Metal ions doped carbon quantum dots: synthesis, physicochemical properties, and their applications. TrAC - Trends Anal Chem 103:87–101

Chapter 5
Graphene-Based Smart Nanomaterials for Photothermal Therapy

Ramakrishna Dadigala, Rajkumar Bandi, Madhusudhan Alle, Bhagavanth Reddy Gangapuram, and Seung-Hwan Lee

5.1 Introduction

Cancer is one of the leading foremost public health complications and widely recognized as one of the most destructive diseases impacting people. According to estimates, almost 18 million new cancer cases and 10 million cancer deaths occur each year (Eskiizmir et al. 2018; Rahman et al. 2019; Wei et al. 2019). As a result, early cancer detection is becoming increasingly relevant. From the past few years, traditional therapies, such as surgery, chemotherapy and radiotherapy, have been developed to treat various cancers (Gai et al. 2018; Estelrich and Busquets 2018), although they are ineffective in completely destroying the cancer cells, they still have several drawbacks, such as cancer cell resistance, immune system depletion, poor water soluble efficiency, poor patient adherence and destroying healthy tissue (Tew et al. 2014). As a result, novel cancer therapy approaches must be developed

R. Dadigala · R. Bandi
Institute of Forest Science, Kangwon National University, Chuncheon, Republic of Korea

M. Alle
Institute of Forest Science, Kangwon National University, Chuncheon, Republic of Korea

Department of Biomedical Science & Institute of Bioscience and Biotechnology, Kangwon National University, Chuncheon, Republic of Korea

B. R. Gangapuram
Department of Chemistry, Palamuru University, Mahabub Nagar, Telangana, India

S.-H. Lee (✉)
Institute of Forest Science, Kangwon National University, Chuncheon, Republic of Korea

Department of Forest Biomaterials Engineering, Kangwon National University, Chuncheon, Republic of Korea
e-mail: lshyhk@kangwon.ac.kr

J.-C. Kim et al. (eds.), *Smart Nanomaterials in Biomedical Applications*, Nanotechnology in the Life Sciences,
https://doi.org/10.1007/978-3-030-84262-8_5

to solve these drawbacks. To combat this issue, researchers have been developing new cancer diagnostic and therapeutic methods like microwave therapy, ultrasound therapy, radio-frequency therapy and phototherapy (Min et al. 2012; Mura et al. 2013; Zhang et al. 2013). Because of their high performance, spatial precision and remote control capacity, phototherapy methodologies have drawn the most interest, due to their more stability and little toxicity under several circumstances (Montazeri 2008; Fan et al. 2017). PTT is an effective novel method for treating diseases, especially cancers, in clinical practice (Khot et al. 2019; Zhi et al. 2020).

In PTT, hyperthermia was induced at the cancer site by irradiating with a particular wavelength (NIR) light, which ablates cancer cells in a localized area while reducing the death of neighbouring cells. Because of its specific wavelengths (650–900 nm), the NIR laser directly transmitted into the tissues without causing harm to normal cells (Khot et al. 2019; Wang and Cheng 2019; Zhi et al. 2020). For cancer PTT, PTAs in the nanosize are needed, since they capture NIR light to destroy cancer cells. Till now many investigators have produced a variety of PTAs (Bao et al. 2016; Hu et al. 2018; Wei et al. 2019). Among several nanomaterials, graphene and its derivatives have been thoroughly researched as PTAs in recent years, like graphene oxide (GO), reduced graphene oxide (rGO), graphene quantum dots and its composites, due to its extraordinary photothermal conversion efficacy, huge surface area, abundance of oxygen-containing functional units and unique optical characteristics (Dinadayalane et al. 2012; Chen et al. 2016b; Mohamadi and Hamidi 2017; Eskiizmir et al. 2018; Ghosal and Sarkar 2018; Lin et al. 2018; Saba and Jawaid 2018; Samantara et al. 2018; Lahir 2019; Nafiujjaman and Nurunnabi 2019; Wang and Cheng 2019; Xie et al. 2019; Barua et al. 2020; Li et al. 2021). In this chapter, we discussed how GO PTAs have been used for thermal ablation of cancer cells.

5.2 Photothermal Therapy

The ability of a material to absorb light or radio-frequency energy and transform it into heat for the purpose of enabling hyperthermia in pernicious cells is known as photothermal therapy. When cancer cells are subjected to NIR light over many minutes, permanent damage to protein and membrane structures occurs, leading to cell death (Fig. 5.1) (Estelrich and Busquets 2018; Fernandes et al. 2020). PTT has several benefits over traditional treatments, including capability to penetrate deep tissue, high sensitivity, low severity and limited nonspecific cell death on surrounding healthy tissue (Khot et al. 2019; Wang and Cheng 2019; Yu et al. 2020; Zhi et al. 2020). The kinetic energy of the PTAs inside the cell raises as the duration of NIR light irradiation improves, as a result of the accumulation of NIR light energy by PTAs and the cell atmosphere heating up causing obstruction of the cytoplasm as well as destroying the cancer cell proteins (Bao et al. 2016; Hu et al. 2018; Wei et al. 2019). PTAs used in PTT will have a high NIR absorbance, high photothermal conversion performance, excellent optical consistency, cancer detection, high

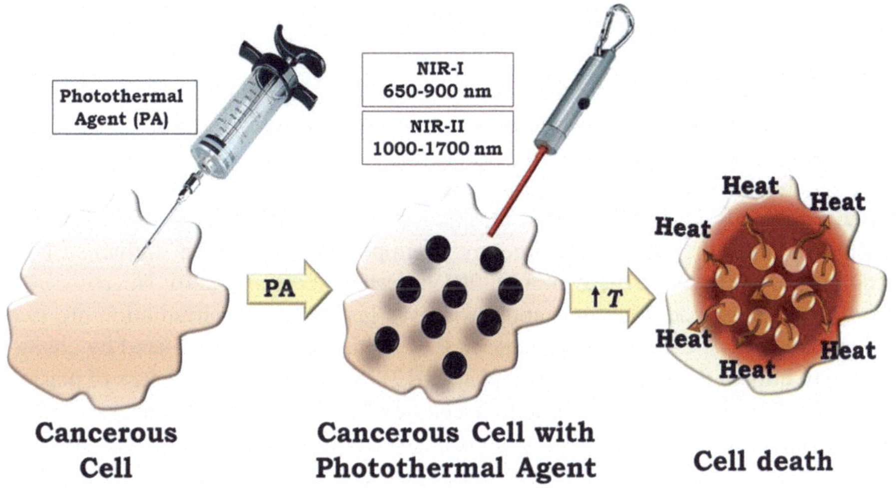

Fig. 5.1 Photothermal therapy process. (Adopted from Estelrich and Busquets 2018)

photostability, appropriate height, consistent shape, good miscibility, good biocompatibility and lower toxicity. Furthermore, ideal PTAs should have broad optical wavelength absorption cross sections, low toxicity, ease of surface modification and complete solubility in biocompatible solutions (Hernández and Galarreta 2019; Wang and Cheng 2019; Fernandes et al. 2020; Yu et al. 2020).

Several PTAs have been increasingly improved by several researchers for possible use in PTT (Bao et al. 2016; Hu et al. 2018; Wei et al. 2019). PTAs are classified into two groups based on their chemical characteristics: inorganic and organic PTAs. Inorganic PTAs include metal-based nanomaterials, coated transition metal di-chalcogenides, carbon materials, quantum dots and black phosphorous quantum dots, while organic PTAs like small dyes, polymer nanoparticles and covalently linked polymer nanoparticles with strong absorption in NIR region (Chen et al. 2016b; Ghosal and Sarkar 2018; Saba and Jawaid 2018; Samantara et al. 2018; Nafiujjaman and Nurunnabi 2019; Rahman et al. 2019). Carbon-based nanomaterials, in particular, have been extensively studied due to its higher NIR absorbance, biologically inertness and low price (Dinadayalane et al. 2012; Shi et al. 2014; Eskiizmir et al. 2018; Lin et al. 2018; Lahir 2019; Barua et al. 2020). Among different carbon-based nanomaterials, graphene-based nanomaterials have been considered as promising PTAs in treatment of cancer (Shi et al. 2014; Chen et al. 2015; Mohamadi and Hamidi 2017; Wang and Cheng 2019; Xie et al. 2019; Li et al. 2021). In recent years, graphene-based nanostructures have been thoroughly analysed for PTT (Dinadayalane et al. 2012; Shi et al. 2014; Shi et al. 2017; Eskiizmir et al. 2018; Saba and Jawaid 2018; Nafiujjaman and Nurunnabi 2019; Xie et al. 2019) Therefore, it has been recognized as an effective therapeutic technique for thermal ablation of cancer cells, and efficient PTAs are important for achieving a sufficient therapeutic impact in cancer PTT.

5.2.1 Mechanism of Photothermal Therapy

PTT has arisen as a recent trend in cancer treatment because of its more effectiveness for killing cancer cells and relatively less risk to healthy tissues. It uses PTAs that assemble in the cancer cells to produce heat form NIR irradiation for killing cancer cells (Hernández and Galarreta 2019; Khot et al. 2019; Wang and Cheng 2019; Fernandes et al. 2020; Yu et al. 2020; Zhi et al. 2020). Type of PTAs, NIR laser strength and NIR contact time are all important aspects that influence PTT efficiency (Bao et al. 2016; Wei et al. 2019; Fernandes et al. 2020). Necrosis over more frequency irradiation and apoptosis over less frequency irradiation are two potential routes of cancer cell damage in PTT, which could be regulated by changing the treatment parameters (Martin et al. 2012). Necrosis will cause loss of plasma membrane reliability and subsequent discharge of intracellular substances, which is the way for death of cancer cells. Apoptosis, like chemotherapy or radiotherapy, can reduce unwanted provocative reactions but also generate resistance to PTT. When used under the right irrigation conditions, PTT will cause apoptosis instead of necrosis (Nafiujjaman and Nurunnabi 2019). PTT's ability to persuade apoptosis begs the intriguing query of how much external or inherent mitochondrial apoptosis path is the primary route to cancer cell damage (Mocan et al. 2014). Because the factors which influence the mode of cancer cell damage are unknown, more research is needed to elaborate the mechanism of PTT-induced cancer cell death.

5.3 Graphene-Based Materials in Photothermal Therapy

Recently, graphene-based nanomaterials have gained extensive consideration from researchers due to their potential use in cancer PTT (Dinadayalane et al. 2012; Chen et al. 2016b; Georgakilas et al. 2016; Ghosal and Sarkar 2018; Lin et al. 2018; Samantara et al. 2018; Bai and Husseini 2019; Nafiujjaman and Nurunnabi 2019; Rahman et al. 2019; Ramezani et al. 2019; Yu et al. 2020). In graphene, carbon atoms form hexagonal honeycomb structure, and it appeared as a sheetlike structure and these sheets are closely bound by Van der Waals interactions. It's a carbon allotrope with a molecular bond length of 0.142 nm and sp^2-bonded carbon atoms. Graphene layers on top of each other from graphite with a 0.335 nm inter-planar gap. Van der Waals interactions hold each sheet of graphene in graphite together, which can be resolved with graphene exfoliation (Dinadayalane et al. 2012; Lin et al. 2018; Saba and Jawaid 2018; Bai and Husseini 2019; Lahir 2019; Ramezani et al. 2019; Barua et al. 2020). It serves as a foundation for other graphitic materials. Graphene nanosheets and its substitutes like GO, rGO and GQDs have been used in PTT as PTAs for thermal ablation of cancer cells from the past few decades (Fig. 5.2) (Shi et al. 2014; Chen et al. 2015; Chen et al. 2016a; Shi et al. 2017; Eskiizmir et al.

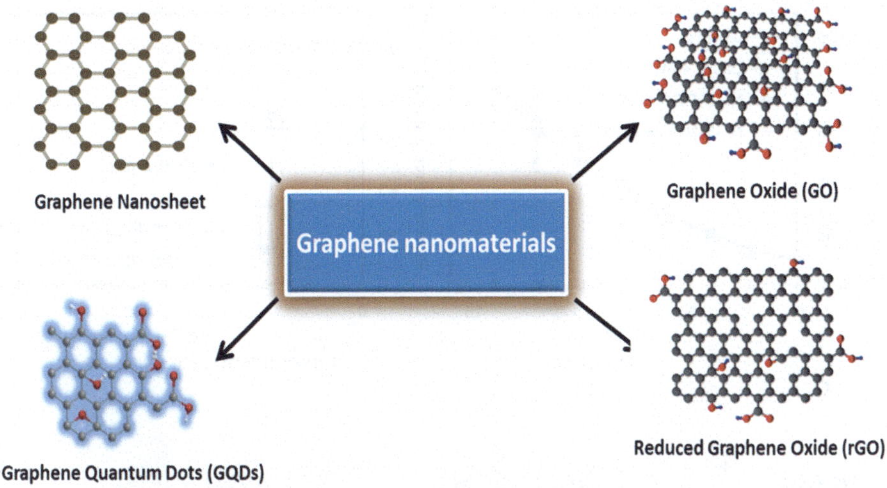

Fig. 5.2 Graphene-based nanomaterials for photothermal therapy

2018; Ghosal and Sarkar 2018; Bai and Husseini 2019; Nafiujjaman and Nurunnabi 2019; Rahman et al. 2019; Ramezani et al. 2019; Barua et al. 2020; Li et al. 2021).

5.3.1 Graphene Nanosheets in Photothermal Therapy

Liu and his colleagues were the first to use nanographene sheets for PTT (Yang et al. 2010). First, they investigated the in vivo behaviour of nanographene sheets (NGS) coated with polyethylene glycol (PEG) using a fluorescent labelling Cy7 (NGS-PEG). After 24 h of NGS-PEG injection, each mouse's cancer was treated with NIR (808 nm) laser. The outside temperature of the cancer in the NGS-PEG-inserted community approached 50 °C, and the cancer was completely eliminated (Fig. 5.3). However, no findings were obtained without introducing NGS-PEG into mouse. Further, Yijuan wang et al. developed a Monte Carlo technique to display a mesoscale model of cancer PTT using NIR and graphene nanosheets (Wang et al. 2020). A cancer cell was embedded within a cube of healthy tissue in a three-dimensional (3D) model. On the surface of a cancer cell, graphene nanosheets were uniformly dispersed (Fig. 5.4). The thermal behaviour of the device was studied quantitatively in relation to the concentration and morphology of graphene nanosheets. To investigate their effect on temperature increase, the interfacial thermal resistance (ITR) around the graphene sheets is measured. It was discovered that a higher concentration of graphene nanosheets resulted in a faster rate of temperature rise and shorter graphene sheets with a larger width/thickness could cause a faster rate of temperature increase in cancer cells. Reducing the ITR between the cancer cell and graphene is another way to get the cancer cell to raise its temperature faster, which can be done by properly functionalizing the graphene.

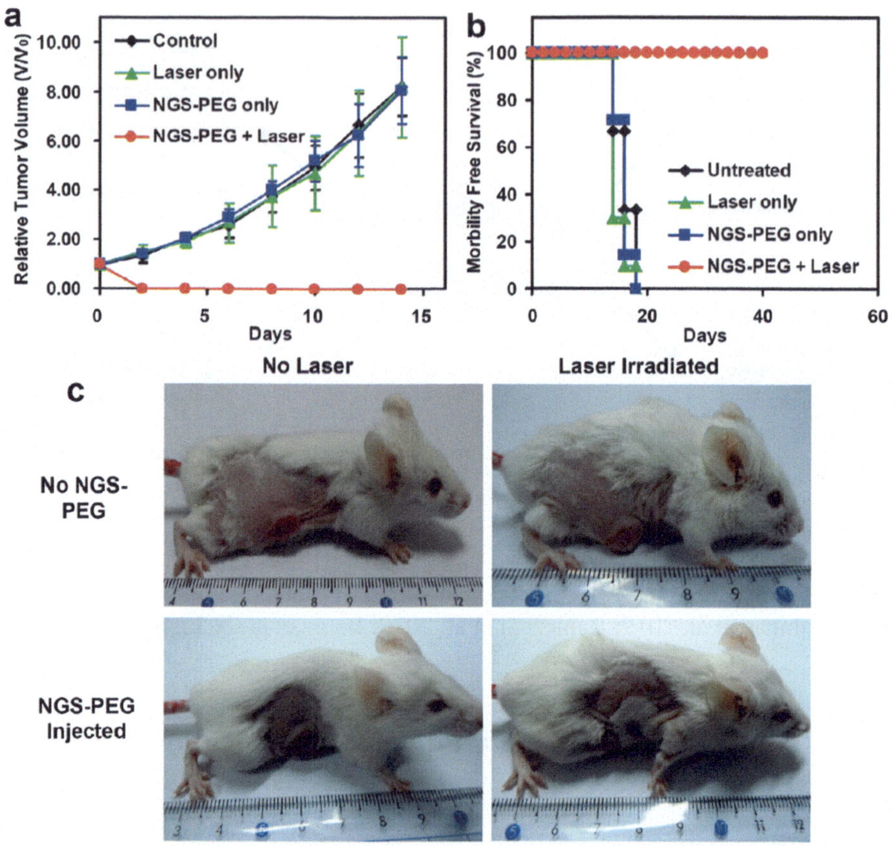

Fig. 5.3 In vivo photothermal therapy using intravenously injected NGS-PEG. (a) Cancer growth curves of different groups after treatment. (b) Survical curves of mice bearing 4T1 cancer after various treatments indicated. NGS-PEG-injected mice after photothermal therapy survived over 40 days without any single death. (c) Reprehensive photos of cancers on mice after various treatments. The laser-irradiated cancer on NGS-injected mouse was completely destructed (Adopted from Yang et al. 2010)

5.3.2 Graphene Oxide and Its Composites in Photothermal Therapy

Graphene oxide is an oxidized form of graphene that has several exceptional physical, chemical, optical, electrical and mechanical properties as well as it contains several oxygen-containing functional groups like epoxy, hydroxyl, carbonyl and carboxylic groups (Mohamadi and Hamidi 2017; Singh et al. 2018; Ghorai 2019; Nizami et al. 2020). Many scientific areas, including biomedical applications, have shown great promise for GO and its composites because of its exceptional properties like large surface area, high water solubility, the presence of oxygen-containing groups, high adsorption ability, good biocompatibility, mechanical strength, low

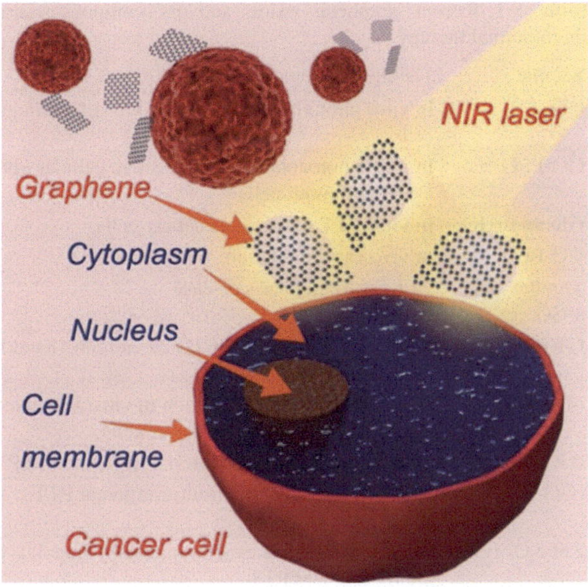

Fig. 5.4 Schematic plot of the photothermal therapy using NIR irradiation and graphene nanosheets (Adopted from Wang et al. 2020)

price and easy functionalization (Mohamadi and Hamidi 2017; Singh et al. 2018; Ghorai 2019; Nizami et al. 2020). Since GO absorbance ranges from UV to NIR, numerous interesting studies have looked at using it for PTT in recent years. Besides that, GO is an appealing nanocarrier for drug delivery because of its fast and efficient cell penetration as well as its ability to shield DNA and peptides from enzymatic cleavage (Georgakilas et al. 2016; Lim et al. 2018; Xie et al. 2019). Table 5.1 shows recent research and key findings of GO and its composites used as PTAs in PTT.

Anticancer agents that react to stimuli are of special importance in the field of cancer therapy. Huabing Chen et al. (Guo et al. 2015) created a pH-responsive GO-cypate nanoplatform with improved photothermal effect in the lysosomal environment. Cypate molecules clumped together on the surface of GO in aqueous solution, generating additional photothermal energy from near-infrared fluorescence (NIRF) via the pH-dependent fluorescence resonance energy transfer (FRET) effect, cause greater photothermal cytotoxicity. Due to this, GO-cypate causes significant cell death, resulting in complementary PTT efficiency with cancer killing after intramuscular injection. As an efficient FRET receptor, the photothermal nanoadduct with broad NIR absorbance will wisely transform transmitted NIRF energy from the donor to cyanine dye with improved PTT. Sei Kwang Hahn et al. (Jung et al. 2014) established a conjugate with nanographene oxide and hyaluronic acid (NGOHA) for photothermal melanoma skin cancer using NIR irradiation. Melanoma skin cancer is among the utmost severe types of skin cancer and leading reason of death. HA is generally used as a transdermal delivery vehicle for chemical and biopharmaceutical products. The extraordinary transdermal distribution of NGOHA in cancer tissues of mice skin was clearly visible using confocal microscopy and

Table 5.1 Recent graphene oxide and its composites used as photothermal agents in photothermal therapy

Sample	Therapeutic remarks	References
GO	In vitro photothermal killing of U251 human glioma cells	Markovic et al. (2011)
GO-PEG	In vitro photoablation and selective cellular uptake in U87MG tumour cells	Robinson et al. (2011)
Glucose-rGO	In vitro PTT for prostate cancer cells	Akhavan et al. (2012)
GO-PEG	In vivo cancer ablation	Yang et al. (2012b)
Thioflavin-s-GO	PTT for Alzheimer's disease	Li et al. (2012b)
GT-rGO	More efficacy of PTT on HT29 and SW48 colon cancer cells	Abdolahad et al. (2013)
rGO-Cu_2O	Under visible light irradiation in vitro selective photocatalytic killing of cancer cells	Hou et al. (2013)
rGO-Au	Gold-coated graphene material for enhanced PTT	Lim et al. (2013)
rGO	Nanomesh structures for ultra-efficient PTT	Akhavan and Ghaderi (2013)
$ZnFe_2O_4$-rGO	Graphene composites with magnetic materials for highly efficient PTT	Akhavan et al. (2014)
HA-rGO	Photothermal killing of transdermal melanoma skin cancer	Jung et al. (2014)
rGO-GO	PTT for cancer under femtosecond laser beam irradiation	Li et al. (2014a)
GO-PEG-AnNR	Enhanced PTT by nanocomposites	Dembereldorj et al. (2014)
Cyanine-GO	Stimuli-mediated PTT for cancer therapy	Guo et al. (2015)
Apt-Au-GO	For breast cancer combining of thermal therapy and inhibitor producing cancericidal effects	Yang et al. (2015)

ex vivo bioimaging (Fig. 5.5), which could be attributed to extremely express HA receptors and comparatively permeable structures across tissues, allowing for better nanoparticle penetration and persistence. ELISA for caspase-3 activity and immune histochemical analysis by TUNEL assay for cancer apoptosis reported the anticancer effect. This method can be used for transdermal treatment of melanoma skin cancer by loading drug with NGO in NGOHA.

5.3.3 Graphene Quantum Dots in Photothermal Therapy

Graphene quantum dots are nanoscale graphene structures with solid quantum properties and edge effects, resulting in extraordinary photoluminescence properties (Bandi et al. 2016; Bandi et al. 2020a; Bandi et al. 2021; Mohammed et al. 2021). GQDs' optical properties are highly dependent on their structure size due to the quantum confinement effect and the varying type and distribution of sp^2 sites in the

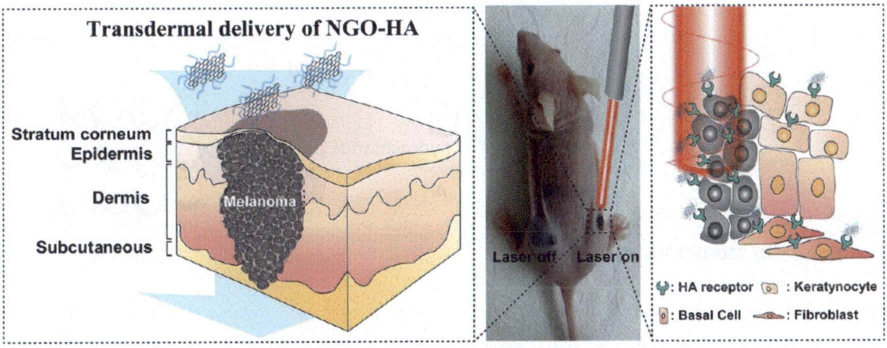

Fig. 5.5 Schematic illustration for the transdermal delivery of NGO-HA conjugates into melanoma skin cancer cells and following photothermal ablation therapy using a NIR irradiation (adopted from Jung et al. 2014)

structure. As a result, manipulating the size and surface chemistry properties of GQDs will alter their energy bandgap. GQDs have exciting biomedical uses, including diagnostics, drug delivery, PTT using NIR laser and in vitro and in vivo bioimaging (Dadigala et al. 2017; Bandi et al. 2018a; Bandi et al. 2018b; Bandi et al. 2020b).

Justin et al. have prepared chitosan-conjugated GQDs with biodegradable lidocaine hydrochloride (CS-GQD-LH) and iontophoresis therapy and then monitored the delivery of both high and small molecular mass drugs into the human skin (Fig. 5.6) (Justin et al. 2015). They discovered that the CS-GQD-LH could passively diffuse low molecular weight drugs into the body and iontophoresis large molecular weight drugs into the body. Tayyebi et al. have stated the synthesis of GQDs under supercritical conditions, which resulted in improved thermal connectivity and photothermal properties (Tayyebi et al. 2018). After 460 s NIR irradiation, the phantoms reached 45^{0}C for GQDs and 40^{0}C for GO, respectively. Ding et al. also published a GQD-based marked anticancer nanocarrier that generates multiple fluorescence signals to monitor DOX transmission, release and response (Ding et al. 2017).

5.3.4 Reduced Graphene Oxide and Its Composites in Photothermal Therapy

Reduced graphene oxide can be made from GO through electrochemical reduction, chemical reduction and thermal reduction. Following the development of rGO, the substance can be functionalized for various applications. By reacting rGO with different chemicals or mixing with other material to make new compounds is a potential way to improve the properties of the compound to fit for biological applications (Hashemi et al. 2017; Liu et al. 2018; Chen et al. 2019). It exhibits high

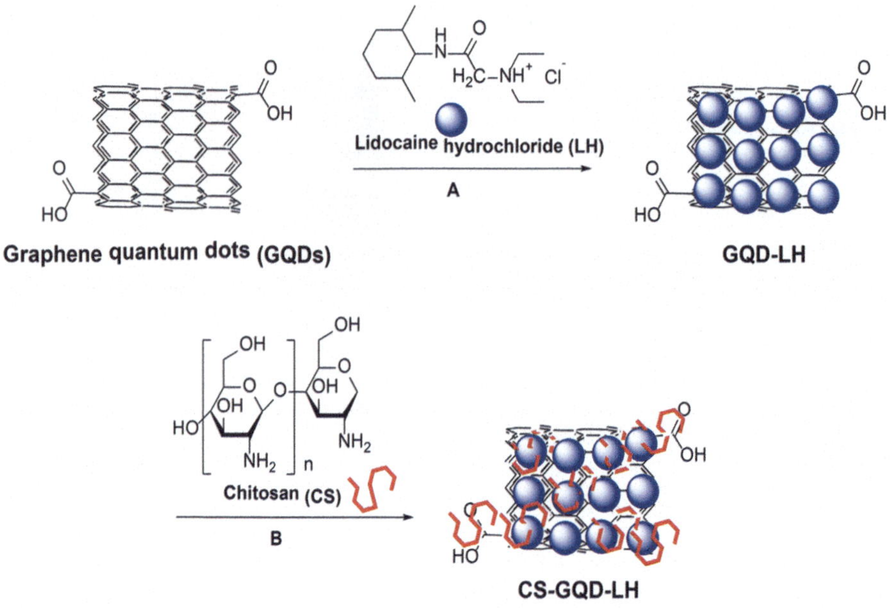

Fig. 5.6 The synthesis procedure of CS-GQD-LH nanocomposite. (**a**) LH to bond to the GQDs via π-π stacking of aromatic rings and hydrogen bonding. (**b**) CS to bond to GQD-LH through hydrogen bonding between the amino groups of CS and carboxyl groups of GQD-LH. (Adopted from Justin et al. 2015)

nanostructured aromatic lattice and have greater π-π and hydrophobic associations with drugs (Darabdhara et al. 2015; Gai et al. 2019). Furthermore, rGO-based nanomaterials have a much greater NIR absorption than GO, resulting in higher photothermal conversion efficiency. The rGO structure's high inclination will contribute to minimal dispersibility and irreversible accumulation (Darabdhara et al. 2015; Liu et al. 2020). To date, an increasing number of studies have concentrated on rGO's biomedical applications. For example, Lima-Sousa et al. synthesized novel hyaluronic acid (HA) functionalized rGO for cancer PTT (Lima-Sousa et al. 2018). Initially, the NIR absorption was used to optimize the green reduction of GO with L-ascorbic acid. Then, rGO was functionalized with a novel HA-dependent amphiphilic polymer, formed by attaching HA onto poly(maleic anhydride-alt-1-octadecene) (PMAO) for enhanced stability and cyto-compatibility, feasible of executing a targeted cancer PTT (Fig. 5.7). The formed HA-rGO shows a superior internalization of cancer cells which overexpressed CD44 receptors, causing low cell sustainability (6%) with rGO dosage (75 μg/mL), NIR laser intensity (1.7 W/cm^2) and irradiation duration (5 min). Overall, HA-rGO seems to be a viable nanomaterial for cancer PTT targeting CD44.

RGD-PEG-rGO nanosheets were created (Robinson et al. 2011) with more NIR light absorption and biocompatibility for efficient PTT to research the impact of PTT with GO-based nanocarriers. The nanosheets were then attached

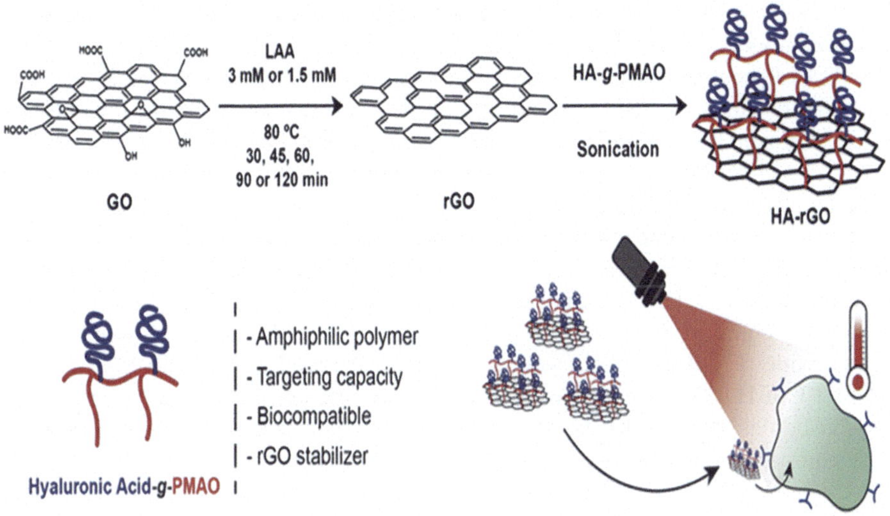

Fig. 5.7 Preparation of HA-rGO and schematic representation of the reduction and functionalization of rGO with HA-g-PMAO and its application in photothermal therapy. (Adopted from Lima-Sousa et al. 2018)

non-covalently by amphiphilic PEGylated polymer chains to improve its stability in biological solutions, resulting in a six-time increase in NIR absorption over covalently and nonreduced PEGylated nano-GO. In addition, adding of Arg-Gly-Asp (RGD) peptide-like substance to nano-rGO allowed for selective cellular absorption and extremely successful photoablation of U87MG cancer cells in vitro. Dai et al. have produced nanosized rGO with noncovalent PEGylation (PEG-rGO) and also, with remaining PEG chains bound to GO, further hydrophobicity of the nano-rGO sheets enhanced after reduction, causing accumulation (Ma et al. 2017). To recover the dispersion, the formed PEG-rGO was again bound with existing PEG using sonication to form a polymer-coated PEG-rGO (2PEG-rGO). The 2PEG-rGO polymers retained equilibrium in biological solutions and buffers without aggregation. Furthermore, the 2-PEG-rGO polymer shows six-time increase in NIR absorption compared to nano-GO and PEGylation nano-GO. This improvement was attributed to the degree of π conjugation promotion in GO after chemical reduction. As a result, the 2PEG-rGO polymer shows great NIR absorbance, which was acceptable for successful photothermal heating of solutions at low 2PEG-rGO polymer concentrations.

It has been noticed that the newly invented rGONM-PEG-Cy7-RGD using r-GO nanomesh grafted with PEG, cyanine-7 and RGD peptide (arginine-glycine-aspartic acid) for in vivo cancer targeting and fluorescence imaging in mouse models (Akhavan and Ghaderi 2013). In vivo fluorescence imaging shows more selective cancer absorption of composite in mice carrying U87MG cancer cells. The ultra-efficient PTT of this material is due to its cancer targeting and tremendous NIR absorption and all mice lived for more than 100 days. Another article by Omid

Akhavan et. al. manufactured superparamagnetic $ZnFe_2O_4$-rGO composites with different amounts of graphene by hydrothermal process and then applied for effective PTT of prostate cancer cells (in vitro) and human glioblastoma cancer cells (in vivo) (Akhavan et al. 2014). About the fact that a low concentration of $ZnFe_2O_4$-rGO composite resulted in 50% of in vitro cancer cell damage within 1 min of NIR irradiation under ideal circumstances and the composite ability to turn into magnetically localized at light irradiated place then made admirable photothermal performance under the same circumstances. Besides that, with and without applied magnetic field, the localization of magnetic nanomaterials inserted into the cancer was investigated. These findings would encourage more magnetic graphene-containing composites to be used in high-efficiency PTT. Green tea-reduced graphene oxide (GT-rGO) sheets were prepared by Abdolahad et al. for high-quality NIR PTT of HT29 (lower metastatic) and SW48 (greater metastatic) colon cancer cells (Abdolahad et al. 2013). SW48 cancer cells photothermally died more than 20% compared to HT29 cancer cells.

5.4 GO PTT with Combination Treatments

Graphene oxide PTT with various cancer treatment methods is a viable area for NIR-based PTT. Moreover, several studies using these products as PTT agents have shown efficacy in in vivo and in vitro cancer treatment. Further, the combination of these treatments such as chemotherapy, photodynamic therapy, gene therapy, fluorescent agent decorated, imaging regulated, low power efficient and bio-functionalized materials was enhanced the therapeutic efficiency (Fig. 5.8) (Chen et al. 2016a). The photothermal cancer therapy using nanosized graphene-based materials is summarized in Table 5.2.

5.4.1 GO PTT with Chemotherapy

Nanosized GO and rGO have a superior drug loading efficacy than other materials due to its large surface area and containing of more functional groups (Cheon et al. 2016). Because of its pH-sensitive release, the anticancer medication DOX (doxorubicin) has been widely used as a basis for treatment of cancer. The release of DOX is increased in acidic environments, as the case in the intracellular environment, owing to its higher water solubility. DOX is mainly loaded into GO or rGO due to molecular piling. For example, Wang et al. have prepared GO coated with mesoporous silica, PEG and IL 31 peptides (GSPI) for chemo-PTT and then GSPI loaded with DOX (Wang et al. 2013b). Under NIR irradiation the interactions between DOX and GSPI were destroyed and then it accelerated the exponential release of DOX at various times and pH values. As a result GSPI shows tremendous therapeutic activity on agonist glioma (Fig. 5.9).

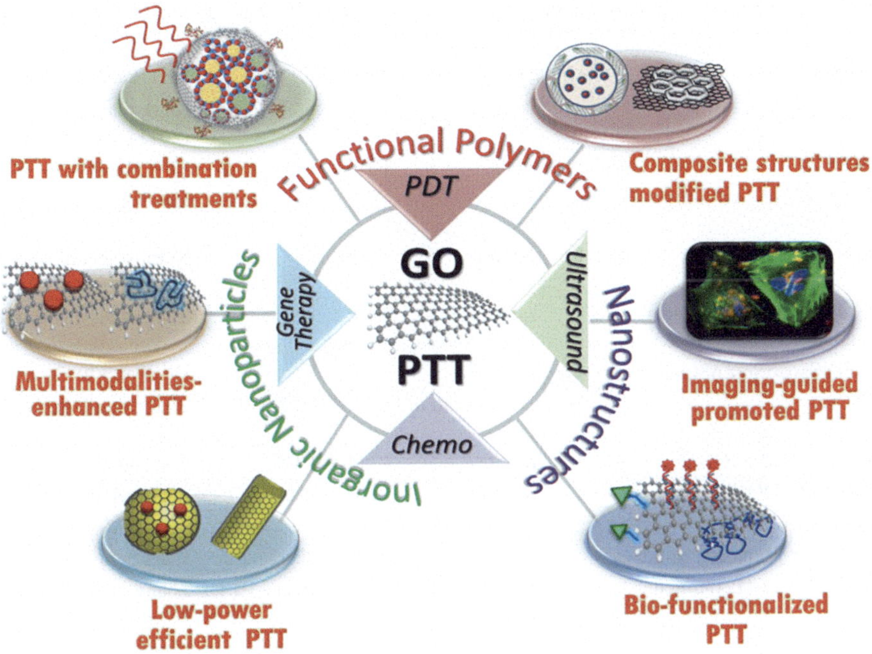

Fig. 5.8 Various functions and nanostructures for enhanced photothermal therapy based on graphene materials. (Adopted from Chen et al. 2016a)

Zhang et al. used PEGylated GO in both in vitro and in vivo synergistic therapy (Zhang et al. 2011b); chemo-PTT was more successful than single photothermal treatment, and synergistic treatment was achieved at high laser strength and very low drug concentration. The photothermal effect takes place when the laser strength is greater. High laser power, on the other hand, is not recommended for in vivo treatment because it may cause safety issues. The combination therapy demonstrated the highest efficacy when the laser strength was set at 2 W/cm^2 and the DOX concentration was increased. This suggests a synergistic effect between DOX and the heat generated by GO. In vivo studies have shown that the combination treatment will totally eliminate cancers in mice. Cheon et al. established rGO amended with serum albumin and DOX for the chemo-PTT of brain cancers (Cheon et al. 2016). In comparison to non-rGO, the receiving nanomaterials showed increased absorption in the NIR region, and the serum albumin-coated rGO caused a temperature rise up to 60 ^{0}C after NIR irradiation. Furthermore, an increase in temperature can be used to activate drug release and cause U87MG cancer cells to die, with 21.8% and 1.76% of sustainable cells in the photothermal and chemo-PTT, respectively. In another study, Roy and colleagues prepared poly(allylamine hydrochloride) to mediated rGO nanosheets for chemo-PTT of breast cancer (Roy et al. 2019). The functionalized rGO irradiating with an NIR laser could cause temperature above 45 ^{0}C; it can ablate the cancer cells. In addition, in in vitro experiments, a computational

Table 5.2 A summary of photothermal cancer therapy using nanosized graphene-based materials

Sample	*In vitro*/*in vivo*	Therapy	References
Graphene NPs	*In vitro*	PTT	Markovic et al. (2011)
GO NPs	*In vitro*	PTT	Li et al. (2012a)
GONSs-PLA-Au composites	*In vitro* and *in vivo*	PTT	Jin et al. (2013)
rGO NSs	*In vitro*	PTT	Robinson et al. (2011)
rGONSs/Fe_3O_4 magnetic nanoparticle composite	*In vitro*	PTT	Yang et al. (2012a)
rGONSs-CdSe/ZnSQDs composite	*In vitro*	PTT	Hu et al. (2012)
GONSs	*In vitro* and *in vivo*	Chemo-PTT	Zhang et al. (2011c)
Silica-coated rGONSs	*In vitro*	Chemo-PTT	Wang et al. (2013b)
rGO-PEG-DOX	*In vivo*	Chemo-PTT	Zhang et al. (2011b)
GO-PEG-Ce6	*In vivo*	PDT-PTT	Tian et al. (2011)
Graphene-phthalocyanine	*In vivo*	PDT-PTT	Jiang et al. (2014)
NGO-PEI-PEG conjugate	*In vitro* and *in vivo*	Gene delivery-PTT	Zhang et al. (2011a)
IO/GO-COOH	*In vitro* and *in vivo*	Imaging-guided PTT	Huang et al. (2015)
rGO-AuNR composites	*In vitro* and *in vivo*	Imaging-guided PTT	Moon et al. (2015)
Apt-AuNPs-GO	*In vitro* and *in vivo*	Low-power efficient PTT	Yang et al. (2015)
AuNR-PEG-GO	*In vitro* and *in vivo*	Low-power efficient PTT	Dembereldorj et al. (2014)
QD-rGO	*In vitro* and *in vivo*	Fluorescent agent-PTT	Hu et al. (2012)

therapeutic modality induced by DOX delivery and photothermal effect caused superior cytotoxicity against MCF-7 cells, with 70% and 6% viable cells in the photothermal and chemo-PTT, respectively. DOX-loaded BSA (bovine serum albumin)-grafted rGO (DOX-BSA-rGO) was recently developed by Yeong Ah Cheon and co-workers for chemo-PTT of brain cancer cells (Cheon et al. 2016). In stimuli-responsive drug delivery and meticulous release applications, the protein-mediated rGO is of great importance. Here heat generated from NIR irradiation breaks the bond between DOX and BSA protein, permitting the release of DOX into cancers (Fig. 5.10). This finding showed that NIR irradiation increased DOX release from DOX-BSA-rGO nanosheets by around twofold. As a result, this DOX-loaded BSA-rGO nanosheet may be an effective chemo-PTT platform for brain cancer cells.

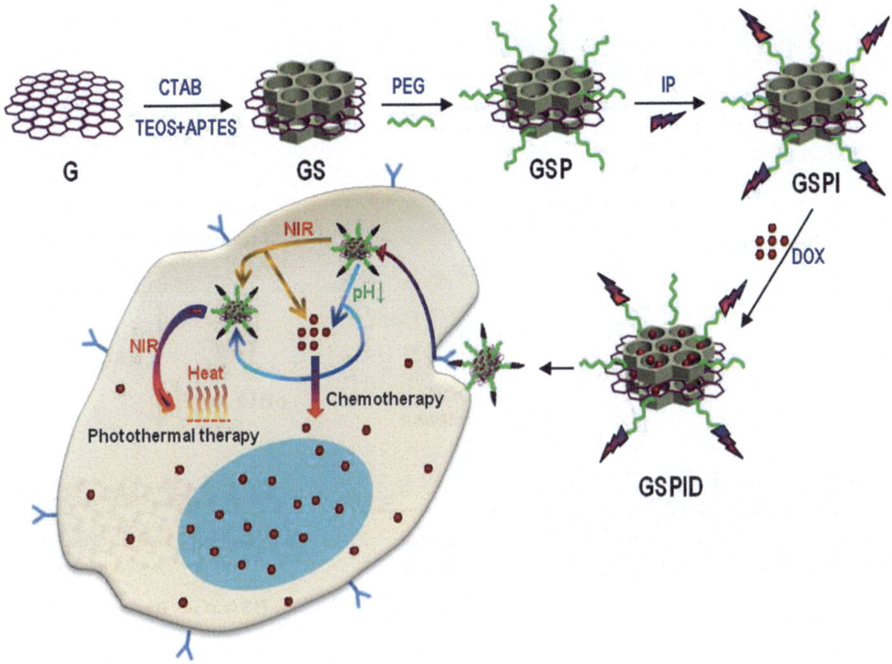

Fig. 5.9 Graphene nanosheet-coated mesoporous silica as a multifunctional drug delivery system for combined chemo-photothermal-targeted therapy of glioma. (Adopted from Wang et al. 2013b)

5.4.2 GO PTT with PDT

Photodynamic therapy generally used the reactive oxygen species (ROS) released by photosensitizers under laser irradiation to trigger cancer cell apoptosis, which is gaining popularity as cancer morbidity rises. There has been some research into the synergistic effects of PDT and PTT. For instance, as shown in Fig. 5.11 (Tian et al. 2011), a PEG-modified GO is combined with a PDT agent (chlorine e6 (Ce6)), for PDT combined PTT to improve the cancer cell ablation. The combination of mild photothermal heating of graphene caused by NIR light and photodynamic treatment with Ce6 delivered by GO-PEG significantly improves PDT efficacy.

5.4.3 GO PTT with Gene Therapy

In gene delivery, GO surface modification may serve as a gene delivery vector for in vitro and in vivo treatment. GO can be effectively bonded with gene delivery agents like polyethylene imine and chitosan by an amide bond, electrostatic interaction and stacking interactions (Zhang et al. 2011a). To facilitate gene-mediated cancer treatment, siRNA and DOX-loaded NGO-PEI-PEG conjugate was used to

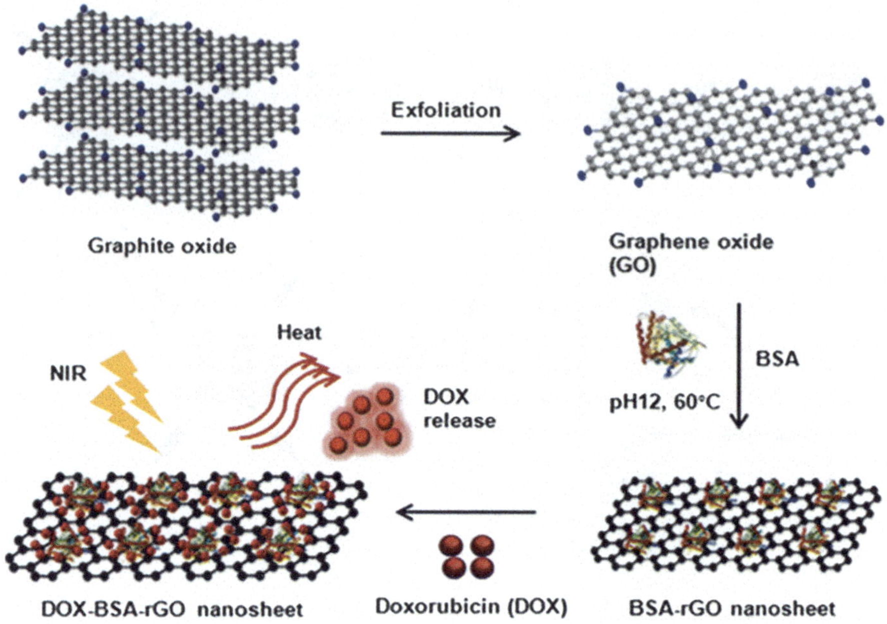

Fig. 5.10 Schematic illustration of synthesis of DOX-BSA-rGO nanosheets and NIR-mediated chemo-photothermal therapy. The graphite oxide was exfoliated to make graphene oxide, and BSA was used for reduction of graphene oxide at pH 12 and 60 °C. Doxorubicin was finally conjugated with BSA-rGO nanosheets. (Adopted from Cheon et al. 2016)

reduce cytotoxicity while also enhancing DOX release and gene delivery ability (Shen et al. 2012). At NIR irradiation with lower energy, cellular uptake of GO was increased because of low photothermal heating that increased permeability of the cell membrane without causing significant cell harm. As a result, GO-based gene transfer agents are efficient under NIR laser to improve cancer therapy.

5.4.4 Fluorescent Agent Decorated GO for PTT

For fluorescent cell imaging and PTT, Liu group developed QD-tagged rGO nanocomposites (Hu et al. 2012). The toxicity of the QDs is reduced by the surface modification, and fluorescence reducing is prevented by accurate meticulous spacer between the QDs and rGO. Incredibly, when QDs-rGO absorbs NIR irradiation, then it increases the temperature and decreases QD light, offering a potential for in situ heat/temperature sensing and an indication of PTT development. In the meantime, aggressive affecting and improved cell internalization via the combining of the QDs-rGO with directing/internalizing ligands have also been accomplished as seen in Fig. 5.12. This nanocomposite's combination of integrations makes it an

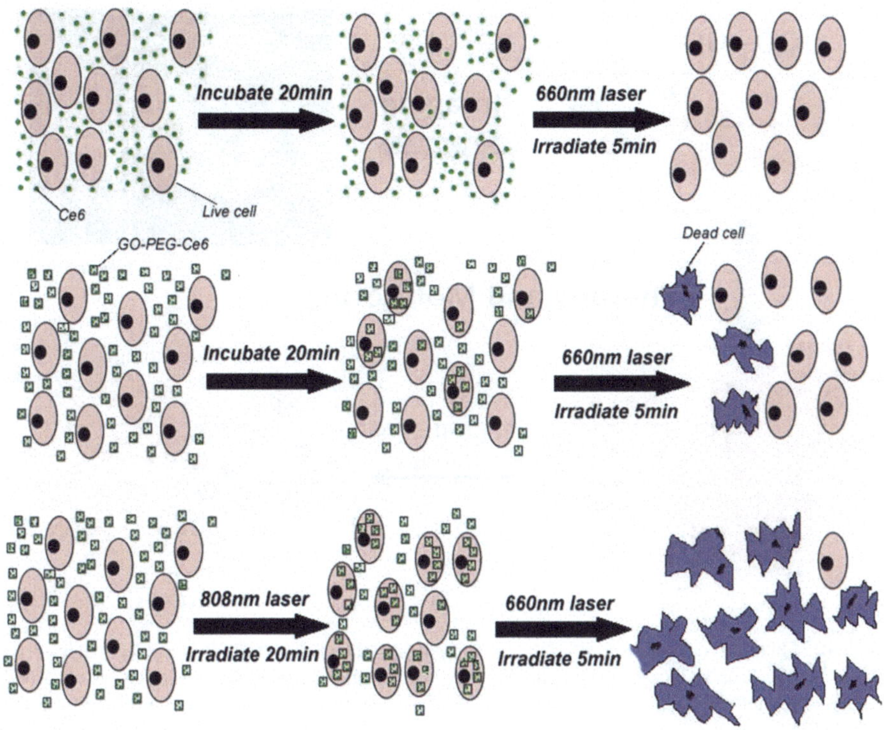

Fig. 5.11 Schemes of the design in photothermally enhanced photodynamic therapy. To combine the photothermal effect, cells incubated with GO-PEG-Ce6 were first exposed to the 808 nm laser (20 min) before PDT treatment, resulting in significant improvement of PDT. (Adopted from Tian et al. 2011)

efficient viable novel method for cancer cell detection, imaging and delivery of drug molecules and PTT.

5.4.5 Imagine-Guided PTT of GO

Bioimaging techniques have been extensively used in various medical remedies and drug delivery processes. Imaging-guided therapy using nanomaterials was in recent times adopted as a novel technique for cancer cure (Yang et al. 2012a; Tian et al. 2013; Song et al. 2014), since imaging offers crucial details like cancer size, location and optimum phototherapy time window. As a result, several studies for imaging-guided PTT have centred on incorporating imaging and remedy components into graphene-based materials (Janib et al. 2010). Taratula et al. prepared a graphene nanoplatform deposited with phthalocyanine for imaging-guided PTT. The combinational therapy led to an improved killing effectiveness of 90% to 95% for

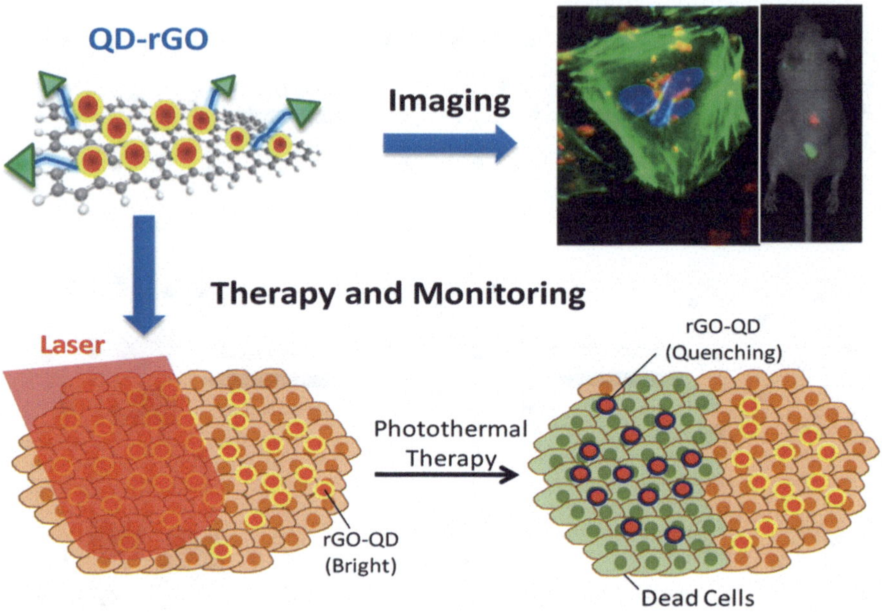

Fig. 5.12 Quantum dot-targeted rGO nanocomposites (left) internalized into targeted cancer cells display bright fluorescence from the QDs (right). NIR irradiation incident on the rGO is absorbed and converted into heat, causing cell death and fluorescence reduction (bottom). (Adopted from Hu et al. 2012)

ovarian cancer cells at low pc and low GO amount (Moon et al. 2015). Furthermore, Huang et al. firstly synthesized carboxylic group-functionalized GO and then iron oxide nanoparticles deposited on it (IONPs/GO COOH) for T2-contrast improvement both in vivo and in vitro as compared to pure IONPs or GO COOH (Huang et al. 2015). Recently, PEG-conjugated rGO was deposited with IONPs for imaging-guided PTT (Ma et al. 2012).

5.4.6 Low-Power Efficient PTT of GO

So far, there has been an emphasis on nanomedicine with optimized imaging and therapeutic modalities for combined disease management and control as a theranostic method to improve therapeutic effectiveness (Kelkar and Reineke 2011; Hu et al. 2012; Yang et al. 2012a; Li et al. 2014b). The existing GO-based PTTs are restricted due to the difficult NIR perception (only several millimetres). As a result, determining how to make use of lower-energy NIR to absorb materials to attain effective photothermal conversion is crucial because it reduces normal tissues damage. To solve this problem, rGO is combined with plasmonic nanomaterials to achieve effective PTT at lower-energy NIR irradiation (Zedan et al. 2013; Lim et al. 2013;

Turcheniuk et al. 2015; Stabile et al. 2020). Gold nanoparticles are the most investigated plasmonic substance in biomedical applications because of its less cytotoxicity, more biocompatibility and exceptional photothermal conversion efficacy (Huang et al. 2008; Yang et al. 2015; Reddy et al. 2015; Chang et al. 2015; Gangapuram et al. 2017; Aioub et al. 2018; Gangapuram et al. 2018; Alle et al. 2020a; Alle et al. 2020b; Bhagavanth et al. 2021; Alle et al. 2021; Kim et al. 2021). Subsequently, the size and morphology of gold nanomaterials can be regulated to control the strengthened photothermal effects of gold nanomaterials.

Chia-Hua Lin et al. (Yang et al. 2015) recently synthesized an aptamer-AuNPs hybridized GO nanocomposite (Apt-AuNPs-GO) for NIR activatable PTT treatment of cancer cells (Fig. 5.13). Because of the unique association between the MUC1-binding aptamer and the MUC1 (type I transmembrane mucin glycoprotein) on cell membrane, the self-assembled Apt-AuNPs-GO nanocomposite could selectively target MUC1-positive human breast cancer cells (MCF-7). Apt-AuNPs-GO also has a high light-to-high heat transfer capability for photoabsorption of NIR light, and it can exert therapeutic effects on MCF-7 cells at ultralow concentrations without harming healthy cells. Besides that, the absence of HSP-70 protein induced by Apt-AuNPs-GO with NIR irradiation can cause cell death that is irreversible. As a result, the degree and length of HSP-70 protein expression are linked to PTT's anti-breast cancer therapeutic efficacy. Hyperthermia and HSP-70 inhibition had a synergistic therapeutic effect, resulting in a substantial increase in breast cancer cell death. Yang and colleagues created the Apt-AuNPs-GO method (Yang et al. 2015).

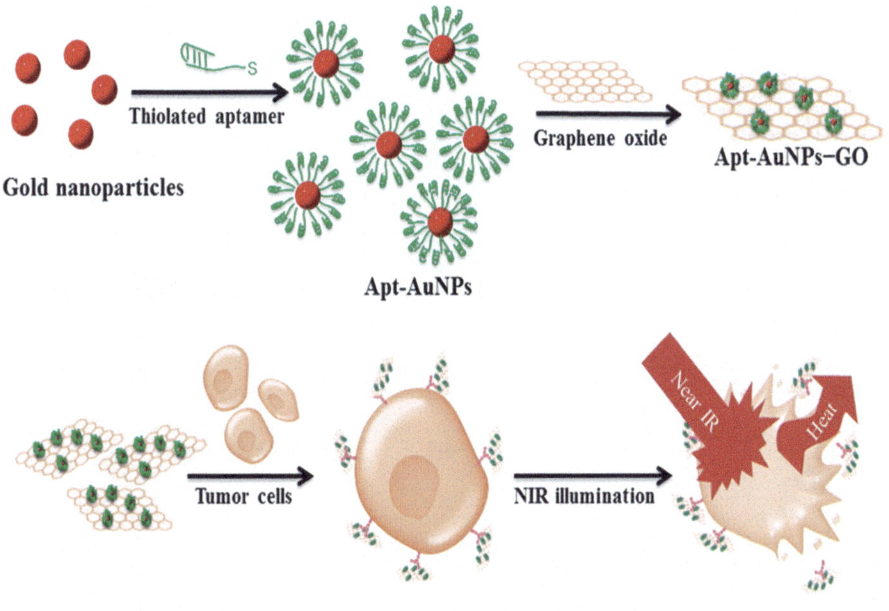

Fig. 5.13 Schematic representation of the preparation of Apt-AuNPs-GO and its application with NIR laser irradiation for photothermal therapy. (Adopted from Yang et al. 2015)

There, GO was fabricated using an enhanced Hummers process, and the single-layered GO had an average size of about 300 nm. By reducing $AuCl_4$ with citrate ions, AuNPs were formed. The aptamer molecules then interacted directly with the AuNPs to form Apt-AuNPs (15 nm), which were used to bind oligo-S-units to the AuNP surfaces. Apt-AuNPs-GO was then generated by combining Apt-AuNPs with GO for 1 h in a buffer solution (5mM, pH 7.4). The stability of this nanostructure was greatly improved due to the strong interaction between gold and aptomer as well as stacking between the multivalent nucleobase of aptamers and GO. Apt-AuNPs-GO showed a photothermal conversion efficiency of 25.4% when exposed to persistent agitation of NIR light at power strength of 3W.

Uuriintuya DembereldorJ et al. prepared gold nanorod-attached PEGylated GO (AuNR-PEG-GO) composites for PTT (Dembereldorj et al. 2014). After abstraction with PEG-GO and the elimination of cetyltrimethylammonium bromide (CTAB) from AuNR-HCl, the cytotoxicity of AuNRs was reduced. A431 epidermoid cancer cells were irradiated with xenon lamp (60 Wcm2) for 5 min to assess the photothermal effect of AuNR-PEG-GO composite. In the presence of composite and NIR light cell, viability decreased by 40%, but only with composite no cell death was observed. In vivo cancer studies have revealed that AuNR-PEG-GO treated with HCL could effectively reduce cancer volumes via a photothermal process (Fig. 5.14).

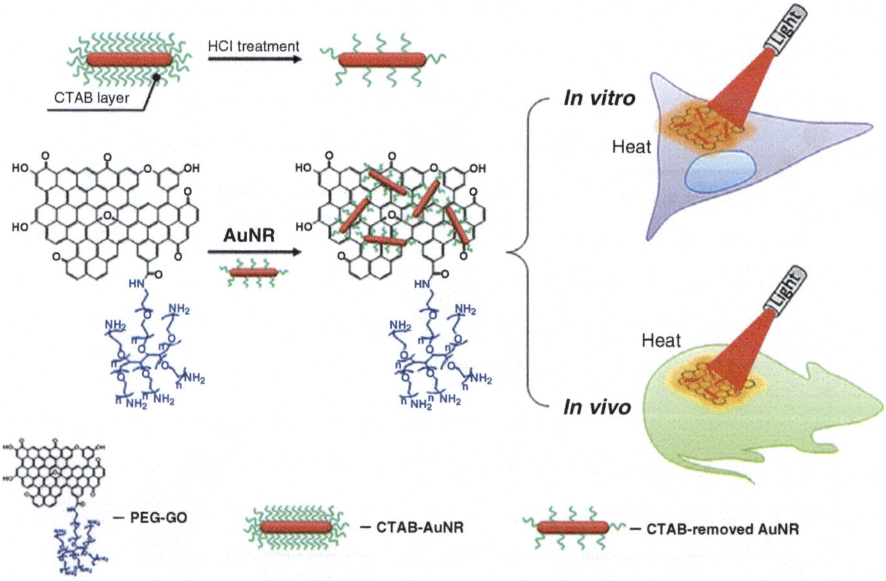

Fig. 5.14 Schematic illustration of the preparation of AuNR-PEG-GO for photothermal therapy. (Adopted from Dembereldorj et al. 2014)

5.4.7 *Bio-functionalized PTT of GO*

The fundamental problem of GO-based materials in terms of biocompatibility is GO amount and surface properties. PTT performance has been improved using GO materials modified with different conjugated molecules. For example, poly(2-dimethyl amino ethyl methacrylate) [poly(PDMAEMA)]-functionalized rGO and indocyanine green (ICG) showed a pH-sensitive release and quenching process that was regulated in vitro by photothermolysis (Wang et al. 2013a; Sharker et al. 2015). In comparison to free ICG, the composite showed significantly enhanced in vitro photothermal efficiency and exceptional removal of malicious cells after 18 days of treatment. Furthermore, PEG is often conjugated to GO for better stability in physiological conditions such as serum and a more positive targeting outcome (Jastrzębska et al. 2012). Biocompatible rGO sheets with appropriate water solubility were acquired by functionalizing reduced sheets with gluconate ions formed during glucose reduction in the presence of Fe catalyst rather than PEGylation (Akhavan et al. 2012), which is a typical method of functionalization. Then, efficiency of the glucose-rGO in PTT of LNCaP prostate cancer cells was studied in vitro. The glucose-rGO sheets outperformed compared to single-wall and multiwall carbon nanotube suspensions and hydrazine-reduced GO, in terms of PTT performance, and also showed some toxicity. The modification of rGO sheets by gluconate ions also avoided their aggregation and showed high PTT efficiency and biocompatibility.

5.5 Conclusions

The heart of the cancer therapy is being able to target and destroy locally diseased cells while causing minimal damage to healthy cells. A feasible approach to solve this issue is PTT with graphene-based materials, which can be used to absorb the irradiated NIR light to convert it into heat to ablate the cancer cells. The key graphene-based nanomaterials produced for cancer PTT were defined in this chapter, including graphene oxide, reduced graphene oxide, graphene quantum dots and their composites, as well as PTT with hybrid therapies. In general, evidence in the literature shows that GO nanosheets as photothermal transfer agents in PTT have drawn a lot of research interest in the last few years, because of their high surface area, strong biocompatibility, ample oxygen supplying functional units, outstanding stability and unique optical properties. In PTT, the GO nanosheets can absorb energy from NIR irradiation and convert it into heat in microenvironment of cancer cells. Then the local hyperthermia will alter the roles of certain structural proteins and enzymes in cancer cells, thus leading to death of cancer cells. Furthermore, other environmental parameters such as cancer localization, NIR laser conditions and aggregation within the cancer cells all influence the therapeutic effectiveness of graphene-mediated PTT. Additionally, combining PTT with other therapies will boost graphene materials' therapeutic potential, lowering the dosage required to eradicate cancer cells and reducing treatment side effects. This chapter briefly presents the graphene-based materials for PTT approaches and addresses current and potential problems in this area.

The PTT controlled by nanomaterials has the ability to improve cancer therapy outcomes. Nonetheless, the majority of these nanomaterials are still in the preclinical stage, and further research is needed to define their short- and long-term fate in the body, biodegradation and protection in order to accelerate their transmission from the lab to the clinic. Furthermore, the invention of simple processing methods and their scaling-up would improve nanomaterials' reproducibility and therapeutic effect.

Acknowledgements This work was supported by the Basic Science Research Program through the National Research Foundation of Korea funded by the Ministry of Education No. 2018R1A6A1A03025582.

References

Abdolahad M, Janmaleki M, Mohajerzadeh S, Akhavan O, Abbasi S (2013) Polyphenols attached graphene nanosheets for high efficiency NIR mediated photodestruction of cancer cells. Mater Sci Eng C 33:1498–1505. https://doi.org/10.1016/j.msec.2012.12.052

Aioub M, Austin LA, El-Sayed MA (2018) Gold nanoparticles for cancer diagnostics, spectroscopic imaging, drug delivery, and plasmonic photothermal therapy. In: Inorganic frameworks as smart nanomedicines. Elsevier, pp 41–91

Akhavan O, Ghaderi E (2013) Graphene nanomesh promises extremely efficient in vivo photothermal therapy. Small 9:3593–3601. https://doi.org/10.1002/smll.201203106

Akhavan O, Ghaderi E, Aghayee S, Fereydooni Y, Talebi A (2012) The use of a glucose-reduced graphene oxide suspension for photothermal cancer therapy. J Mater Chem 22:13773–13781. https://doi.org/10.1039/c2jm31396k

Akhavan O, Meidanchi A, Ghaderi E, Khoei S (2014) Zinc ferrite spinel-graphene in magneto-photothermal therapy of cancer. J Mater Chem B 2:3306–3314. https://doi.org/10.1039/c3tb21834a

Alle M, Reddy GB, Kim TH, Park SH, Lee SH, Kim JC (2020a) Doxorubicin-carboxymethyl xanthan gum capped gold nanoparticles: microwave synthesis, characterization, and anti-cancer activity. Carbohydr Polym 229:115511. https://doi.org/10.1016/j.carbpol.2019.115511

Alle M, Lee S-H, Kim J-C (2020b) Ultrafast synthesis of gold nanoparticles on cellulose nanocrystals via microwave irradiation and their dyes-degradation catalytic activity. J Mater Sci Technol 41:168–177. https://doi.org/10.1016/j.jmst.2019.11.003

Alle M, Park SC, Bandi R, Lee S-H, Kim J-C (2021) Rapid in-situ growth of gold nanoparticles on cationic cellulose nanofibrils: Recyclable nanozyme for the colorimetric glucose detection. Carbohydr Polym 253:117239. https://doi.org/10.1016/j.carbpol.2020.117239

Bai RG, Husseini GA (2019) Graphene-based drug delivery systems. In: Biomimetic nanoengineered materials for advanced drug delivery. Elsevier, pp 149–168

Bandi R, Alle M, Park CW, Han SY, Kwon GJ, Kim NH, Kim JC, Lee SH (2021) Cellulose nanofibrils/carbon dots composite nanopapers for the smartphone-based colorimetric detection of hydrogen peroxide and glucose. Sensors Actuators B Chem 330:129330. https://doi.org/10.1016/j.snb.2020.129330

Bandi R, Dadigala R, Gangapuram BR, Guttena V (2018a) Green synthesis of highly fluorescent nitrogen – doped carbon dots from Lantana camara berries for effective detection of lead(II) and bioimaging. J Photochem Photobiol B Biol 178:330–338. https://doi.org/10.1016/j.jphotobiol.2017.11.010

Bandi R, Dadigala R, Gangapuram BR, Sabir FK, Alle M, Lee S-H, Guttena V (2020a) N-Doped carbon dots with pH-sensitive emission, and their application to simultaneous fluorometric determination of iron(III) and copper(II). Microchim Acta 187:30. https://doi.org/10.1007/s00604-019-4017-1

Bandi R, Devulapalli NP, Dadigala R, Gangapuram BR, Guttena V (2018b) Facile conversion of toxic cigarette butts to N,S-codoped carbon dots and their application in fluorescent film, security ink, bioimaging, sensing and logic gate operation. ACS Omega 3:13454–13466. https://doi.org/10.1021/acsomega.8b01743

Bandi R, Gangapuram BR, Dadigala R, Eslavath R, Singh SS, Guttena V (2016) Facile and green synthesis of fluorescent carbon dots from onion waste and their potential applications as sensor and multicolour imaging agents. RSC Adv 6:28633–28639. https://doi.org/10.1039/C6RA01669C

Bandi R, Gavash KH, Dadigala R, Reddy GB, Rao VJ, Guttena V (2020b) One step synthesis of hydrophobic carbon dots powder with solid state emission and application in rapid visualization of latent fingerprints. Opt Mater (Amst) 109:110349. https://doi.org/10.1016/j.optmat.2020.110349

Bao Z, Liu X, Liu Y, Liu H, Zhao K (2016) Near-infrared light-responsive inorganic nanomaterials for photothermal therapy. Asian J Pharm Sci 11:349–364. https://doi.org/10.1016/j.ajps.2015.11.123

Barua S, Geng X, Chen B (2020) Graphene-based nanomaterials for healthcare applications. Elsevier Inc.

Bhagavanth RR, Dadigala R, Bandi R, Seku K, Koteswararao D, Mangatayaru KG, Shalan AE (2021) Microwave-assisted preparation of a silver nanoparticles/N-doped carbon dots nanocomposite and its application for catalytic reduction of rhodamine B, methyl red and 4-nitrophenol dyes. RSC Adv 11:5139–5148. https://doi.org/10.1039/d0ra10679h

Chang G, Wang Y, Gong B, Xiao Y, Chen Y, Wang S, Li S, Huang F, Shen Y, Xie A (2015) Reduced graphene oxide/amaranth extract/AuNPs composite hydrogel on tumor cells as integrated platform for localized and multiple synergistic therapy. ACS Appl Mater Interfaces 7:11246–11256. https://doi.org/10.1021/acsami.5b03907

Chen X, Li C, Wang X, Zhao X (2019) Infrared heating of reduced graphene oxide nanosheets as photothermal radiation therapeutic agents for tumor regressions. Mater Res Express:6. https://doi.org/10.1088/2053-1591/ab13c3

Chen Y-W, Su Y-L, Hu S-H, Chen S-Y (2016a) Functionalized graphene nanocomposites for enhancing photothermal therapy in tumor treatment. Adv Drug Deliv Rev 105:190–204. https://doi.org/10.1016/j.addr.2016.05.022

Chen Y, Tan C, Zhang H, Wang L (2015) Two-dimensional graphene analogues for biomedical applications. Chem Soc Rev 44:2681–2701. https://doi.org/10.1039/C4CS00300D

Chen YW, Su YL, Hu SH, Chen SY (2016b) Functionalized graphene nanocomposites for enhancing photothermal therapy in tumor treatment. Adv Drug Deliv Rev 105:190–204. https://doi.org/10.1016/j.addr.2016.05.022

Cheon YA, Bae JH, Chung BG (2016) Reduced graphene oxide nanosheet for chemo-photothermal therapy. Langmuir 32:2731–2736. https://doi.org/10.1021/acs.langmuir.6b00315

Dadigala R, Bandi R, Gangapuram BR, Guttena V (2017) Carbon dots and Ag nanoparticles decorated g-C 3 N 4 nanosheets for enhanced organic pollutants degradation under sunlight irradiation. J Photochem Photobiol A Chem 342:42–52. https://doi.org/10.1016/j.jphotochem.2017.03.032

Darabdhara G, Das MR, Turcheniuk V, Turcheniuk K, Zaitsev V, Boukherroub R, Szunerits S (2015) Reduced graphene oxide nanosheets decorated with AuPd bimetallic nanoparticles: a multifunctional material for photothermal therapy of cancer cells. J Mater Chem B 3:8366–8374. https://doi.org/10.1039/c5tb01704a

Dembereldorj U, Choi SY, Ganbold EO, Song NW, Kim D, Choo J, Lee SY, Kim S, Joo SW (2014) Gold nanorod-assembled pegylated graphene-oxide nanocomposites for photothermal cancer therapy. Photochem Photobiol 90:659–666. https://doi.org/10.1111/php.12212

Dinadayalane TC, Leszczynska D, Leszczynski J (2012) Graphene: properties, biomedical applications and toxicity. RSC Nanosci Nanotechnol:1–26. https://doi.org/10.1039/9781849735476-00001

Ding H, Zhang F, Zhao C, Lv Y, Ma G, Wei W, Tian Z (2017) Beyond a carrier: graphene quantum dots as a probe for programmatically monitoring anti-cancer drug delivery, release, and response. ACS Appl Mater Interfaces 9:27396–27401. https://doi.org/10.1021/acsami.7b08824

Eskiizmir G, Baskın Y, Yapıcı K (2018) Graphene-based nanomaterials in cancer treatment and diagnosis. In: Fullerenes, graphenes and nanotubes. Elsevier, pp 331–374

Estelrich J, Busquets M (2018) Iron oxide nanoparticles in photothermal therapy. Molecules 23:1567. https://doi.org/10.3390/molecules23071567

Fan W, Yung B, Huang P, Chen X (2017) Nanotechnology for Multimodal synergistic cancer therapy. Chem Rev 117:13566–13638. https://doi.org/10.1021/acs.chemrev.7b00258

Fernandes N, Rodrigues CF, Moreira AF, Correia IJ (2020) Overview of the application of inorganic nanomaterials in cancer photothermal therapy. Biomater Sci 8:2990–3020. https://doi.org/10.1039/d0bm00222d

Gai L-X, Wang W-Q, Wu X, Su X-J, Yang F-C (2019) NIR absorbing reduced graphene oxide for photothermal radiotherapy for treatment of esophageal cancer. J Photochem Photobiol B Biol 194:188–193. https://doi.org/10.1016/j.jphotobiol.2019.03.014

Gai S, Yang G, Yang P, He F, Lin J, Jin D, Xing B (2018) Recent advances in functional nanomaterials for light–triggered cancer therapy. Nano Today 19:146–187. https://doi.org/10.1016/j.nantod.2018.02.010

Gangapuram BR, Bandi R, Alle M, Dadigala R, Kotu GM, Guttena V (2018) Microwave assisted rapid green synthesis of gold nanoparticles using Annona squamosa L peel extract for the efficient catalytic reduction of organic pollutants. J Mol Struct 1167:305–315. https://doi.org/10.1016/j.molstruc.2018.05.004

Gangapuram BR, Bandi R, Dadigala R, Kotu GM, Guttena V (2017) Facile green synthesis of gold nanoparticles with carboxymethyl gum karaya, selective and sensitive colorimetric detection of copper (II) ions. J Clust Sci 28:2873–2890. https://doi.org/10.1007/s10876-017-1264-3

Georgakilas V, Tiwari JN, Kemp KC, Perman JA, Bourlinos AB, Kim KS, Zboril R (2016) Noncovalent functionalization of graphene and graphene oxide for energy materials, biosensing, catalytic, and biomedical applications. Chem Rev 116:5464–5519. https://doi.org/10.1021/acs.chemrev.5b00620

Ghorai TK (2019) Graphene oxide-based nanocomposites and biomedical applications. Elsevier Ltd.

Ghosal K, Sarkar K (2018) Biomedical applications of graphene nanomaterials and beyond. ACS Biomater Sci Eng 4:2653–2703. https://doi.org/10.1021/acsbiomaterials.8b00376

Guo M, Huang J, Deng Y, Shen H, Ma Y, Zhang M, Zhu A, Li Y, Hui H, Wang Y, Yang X, Zhang Z, Chen H (2015) PH-responsive cyanine-grafted graphene oxide for fluorescence resonance energy transfer-enhanced photothermal therapy. Adv Funct Mater 25:59–67. https://doi.org/10.1002/adfm.201402762

Hashemi M, Omidi M, Muralidharan B, Smyth H, Mohagheghi MA, Mohammadi J, Milner TE (2017) Evaluation of the photothermal properties of a reduced graphene oxide/arginine nanostructure for near-infrared absorption. ACS Appl Mater Interfaces 9:32607–32620. https://doi.org/10.1021/acsami.7b11291

Hernández Y, Galarreta BC (2019) Noble metal-based plasmonic nanoparticles for SERS imaging and photothermal therapy. In: Nanomaterials for Magnetic and optical hyperthermia applications. Elsevier, pp 83–109

Hou C, Quan H, Duan Y, Zhang Q, Wang H, Li Y (2013) Facile synthesis of water-dispersible Cu2O nanocrystal-reduced graphene oxide hybrid as a promising cancer therapeutic agent. Nanoscale 5:1227–1232. https://doi.org/10.1039/c2nr32938g

Hu JJ, Cheng YJ, Zhang XZ (2018) Recent advances in nanomaterials for enhanced photothermal therapy of tumors. Nanoscale 10:22657–22672. https://doi.org/10.1039/c8nr07627h

Hu S-H, Chen Y-W, Hung W-T, Chen I-W, Chen S-Y (2012) Quantum-dot-tagged reduced graphene oxide nanocomposites for bright fluorescence bioimaging and photothermal therapy monitored in situ. Adv Mater 24:1748–1754. https://doi.org/10.1002/adma.201104070

Huang G, Zhu X, Li H, Wang L, Chi X, Chen J, Wang X, Chen Z, Gao J (2015) Facile integration of multiple magnetite nanoparticles for theranostics combining efficient MRI and thermal therapy. Nanoscale 7:2667–2675. https://doi.org/10.1039/C4NR06616B

Huang X, Jain PK, El-Sayed IH, El-Sayed MA (2008) Plasmonic photothermal therapy (PPTT) using gold nanoparticles. Lasers Med Sci 23:217–228. https://doi.org/10.1007/s10103-007-0470-x

Janib SM, Moses AS, MacKay JA (2010) Imaging and drug delivery using theranostic nanoparticles. Adv Drug Deliv Rev 62:1052–1063. https://doi.org/10.1016/j.addr.2010.08.004

Jastrzębska AM, Kurtycz P, Olszyna AR (2012) Recent advances in graphene family materials toxicity investigations. J Nanoparticle Res 14:1320. https://doi.org/10.1007/s11051-012-1320-8

Jiang B-P, Hu L-F, Wang D-J, Ji S-C, Shen X-C, Liang H (2014) Graphene loading water-soluble phthalocyanine for dual-modality photothermal/photodynamic therapy via a one-step method. J Mater Chem B 2:7141–7148. https://doi.org/10.1039/C4TB01038H

Jin Y, Wang J, Ke H, Wang S, Dai Z (2013) Graphene oxide modified PLA microcapsules containing gold nanoparticles for ultrasonic/CT bimodal imaging guided photothermal tumor therapy. Biomaterials 34:4794–4802. https://doi.org/10.1016/j.biomaterials.2013.03.027

Jung HS, Kong WH, Sung DK, Lee M, Beack SE, Keum DH, Kim KS, Yun SH, Hahn SK (2014) Nanographene oxide–hyaluronic acid conjugate for photothermal ablation therapy of skin cancer. ACS Nano 8:260–268. https://doi.org/10.1021/nn405383a

Justin R, Román S, Chen D, Tao K, Geng X, Grant RT, MacNeil S, Sun K, Chen B (2015) Biodegradable and conductive chitosan-graphene quantum dot nanocomposite microneedles for delivery of both small and large molecular weight therapeutics. RSC Adv 5:51934–51946. https://doi.org/10.1039/c5ra04340a

Kelkar SS, Reineke TM (2011) Theranostics: combining imaging and therapy. Bioconjug Chem 22:1879–1903. https://doi.org/10.1021/bc200151q

Khot MI, Andrew H, Svavarsdottir HS, Armstrong G, Quyn AJ, Jayne DG (2019) A review on the scope of photothermal therapy–based nanomedicines in preclinical models of colorectal cancer. Clin Colorectal Cancer 18:e200–e209. https://doi.org/10.1016/j.clcc.2019.02.001

Kim TH, Alle M, Park SC, Zhao F, Long W, Samala S, Kim J-C (2021) Self-assembly prepared using an ion pair of poly(ethylene imine) and (phenylthio) acetic acid as a drug carrier for oxidation, temperature, and NIR-responsive release. Chem Eng J 415:128954. https://doi.org/10.1016/j.cej.2021.128954

Lahir Y (2019) Graphene and graphene-based nanomaterials are suitable vehicles for drug delivery. Elsevier Inc.

Li JL, Bao HC, Hou XL, Sun L, Wang XG, Gu M (2012a) Graphene oxide nanoparticles as a nonbleaching optical probe for two-photon luminescence imaging and cell therapy. Angew Chemie – Int Ed 51:1830–1834. https://doi.org/10.1002/anie.201106102

Li JL, Hou XL, Bao HC, Sun L, Tang B, Wang JF, Wang XG, Gu M (2014a) Graphene oxide nanoparticles for enhanced photothermal cancer cell therapy under the irradiation of a femtosecond laser beam. J Biomed Mater Res A 102:2181–2188. https://doi.org/10.1002/jbm.a.34871

Li M, Yang X, Ren J, Qu K, Qu X (2012b) Using graphene oxide high near-infrared absorbance for photothermal treatment of Alzheimer's disease. Adv Mater 24:1722–1728. https://doi.org/10.1002/adma.201104864

Li X-D, Liang X-L, Yue X-L, Wang J-R, Li C-H, Deng Z-J, Jing L-J, Lin L, Qu E-Z, Wang S-M, Wu C-L, Wu H-X, Dai Z-F (2014b) Imaging guided photothermal therapy using iron

oxide loaded poly(lactic acid) microcapsules coated with graphene oxide. J Mater Chem B 2:217–223. https://doi.org/10.1039/C3TB21281E
Li Z, Lei H, Kan A, Xie H, Yu W (2021) Photothermal applications based on graphene and its derivatives: a state-of-the-art review. Energy 216:119262. https://doi.org/10.1016/j.energy.2020.119262
Lim D-K, Barhoumi A, Wylie RG, Reznor G, Langer RS, Kohane DS (2013) Enhanced photothermal effect of plasmonic nanoparticles coated with reduced graphene oxide. Nano Lett 13:4075–4079. https://doi.org/10.1021/nl4014315
Lim JH, Kim DE, Kim EJ, Ahrberg CD, Chung BG (2018) Functional graphene oxide-based nanosheets for photothermal therapy. Macromol Res 26:557–565. https://doi.org/10.1007/s13233-018-6067-3
Lima-Sousa R, de Melo-Diogo D, Alves CG, Costa EC, Ferreira P, Louro RO, Correia IJ (2018) Hyaluronic acid functionalized green reduced graphene oxide for targeted cancer photothermal therapy. Carbohydr Polym 200:93–99. https://doi.org/10.1016/j.carbpol.2018.07.066
Lin J, Huang Y, Huang P (2018) Graphene-based nanomaterials in bioimaging. In: Biomedical applications of functionalized nanomaterials. Elsevier, pp 247–287
Liu X, Wu X, Xing Y, Zhang Y, Zhang X, Pu Q, Wu M, Zhao JX (2020) Reduced graphene oxide/mesoporous silica nanocarriers for ph-triggered drug release and photothermal therapy. ACS Appl Bio Mater 3:2577–2587. https://doi.org/10.1021/acsabm.9b01108
Liu Z, Zhang J, Tian Y, Zhang L, Han X, Wang Q, Cheng W (2018) Targeted delivery of reduced graphene oxide nanosheets using multifunctional ultrasound nanobubbles for visualization and enhanced photothermal therapy. Int J Nanomedicine 13:7859–7872. https://doi.org/10.2147/IJN.S181268
Ma L, Wang G, Dai J (2017) Preparation of functional reduced graphene oxide and its influence on the properties of polyimide composites. J Appl Polym Sci 134:1–8. https://doi.org/10.1002/app.45119
Ma X, Tao H, Yang K, Feng L, Cheng L, Shi X, Li Y, Guo L, Liu Z (2012) A functionalized graphene oxide-iron oxide nanocomposite for magnetically targeted drug delivery, photothermal therapy, and magnetic resonance imaging. Nano Res 5:199–212. https://doi.org/10.1007/s12274-012-0200-y
Markovic ZM, Harhaji-Trajkovic LM, Todorovic-Markovic BM, Kepić DP, Arsikin KM, Jovanović SP, Pantovic AC, Dramićanin MD, Trajkovic VS (2011) In vitro comparison of the photothermal anticancer activity of graphene nanoparticles and carbon nanotubes. Biomaterials 32:1121–1129. https://doi.org/10.1016/j.biomaterials.2010.10.030
Martin SJ, Henry CM, Cullen SP (2012) A perspective on mammalian caspases as positive and negative regulators of inflammation. Mol Cell 46:387–397. https://doi.org/10.1016/j.molcel.2012.04.026
Min Y, Mao CQ, Chen S, Ma G, Wang J, Liu Y (2012) Combating the drug resistance of cisplatin using a platinum prodrug based delivery system. Angew Chemie – Int Ed 51:6742–6747. https://doi.org/10.1002/anie.201201562
Mocan T, Matea CT, Cojocaru I, Ilie I, Tabaran FA, Zaharie F, Iancu C, Bartos D, Mocan L (2014) Photothermal treatment of human pancreatic cancer using PEGylated multi-walled carbon nanotubes induces apoptosis by triggering mitochondrial membrane depolarization mechanism. J Cancer 5:679–688. https://doi.org/10.7150/jca.9481
Mohamadi S, Hamidi M (2017) The new nanocarriers based on graphene and graphene oxide for drug delivery applications. In: Nanostructures for drug delivery. Elsevier, pp 107–147
Mohammed A, Gugulothu Y, Bandi R, Dadigala R, Utkoor UK (2021) Ultraspeed synthesis of highly fluorescent N-doped carbon dots for the label-free detection of manganese (<scp>VII</scp>). J Chinese Chem Soc jccs.202000545. https://doi.org/10.1002/jccs.202000545
Montazeri A (2008) Health-related quality of life in breast cancer patients: a bibliographic review of the literature from 1974 to 2007. J Exp Clin Cancer Res 27:1–31. https://doi.org/10.1186/1756-9966-27-32

Moon H, Kumar D, Kim H, Sim C, Chang J-H, Kim J-M, Kim H, Lim D-K (2015) Amplified photoacoustic performance and enhanced photothermal stability of reduced graphene oxide coated gold nanorods for sensitive photoacoustic imaging. ACS Nano 9:2711–2719. https://doi.org/10.1021/nn506516p

Mura S, Nicolas J, Couvreur P (2013) Stimuli-responsive nanocarriers for drug delivery. Nat Mater 12:991–1003. https://doi.org/10.1038/nmat3776

Nafiujjaman M, Nurunnabi M (2019) Graphene and 2D Materials for Phototherapy. In: Biomedical Applications of Graphene and 2D Nanomaterials. Elsevier, pp 105–117

Nizami MZI, Takashiba S, Nishina Y (2020) Graphene oxide: A new direction in dentistry. Appl Mater Today 19:100576. https://doi.org/10.1016/j.apmt.2020.100576

Rahman M, Kazmi I, Beg S, Hafeez A, Afzal M, Kumar V, Anwar F, Ahmad FJ (2019) Functionalized graphene-based nanomaterials for drug delivery and biomedical applications in cancer chemotherapy. In: Nanoparticles in pharmacotherapy. Elsevier, pp 429–460

Ramezani M, Alibolandi M, Nejabat M, Charbgoo F, Taghdisi SM, Abnous K (2019) Graphene-based hybrid nanomaterials for biomedical applications. Elsevier Inc.

Reddy GB, Ramakrishna D, Madhusudhan A, Ayodhya D, Venkatesham M, Veerabhadram G (2015) Catalytic reduction of p-nitrophenol and hexacyanoferrate (III) by borohydride using green synthesized gold nanoparticles. J Chinese Chem Soc 62:420–428. https://doi.org/10.1002/jccs.201400513

Robinson JT, Tabakman SM, Liang Y, Wang H, Sanchez Casalongue H, Vinh D, Dai H (2011) Ultrasmall reduced graphene oxide with high near-infrared absorbance for photothermal therapy. J Am Chem Soc 133:6825–6831. https://doi.org/10.1021/ja2010175

Roy S, Sarkar A, Jaiswal A (2019) Poly(allylamine hydrochloride)-functionalized reduced graphene oxide for synergistic chemo-photothermal therapy. Nanomedicine 14:255–274. https://doi.org/10.2217/nnm-2018-0320

Saba N, Jawaid M (2018) Energy and environmental applications of graphene and its derivatives. Elsevier Ltd.

Samantara AK, Acharya C, Satpathy D, Panda CR, Bhaskara PK, Sasmal A (2018) Functionalized graphene: an unique platform for biomedical application. Elsevier Inc.

Sharker SM, Lee JE, Kim SH, Jeong JH, In I, Lee H, Park SY (2015) pH triggered in vivo photothermal therapy and fluorescence nanoplatform of cancer based on responsive polymer-indocyanine green integrated reduced graphene oxide. Biomaterials 61:229–238. https://doi.org/10.1016/j.biomaterials.2015.05.040

Shen H, Zhang L, Liu M, Zhang Z (2012) Biomedical applications of graphene. Theranostics 2:283–294. https://doi.org/10.7150/thno.3642

Shi S, Chen F, Ehlerding EB, Cai W (2014) Surface engineering of graphene-based nanomaterials for biomedical applications. Bioconjug Chem 25:1609–1619. https://doi.org/10.1021/bc500332c

Shi Y, Liu M, Deng F, Zeng G, Wan Q, Zhang X, Wei Y (2017) Recent progress and development on polymeric nanomaterials for photothermal therapy: a brief overview. J Mater Chem B 5:194–206. https://doi.org/10.1039/C6TB02249A

Singh DP, Herrera CE, Singh B, Singh S, Singh RK, Kumar R (2018) Graphene oxide: an efficient material and recent approach for biotechnological and biomedical applications. Mater Sci Eng C 86:173–197. https://doi.org/10.1016/j.msec.2018.01.004

Song X, Gong H, Yin S, Cheng L, Wang C, Li Z, Li Y, Wang X, Liu G, Liu Z (2014) Ultra-small iron oxide doped polypyrrole nanoparticles for in vivo multimodal imaging guided photothermal therapy. Adv Funct Mater 24:1194–1201. https://doi.org/10.1002/adfm.201302463

Stabile J, Najafali D, Cheema Y, Inglut CT, Liang BJ, Vaja S, Sorrin JA, Huang H-C (2020) Engineering gold nanoparticles for photothermal therapy, surgery, and imaging. In: Nanoparticles for biomedical applications. Elsevier, pp 175–193

Tayyebi A, Akhavan O, Lee B-K, Outokesh M (2018) Supercritical water in top-down formation of tunable-sized graphene quantum dots applicable in effective photothermal treatments of tissues. Carbon N Y 130:267–272. https://doi.org/10.1016/j.carbon.2017.12.057

Tew WP, Muss HB, Kimmick GG, Von Gruenigen VE, Lichtman SM (2014) Breast and ovarian cancer in the older woman. J Clin Oncol 32:2553–2561. https://doi.org/10.1200/JCO.2014.55.3073

Tian B, Wang C, Zhang S, Feng L, Liu Z (2011) Photothermally enhanced photodynamic therapy delivered by nano-graphene oxide. ACS Nano 5:7000–7009. https://doi.org/10.1021/nn201560b

Tian Q, Hu J, Zhu Y, Zou R, Chen Z, Yang S, Li R, Su Q, Han Y, Liu X (2013) Sub-10 nm Fe 3 O 4 @Cu 2– x S core–shell nanoparticles for dual-modal imaging and photothermal therapy. J Am Chem Soc 135:8571–8577. https://doi.org/10.1021/ja4013497

Turcheniuk K, Boukherroub R, Szunerits S (2015) Gold–graphene nanocomposites for sensing and biomedical applications. J Mater Chem B 3:4301–4324. https://doi.org/10.1039/C5TB00511F

Wang X, Cheng L (2019) Multifunctional two-dimensional nanocomposites for photothermal-based combined cancer therapy. Nanoscale 11:15685–15708. https://doi.org/10.1039/c9nr04044g

Wang Y-W, Fu Y-Y, Peng Q, Guo S-S, Liu G, Li J, Yang H-H, Chen G-N (2013a) Dye-enhanced graphene oxide for photothermal therapy and photoacoustic imaging. J Mater Chem B 1:5762. https://doi.org/10.1039/c3tb20986e

Wang Y, Leng S, Huang J, Shu M, Papavassiliou DV (2020) Modeling of cancer photothermal therapy using near-infrared radiation and functionalized graphene nanosheets. Int j numer method biomed eng 36:1–10. https://doi.org/10.1002/cnm.3275

Wang Y, Wang K, Zhao J, Liu X, Bu J, Yan X, Huang R (2013b) Multifunctional mesoporous silica-coated graphene nanosheet used for chemo-photothermal synergistic targeted therapy of glioma. J Am Chem Soc 135:4799–4804. https://doi.org/10.1021/ja312221g

Wei W, Zhang X, Zhang S, Wei G, Su Z (2019) Biomedical and bioactive engineered nanomaterials for targeted tumor photothermal therapy: a review. Mater Sci Eng C 104:109891. https://doi.org/10.1016/j.msec.2019.109891

Xie M, Zhang F, Peng H, Zhang Y, Li Y, Xu Y, Xie J (2019) Layer-by-layer modification of magnetic graphene oxide by chitosan and sodium alginate with enhanced dispersibility for targeted drug delivery and photothermal therapy. Colloids Surf B Biointerfaces 176:462–470. https://doi.org/10.1016/j.colsurfb.2019.01.028

Yang K, Hu L, Ma X, Ye S, Cheng L, Shi X, Li C, Li Y, Liu Z (2012a) Multimodal imaging guided photothermal therapy using functionalized graphene nanosheets anchored with magnetic nanoparticles. Adv Mater 24:1868–1872. https://doi.org/10.1002/adma.201104964

Yang K, Wan J, Zhang S, Tian B, Zhang Y, Liu Z (2012b) The influence of surface chemistry and size of nanoscale graphene oxide on photothermal therapy of cancer using ultra-low laser power. Biomaterials 33:2206–2214. https://doi.org/10.1016/j.biomaterials.2011.11.064

Yang K, Zhang S, Zhang G, Sun X, Lee S-T, Liu Z (2010) Graphene in mice: ultrahigh in vivo tumor uptake and efficient photothermal therapy. Nano Lett 10:3318–3323. https://doi.org/10.1021/nl100996u

Yang L, Tseng Y-T, Suo G, Chen L, Yu J, Chiu W-J, Huang C-C, Lin C-H (2015) Photothermal therapeutic response of cancer cells to aptamer–gold nanoparticle-hybridized graphene oxide under NIR illumination. ACS Appl Mater Interfaces 7:5097–5106. https://doi.org/10.1021/am508117e

Yu C, Xu L, Zhang Y, Timashev PS, Huang Y, Liang X-J (2020) Polymer-based nanomaterials for noninvasive cancer photothermal therapy. ACS Appl Polym Mater 2:4289–4305. https://doi.org/10.1021/acsapm.0c00704

Zedan AF, Moussa S, Terner J, Atkinson G, El-Shall MS (2013) Ultrasmall gold nanoparticles anchored to graphene and enhanced photothermal effects by laser irradiation of gold nanostructures in graphene oxide solutions. ACS Nano 7:627–636. https://doi.org/10.1021/nn304775h

Zhang L, Lu Z, Zhao Q, Huang J, Shen H, Zhang Z (2011a) Enhanced chemotherapy efficacy by sequential delivery of sirna and anticancer drugs using pei-grafted graphene oxide. Small 7:460–464. https://doi.org/10.1002/smll.201001522

Zhang W, Guo Z, Huang D, Liu Z, Guo X, Zhong H (2011b) Synergistic effect of chemo-photothermal therapy using PEGylated graphene oxide. Biomaterials 32:8555–8561. https://doi.org/10.1016/j.biomaterials.2011.07.071

Zhang W, Guo Z, Huang D, Liu Z, Guo X, Zhong H (2011c) Synergistic effect of chemo-photothermal therapy using PEGylated graphene oxide. Biomaterials 32:8555–8561. https://doi.org/10.1016/j.biomaterials.2011.07.071

Zhang Z, Wang J, Chen C (2013) Near-infrared light-mediated nanoplatforms for cancer thermo-chemotherapy and optical imaging. Adv Mater 25:3869–3880. https://doi.org/10.1002/adma.201301890

Zhi D, Yang T, O'Hagan J, Zhang S, Donnelly RF (2020) Photothermal therapy. J Control Release 325:52–71. https://doi.org/10.1016/j.jconrel.2020.06.032

Further Reading

Eskiizmir G, Baskın Y, Yapıcı K (2018) Graphene-based nanomaterials in cancer treatment and diagnosis. In: Fullerenes, graphenes and nanotubes. Elsevier, pp 331–374

Barua S, Geng X, Chen B (2020) Graphene-based nanomaterials for healthcare applications. Elsevier Inc.

Chapter 6
A New Era of Cancer Treatment: Carbon Nanotubes as Drug Delivery Tools

Sayan Deb Dutta, Keya Ganguly, Rajkumar Bandi, and Madhusudhan Alle

6.1 Introduction

Life-threatening diseases like cancer continue to increase despite substantial progress in modern medicine. Cancer treatment is limited to surgery, chemotherapy, hormone therapy, immune therapy, and radiation therapy (Faubert et al. 2020). These conventional treatment methods' significant drawbacks are their inability to distinguish normal cells from the tumor cells, leading to multiple side effects (Kilgallen et al. 2020). Exploring the efficient ways of targeted drug delivery and early detection of cancer cells is of utmost importance to revolutionize the existing cancer treatment (Ingle et al. 2020). The use of carbon-based nanomaterials in the range of 5–200 nm has emerged as a potential medical weapon for cancer therapy (Adrita et al. 2020). The commonly used carbon nanomaterials include carbon nanotubes (CNTs), nanodiamonds (ND), and graphene-based materials like graphene oxide (GO), reduced graphene oxide (rGO), and few-layer graphene (FLG), among others (Simon et al. 2019). Several recognized advantages of nanomaterial-based cancer treatment are increased systemic circulation lifetime, enhanced serum solubility, sustained drug release, and multiple drug molecules' codelivery (Kaur et al. 2019). Remarkably, the fabrication of CNT-based nanocarrier has shown

S. D. Dutta · K. Ganguly
Department of Biosystems Engineering, Kangwon National University, Chuncheon, Republic of Korea

R. Bandi
Institute of Forest Science, Kangwon National University, Chuncheon, Republic of Korea

M. Alle (✉)
Institute of Forest Science, Kangwon National University, Chuncheon, Republic of Korea

Department of Biomedical Science & Institute of Bioscience and Biotechnology, Kangwon National University, Chuncheon, Republic of Korea

J.-C. Kim et al. (eds.), *Smart Nanomaterials in Biomedical Applications*, Nanotechnology in the Life Sciences,
https://doi.org/10.1007/978-3-030-84262-8_6

tremendous promising results to overcome cancer treatment challenges, especially drug delivery.

CNTs consist primarily of a series of condensed benzene rings rolled up into a tubular structure and belong to fullerenes' family, the third allotropic form of carbon. CNTs exhibit a unique combination of desired properties such as biocompatibility and excellent mechanical properties (Yadav and Mohite 2020). However, the insolubility of CNTs in the aqueous medium was a significant challenge for its utility in biomedical applications (Naqvi et al. 2020). Nevertheless, a wide range of applications of CNTs emerged, including the delivery of drugs after the functionalization of the CNTs was reported to increase its solubility drastically. Moreover, the solubility of CNTs in different solvents can be modulated depending on the type of its surface functionalization (covalent or non-covalent) (Panigrahi and Nayak 2020). The various applications of CNTs as a drug delivery tool are depicted in Fig. 6.1. This chapter discusses the structure, synthesis, functionalization, and potential applications of CNTs in cancer theranostics.

6.2 Classification and Structure of CNTs

As mentioned earlier, CNTs are allotropes of carbon, comprising graphite sheets rolled into a tube of either single or multiple layers (Tiwari et al. 2016). The surface of the layers displays a honeycomb lattice structure originating from the *sp*[2]

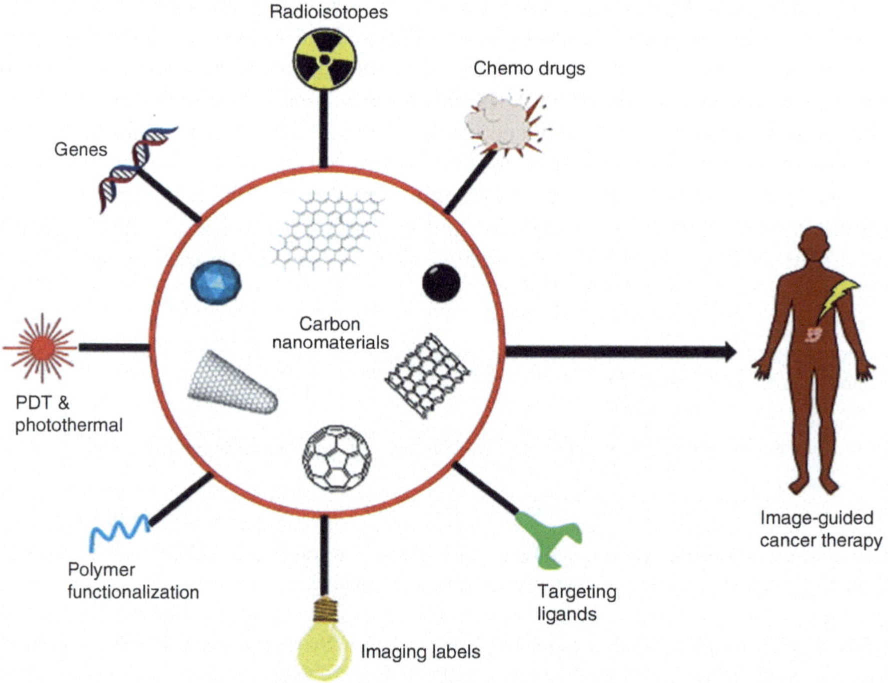

Fig. 6.1 Schematic presentation for the application of carbon-based nanomaterials for cancer theranostics

hybridization of graphite, where each atom is connected uniformly to three carbons (120°) in the *XY* plane. Additionally, the Z-axis harbors a weak π bond. Based on the number of layers, CNTs are classified into two major types: single-walled carbon nanotubes (SWCNTs) and multiple-walled carbon nanotubes (MWCNTs) (Wang et al. 2007; Li et al. 2002; Takamori et al. 2007).

6.2.1 Classification of CNTs

6.2.1.1 Single-Walled Carbon Nanotubes

SWCNTs are a single roll of a graphene sheet with a length to diameter ratio of ~1000, making them almost one dimensional in structure. Generally, SWCNTs have a diameter of ~1 nm; however, nanotubes with a diameter of 0.4 nm have also been synthesized successfully (Wang et al. 2000). Nanotubes wider than 2.5 nm are rarely reported. Contrary, there is no such restriction on the nanotube length, which usually depends upon the adopted preparation method. Generally, the reported lengths of SWCNTs are in the range of micrometer to the millimeter. The structural feature of these nanotubes is extremely valuable in rendering the unique functions of these nanotubes. As reported, all the carbon atoms in SWCNTs are arranged in hexagonal rings except for the nanotube tips, where the carbon atoms are most likely placed in the pentagonal rings. Therefore, the nanotubes' chemical reactivity is thermodynamically favored at the ends of the nanotubes stretched with the pentagonal rings. Moreover, despite the involvement of the carbon atoms to form aromatic rings, the C=C bond angles are not always planar. Indicating the involvement of a certain degree of sp^3 character, which is inversely proportional to the tubes' radius of curvature, giving them overlapping energy bands and versatile electronic behavior. This feature of SWCNTs is believed to make the surface more reactive than planar graphene. There is potentially more than one way to roll a graphene sheet into a single-walled nanotube. These include rolling the sheet onto perpendicular, parallel, or helical symmetry planes to the nanotube axis, making SWCNTs zigzag type (angle of helicity = 0°), armchair-type (angle of helicity = 30°), and helical type, respectively (Rao et al. 2003; Hussain et al. 2020).

6.2.1.2 Multi-walled Carbon Nanotubes

Multi-walled carbon nanotubes involve numerous ways to fold and arrange graphene sheets into filamentary morphology and have diameters of up to 100 nm. Concentric-type MWCNTs (c-MWCNTs) are relatively simple to synthesize, where SWCNTs with increasing diameter are coaxially arranged. MWCNTs can also be synthesized by rolling a graphene sheet onto itself, resembling a scroll of paper. The interactions between adjacent walls of MWCNTs are primarily caused by weak van der Waals forces resulting in a dynamic shear property of MWCNTs (Filleter et al.

2014). Notable differences have been recognized between SWCNTs and MWCNTs in cancer therapy. MWCNTs are more useful for the thermal treatment of cancer than SWCNTs due to their higher vibrational energy dissipation capacity upon near-infrared light exposure (Hirsch et al. 2003). However, the efficiency of SWCNTs in drug delivery is more profound than MWCNTs owing to SWCNTs' unique features, including their one-dimensional structure and higher surface area (Feazell et al. 2007). Additionally, longer blood circulation time of anticancer drug conjugated to SWCNTs was also reported. Figure 6.2 represents the different structures of CNTs and their corresponding AFM images.

6.3 Synthesis of CNTs

Numerous methods are used for the synthesis of CNTs from different carbon sources. Synthesis methods include arc discharge, laser ablation, sonochemical techniques, hydrothermal techniques, chemical vapor deposition, and electrolysis. A few of the commonly used synthesis processes are described here.

6.3.1 Arc Discharge

Arc discharge method utilizes temperature above 1700 °C for CNT synthesis. This method uses a DC arc discharge between two graphite electrodes with diameters between 6 and 12 mm in a chamber filled with either helium, hydrogen, or methane at subatmospheric pressure to synthesize CNTs, with certain differences in the synthesis of SWCNTs and MWCNTs (Fig. 6.3a). A catalyst precursor such as a transition metal catalyst is often used for the synthesis of SWCNTs. A composite anode facilitates SWCNTs' growth in the presence of hydrogen or argon atmosphere. The anode comprises graphite and a metal such as Ni, Co, Fe, Ag, Pd, Pt, Fe-No, Ni-Cu, Co-Cu, Ni-Ti, etc. Contrary, MWCNTs are produced in the absence of a catalyst precursor. The yield and purity of the synthesized product largely depend upon the gas pressure in the reaction vessel. Besides, a higher yield of MWCNTs can be obtained by arc discharges in an organic atmosphere of ethanol, acetone, or hexane. Pulsed currents are also efficient in the production of vertically oriented MWCNTs. Synthesis of MWCNTs by arc discharge in liquid nitrogen shows promising results for their large-scale production (Prasek et al. 2011). This technique is reported to cause the least structural defects compared to other methods.

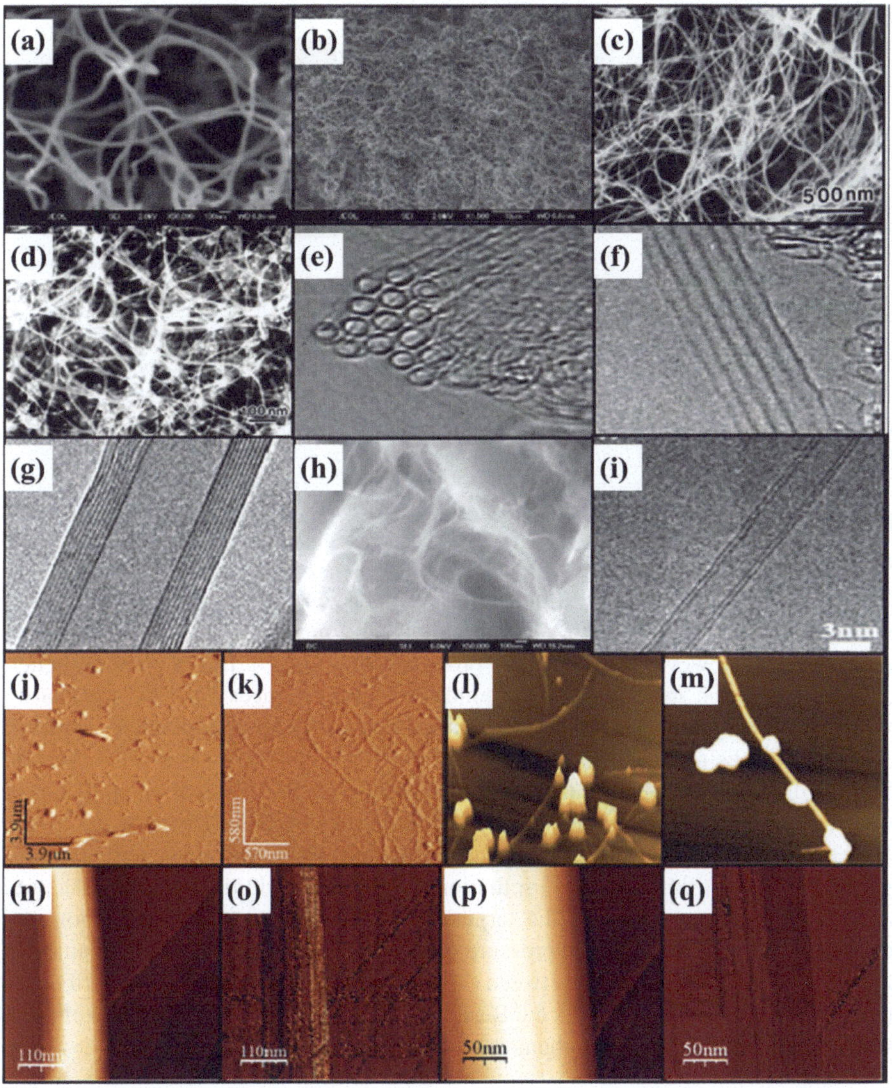

Fig. 6.2 Structure of various carbon nanotubes (CNTs). (**a** and **b**) FE-SEM morphology of MWCNTs with corresponding low and high magnification, (**c** and **d**) morphology of SWCNTs with corresponding low and high magnification, (**e** and **f**) FE-TEM morphology of the cross section of a bundle of SWCNTs, (**g**) an individual tube of SWCNTs, (**h** and **i**) morphology of the DWCNTs, (**j-k**) AFM surface topology of MWCNTs, (**l**) AFM morphology of the SWCNTs with corresponding high (**m**) magnification image, (**n**) STM photograph of the MWCNTs with corresponding high magnification images indicating the line of growth (**o, p,** and **q**). (Adopted from Dutta et al. 2020)

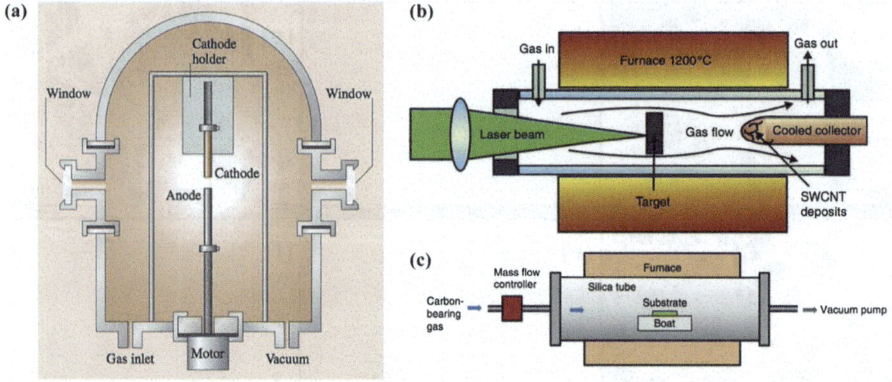

Fig. 6.3 Synthesis of CNTs. (**a**) Arc discharge process, (**b**) laser ablation method, and (**c**) chemical vapor deposition method (adopted from Dutta et al. 2020)

6.3.2 Laser Ablation

In the laser ablation process, a pulsed laser vaporizes a graphite target in a high-temperature reactor (Fig. 6.3b). The principles and mechanisms are similar to the arc discharge; however, the energy is provided by a laser hitting a graphite pellet. The laser ablation method yields around 70% and produces SWCNTs primarily with controllable diameter, highly determined by the reaction temperature. Uniform SWCNTs can be made in the presence of transition metal catalysts like Co, Ni, and Fe. The diameter distribution can be tuned by altering the chemical composition of the target and the gaseous atmosphere. The properties of the CNTs prepared by the laser deposition process strongly depend on the laser properties like energy fluence, peak power, repetition rate, and oscillation wavelength. Additionally, the structural and chemical composition of the target material, the chemical composition and chamber pressure, the flow and pressure of the buffer, the substrate and the ambient temperature, and the distance between the target and the substrate also contribute to the property of the CNTs produced. Besides, the relationship between the excitation wavelength and the growth mechanism of CNTs is recently being studied extensively.

6.3.3 Chemical Vapor Deposition

The chemical vapor deposition (CVD) technique is the most standard method used to produce CNTs. In this method, CNTs are generated onto the surface of a catalyst by cleaving a carbon atom-containing gas that flows continuously through the catalyst nanoparticle (Fig. 6.3c). Usually, a mixture of hydrocarbon as the carbon source (methane, ethane, ethylene, xylene, acetylene, isobutene, ethanol), a metal catalyst, and an inert gas is introduced into the reaction chamber at a temperature of 700–900

°C and atmospheric pressure. CNTs are formed on the substrate by the decomposition of hydrocarbon. The catalyst used in the CVD process helps in the decomposition of the carbon source. The frequently used catalysts are the transition metals like Fe, Co, and N. Alternatively, gaseous carbon sources are also utilized. Other CVD techniques are water-assisted CVD, oxygen-assisted CVD, hot-flame CVD, microwave plasma CVD, or radiofrequency CVD. CVD is considered to be an economically viable process for large-scale and pure CNT production. The main advantage of CVD is the easy to control of the reaction course and the high purity of the obtained material.

6.3.4 Vapor Phase Growth

In this method, CNT synthesis is aided by a reaction gas and an organometallic catalyst in a reaction furnace without the involvement of a substrate. This is typically a four-step method, involving the use of a laser or arc discharge source, hydrocarbon gas, catalyst, and deposition of carbon nanotubes in the reaction chamber. Figure 6.4 indicates the vapor phase growth mechanism of CNTs' formation. The vapor phase growth is mostly used commercially for bulk manufacturing of CNTs.

6.4 Functionalization of CNTs

The pristine CNT is mostly hydrophobic due to the presence of sp^2 carbon nanostructure. Therefore proper functionalization is required to make it water-soluble for biomedical applications. There are two main strategies for CNTs' functionalization, namely, covalent and non-covalent type. Table 6.1 depicts an overview of the various functionalization techniques of CNTs for cancer theranostics. Generally,

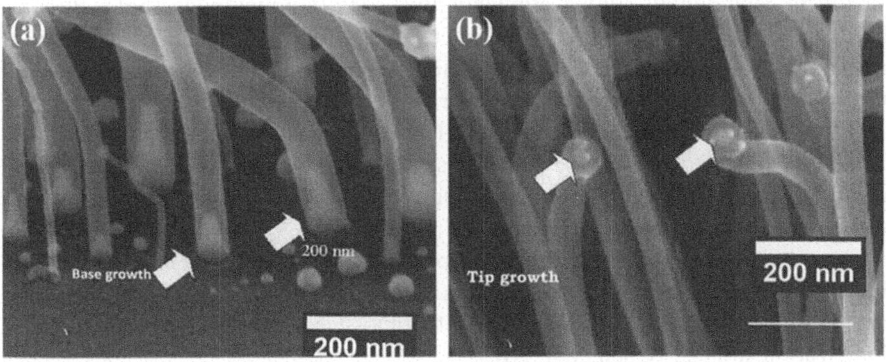

Fig. 6.4 FE-SEM images of CNTs indicating the (**a**) tube growth and (**b**) tip growth during vapor phase growth (adopted from Dutta et al. 2020)

Table 6.1 An overview of functionalization strategies for CNTs as drug delivery vehicle for cancer therapy

Type of interactions	Functionalization molecules	Purpose(s)	Toxicity
Covalent	Polyethylenimine (PEI)	Stabilization and gene delivery	Mild toxicity and immune response
	Radioisotopes and chelators	Imaging label	Radiation exposure
	Gadolinium (Gd)	Imaging label	Toxicity from premature release
	Chemo drugs	Tumor therapy	Toxicity from premature release
	Antibodies, proteins, or peptides	Tumor targeting	Potential immune response
Non-covalent	Polyethylenimine (PEI)	Stabilization and gene delivery	Mild toxicity and immune response
	M13 phage	Stabilization and tumor targeting	Potential immune response
	DNA	Stabilization	Potential immune response
	DNA/siRNA	Cancer therapy	Gene inhibition in normal tissues
	Nanostructures (IONP, QDs, gold NPs, etc.)	Imaging label, PDT, and PTT	MPS capture and potential metal contamination and serious toxicity
	Gadolinium (Gd)	Imaging label	Toxicity from premature release
	Photosensitizers	PDT	Damage to normal tissues by ROS

Adopted from Chen et al. (2015)

covalent modification at the surface of CNTs involves the introduction of hydrophilic functional moieties, such as hydroxyl groups (-OH), carboxylic groups (-COOH), and amino groups (-NH_2) along with some biocompatible polymers, such as polyethylene glycols (PEG), ethylene glycols (EG), polyethylenemine (PEI), or drug/gene cargos (Yang et al. 2014). The covalently functionalized CNTs are structurally stable and can be easily dispersed in any polar solvents. However, problems regarding covalent modification are the rapid change of functional properties of CNTs, such as photothermal properties. Thus, non-covalent modification is comparatively mild and retains the original structural properties of CNTs, therefore making it more suitable for drug delivery applications. Non-covalent interaction can be achieved by electrostatic interactions, π–π stacking, and hydrogen bonding, among others. A lot of knowledge and expertise should be needed before choosing the appropriate functionalization method for CNt-based materials. CNT-based nanohybrids are the most effective cancer theranostics if chosen as the proper functionalization strategy (Chen et al. 2015).

6.5 Applications

A wide variety of drug delivery system are currently available that promotes the pharmacological profile and therapeutic properties of drugs. Functionalized CNTs (fCNTs) are highly emerging nanovectors for the delivery of cancer therapeutic drug owing to their capacity to penetrate into the cells via two established mechanisms, via passive diffusion across the lipid bilayer without causing membrane damage or via the process of endocytosis (Pantarotto et al. 2004; Cai et al. 2005; Kam and Dai 2005; Shi Kam et al. 2004). The conjugation of drugs to CNTs is achieved through several ways of surface functionalization which includes covalent linkages, hydrophobic interactions, π-stacking interaction, and capillary-induced filling, among others (Lay et al. 2011). This section discusses a few selected CNT-loaded drug as a nanocarrier.

6.5.1 CNTs-Paclitaxel Conjugates

One of the initial studies of CNTs-based in vivo tumor treatment in mice came up with the use of CNT-paclitaxel complex. Paclitaxel (PTX) conjugated to the branched polyethylene glycol (PEG) chains via ester bonding on the surface of SWCNTs has been used in the synthesis of water-soluble SWCNTs-PTX conjugates to suppress tumor growth in murine 4T1 breast cancer model. Moreover, the SWCNT-PTX complex was found to have tenfold higher uptake rate by the targeted tumor, much likely due to enhanced cell permeability and retention. Additionally, this SWCNT-PTX formulation was much safer compared to Cremophor containing Taxol owing to the higher efficiency of SWCNTs as the drug carrier. The reported amount of SWCNTs required to deliver 5mg/kg PTX is ~4 g/kg, contrary to ~420 mg/kg Cremophor in the case of Taxol for the same PTX dose (Liu et al. 2008). Dual-functionalized MWCNTs with ethylenediamine and phenylboronic acid groups showed promising results of PTX loading through non-covalent π-π stacking within the CNTs interior to target colon cancer cells. Higher water dispersity and biocompatibility were noted as a result of the dual functionalization and drug loading of 30.85% achieved (Rathod et al. 2019). CNT-PTX conjugates can also be used for ultrasound imaging of certain tumors. Due to low xenotoxicity, PTX is commonly used in clinical and preclinical application for ovarian and breast cancer (Högberg et al. 2001). A study conducted by Zhang et al. reported that PEGylated folic acid (FA)-CNT conjugates have superior potential for in vivo high-resolution ultrasound imaging of breast cancers on live Kunming mice (Fig. 6.5). Therefore, functionalized CNTs in combination of PTX can be used as multifunctional toolkit for anticancer therapy as well as bioimaging applications (Zhang et al. 2019).

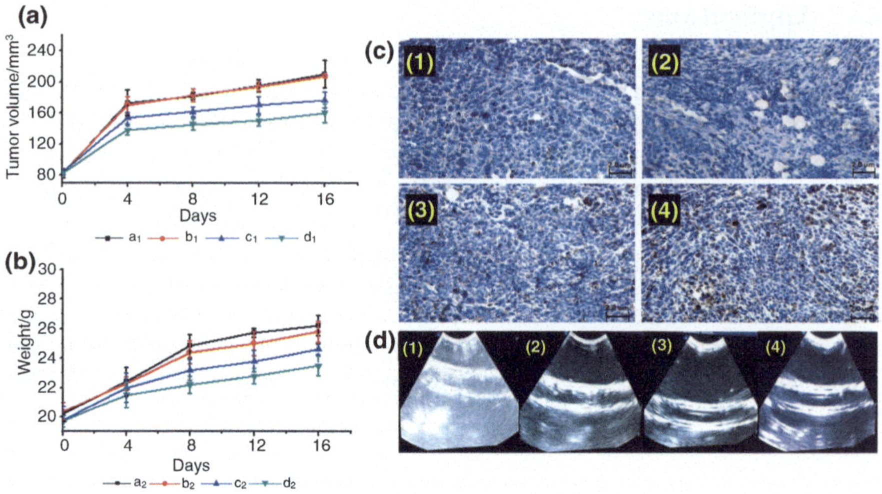

Fig. 6.5 (**a–b**) Tumor volume and weight measurements after administration of the Span-PEG (**a1**), Span-PEG-FA-CNTs (**b1**), Span-PEG-CNT@PTX (**c1**), and Span-PEG/FA-CNT@PTX, respectively, (**c**) in vivo apoptosis study in different formulations, and (**d**) in vivo ultrasound imaging of tumor after administration of formulated drugs. (Adopted from Zhang et al. 2019)

6.5.2 *CNT-Doxorubicin Conjugates*

Folic-acid (FA)-modified MWCNTs have shown promising results in the development of pH-responsive Dox delivery system, including exceptional colloidal stability and biocompatibility. The FA-CNTs often exhibited a substantially high drug loading and encapsulation efficiency of 70.4% besides drug release in an acidic environment. Moreover, FA-CNTs loaded with Dox have been shown to inhibit the tumor cells overexpressing FA receptors (Yan et al. 2018). Non-covalent MWCNT-Dox supramolecular complex is also a promising candidate to kill breast cancer cells. In this regard, MWCNTs coated with tri-block copolymer (Pluronic F127) were found to possess notably higher cytotoxic activity in comparison to Dox alone and Dox-pluronic complexes (Ali-Boucetta et al. 2008). DOX-loaded CNTs incorporated into the poly(lactic-co-glycolic acid) (PLGA) electrospun nanofibers have high antitumor efficiency against HeLa cells. Such electrospun mats render a uniform distribution and trigger sustained and prolonged release of the drug, making PLGA/Dox-CNTs' electrospun mats a long-term drug-releasing platform for potential chemotherapy (Yu et al. 2015). CNTs can also be used for encapsulating nanoparticle-DOX conjugates for effective delivery in tumor cells. Seyfoori et al. reported a pH-responsive chitosan nanogel coated with $MnFe_2O_4$ nanoparticles and grafted it on the surface of MWCNTs to enhance the delivery of Dox in U-87 glioblastoma cells (Seyfoori et al. 2019). The chitosan/$MnFe_2O_4$-grafted MWCNT

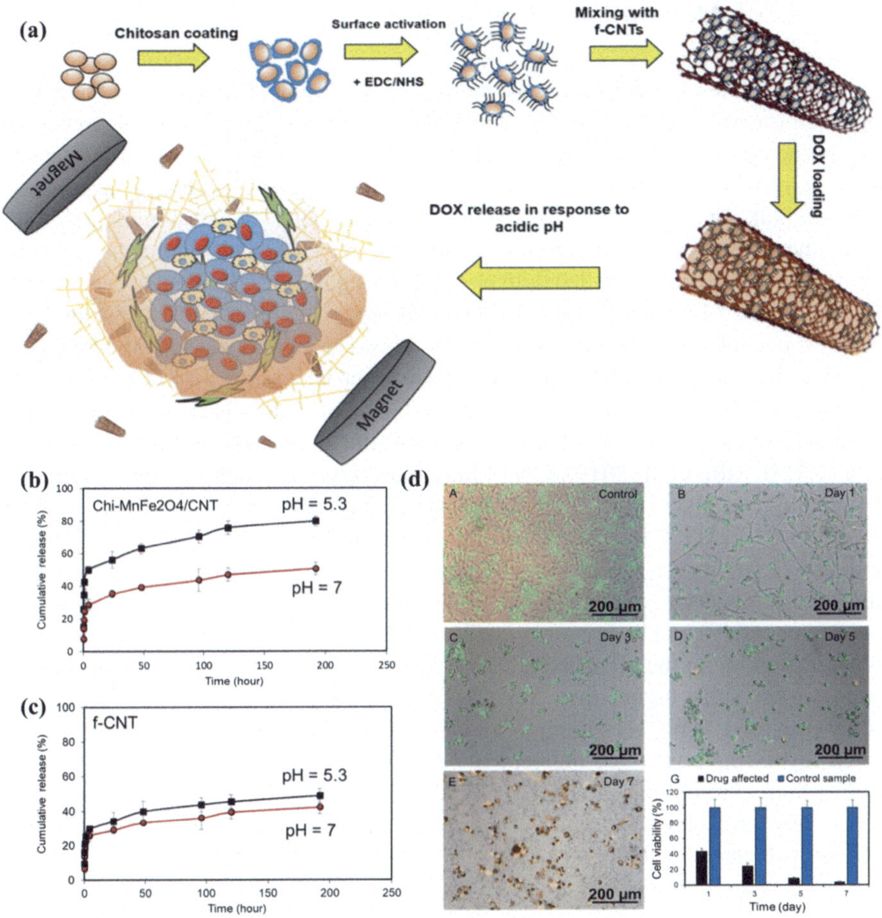

Fig. 6.6 (**a**) Schematic illustration of chitosan/$MnFe_2O_4$ (Chi-$MnFe_2O_4$) nanoparticle synthesis and grafting onto the surface of CNTs, (**b-c**) cumulative release profile of Chi-$MnFe_2O_4$ and Chi-$MnFe_2O_4$/CNT at different pH, and (**d**) cytotoxicity evaluation of Chi-$MnFe_2O_4$/CNTs on U-87 glioblastoma cells. (Adopted from Seyfoori et al. 2019)

exhibited greater pH responsiveness in acidic environment compared to MWCNTs alone (Fig. 6.6). CNTs can also be used as strong photothermal agent owing to its strong absorption at near-infrared region-I (NIR-1). A study conducted by Oh et al. showed NIR-I responsive SWCNT-Dox conjugates as promising treatment strategy for breast cancers. Therefore, the CNT-based nanomaterials could be used to encapsulate Dox for minimizing the high cytotoxic effect of Dox.

6.5.3 CNT-Methotrexate Conjugates

A comparative study of drug release between carboxylated MWCNTs conjugated with methotrexate (MTX) and free MTX was reported to favor slow release of the drug in the conjugated form (Karimi et al. 2018). Furthermore, covalently attached MTX onto MWCNTs through cleavable linkers has been reported to increase its cellular uptake and cytotoxic efficiency to the target cells. The design of such linkers is the critical factor determining the efficiency of such conjugate systems (Samorì et al. 2010). Controlled delivery of MTX against MCF-7 breast cancer cells was achieved by CNT-loaded MTX-injected hydrogel of chitosan and *β*-glycerophosphate (Saeednia et al. 2019). Besides, hybrid system of MWCNTs coated with chitosan was fabricated as a pH-responsive carrier for MTX delivery to lung cancer cells. The MTX release from such hybrid system is favored at an acidic pH of H1299 cancer cells (5.0), as compared to the neutral pH of the noncancerous cells (7.4) (Cirillo et al. 2019). CNTs functionalized with interpolymer complexes of polyethylene glycol (PEG) and polyacrylic acid (PAA) have also been used for the delivery of MTX to increase the biocompatibility of CNTs itself (Azqhandi et al. 2017). Moreover, CNTs functionalized with both folic acid and MTX showed enhanced anticancer activity against folate receptor-positive (FR) lung epithelial carcinoma (A549) and breast cancer cells (MCF-7) (Das et al. 2013). The endolysosomal trafficking in the presence of folate-modified MWCNT-MTX conjugates was found significantly higher than MWCNTs alone (Fig. 6.7). In another study, the amine-modified MWCNTs were conjugated with MTX via esterification reaction which favors the controlled delivery of MWCNT-MTX conjugates inside cells as well as high cytotoxicity in cancer cells (Joshi et al. 2017). Therefore, functionalized CNTs have several merits over pure CNTs and could be used as a potential delivery vehicle for MTX.

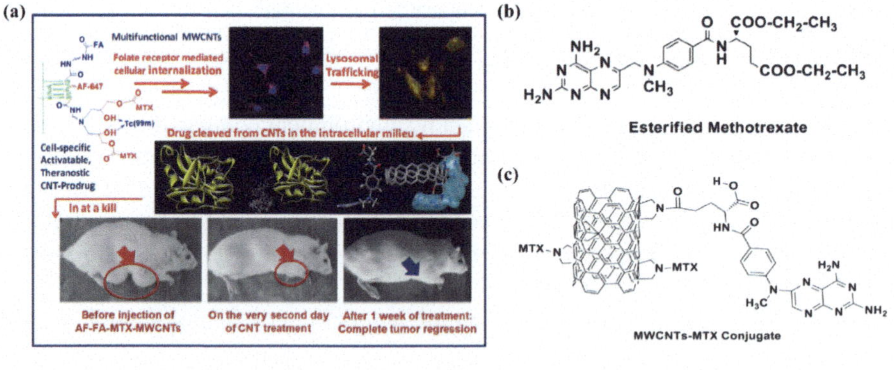

Fig. 6.7 (**a**) Schematic illustration for the synthesis of MTX-loaded folic acid-modified MWCNTs with enhanced anticancer properties (Das et al. 2013), (**b–c**) synthesis of esterified MTX and its loading onto aminated MWCNTs. (Adopted from Joshi et al. 2017)

6.5.4 Biomolecule-Functionalized CNTs

Targeted delivery of doxorubicin to HeLa cancer cells is often enhanced by the synthesis of programmable drug delivery system, DNA-functionalized CNTs. For instance, the incorporation of GC/CG-rich stem loop DNA and specific aptamer facilitates selective binding of the intercalating drug and the cell surface receptors, respectively (Hu and Niemeyer 2020). CNT-DNA conjugates are structurally stable. The DNA helix can bind tightly with SWCNTs either by covalently or non-covalently. An overview of various functionalization strategies of CNTs by using different biomolecules is shown in Fig. 6.8.

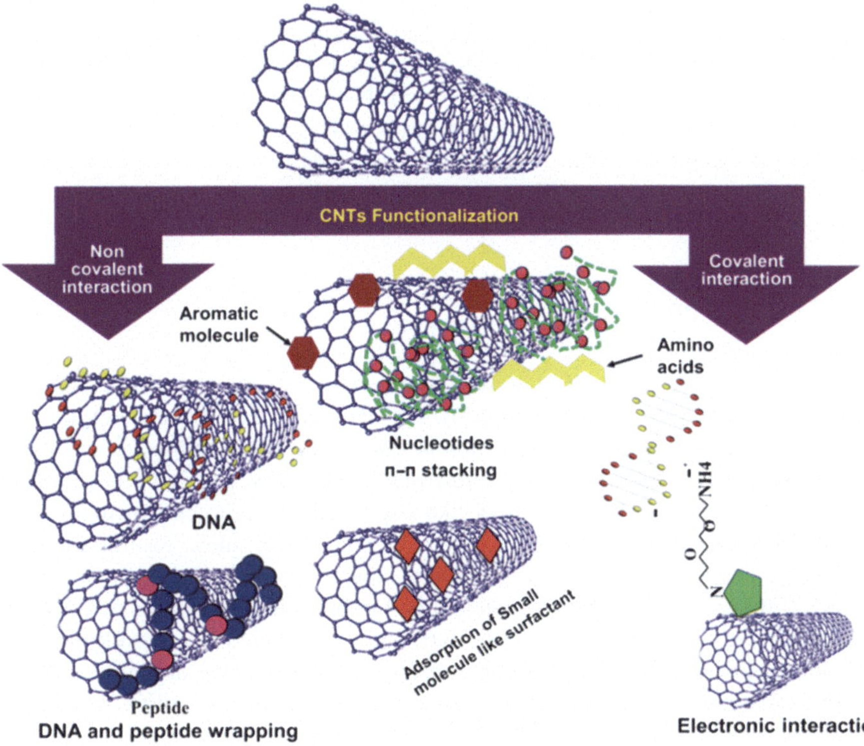

Fig. 6.8 Schematic illustration for functionalized CNTs using various biomolecules. (Adopted from Maheshwari et al. 2019)

6.5.5 Molecular Dynamic Simulation Studies

Simulation studies related to the interaction of drugs or a combination of drugs with functionalized CNTs have tremendously improved the selection of potential CNT-drug complexes. CNTs as a promising anticancer drug carrier are successfully determined through many computer simulations. A simulation study addressing the DOX adsorption efficiency on SWCNTs and MWCNTs at physiological pH of 7.0 revealed stronger adsorption of the drug than it is in acidic pH. This shows the suitability of CNTs as nanovehicles to target acidic cancerous tissues (Maleki et al. 2020). In a similar study, the adsorption of DOX onto covalently functionalized CNTs was studied by MD simulation and quantum mechanics calculations. The outcome of the simulations indicated carboxyl and amine-f-CNTs should primarily be the choice as a pH-sensitive drug carrier favored by the electrostatic interactions by the chemical moieties (Kordzadeh et al. 2019). Moreover, molecular dynamic (MD) simulation studies of DOX and PTX co-loading onto SWCNTs have shown favorable π-π stacking of DOX through conjugated aromatic rings, whereas PTX interacts with SWCNTs through x-π (x = C-H, N-H, and C=O) interactions along with π-π stacking. Upon functionalization of the SWCNTs with chitosan, the affinity of these drugs for the surface of SWCNTs was measured to decrease indicating a desired drug release strategy (Karnati and Wang 2018).

6.6 Conclusion

CNTs are invaluable nanomaterials for the targeted delivery of anticancer drugs. The exceptional physicochemical properties of CNTs based on their type, synthesis process, and numerous covalent and non-covalent functionalizations render them their unique cargo loading capacity. Functionalized CNTs have been used in the delivery of the commonly used drugs like paclitaxel, doxorubicin, methotrexate, DNA conjugates, peptide conjugates, and even small molecules. Several attempts have been made to improve the drug loading capacity of CNTs by enhancing their aqueous solubility and cell penetration ability. Despite such advantages of CNTs as nanocarriers, certain limitations prevail to be addressed which includes eliminating their limited biocompatibility, potential to trigger oxidative stress at higher concentrations, improved real-time imaging techniques, and biodegradability, which can help in the fabrication of smart CNT-based materials with improved pharmacokinetic behavior.

Acknowledgment This research was supported by Basic Science Research Program through the National Research Foundation of Korea (NRF) funded by the Ministry of Education (No. 2018R1A6A1A03025582).

References

Adrita SH, Tasnim KN, Ryu JH, Sharker SM (2020) Nanotheranostic carbon dots as an emerging platform for cancer therapy. J Nanotheranostics 1:59–78. https://doi.org/10.3390/jnt1010006

Ali-Boucetta H, Al-Jamal KT, McCarthy D, Prato M, Bianco A, Kostarelos K (2008) Multiwalled carbon nanotube–doxorubicin supramolecular complexes for cancer therapeutics. Chem Comm 12:459–461. https://doi.org/10.1039/B712350G

Azqhandi MHA, Farahani BV, Dehghani N (2017) Encapsulation of methotrexate and cyclophosphamide in interpolymer complexes formed between poly acrylic acid and poly ethylene glycol on multi-walled carbon nanotubes as drug delivery systems. Mater Sci Eng C 79:841–847. https://doi.org/10.1016/j.msec.2017.05.089

Cai D, Mataraza JM, Qin Z-H, Hunag Z, Huang J, Chiles TC, Carnhan D, Kempa K, Ren Z (2005) Highly efficient molecular delivery into mammalian cells using carbon nanotube spearing. Nat Methods 2:449–454. https://doi.org/10.1038/nmeth761

Chen D, Dougherty CA, Zhu K, Hong H (2015) Theranostic applications of carbon nanomaterials in cancer: Focus on imaging and cargo delivery. J Control Release 210:230–245. https://doi.org/10.1016/j.jconrel.2015.04.021

Cirillo G, Vittorio O, Kunhardt D, Kunhardt D, Valli E, Voli F, Farfalla A, Curcio M, Spizzirri UG, Hampel S (2019) Combining carbon nanotubes and chitosan for the vectorization of methotrexate to lung cancer cells. Materials 12:2889. https://doi.org/10.3390/ma12182889

Das M, Datir SR, Singh RP, Jain S (2013) Augmented anticancer activity of a targeted, intracellularly activatable, theranostic nanomedicine based on fluorescent and radiolabeled, methotrexate-folic acid-multiwalled carbon nanotube conjugate. Mol Pharm 10:2543–2557. https://doi.org/10.1021/mp300701e

Dutta SD, Patel DK, Lim K-T (2020) Carbon nanotube-based nanohybrids for agricultural and biological applications. In: Multifunctional hybrid nanomaterials for sustainable agri-food and ecosystems. Elsevier, pp 505–535. https://doi.org/10.1016/B978-0-12-821354-4.00021-2

Faubert B, Solmonson A, DeBerardinis RJ (2020) Metabolic reprogramming and cancer progression. Science 368. https://doi.org/10.1126/science.aaw5473

Feazell RP, Nakayama-Ratchford N, Dai H, Lippard SJ (2007) Soluble single-walled carbon nanotubes as longboat delivery systems for platinum (IV) anticancer drug design. J Am Chem Soc 129:8438–8439. https://doi.org/10.1021/ja073231f

Filleter T, Beese AM, Roenbeck M, Wei X, Espinosa HD (2014) Tailoring the mechanical properties of carbon nanotube fibers. In: Nanotube superfiber materials. Elsevier, pp 61–85. https://doi.org/10.1016/B978-1-4557-7863-8.00003-7

Hirsch LR, Stafford RJ, Bankson JA, Sershen SR, Rivera B, Price RE, Hazle JD, Halas NJ, West JL (2003) Nanoshell-mediated near-infrared thermal therapy of tumors under magnetic resonance guidance. PNAS USA 100:13549–13554. https://doi.org/10.1073/pnas.2232479100

Högberg T, Glimelius B, Nygren P (2001) A systematic overview of chemotherapy effects in ovarian cancer. Acta Oncol 40:340–360. https://doi.org/10.1080/02841860119644

Hu Y, Niemeyer CM (2020) Designer DNA–silica/carbon nanotube nanocomposites for traceable and targeted drug delivery. J Mater Chem B 8:2250–2255. https://doi.org/10.1039/C9TB02861G

Hussain A, Ding E, McLean B, Mustonen K, Ahmed S, Tavakkoli M, Page AJ, Zhang Q, Kotakoshi J, Kauppinen EI (2020) Scalable growth of single-walled carbon nanotubes with highly uniform structure. Nanoscale. 12:12263–12267. https://doi.org/10.1039/D0NR01919D

Ingle AP, Golińska P, Yadav A, Razzaghi-Abyaneh M, Patel M, Patel R, Pllekhanova Y, Reshetilov A, Rai M (2020) Nanotechnology: a new era in the revolution of drug delivery, diagnosis, and treatment of diseases. In: Nanobiotechnology in diagnosis, drug delivery, and treatment, pp 1–24. https://doi.org/10.1002/9781119671732.ch1

Joshi M, Kumar P, Kumar R, Sharma G, Singh B, Katare OP, Raza K (2017) Aminated carbon-based "cargo vehicles" for improved delivery of methotrexate to breast cancer cells. Mater Sci Eng C 75:1376–1388. https://doi.org/10.1016/j.msec.2017.03.057

Kam NWS, Dai H (2005) Carbon nanotubes as intracellular protein transporters: generality and biological functionality. J Am Chem Soc 127:6021–6026. https://doi.org/10.1021/ja050062v

Karimi A, Erfan M, Mortazavi SA, Ghorbani-Bidkorbeh F, Kobarfard F, Shirazi FH (2018) Functionalisation of carbon nanotubes by methotrexate and study of synchronous photothermal effect of carbon nanotube and anticancer drug on cancer cell death. IET nanobiotechnol 13:52–57. https://doi.org/10.1049/iet-nbt.2018.5085

Karnati KR, Wang Y (2018) Understanding the co-loading and releasing of doxorubicin and paclitaxel using chitosan functionalized single-walled carbon nanotubes by molecular dynamics simulations. Phys Chem Chem Phys 20:9389–9400. https://doi.org/10.1039/C8CP00124C

Kaur J, Gill GS, Jeet K (2019) Applications of carbon nanotubes in drug delivery: a comprehensive review. In: *Characterization and biology of nanomaterials for drug delivery*. Elsevier, pp 113–135. https://doi.org/10.1016/B978-0-12-814031-4.00005-2

Kilgallen AB, Štibler U, Printezi MI, Putker M, Punt CJA, Sluijter JPG, May AM, Laake LWV (2020) Comparing conventional chemotherapy to chronomodulated chemotherapy for cancer treatment: protocol for a systematic review. JMIR Res Protoc 9:e18023. https://doi.org/10.2196/18023

Kordzadeh A, Amjad-Iranagh S, Zarif M, Modarress H (2019) Adsorption and encapsulation of the drug doxorubicin on covalent functionalized carbon nanotubes: a scrutinized study by using molecular dynamics simulation and quantum mechanics calculation. J Mol Graph Model 88:11–22. https://doi.org/10.1016/j.jmgm.2018.12.009

Lay CL, Liu J, Liu Y (2011) Functionalized carbon nanotubes for anticancer drug delivery. Expert Rev Med Devices 8:561–566. https://doi.org/10.1586/erd.11.34

Li S, Li H, Wang X, Song Y, Liu Y, Jiang L, Zhu D (2002) Super-hydrophobicity of large-area honeycomb-like aligned carbon nanotubes. J Phys Chem B 106:9274–9276. https://doi.org/10.1021/jp0209401

Liu Z, Chen K, Davis C, Sherlock S, Cao Q, Chen X, Dai H (2008) Drug delivery with carbon nanotubes for in vivo cancer treatment. Cancer Res 68:6652–6660. https://doi.org/10.1158/0008-5472.CAN-08-1468

Maheshwari N, Tekade M, Soni N, Ghode P, Sharma MC, Deb PK, Tekade RK (2019) Functionalized carbon nanotubes for protein, peptide, and gene delivery. In: Biomaterials and bionanotechnology. Elsevier, pp 613–637. https://doi.org/10.1016/B978-0-12-814427-5.00016-0

Maleki R, Afrouzi HH, Hosseini M, Toghraie D, Piranfar A, Rostami S (2020) pH-sensitive loading/releasing of doxorubicin using single-walled carbon nanotube and multi-walled carbon nanotube: a molecular dynamics study. Comput Meth Programs Biomed 186:105210. https://doi.org/10.1016/j.cmpb.2019.105210

Naqvi STR, Rasheed T, Hussain D, Haq MN, Majeed S, Shafi S, Ahmed N, Nawaz R (2020) Modification strategies for improving the solubility/dispersion of carbon nanotubes. J Mol Liq 297:111919. https://doi.org/10.1016/j.molliq.2019.111919

Panigrahi BK, Nayak AK (2020) Carbon nanotubes: an emerging drug delivery carrier in cancer therapeutics. Curr Drug Deliv 17:558–576. https://doi.org/10.2174/1567201817999200508092821

Pantarotto D, Singh R, McCarthy D, Erhardt M, Briand J-P, Prato M, Kostarelos K, Bianco A (2004) Functionalized carbon nanotubes for plasmid DNA gene delivery. Angew Chem Int Ed Engl. 116:5354–5358. https://doi.org/10.1002/anie.200460437

Prasek J, Drbohlavova J, Chomoucka J, Hubalek J, Jasek O, Adam V, Kizek R (2011) Methods for carbon nanotubes synthesis. J Mater Chem 21:15872–15884. https://doi.org/10.1039/C1JM12254A

Rao SG, Huang L, Setyawan W, Hong S (2003) Large-scale assembly of carbon nanotubes. Nature 425:36–37. https://doi.org/10.1038/425036a

Rathod V, Tripathi R, Joshi P, Jha PK, Bahadur P, Tiwari S (2019) Paclitaxel encapsulation into dual-functionalized multi-walled carbon nanotubes. Aaps PharmSciTech 20:51. https://doi.org/10.1208/s12249-018-1218-6

Saeednia L, Yao L, Cluff K, Asmatulu R (2019) Sustained releasing of methotrexate from injectable and thermosensitive chitosan–carbon nanotube hybrid hydrogels effectively controls tumor cell growth. ACS Omega 4:4040–4048. https://doi.org/10.1021/acsomega.8b03212

Samorì C, Ali-Boucetta H, Sainz R, Guo C, Toma FM, Fabbro C, Ros TD, Prato M, Kostarelos K, Bianco A (2010) Enhanced anticancer activity of multi-walled carbon nanotube–methotrexate conjugates using cleavable linkers. Chem Commun 46:1494–1496. https://doi.org/10.1039/B923560D

Seyfoori A, Sarfarazijami S, Seyyed Ebrahimi S (2019) pH-responsive carbon nanotube-based hybrid nanogels as the smart anticancer drug carrier. Artif Cells Nanomed Biotechnol 47:1437–1443. https://doi.org/10.1080/21691401.2019.1596939

Shi Kam NW, Jessop TC, Wender PA, Dai H (2004) Nanotube molecular transporters: internalization of carbon nanotube– protein conjugates into mammalian cells. J Am Chem Soc 126:6850–6851. https://doi.org/10.1021/ja0486059

Simon J, Flahaut E, Golzio M (2019) Overview of carbon nanotubes for biomedical applications. Materials 12:624. https://doi.org/10.3390/ma12040624

Takamori H, Fujigaya T, Yamaguchi Y, Nakasima N (2007) Simple preparation of self-organized single-walled carbon nanotubes with honeycomb structures. Adv Mater 19:2535–2539. https://doi.org/10.1002/adma.200602704

Tiwari SK, Kumar V, Huczko A, Oraon HR, Adhikari AD, Nayak GC (2016) Magical allotropes of carbon: prospects and applications. Crit Rev Solid State 41:257–317. https://doi.org/10.1080/10408436.2015.1127206

Wang M, Qiu X, Zhang X (2007) Mechanical properties of super honeycomb structures based on carbon nanotubes. Nanotechnol 18:075711. https://doi.org/10.1088/0957-4484/18/7/075711

Wang N, Tang Z-K, Li G-D, Chen JS (2000) Single-walled 4 Å carbon nanotube arrays. Nature 408:50–51. https://doi.org/10.1038/35040702

Yadav AR, Mohite SK (2020) Carbon nanotubes as an effective solution for cancer therapy. Res J Pharma Dosage Forms Technol 12:301–307. https://doi.org/10.5958/0975-4377.2020.00050.6

Yan Y, Wang R, Hu Y, Sun R, Song T, Shi X, Yin S (2018) Stacking of doxorubicin on folic acid-targeted multiwalled carbon nanotubes for in vivo chemotherapy of tumors. Drug Deliv 25:1607–1616. https://doi.org/10.1080/10717544.2018.1501120

Yang X, Ebrahimi A, Li J, Cui Q (2014) Fullerene–biomolecule conjugates and their biomedicinal applications. Int J Nanomed 9:77. https://doi.org/10.2147/IJN.S52829

Yu Y, Kong L, Li L, Li N, Yan P (2015) Antitumor activity of doxorubicin-loaded carbon nanotubes incorporated poly (lactic-co-glycolic acid) electrospun composite nanofibers. Nanoscale Res Lett 10:343. https://doi.org/10.1186/s11671-015-1044-7

Zhang J, Song L, Zhou S, Hu M, Jiao Y, Teng Y, Wang Y, Zhang X (2019) Enhanced ultrasound imaging and anti-tumor in vivo properties of Span–polyethylene glycol with folic acid–carbon nanotube–paclitaxel multifunctional microbubbles. RSC Adv. 9:35345–35355. https://doi.org/10.1039/C9RA06437K

Chapter 7
Hydrogels as Smart Drug Delivery Systems: Recent Advances

R. Jalababu, M. Kiranmai Reddy, K. V. N. Suresh Reddy, and Kummari S. V. Krishna Rao

Abbreviations

IPN	Interpenetrating polymeric network
CST	Critical solution temperature
NiPAAm	N-isopropylacrylamide
PBA	Phenylboronic acid

7.1 Introduction

Compounds having three-dimensional interpenetrated networks of physical or chemical cross-linked polymers are known as hydrogels. Initially, synthetic polymers were used to prepare the hydrogels, and now it is extended to natural polymers as they contain various functional groups to react with free radicals of another polymeric group. Hydrogels resemble natural tissues and have numerous additional features like biocompatibility, nontoxicity, and biodegradability and absorb a large amount of water and other biological fluids due to their hydrophilic nature (Sharpe et al. 2014). Hydrogels have been using as therapeutic drug delivery systems for the

R. Jalababu
Department of Chemistry, S.C.I.M. Government College, Tanuku, Andhra Pradesh, India

M. K. Reddy
Department of Environmental Science, GITAM Institute of Science, GITAM (Deemed to be University), Visakhapatnam, Andhra Pradesh, India

K. V. N. S. Reddy (✉)
Department of Chemistry, GITAM Institute of Science, GITAM (Deemed to be University), Visakhapatnam, Andhra Pradesh, India

K. S. V. K. Rao
Department of Chemistry, Yogi Vamana University, Kadapa, Andhra Pradesh, India

J.-C. Kim et al. (eds.), *Smart Nanomaterials in Biomedical Applications*, Nanotechnology in the Life Sciences,
https://doi.org/10.1007/978-3-030-84262-8_7

past few decades. Especially in the late twentieth century and early twenty-first century, many hydrogel fabrication methods were developed for drug delivery as conventional administration of oral drugs caused side effects due to high-level drug concentrations in the bloodstream within a short period. The benefits of the hydrogel are releasing the drug at a predetermined rate and enhancing drug bioavailability. One more attractive hydrogel feature is its spongelike structure, which is regulated by a cross-linked gel matrix. At present, hydrogels are prepared by coupling the stimuli-responsive polymers with nanoparticles and nanorods and making them dual and multi-responsive hydrogels.

7.2 Types of Hydrogels

Hydrogels are self-regulating controlled drug delivery systems and release the drug at a controlled rate and maintain the blood's optimum drug for an extended period. These hydrogels respond to external stimuli like temperature, magnetic field, light, pH, electric field, biological conditions, and ionic power (Fig. 7.1). It forms the basis for the usage of hydrogels in a wide range of applications. Mostly, great importance has been given to design stimuli-responsive hydrogels during the past few decades. The stimuli-responsive hydrogels change their behavior with respect to external stimuli and find their importance in drug release applications (Almeida et al. 2012). Here in this chapter, we explore the basic principles and applications of the following stimuli-responsive hydrogels:

- pH-responsive hydrogels
- Temperature-responsive hydrogels

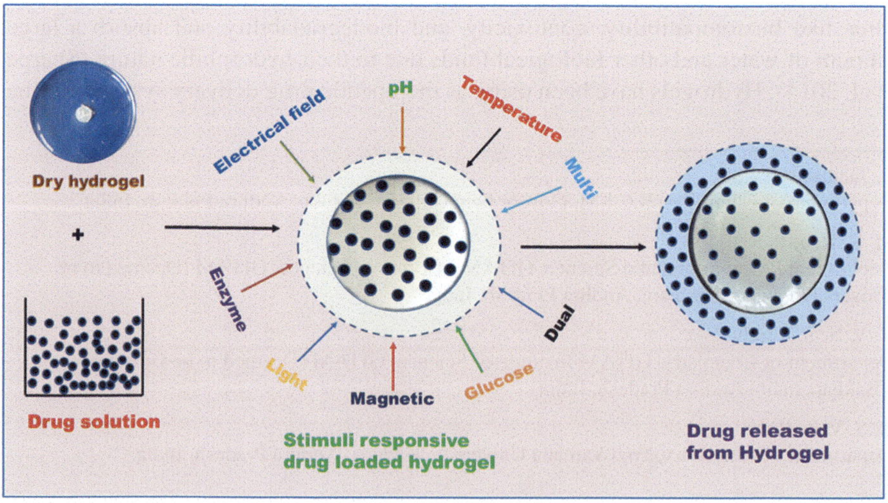

Fig. 7.1 Stimuli responsiveness of hydrogels at various conditions

- Electrical field–responsive hydrogels
- Enzyme-responsive hydrogels
- Light-responsive hydrogels
- Magnetic-responsive hydrogel
- Glucose-responsive hydrogels
- Dual-responsive hydrogels
- Multi-responsive hydrogels

7.2.1 pH-Responsive Hydrogels

Researchers pay much attention to the pH-responsive hydrogels; these are important drug delivery systems because of their distinctive properties and useful drug delivery applications. Every part of the human body shows variation in pH. The pH-responsive hydrogels could be constructed to self-control and target drug delivery to a specific site varying in the body's range of pH areas. The researchers have designed various advanced pH-responsive hydrogels using different natural and synthetic polymers. The pH-responsive hydrogels mainly contain acidic and basic groups for stimuli responsiveness in different pH conditions (El-Sawy et al. 2020). Different acidic and basic polymers and their responsive, functional groups are listed in Table 7.1 (Das and Pal 2015, Qiu and Park 2001, Liu et al. 2008).

Because of their acidic or basic structures, hydrogels accept or release protons concerning the surrounding pH, as shown in Fig. 7.2. The volume of the polymer in solution changes as per the pKa values of the ionizable polymers. When hydrogel has acidic groups (-COOH, $-SO_3H$) in its structure, at pH 7.40, deprotonation of acidic groups results in anions' formation. The repulsions between this anion–anion increase the pore size by enlarging polymer chains, resulting in more drug molecules being delivered from hydrogels at pH 7.40. Whereas at pH 1.20, acidic groups involved in forming hydrogen bonds between the polymer chains strictly obstructed the release of drug molecules from the hydrogel network, and less % drug release is observed at pH 1.20 in Fig. 7.2a. When hydrogel contains alkaline functional groups ($-NH_2$) at lower pH, protonation of these groups leads to electrostatic repulsion between the chains and releases more drugs. At alkaline pH, hydrogen bonds are formed, which strictly restrict the release of drug particles and show a slow-release rate, as shown in Fig. 7.2b.

The hydrogels bear acidic or basic structures that accept or release protons in response to the surrounding pH, which becomes the cause for the hydrogel's pH-sensitive behavior. The hydrodynamic volume of the polymer changes according to the difference in pKa values of the ionizable polymers. Apart from the ionizable values, the ratio of cross-linking concentration, hydrophobicity, and polymer composition also affects the hydrogels' pH-sensitive behavior (Aguila et al. 2007). Using various carbohydrate polymers, researchers have synthesized advanced pH-responsive hydrogels to be used with assurance in drug delivery and other biomedical applications.

Table 7.1 Different acidic and basic polymers and their responsive functional groups

S. no	Name of the polymer	Polymer type	Polymer nature	Responsive functional group
1	Alginic acid	Natural	Acidic	Carboxylic group
2	Hyaluronic acid	Natural	Acidic	Carboxylic group
3	Pectin	Natural	Acidic	Carboxylic group
4	Xanthan gum	Natural	Acidic	Carboxylic group
5	Carboxymethyl cellulose	Natural	Acidic	Carboxylic group
6	Carrageenan	Natural	Acidic	Sulfate group
7	Dextran sulfate	Natural	Acidic	Sulfate group
8	Chondroitin sulfate	Natural	Acidic	Sulfonic/carboxylic group
9	Heparin	Natural	Acidic	Sulfonic/carboxylic group
10	Poly(Methacrylic acid)	Synthetic	Acidic	Carboxylic group
11	Poly(L-glutamic acid)	Synthetic	Acidic	Carboxylic group
12	Poly(Acrylic acid)	Synthetic	Acidic	Carboxylic group
13	Poly(Lactic acid)	Synthetic	Acidic	Carboxylic group
14	2-Acrylamido-2-methyl propane sulfonic acid	Synthetic	Acidic	Sulfonic acid group
15	Chitosan	Natural	Basic	Amine group
16	Poly(L-lysine)	Natural	Basic	Amine group
17	Poly(Ethylene imine)	Synthetic	Basic	Amine group
18	Poly(N,N′-diethylamino) ethyl methacrylate)	Synthetic	Basic	Tertiary amine group
19	Poly(N,N′-dimethylamino) ethyl methacrylate)	Synthetic	Basic	Tertiary amine group
20	Poly(2-Vinylpyridine)	Synthetic	Basic	Tertiary amine group

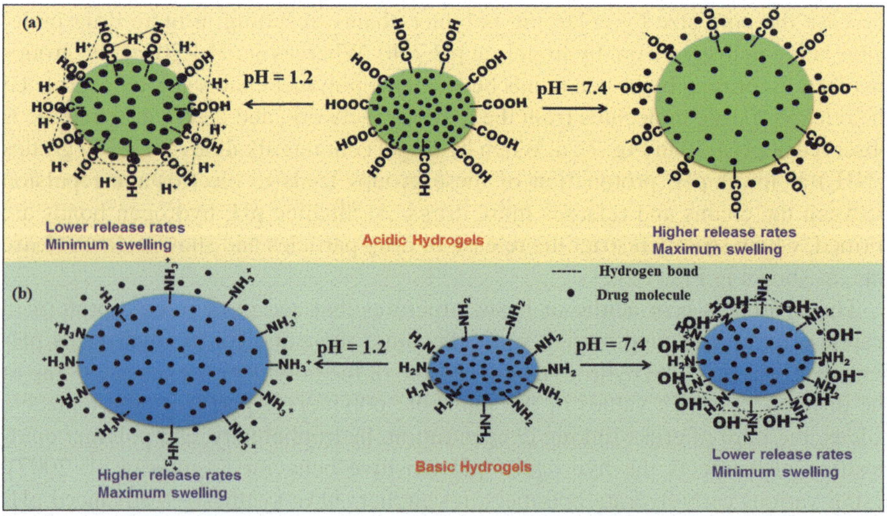

Fig. 7.2 Swelling and drug release pattern of pH-responsive hydrogels. (**a**) Acidic hydrogels. (**b**) Basic hydrogels

Suhail et al. (2021) have developed chondroitin sulfate-co-poly(acrylic acid) pH-responsive hydrogels ((CSPAA-hydrogels) for sustained release of diclofenac sodium. The CSPAA-hydrogels were synthesized from chondroitin sulfate (CS) polymer, acrylic acid (AA) monomer, and ethylene glycol dimethyl acrylate (EGDMA) cross linker by free radical polymerization method. For all CSPAA hydrogels' formulations, in vitro drug release studies were done at pH 1.20 and pH 7.40. The more significant amount of diclofenac sodium was released at pH 7.40 when compared with pH 1.20. At alkaline pH conditions, deprotonation of -COOH groups of chondroitin sulfate and acrylic acid causing anion-anion inter-ionic repulsions leads to more significant swelling and more release drug. In this study, the maximum release of diclofenac sodium from CSPAA pH-sensitive hydrogels is observed at pH 7.4 up to 72 h.

Krishna Rao et al. (2006) reported a novel pH-responsive interpenetrating polymeric network (IPN) microgels (MGs) using chitosan, acrylamide-g-poly(vinyl alcohol), and hydrolyzed acrylamide-g-poly(vinyl alcohol), which are cross-linked with glutaraldehyde for sustained release of cefadroxil antibiotic. The drug release studies of chitosan with all other formulation in both pH 1.20 and pH 7.4 media were carried out. Almost 100% release of cefadroxil occurred at 10 h for CS-25 formulation in pH 1.20 media, whereas only 60% cefadroxil release was detected at 10 h in pH 7.4 media. At pH 1.2 protonation of the amino group of chitosan is involved, due to inter-ionic repulsions between cation and cation leading to an increase in pore size by the enlargement of polymer chains and showing higher drug release. Whereas at pH 7.40, alkaline groups (-NH_2) involved in forming hydrogen bonds between the chitosan polymer chains strictly hindered the release of drug molecules from CS microgels, hence, a lower degree of cefadroxil drug molecule delivery from CS microgels at pH 1.20.

7.2.2 *Temperature-Responsive Hydrogels*

Temperature is vital in controlling the drug release from the polymer network. The specific temperature at which the thermoresponsive hydrogels display transitional properties in the aqua solutions is critical solution temperature (CST). These hydrogels are temperature sensitive with solution-to-gel phase transitions at a specific temperature and can swell and deswell by varying the fluid temperature. Thermosensitive hydrogels are mainly divided into negative thermoresponsive hydrogels and positive thermoresponsive hydrogels.

The negative thermoresponsive hydrogel has a lower critical solution temperature (LCST). The LCST-type hydrogels exist in one phase (miscible) when the temperature is less than CST and present in two phases (gel) above the CST. The swelling nature of hydrogels is tuned by changing the temperature. When the temperature is less than the LCST, the water interacts with the polymer chain's hydrophilic portion, and swelling occurs (Ullah et al. 2015).

When the temperature is raised above the LCST, hydrophobic interaction among the polymer chains predominates, and the shrinking of the hydrogel occurs. Hydrophobic interactions lead to the transition of polymer chains from coiled form to the dehydrated globular state. Besides intra- and intermolecular hydrophobic interactions between the existing hydrogel networks, the polymer's water molecules participate in salvation and hydrogen bonding and lead to the hydrophilicity change (Qureshi et al. 2019).

The positive thermoresponsive hydrogel has an upper critical solution temperature (UCST). The UCST-type hydrogels exist in two phases when the temperature is lower than CST and present in one phase (miscible) at a temperature higher than CST. When below the UCST, the hydrogels shrink and release the drug from the hydrogel network. At a temperature above UCST, swelling occurs due to the formation of hydrogen bonding (Ward and Georgiou 2011). Schematic presentation of temperature-responsive hydrogel swelling and drug release based on LCST and UCST is shown in Fig. 7.3

Hydrogel's thermoresponsive behavior is caused by hydrophobic or a combination of hydrophilic and hydrophobic groups in them. These thermoresponsive hydrogel factors are easily controllable and can be used to perform both in vitro and in vivo calculations; they find their extensive usefulness in controlled drug delivery systems. The familiar thermoresponsive polymers and their phase transition temperatures are shown in Table 7.2 (Qureshi et al. 2019).

Isiklan and Kucukbalci (2015) synthesized sodium alginate and poly(N-isopropylacrylamide/acrylic acid) [SA-g-PIPAA/PAA] temperature and

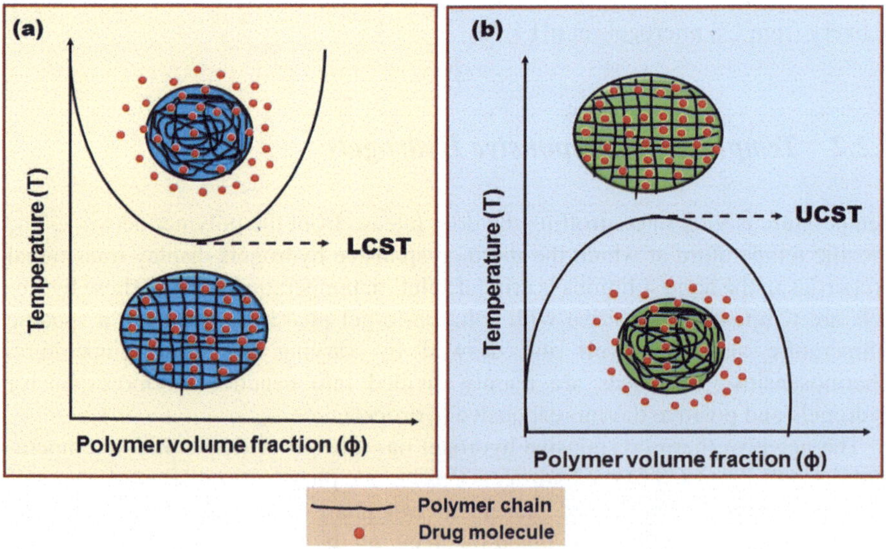

Fig. 7.3 Swelling and drug release pattern of temperature-responsive hydrogels. (**a**) Lower critical solution temperature (LCST) phase transition. (**b**) Upper critical solution temperature (UCST) phase transition

Table 7.2 Types of temperature-responsive polymers with their CST

S. no	Name of the polymer	CST type	Phase transition temperature
1	Poly(N, N-diethylacrylamide)	LCST	25–32 °C
2	Poly(N-vinyl caprolactam)	LCST	32–34 °C
3	Poly(N-isopropylacrylamide)	LCST	32–34 °C
4	Poly(Methyl vinyl ether)	LCST	37 °C
5	Poly(2-(Dimethylamino)ethyl methacrylate)	LCST	38 °C
6	Poly(N-cyclopropylacrylamide)	LCST	58 °C
7	Poly(N-ethyl acrylamide)	LCST	74 °C
8	Poly(Acrylic acid)	UCST	25 °C
9	Poly(Acrylamide)	UCST	25 °C

pH-responsive networks for controlled release of indomethacin (IND). To study temperature responsiveness of SA-g-PIPAA/PAA networks, the release of IND was observed higher at 37 °C than 25 °C. These findings might explain the temperature above LCST (> 34 °C) of poly(N-isopropylacrylamide) subsequently polymer chains will aggregate, will get more hydrophobic, and will collapse, which leads to squeezing out IND molecules from network. Drug release experiments were carried out at 25 °C, 37 °C, and 47 °C and the results observed that increasing IND from network increases temperature.

As per the study in the literature, chitosan-poly(N-vinyl pyrrolidone)-poly(N-isopropylacrylamide) temperature-responsive interpenetrating polymeric networks (IPNs) with different concentrations of glutaraldehyde cross-linker release more amount of amoxicillin trihydrate drug (82%) at 37 °C and 60% of the drug at 25 °C (Ekici and Saraydin 2017). Poly(N-vinyl pyrrolidone) and poly(N-isopropylacrylamide) are the LCST polymers present in interpenetrating polymeric networks, which are responsible for thermal responsiveness.

Hill and co-workers (2019) first developed new thermoresponsive a fibrous hydrogel network by self-assembly of a single α-helical coiled-coil Q protein. The Q gel is reversible thermoresponsiveness below and above its UCST, and it is determined to be at ~16.2 °C. The Q gel exhibits UCST-type behavior through gel-sol transition above the UCST. At physiological temperature (> UCST), curcumin-loaded gels showed a slow release of curcumin over 17–18 days at 37 °C.

7.2.3 Electric-Responsive Hydrogels

Hydrogels made up of electroactive polymers (natural or synthetic polyelectrolytes) can change their size when an electric field is applied. Under the electric field, hydrogel undergoes bending or swelling (Murdan 2003). It is observed that the drug release is more in the presence of an electric field when compared to its passive state (in the absence of an electric field), and the release rate can be controlled by applied electric strength. There could be three mechanisms by which electric field functions:

1. The advancement of stress gradient in the hydrogel
2. Local pH changes around the electrodes
3. Electroosmosis of water coupled to electrophoresis

The gel's experimental conditions and nature lead to any one of the above mechanisms and decide the gel swelling (Merino et al. 2015). The drug delivery took place by getting attracted to the oppositely charged electrode, then contraction of hydrogel results in the release of drug through electric stimulation. Figure 7.4 describes the shrinkage of gel and release of the drug, which depends on the change of pH (increase in pH at the cathode and decreased pH at the anode) and stress gradient. A change in pH due to electric current led to distort hydrogen bonds between polymer chains to mortify or bend the polymer to release drug (James et al. 2014). Application of electric field accelerates the ionization of ionizable groups (-COOH) present in hydrogel. As per the study in the literature, the ionizable group (-COOH) present in the hybrid hydrogel of cellulose and sodium alginate tends to ionize into $-COO^-$ and H^+ ions (Shi et al. 2014). Here H^+ ions present (counterions) move toward the cathode surface. The decrease in H^+ ions leads to the greater number of $-COO^-$ ions in hydrogel matrix, thus enhancing the hydrogel's swelling behavior

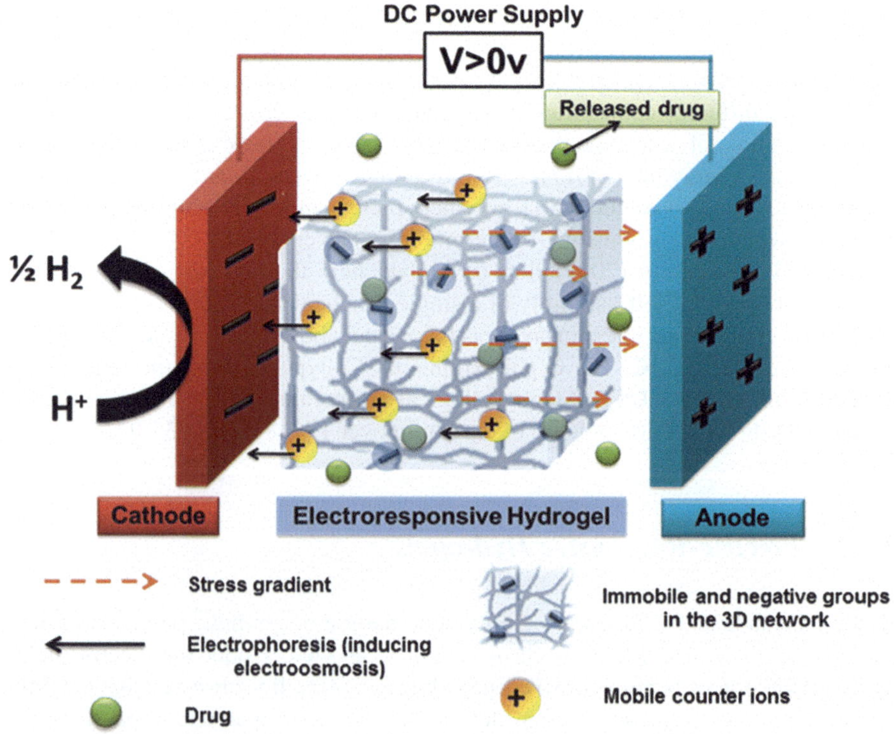

Fig. 7.4 Pictorial representation of drug release mechanism of electro-responsive hydrogel. (Adapted from Merino et al. 2015)

under electric stimulus due to repulsive forces exerted between carboxylate anions and improving the electric response sensitivity. At a lower voltage, the low concentration of drug release is observed, and at a higher voltage high concentration of release of drug is notified. Electro-stimulation of hydrogel also involves stress gradient and electrophoresis (Merino et al. 2015). Recently electro-responsive method is widely used in biomedical applications (Mariouras and Vamvakaki 2016).

The electroactive polymers are used in many fields like artificial muscles, hands, or fingers for mechanical gripping, tissue engineering scaffold, actuators, etc. (Saikia et al. 2015). The electro-responsive hydrogels have wide applications in sound dampening, artificial muscle actuators, drug delivery, and energy transductions. Polymer [poly(2-acrylamido-2-methylpropane sulfonic acid-co-n-butylmethacrylate)] acted as efficient electro-responsive toward drug delivery of edrophonium hydrochloride and hydrocortisone (Zhao et al. 2011). Similarly, application of polymer [poly(3,4-ethylenedioxythiophene)/poly (styrenesulfonate)] in gelatin methacryloyl as hydrogel with electric response was used to treat skin cancer (Oktay and Alemdar 2018). Ibuprofen and amoxicillin exhibited "on and off" release phenomena when subjected to electric voltage (Qu et al. 2018); these were drafted with chitosan-graft-polyaniline and oxidized dextran polymer-made hydrogels that showed high efficiency in drug release.

7.2.4 Enzyme-Responsive Hydrogels

Enzyme acts as a catalyst in all living cells; at molecular levels, the metabolism of cells and tissues is regulated by enzymes, but their irregular actions lead to diseases such as inflammation, cancer, Alzheimer's, cardiovascular, and osteoarthritis (Quershi et al. 2019, Wang et al. 2018a; Chandrawati 2016). Overexpressed enzymes change the biomaterial (drug-loaded) properties by their action, thereby attaining continuous and regulated drug release at the site of action. The materials selected for enzymatic response must have the following requirements:

1. The materials are integrated with identified elements or substrates that can be recognized by those enzymes selectively (use of peptide chains/linkers or polymer-peptide by conjugation with specific amino acid sequences toward enzymes of interest).
2. These integrated substrates are accessible by enzymes for successful enzyme-catalyzed reactions' kinetics.
3. These enzyme-catalyzed reactions' kinetics must lead to changes in material properties. These can include degradation or morphological transformation of biomaterials.

Figure 7.5a describes how the overexpressed enzyme has participated in drug release when the enzyme-sensitive cross-linked hydrogel (loaded with the required drug) is injected into the body. Here the drug from hydrogel is released by breaking the peptide cross-linkers of polymer strands by the action of enzymes.

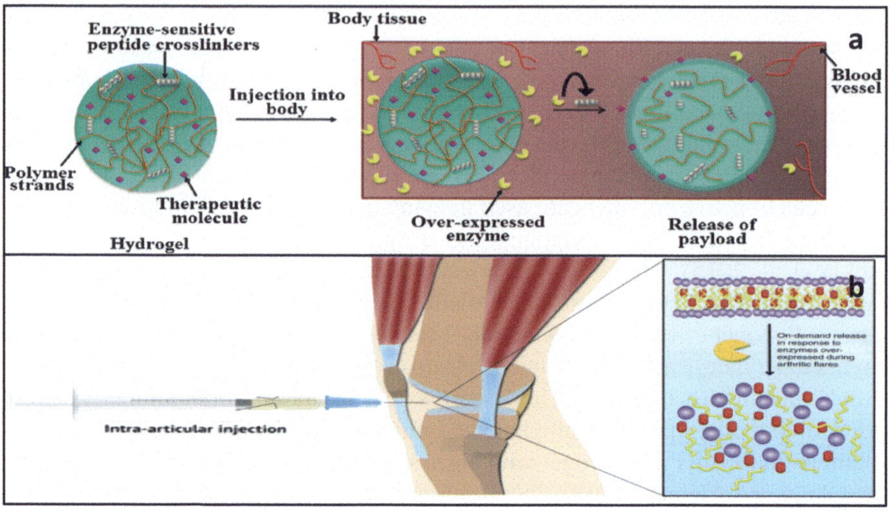

Fig. 7.5 (**a**) Schematic representation of drug release pattern of enzyme-responsive hydrogel (Adapted and reproduced from reference Quershi et al. 2019 with the permission from Elsevier). (**b**) Schematic representation of drug release in response to enzyme overexpressed during arthritic flares. (Adapted from Joshi et al. 2018)

In one study, a network of macromolecules of cellulose coupled with polymeric materials is used for enzyme-responsive hydrogels (Billah et al. 2019). Therapeutic molecules can be compressed through covalent attachment or physical compression into polymeric material delivered either by degradation or swelling of material upon enzyme action when it reaches the target site. Compared to physical or chemical conventional stimuli (pH, temperature, ligand-receptor interactions, ionic strength, etc.), enzymatic response to biology's signals is highly selective and includes catalytic amplification to facilitate fast response (Haija and Ulijin 2014, Spruell and Hawker 2011).

Many innovations were made in the field of enzyme-sensitive systems. For pulmonary drug delivery, poly(ethylene glycol) was used to prepare enzyme-sensitive hydrogel microparticles (Secret et al. 2014). Solution polymerization was utilized to develop PEG microparticles from poly(ethylene glycol) diacrylate with the inclusion of a peptide Gly-Leu-Lys for metalloproteinase enzyme. Figure 7.5b discusses the overexpressed enzymatic response toward drug delivery when it is injected during arthritic flares (Joshi et al. 2018). It was observed that the development of TG-18 hydrogel acts as a drug reservoir and releases the drug in response to overexpressed enzymes in inflammatory joints. The current drug delivery system suffers from low local concentration and short retention time at the oral disease target site. To overcome this, Guo et al. (2019) prepared diacrylate polyethylene glycol-based scaffolds with a cysteine-terminated peptide cross-linker cleaved by the matrix to overcome this problem metalloproteinase 8 (MM8) (Guo et al. 2019). The use of enzyme-sensitive MMP-8 with hydrogel has good results in drug delivery at the target site, proven by HPLC and FTIR testing. The use of MMP-9 (matrix mettaloproteinase-9) overexpression enzyme in the extracellular matrix of tumor tissues can trigger chemically modulating the drug delivery from the carrier (Kulkarni et al. 2015).

7.2.5 Photoresponsive Hydrogels

Light stimulation is a convenient drug release parameter due to its high spatial resolution, non-invasive nature, comfortability, and convenience. Light can penetrate deep into the skin and quickly release the compounds (drug) at a specific site. Among different stimuli-responsive hydrogels, photoresponsive hydrogels became promising drug delivery systems owing to their invasive and remote control options. The mechanism of light-responsive hydrogel consists of a polymeric network and a photoreactive moiety (consists of a photochromic chromophore as a functional part).

The study of Alfimove et al. (2003) stated that photochromic molecules first receive optical signals, and then the chromophores in the photoreceptor convert the photoirradiation to a chemical signal through a photoreaction such as a cleavage, thermal, and isomerization. Later the signal is transferred to the functional part of hydrogel and controls its properties. The change in chromophores upon photoirradiation strongly depends on the molecular structure. The functioning of hydrogels is broadly categorized into photochemical, photoisomerization, and photothermal methods.

7.2.5.1 Photochemical

Under this category, the drug is released by breaking the covalent bond by light irradiation. These hydrogels are prepared by incorporating the photoresponsive moieties (o-nitrobenzyl, coumarin, and pyrene derivatives) in the polymer network. Initially, UV light is used to irradiate; now, it extended to near IR due to its high penetration through tissue. The most popular photoresponsive moiety for photodegradable hydrogel is the o-nitrobenzyl group, in which irradiation with UV produces carboxylic acid o-nitroso benzaldehyde and releases the drug at the site of action. In hydrogel preparation, the o-nitrobenzyl group is used as a cross-linker between the polymeric chains of poly(ethylene glycol) and polyacrylamide hydrogels (Linsely and Wu 2017).

7.2.5.2 Photoisomerization

Here drug release is controlled by reversible conformational changes of particular molecules by irradiation with UV and visible light. Azobenzene is a frequently used chemical moiety for this kind of mechanism; upon irradiation with UV light,

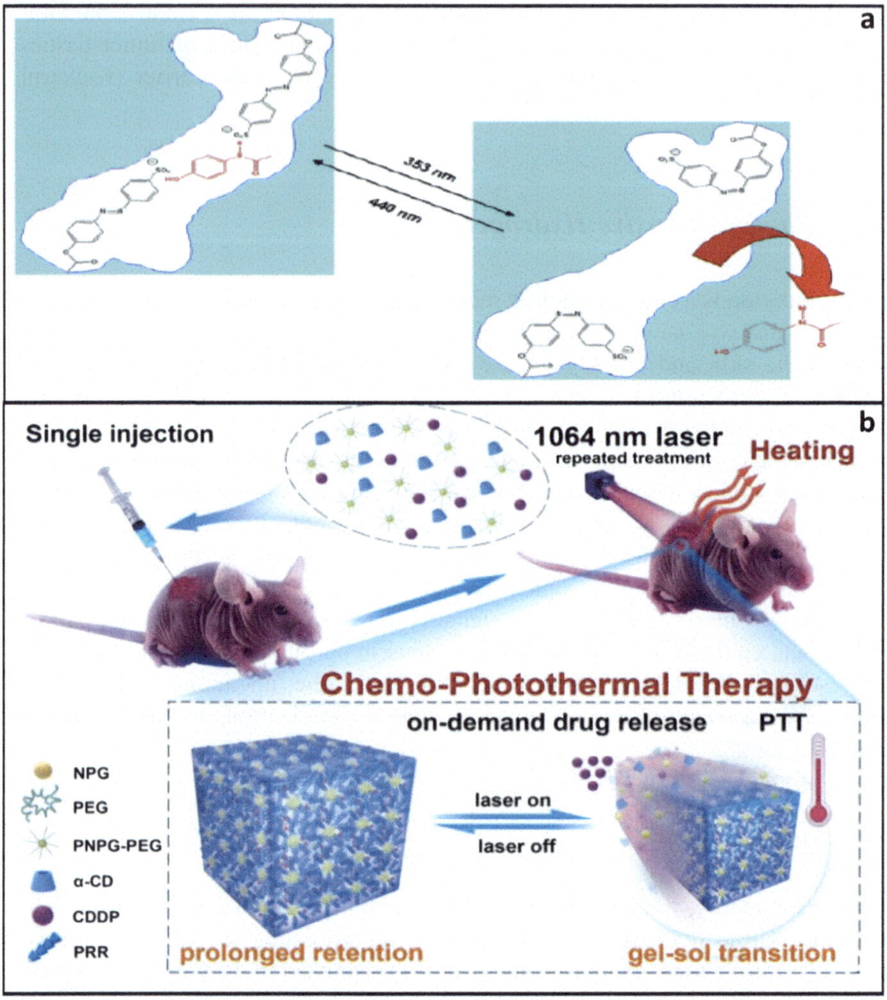

Fig. 7.6 (**a**) Schematic representation of photoregulated substrate release and uptake with molecularly imprinted photoresponsive hydrogel (adapted and reproduced from reference Gong et al. 2008 with the permission from American Chemical Society). (**b**) Schematic presentation of NIR-triggered cisplatin through chemo-photothermal therapy. (Adapted from Ruan et al. 2019)

transconfiguration converts into cis form and back to cis form by interacting in visible light, as shown in Fig. 7.6a (Timko et al. 2010). Cis form shows high polarity than transform, which is used to control the hydrophobic interactions. In one study, photoresponsive molecular etched hydrogel drug delivery systems are prepared. When hydrogels are prepared in the presence of paracetamol, azobenzene-based molecular sites were formed for the target molecule (Gong et al. 2008). While it is irradiated with UV/Vis light, photoisomerization of the azobenzene moieties stimulates structural change as shown in Fig. 7.6a of the recognized sites, and uptake or release of paracetamol was controlled.

Chiang and Chu (2015) developed light-sensitive hybrid alginate hydrogel. The preparation of β-cyclodextrin-grafted alginate (β-CD-Alg) hydrogel was made photosensitive by permeating diazobenzene-modified PEGs [poly(ethylene glycol)) into it. The exposure of UV light on the hydrogel disrupts gel structure, resulting in the drug's release. This photo-triggered structure degradation of the hybrid gel was used in wound healing and other biomedical applications.

7.2.5.3 Photothermal Mechanism

This hydrogel contains thermosensitive molecules and generates heat upon irradiation with light, and the drug is released. It is made with two key components, i.e., one component converts the light energy into heat (chromophore), and the second molecule responds to heat released (thermo-responsive). Gold nanoparticles, carbon nanotubes, and graphene oxide nanoparticles absorb NIR light, and methylene blue, cardio green, and riboflavin absorb both visible light and NIR light. N-Isopropylacrylamide (NiPAAm) is a commonly used thermo-responsive monomer. The functioning of NiPAAm is discussed in thermo-responsive hydrogels. When these hydrogels are irradiated with light, photothermal agents present in hydrogels mediate the photothermal conversation and accelerate the gel-sol formation of thermo-responsive monomers, and the drug is released. However, prolonged irradiation causes the leaching out of photothermal agents and leads to decreased photothermal effect and drug delivery control.

Leaching of nanoparticles from hydrogels overcome by preparing hydrogel from polymer poly(N-phenylglycine), poly(ethylene glycol), and α-cyclodextrin has been reported as photothermal hydrogel in the literature. In this study poly(N-phenylglycine) not only ties the poly(ethylene glycol) chains and but also serves as NIR-absorbing facilitators (Ruan et al. 2019). The polymer PNPG absorbs NIR-II light, facilitates photothermal conversion, makes prolonged retention time, and permits repeatable treatment. The polymer building blocks were ruptured after treating with NIR-II laser, which releases the desired drug to the target site, minimizing off-target toxicity, as shown in Fig. 7.6b. Under NIR-II irradiation, the hydrogel is converted into a sol and released the chemotherapeutic drug, cisplatin (cis-diamminedichloroplatinum(II)), in an easy way and treats tumor cells.

7.2.6 *Magnetic-Responsive Hydrogel*

Hydrogels consist of a polymer matrix made up of considerable quantities of water. The matrix, which is of 3D structured polymer networks, has solid and liquid characteristics resembling soft tissue living entities (Li et al. 2012, Shi et al. 2020). These hydrogels have a specific property like swelling, wetting, and elastic in nature, due to which it has been used in many fields. The hydrogel's primary assembly is cross-linking of hydrophilic polymers with different ways with covalent bonds, interaction with hydrophobic compounds, formation with crystals, etc. (Duan and Zang 2017). However, the major obstacle of hydrogels is their functionality, which needs to be improved. For many decades scientists are trying different methodologies to strengthen the functionality by adopting various strategies such as assembly of supramolecules, alteration in polymeric chains, a collaboration of other networks, etc. (Kowalczyk et al. 2014, Matsuda et al. 2019). Nevertheless, these technologies also have some pitfalls, so researchers focused on electromagnetic technology. The use of a magnetic field with some magnetic materials improved the functionality of the hydrogels.

The magnetic field-sensitive hydrogel is prepared with a combination of water and magnetic particles (at nano- or microscale). Most magnetic field hydrogels are made up of nanoparticles called superparamagnetic iron oxide nanocrystals (SPIONs), which act as a matrix (Li et al. 2013). When a magnetic field is combined with this desired material, a new magnetically induced system is developed, controlled by the *ON* and *OFF* mode of the magnetic field (Liu et al. 2006), as shown in Fig. 7.7 (Qin et al. 2009). When these hydrogels are subjected to an external magnetic field, it gets twisted, and the drug can be released, which can be identified easily through responsive behavior, as shown in Fig. 7.7. A unique feature of a magnetic field is fast response, remote operation, and highly controlled.

In the past few decades, the application of magnetic fields has gained importance in hydrogels' preparation. Its quick response, remote monitoring, deep penetration, non-contact, and controlled magnetic-responsive hydrogels gained importance in biomedical applications. Compared to the conventional drug release process, magnetic field-sensitive hydrogels are more advantageous in triggering drug release with minimal side effects (Frachini and Petri 2019). In one study, the drug is administered through the magnetic hydrogels in a transdermal patch. The magnetic nanoparticles present in the hydrogels are suitable for the magnetic stimulating application. Magneto-responsive hydrogel of ethylene-vinyl acetate loaded with

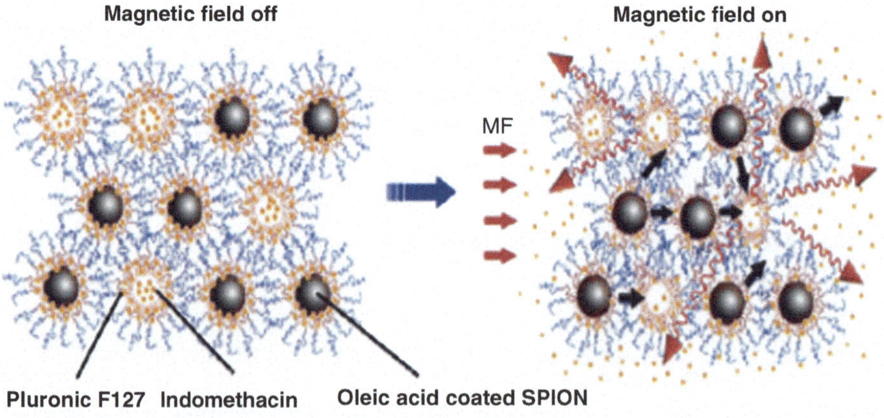

Fig. 7.7 Illustration of magnetic field hydrogel release of the magnetic field–induced drug. (Adapted from Qin et al. 2009)

Table 7.3 Examples of different magnetic field-sensitive drug delivery systems

S.no	Name of the polymer	Drug	Application	References
1.	Poly(Ɛ-carprolactone)	Doxorubicin hydrochloride	Anticancer	Wang et al. (2018)
2.	Starch-g-poly (methylmethacrylate-co-PEG-acrylamide)	Doxorubicin	Anticancer	Zohreh et al. (2016)
3.	Chitosan	Paclitaxel	Anticancer	Mangaiyarkarasi et al. (2016)
4.	Carboxymethyl cellulose	Curcumin		Reddy et al. (2011)
5.	Poly(Vinyl alcohol)	Theophylline	Pulmonary	Bertoglio et al. (2010)

insulin was successfully prepared, and insulin is easily released into the body. In the absence of a magnetic field (OFF), the insulin (drug) diffusion depends on drug and carrier interactions. If the interactions among insulin (drug) molecule and matrix are weak, then drug molecules tend to diffuse rapidly to the medium (burst effect). However, if the interactions between drug molecules and matrix are strong, the release tends to be slow. If the magnetic field is ON, the moments of magnetic nanoparticles (MNP) become aligned with the magnetic field rather than randomly oriented. The alternating magnetic field from ON to OFF mode accelerates MNP cases' movement vibrations in the cross-linked polymer chains, stimulating the drug release to the medium (Finotelli et al. 2010). Magnetic responsive hydrogels made of polymers for different drugs are summarized in Table 7.3.

7.2.7 Glucose-Responsive Hydrogels

In healthy humans, blood glucose levels in the body are regulated by insulin secreted by β cells. The deficiency of insulin secretion from β cells leads to "diabetes mellitus" (Johan et al. 2003). Diabetes mellitus is controlled by maintaining optimum blood glucose concentration by administering sufficient insulin quantity at a definite time into the body (Sharifzadeh and Hosseinkhani 2017). Researchers have paid more attention to develop glucose-sensitive, self-controlled smart hydrogel systems to deliver insulin. These glucose-responsive hydrogels can release a required amount of insulin with the blood glucose level variation. These are categorized into three main types based on stimulation: i.e., glucose oxidase (GOS) hydrogels, phenylboronic acid (PBA) hydrogels, and glucose-binding molecule-based hydrogels (Wang et al. 2019a, b).

7.2.7.1 *Glucose Oxidase (GOS)-Stimulated Systems*

The glucose oxidase (GOS)-stimulated hydrogels release insulin by oxidation of glucose which GOS catalyzes. GOS is a flavin-containing glycoprotein that catalyzes the oxidation of glucose in the presence of oxygen to produce gluconic acid and hydrogen peroxide (H_2O_2) as a by-product (Wilson and Turner 1992). GOS is highly selective toward catalytic oxidation of glucose into gluconic acid, thereby decreasing the system's overall pH. Generally, hydrogels are classified into cationic hydrogels and anionic hydrogels based on pH sensitivity. The cationic hydrogels are ionized and swell at low pH, while anionic hydrogels are swelled at high pH conditions. When glucose levels rise, the GOS oxidizes the glucose molecules by producing gluconic acid. So, gluconic acid drops the local environment's pH, and excess of H^+ ions accumulates more positive charge on the polymer backbone, resulting in swelling of hydrogel due to the expansion of polymer chains (Hassan et al. 1997). As hydrogel swells, it increases hydrogel's mesh size and permits imbibed insulin to diffuse out the hydrogel as shown in Fig. 7.8 (Wang et al. 2019a).

When insulin is caused to decrease glucose levels, the amount of gluconic acid inside the hydrogel decreases and the insulin release slows down. The insulin release rate depends on the glucose concentrations, and the glucose-stimulated swelling and shrinking of the hydrogel are shown to be reversible. The enzymatic conversions of glucose to gluconic acid and hydrogen peroxide to oxygen are shown below (Wu et al. 2011):

$$\text{Glucose} + O_2 + H_2O \xrightarrow{\text{GOS}} \text{Gluconic acid} + H_2O_2$$

$$2H_2O_2 \xrightarrow{\text{CAT}} 2H_2O_2 + O_2$$

H_2O_2 is a by-product in the GOS-catalyzed reaction, and the inhibitory effect of H_2O_2 in the reaction was overawed by co-immobilization of catalase (CAT) with

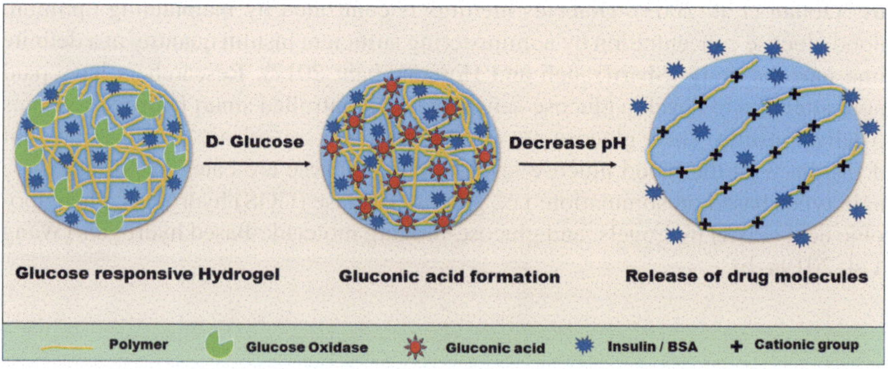

Fig. 7.8 Glucose-responsive mechanism of GOS-based hydrogel

GOS to convert H_2O_2 into O_2 and H_2O, which enhanced the availability of the reactants and increased the swelling rate.

Zou et al. (2015) have prepared pH-responsive cationic chitosan hydrogel (HTCC) for glucose-responsive controlled drug delivery system. In a pH-responsive GOS immobilized HTCC hydrogel matrix, the enzymatic oxidation of GOS converts the glucose into gluconic acid, and a decrease in pH of a solution from 4.7 to 6.6 is observed. Thus, lowering pH values leads to a significant cumulative release of bovine serum albumin (BSA) from the HTCC hydrogel matrix. The release kinetics data of BSA under different pH values and glucose concentrations is substituted in Korsmeyer-Peppas kinetic model, which follows a Fickian release.

Li et al. (2017) developed glucose-responsive peptide hydrogel for controlled delivery of insulin. The glucose-sensitive peptide hydrogel is encapsulated with insulin and GOS that can oxidize glucose into gluconic acid and decrease the local pH. When the matrix's pH is below the pKa of the peptide chains, there would be electrostatic repulsion between them and the consequent unfolding of individual chains. Ultimately the hydrogel disassembled and loaded insulin in the hydrogel matrix were diffused. The in vitro and in vivo studies illustrated that these peptide hydrogels are biocompatible, injectable, and active in controlling blood glucose levels for a long time.

7.2.7.2 Phenylboronic Acid (PBA)-Stimulated Systems

Phenylboronic acid (PBA) is a synthetic, highly stable, and biocompatible compound present in the physiological environment, which has been most extensively studied and used to design glucose-responsive networks to deliver insulin. PBA exists in equilibrium between a neutral triangle form and an anionic tetrahedral form. The neutral structure is hydrophobic; however, the anionic structure is hydrophilic. While the two structures can react with 1,2-cis-diols of glucose reversibly, the binding of the neutral structure with glucose is unstable as it is more labile to hydrolysis. PBA's anionic state can form stable complex with cis-diol of glucose, which can move equilibrium toward increasing hydrophilic form and lowering the hydrophobic form (Ma and Shi 2014). The hydrophilic state of the PBA-glucose complex that undergoes swelling results in the release of encapsulated drug from the cross-linked networks into the solution.

The PBA and its derivatives are Lewis acids with a pKa value in 8.20–8.86. Due to the PBA components' high pKa value, these PBA-based glucose-responsive systems are lower ionization at physiological conditions (pH 7.4), but more ionized in alkaline medium only. At physiological conditions, PBA moieties are lower binding capacity toward glucose molecules because of poor solubility in an aqueous medium. Initially, two methods were proposed to drop PBA's pKa and enhance the binding affinity of glucose at physiological pH. The first method was substituting electron-withdrawing groups, for example, fluoro, nitro, and carbonyl group at meta position with respect to boronic acid on PBA (Matsumoto et al. 2003). The second method was the incorporation of the amino group onto PBA having polymer (Wu et al. 2010) or onto phenyl ring close to the boronic acid (Kim et al. 2009).

In recent times, other new methods were established to reduce the pKa of PBA and raise glucose binding sensitivity at physiological conditions. The first method was a partial modification of the PBA chain by incorporating carboxylic groups on a polymer backbone prominent to PBA-carboxylic interactions in the network and thus lowering its pKa value (Wang et al. 2009). The second method for enhancing glucose responsiveness and lowering the pK_a was both complexation of PBA polymer with glycopolymer and coordination between PBA polymer and polyol-containing compounds (Yang et al. 2013). The third method was synthesis of new type of substituted PBA benzoboroxole compounds, which influenced more water solubility and a low pKa value compared to that of simple PBA (Dowlut and Hall 2006).

Zhang et al. (2008) introduced a new type of hydrogels to enhance swelling and glucose response rate. They were prepared in normal type of poly(N-isopropylacrylamide-co-3-acrylamidophenylboronic acid) (poly(NIPAM-co-AAPBA)] cross-linked hydrogels and giant hydrogels, which are synthesized by grafting (poly(NIPAM-co-AAPBA)] side chain on normal hydrogels. The giant hydrogels have additional grafted side chains with PBA functional groups and can faster swelling and deswelling rates than normal hydrogels. The grafted giant hydrogels have shown faster glucose response than that of normal cross-linked hydrogels at physiological conditions. The rapid glucose response hydrogels are smart carriers for drug delivery systems.

Dong et al. (2016) developed injectable, self-healing, and glucose-responsible polymeric hydrogels. These hydrogels were synthesized by cross-linking glucose with PBA, and this PBA-glucose complex was fixed on the polymer chain. This glucose-responsible hydrogel is the effective release of rhodamine B model drug in in vitro conditions.

7.2.7.3 Glucose-Binding Molecule-Based Systems

Glucose-responsive hydrogels can be prepared from the glucose-binding proteins. Among glucose-binding proteins, lectins are the main kind of proteins, which can strongly bind with carbohydrates and act as natural receptors in glucose-responsive networks. Among these lectins, concanavalin A (Con A) is the most extensively used non-sugar protein with four binding sites, which is extracted from the jack bean plant *Canavalia ensiformis* (Michael et al. 2019). Con A molecules show reversible and strong attraction with unchanged -OH groups of pyranose ring-containing monosaccharides, polysaccharides, and glycopolymers (Lina et al. 2019). Four Con A molecules and pyranose rings were assembled to form tetramer at physiological conditions. Con A can form glucose-responsive hydrogels; it has a more affinity for free glucose than for glycosylated polysaccharides and glycopolymers. In the hyperglycemia condition, glucose molecules can competitively bind to Con A; a gel-sol transformation occurs in presence of free glucose; this offers a switch

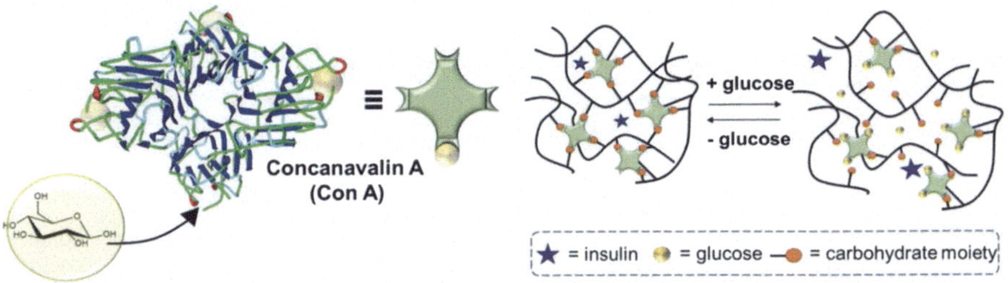

Fig. 7.9 Concanavalin A-based cross-linked networks with response to free glucose via a competitive binding equilibrium. (Adapted from Michael and Webber 2019)

for controlling the release of the insulin from immobilized hydrogels as per Fig. 7.9 (Michael et al. 2019). Con A's major drawbacks are poor biocompatibility and instability, which result in less glucose sensitivity and adverse cytotoxicity and immunogenicity (Yin et al. 2010). To avoid these problems, Con A has been covalently cross-linked with the hydrogel matrix against escape from human bodies (Miyata et al. 2004). Recent studies reported the glucose-binding capacities of glucose transporter (Wang et al. 2017) and aptamer (Yan et al. 2013) for the glucose-responsive systems for insulin delivery applications.

Lina et al. (2019) prepared covalently adopted pullulan-based glucose-sensitive hydrogels consisting of carboxylated pullulan and Con A (Con A-CPUL) by a 1-ethyl-3-(3-dimethylaminopropyl) carbodiimide/N-hydroxysuccinimide (EDC/NHS) activation method. The Con A-CPUL hydrogels were loaded with insulin by absorption method. The in vitro controlled release of insulin from the glucose sensitivity of Con A-CPUL hydrogels was studied. These hydrogels might be used in closed-loop systems and controlled release of insulin for diabetes therapy.

7.2.8 Dual-Responsive Hydrogels

Dual-responsive hydrogels respond to two different stimulus conditions and are demonstrated as a promising material in recent years. Recently we reported a designed dual-responsive NaAla-g-PAPA-cl-(PNIPAAm-co-PAAm) hydrogels from NaAla-g-PAPA polymer NIPAM and AAm for the releasing of anticancer drug (Jalababu et al. 2018). Here the mechanical strength of sodium alginate is strengthened by grafting the N-acryloyl-L-phenylalanine on the backbone of the sodium alginate and further made it dual responsive through polymerization of this grafted polymer with poly(N-isopropylacrylamide) [PNIPAAm] (thermosensitive polymer). The -COOH groups present on sodium alginate made the hydrogel to respond to pH. It is observed that hydrogels released more amount of drug at pH 7.4 than pH 1.2, and released data indicate that % cumulative release of drugs is more at 37 °C than 25 °C. The mechanism behind the pH and temperature responsiveness is discussed in earlier sections of this chapter.

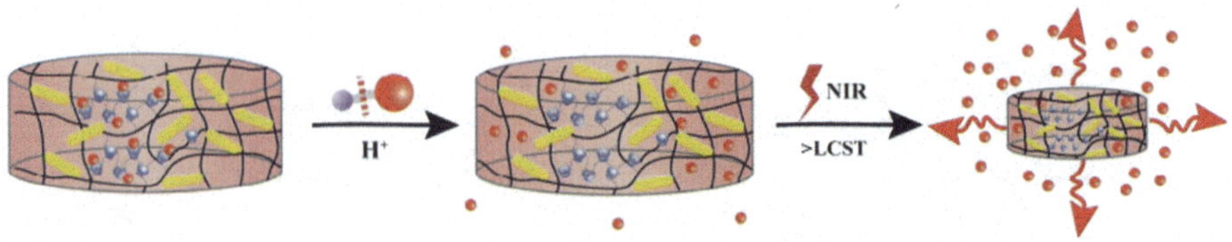

Fig. 7.10 Illustration of dual-responsive pH/NIR for release of drug in cancer therapy. (Adapted from Xu et al. 2017)

In another study, a hydrogel that can respond to both pH and NIR (near infrared) is designed to release adamantane-modified doxorubicin prodrug using N-isopropylacrylamide (NIPAm) and methacrylated β-cyclodextrin-based macromer (MPCD) hydrogel in the presence of gold nanorods (Xu et al. 2017). Here, doxorubicin is conjugated with the adamantyl group through the hydrazone bond (acid-responsive), which is in acidic tumor condition, release the drug by breaking the hydrazone bond. Presence of gold nanorods and PNIPAm made the hydrogel to respond to NIR light and accelerate the drug release on demand, as shown in Fig. 7.10.

Wang et al. (2019a) designed magnetic-thermosensitive hydrogels by integrating vinyl $Fe_3O_4@SiO_2$ in nanoparticles in the hydrogel made up with N-isopropylacrylamide-co-acrylamide copolymers for controlled release of chelerythrine. Here, magnetic vinyl-$Fe_3O_4@SiO_2$ shows good cross-linking ability and water dispersibility and accelerates the drug release in magnetic field presence. The drug release is controlled by applying the magnetic field every hour in 2 h through a switch on/switch off mode. When the magnetic field is applied, it increases the hydrogel's temperature and releases more amount of the drug. The presence of N-isopropylacrylamide-co-acrylamide made the hydrogel thermoresponsive; drug release experiment was conducted at 25, 37, and 45 °C, and more amount of drug release is observed at 45 °C (>LCST) when compared to 25 and 37 °C (<LCST). Here LCST of hydrogel was found as 41 °C. When LCST is more than operating temperature due to the hydrogel's shrinking, more amount of drug is released (Sun et al. 2021). Different dual-responsive hydrogels and their applications are presented in Table 7.4

7.2.9 *Multi-responsive hydrogels*

Multi-stimuli hydrogels have been created as responsive drug delivery systems in different environmental conditions. Various factors affect drug delivery in the human body due to variations in physiological or pathological conditions such as pH, temperature, enzymes, and glucose concentrations. Multi-stimuli polymers are required for the synthesis of smart multiple responsive hydrogel networks (Siangsanoh et al. 2018).

Table 7.4 Examples of various dual-responsive hydrogels

S. no	Dual response	Applications	Authors
1	pH-magnetic	Cancer therapy	Zhou et al. (2018)
2	Light-reductant	Drug delivery	Li et al. (2015)
3	Light-magnetic	Antitumor therapy	Pham et al. (2020)
4	Glucose-pH	Wound healing	Zhao et al. (2017)
5	Light-temperature	Photodynamic therapy	Belali et al. (2018)
6	Enzyme-pH	Drug delivery	Kumar and Nayak (2019)
7	Oxidation-thermo	Drug delivery	Cheng et al. (2015)

Zhang et al. (2017) have developed novel multi-responsive (pH, glucose & temperature) biocoating nanogels for specific site delivery of anticancer drug doxorubicin. Biocoating nanogels were synthesized in four steps. In the first step, nanogels were prepared utilizing 90% N-isopropylacrylamide, 5% glycosyloxyethyl methacrylate, and 5% N,N′-methylenebis(acrylamide) via free radical precipitation polymerization method. In the second step, concanavalin A (Con A) was connected to the nanogel surface through its interaction with the glucosyl groups of glycosyloxyethyl methacrylate. In the third step, glycogen was subjected to the nanogels, glued to the Con A by sugar-lectin-based biorecognition. Finally, in the fourth step, the Con A and glycogen coverage nanogels are coated over with transferrin, a similar glycoprotein–lectin binding mechanism; these coated nanogels are called NG-CGT. These NG-CGT nanogels are loaded with doxorubicin, which is pH-, glucose-, and temperature-responsive in doxorubicin delivery. Schematic representation of the synthesis of biocoating NG-CGT nanogels and Dox loading/release process is shown in Fig. 7.11

7.2.9.1 pH Responsivity of NG-CGT Nanogels

Dox release studies of pH stimuli NG-CGT nanogels were studied over 4.00–7.40 pH range. At physiological pH (7.40), 1.8 ppm after 20 h and 2.1 ppm after 2 days, Dox drug was released from NG-CGT nanogels. When the solution's pH was reduced to 4.00, the Dox release rate intensely raised; after 20 h, 7.2 ppm of Dox was released, whereas 6.5 ppm of Dox was released at pH 5 in the same conditions. The pH-responsive behavior of NG-CGT nanogels can be attributed by the collapse of the biocaoting of nanogel at low pH, which results in opening of the pores of nanogels and release of the encapsulated Dox from endocytosed nanogels.

7.2.9.2 Glucose Responsivity of NG-CGT Nanogels

Zhang et al. (2017) evaluated glucose-responsive nature of NG-CGT nanogels for Dox release. When the glucose concentration rose from 400 to 1200 ppm, the Dox release quantity also increased from 1.8 to 7.6 ppm. The resultant drug release

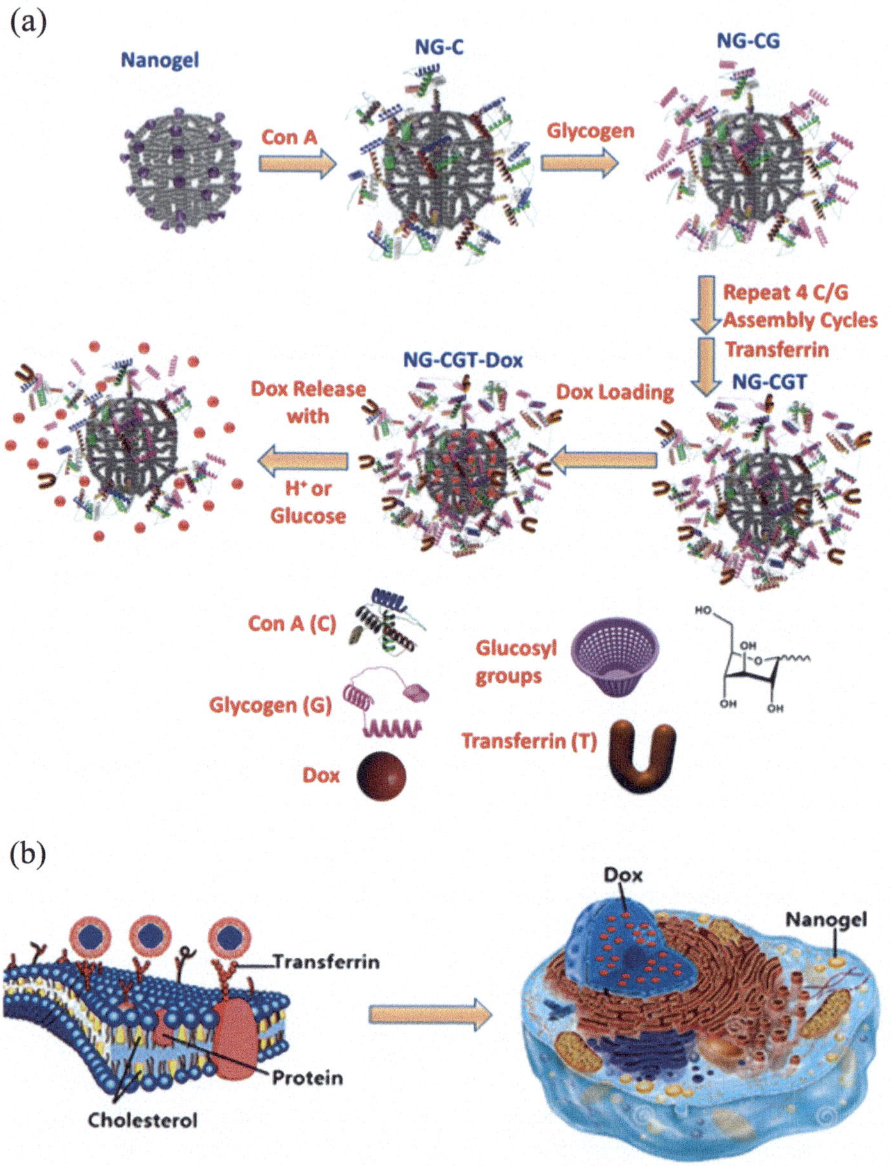

Fig. 7.11 (**a**) Schematic representation of synthesis of biocoating NG-CGT nanogels and Dox loading/release process. (**b**) The targeting/delivery process of Dox from nanogel. (Adapted from Zhang et al. 2017)

studies explained that glucose could bind to Con A via multiple hydrogen bond interactions. In presence of glucose, the biocoating of the nanogel becomes more porous, and Dox is released from the nanogel network. When the concentration of glucose increases, the amount of Dox also increases.

7.2.9.3 Temperature Responsivity of NG-CGT Nanogels

The thermal responsive nature of NG-CGT nanogels is explained that lower critical solution temperature (LCST) of NG-CGT nanogels (36 °C) is higher than that of pure N-isopropylacrylamide (32 °C). The LCST of NG-CGT gels is greater than body temperature (37 °C), making them an ideal system for Dox delivery at physiological temperature. At more than LCST, significant changes in volume and size due to the collapse of nanogels could squeeze out the drug from NG-CGT nanogels. Finally, the temperature response of NG-CGT nanogel Dox release was also determined. Doxorubicin was released to the nucleus of the human liver cancer cell line (HepG2) within 3 h. Doxorubicin-loaded NG-CGT nanogels exhibit a healthy growth inhibition ability toward HepG2 (Zhang et al. 2017).

Ding et al. (2019) synthesized UCST-type poly(acrylic acid-co-acrylamide) [P(AA-co-AM)] hydrogels, which combine with SiO_2 and polypyrrole (PPy) for the construction of smart drug delivery hydrogels [SiO_2@PPy@P(AA-co-AM)] for light-, thermo-, and pH-responsive release of 5-fluorouracil (5-FU) anticancer drug.

7.2.9.4 Photo and Thermal Responsive Ability of SiO_2@PPy@P(AA-co-AM) Hydrogels

In SiO_2@PPy@P(AA-co-AM) hydrogels, PPy acts as a photothermal transducer (conversion of light into heat) by the NIR irradiation and leads to the thermal swelling of hydrogels and release of the drug molecules from the hydrogel. The cumulative release of 5-FU from SiO2@PPy@P(AA-co-AM) is estimated only 56.6 % after 6 h with exposure to NIR irradiation. The cumulative release of 5-FU from the same hydrogels is 83.9% with the NIR irradiation after the same time (6 h). Higher cumulative release of 5-FU with NIR irradiation could be attributed by light to thermal conversion ability of PPy present in the hydrogel and the characteristic UCST performance of P(AA-co-AM). The hydrogel's thermal change leads to rise the temperature and rupture of the hydrogen bonds between acrylic acid and acrylamide units in P(AA-co-AM). Schematic representation of SiO_2@PPy@P(AA-co-AM) hydrogel preparation and multi-responsive release of 5-FU through the action of light, thermo, and pH is shown in Fig. 7.12.

7.2.9.5 pH-Responsive Ability of SiO_2@PPy@P(AA-co-AM) Hydrogels

SiO_2@PPy@P(AA-co-AM) hydrogels' pH-responsive performance was estimated by carrying the drug release experiments at 1.2, 5.8, and 7.4 pH values. The cumulative release of 5-FU is computed as 17.4%, 46.3%, and 56.6% at pH 1.20, 5.80, and 7.40, respectively, after 6 h. At pH 1.2, the formation of hydrogen bonds among the -COOH groups of acrylic acids in P(AA-co-AM) and the shrinkage of the hydrogels obstruct the release of 5-FU from the SiO2@PPy@P(AA-co-AM) hydrogels. At pH 7.4, ionization of -COOH groups on acrylic acid into -COO leads to

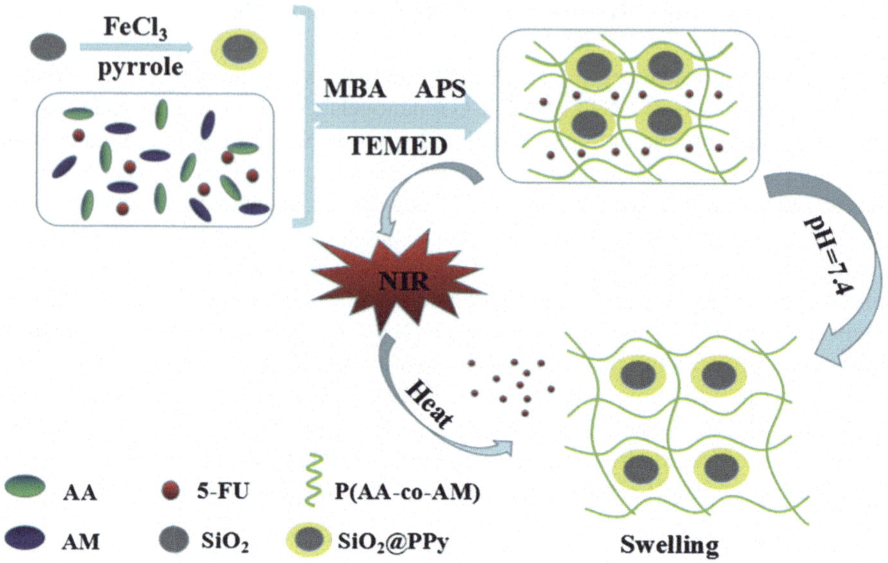

Fig. 7.12 Schematic representation of SiO_2@PPy@P(AA-co-AM) hydrogel preparation and multi-responsive release of 5-FU through the action of light, thermo, and pH. (Adapted from Ding et al. 2019)

increasing swelling ratio of hydrogels due to anion-anion repulsion. The cumulative release of 5-FU from SiO_2@PPy@P(AA-co-AM) hydrogels is higher at pH values.

Wang et al. (2010) designed and developed a novel pH, thermo, and glucose triple responsive (2-dimethylamino)ethyl methacrylate and 3-acrylamidophenylboronic acid hydrogel by free radical polymerization method. Release behavior of bovine serum albumin from hydrogel matrix affected by pH, temperature, and glucose concentration at physiological pH was studied.

Huang et al. (2012) fabricated magnetic and triple responsive characteristics of Fe_3O_4/poly(3-acrylamidophenylboronic acid-co-(2-dimethylamino)ethyl methacrylate) hydrogels and semi-interpenetrated by b-cyclodextrin-epichlorohydrin via radical polymerization. These magnetic hydrogels' swelling behavior was affected by pH, glucose concentration, temperature, and magnetic field. The hydrogels efficiently target the site and release quercetin in a controlled manner by adjusting temperature, pH value, magnetic field, and glucose concentration.

7.3 Conclusion

Researchers pitched more interest toward smart hydrogels in drug delivery and other biomedical applications because of altering their network structure and swelling capacity in response to specific stimuli, either exogenous stimuli (temperature, magnetic field, light, or electric) or endogenous stimuli (pH or enzyme). Initially, single responsive hydrogels were designed to release the drug at specific stimuli conditions using both natural and synthetic polymers, and now studies are extended to design dual and multi-responsive hydrogels. This chapter summarized the basic principle behind each various stimuli-responsive hydrogel and their drug delivery applications. The presence of acidic or alkaline groups in hydrogel made them pH-responsive. Most of the thermoresponsive hydrogels are prepared by inserting N-isopropylacrylamide, a thermoresponsive monomer in the hydrogel network. Electro-responsive hydrogels made with electroactive polymers release the drug by changing the pH of the hydrogel's atmosphere. Enzyme-responsive hydrogels are very selective and designed by using specific cross-linkers in hydrogel networks and release the drug as per the enzyme action. Photoresponsive hydrogels are prepared by inserting active photo moieties in hydrogel networks. Hydrogels are made to respond to the magnetic field by inserting the magnetic nanoparticles into the hydrogel networks. Glucose-responsive hydrogels control the glucose through different mechanisms, i.e., by oxidation of glucose, by inserting the phenylboronic acid (PBA) in hydrogels, and by coupling with glucose-binding materials. Upcoming research attempts may emphasize the advancement of rational hydrogels in tissue engineering and regenerative medicine.

References

Aguila M, Elvir C, Gallard A, Vazque B, Roman J (2007) In: Ashammakhi N, Reis R, Chiellini E (eds) Smart polymers & their applications as biomaterials, Topics in tissue engineering, vol 3

Alfimove MV, Fedorova OA, Gromov SP (2003) Photo switchable molecular receptors. J Photochem Photobiol A 158(2-3):183–198

Almeida H, Amaral MH, Lobao P (2012) Temperature & pH stimuli-responsive polymers and their applications in controlled and self-regulated drug delivery. J Appl Pharm Sci 2(6):1–10

Belali S, Savoie H, O'Brien JM, Cafolla AA, O'Connell B, Karimi AR, Boyle RW, Senge MO (2018) Synthesis and Characterization of Temperature-sensitive and Chemically Cross-linked Poly(Nisopropylacrylamide)/Photosensitizer Hydrogels for Applications in Photodynamic Therapy. Biomacromolecules 19(5):1592–1601

Bertoglio P, Jacobo SE, Daraio ME (2010) Preparation and characterization of PVA films with magnetic nanoparticles the effect of particle loading on drug release behaviour. J Appl Polym Sci 115(3):1859–1865

Billah SMR, Mondal MIH, Somoal SH, Pervez MN, Haque MO (2019) Enzyme – responsive hydrogels. In: Cellulose based superabsorbent hydrogels, pp 309–330

Chandrawati R (2016) Enzyme-responsive polymer hydrogels for therapeutic delivery. Exp Biol Med 241(9):972–979

Cheng X, Jin Y, Sun T, Qi R, Fan B, Li H (2015) Oxidation and thermo-responsive poly (N-isopropylacrylamide-co-2-hydroxyethyl acrylate) hydrogels cross-linked via diselenides for controlled drug delivery. RSC Adv 5(6):4162–4170

Chiang CY, Chu CC (2015) Synthesis of photo responsive hybrid alginate hydrogel with photo-controlled release behavior. Carbohydr Polym 119:18–25

Das D, Pal S (2015) Modified biopolymer-dextrin based crosslinked hydrogels: application in controlled drug delivery. RSC Adv 5(32):25014–25050

Ding C, Guo Z, Xiong J, Wu D, Tao Y, Qin Y, Kong Y (2019) Rational design of a multi-responsive drug delivery plat form based on SiO_2@PPy@poly (acrylic acid-co-acrylamide). React Funct Polym 137:88–95
Dong Y, Wang W, Veiseh O, Appel EA, Xue K, Webber MJ, Tang B, Yang XW, Weir W, Langer R, Anderson DG (2016) Injectable and glucose-responsive hydrogels based on boronic acid–glucose complexation. Langmuir 32(34):8743–8747
Dowlut M, Hall DG (2006) An improved class of sugar-binding boronic acids, soluble and capable of complexing glycosides in neutral water. J Am Chem Soc 128(13):4226–4227
Duan JJ, Zang LN (2017) Robust and smart hydrogels based on natural polymers. Chin J Polym Sci 35:1165–1180
Ekici S, Saraydin D (2017) Temperature-sensitive ternary interpenetrating polymeric networks for potential gastrointestinal drug release. J Drug 1(3):1–6
El-Sawy NM, Raafat AI, Badawy NA, Mohamed AM (2020) Radiation development of pH-responsive (xanthan-acrylic acid)/MgO nanocomposite hydrogels for controlled delivery of methotrexate anticancer drug. Int J Biol Macromol 142:254–264
Finotelli PV, Silva DD, Penna MS, Rossi AM, Farina M, Andrade LR, Takeuchi AY, Leao MHR (2010) Microcapsules of alginate/chitosan containing magnetic nanoparticles for controlled release of insulin. Colloids Surf B Biointerfaces 81(1):206–211. https://doi.org/10.1016/j.colsurfb.2010.07.008
Frachini ECG, Petri DFS (2019) Magneto-responsive hydrogels: preparation, characterization, biotechnological and environmental applications. Rev J Braz Chem Soc 30(10):2010–2028
Gong CB, Wong KL, Lam (2008) Photo responsive molecularly imprinted hydrogels for the photo regulated release and uptake of pharmaceuticals in the aqueous media. Chem Mater 20(4):1353–1358
Guo J, Sun H, Lei W, Tang Y, Hong S, Yang H, Tay FR, Huang C (2019) MMP-8-responsive polyethylene glycol hydrogel for intraoral drug delivery. J Dent Res 98(5):564–571
Haija YMA, Ulijin RV (2014) Enzyme-responsive hydrogels for biomedical application. In: Connon CJ, Hamley IW (eds) Hydrogels in cell-based therapies. The Royal Society of Chemistry, Cambridge UK, pp 112–134
Hassan CM, Doyle FJ III, Peppas NA (1997) Dynamic behavior of glucose-responsive poly (methacrylic acid-g-ethylene glycol) hydrogels. Macromolecules 30(20):6166–6173
Hill KL, Meleties M, Katyal P, Xie X, Delgado-Fukushima E, Jihad T, Liu C-F, O'Neill S, Tu SR, Renfrew PD, Bonneau R, Wadghiri YZ, Montclare JK (2019) Thermoresponsive protein-engineered coiled-coil hydrogel for sustained small molecule release. Biomacromolecules 20(9):3340–3351
Huang Y, Liu M, Chen J, Gao C, Gong Q (2012) A novel magnetic triple-responsive composite semi-IPN hydrogels for targeted and controlled drug delivery. Eur Polym J 48(10):1734–1744
Işıklan N, Küçükbalcı G (2015) Synthesis and characterization of pH- and temperature sensitive materials based on alginate and poly (N-isopropyl acrylamide/acrylic acid) for drug delivery. Polym Bull 73:1321–1342
Jalababu R, Satyaveni S, Suresh Reddy KVN (2018) Synthesis and characterization of dual responsive sodium alginate-g-acryloyl phenylalanine-poly N-isopropyl acrylamide smart hydrogels for the controlled release of anticancer drug. J Drug Deliv Sci Technol 44:190–204
James HP, John R, Alex A, Anoop KR (2014) Smart polymers for the controlled delivery of drugs-a concise overview. Acta Pharma Sin B 4(2):120–127
Johan H, Koeslag Peter T, Saunders TE (2003) A reappraisal of the blood glucose homeostat which comprehensively explains the type 2 diabetes mellitus–syndrome X complex. J Physiol 549(2):333–346
Joshi N, Yan J, Levy S, Bhagchandani S, Slaughter KV, Sherman NE, Amirault J, Wang Y, Riegel L, He X, Rui TS, Valic M, Vemula PK, Miranda OR, Levy O, Gravallese EM, Aliprantis AO, Ermann J, Karp JM (2018) Towards an arthritis flare-responsive drug delivery system. Nat Commun 9:1275

Kim KT, Cornelissen J, Nolte R, Hest JV (2009) Polymeric monosaccharide receptors responsive at neutral pH. J Am Chem Soc 131(31):13908–13909
Kowalczyk A, Fau M, Karbarz M, Donten M, Stojek Z, Nowicka AM (2014) Hydrogel with chains functionalized with carboxyl groups as Universal 3D platform in DNA biosensors. Biosens Bioelectron 54:222–228
Krishna Rao KSV, Vijaya Kumar Naidu B, Subha MCS, Sairam M, Aminabhavi TM (2006) Novel chitosan-based pH-sensitive interpenetrating network microgels for the controlled release of cefadroxil. Carbohydr Polym 66(3):333–344
Kulkarni PS, Haldar MK, Nahire RR, Katti P, Ambre AH, Muhonen WW, Shabb JB, Padi SKR, Singh RK, Borowicz PP, Shrivastava DK, Katti KS, Reindl K, Guo B, Mallik S (2015) MMP-9 Responsive PEG cleavable nanovesicles for efficient delivery of chemotherapeutics to pancreatic cancer. Mol Pharm 11(7):2390–2399
Kumar BA, Nayak RR (2019) Supramolecular phenoxy-alkyl maleate-based hydrogels and their enzyme/pH -responsive curcumin release. New J Chem 43:5559–5567
Li Y, Huang G, Zhang X, Li B, Chen Y, Lu T, Lu TJ, Xu F (2012) Magnetic hydrogels and their potential biomedical applications. Adv Funct Mater 23(6):660–672. https://doi.org/10.1002/adfm.201201708
Li YH, Huang GY, Zhang XH, Li BQ, Chen YM, Lu TL, Lu TJ, Xu F (2013) Magnetic hydrogels and their potential biomedical applications. Adv Funct Mater 23(6):660–672
Li J, Ma L, Chen G, Zhou Z, Li Q (2015) A high water-content and high elastic dual-responsive polyurethane hydrogel for drug delivery. J Mater Chem B 3(42):8401–8409
Li X, Fu M, Wu J, Zhang C, Deng X, Dhinakar A, Huang W, Qian H, Ge L (2017) pH-sensitive peptide hydrogel for glucose-responsive insulin delivery. Acta Biomater 51:294–303
Lina K, Yia J, Maob X, Wua H, Zhangb LM, Yang L (2019) Glucose-sensitive hydrogels from covalently modified carboxylated pullulan and concanavalin A for smart controlled release of insulin. React Funct Polym 139:112–119
Linsely CS, Wu BM (2017) Recent advances in light responsive on-demand drug delivery systems. Ther Deliv 8(2):89–107
Liu TY, Hu SH, Liu KH, Liu DM, Chen SY (2006) Preparation and characterization of smart magnetic hydrogels and its use for drug release. J Magn Magn Mater 204(1):e397–e399
Liu Z, Jiao Y, Wang Y, Zhou C, Zhang Z (2008) Polysaccharides-based nanoparticles as drug delivery systems. Adv Drug Deliv Rev 60(15):1650–1662
Ma R, Shi L (2014) Phenylboronic acid-based glucose-responsive polymeric nanoparticles: synthesis and applications in drug delivery. Polym Chem 5(5):1503–1518
Mangaiyarkarasi R, Chinnathambi S, Karthikeyan S, Aruna P, Ganesan S (2016) Paclitaxel conjugated Fe_3O_4@LaF_3:Ce^{3+}, Tb^{3+} nanoparticles as bifunctional targeting carriers for Cancer theranostics application. J Magn Mat 399:207–215. https://doi.org/10.1016/j.jmmm.2015.09.084
Mariouras T, Vamvakaki M (2016) Field responsive materials: photo-, electro-, magnetic- and ultrasound-sensitive polymers. Polym Chem 8:74–96
Matsuda T, Kawakami R, Namba R, Nakajima T, Gong JP (2019) Mechanoresponsive self-growing hydrogels inspired by muscle training. Science 363(6426):504–508
Matsumoto A, Ikeda S, Harada A, Kataoka K (2003) Glucose-responsive polymer bearing a novel phenylborate derivative as a glucose-sensing moiety operating at physiological pH conditions. Biomacromolecules 4(5):1410–1416
Merino S, Martin C, Kostarelos K, Prato M, Vazques E (2015) Nanocomposite hydrogels: 3D polymer – nanoparticle synergies for on – demand drug delivery. ACS Nano 9(5):4686–4697
Michael AV, Webber MJ (2019) Biologically inspired and chemically derived methods for glucose-responsive insulin therapy. Adv Healthcare Mater 8(12):1801466
Miyata M, Jikihara A, Nakamae K, Hoffman AS (2004) Preparation of reversibly glucose-responsive hydrogels by covalent immobilization of lectin in polymer networks having pendant glucose. J Biomater Sci Polym 15(9):1085–1098
Murdan S (2003) Electro-responsive drug delivery from hydrogels. JCR 92(1-2):1–17

Oktay S, Alemdar N (2018) Electrically controlled release of 5-fluorouracil from conductive gelatin methacryloyl-based hydrogels. J Appl Polym Sci 136(1):46914
Pham S, Choi Y, Choi J (2020) Stimuli-responsive nanomaterials for application in antitumor terapy and drug delivery. Pharmaceutics 12(7):630
Qin J, Asempah L, Laurent S, Fornara A, Muller RN, Muhammed M (2009) Injectable superparamagnetic ferrogels for controlled release of hydrophobic drugs. Adv Mater 21(13):1354–1357
Qiu Y, Park K (2001) Environment-sensitive hydrogels for drug delivery. Adv Drug Deliv Rev 53(3):321–339
Qu J, Zhao X, Ma PX, Guo B (2018) Injectable antibacterial conductive hydrogels with dual response to an electric field and pH for localized "smart" drug release. Acta Biomater 72:55–69
Quershi D, Nayak SK, Maji S, Anis A, Kim D, Pal K (2019) Environment sensitive hydrogels for drug delivery applications. Eur Polym J 120:109220
Reddy NN, Mohan YM, Varaprasad K, Ravindra S, Joy PA, Raju KM (2011) Magnetic and electric responsive hydrogel magnetic nanocomposites for drug-delivery application. J Appl Poym Sci 122(2):1364–1375
Ruan C, Liu C, Hu H, Guo XL, Jiang BP, Liang H, Shen XC (2019) NIR-II light-modulated thermosensitive hydrogel for light-triggered cisplatin release and repeatable chemo-photothermal therapy. Chem Sci 10:4699–4706
Saikia AK, Aggarwal S, Mandal UK (2015) Electrically induced swelling and methylene blue release behavior of poly (N-isopropylacrylamide-co-acrylamido-2-methylpropyl sulphonic acid) hydrogels. Colloid Polym Sci 293:3533–3544
Secret E, Kelly SJ, Crannell KE, Andrew JS (2014) Enzyme-responsive hydrogel microparticles for pulmonary drug delivery. ACS Appl Mater Interfaces 6(13):10313–10321
Sharifzadeh G, Hosseinkhani H (2017) Biomolecule-responsive hydrogels in medicine. Adv Healthcare Mater 6(24). https://doi.org/10.1002/adhm.201700801
Sharpe LA, Daily AM, Horava SD, Peppas NA (2014) Therapeutic applications of hydrogels in oral drug delivery. Expert Opin Drug Deliv 11(6):901–915
Shi X, Zheng Y, Wang G, Lin Q, Fan H (2014) pH-and electro-responsive characteristics of bacterial cellulose nanofiber/sodium alginate hybrid hydrogels for the dual controlled drug delivery. RSC Adv 4(87):47056–47065
Shi W, Huang J, Fang R, Liu M (2020) Imparting functionality to the hydrogel by magnetic-field-induced nano-assembly and macro-response. ACS Appl Mater Interfaces 12(5):5177–5194
Siangsanoh C, Ummartyotin S, Sathirakul K, Rojanapanthu P, Treesuppharat W (2018) Fabrication and characterization of triple-responsive composite hydrogel for targeted and controlled drug delivery System. J Mol Liq 256:90–99
Spruell JM, Hawker CJ (2011) Triggered structural and property changes in polymeric nanomaterials. J Chem Sci 2:18–26
Suhail M, Wu P-C, Usman Minhas M (2021) Development and characterization of pH-sensitive chondroitin sulfate- co-poly (acrylic acid) hydrogels for controlled release of diclofenac sodium. J Saudi Chem Soc 25(4):101212
Sun L, Mo Z, Li Q, Zheng D, Qiu X, Pan X (2021) Facile synthesis and performance of pH/temperature dual-response hydrogel containing lignin-based carbon dots. Int J Biol Macromol 175:516–525
Timko BP, Dvir T, Kohane DS (2010) Remotely triggerable drug-delivery systems. Adv Mater 22(44):4925–4943
Ullah F, Othma MBH, Javed F, Ahmada Z, Ad AH (2015) Classification, processing and application of hydrogels: a review. Materi Sci Eng C 57:414–433
Wang B, Ma R, Liu G, Li Y, Liu X, An Y, Shi L (2009) Glucose-responsive micelles from self-assembly of poly (ethylene glycol)-b-Poly (acrylic acid-co-acrylamidophenylboronic acid) and the controlled release of insulin. Langmuir 25(21):12522–12528
Wang L, Liu M, Gao C, Ma L, Cui D (2010) A pH-, thermo-, and glucose-, triple-responsive hydrogels: synthesis and controlled drug deliver. React Funct Polym 70(3):159–167

Wang C, Ye Y, Sun W, Yu J, Wang J, Lawrence DS, Buse JB, Gu Z (2017) Red blood cells for glucose-responsive insulin delivery. Adv Mater 29(18):1606617

Wang J, Zhang H, Wang F, Ai X, Huang D, Liu G, Mi P (2018a) Enzyme-responsive polymers for drug delivery and molecular imaging. In: Stimuli responsive nanocarriers for drug delivery applications, 1st edn. Elsevier, Publisher Woodhead Publishing., pp 101–119

Wang Y, Wei G, Zhang X, Huang X, Zhao J, Guo X, Zhou S (2018b) Multistage targeting strategy using magnetic composite nanoparticles for synergism of photothermal therapy and chemotherapy. Small 14(12):e1702994. https://doi.org/10.1002/smll.201702994

Wang J, Wang Z, Yu J, Kahkoska AR, Buse JB, Gu Z (2019a) Glucose-responsive insulin and delivery systems: innovation and translation. Adv Mater 32(13):1902004

Wang Y, Zong M, Wang M, Wang L, Liu Y, Wang B, Li Y (2019b) Chelerythrine loaded composite magnetic thermosensitive hydrogels as s novel anticancer drug-delivery system. J Drug Deliv Sci Technol 54: Article 101293.

Ward MA, Georgiou TK (2011) Thermo-responsive polymers for biomedical applications. Polymers 3:215–1242

Wilson R, Turner APF (1992) Glucose oxidase: an ideal enzyme. Biosens Bioelectron 7(3):165–185

Wu WT, Mitra N, Yan E, Zhou SQ (2010) Multifunctional hybrid nanogel for integration of optical glucose sensing and self-regulated insulin release at physiological pH. ACS Nano 4(8):4831–4839

Wu Q, Wang L, Yu HJ, Wang JJ, Chen ZF (2011) Organization of glucose-responsive systems and their properties. Chem Rev 111(12):7855–7875

Xu X, Huang Z, Huang Z, Zhang X, He S, Sun X, Shen Y, Yan M, Zhao C (2017) Injectable. NIR/pH responsive nanocomposite hydrogel as long-acting implant for chemo-photothermal synergistic cancer therapy. ACS Appl Mater Interfaces 9(24):20361–20375

Yan L, Zhu Z, Zou Y, Huang Y, Liu D, Jia S, Xu D, Wu M, Zhou Y, Zhou S, Yang CJ (2013) Target-responsive "sweet" hydrogel with glucometer readout for portable and quantitative detection of non-glucose targets. J Am Chem Soc 135(10):3748–3751

Yang H, Sun X, Liu G, Ma R, Li Z, An Y, Shi L (2013) Glucose-responsive complex micelles for self-regulated release of insulin under physiological conditions. Soft Matter 9(35):8589–8599

Yin R, Han J, Zhang J, Nie J (2010) Glucose-responsive composite microparticles based on chitosan, concanavalin A and dextran for insulin delivery. Colloids Surf B 76(2):483–488

Zhang S, Chu LY, Xu D, Zhang J, Ju XJ, Xie R (2008) Poly(N-isopropylacrylamide)-based comb-type grafted hydrogel with rapid response to blood glucose concentration change at physiological temperature. Polym Adv Technol. 19(8):937–943

Zhang Q, Colazo J, Berg D, Mugo SM, Serpe JM (2017) Multi responsive nanogels for targeted anticancer drug delivery. Mol Pharm 14:2624–2628

Zhao X, Kim J, Cezar CA, Huebsch N, Lee K, Bouhadir K, Mooney DJ (2011) Active scaffolds for on-demand drug and cell delivery. Proc Natl Acad Sci USA 108(1):67–72

Zhao L, Niu L, Liang H, Tan H, Liu C, Zhu F (2017) pH and glucose dual-responsive Injectable hydrogels with insulin and fibroblasts as bioactive dressings for diabetic wound healing. ACS Appl Mater Interfaces 9(43):37563–37574

Zhou X, Wang L, Xu Y, Du W, Cai X, Wang F, Ling Y, Chen H, Wang Z, Hu B, Zheng Y (2018) A pH and magnetic dual-response hydrogel for synergistic chemi-magnetic hyperthermia tumor therapy. RSC Adv 8(18):9812–9821

Zohreh N, Hosseini SH, Pourjavadi A (2016) Hydrazine-modified starch coated magnetic nanoparticles as an effective pH-responsive nanocarrier for doxorubicin delivery. J Indus Eng Chem 39:203–209. https://doi.org/10.1016/j.jiec.2016.05.029

Zou X, Zhao X, Ye L (2015) Synthesis of cationic chitosan hydrogel and its controlled glucose-responsive drug release behavior. Chemical Engineering Journal 273:92–100

Chapter 8
Recent Trends in Preparation and Biomedical Applications of Nanocellulose-Based Hydrogels

Rajkumar Bandi, Ramakrishna Dadigala, Madhusudhan Alle, and Seung-Hwan Lee

8.1 Introduction

A hydrogel is a soft and porous 3D network substance made up of a crosslinked polymer chains that can grasp water up to 99.9%. They can be manufactured using natural or synthetic polymers and possess fascinating properties like high biocompatibility and outstanding mechanical properties (Miyata 2010). Hydrogels can be prepared in such a way to become sensitive to environmental factors such as light, pH, temperature, ionic strength, pressure, and electric and magnetic fields (Namdeo et al. 2009; Mendoza et al. 2018a). Therefore, hydrogels can be engineered for scientific and industrial applications. Hydrogels hold a great promise in biomedical applications. This is because of their ability to mimic biological, chemical, and physical properties present in animal body that are required for several biotic processes (Caliari and Burdick 2016). With this promising feature, hydrogels are finding applications as hygienic products, wound treatment, contact lenses, drug delivery, scaffolds, and diagnostic devices (Zhong et al. 2010; Willcox et al. 2010; Camci-Unal et al. 2014; Jung et al. 2017). Each implementation has its own set of specifications. Cells and organoids culture, for instance, need a mutable 3D support covered by adherence proteins that has a sol-gel conversion to allow the operation of biological sample at altered phases (Haycock 2011). Controlling mechanical properties and facile processing into complex shapes are needed for engineered

R. Bandi · R. Dadigala · S.-H. Lee (✉)
Institute of Forest Science, Kangwon National University, Chuncheon, Republic of Korea
e-mail: lshyhk@kangwon.ac.kr

M. Alle
Institute of Forest Science, Kangwon National University, Chuncheon, Republic of Korea

Department of Biomedical Science & Institute of Bioscience and Biotechnology, Kangwon National University, Chuncheon, Republic of Korea

J.-C. Kim et al. (eds.), *Smart Nanomaterials in Biomedical Applications*, Nanotechnology in the Life Sciences,
https://doi.org/10.1007/978-3-030-84262-8_8

cartilage and skin regeneration (Jayakumar et al. 2011; Fu et al. 2017). Drug delivery systems are designed to keep therapeutic agents safe until they hit their intended destination, triggering a regulated kinetics of drug release (Fenton et al. 2018).

Nevertheless, the widespread usage of hydrogels for biomedical industry has been hampered by a few key factors. When making hydrogels, eliminating the toxic polyfunctional crosslinking agents is very slow, and remnant can linger in the medium (Caló and Khutoryanskiy 2015). Furthermore, the polymers employed are mostly nonbiodegradable and nonrecyclable synthetic origins, posing a threat to the sustainability of the environment (Corkhill et al. 1989; Gyles et al. 2017). As a result, natural materials that can be easily engineered into biocompatible hydrogels are highly needed.

For water-based gels, cellulose has arisen as a convincing natural, nontoxic, and green material. Although starch and dextran are engineered by nature to store and discharge energy slowly, cellulose is used by plants and bacteria for structural purposes (Neely 1961; Copeland et al. 2009). Glucose is the monomer of both of these polysaccharides. The molecular weight, shape, and bonds that connect their monomers are the key differences between them. Amylose has only α-1,4 glycosidic bonds, while amylopectin has a mixture of α-1,4 and α-1,6 glycosidic bonds and some α-1,3 bonds additionally present in dextran. In cellulose, there is only β-1,4 linkage (Hoover 2001). The chain structure of glycosidic polymers and their hydrogen bonding power, as well as their hierarchical like arrangement, affect these structures' solubility in water, mechanical properties and its configuration; these are all governed by different above-described bonds (Dufresne 2013). Cellulose (colloid) is insoluble in water, while dextran and starch dissolve into polymeric solutions. Cellulose gels have a statistical network of meshes of varying diameters and rigidities, as well as various crosslinking densities (Moberg et al. 2017; Curvello et al. 2019). Nanocellulose hydrogels have been shown to be useful for a variety of biomedical applications due to their biocompatibility, nontoxicity, and stability.

This book chapter accumulates the recent progress of nanocellulose hydrogels in various fields of applications and engineering of their properties. It recognizes the distinct properties of nanocellulose gels as they interact with biomolecules and cells. At first, this book chapter discusses about how nanocellulose hydrogels are formed and their chemical and physical properties in relation to biomedical applications. Next, recent discoveries and important concerns related to the use of nanocellulose-based hydrogels in biomedical applications are detailed. The aim of this book chapter is to wholly explore the biomedical applications of nanocellulose hydrogels.

8.2 Formation of Nanocellulose-Based Hydrogels

8.2.1 Characteristics of Nanocellulose

In 1951, Beng Randy first defined nanocellulose which appeared as bunch of cellulose molecules making micelles in aqueous solutions (Rånby 1951). Several researchers have explored the preparation and properties of nanocellulose in the last decades and to mention some specific inventions are De France, Dufresne, and Isogai (Saito and Isogai 2004; Dufresne 2013; De France et al. 2017a). Nanocellulose can be extracted from plants and bacteria. Extraction from plants first involves chemical or enzymatic treatments to eliminate lignin and hemicellulose, finally mechanical treatment to decompose the cellulosic hierarchical structure (Kwon et al. 2020). The term "nanocellulose" is broadly used to refer the three major types: cellulose nanofibrils (CNF), cellulose nanocrystals, (CNC) and bacterial nanocellulose (BNC) (Alle et al. 2020a). TEMPO-mediated oxidation is most popular method for CNF extraction from wood pulp, which was established by Isogai, and by this process several nanometers to microns length fibers can be produced (Saito and Isogai 2004). The most common method of CNC preparation involves the hydrolysis of cellulose pulp using strong acids such as sulfuric acid and phosphoric acid. This process completely dissolves the amorphous regions of the fibers and leaves the crystalline parts (Habibi et al. 2010). CNC will appear as rod-like structures with width of 5 nm and lengths of 20–100 nm. However, there will be some variations in the sizes of both CNC and CNF depending on the cellulose source and treatment process. Apart from biomedical applications, CNF and CNC also have wide applications including sensors (Alle et al. 2020d; Bandi et al. 2021) and catalytic support (Bandi et al. 2020; Alle et al. 2020c, 2021). BNC can be produced from various bacterial strains like *Gluconacetobacter xylinus*, *Agrobacterium*, *Salmonella*, *Komagataeibacter*, *Sarcina*, *Rhodobacter*, *Rhizobium*, *and Escherichia* as extracellular secretion. Among all, *Gluconacetobacter xylinus* is widely used bacteria for BNC production (Iguchi et al. 2000; Müller et al. 2013). Irrespective of their source, all nanocellulose types can make hydrogels. This book chapter mainly focuses on plant-based nanocellulose.

8.2.2 Mechanism of Hydrogel Formation

In comparison with other material hydrogels, the nanocellulose-based hydrogels have a distinct difference, i.e., the nanocellulose hydrogels is water insoluble whereas other hydrogels are water soluble. This leads to the formation of a colloidal suspension when CNF or CNC is added to water (Mendoza et al. 2018b). However, the other gels form a solution in water. This property causes two major implications. The first one is, compared to other gels, the critical dimension of the constitutive unit and order of magnitude higher for nanocellulose (De France et al. 2017a). The second one is that the mode of gel networking varies. In nanocellulose hydrogels, the formation of gel network primarily involves the entanglement of fibrils together with electrostatic stabilization, and further strengthening of the gel structure can be achieved by chemical crosslinking (Shen et al. 2015).

Electrostatic interactions, hydrogen bond, van der Waal, hydrophobic, ionic interactions, and chain entanglements come under the category of physical interactions. These are reversible and as formed hydrogel can be disrupted easily upon the disintegration of the interacting force between fibers (Dufresne 2013). For instance,

CNF produced from TEMPO-mediated oxidation comprises anionic –COOH groups, and CNC produced from sulfuric acid treatment possesses anionic $-SO_3$ groups. Due to these charged groups, CNF and CNC can be crosslinked by interactions that are formed as a balance between van der Waal attraction and electrostatic repulsion (Saito and Isogai 2004; Oksman et al. 2006). In some cases, external crosslinkers were used for hydrogel formation by providing hydrophobic or electrostatic interactions and covalent crosslinking. For instance, sodium citrate was used as an electrostatic crosslinker to form CNF hydrogels in an acetic acid medium (Shu et al. 2001). Crosslinking networks formed by chemical/covalent bonding are durable and stable. Covalent crosslinking of cellulose fibers can be achieved by various chemical crosslinkers such as epichlorohydrin (ECH), glutaraldehyde, citric acid, metal ions, and succinic anhydride (Demitri et al. 2008; Navarra et al. 2015).

Both physical and chemical crosslinking can produce a mesh-like structure with fibers/chains through macrostructures which can be seen in Fig. 8.1. The point at which two separate chains are attached is called reticulation point. An individual chain also connects to itself which leads to the formation of a closed loop. Free-end sections of chains are termed as dangling ends. The actual difference between physical and chemical crosslinking can be found when looking into the hydrogel microstructure. Physical crosslinking of two or more chains leads to the formation of helical structures. Likewise, entanglement of fibers will result in crystalline microstructures that differ in complexity and size (Mendoza et al. 2018c). These kinds of interactions are generally observed in natural polymer containing hydrogels

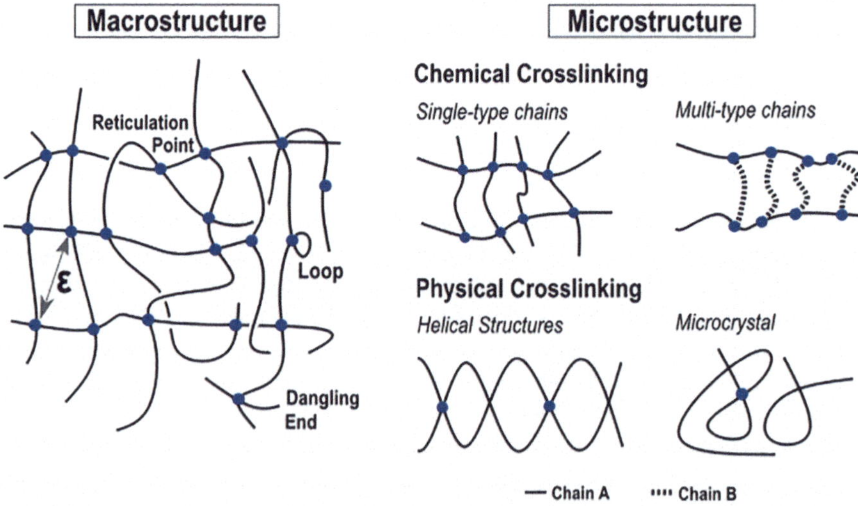

Fig. 8.1 The hydrogel gel macrostructure is formed by a well-defined mesh, where fibers connect at the reticulation point; the self-connection of fibers forms loops. Free terminal segments are named dandling ends. The pore size (ε) is directly proportional to the fiber concentration and inversely proportional to the crosslinking density. The microstructure of hydrogels is based in organized mesh for those chemically crosslinked and differs by the types of chains/fibers. Helical and microcrystalline microstructures are found in physically crosslinked hydrogels. (Adopted from Curvello et al. 2019)

(Mendoza et al. 2018a). In contrast to physical, the microstructures of chemical crosslinking generally vary in interaction between individual or different type polymers and form a uniform or composite gel, respectively. The crosslinking density (CD) defines the porosity of the hydrogel which is independent of the type of crosslinking (Bae et al. 2013). Various crosslinking densities create flexible spaces inside the hydrogel structures, and CD represent modifications in pore size, propagation of hydrogel, and correspondence length. Dissemination of various molecular substances such as drugs, proteins, and gases present in the hydrogel is highly influenced by the pore size (Li et al. 2016). Furthermore, the number of interconnected chains affects the hydrogel's swelling ratio, shear modulus, and its properties (Bae et al. 2013; Li et al. 2016). Therefore, an appropriate crosslinking method used for hydrogel preparation will determine its structure and application.

8.2.3 Crosslinking Nanocellulose

Due to the strong intra- and intermolecular hydrogen bonding, it is difficult to dissolve cellulose chains in a solvent and encourage crosslinking for hydrogel construction. Dissolution of cellulose is controlled by entropy which favors solubility and hydrogen bonding interactions which opposes it. Hence, for efficient cellulose dissolution, hydrogen bonds breakdown is essential (Lindman et al. 2017).

For the dissolution of cellulose in water or organic solvents, various chemical techniques and modification protocols have been established (Heinze and Koschella 2005; Raghuwanshi et al. 2018). Cai et al. investigated the ability of NaOH/urea, and LiOH/urea aqueous solutions can dissolve cellulose below −10 °C (Cai and Zhang 2005). Cellulose was successfully dissolved by both alkali and urea hydrates with urea delivering better results at a concentration of 12%. After dissolution, an optically transparent hydrogel can be prepared by crosslinking the regenerated cellulose with epichlorohydrin ECH at −3 °C (Chang et al. 2010). Liu et al. prepared a series of nanocellulose hydrogels by crosslinking TEMPO-oxidized CNF with three different hemicelluloses, namely, xylan, xyloglucan (XG), and galactoglucomannan (GGM). For this, pre-sorption and in situ hemicellulose sorption methods were used and compared. In pre-sorption, hemicellulose adsorption on CNF surface was done via hot mixing and swelling (by water addition). In in situ sorption, the hydrogel structure was defined by simultaneously carrying out the hemicellulose adsorption and swelling water. Out of the three hemicelluloses, XG crosslinker exhibited the greatest sorption potential onto CNF using pre-sorption method (Liu et al. 2016).

In an another study, Zander et al. used metal ions as crosslinkers to create hydrogels with improved strength and stiffness as well as thermal and optical properties (Zander et al. 2014). Metal ions like Ca^{2+}, Zn^{2+}, Cu^{2+}, Al^{3+}, and Fe^{3+} were used for hydrogel preparation. Rapid gelation was observed upon adding metal ions to the CNF suspension. The repulsive charges can be screened by cation-carboxylate interactions causing in the CNF gelation through ionic crosslinking (Dong et al. 2013). In a similar work, Yang et al. created polyacrylamide hydrogels reinforced

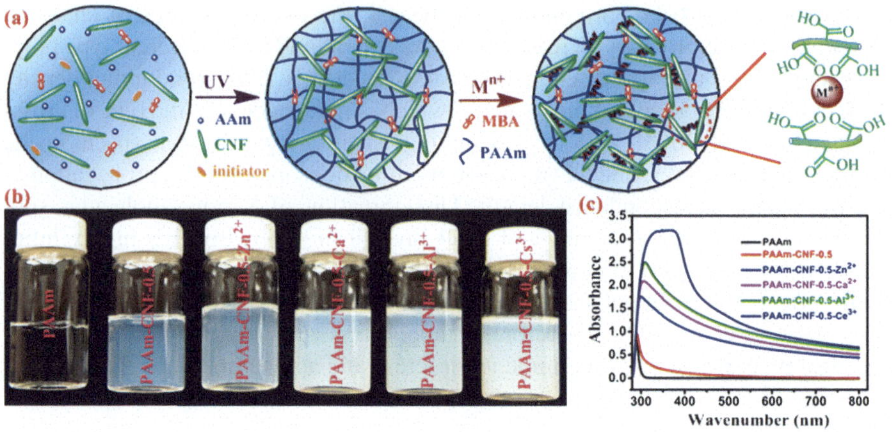

Fig. 8.2 (**a**) Schematic illustration of ionic hydrogel synthetic process that includes the in situ polymerization to form composite hydrogels followed by immersion in a cation solution to produce ionic coordination. (**b**) Ionic coordination leads to CNF aggregation in ionic gels of increased opacity and (**c**) corresponding absorbance in UV/vis spectra. (Adopted from Yang et al. 2017)

by CNF and submerged in cation solution (Ca^{2+}, Zn^{2+}, Al^{3+}, and Ce^{3+}) (Fig. 8.2) to form ionic-covalent crosslinked hydrogels (Yang et al. 2017). Metal ions form ionic bonds with CNF, while polyacrylamide contributes to covalent bonds. The hydrogel tensile strength and hardness improved with the type of metal ions in the order of $Zn^{2+}<Ca^{2+}<Al^{3+}<Ce^{3+}$. Hydrogels with improved mechanical properties resulted from the strong ionic-covalent bonds.

Wide variety protocols are available for the preparation of nanocellulose hydrogels, and the abovementioned examples are some of the representative studies. The hydrogel network structure is mainly determined by its preparation, and the pore size distribution is determined by the mesh size. The amount of crosslinking is controlled by fiber concentration, and length, as well as crosslinking points uniformity, affects both structure and shape of the hydrogel. Nanocellulose hydrogels normally appear as translucid, but at greater solid content it appeared as turbid, and it is colorful in the existence of additives like ions, polymers, copolymers, and proteins. Selecting the utmost suitable methods for nanocellulose hydrogels preparation is governed by aimed end prerequisites and its application (Curvello et al. 2019).

8.3 Biomedical Applications

8.3.1 Drug Delivery

Drug delivery systems are a group of bioengineered devices that help therapeutic agents get to their target sites, such as tissues and organs. For the specific and regulated release of drugs and biomolecules, such a device includes both carrier vesicles and coating treatments (Allen and Cullis 2004; Madhusudhan et al. 2014; Alle et al. 2020b). These changes are mostly intended to improve pharmacokinetics and optimize drug biodistribution throughout the human body. Micelles are the most commonly studied drug delivery method; they offer a hydrophobic or hydrophilic state that improves the solubility of drugs in biological fluids (Langer 1990; Kim et al. 2021). Drugs encapsulated in synthetic or natural shells, on the other hand, are protected from premature decay and have fewer side effects on non-targeted tissues (Zelikin et al. 2016). Until reaching the target location in the body or organ, the nanocellulose hydrogel (as micelle or encapsulation) must be sufficiently stable for a period of time and unaffected by changes in pH, ionic strength, or temperature. These factors, on the other hand, are critical for the safe release and degradation of drugs while avoiding unintended organ damage (Hoare and Kohane 2008). Because of their biodegradability, biocompatibility, low toxicity, and high affordability, biopolymers have been widely researched as materials for drug delivery systems (Lee et al. 2019).

8.3.1.1 CNF in Drug Delivery

Laurén et al. prepared an injectable CNF hydrogel for in vivo drug delivery (Laurén et al. 2014). They employed technetium-99m-NFC labelling to monitor the in vivo location of the hydrogel (Fig. 8.3). The hydrogel was applied for in vivo drug delivery where the localization of hydrogel and drug delivery can be tracked or monitored as a function of time. The hydrogel proved great potential for regulated release or localized delivery of large molecules. Interestingly, after function this biocompatible CNF hydrogel can be potentially break down into glucose via locally administrating cellulose-degrading enzymes.

In an interesting research by Paukkonen et al., the influence of freeze-drying and succeeding rehydration on rheological characteristics and drug release behavior of anionic nanofibrillar cellulose (ANFC) hydrogels were studied (Paukkonen et al. 2017). ANFC were prepared by the TEMPO oxidation. It was found that the release behaviors of loaded molecules were alike before and after freeze-drying indicating that the freeze-drying has no significant effect. This is highly advantageous for manufacturing pharmaceutical formulations because the dry state of the aerogel will extend the life span of hydrolysis-sensitive molecules.

In recent times, CNF-based stimuli-responsive hydrogels gained significant attention with their on-demand drug releasing nature in response to specific stimuli like pH, temperature, and ionic strength. For example, Masruchin et al. described the fabrication of dual responsive composite hydrogels by combining TEMPO-oxidized CNF and thermally responsive PNIPAM intended for drug release (Masruchin et al. 2018). They observed that the pH response nature of the hydrogels was attained by regulating the carboxyl charge content of the CNF during TEMPO oxidation. The hydrogel swelling ratio can be altered by regulating the temperature, and the hydrogels were enlarged and transparent below the LCST and contracted and opaque above the LCST. In a study by Zhang et al., pH-responsive gel

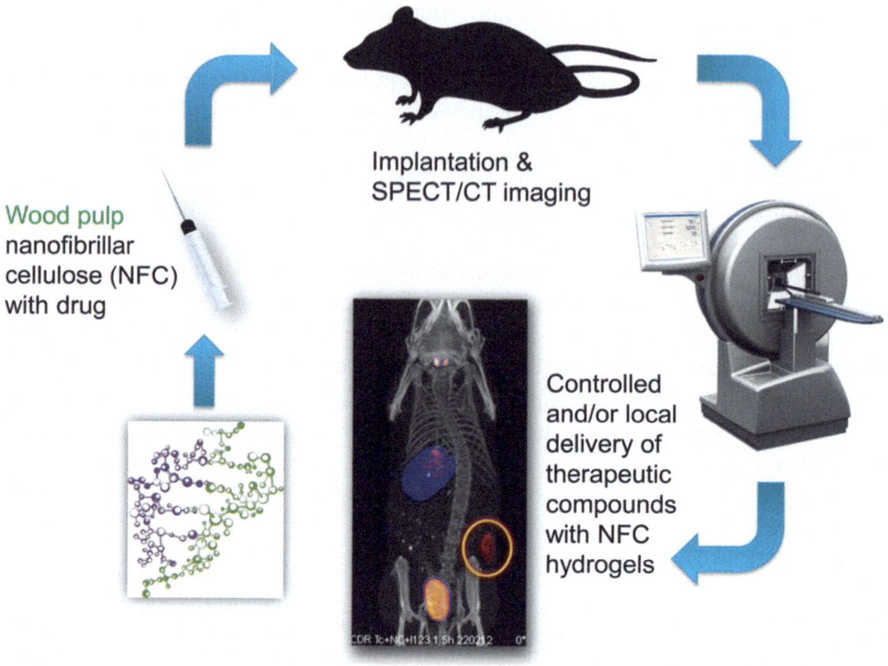

Fig. 8.3 Flow chart of technetium-99m-labeled CNF hydrogels for in vivo drug release. (Adopted from Laurén et al. 2014)

macrospheres were prepared by crosslinking the mixture of sodium alginate and TEMPO-oxidized CNF using $CaCl_2$ solution and evaluated for potential intestinal delivery of probiotics (Zhang et al. 2018). As formed gel macrospheres exhibited decent stability in simulated gastric fluid, this is confirming that the embedded probiotics are best protected in an acidic environment. In contrary, the gel macrospheres can swell and release the encapsulated probiotics in simulated intestinal fluid. Therefore, the gel macrospheres can be considered as a promising strategy for intestinal-targeted delivery of probiotics. Using a LbL process, Paulraj et al. described the manufacturing of bioinspired microcapsules using apple pectin (AP), xyloglucan (XyG), and cationic CNF (Paulraj et al. 2018). The capsules were found to have stimuli-responsive (ON/OFF) penetrability and biocompatibility. Furthermore, the nontoxic microcapsules were found to promote cell growth in a live cell staining experiment. Lastly, the researchers speculated that the microcapsules have potential bio-applications like colon targeted delivery in the intestine.

8.3.1.2 CNC in Drug Delivery

Li et al. studied the long-time delivery of basic fibroblast growth factor from CNC-reinforced collagen scaffolds under in vitro and in vivo conditions (Li et al. 2015). First, they prepared the eco-friendly gelatin microspheres (GMs) by incorporating basic fibroblast growth factor (bFGF). These GMs were then combined with a porous collagen scaffold reinforced with 5 wt% CNC. In vitro release of bFGF from these scaffolds was studied by ELISA. Further studies with human umbilical vein endothelial cells revealed the high biocompatibility of scaffolds, and then significant improvement in the cell proliferation was observed with scaffolds containing bFGF. In in vivo studies by internal embedding in rats, formation of a more amount of mature (CD31+ and αSMA+) blood vessels was observed. These results designated the potential of collagen/CNCs/bFGF-GMs scaffold in promoting angiogenesis. Mauricio et al. employed the covalent microhydrogel produced from starch and CNW as a drug delivery system (Mauricio et al. 2015). The authors observed that the drug released was regulated by the changes in the amount of CNW in starch particle. CNW act as a delayed aspect for drug delivery, and the rate was found to be 2.9 times slower with the addition of CNW. Thus, the CNW helped in sustained release for long-term effects. In an interesting study by Lin et al., cationic CNC-crosslinked alginate hydrogels with a two layered structure were synthesized and loaded with two drugs (Lin et al. 2016). The outer layer contains plain alginate, loaded with an antibiotic, and it exhibited rapid drug delivery, while the internal layer was loaded with a growth factor and exhibited sustained delivery behavior due to the "nano-obstruction effect" and "nano-locking effect" prompted through CCNC. This way we can achieve the complexing drugs co-delivery with rapid delivery of one drug and sustained delivery of one more drug; it will be helpful in providing a synergistic effect in biomedical field (Fig. 8.4). Supramaniam et al. used iron oxide-modified magnetic CNC incorporated alginate hydrogels as drug delivery system for discharge of ibuprofen (Supramaniam et al. 2018). They observed that the incorporation of mCNC in hydrogel beads improved the integrity and swelling nature. Furthermore, it decreased the rate of in vitro drug release. Ahlen et al. studied the influence of CNC in chitosan-PAA nanoparticle incorporated PVA hydrogel eye lenses on optical drug delivery (Åhlén et al. 2018). They found that PVA lenses charged by the CNC showed greater delivery potential relative to

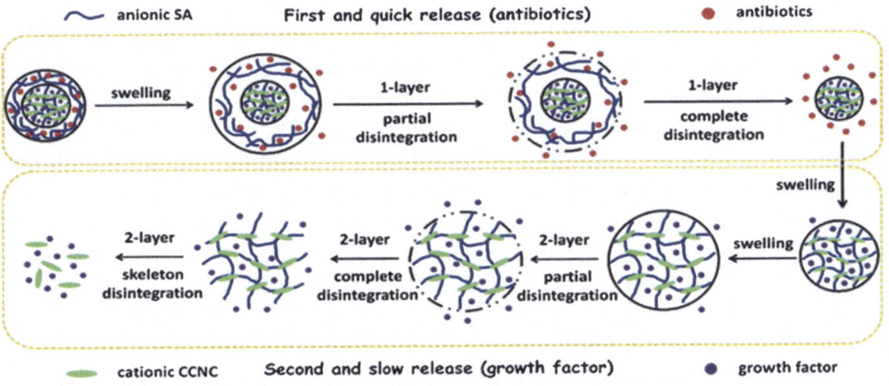

Fig. 8.4 Proposed complexing drug release model for the double-membrane hydrogel with the formation of cationic CCNC and anionic alginate. (Adopted from Lin et al. 2016)

CNC-less lenses. The in situ gelation of nanoparticles and CNC can be attributed to it by interlocking the CNC's particles and allowing the gel to restrain in the PVA network. It will prevent it from leaching. The new controlled drug release optical lenses should also be assumed to be integrated with working CNCs.

8.3.2 Wound Dressing

The human skin serves as a physical-chemical barrier and a defensive mask against environmental aggressors. Skin accidents, on the other hand, are common in everyday life, necessitating careful skin care in order to prevent long-term injuries and ensure a speedy recovery (Kamoun et al. 2017). Wound dressing is a quick and easy way to treat skin injuries. For a substance to be called an "ideal" wound dressing component, it must possess a number of characteristics. An ideal wound dressing should be nontoxic and nonallergenic, be able to sustain a moist environment at the wound surface, have decent gas permeability, have antimicrobial activity, absorb excess exudate and contaminants, facilitate healing, avoid more infection, and be possible to separate without causing further damage to the wound (Caló and Khutoryanskiy 2015; Hamedi et al. 2018). Hydrogels based on CNCs and CNFs follow the majority of the above criteria and have gained a lot of attention in recent years for wound dressing applications.

Liu et al. stated the synthesis of all polysaccharide-based hydrogels in which hemicelluloses serve as linkers for NFC and studied their wound healing ability (Liu et al. 2016). Xylan, xyloglucan (XG), and galactoglucomannan (GGM) were used as crosslinkers at various amounts to alter the mechanical and structural properties of hydrogels. The XG was discovered to have the highest adsorption potential on CNFs as well as the highest reinforcing effect. As a consequence, it exhibited the highest effectiveness in promoting fibroblast cell (NIH 3T3) growth and proliferation. These composite hydrogels have presented promise in wound healing applications, where they can offer subsidiary networks and uphold cell adhesion, expansion, and propagation. Huang et al. prepared injectable hydrogel with on-demand dissolvability and self-healing for deep burn wound (Huang et al. 2018b). For this, hydrogel was prepared by crosslinking carboxymethyl chitosan and dialdehyde cellulose nanocrystals via Schiff's base formation. Injected into abnormal and extreme burn wound litters, this hydrogel can self-heal and form an integrated hydrogel which occupies the wound area and prevents it against external surroundings. On demand, the hydrogel can be dissolved with amino acid solutions which can avoid the pain and discomfort during the dressing shift (Fig. 8.5). Basu et al. prepared Ca ion-crosslinked nanofibrillated cellulose (NFC) hydrogels and evaluated their applicability for wound dressing healing (Basu et al. 2017). The hydrogels exhibited good mechanical stability and water retention property that helps in maintaining the wet surroundings for diverse types of wounds. The hydrogels show no effect on dermal fibroblasts monolayer cultures and maintained inert nature to inflammatory response in terms of cytokine secretion and ROS production. With all these

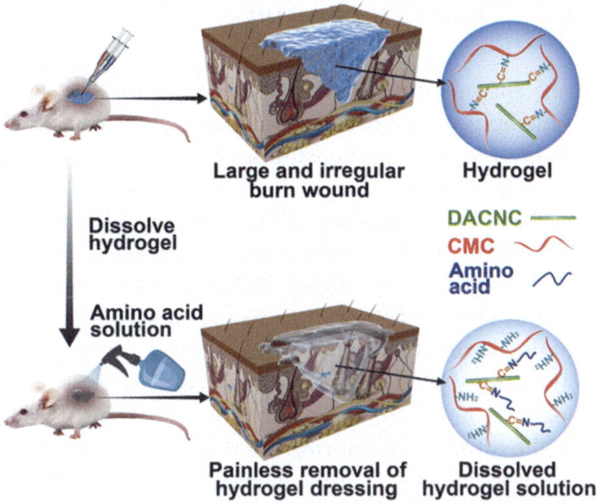

Fig. 8.5 Flow chart of on-demand dissolvable self-healing hydrogels for deep partial thickness burn wound healing. (Adopted from Huang et al. 2018b)

properties, the authors proposed that the Ca^{2+}-crosslinked hydrogels can best serve for wound dressing application, and adding additional functionality will lead to improved properties for specific wound healing. Further, the same author group studied the in vitro antibacterial potential of Ca^{2+}- and Cu^{2+}-crosslinked hydrogels on *Staphylococcus epidermidis* and *Pseudomonas aeruginosa* bacterial strains (Basu et al. 2018). The results revealed that the Ca-crosslinked hydrogels can retard the growth of *S. epidermidis* and inhibit *P. aeruginosa* biofilm production, while the Cu-crosslinked one prohibited the growth of S. *epidermidis* and bacteriostatic toward *P. aeruginosa*. Furthermore, the two hydrogels displayed barrier properties toward both the bacterial strains. Xu et al. prepared the CNF hydrogel by double-crosslinking TOCNF, first in situ crosslinking by Ca ion and second chemical cross-linking by 1,4-butanediol (Xu et al. 2018). With this double-crosslinking strategy, the mechanical strength of 3D-printed scaffolds with 1% hydrogels can be tuned from 3 to 8 kPa. Further, the cell tests were non-hazardous to fibroblast cells, and the cell proliferation is promoted with the increase in rigidity within the tunable range of 3–8 kPa. Liu et al. prepared a new multi-responsive hydrogels based on TEMPO-oxidized CNF and polydopamine (PDA). The PDA acts as a reinforcing agent, photothermal agent, and drug carrier (Liu et al. 2018). PDA nanoparticles loaded with tetracycline hydrochloride (TH) drug were added to TEMPO-oxidized CNF and crosslinked Ca^{2+} ions to form hydrogel. The hydrogel exhibited pH/NIR-responsive nature and released drug in on demand manner when exposed to NIR radiation or low pH. No burst discharge was detected in the starting, and the discharge was continuous for 24 h and reached extreme drug discharge rate 77%. Therefore, hydrogel was effectively used in in vivo skin wound dressing test.

8.3.3 Tissue Engineering

Tissue engineering was first described in 1988 as the "application of the principles and methods of engineering and life sciences toward fundamental understanding of structure-function relationship in normal and pathological mammalian tissues and the development ofbiological substitutes for the repair or regeneration of tissue or organ function" (Chapekar 2000; Caló and Khutoryanskiy 2015). CNCs and CNFs-based hydrogels have been extensively used in tissue engineering for the past few years because of their highly hydrated 3D porous structure that mimics biological tissue, as well as the excellent mechanical properties that CNCs and CNFs provide (Domingues et al. 2014).

By using UPy-combined CNCs (CNC-UPy) as pickering stabilizer, Liu et al. prepared a high internal phase emulsion (HIPE) for preparation of macroporous hybrid hydrogel. Monitoring the CNC-UPy amounts change the porous structures, swelling behaviors, and mechanical strength of the resulting macroporous hybrid hydrogels (Liu et al. 2017). The hybrid hydrogels were found to be cytocompatible and had good cell adhesion. Consequently, the researchers thought that these hydrogels demonstrated high potential as active natural scaffolds for tissue engineering. Huang et al. used a simple in situ hydroxyapatite (HAP) coating on CNC medium and create a highly porous and lightweight synthetic bone scaffold (Huang et al. 2018a). Further crosslinking of CNC/HAP scaffold with poly(methyl vinyl ether-alt-maleic acid) (PMVEMA) and PEG was found to greatly increase the water stability and mechanical property (up to 41.8 MPa). Furthermore, the scaffold had excellent biocompatibility and capacity to stabilize bovine serum albumin (BSA) protein, indicating that it could be a promising bone scaffold.

Aldehyde-functionalized CNCs (CHO-CNCs)-reinforced CMC and dextran-based injectable hydrogels were prepared by Yang et al. They observed that these hydrogels exhibited significantly greater elastic moduli as well as greater dimensional stability in long-term swelling tests. In addition, it does not show substantial toxicity in NIH 3T3 fibroblast cells. Based on these findings, the researchers advised that these hydrogels can be useful in tissue engineering (Yang et al. 2013). Based on hydrazine-crosslinked POEGMA and magnetically aligned CNCs, the same author group further established a novel injectable nanocomposite hydrogels. By the application of external magnetic field on CNC-POEGMA-coextruded mixture, the CNC could be aligned and resulting in hydrogel nanocomposites with desired degree of arrangement (Fig. 8.6). It revealed that fabricated hydrogels with aligned CNCs displayed greater swelling, minor shear moduli, and greatly increased myotube alignment compared to hydrogels composed of isotropic CNCs (De France et al. 2017b).

As shown in Fig. 8.7, Domingues et al. prepared a series of injectable hydrogels containing adipic acid dihydrazide-modified HA (ADA-HA) and aldehyde-modified HA (a-HA) reinforced with different amounts of aldehyde-modified CNCs (a-CNCs) (Domingues et al. 2015). In contrast to unfilled hydrogels, CNC-reinforced HA hydrogels had a more organized and compact network structure, which resulted in

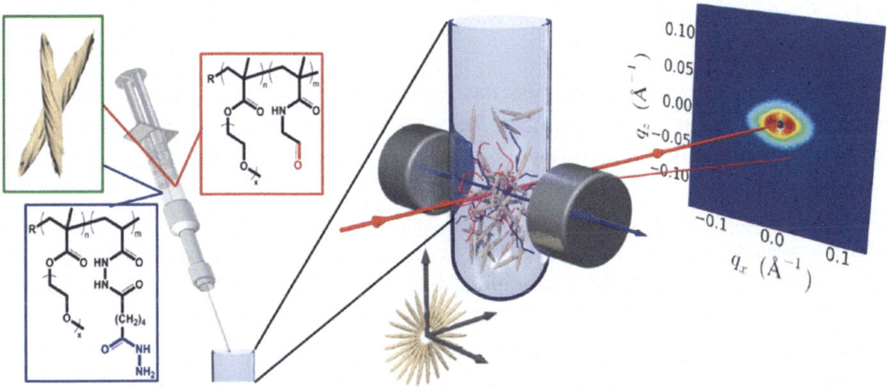

Fig. 8.6 Schematic representation of injectable CNC-POEGMA nanocomposite hydrogel. (Adopted from De France et al. 2017b)

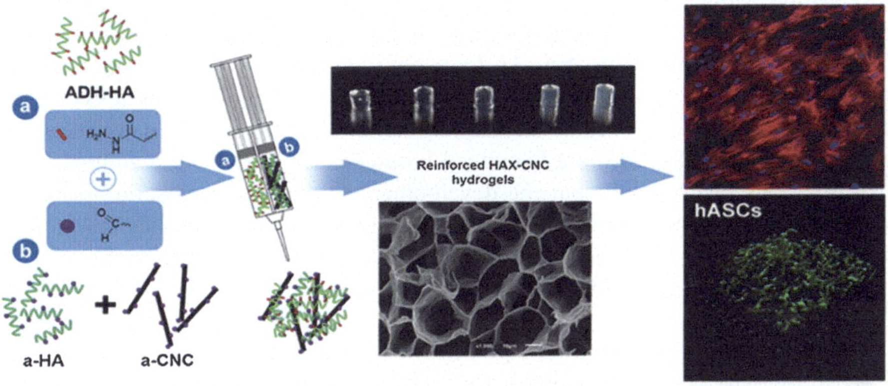

Fig. 8.7 Schematic representation of hyaluronic acid-based hydrogels (HAX) reinforced with CNCs, photograph of the HAX-CNC hydrogels and the corresponding cross-sectional SEM picture, as well as actin filaments staining image (top right) and live/dead staining image (bottom right) of HAX-CNC hydrogels. (Adopted from Domingues et al. 2015)

stiffer hydrogels with improved storage modulus (improvement was up to 135%). The biological efficiency of CNC-reinforced HA hydrogels was also evaluated using human adipose-derived stem cells (hASCs). HA-CNCs nanocomposite hydrogels presented preferential cell supportive properties in in vitro conditions, and hASCs encapsulated in nanocomposite hydrogel presented the capability to expand throughout the volume of gels and exhibited pronounced proliferative activity.

8.4 Conclusions

In recent years, substantial progress has been made in the fabrication, moderation, and implementation of nanocellulose because of their excellent mechanical property, unique structure, more surface area, and its biocompatibility. CNFs and CNCs have been recently proved advantageous substances for hydrogel preparation. In this book chapter, the authors presented few traditional methods for CNFs and CNCs production from lignocellulose and outlined current developments of its hydrogels as well as their biomedical applications. Hydrogels based on nanocellulose are now regarded as one-of-a-kind practical materials that can be employed in various biomedical applications. However, the majority of these hydrogel technologies in the biomedical sector are still in the early stages of development and laboratory testing. In vitro therapy has been the subject of only a few published clinical trials. However, for real-world applications, toxicology testing and sustainable evaluations of in vivo toxicity and biocompatibility for hydrogels are critical. As a result, more similar trials should be performed on nanocellulose-based hydrogels for real-world clinical use. Furthermore, large-scale manufacturing of nanocellulose-based hydrogels remains an obstacle. In recent times, the new 3D printing technology has been explored as a modern route to produce nanocellulose-based hydrogels with customized hierarchical structures, which could benefit bulk manufacturing and allow the creation of more beneficial biomedical materials. The authors hope that this book chapter will spur the development of practical nanocellulose-based hydrogels for a variety of biomedical applications. However, further research is required to identify biocompatibility and long-term toxicology of these hydrogels, and the current findings point to a promising future.

Acknowledgments This research was supported by Basic Science Research Program through the National Research Foundation of Korea (NRF) funded by the Ministry of Education (No. 2018R1A6A1A03025582).

References

Åhlén M, Tummala GK, Mihranyan A (2018) Nanoparticle-loaded hydrogels as a pathway for enzyme-triggered drug release in ophthalmic applications. Int J Pharm 536:73–81. https://doi.org/10.1016/j.ijpharm.2017.11.053

Alle M, Bandi R, Lee S-H, Kim J-C (2020a) Recent trends in isolation of cellulose nanocrystals and nanofibrils from various forest wood and nonwood products and their application. In: Nanomaterials for agriculture and forestry applications. Elsevier, pp 41–80

Alle M, Reddy GB, Kim TH, Park SH, Lee SH, Kim JC (2020b) Doxorubicin-carboxymethyl xanthan gum capped gold nanoparticles: microwave synthesis, characterization, and anti-cancer activity. Carbohydr Polym 229:115511. https://doi.org/10.1016/j.carbpol.2019.115511

Alle M, Lee SH, Kim JC (2020c) Ultrafast synthesis of gold nanoparticles on cellulose nanocrystals via microwave irradiation and their dyes-degradation catalytic activity. J Mater Sci Technol 41:168–177. https://doi.org/10.1016/j.jmst.2019.11.003

Alle M, Park SC, Bandi R, Lee S, Kim J-C (2020d) Rapid in-situ growth of gold nanoparticles on cationic cellulose nanofibrils: recyclable nanozyme for the colorimetric glucose detection. Carbohydr Polym 253:117239. https://doi.org/10.1016/j.carbpol.2020.117239

Alle M, Bandi R, Sharma G, Lee S, Kim J (2021) Shape recoverable, Au nanoparticles loaded nanocellulose foams as a recyclable catalyst for the dynamic and batch discoloration of dyes. Carbohydr Polym 258:117693. https://doi.org/10.1016/j.carbpol.2021.117693

Allen TM, Cullis PR (2004) Drug delivery systems: entering the mainstream. Science (80-) 303:1818–1822

Bae KH, Wang LS, Kurisawa M (2013) Injectable biodegradable hydrogels: Progress and challenges. J Mater Chem B 1:5371–5388

Bandi R, Alle M, Park C-W, Han S-Y, Kwon G-J, Kim J-C, Lee S-H (2020) Rapid synchronous synthesis of Ag nanoparticles and Ag nanoparticles/holocellulose nanofibrils: Hg(II) detection and dye discoloration. Carbohydr Polym 240:116356. https://doi.org/10.1016/j.carbpol.2020.116356

Bandi R, Alle M, Park C, Han S, Kwon G, Kim N-H, Kim J, Lee S (2021) Cellulose nanofibrils/carbon dots composite nanopapers for the smartphone-based colorimetric detection of hydrogen peroxide and glucose. Sensors Actuators B Chem 330:129330. https://doi.org/10.1016/j.snb.2020.129330

Basu A, Lindh J, Ålander E, Strømme M, Ferraz N (2017) On the use of ion-crosslinked nanocellulose hydrogels for wound healing solutions: physicochemical properties and application-oriented biocompatibility studies. Carbohydr Polym 174:299–308. https://doi.org/10.1016/j.carbpol.2017.06.073

Basu A, Heitz K, Strømme M, Welch K, Ferraz N (2018) Ion-crosslinked wood-derived nanocellulose hydrogels with tunable antibacterial properties: candidate materials for advanced wound care applications. Carbohydr Polym 181:345–350. https://doi.org/10.1016/j.carbpol.2017.10.085

Cai J, Zhang L (2005) Rapid dissolution of cellulose in LiOH/urea and NaOH/urea aqueous solutions. Macromol Biosci 5:539–548. https://doi.org/10.1002/mabi.200400222

Caliari SR, Burdick JA (2016) A practical guide to hydrogels for cell culture. Nat Methods 13:405–414. https://doi.org/10.1038/nmeth.3839

Caló E, Khutoryanskiy VV (2015) Biomedical applications of hydrogels: a review of patents and commercial products. Eur Polym J 65:252–267. https://doi.org/10.1016/j.eurpolymj.2014.11.024

Camci-Unal G, Annabi N, Dokmeci MR, Liao R, Khademhosseini A (2014) Hydrogels for cardiac tissue engineering. NPG Asia Mater 6:e99–e99. https://doi.org/10.1038/am.2014.19

Chang C, Zhang L, Zhou J, Zhang L, Kennedy JF (2010) Structure and properties of hydrogels prepared from cellulose in NaOH/urea aqueous solutions. Carbohydr Polym 82:122–127. https://doi.org/10.1016/j.carbpol.2010.04.033

Chapekar MS (2000) Tissue engineering: challenges and opportunities. J Biomed Mater Res 53:617–620. https://doi.org/10.1002/1097-4636(2000)53:6<617::AID-JBM1>3.0.CO;2-C

Copeland L, Blazek J, Salman H, Tang MC (2009) Form and functionality of starch. Food Hydrocoll 23:1527–1534. https://doi.org/10.1016/j.foodhyd.2008.09.016

Corkhill PH, Hamilton CJ, Tighe BJ (1989) Synthetic hydrogels VI. Hydrogel composites as wound dressings and implant materials. Biomaterials 10:3–10. https://doi.org/10.1016/0142-9612(89)90002-1

Curvello R, Raghuwanshi VS, Garnier G (2019) Engineering nanocellulose hydrogels for biomedical applications. Adv Colloid Interf Sci 267:47–61. https://doi.org/10.1016/j.cis.2019.03.002

De France KJ, Hoare T, Cranston ED (2017a) Review of hydrogels and aerogels containing Nanocellulose. Chem Mater 29:4609–4631. https://doi.org/10.1021/acs.chemmater.7b00531

De France KJ, Yager KG, Chan KJW, Corbett B, Cranston ED, Hoare T (2017b) Injectable anisotropic nanocomposite hydrogels direct in situ growth and alignment of Myotubes. Nano Lett 17:6487–6495. https://doi.org/10.1021/acs.nanolett.7b03600

Demitri C, Del Sole R, Scalera F, Sannino A, Vasapollo G, Maffezzoli A, Ambrosio L, Nicolais L (2008) Novel superabsorbent cellulose-based hydrogels crosslinked with citric acid. J Appl Polym Sci 110:2453–2460. https://doi.org/10.1002/app.28660
Domingues RMA, Gomes ME, Reis RL (2014) The potential of cellulose nanocrystals in tissue engineering strategies. Biomacromolecules 15:2327–2346
Domingues RMA, Silva M, Gershovich P, Betta S, Babo P, Caridade SG, Mano JF, Motta A, Reis RL, Gomes ME (2015) Development of injectable hyaluronic acid/cellulose nanocrystals bionanocomposite hydrogels for tissue engineering applications. Bioconjug Chem 26:1571–1581. https://doi.org/10.1021/acs.bioconjchem.5b00209
Dong H, Snyder JF, Williams KS, Andzelm JW (2013) Cation-induced hydrogels of cellulose nanofibrils with tunable moduli. Biomacromolecules 14:3338–3345. https://doi.org/10.1021/bm400993f
Dufresne A (2013) Nanocellulose: a new ageless bionanomaterial. Mater Today 16:220–227
Fenton OS, Olafson KN, Pillai PS, Mitchell MJ, Langer R (2018) Advances in biomaterials for drug delivery. Adv Mater 30:1705328. https://doi.org/10.1002/adma.201705328
Fu N, Zhang X, Sui L, Liu M, Lin Y (2017) Application of scaffold materials in cartilage tissue engineering. Humana Press, Cham, pp 21–39
Gyles DA, Castro LD, Silva JOC, Ribeiro-Costa RM (2017) A review of the designs and prominent biomedical advances of natural and synthetic hydrogel formulations. Eur Polym J 88:373–392. https://doi.org/10.1016/j.eurpolymj.2017.01.027
Habibi Y, Lucia LA, Rojas OJ (2010) Cellulose nanocrystals: chemistry, self-assembly, and applications. Chem Rev 110:3479–3500. https://doi.org/10.1021/cr900339w
Hamedi H, Moradi S, Hudson SM, Tonelli AE (2018) Chitosan based hydrogels and their applications for drug delivery in wound dressings: a review. Carbohydr Polym 199:445–460. https://doi.org/10.1016/j.carbpol.2018.06.114
Haycock JW (2011) 3D cell culture: a review of current approaches and techniques. In: Methods in molecular biology. Humana Press, Clifton, N.J, pp 1–15
Heinze T, Koschella A (2005) Solvents applied in the field of cellulose chemistry: a mini review. Polímeros 15:84–90. https://doi.org/10.1590/s0104-14282005000200005
Hoare TR, Kohane DS (2008) Hydrogels in drug delivery: Progress and challenges. Polymer (Guildf) 49:1993–2007. https://doi.org/10.1016/j.polymer.2008.01.027
Hoover R (2001) Composition, molecular structure, and physicochemical properties of tuber and root starches: a review. Carbohydr Polym 45:253–267. https://doi.org/10.1016/S0144-8617(00)00260-5
Huang C, Hao N, Bhagia S, Li M, Meng X, Pu Y, Yong Q, Ragauskas AJ (2018a) Porous artificial bone scaffold synthesized from a facile in situ hydroxyapatite coating and crosslinking reaction of crystalline nanocellulose. Materialia 4:237–246. https://doi.org/10.1016/j.mtla.2018.09.008
Huang W, Wang Y, Huang Z, Wang X, Chen L, Zhang Y, Zhang L (2018b) On-demand dissolvable self-healing hydrogel based on Carboxymethyl chitosan and cellulose nanocrystal for deep partial thickness burn wound healing. ACS Appl Mater Interfaces 10:41076–41088. https://doi.org/10.1021/acsami.8b14526
Iguchi M, Yamanaka S, Budhiono A (2000) Bacterial cellulose – a masterpiece of nature's arts. J Mater Sci 35:261–270. https://doi.org/10.1023/A:1004775229149
Jayakumar R, Prabaharan M, Sudheesh Kumar PT, Nair SV, Tamura H (2011) Biomaterials based on chitin and chitosan in wound dressing applications. Biotechnol Adv 29:322–337. https://doi.org/10.1016/j.biotechadv.2011.01.005
Jung IY, Kim JS, Choi BR, Lee K, Lee H (2017) Hydrogel based biosensors for in vitro diagnostics of biochemicals, proteins, and genes. Adv Healthc Mater 6:1601475. https://doi.org/10.1002/adhm.201601475
Kamoun EA, Kenawy ERS, Chen X (2017) A review on polymeric hydrogel membranes for wound dressing applications: PVA-based hydrogel dressings. J Adv Res 8:217–233
Kim TH, Alle M, Park SC, Zhao F, Long W, Samala S, Kim J-C (2021) Self-assembly prepared using an ion pair of poly(ethylene imine) and (phenylthio) acetic acid as a drug carrier for

oxidation, temperature, and NIR-responsive release. Chem Eng J 415:128954. https://doi.org/10.1016/j.cej.2021.128954

Kwon G-J, Han S-Y, Park C-W, Park J-S, Lee E-A, Kim N-H, Alle M, Bandi R, Lee S-H (2020) Adsorption characteristics of ag nanoparticles on cellulose Nanofibrils with different chemical compositions. Polymers (Basel) 12:164. https://doi.org/10.3390/polym12010164

Langer R (1990) New methods of drug delivery. Science (80-) 249:1527–1533. https://doi.org/10.1126/science.2218494

Laurén P, Lou YR, Raki M, Urtti A, Bergström K, Yliperttula M (2014) Technetium-99m-labeled nanofibrillar cellulose hydrogel for in vivo drug release. Eur J Pharm Sci 65:79–88. https://doi.org/10.1016/j.ejps.2014.09.013

Lee S-H, Kim H-J, Kim J-C (2019) Nanocellulose applications for drug delivery: a review. J For Environ Sci 35:141–149

Li W, Lan Y, Guo R, Zhang Y, Xue W, Zhang Y (2015) In vitro and in vivo evaluation of a novel collagen/cellulose nanocrystals scaffold for achieving the sustained release of basic fibroblast growth factor. J Biomater Appl 29:882–893. https://doi.org/10.1177/0885328214547091

Li J, Yan Y, Liang Z, Zhang T (2016) Experimental and numerical study of adhesively bonded cfrp scarf-lap joints subjected to tensile loads. J Adhes 92:1–17. https://doi.org/10.1080/00218464.2014.987343

Lin N, Gèze A, Wouessidjewe D, Huang J, Dufresne A (2016) Biocompatible double-membrane hydrogels from cationic cellulose nanocrystals and anionic alginate as complexing drugs Codelivery. ACS Appl Mater Interfaces 8:6880–6889. https://doi.org/10.1021/acsami.6b00555

Lindman B, Medronho B, Alves L, Costa C, Edlund H, Norgren M (2017) The relevance of structural features of cellulose and its interactions to dissolution, regeneration, gelation and plasticization phenomena. Phys Chem Chem Phys 19:23704–23718

Liu J, Chinga-Carrasco G, Cheng F, Xu W, Willför S, Syverud K, Xu C (2016) Hemicellulose-reinforced nanocellulose hydrogels for wound healing application. Cellulose 23:3129–3143. https://doi.org/10.1007/s10570-016-1038-3

Liu S, Jin M, Chen Y, Gao H, Shi X, Cheng W, Ren L, Wang Y (2017) High internal phase emulsions stabilised by supramolecular cellulose nanocrystals and their application as cell-adhesive macroporous hydrogel monoliths. J Mater Chem B 5:2671–2678. https://doi.org/10.1039/c7tb00145b

Liu Y, Sui Y, Liu C, Liu C, Wu M, Li B, Li Y (2018) A physically crosslinked polydopamine/nanocellulose hydrogel as potential versatile vehicles for drug delivery and wound healing. Carbohydr Polym 188:27–36. https://doi.org/10.1016/j.carbpol.2018.01.093

Madhusudhan A, Reddy G, Venkatesham M, Veerabhadram G, Kumar D, Natarajan S, Yang M-Y, Hu A, Singh S (2014) Efficient pH dependent drug delivery to target cancer cells by gold nanoparticles capped with Carboxymethyl Chitosan. Int J Mol Sci 15:8216–8234. https://doi.org/10.3390/ijms15058216

Masruchin N, Park BD, Causin V (2018) Dual-responsive composite hydrogels based on TEMPO-oxidized cellulose nanofibril and poly(N-isopropylacrylamide) for model drug release. Cellulose 25:485–502. https://doi.org/10.1007/s10570-017-1585-2

Mauricio MR, da Costa PG, Haraguchi SK, Guilherme MR, Muniz EC, Rubira AF (2015) Synthesis of a microhydrogel composite from cellulose nanowhiskers and starch for drug delivery. Carbohydr Polym 115:715–722. https://doi.org/10.1016/j.carbpol.2014.07.063

Mendoza L, Batchelor W, Tabor RF, Garnier G (2018a) Gelation mechanism of cellulose nanofibre gels: a colloids and interfacial perspective. J Colloid Interface Sci 509:39–46. https://doi.org/10.1016/j.jcis.2017.08.101

Mendoza L, Batchelor W, Tabor RF, Garnier G (2018b) Gelation mechanism of cellulose nanofibre gels: a colloids and interfacial perspective. J Colloid Interface Sci 509:39–46. https://doi.org/10.1016/j.jcis.2017.08.101

Mendoza L, Gunawardhana T, Batchelor W, Garnier G (2018c) Effects of fibre dimension and charge density on nanocellulose gels. J Colloid Interface Sci 525:119–125. https://doi.org/10.1016/j.jcis.2018.04.077

Miyata T (2010) Preparation of smart soft materials using molecular complexes. Polym J 42:277–289

Moberg T, Sahlin K, Yao K, Geng S, Westman G, Zhou Q, Oksman K, Rigdahl M (2017) Rheological properties of nanocellulose suspensions: effects of fibril/particle dimensions and surface characteristics. Cellulose 24:2499–2510. https://doi.org/10.1007/s10570-017-1283-0

Müller A, Ni Z, Hessler N, Wesarg F, Müller FA, Kralisch D, Fischer D (2013) The biopolymer bacterial Nanocellulose as drug delivery system: investigation of drug loading and release using the model protein albumin. J Pharm Sci 102:579–592. https://doi.org/10.1002/jps.23385

Namdeo M, Bajpai SK, Kakkar S (2009) Preparation of a magnetic-field-sensitive hydrogel and preliminary study of its drug release behavior. J Biomater Sci Polym Ed 20:1747–1761. https://doi.org/10.1163/156856208X386372

Navarra MA, Dal Bosco C, Moreno JS, Vitucci FM, Paolone A, Panero S (2015) Synthesis and characterization of cellulose-based hydrogels to be used as gel electrolytes. Membranes (Basel) 5:810–823. https://doi.org/10.3390/membranes5040810

Neely WB (1961) Dextran: structure and synthesis. In: Advances in carbohydrate chemistry. Academic Press, pp 341–369

Oksman K, Mathew AP, Bondeson D, Kvien I (2006) Manufacturing process of cellulose whiskers/polylactic acid nanocomposites. Compos Sci Technol 66:2776–2784. https://doi.org/10.1016/j.compscitech.2006.03.002

Paukkonen H, Kunnari M, Laurén P, Hakkarainen T, Auvinen VV, Oksanen T, Koivuniemi R, Yliperttula M, Laaksonen T (2017) Nanofibrillar cellulose hydrogels and reconstructed hydrogels as matrices for controlled drug release. Int J Pharm 532:269–280. https://doi.org/10.1016/j.ijpharm.2017.09.002

Paulraj T, Riazanova AV, Svagan AJ (2018) Bioinspired capsules based on nanocellulose, xyloglucan and pectin – the influence of capsule wall composition on permeability properties. Acta Biomater 69:196–205. https://doi.org/10.1016/j.actbio.2018.01.003

Raghuwanshi VS, Cohen Y, Garnier G, Garvey CJ, Russell RA, Darwish T, Garnier G (2018) Cellulose dissolution in ionic liquid: ion binding revealed by neutron scattering. Macromolecules 51:7649–7655. https://doi.org/10.1021/acs.macromol.8b01425

Rånby BG (1951) Fibrous macromolecular systems. Cellulose and muscle. The colloidal properties of cellulose micelles. Discuss Faraday Soc 11:158–164. https://doi.org/10.1039/DF9511100158

Saito T, Isogai A (2004) TEMPO-mediated oxidation of native cellulose. The effect of oxidation conditions on chemical and crystal structures of the water-insoluble fractions. Biomacromolecules 5:1983–1989. https://doi.org/10.1021/bm0497769

Shen X, Shamshina JL, Berton P, Gurau G, Rogers RD (2015) Hydrogels based on cellulose and chitin: fabrication, properties, and applications. Green Chem 18:53–75

Shu XZ, Zhu KJ, Song W (2001) Novel pH-sensitive citrate cross-linked chitosan film for drug controlled release. Int J Pharm 212:19–28. https://doi.org/10.1016/S0378-5173(00)00582-2

Supramaniam J, Adnan R, Mohd Kaus NH, Bushra R (2018) Magnetic nanocellulose alginate hydrogel beads as potential drug delivery system. Int J Biol Macromol 118:640–648. https://doi.org/10.1016/j.ijbiomac.2018.06.043

Willcox MDP, Phillips B, Ozkan J, Jalbert I, Meagher L, Gengenbach T, Holden B, Papas E (2010) Interactions of lens care with silicone hydrogel lenses and effect on comfort. Optom Vis Sci 87:839–846. https://doi.org/10.1097/OPX.0b013e3181f3e2fc

Xu C, Zhang Molino B, Wang X, Cheng F, Xu W, Molino P, Bacher M, Su D, Rosenau T, Willför S, Wallace G (2018) 3D printing of nanocellulose hydrogel scaffolds with tunable mechanical strength towards wound healing application. J Mater Chem B 6:7066–7075. https://doi.org/10.1039/c8tb01757c

Yang X, Bakaic E, Hoare T, Cranston ED (2013) Injectable polysaccharide hydrogels reinforced with cellulose nanocrystals: morphology, rheology, degradation, and cytotoxicity. Biomacromolecules 14:4447–4455. https://doi.org/10.1021/bm401364z

Yang J, Xu F, Han CR (2017) Metal ion mediated cellulose Nanofibrils transient network in covalently cross-linked hydrogels: mechanistic insight into morphology and dynamics. Biomacromolecules 18:1019–1028. https://doi.org/10.1021/acs.biomac.6b01915

Zander NE, Dong H, Steele J, Grant JT (2014) Metal cation cross-linked nanocellulose hydrogels as tissue engineering substrates. ACS Appl Mater Interfaces 6:18502–18510. https://doi.org/10.1021/am506007z

Zelikin AN, Ehrhardt C, Healy AM (2016) Materials and methods for delivery of biological drugs. Nat Chem 8:997–1007. https://doi.org/10.1038/nchem.2629

Zhang H, Yang C, Zhou W, Luan Q, Li W, Deng Q, Dong X, Tang H, Huang F (2018) A pH-responsive gel macrosphere based on sodium alginate and cellulose nanofiber for potential intestinal delivery of probiotics. ACS Sustain Chem Eng 6:13924–13931. https://doi.org/10.1021/acssuschemeng.8b02237

Zhong SP, Zhang YZ, Lim CT (2010) Tissue scaffolds for skin wound healing and dermal reconstruction. WIREs Nanomed Nanobiotechnol 2:510–525. https://doi.org/10.1002/wnan.100

Further Reading

Alle M, Bandi R, Lee S-H, Kim J-C (2020) Recent trends in isolation of cellulose nanocrystals and nanofibrils from various forest wood and nonwood products and their application. In: Nanomaterials for agriculture and forestry applications. Elsevier, pp 41–80

De France KJ, Hoare T, Cranston ED (2017) Review of hydrogels and aerogels containing Nanocellulose. Chem Mater 29:4609–4631

Chapter 9
Microneedle Array Patches for the Delivery of Therapeutic Agents

Shreya Shashank Chauhan, Venessa Maurice Lobo, Samruddhi Nandkumar Borate, Saili Sachin Jagade, and Venkata Vamsi Krishna Venuganti

9.1 Introduction

Solid oral dosage forms are the most convenient form of drug administration. The therapeutic compounds that are amenable to the development of solid oral dosage forms require ideal physicochemical properties. The drug compounds prone to degradation in the gastrointestinal tract show poor systemic absorption and undergo the first-pass metabolism that would require alternative routes of administrations. The compounds with nonideal physicochemical properties are administered through parenteral routes using injectable dosage forms. Several disadvantages are associated with the administration of injectable preparations, including needle injuries, pain, syringe phobia, patient noncompliance, requirement of a medical practitioner for administration, and requirement of sterile liquid preparations, among others. Some of these disadvantages are attributed to the design and length of the conventional syringe needle ranging from 2 to 3 cm. It was found that reducing the needle dimensions to a length of less than one millimeter would reduce the pain and injury associated with hypodermic syringe needles (Gill and Prausnitz 2007). The length of 500–600 μm would allow needle penetration to a depth of about 150–250 μm inside skin. This penetration depth would only reach the dermal-epidermal junction or a little inside the dermis (Henry et al. 1998). Therefore, the pain and injury associated with the deeper needle penetration is avoided. The microneedles (MN) are arranged in an array format to improve the applicability and allow prolonged administration of drugs. Over the past decade, microneedle devices have evolved with respect to their designs, materials of construction, and applications. One of the

S. S. Chauhan · V. M. Lobo · S. N. Borate · S. S. Jagade · V. V. K. Venuganti (✉)
Department of Pharmacy, Birla Institute of Technology and Science (BITS) Pilani, Hyderabad, Telangana, India
e-mail: vamsi@hyderabad.bits-pilani.ac.in

J.-C. Kim et al. (eds.), *Smart Nanomaterials in Biomedical Applications*, Nanotechnology in the Life Sciences,
https://doi.org/10.1007/978-3-030-84262-8_9

applications explored extensively is drug delivery through accessible barriers, including skin and ocular route. MN have been investigated to deliver varied vaccines and therapeutics including influenza vaccine, chemotherapeutics, antibiotics, and vitamins, among others. This chapter provides an account of the delivery of different vaccines and therapeutics using microneedle array patches.

9.1.1 Microneedle Designs, Drug Loading, and Release Mechanisms

Polymers are the most widely investigated materials for the preparation of MNs. The desired quality parameters of MN can be optimized by fabricating primary master molds and secondary production molds. The master molds would be made of stainless steel or silicon (among other materials), and secondary molds are generally made of polydimethylsiloxane (PDMS). The secondary molds can be utilized to fabricate multiples of MN patches by using varied combinations and concentrations of materials. Polymers can be selected based on the purpose of the MN, the type of drug to be delivered, site of patch application, biocompatibility, biodegradability, and non-immunogenicity. The shape and other geometrical aspects of the needles, like tip diameter, aspect ratio, length of the needle, and base width, significantly affect MN performance capabilities. The desired performance characteristics like physical/ biological stimuli-responsiveness, degradation/swelling behaviors, and mechanical strength can be optimized. Polymeric MN exhibit ease of fabrication on scale-up and provide high drug loading capacity. Polymeric MN can be broadly characterized into six types, namely, solid, hollow, biodegradable, dissolvable, bio-responsive, and coated (Fig. 9.1) (Azmana et al. 2020).

9.1.1.1 Solid Polymeric Microneedles

MN can be categorized as solid or hollow MN based on the design factor. The solid MN are effective in perforation of the uppermost layer of the epidermis, stratum corneum, to painlessly deliver drugs inside the skin. Solid MN create microchannels after insertion, and upon removal, drug-loaded formulation is applied to enhance the permeation. The solid MN provides the required strength and flexibility to the needles (Lee et al. 2013). Polymers, such as polyvinylpyrrolidone (PVP), polyvinyl alcohol (PVA), polylactic acid (PLA), and polyglycolic acid (PGA), and their copolymers, poly-lactic-co-glycolic acid (PLGA), have been used to prepare solid MN (Lee et al. 2013). Donnelly et al. (2011) used micromolding technique to fabricate solid MN using mucoadhesive copolymer Gantrez® AN-139. PLA-based solid MNs with different aspect ratios could be prepared using commercially available acupuncture MN (Cha et al. 2014). Li et al. (2017) made PLA MN of different lengths for pretreatment of female BALB/c mice skin to create microchannels before applying insulin for diabetes treatment. The MN system was compared

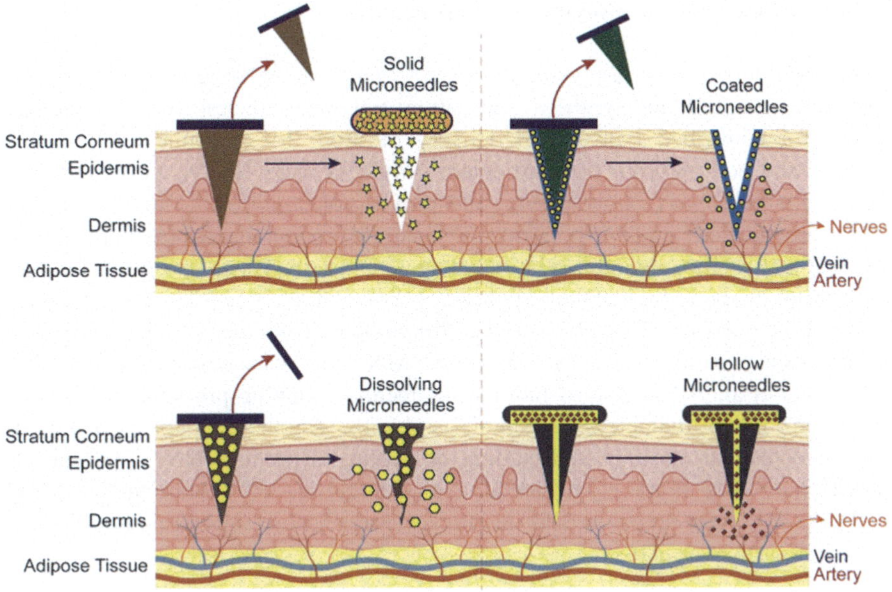

Fig. 9.1 Representation of different types of MN including solid, coated, dissolving, and hollow MN. (Adopted from Dsouza et al. 2020)

against other conventional methods and insulin delivery without MN pretreatment. The bigger microchannels created because of the insertion of MN resulted in greater drug permeation (Wang et al. 2017).

9.1.1.2 Hollow Polymeric Microneedles

Hollow MN consists of a hollow channel where the drug solution is allowed to diffuse. The drug solution is directly delivered to the dermal layer upon insertion of MN in the skin. Compounds having high molecular weights like proteins and oligonucleotides are generally delivered using hollow MN. Maintenance of a steady flow is of importance here, as it will eventually affect the absorption pattern of the drug. An increase in the bore size of the needle will increase the space available for drug loading and also increase flow rate but will adversely affect the needle strength and sharpness. Hollow MN were fabricated using techniques including UV photolithography, deep X-ray photolithography, laser micromachining, and reactive ion etching (RIE) (Donnelly et al. 2010; Kim et al. 2020b). Polymers such as poly(methyl methacrylate) (PMMA) and SU-8 are often combined with lithography techniques to obtain polymeric hollow MN. McAllister et al. were the first group to fabricate polymer hollow MN (McAllister et al. 2003). Poly(dimethylsiloxane) (PDMS) replication method was employed to produce biodegradable needles from an RIE-fabricated silicon master mold. A backside exposure procedure to fabricate MN array from SU-82050 and SU-82005 negative photoresist by UV photolithography was proposed by Huang and Fu (2007).

9.1.1.3 Biodegradable Polymeric Microneedles

Biodegradable polymers are generally used to fabricate MN meant for prolonged delivery of drugs. Many synthetic and natural polymers are researched in the last decade for controlled transdermal delivery of drugs for systemic applications. Polyglycolic acid (PGA), polycarbonate, polylactic acid (PLA), poly(lactic-co-glycolic) acid (PLGA), polystyrene, and polycaprolactone (PCL) are commonly used synthetic biodegradable polymers. The natural biodegradable polymers include silk, chitosan, chitin, alginate, gelatin, collagen, and zein. These are eliminated naturally from the body by metabolic pathways (Wang et al. 2017). The rate of degradation of the polymers controls the rate of drug release from the polymer matrix and the duration of drug action. A 3D-printed MN array was designed using PLA (Luzuriaga et al. 2018). A post-fabrication chemical etching protocol enabled the formation of tip diameters as small as 1 μm. Another group used micro-electromechanical masking and etching procedure to produce bevel- and chisel-tipped sharp MN (Park et al. 2005). Doxorubicin-loaded gelatin methacryloyl (GelMA) MN were prepared, and its efficacy was checked against melanoma cell line A375 (Luo et al. 2018).

9.1.1.4 Dissolvable Polymeric Microneedles

The MN will dissolve in the interstitial fluid (ISF) upon insertion in the skin and subsequently release the drug. ISF is present in more amounts in the dermis than in the outer stratum corneum layer. The hydrophilic polymers that form the MN are easily dissolved without leaving behind any bio-hazardous waste. Water-soluble polysaccharides including sodium chondroitin sulfate, sodium alginate, HPMC (hydroxypropyl methylcellulose), HPC (hydroxypropyl cellulose), CMC (carboxymethylcellulose), amylopectin, dextran, and HA (hyaluronic Acid) can form dissolvable MN. Besides gelatin, poly(methyl vinyl ether-maleic anhydride) (PMVE/MA), poly(vinylpyrrolidone-co-methacrylic acid) (PVP-MAA), poly-γ-glutamic acid, polyvinylpyrrolidone (PVP), polyvinyl pyrrolidone-polyvinyl alcohol (PVP-PVA), and PVP-cyclodextrin (PVP-CD) are also employed (Park et al. 2005). The amount of drug dissolved in the polymer matrix before MN casting is critical in the fabrication process. Sufficient mechanical strength to allow skin penetration is essential for effective drug delivery at the targeted site. Dissolving MN are currently being researched extensively for local delivery to the skin, for systemic absorption and ocular delivery to treat various diseases related to the inner and outer eye, among many others. Bhatnagar et al. (2018) fabricated rapidly dissolving PVA-PVP MNs for the ocular delivery of besifloxacin through the cornea. Complete dissolution of needle structure was seen within 5 minutes of insertion in excised human cornea. Dissolving MN were fabricated using HPMC as MN tip material and CMC as the base material. Donepezil hydrochloride (DPH)-loaded MN completely dissolved in the skin within 15 mins (Kim et al. 2016). Chemical combination of different polymers by copolymerization of monomers or simple mixing is also reported to optimize performance by the resultant MN.

9.1.1.5 Bio-responsive Polymeric Microneedles

Bio-responsive MN were fabricated to respond to the biological signals of the body to release the entrapped drug. Physiological signals, including blood glucose, pH, or enzymes, act as triggers for the drug release. pH-sensitive MN can be triggered by the acidic nature of skin (pH of healthy skin being 4.0 to 7.0). Crosslinked PVA, methacrylate hyaluronic acid (MeHA), and crosslinked alginate are used to prepare pH-sensitive MN (Du and Sun 2020; Ullah et al. 2019). Hypoxia-sensitive hyaluronic acid was used to make glucose level-sensitive MN containing glucose-responsive vesicles loaded with insulin and glucose oxidase. In hyperglycemic conditions, the vesicle would rapidly degrade, resulting in the discharge of insulin (Yu et al. 2015).

9.1.1.6 Coated Polymeric Microneedles

Coated polymeric MN were prepared where drug solution or drug dispersion is layered onto the surface of the MN. The drug concentration in the coat is governed by the thickness of the layer and length of the needle. Spray coating, spin coating, and/or dip coating are the generally employed methods for fabrication. However, newer techniques such as inkjet printing are emerging (Uddin et al. 2020). Uniform coating on the MN surface is a formidable challenge. Chen et al. have designed a new adjustable apparatus that can be "lifted and lowered" into the coating solutions with the attached polymer patches. PLA MN showed drug delivery efficiency of 90% using this method (Chen et al. 2017a, b). Baek et al. loaded lidocaine on poly(L-lactide) (PLLA) microneedle arrays, which released rapidly in phosphate buffer saline and was found to be stable for 3 weeks (Baek et al. 2017). More recently, polyvinyl alcohol (PVA) and chitosan-coated polyvinyl pyrrolidone (PVP) MN were constructed for delivering doxorubicin and gold core silica shell (AuMSS) (Moreira et al. 2020).

9.1.2 Drug Loading and Release Mechanisms

Techniques for efficient drug loading in MN and its consequent release have considerably evolved over the years. In general, the drug compounds are loaded in MN by coating or entrapment in the matrix depending on the MN design and material of construction. The earliest references to MN fabrication show the development of silicon MN with etching (dry or wet) techniques to form sharper tips that could easily penetrate the stratum corneum. However, concerns related to bio-incompatibility and higher fabrication costs led to the emergence of metal and ceramic microneedle, made in solid or hollow designs. The drug solution or dispersion was either loaded into the hollow MNs or coated over the solid MN. Hollow metal MN were contrived by a novel micromachining technique from copper, titanium, and nickel to deliver insulin to diabetic rats (Davis et al. 2005). Hollow MN were also used to pretreat the skin to create microchannels and augment the delivery of applied drugs (Kalluri et al. 2011). Influenza virus vaccine in polymer solution was coated on titanium MN that showed increased stability and immune response (Choi et al. 2013). DOX-encapsulated PLGA nanoparticles were coated on stainless steel MN using dip-coating method to prevent drug loss from the tumor to the systemic circulation (Ma et al. 2015). Silica glass and sugars such as maltose, mannitol, and trehalose are other examples of materials used for MN fabrication. Maltose-based MN were used for the delivery of doxorubicin to the dermis through the microchannels created (Nguyen and Banga 2017). Coating of drug solutions or dispersions on the surface of polymeric MN can be achieved by various techniques. BCG vaccine solution was poured into the PDMS mold followed by the addition of sodium alginate solution under vacuum (Arshad et al. 2020). Solid MN dip-coated with lidocaine hydrochloride aqueous solution were evaluated for prolonged local analgesic action (Zhang et al. 2012).

Encapsulation of drugs inside MN provides advantages, including higher drug loading and precise dose control (Chen et al. 2020a; Ito et al. 2011; Migalska et al. 2011). Encapsulation of fragile molecules like proteins and vaccines into the polymeric MN spares them the harsh conditions they otherwise would have been subjected to with coating processes. BSA-loaded poly(ethylene glycol) diacrylate (PEGDA) MN were fabricated by crosslinking after UV exposure (Kochhar et al. 2012). Exenatide-loaded dissolving sodium hyaluronate MN exhibited sufficient mechanical strength to penetrate the skin. On coming in contact with interstitial fluid, MN dissolved rapidly to release drug under 2 minutes (Zhu et al. 2014). Drugs concentrated only in the tips of the MN can be prepared by micromolding and subsequent centrifugation. MN prepared with HPMC-ethanol-water as tip material and CMC-water for MN base showed desired mechanical strength (Kim et al. 2016). Powder-carrying MN (PCMs) is an attractive approach for delivering moisture-sensitive drugs to eliminate the step of reconstitution and possible degradation. S. Kim et al. loaded high dose insulin in powder form by micro-shell fabrication using solvent casting technique with CMC as the polymer. Briefly, CMC solution was filled in PDMS molds under centrifugation, and upon drying, the shrunken volume was refilled with insulin powder. Again, the CMC solution was poured over it to form a protective layer. Thus, the array was formed (Kim et al. 2020c). Both hydrophilic and hydrophobic drugs can be administered transdermally by encapsulating them in nanoparticle preparations, which in turn can be loaded into the MN (Chen et al. 2020a). Tetanus toxoid-loaded nanoparticles prepared by ionotropic gelation method with crosslinking by poly(sodium-4-styrene sulfonate) (PSS) were encapsulated in chitosan MN (Siddhapura et al. 2016). Similarly, Lan et al. fabricated PD-1 inhibitor and cisplatin-loaded pH-responsive solid lipid nanoparticles that were delivered through MN directly to the tumor site. A positive synergistic effect was seen in tumor reduction with controlled side effects (Lan et al. 2020). Thus, various techniques are emerging to enhance drug loading and subsequent release to target tissues with the minimally invasive MN application technique.

9.2 Microneedle Array Patches to Deliver Vaccines

9.2.1 *Influenza*

The most used strains to deliver influenza vaccine in microneedle patches have been whole inactivated influenza virus, influenza hemagglutinins, influenza subunit, and split vaccine. Various vaccine strains produce varied effects in these patches.

Presently, two influenza MN delivery systems have been approved by US-FDA (Kolli 2015). Intanza®, IDflu®, and Fluzone Intradermal® made by Becton Dickinson/Sanofi-Pasteur contain a quadrivalent influenza vaccine. MicronJet®, developed by Nanopass, contains four hollow silicon MN (Kolli 2015). The product has shown improved immunogenicity to a full dose of influenza and zoster vaccines.

MN for delivery of influenza vaccine were initially evaluated pre-clinically by affixing the micron-scale needle tips to a traditional syringe (Alarcon et al. 2007). Metal (Kim et al. 2010) and dissolvable patches (Raphael et al. 2010; Sullivan et al. 2010) containing influenza vaccine strains have been developed. The administration of influenza vaccines by these MN patches proved to produce a wide range of immune responses. Studies to support these patches' efficiency showed high responsiveness of $CD4^+$ and $CD8^+$ cells in spleen and draining lymph nodes (Kim et al. 2010; Koutsonanos et al. 2012). These MN patches not only presented an increase in specific influenza IgG A antibody levels but were also capable of activating memory B-cells. Simultaneously, dose-sparing effects were reported in the influenza vaccine delivery with the MN patches compared to IM administration. Some polymers that could be used to contrive these patches are polyvinylpyrrolidone, hyaluronic acid (Choi et al. 2018), carboxymethylcellulose, polycaprolactone (Choi et al. 2019), and polyvinyl alcohol (Littauer et al. 2018). Single-layer or layer-by-layer dissolvable MN can be prepared to improve the vaccine payload.

The MN patch designs were bettered with adjuvants and stabilizers to improve vaccine delivery compared to IM administration. For instance, studies were aimed to enhance the efficacy of the immune response by incorporating adjuvants like cytokines, ligands, and peptides. In one such report, the addition of immunomodulatory cytokines like granulocyte-macrophage colony-stimulating factor (GM-CSF) in the dissolvable patch with influenza subunit vaccine produced robust, long-lived, and better antibody response. The improved vaccine effectiveness occurs as cytokines shape immune responses by modulating the cellular targets, thereby making them attractive adjuvants in a microneedle patch. (Littauer et al. 2018). Likewise, the effectiveness of toll-like receptors TLR7 and TLR3 as prospective ligand adjuvants were analyzed using imiquimod and poly(I:C), respectively, in an MN patch. These TLRs displayed a significant increase in IgG2a and induction of functional antibodies (Weldon et al. 2012).

The adequacy of peptides as adjuvants was demonstrated for the M2 extracellular domain (M2e) peptides found among influenza strains. A better immune response was observed when the peptide adjuvants were fused with natural TLR5 ligand flagellin (FliC) (Zhu et al. 2017). It was also pointed out that to boost the influenza vaccinations through dissolvable MN patches, co-immunization of the influenza split vaccine with M2e peptides produced cross-protection with the production of humoral and cellular immunity (Kim et al. 2019).

Similarly, complementing the microneedle formulation with stabilizers like trehalose, sucrose, glucose, inulin, chicory, and dextran helped prevent loss of influenza antigen activity. Additionally, other excipients that can be used as stabilizers comprise bovine serum albumin, arginine, heptagluconate, potassium buffers, and maltose. Trehalose was found to be the most attractive stabilizer as it protects and retains the vaccine strains' activity in the microneedle patch by preventing particle aggregation while also providing excellent thermal stability (Kim et al. 2010).

The use of a microneedle patch to deliver the influenza vaccine to geriatric patients undergoing chronic statin therapy was evaluated. This study is of significance as elderly patients exhibit an age-related decrease in the potency of antibody response. If they are exposed to chronic statin therapy, an even lower response to conventional IM influenza vaccination is observed. Therefore, to understand and mitigate the immunosuppressive effect from statins, a dissolvable influenza vaccine MN patch made from PVA was subjected to mice models administered with atorvastatin. It was reported that age-related decline in antibody response is lower after dermal vaccination through MNs than in systemic immunization group. Statin therapy dampened antibody titers in both MN and systemic groups. However, co-delivery of adjuvant with vaccines in MN could boost the humoral responses (Vassilieva et al. 2017).

Other innovations in microneedle designs for better delivery of vaccine includes a novel system called insertion-responsive MN (IRMNs) (Choi et al. 2018). Herein, the tips instantly separate after microneedle application. This study is particularly attractive to animal vaccinations. The studies carried out in guinea pigs suggested that the novel MN patch was capable of inducing antibody production, thus reducing the extent and duration of viral shedding when compared to intramuscular injections. In another study ceramic solid nanoporous microneedle arrays (npMNAs) have been explored that promised influenza vaccine delivery by using a mere dose of 3 μg in comparison to its corresponding IM dose of 15 μg. The animal model of female BALB/c mice showed an enhanced immunogenic effect on applying the patch behind the ears (Schepens et al. 2019).

Although influenza MNs exhibit dose-sparing effects, the extent to which the lowest dose of vaccine is required to produce adequate immunization was examined in a Nanopatch developed from an array of gold-coated silicon projections. This Nanopatch was deemed to be different from current microneedle devices as it had a high packing density of projections and was intended to deposit influenza antigen directly to the antigen-presenting cells (APCs) spread across the dermal layers (Fernando et al. 2010).

A few clinical trials have been reported for the delivery of influenza antigens using MN. In one such clinical trial, a trivalent influenza hemagglutinins vaccine was incorporated in a dissolvable MN. The MNs were compared to the influenza vaccine administered subcutaneously. The efficacy was deduced as a secondary endpoint by anti-hemagglutinin antibody in the sera. The MN application enhanced the production of antibodies against the B strain when compared to the subcutaneous route (Hirobe et al. 2015). However, this clinical trial did not assess MN's self-administration or with a negative control group. Therefore, these shortcomings were evaluated in another clinical trial wherein a dissolvable self-administered MN patch was developed and investigated. A strong antibody generation was observed; the patches were reliable to self-administer and demonstrated stability at 40 °C for 12 months (Rouphael et al. 2017).

In a different clinical trial, inactivated influenza vaccine (IIV) was evinced when added in a dissolvable MN patch. This clinical study focused on the acceptability of the MN patches by the patients and compared to intramuscular influenza vaccine delivery. Overall, from these clinical trials, it is inferred that MN patches were more effective than subcutaneous/intramuscular injections. They produced better immunogenicity, robustness, patient acceptance, and compliance (Frew et al. 2020).

9.2.2 Hepatitis

Most works for the development of MN-administered hepatitis B and hepatitis C virus vaccine have used DNA vaccines. Various authors have tried to determine the MN patch utility with and without adjuvants.

The commercially available conventional HBV vaccines administered use alum adjuvants. They are associated with injection site inflammation indicating perturbations of nodules and granuloma formation in humans and animals. Additionally, the HBV vaccines require cold-chain storage and lose their stability at higher temperatures. Therefore, for better immunization methods, a preventive and therapeutic design of MNs was suggested. This MN patch was made from silicon wafers that deliver plasmid DNA encoding hepatitis B virus (HBsAg) with an adjuvant Flt3L. The adjuvant used is an FMS-like tyrosine-kinase-3 ligand, a known hematopoietic growth factor activating dendritic cells (DCs) and natural killer cells (NK) to orchestrate primary and memory immune responses. Their studies showed potent cytotoxic T-cell lymphocytes (CTL), IFN-γ, and IL-12 activation in the MN patch group compared to IM groups (Zhou et al. 2010).

A few dissolving MN patches have been proposed to use cationic liposomes to construct a vaccine adjuvant delivery system (VADS) (Qiu et al. 2016; Wang et al. 2015). A novel cationic liposome encapsulating hepatitis B DNA vaccine was developed. It was adjuvanted with CpG oligodeoxynucleotide (CpG ODN) and fabricated in a PVP-K17 dissolving MN patch (Qiu et al. 2016). This patch was administered transcutaneously. It showed significant improvement in IgG A /IgG1 ratios and a balance in Th1/Th2 immune response compared to IM administration. The authors were also able to corroborate for a gradual release with a long duration of gene expression in the skin through the MN patch. In another case, mannose-PEG-cholesterol liposomes adjuvanted with lipid A (TLR2 ligand) were developed. This combination was loaded along HBsAg in an MN patch. The MN patch was supplemented onto the oral mucosa. The immune responses were evidenced with high levels of HbsAg-specific IgG in the sera and IgA in the salivary, intestinal, and vaginal secretions. There was a mixed Th1/Th2 response, an increase in $CD8^+$ T cells, and enhanced IgG2a and IFN-γ (Wang et al. 2015).

To improve MN patch thermostability and retention of HBV vaccine antigenicity, formulators proposed a dissolvable MN composed of polymers like hydroxyethyl starch (HES) and chondroitin sulfate. Along with these polymers, adjuvant saponin QS-21 was employed. The retention of antigenic activity was maintained at body temperature and 45 °C. A loss of 10% HBV antigenic activity in the patch at 50 °C was observed by the end of 6 months (Poirier et al. 2017). Recent studies have been aimed to prepare MN patches without any adjuvants in vaccine delivery as specific adjuvants resulted in skin irritation. Scientists attempted to coat HBsAg on a PLA patch or used a third-generation HBV vaccine (L-HBsAg) on a PVA patch (Na et al. 2020; Nguyen et al. 2019). Both these formulations showed promise in avoiding dermal toxicity. While the PLA patch exhibited moderate immunogenicity, the PVA patch had an improved antibody-inducing effect due to a third-generation L-HBsAg antigen. The PVA patch also incorporated 15% w/w trehalose as a stabilizer entailing better thermal stability to the MN patch vaccine.

There have been two pre-clinical evaluations for MN delivery of HBV vaccine. In the first study, the authors' constructed a solid metallic and a dissolvable polymeric MN patch delivering adjuvant-free HBV vaccine. Their examinations conducted in female mice and rhesus macaques demonstrated that these MNs were immunogenic and generated HBV surface antibody levels comparable to human seroprotection (Perez Cuevas et al. 2018). Their main goal was to ascertain the immunogenicity when aluminum was not present as an adjuvant. While the adjuvanted IM delivery produced higher antibody titers, the MN patches were also capable of stimulating cellular and humoral responses.

Later, the same authors (Choi et al. 2019) performed the pre-clinical evaluation of dissolvable MN containing adjuvant-free monovalent HBV vaccine and compared it to standard adjuvanted IM and adjuvant-free IM in rhesus macaques. The antibody levels produced were lower with the MN patch, but they did not differ significantly. After a booster dose, there was a more significant induction of antigen-specific cellular immune response along with high levels of inflammatory cytokines and chemokines. Their results suggested that the MN patch elicited antibodies specific to HBsAg that may be prognostic of human seroprotection.

There have been only two studies that have evaluated the efficacy of hepatitis C vaccine in a microneedle array. A stainless-steel MN patch was designed by coating plasmid encoding hepatitis C virus non-structured 3/4A protein. This study was compared with IM in mice and gene gun in humans. The MNs showed comparable priming of antigen-specific cytotoxic T-cell lymphocytes (CTLs). The CTL priming for MNs was homogenous with gene gun doses. Therefore, the immune response generated using MNs could be sufficient for DNA vaccination (Gill et al. 2010).

The pre-clinical results of the quadrivalent HCV virus-like particles (VLP) vaccine in pigs were evaluated using a microneedle injector intradermally. The immune responses were long-lived multi-genotype-specific, high neutralizing antibody (NAb), strong T cell, and acceptable granzyme B with normal Th1/Th2 cytokine levels. Large animal models have provided robust immune responses with their methodology (Christiansen et al. 2019).

9.2.3 Human Immunodeficiency Virus (HIV)

HIV-1 vaccine strain was evaluated for MN-mediated delivery. Dermal immunization is an excellent strategy to administer HIV-1 vaccine due to an abundance of dendritic cells (DCs) and Langerhans cells (LCs). Scientists have focused on the skin's rapid dissolution kinetics and the induction of polyfunctional $CD8^+$ T cells that occur through MN mediated HIV-1 vaccine delivery. Various adjuvants were employed to intensify the immune response generated from the MNs.

In one such work, a dissolving MN patch was constructed encompassing a trimeric recombinant HIV-1 CN54 clade C gp140 envelope protein. It was accompanied with a TLR4 agonist monophosphoryl lipid (MPLA) as an adjuvant loaded in a mucoadhesive copolymer, Gantrez®. With this MN design, a balanced Th1/Th2 profile was observed with high IgA levels in both serum and mucosa, especially in the vaginal mucosa. When boosted with a topical mucosal or intranasal vaccine dose, the MN administration further elicited IgG antibodies and lymphocyte proliferation. However, $CD8^+$ and CTL responses were not addressed (Pattani et al. 2012).

Similarly, another dissolvable MN patch was fabricated with live recombinant human adenovirus type 5 (rAdHu5) using the polymer Na-CMC. This study demonstrated thermostability of live vaccines in an MN patch. The design was used to exploit dendritic cells (DCs) to induce potent cellular immunity. The antigen combination in the MN patches maintained bioactivity, enabled painless skin administration, and evoked polyfunctional cytolytic $CD8^+$ T cell responses (Bachy et al. 2013). With the same polymer composition in another microneedle patch, a different HIV vaccine antigen was incorporated. The live recombinant human adenovirus type 5 vector (AdHu5) encoding HIV-1 gag was utilized in this case. The aim was to deliver a comprehensive phenotype study profiling of $CD8^+$ T cell and antigen precursors against the intradermal injection (Becker et al. 2015). This model was further evaluated in another study to demonstrate rapid dissolution kinetics in the skin. After dermal application of MN patch, long-lived antigen-specific $CD8^+$ T cells were observed in the vaginal and respiratory mucosa. The MNs also showed a polyfunctional $CD8^+$ T cells profile with greater levels of IFN-γ and granzyme B (Zaric et al. 2015).

The first study for oral administration of HIV-1 vaccine was reported by Ma et al. (2014). Two HIV-1 antigens, a virus-like particle and a DNA vaccine expressing gp160, were evaluated. These antigens were coated onto stainless steel MN sheets. The MN application to the oral mucosa in white rabbits produced greater serum IgG and mucosal IgA production.

Innovation in the MN system for delivery of HIV-1 vaccine was formulated in the form of stainless steel-coated HIV-1 p24 gag peptide conjugated with polypropylene sulfide nanoparticles (Caucheteux et al. 2016). The immune response was achieved by targeting resident immature DCs. At very low doses, nanoparticles coated MN patches provided $CD4^+$, $CD8^+$, and IFN-γ response. Recently, solid pyramidal MNs made from silk fibroin protein have been reported. The tips encapsulated a stabilized HIV trimer immunogen along with some TLR adjuvants on a dissolving polymer base. These MN promoted increased B-cell responses with high IgG titers. A sustained release of the vaccine from the MNs resulted in priming maximal antibody responses (Boopathy et al. 2019).

9.2.4 *Measles*

Researchers from the Georgia Institute of Technology have been developing measles microneedle patches (Joyce et al. 2018; Moon et al. 2013). They initially began the patch development by using the live attenuated measles vaccine coated on stainless steel arrays. The formulation was optimized using excipients like CMC, trehalose, and Lutrol F68. The immunogenicity of patch was tested in an animal model of female cotton rats *(Sigmodon hispidus)*. It was found that measles vaccination using MN can induce an antibody response analogous to the response after standard subcutaneous administration. Following these results, the group corroborated their pre-clinical studies in rhesus macaques. A dissolvable MN patch composed of sucrose, threonine, CMC, and PVA encapsulating a standard dose of measles vaccine was formulated. It was observed that more than 90% of the dose could be delivered. The patch maintained full potency after storage at 25 °C for 4 months. Compared to the standard subcutaneous injection, the patches produced equivalent neutralizing antibodies (Nab) and detectable IgM responses with an added advantage that reconstitution is not required with MNs. It was found that immune response was similar in rhesus macagues and human subjects after MN application.

Furthermore, human safety and immunogenicity was investigated. Two live attenuated vaccines (measles and rubella) were loaded on a single MN patch administered to infant rhesus macaques (Joyce et al. 2018). This study is the first report to show that protective titers of Nab against both measles and rubella can be developed. At the same time, the MN patch showed high-temperature stability for at least a month.

9.2.5 *Polio*

There are three virulent poliovirus strains (Type 1, Type 2, and Type 3). Two vaccines are commercially available to combat the virus, the oral polio vaccine (OPV) and the inactivated polio vaccine (IPV). While OPV provides a convenient administration mode, there is a constant risk associated with the reversion of the vaccine strains. This reversion leads to a more neuro-virulent profile resulting in vaccine-associated paralysis. On the other hand, IPV is more stable. The conventional mode of administration of IPV is through IM. It is also essential to understand that out of the two antigenic forms of IPV (C-antigen and D-antigen), the D-antigen expresses itself in the native infectious poliovirus and plays a crucial role in stimulating an immune response. Researchers have tried to develop IPV vaccine incorporated MN with a favorable D-antigen immunogenic response.

The layer-by-layer assembly of IPV and N-trimethyl chitosan chloride (TMC) as the polymer on pH-sensitive MNs (Van Der Maaden et al. 2015). Silicon microneedle arrays were modified with positively charged pyridine groups, followed by the coating of negatively charged IPV and positively charged TMC. IPV and TMC form the layers on the MN via electrostatic interactions. Ex vivo human skin studies showed that IPV-TMC MNs were successfully delivered in the epidermis and dermis and could release the vaccine within 5 minutes of application. The immunization study in rats indicated the production of IPV-specific robust C-antigen antibodies by the MNs. However, compared to IM, the D-antigen antibodies produced were lower. The authors speculated that a possible occurrence of D-antigen conversion to the C-antigen on the MN surface as IPV is very sensitive to surface-induced denaturation in the coating process.

Similarly, a dissolvable IPV-MN patch composed of sucrose and threonine was prepared. Moreover, an MN patch with a booster dose of IPV composed of maltodextrin was constructed. It was found that female infant rhesus macaques could produce neutralizing antibodies (Nab) with at least half the dose seroconverted against poliovirus type 1 (Edens et al. 2015). After the booster dose, a 100% seroconversion was observed. A better response was observed for poliovirus type 2, with increased Nab titers and 100% seroconversion. Consequent administration of the booster dose only further increased the immune response. However, a weak poliovirus type 3 response was observed. The authors conclude that the inadequate response is due to the erroneous pre-production of the type 3 polio strain's analytical method. The flawed analytical approach caused a lower dose production and resulted in the nonspecific D-antigen conformation of IPV. Overall, this MN patch was found to generate a sufficient immune response in infant rhesus macaques.

9.2.6 Dengue

FDA approved the first dengue vaccine, Dengvaxia™, against dengue fever in 2019 by Sanofi Pasteur. The vaccine has limitations of age-restricted use, while WHO has also recommended using it in patients previously infected. Alternative methods have been proposed by fabricating dengue vaccine coated on MN (Muller et al. 2019; Turvey et al. 2019).

Microneedle arrays prepared using solid poly(L-lactide) were drop-casted with saccharide-formulated virus (two live-attenuated serotypes were assessed, DENV-2 and DENV-4) (Turvey et al. 2019). The MNs could significantly preserve the serotypes at ambient temperature attributed to the presence of CMC and trehalose in the formulation. The authors found that monosaccharides like xylose and mannose and surfactants like tween 80 and Lutrol F68 deterred the viability of the vaccine in DENV-MNs to the levels of the unformulated live attenuated vaccine. Conversely, the use of polyvinylpyrrolidone or maltodextrin was encouraged to improve DENV-MNs stability. Their studies also indicated that DENV-MNs could generate comparable antibody responses to that of subcutaneous vaccine administration.

However, there are multiple challenges associated with using a live attenuated vaccine strain. Therefore, an MN model was proposed by using DENV surface glycoprotein E. This protein is an ideal candidate for a tetravalent subunit vaccine dry-coated on an ultra-high-density Nanopatch. The Nanopatch is effective in producing neutralizing antibody responses against all 4 DENV serotypes. This unique microneedle patch could achieve an enhanced immune response with fractional doses compared to IM or SC administration. The use of an adjuvant Quil-A further boosted the immune response as it overcame the weak immunogenic nature of protein E. However, Quil-A is not a suitable adjuvant for human use, and alternate adjuvants for improving protein E's immunogenicity are needed to be screened. The authors emphasized the need to further evaluate DENV serotype antigen ratios and epitope bias (Muller et al. 2019).

9.2.7 Rotavirus

Rotavirus causes severe diarrhea among children below 5 years of age. The conventional licensed oral vaccines RotaTeq® and Rotarix™ are likely to cause gastroenteritis and intussusception in infants. Therefore, scientists have explored the possibility of an inactivated rotavirus vaccine (IRV) administered intradermally. The MN projections coated with IRV could induce the antigen-presenting cells (APCs) and elicit a better immune response. The IgG and Nab titers specific to rotavirus were seen to be higher in MN as compared to intramuscular administration. Additionally, IM required a dose of 5 μg to produce an immune response, whereas MN required a reduced dose of 0.5 μg to provide a similar immune response (Moon et al. 2013).

To further evaluate the immunogenicity and protective efficacy of IRV administered with MNs, scientists provided proof of concept in gnotobiotic pigs. The animal model of gnotobiotic pigs does not just mimic but also exhibits heterogeneity in immune responses similar to humans. The delivery of IRV intradermally was achieved using an MN device, MicronJet600®, which is commercially used to deliver the influenza vaccine. The results of MN antibody titers were compared with intramuscularly administered IRV with aluminum hydroxide adjuvant. Both MN and IM delivery induced similar levels of IgA, IgG, and Nab titers. They could also develop a protective response against the rotavirus antigen shedding in stools and reduce the cumulative diarrhea scores of the gnotobiotic pigs (Wang et al. 2016).

9.2.8 Ebola Virus

There have been two reports of Ebola MN patches incorporating different vaccine types. The first study presented a unique Ebola immunization method using a DNA vaccine administered with an MN patch. The DNA vaccine was coated on polylactic-co-glycolic acid – polylysine/poly-γ-glutamic acid (PLGA-PLL/γPGA) nanoparticles incorporated into the MN patch. The data showed safety, immunogenicity, and stability and did not require cold-storage conditions (Yang et al. 2017).

The second study used the EBOV glycoprotein (GP) subunit vaccine on metallic MN and compared the immune responses to IM delivery. Higher and long-lasting antibody response was observed as compared to IM against the glycoprotein. Furthermore, when the vaccine was co-delivered with a saponin-based adjuvant, Matrix-M, the antibody responses induced were higher in the arrays than IM (Liu et al. 2018).

9.2.9 Japanese Encephalitis

A flavivirus causes Japanese encephalitis. Authors Dean et al. (2005) first showcased the utility of a live attenuated JE vaccine ChimeriVax™-JE delivered using MN. They chose an animal model of non-human primates, cynomolgus monkeys, that successfully demonstrated longer mean Nab titers sevenfold greater than subcutaneous injection.

9.2.10 Herpes Simplex Virus

Coated Nanopatches made from silicon with an HSV-2-gD2 plasmid DNA vaccine were constructed. The MN patch captured thousands of antigen-presenting cells. The total IgG anti-gD titers were significantly higher than the IM injection while using 1/10th the dose. There was no statistical difference in protective immune responses produced between MN and IM on the HSV-2 virus challenge. Further investigation is needed to understand if the Nanopatch can produce gD2-specific $CD4^+$ T cells (Chen et al. 2010).

9.2.11 Coronavirus

Sars-CoV-2 causes the novel coronavirus infection, and it has been found that spike (S) protein is pivotal for viral transmission and infection. When incorporated in a dissolving MN patch, Kim et al. (2020a) demonstrated that the trimeric recombinant S1 subunit vaccine could induce robust humoral response and elicit Nab titers compared to subcutaneous delivery. A dose sparing effect was observed, and currently, studies are underway to understand the MN thermostability (Kim et al. 2020a).

9.2.12 Tuberculosis – BCG Vaccine

Bacillus Calmette-Guérin (BCG) is the only approved live vaccine for humans extensively used against tuberculosis. The reconstituted BCG vaccine is injected into neonates or newborns intradermally as per WHO recommendations. However, its efficacy in averting adult pulmonary diseases is still uncertain. Therefore, to overcome these shortcomings, scientists have focused on developing better alternatives to boost BCG vaccination in adolescent life and improve administration mode.

One such work was presented by developing stainless steel MN dip-coated with polymeric CMC solution containing BCG vaccine. Other excipients used were Lutrol F-68 and trehalose (Hiraishi et al. 2011). The bacteria can undergo osmotic stress, threatening loss of efficacy during drying when delivered through the solid-state MN. Minimal osmotic pressure was achieved by using a salt-free suspension that retained activity and shelf-life stability. The addition of 15% w/v trehalose imparted better stabilization as compared to other cryoprotectants. The study carried out in guinea pigs showed a comparable increase in IFN-γ, TNF-α and bi-functional CD4$^+$ T-cell induction after 3–12 weeks by MN and the hypodermic needles. The authors focused on successfully providing evidence on the feasibility and applicability of BCG-coated MN. However, a loss of 25% viability was observed even in the presence of trehalose and surfactant.

In another study, the loss of BCG viability and stability issues were overcome by preparing a dissolvable MN patch (Chen et al. 2017a, b). The dissolvable MN patch composed of hyaluronic acid had a deep cave in its basal portion. In this cave, freeze-dried BCG powder was packed. Following microneedle application on mice skin, the arrays melted and exposed the powder. The powder was sucked and dissolved slowly in the interstitial fluid. It diffused within a day across the epidermis against the interstitial fluid influx. This novel MN design showed good BCG viability, mechanical MN strength, and thermostability at room temperature for about 60 days. Compared to traditional BCG vaccination, this novel MN produced little skin irritation while efficiently eliciting innate, humoral, and cellular immunity. Another BCG-encapsulated dissolvable MN patch was fabricated with sodium alginate and trehalose (Arshad et al. 2020).

There were two separate studies published by Yan et al. (2017, 2018) using different strategies to deliver the BCG vaccine through MN. Polysaccharide nucleic acid fractions of BCG termed as BCG-PSN were used in the earlier study. They encapsulated the BCG-PSN powder in the MN cave and ascertained its immune response. BCG-PSN is traditionally used as an immunomodulator for the treatment of various diseases. After a week of treatment, these BCG-PSN-MNs on female

BALB/c mice produced more IFN-γ and TFN-α from the peripheral $CD4^+$ T cells than BCG-PSN-IM injection group. Later, a plasmid DNA vaccine encoding the protein Ag85B of *M. tuberculosis* was used. The dose selected for both MN and IM groups showed low immunization capabilities. The MN needed to increase the loading capabilities using a proper adjuvant. However, the immune protection provided by the Ag85B-MNs showed promise of a new strategy against TB.

However, no studies have so far been able to verify the BCG-microneedle protection elicited against the bacterium when challenged by *M. tuberculosis*.

9.2.13 Anthrax

There have been two studies that used freeze-dried *Bacillus anthracis* protective antigen (rPA) (Mikszta et al. 2005, 2006). In the first study, authors compared the delivery of rPA through cutaneous, intramuscular, and nasal routes. For the cutaneous route, a specialized micro-enhancer array (MEA) device loaded with rPA was compared to IM injection. It was found that rPA was 90% seroconverted after a single dose through MEA delivery in female mice and white rabbits. The IM route showed only 20% seroconversion after a single dose, whereas the nasal route required at least three doses to produce a comparable immune response.

In the second study, the authors compared aluminum adjuvanted MN with the intramuscular injection and accounted for serum IgG and toxin-Nab titers in female New Zealand white rabbits. The MNs showed a dose sparing effect, which was evident during the initial 2 weeks following the primary and booster dose. In an aerosol spore challenge, a dose of 10 g of rPA exhibited 71% seroprotection through IM. In contrast, the dose showed a 100% seroprotection in MN.

In a pre-clinical study, rhesus macaques were vaccinated with four different recombinant protein vaccines using stainless steel MN attached to a syringe (Morefield et al. 2008). There were physical, chemical, and biological interactions between vaccines when administered as a combination, even though the total number of injections was reduced. The different recombinant proteins tested were protective antigen rPA (anthrax), botulinum neurotoxin type A (botulism), staphylococcal enterotoxin B (staphylococcus aureus), and recombinant fusion proteins of bacterial antigens CaF1 and LcrV (bubonic plague). Protective antibodies were found in rhesus macaques when they were challenged by lethal aerosol. This method provided an alternative for multiple vaccination while avoiding physical and chemical incompatibilities.

9.2.14 Whooping cough (Pertussis)

Bordetella pertussis is the causative bacterium of whooping cough. Zhu and group (Zhu et al. 2019) fabricated a dissolvable MN patch composed of PVA/sucrose/water in 8:6:15 ratio and evaluated it in female NIH mice to deliver pertussis toxin

(PT) vaccine. The conventional vaccination requires massive doses, while it also carries the risk of neurological damage and death in infants. This dissolvable MN patch could successfully deliver the PT vaccine intradermally and resulted in the quick formation of PT-specific IgG after first immunization compared to subcutaneous injection. Immune responses were produced primarily through Th1 and Th17 pathways. A dose sparing effect was observed with the MN's administration.

9.2.15 Gonorrhea

There is no vaccine approved to treat *Neisseria gonorrhoeae*. The immunogenicity of the experimental vaccine nanoparticle composed of formalin-fixed whole-cell-inactivated gonococcal microparticle encapsulated in a biodegradable crosslinked albumin matrix loaded on a dissolvable MN patch was tested (Gala et al. 2018). The MN patch was composed of the vaccine nanoparticle, trehalose, maltose, PVA, and HPMC. The immune response observed after MN application was compared to the gonococcal antigen solution or empty MN application. The immune response showed increased antigen-specific IgG titers, specific $CD4^+$, and $CD8^+$ T-cell lymphocytes. The mice also showed a notable increase in antigen-specific IgG titers at the end of 2 weeks.

9.2.16 Shigellosis

Shigella flexneri is the enteropathogen causing shigellosis. MN composed of aqueous blunge of 30% poly(methyl vinyl ether-co-maleic anhydride) commercially known as Gantrez® were prepared (Pastor et al. 2019). The MNs were loaded with an antigenic complex of outer membrane vesicle-based subunit vaccine of *Shigella*. The vaccine was prepared from the heat-inactivated *Shigella flexneri ΔtolR* strain. Ex vivo and in vivo studies of the MN patches indicated that the liberated vaccine reached the gastrointestinal target site after 6 h of administration. The in vivo studies in female mice demonstrated that a single 200 μg dose could produce *S. flexneri*-specific IgG and mucosal IgA antibodies. In the *S. flexneri* challenge study, the vaccine gave adequate protection after four weeks of immunization.

9.2.17 Malaria

Researchers have used silicon MN to deliver malarial recombinant vaccines that provide multi-stage malaria immunity to tackle the infection caused by *Plasmodium* species. Silicon MN containing the recombinant poxvirus modified vaccinia virus Ankara (MVA) expressing the *Plasmodium berghei* antigen were investigated

(Carey et al. 2011, 2014). The results showed that MN with small pore volume could induce a $CD8^+$ T-cell immune response with equivalent or a more significant proportion of memory T cells than conventional needle and syringe. The MN did not produce an inflammatory response on administration and exhibited a significant reduction in inflammatory markers in draining lymph nodes compared to intradermal injection.

Using the same microneedle design, an adenovirus-based malaria vaccine (HAdV5) expressing the C-terminal region of *Plasmodium yoelii* was fabricated. The results showed equivalent or increased antibody responses as compared to intradermal delivery. However, there was a reduction in anti-vector responses (Carey et al. 2014). Therefore, repeated use of the same adenovirus vector was required to produce adequate anti-vector titers. A combination of an intradermal and MN immunization provided the highest immunization against the blood-stage malaria challenge (Fig. 9.2). In a similar study, a recombinant simian adenovirus-vectored vaccine (ChAd63.ME-TRAP) was utilized. The MN-delivered vaccine showed protective efficacy similar to intradermal administration when challenged with live *Plasmodium berghei* sporozoites. The authors correlated the results to the "total array volume" to better understand the transgenic expression (Pearson et al. 2015).

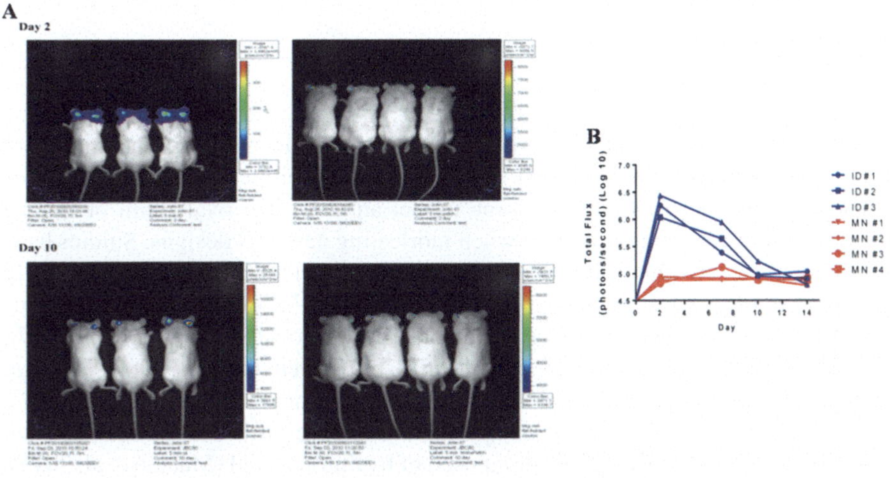

Fig. 9.2 Antigen expression levels by HAdV5 when delivered by intradermal or MN route. (**a**) In vivo imaging of luciferase expression following delivery of HAdV5-expressing luciferase to BALB/c mice by the ID (n = 3, left panels) or microneedle array type D (n = 4, right panels). Photon emissions in each group are shown at day 2 or day 10 (5-minute exposure); the emissions were measured using an RLU scale. (**b**) Total flux (as defined by photons emitted per second) was assessed by injection of D-luciferin at day 2, 7, 10, and 14 in each animal and is graphed on a log scale. (Adopted from Carey et al. 2014)

9.2.18 Leishmaniasis

Stainless steel MN were coated with a plasmid DNA cocktail encoding *Leishmania infantum* histones (Moreno et al. 2017). The protective immunity produced by these MN patches was compared with SC or intradermal injection in mice. Compared to the other routes, the MN application showed increased ratios of IFN-γ, TGF-β, TNF-α, and various interleukins in the spleen and lymph nodes. MN observed a polarized Th1 cytokine response indicated with increased levels of chemokines (CXCL9, CXCL10, and CCL2). The MN or any other route could not confer protection when the BALB/c mice were challenged with *Leishmania major*. The authors have highlighted that the dose was too low to produce an adequate immune response; therefore, strategies to increase the payload are necessary.

On the other hand, a dissolving MN patch was contrived to contain the *Leishmania* recombinant antigen LiHyp1 incorporated in cationic liposomes with the adjuvant, TLR9 agonist, CpG (Lanza et al. 2020). When checked for its efficacy in mice with or without the adjuvant either by SC injection or MN, the antibody responses were significantly high with MN application. In contrast, the loading of cationic liposomes caused the elimination of the immune response. The MN design with just the vaccine and TLR9 agonist could produce a protective immune response when challenged by *Leishmania donovani* infection.

9.2.19 Model Antigens

MN loaded with model antigens provide the basis for understanding the microneedle system's penetration depth and immunization capabilities that can be used for transcutaneous vaccine delivery. Some model antigens like ovalbumin or bovine serum albumin are commonly used when fabricating new MN designs. Studies have exhibited that coated microneedle shows dose-dependent immune response mostly independent of penetration depth (Widera et al. 2006). MN showed improved antibody titers when embedded with groves and loaded with ovalbumin (Han et al. 2009).

Dissolving MN composed of chondroitin sulfate delivered the model antigen within three minutes, producing a robust immune response (Naito et al. 2012). Similarly, soluble antigens like ovalbumin and insoluble particulate antigen like Ad, when loaded onto dissolving MN composed of hyaluronic acid (Matsuo et al. 2012), were capable of eliciting an immune response. The layer-by-layer poly(lactide-co-glycolide) MN with cationic polymer and anionic lipid vesicles could successfully coat the lipid nanocapsules onto the MN with a favorable immune response (Demuth et al. 2012). In another study, an aqueous fusion of 20% w/v PMVE/MA encapsulating the model ova antigen was capable of targeting the skin dendritic cells to generate an immune response (Zaric et al. 2015). Coating MN with liposomes entrapping model antigen BSA was an effective vaccine adjuvant delivery system

(VADS) strategy (Zhen et al. 2015). Similarly, zein MN patch showed effective vaccine delivery (Bhatnagar et al. 2017).

Researchers have used a model heat-inactivated bacteria *Pseudomonas aeruginosa* (strain PA01) formulated in aqueous blends of 30% Gantrez® dissolving MNs for understanding the immune effects produced by bacterial vaccine. These MNs dissolved within five minutes of insertion. The challenge study with live *Pseudomonas* exhibited a lower bacterial load in the vaccinated group (Rodgers et al. 2018).

9.3 Microneedle Array Patches to Deliver Anti-cancer Agents

9.3.1 Breast Cancer

Since the administration of chemotherapeutics intravenously is associated with severe toxicity, dissolvable polymeric MNs have been investigated to deliver chemotherapeutics safely. Single and dual drug-loaded PVA-PVP dissolving MN encapsulating doxorubicin (DOX) and docetaxel (DTX) were fabricated (Bhatnagar et al. 2019). The efficacy between intratumoral injection and drug-loaded MNs was compared. Anti-tumor effectiveness was determined in the athymic breast cancer mouse model. The MN group showed a 100% survival rate despite the amount of drug loaded in MN was equivalent to the intratumoral injections. It was concluded that the dual drug-loaded (DOX + DTX) MNs were more effective at tumor inhibition than individual drugs.

Zein, a prolamine extracted from maize, has been used to prepare MNs loaded with tamoxifen and gemcitabine to treat breast cancer (Bhatnagar et al. 2017). These agents were either entrapped or coated onto zein MNs (ZMN). It was observed that 33% of the tamoxifen released in the epidermis when coated onto ZMN. In contrast, only 0.23% tamoxifen released in the epidermis when it was entrapped in ZMN. When poke and patch method was employed, 16.95% drug release was observed. In the case of gemcitabine, the poke and patch method was found to be more effective, resulting in the highest permeation of 41% of the loaded amount. Photosensitizer IR820 was used along with cisplatin to produce synergistic chemo-photodynamic therapy (Fu et al. 2020). The molecules were loaded into PVP/VA matrix for the MN fabrication. Anti-tumor efficacy was evaluated on female BALB/c mice by subcutaneously injecting 4 T1 cells. The treatment groups consisted of cisplatin MNs, IR820 MNs, and CDDP-IR820 MNs for comparison. Partial tumor inhibition was displayed in the mice treated with only cisplatin MNs and IR820 MNs. The CDDP-IR820 MNs showed excellent in vivo therapeutic effect with 90% tumor inhibition after laser irradiation (Fig. 9.3).

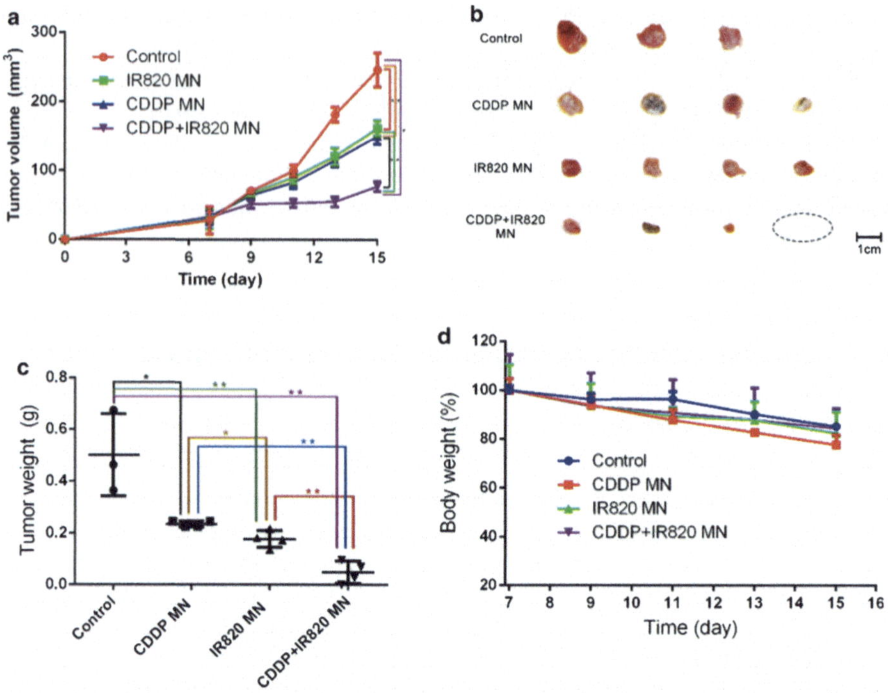

Fig. 9.3 The therapeutic effects of the MN patches against breast cancer (*represents statistic difference, $p < 0.05$). (**a**) The tumor growth profiles; (**b**) pictures of the tumor samples after the treatment; (**c**) the statistical analysis of the tumor weights after the treatment (*indicates $p < 0.05$, **means $p < 0.01$); (**d**) mice body weights throughout the treatment process. (Adopted from Fu et al. 2020)

9.3.2 Skin Cancer

5-Aminolevulinic acid (5-ALA)-coated MN arrays were fabricated for photodynamic therapy against skin cancer (Jain et al. 2016). The efficacy of 5-ALA-coated MNs to treat nodular subcutaneous tumors in mice was determined by subcutaneously injecting A-20 cells (B-lymphoma cells). After topical 5-ALA cream application, the tumor volume increased after 7 days resulting in a tumor burden of 507.8%. On the other hand, coated MNs showed significant tumor suppression with a burden of 43.3%. It should be noted that the coated MNs had lesser drug loading as well as a lesser time of application than the other treatment groups. Tip-loaded fast-dissolving MNs of sodium hyaluronate with 5-ALA were prepared (Zhao et al. 2018). The ALA-MN exhibited a tumor volume decrease of 44%, with the highest inhibition efficacy of 97%. It was concluded that the PDT efficacy was highest in the ALA-MN group (despite lower drug loading) as compared to ALA injection.

A dissolving HA-based microneedle patch was prepared for 5-ALA to target precancerous deep skin lesions (Champeau et al. 2020). The MN patch was effective in treating lesions of basal cell carcinoma, and invasive squamous cell carcinoma as the stratum corneum barrier was eliminated. The ex vivo rat skin studies showed effective permeation of 5-ALA with no discernible cytotoxicity. To treat a less intense form of skin lesions caused by *Propionibacterium acnes (P. acnes)*, a poly-ionic liquid-based MN containing salicylic acid (SA) was produced (Zhang et al. 2020). They were prepared by photocrosslinking of imidazolium-type ionic liquid with MN micromolds followed by anion exchange with SA anions. Improved therapeutic efficiency was seen in the in vitro studies performed on *P. Acnes* strains and in vivo studies conducted on mice models (Fig. 9.4a). Figure 9.4b and c shows that the ear thickness and redness caused by infection decreased significantly in 72 h when treated with SA-PIL-MN. Similarly, Fig. 9.4d depicts a significant increase in the % inhibition of acne on treatment with SA- PIL- MN.

The importance of inhibiting checkpoints that control T-cell count has generated interest over the past few years. The synergistic activity of anti-PD-1/PD-L1 and 1-methyltryptophan to combat melanoma delivered using a core-shell microneedle system (CSMN) was studied (Yang et al. 2020). A charged shell (chitosan or alginate) enclosing a hydrophilic core (polyvinyl alcohol) was used to deliver PD-L1 and 1-MT (1-methyl-D, L-tryptophan). The anti-tumor efficacy was evaluated in female mice with intratumoral injection of PD-L1/1-MT solution compared with PD-L1/1-MT MNs. Anti-tumor efficiency was found to be better with PD-L1/1-MT

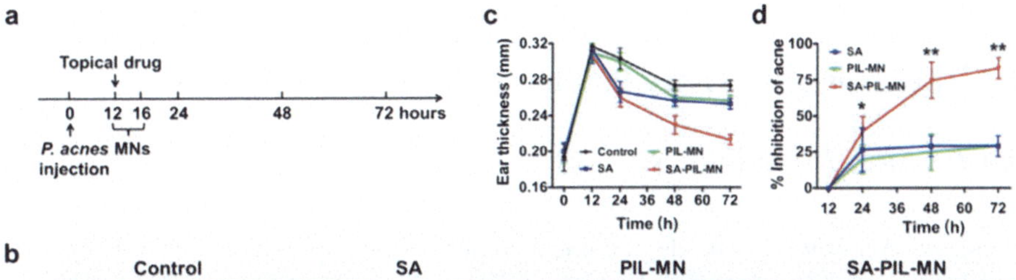

Fig. 9.4 (**a**) Skin healing evaluation of PIL-based MN patches in mouse *P. acnes*-infected model, (**b**) pictorial depiction of the skin acne therapy in a mouse model via topical SA, PIL-MN, and SA-PIL-MN treatment, (**c**) skin thickness changes, and (**d**) acne inhibition percentages of the left ears after subcutaneous injection of *P. acnes* in four groups, while the untreated right ear of each mouse was considered as the initial control group (0 h). Data in Fig. 4c and d are presented as mean ± SD, n = 3 for each group; ∗ represents $P < 0.05$, ∗∗ represents $P < 0.01$. (Adopted from Zhang et al. 2020)

MNs than the intratumoral injection. This can be attributed to the fact that chitosan-MN led to continuous drug permeation from the skin to the tumor site.

PVPVA MN were loaded with imiquimod to enhance intradermal delivery to treat nodular basal cell carcinoma (BCC) (Sabri et al. 2020). The studies performed on porcine skin were used to determine the effectiveness of these imiquimod PVPVA MNs compared to the commercial topical Aldara™ cream. Since nodular BCC are typically found 400 μm below the skin surface, MNs can be used as a minimally invasive technique to penetrate BCC tumor to deliver the drug.

Chen et al. (2016) combined photothermal therapy with chemotherapy to target superficial tumors. Polycaprolactone MN containing a near-infrared (NIR)-light-activatable lanthanum hexaboride and doxorubicin was developed. The tumor growth was significantly inhibited after MNs application.

pH-responsive polyelectrolyte multilayers (PEM)-coated polycaprolactone (PCL) MNs were used to treat skin lesions (Li et al. 2019). The multilayers consisted of two parts, gene-loaded layers (p53 DNA/polyethyleneimine) sandwiched between transition layers of pH-sensitive (PLL-DMA/polyethyleneimine). Two types of MN patches were fabricated; the first tr-MN patch contained transition layers of PLL-DMA/PEI and gene-loaded layers of model DNA/PEI; the second ntr-MN patch contained only a gene-loaded layer of model DNA/PEI. It was concluded that tr-MNP showed greater tumor inhibition and greater DNA release (33%) in simulated conditions, compared to only 4% release from ntr-MNP.

9.4 Microneedle Array Patches to Deliver Antibiotics

9.4.1 Skin Infections

Gantrez® AN 169 BF, a biodegradable acid anhydride copolymer, was used against skin infections as an antimicrobial agent. The MN made of Gantrez® AN 169 BF were evaluated against *E. coli*, *B. subtilis*, *E. faecalis*, *C. albicans*, *P. aeruginosa*, and *S. aureus* cultures. The zone of inhibition was found to be most significant for *B. subtilis* culture (45 mm), with *S. aureus* showing the least zone of inhibition (25 mm) (Boehm et al. 2012). Similarly, an evaluation of the polystyrene MN coated with Au and ZnO nanobrushes for their antibacterial activity was performed (Chew et al. 2019). Agar diffusion assay assessed the antimicrobial property in Gram-negative, *S. aureus*, and Gram-positive, *Salmonella*, models. The ZnO-Au-polystyrene MN were successful in reducing bacterial colony number by approximately 85.6%. In a recent study, the antibacterial activity of graphene oxide (GO)-based PVP MN was reported (Chen et al. 2020b). Among the dissolvable polymers used, PVP MN showed 95.8% inhibition of *S. aureus* and *E. coli* compared with 60% inhibition by HA and CMC MN.

Skin infections can be a result of biofilm formation due to the community-like behavior of skin-associated microbiome. The biofilm formation can lead to antibiotic resistance. Dissolvable PVP MN carrying chloramphenicol (CMP)-loaded gelatin nanoparticles to treat *Vibrio vulnificus* biofilms were developed (Xu et al. 2019). The active sites of infections were found to have an increased expression of the gelatinase enzyme. Thus, in situ drug release was triggered by employing drug-loaded gelatin particles. Biofilms are known to produce extracellular polymeric substances that obstruct the penetration of antibiotics, due to which the CMP-loaded GNPs were combined with MN array patch. The MN application showed a decrease in CFU/ml by 55% at 4 h and 63% at 8 h than the CMP solution. Thus, the integration of CMP-loaded GNPs with MN array could help in combating *Vibrio vulnificus* biofilms.

For the treatment of biofilms of *Staphylococcus aureus and Pseudomonas aeruginosa* at the burn and chronic wounds, doxycycline-loaded PVP-PVA dissolving MN were designed (Permana et al. 2020). The nanoparticles of doxycycline were prepared by employing biodegradable polymers, PLGA, and PCL, and further coated with chitosan. In vitro drug release studies of free DCL compared to the DCL NPs suggested that the drug release was bacteria sensitive. The colony biofilm model illustrated that 90% of bacterial biofilms were inhibited when incubated for a day.

In another study, the synergistic antibacterial effectiveness of chitosan-Zn^{2+} MNs was explored. When compared against chitosan MNs and the control group, the application of chitosan-Zn^{2+} MNs resulted in lower CFU in the presence of *E. coli.* (Gram-negative bacteria) and *S. aureus* (Gram-positive bacteria) (Yi et al. 2020).

9.4.2 Dermal Wounds

MN consisting of green tea extracts and hyaluronic acid have been investigated to accelerate dermal wound healing (Park et al. 2014). The antibacterial activity studies performed using agar plating assay against Gram-negative bacteria (*E. coli*, *S. typhimurium*, and *P. putida*) and Gram-positive bacteria (*B. subtilis* and *S. aureus*) showed up to 95% CFU reduction at 70% green tea extracts concentration. In vivo studies showed a dose-dependent decrease in CFU.

The effectiveness of carvacrol (CRL) polycaprolactone nanoparticles was evaluated after entrapment in PVA/PVP dissolving MN to treat chronic wounds (Mir et al. 2020). The ex vivo dermato-kinetic studies showed a higher CRL concentration in the dermal layers while using CRL nanoparticles MN compared to free CRL MN. In the porcine skin infection model, both CRL-MN and CRL nanoparticles MNA showed more than 90% decrease in microbial burden at the wounded site infected with *S. aureus*. However, in *P. aeruginosa*-infected site, CRL nanoparticles MNA showed a greater (87%) reduction in bioburden compared with CRL-MN (40%).

Similarly, chitosan MN encapsulated with VEGF were studied for wound healing in *S. aureus*- and *E. coli*-infected rat model (Chi et al. 2020). VEGF-chitosan MN showed a greater wound closure rate compared with blank chitosan MN and chitosan film (Fig. 9.5). It was suggested that the VEGF-PNIPAM-loaded chitosan MN promote wound healing by collagen deposition and angiogenesis.

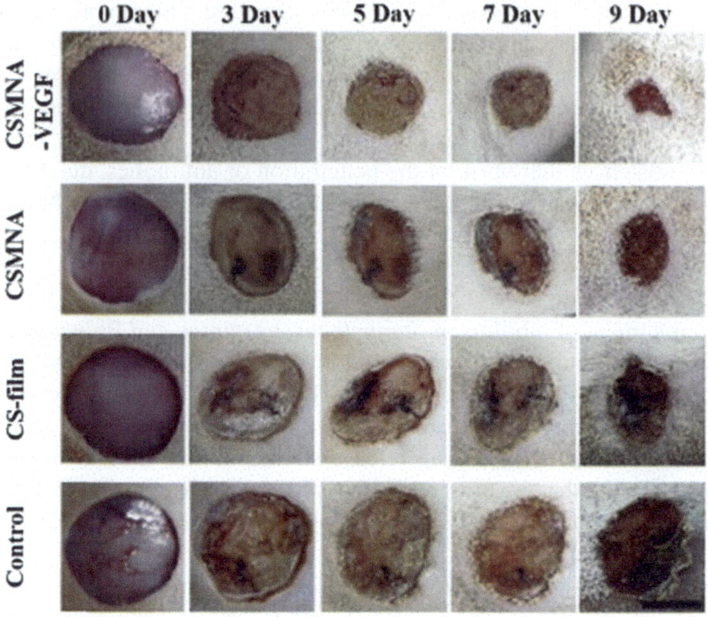

Fig. 9.5 Wound closure and H&E staining: images of the wounds treated with CSMNA-VEGF, CSMNA, CS-film, and PBS solution (control). (Adopted from Chi et al. 2020)

9.4.3 Acne

Reactive oxygen species (ROS)-responsive PVA MN array (MNA) loaded with clindamycin for the treatment of acne vulgaris caused by *P. acne* was developed (Zhang et al. 2018). Hyaluronic acid/diatomaceous earth was used as the MN base that further helped in healing owing to their absorptive properties. The in vitro antibacterial studies were conducted in *P. acnes*-induced inflammation mouse model. Studies revealed shrinkage in swollen skin by 90% and the onset of disappearance by day five.

9.4.4 Warts

Warts are skin infections caused by human papillomavirus (HPV). Bleomycin-coated PLA-PLGA microneedle patch was attempted to treat warts (Lee et al. 2017). Drug release studies in porcine skin model enhanced bleomycin permeation with the increase in MN insertion time from 10 to 15 min. The pharmacokinetics studies performed in the rat model showed that the C_{max} was attained in 50 min after administration of MN. The $t_{1/2}$ of bleomycin-coated MN was 36 min compared with 8 min after SC injection.

9.4.5 *Neonatal Sepsis*

Neonatal sepsis has been a leading cause of concern in infants. Currently, gentamicin IM injection is the first line of treatment. Gonzalez-Vazquez and group (González-Vázquez et al. 2017) developed MN made of hyaluronate and PVP containing gentamicin. In vitro studies in neonatal porcine skin model showed 10% and 75% gentamicin permeation 3 and 24 h after MN application, respectively. The pharmacokinetics studies revealed constant plasma concentration after MN application compared with IM administration (Fig. 9.6).

9.4.6 *Fungal Infections*

Antifungal agents including amphotericin B (Boehm et al. 2013), itraconazole (Boehm et al. 2016), miconazole (Boehm et al. 2014), and voriconazole (Boehm et al. 2015) were entrapped in MN using a piezoelectric inkjet printing technique.

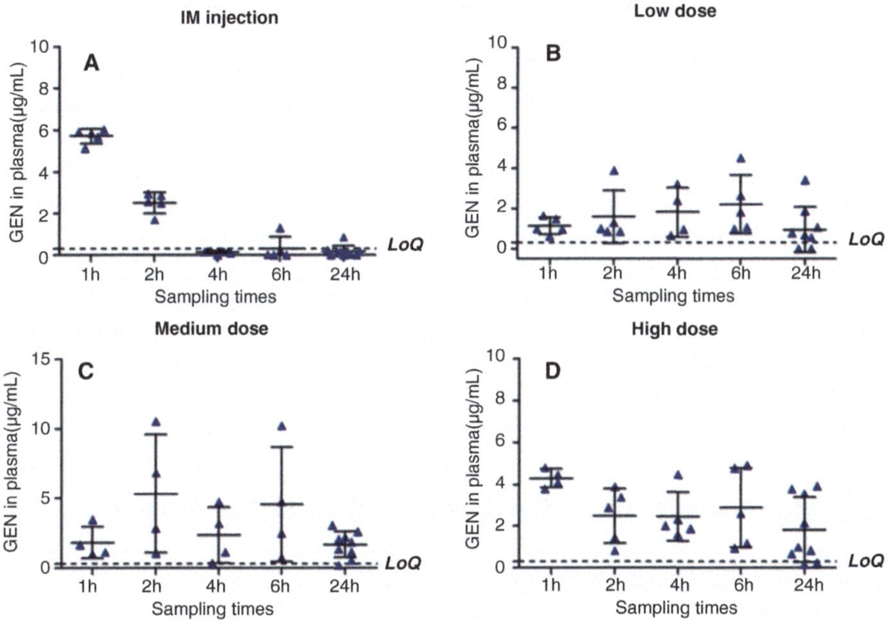

Fig. 9.6 (**a**) In vivo plasma profile of gentamicin following IM injection of gentamicin sulfate dissolved in sterilized water for injections (dose was 7.5 mg/kg). Means ± S.D., n = 5 at 1 h, 2 h, 4 h, and 6 h; n = 10 at 24 h. (**b**) Transdermal delivery by using one dissolving MN (low dose). Means ± S.D., n = 5 at 1 h, 2 h, and 6 h; n = 4 at 4 h; n = 10 at 24 h. (**c**) Transdermal delivery by using two dissolving MNs (medium dose). Means ± S.D., n = 4 at 1 h, 2 h, 4 h, and 6 h; n = 9 at 24 h. (**d**) Transdermal delivery by using four dissolving MNs (high dose). Means ± S.D., n = 4 at 1 h; n = 5 at 2 h, 4 h, and 6 h; n = 10 at 24 h. Dashed lines represent limit of quantification for GEN (LoQ = 0.30 µg/mL). (Adopted from González-Vázquez et al. 2017)

Agar plating assay of amphotericin B-coated Gantrez AN 169 BF MN resulted in 100% growth inhibition of *Candida parapsilosis*. Similarly, miconazole Gantrez AN 169 BF MN also exhibited antifungal activity against *Candida albicans*. In the case of itraconazole MN, an additional polymer, poly(methyl vinyl ether–co–maleic anhydride), coconut oil, and benzyl alcohol were incorporated for faster drug release. The antimicrobial activity showed similar zones of inhibition by itraconazole PGA and PMVA/MA MN containing coconut oil and benzyl alcohol. PGA MN coated with voriconazole through inkjet printing exhibited a zone of inhibition of 35 mm in diameter.

9.4.7 Corneal Infections

Bhatnagar et al. aimed at determining the release kinetics and antibacterial effectiveness of PVP-PVA MN loaded with besifloxacin HCl (Bhatnagar et al. 2018). In vitro studies indicated 100% release at 25 min. The drug-loaded MN showed 55% permeation of besifloxacin across excised cornea compared with 15% after application of besifloxacin solution at the end of 24 hours. The agar plate assay after *S. aureus* inoculation showed a similar zone of inhibition for besifloxacin MN and besifloxacin solution. In an ex vivo corneal infection (*S. aureus*) model, the CFU/ml was lowest for besifloxacin MN. The overall results suggested besifloxacin-loaded MNs to be a powerful alternative for the clinically existing eye drop formulations.

Likewise, amphotericin B-loaded MN were developed to treat fungal keratitis (Roy et al. 2019). The microneedle ocular patch loaded with amphotericin B imitated the curvature of the contact lens. To further improve the effective concentration within the corneal tissue, liposomal solution was incorporated into MN. Ex vivo studies performed with human excised cornea revealed significant differences in corneal drug retention after MN application against free liposomal formulation. The amphotericin B MN has significantly reduced the CFU of *Candida albicans* in ex vivo corneal infection model and in vivo rabbit model in comparison with the liposomal formulation (Fig. 9.7).

9.5 Microneedle Array Patches to Deliver Other Agents

9.5.1 Ocular Drug Delivery

Apart from delivering vaccines, anti-cancer agents, and antibiotics, MN array patches have been investigated to deliver drugs to the ocular region, to deliver various hormones, proteins, and immunosuppressants. MN for drug delivery to the eye were first tested by Jiang et al. (2007), wherein the ability of a drug-loaded polymeric solution coated on metallic MN to deliver drug via intrascleral and intracorneal routes was assessed. Three agents were tested, a model drug (pilocarpine), a protein (bovine serum albumin), and plasmid DNA. The strength of the MN to

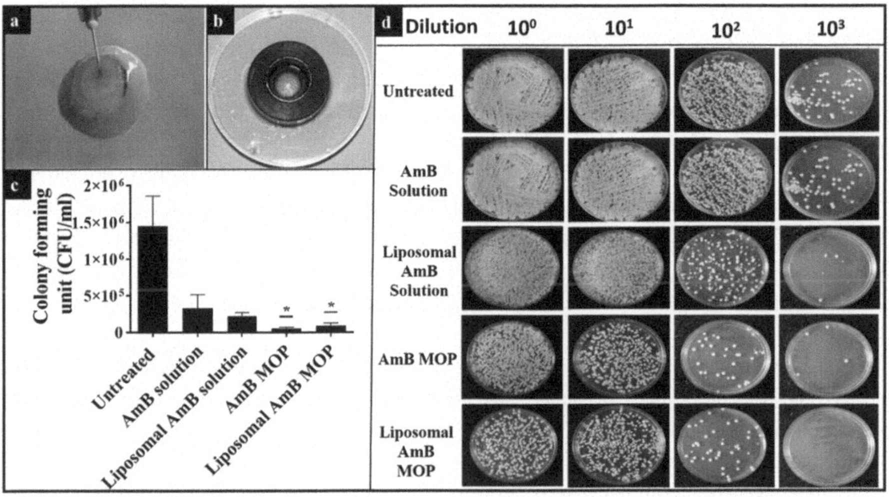

Fig. 9.7 Ex vivo study in *C. albicans*-infected excised cornea. (**a**) Photograph of intrastromal injection of *C. albicans* in excised cornea using 26G needle; (**b**) photograph of experimental setup for the application of AmB and liposomal AmB solutions; (**c**) colony forming units after 48 h of plating of corneal homogenate treated with different formulations; (**d**) photographs of *C. albicans* colonies formed after 48 h of plating corneal homogenate treated with different formulations. Asterisk (*) represents that the values are significantly different compared with all other groups. (Adopted from Roy et al. 2019)

penetrate the sclera was tested in vitro on human cadaveric scleral tissue, and in vivo studies were performed in rabbits to determine drug delivery using fluorescein and pilocarpine as the model molecules. Safety examinations were done to check whether the microneedle insertion site had any persistent perforations or inflammatory responses. The results obtained indicated that the MNs were of optimum strength and sharpness, delivered drug with an improved efficacy with no inflammatory response compared to the topical application of pilocarpine.

This study made way to studies on a safe, effective, and minimally invasive treatment of ocular diseases, including age-related macular degeneration (AMD), uveitis, and diabetic retinopathy. Another study investigated the use of suprachoroidal space for drug delivery to the back of the eye (Patel et al. 2011). Nano- and microparticles were loaded into the hollow MN of 1000 μm length. Sulforhodamine B dye was used as an indicative agent, and FluoSpheres and barium sulfate particles were used as nano- and microparticles, respectively. After a single MN application, the agents were seen to spread circumferentially around the eye and were able to target the suprachoroidal space. Furthermore, it was found that elevated intraocular pressure (IOP) caused a 100% success rate at the lowest infusion pressures and shortest MN lengths instead of normal IOP. It was concluded that MN insertion might be more effective at higher IOPs.

The use of MN for targeted delivery in the eye was investigated (Chun et al. 2015). It was hypothesized that by controlling the rheological properties of the injected formulation into the suprachoroidal space, the drug can be localized at the site or distributed broadly throughout the circumference of the eye. Single hollow MN was fabricated, wherein non-Newtonian shear-thinning polymer solutions were employed. Non-Newtonian moderately shear-thinning solutions were used when a larger circumferential spread was required. Solutions prepared in saline base spread over 29–42% of the suprachoroidal space immediately, whereas moderately non-Newtonian solutions made with hyaluronic acid showed 100% spreading. Highly non- Newtonian formulations localized the spread to 8.3–20% of the suprachoroidal space.

A complex MN design was reported (Hyun et al. 2018), wherein a microneedle pen (MNP) device was developed. The possibility of precisely controlling the depth of penetration of MN into the scleral tissue with a spring-controlled pen was explored. Drug delivery to the posterior eye by partial penetration of sclera by MN is an attractive approach for diseases concerning the outer layers of the retina including retinal detachment, uveitis, and exophytic retinoblastoma and in cases where multiple intravitreal dosing is required. The MNP developed has a SU-8 MN at the tip, which is dip-coated with the agents of choice. CMC salt and deionized water were used as viscosity enhancers, and rhodamine B dye was used for studies on the porcine sclera. Enhanced permeation and diffusion across the scleral tissue were observed and the indicative molecule could traverse the whole scleral tissue within 24 h. This time is reduced to 6 hrs when MN with "medium" and "deep" spring systems were used. Many such applicator-based MN treatments are studied by different groups wherein electrically driven, spring-loaded, pen-type syringe for hollow MNs and applicators using an electromagnet are made. Such devices can be brought into clinical use.

9.5.2 *Hormone Replacement Therapy*

Human growth hormone was entrapped in dissolving CMC and trehalose-based MN for the treatment of short stature and other endocrine disorders among children (Lee et al. 2011). Subcutaneous injection of hGH was considered positive control. CMC + trehalose MN showed a C_{max} fourfold higher and bioavailability fivefold higher than the positive control.

Powdered insulin was incorporated into CMC bubble MN (BMNs) (Lee et al. 2014). The micro-shells were fabricated such that they had a micro-cavity carrying insulin powder, which can be implanted into the dermis with the help of a customized applicator. The CMC matrix casing will dissolve, releasing the encapsulated insulin, thereby eliminating the reconstitution step. Optimization of the CMC concentration will help encapsulate the required amount of drug without compromising the mechanical strength. Compared to the dissolving MN, powder-carrying MN showed a prolonged release with T_{max} of 3.40 ± 0.60 and 5.40 ± 1.17, respectively.

9.5.3 Supplemental Agents

The use of MN for delivering supplemental drugs was studied (Abiandu and Ita 2019; Maurya et al. 2018). Iron replenishment therapy in anemic children and adults was tried by incorporating ferric pyrosulfate in dissolving MN made from hyaluronic acid. IDA-induced rats were treated with FPP-loaded MN patches, and recovery was tracked by measuring hematological and biochemical parameters. A 1.5 mg/dl recovery in the Hb levels per week was observed, followed by a steady boost in RBC counts and HCT levels (Maurya et al. 2018). A similar study was performed to treat hypokalemia with KCl transdermal delivery using solid MNs (Abiandu and Ita 2019). MN rollers were employed to increase permeation in porcine skin, and KCl delivery was measured by transdermal flux obtained from the permeation studies. A statistically significant increase was seen with the use of MN rollers in contrast to passive diffusion.

9.5.4 Arthritis

Dissolving MN carrying nanostructured lipid carriers (NLCs) with encapsulated hydrophobic drug aconitine significantly increased its plasma concentration (Guo et al. 2019). PVP-based dissolving MN were fabricated by UV crosslinking, in which a suspension of ACO-NLC was encapsulated. The resulting formulation was characterized for stability and efficacy. The T_{max} of ACO-NLC against ACO-NLC-MN was 8.00 ± 1.58 and 6.40 ± 2.61 h, respectively. Similarly, C_{max} was found to be 2570 ± 1630 and 5337 ± 625 ng/ml, respectively, indicating a greater efficacy.

Another drug, polydactin, was incorporated into the hydroxyl-β-cyclodextrin (HP-β-CD) complex, and, together, they were encapsulated in dissolving MN for treating acute gout arthritis (Chen et al. 2020c). Different polymers were studied as matrix materials among which PVP was found to show a better permeation profile. A fourfold increase in permeation was demonstrated after applying PD/ HP-β-CD dissolving MN upon increasing the dose of polydactin from 300 to 2400 μg.

Immunosuppressant, tacrolimus (TAC), and an NSAID diclofenac (DIC) were loaded in dissolving MN, in a combination approach (Yu et al. 2020). DIC was loaded on the tip and TAC in the interlayer. Upon skin insertion, the DIC reached the deeper layer into the articular cavity to treat arthritis, and TAC was retained within the epidermis for treating psoriatic lesions. The insertion depths were recorded using optical coherence tomography (OCT), and mechanical strength was determined using a texture analyzer. In vitro and in vivo skin permeation studies were performed along with in vivo visualization of drug distribution in the articular cavity comparing MN application and intra-articular injection (Fig. 9.8). The fluorescence intensity v/s time graph shows a 24 h retention of RhB in the articular cavity administered with MN instead of intra-articular injection for 12 h. The PASI (Psoriasis Area and Severity Index) and TEWL (transepidermal water loss) values were reduced significantly, which indicates mitigation and recovery of skin lesions.

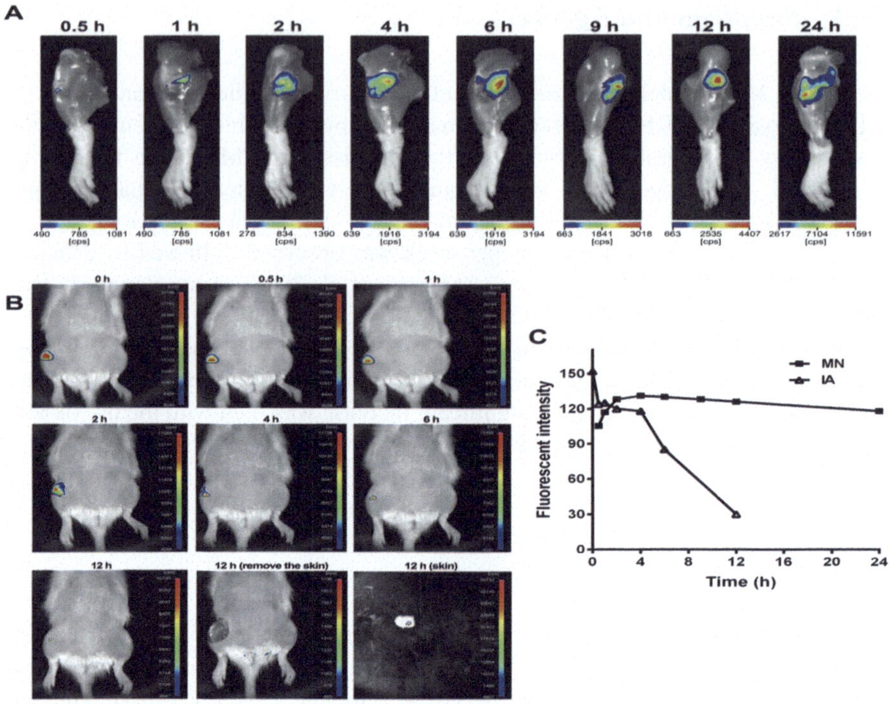

Fig. 9.8 In vivo imaging of rat knees after treated with layered MNs and intra-articular injection. In vivo images of RhB delivery from layered MNs into the articular cavity of rats throughout 24 h (**a**); RhB in the articular cavity of rats via intra-articular injection throughout 12 h (**b**); and the fluorescence intensity with time (**c**). In the layered MNs treated group, the treated skin covering the articular cavity was removed prior to imaging at each time point, while the treated skin in the intra-articular injection group was removed at 12 h post-injection. (Adopted from Yu et al. 2020)

9.5.5 Protein Delivery

Layer-by-layer deposition of proteins and DNA as polyelectrolytes multilayers (PEMs) on the surface of stainless steel MN was demonstrated (Saurer et al. 2010). Precise control on the amount of DNA or protein loading was obtained by controlling the thickness of the film. The PEMs were coated on the surface of MNs by an abbreviated dip coating method, with a prior piercing of the needles with a Parafilm® sheet, to avoid deposition of PEM on the baseplate. The MNs were characterized for film erosion, cell transfection assays, and release studies. A sustained release of DNA was achieved over 48 h.

Another way of delivering proteins or peptides through MN is by coating on tips of dense solid MN (Kapoor et al. 2020). A solid microstructured transdermal system (sMTS) is formulated for an investigational peptide A showing physical and chemical stability challenges. The injection-molded MN array was dip-coated with peptide A formulations. The absolute bioavailability was seen comparably to subcutaneous injection (88% and 74%, respectively) with significantly improved stability profiles.

9.5.6 *Brain Targeting*

Microneedle patches have also been used to deliver drugs to treat neuronal disorders such as depression and deliver anesthetic or anti-nociceptive drugs. Hydrogel-forming MN were fabricated to deliver esketamine to treat patients with depressive crisis states (Courtenay et al. 2020). Gantrez® and PEG 400 were used as a building material. The MNs showed plasma concentrations significantly elevated to 0.15–0.3 μg/ml over 24 h.

Similarly, fentanyl MN were fabricated for anti-nociceptive activity using hyaluronic acid as the casting material (Maurya et al. 2019). A relatively less amount of drug was loaded on the soluble patches than the amount required for systemic activity. The MN patches were tested on rat models (hot plate analgesia model) against the effects of dermal patches. A faster onset of analgesic activity was shown with the MN patch (30 mins) than the dermal patch (6 h). The paw withdrawal latency time significantly increased from 13.94 ± 1.76 s to 36.72 ± 2.09 s (Fig. 9.9). Thus, the mu-opioid receptors present on the peripheral somatic sites can be a target to achieve pain relief.

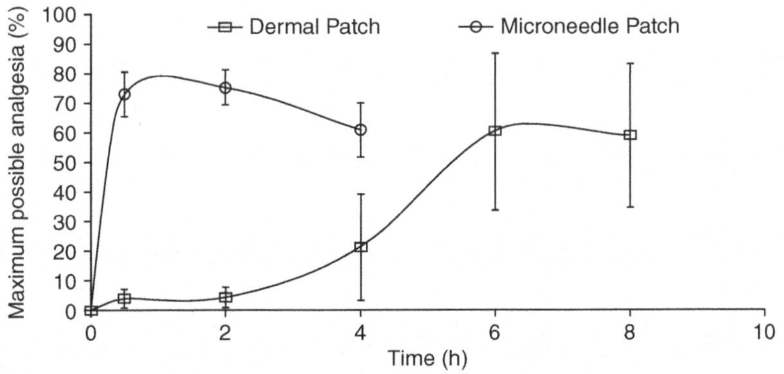

Fig. 9.9 Maximum possible analgesia (MPA) (%) in response to the thermal stimulus after application of fentanyl MN vs fentanyl dermal patch to the plantar surface of the hind paw in a rat model (hot plate analgesia). The MPA 0.5 h for fentanyl MN patch vs 6 h for fentanyl dermal patch. The paw withdrawal latency was measured immediately following removal of the test patches (microneedle or dermal). The control in the experiment is the untreated paw. The data points represented in the graph are an average of n = 6 ± S.D. (adopted from Maurya et al. 2019)

Another potent mu-opioid receptor antagonist, naltrexone (NTX), was used to treat opiate and alcohol dependence. The efficacy and bioavailability of NTX in an MN patch are tested against a conventional transdermal patch (Wermeling et al. 2008). A steady-state plasma concentration of NTX ($\approx$ 2.5 ng/ml) was observed within 48 h of MN patch application instead of negligible amounts after transdermal patch application. The MN patch was compared against oral tablet administration of NTX as well. A clinical study was performed on six healthy male and female volunteers. Serial blood samples and vital signs were recorded before and after patch application. Skin irritation and resealing kinetics study was also performed. A zero-order delivery was found with the maximum concentrations varying between 1.6 and 8.1 ng/ml.

Lidocaine, a hydrophobic drug, was coated as a solid dispersion on MN patches made from polyethylene glycol (PEG) (Ma and Gill 2015). Physical mixtures of lidocaine and PEG were analyzed for thermal behavior and viscosity changes. The results were used for the optimization process, and uniform coatings were obtained on microneedle patches by a novel molten coating process. In vitro studies on porcine skins proved a significantly higher delivery of lidocaine in just 3 mins than topical application of commercially available creams for 1 h.

Sumatriptan succinate for the treatment of migraine has been loaded in PVP-based dissolving MN (Ronnander et al. 2020). It was found that 30% w/w PVP DMN loaded with sumatriptan showed steady-state flux of 395 mg/cm^2/h over a period of 7 h. Further studies were made by the same group wherein iontophoresis was used to assist MN delivery (Ronnander et al. 2019). Permeation studies revealed that under mild electric current (0.5 mA/cm^2), cumulative amount of drug permeated in 6 h (Q_{6h}) and flux (Jss) increased by twofold.

Vinpocetine, a neuroprotective agent, shows poor bioavailability and a short half-life when given as oral tablet form. Ultradeformable liposomes were formulated with vinpocetine, loaded into fast-dissolving MNs (Srivastava and Thakkar 2020). The PVA-PVP MN showed threefold increase in the relative bioavailability compared with VPN suspension.

9.5.7 Miscellaneous

Adipose tissue transformation from WAT to BAT by "browning agent" is a novel way to treat obesity. Rosiglitazone nanoparticles were loaded into the tips of the dissolvable MN patch, which helped its delivery to transcutaneous adipocytes (Zhang et al. 2017). MeHA was selected as a matrix polymer due to its excellent biocompatibility and mechanical properties. Crosslinking of the polymer was done by using N, N′-methylenebisacrylamide, which helped in obtaining a sustained release. In vivo studies on mice were performed where biomarkers such as PPARγ target aP2 and CL 316243 were recorded. The metabolic rates of the mice were recorded while on treatment, and the treatment group mice were found to lose weight. The oxygen consumption was also increased due to the browning effect.

9.6 Summary

MN devices have shown tremendous potential in enhancing the availability of drugs in localized tissues. This is evident from the exponential increase in the number of publications reported yearly over the past decade. Especially, polymeric MN overcame the limitations of silica and metal MN in enhancing the drug payload and stability. Furthermore, polymeric MN allowed application in sensitive tissue such as cornea and sclera. MN have been increasingly investigated to deliver a wide variety of therapeutics, including small molecules with varied physicochemical properties, peptides, proteins, vaccine agents, and nucleic acids, as discussed in this chapter. Moving forward, there is a need to identify materials, including polymers, that provide a greater safety profile and address the challenges associated with the repeat application of MN.

References

Abiandu I, Ita K (2019) Journal of drug delivery science and technology transdermal delivery of potassium chloride with solid microneedles. J Drug Deliv Sci Technol 53(August):101216. https://doi.org/10.1016/j.jddst.2019.101216

Alarcon JB, Hartley AW, Harvey NG, Mikszta JA (2007) Preclinical evaluation of microneedle technology for intradermal delivery of influenza vaccines. Clin Vaccine Immunol 14(4):375–381. https://doi.org/10.1128/CVI.00387-06

Arshad MS, Fatima S, Nazari K, Ali R, Farhan M, Muhammad SA, Abbas N, Hussain A, Kucuk I, Chang MW, Mehta P, Ahmad Z (2020) Engineering and characterisation of BCG-loaded polymeric microneedles. J Drug Target 28(5):525–532. https://doi.org/10.1080/1061186X.2019.1693577

Azmana M, Mahmood S, Rebhi A, Kumar U (2020) Journal of drug delivery science and technology transdermal drug delivery system through polymeric microneedle: a recent update. J Drug Deliv Sci Technol 60(June):101877. https://doi.org/10.1016/j.jddst.2020.101877

Bachy V, Hervouet C, Becker PD, Chorro L, Carlin LM, Herath S, Papagatsias T, Barbaroux JB, Oh SJ, Benlahrech A, Athanasopoulos T, Dickson G, Patterson S, Kwon SY, Geissmann F, Klavinskis LS (2013) Langerin negative dendritic cells promote potent CD8+ T-cell priming by skin delivery of live adenovirus vaccine microneedle arrays. Proc Natl Acad Sci U S A 110(8):3041–3046. https://doi.org/10.1073/pnas.1214449110

Baek SH, Shin JH, Kim YC (2017) Drug-coated microneedles for rapid and painless local anesthesia. Biomed Microdevices 19(1). https://doi.org/10.1007/s10544-016-0144-1

Becker PD, Hervouet C, Mason GM, Kwon SY, Klavinskis LS (2015) Skin vaccination with live virus vectored microneedle arrays induce long lived CD8+ T cell memory. Vaccine 33(37):4691–4698. https://doi.org/10.1016/j.vaccine.2015.04.046

Bhatnagar S, Chawla SR, Kulkarni OP (2017) Zein microneedles for transcutaneous vaccine delivery: fabrication, characterization, and in vivo evaluation using ovalbumin as the model antigen. https://doi.org/10.1021/acsomega.7b00343

Bhatnagar S, Saju A, Cheerla KD, Gade SK, Garg P, Venuganti VVK (2018) Corneal delivery of besifloxacin using rapidly dissolving polymeric microneedles. Drug Deliv Transl Res 8(3):473–483. https://doi.org/10.1007/s13346-017-0470-8

Bhatnagar S, Bankar NG, Kulkarni MV, Venuganti VVK (2019) Dissolvable microneedle patch containing doxorubicin and docetaxel is effective in 4T1 xenografted breast cancer mouse model. Int J Pharm 556(December 2018):263–275. https://doi.org/10.1016/j.ijpharm.2018.12.022

Boehm RD, Miller PR, Singh R, Shah A, Stafslien S, Daniels J, Narayan RJ (2012) Indirect rapid prototyping of antibacterial acid anhydride copolymer microneedles. Biofabrication 4(1). https://doi.org/10.1088/1758-5082/4/1/011002

Boehm RD, Miller PR, Schell WA, Perfect JR, Narayan RJ (2013) Inkjet printing of amphotericin B onto biodegradable microneedles using piezoelectric inkjet printing. JOM 65(4):525–533. https://doi.org/10.1007/s11837-013-0574-7

Boehm RD, Miller PR, Daniels J, Stafslien S, Narayan RJ (2014) Inkjet printing for pharmaceutical applications. Mater Today 17(5):247–252. https://doi.org/10.1016/j.mattod.2014.04.027

Boehm RD, Daniels J, Stafslien S, Nasir A, Lefebvre J, Narayan RJ (2015) Polyglycolic acid microneedles modified with inkjet-deposited antifungal coatings. Biointerphases 10(1):011004. https://doi.org/10.1116/1.4913378

Boehm RD, Jaipan P, Skoog SA, Stafslien S, VanderWal L, Narayan RJ (2016) Inkjet deposition of itraconazole onto poly(glycolic acid) microneedle arrays. Biointerphases 11(1):011008. https://doi.org/10.1116/1.4941448

Boopathy AV, Mandal A, Kulp DW, Menis S, Bennett NR, Watkins HC, Wang W, Martin JT, Thai NT, He Y, Schief WR, Hammond PT, Irvine DJ (2019) Enhancing humoral immunity via sustained-release implantable microneedle patch vaccination. Proc Natl Acad Sci U S A 116(33):16473–16478. https://doi.org/10.1073/pnas.1902179116

Carey JB, Pearson FE, Vrdoljak A, McGrath MG, Crean AM, Patrick T, Walsh PT, Doody T, O'Mahony C, Hill AVS, Moore AC (2011) Microneedle array design determines the induction of protective memory CD8 + T cell responses induced by a recombinant live malaria vaccine in mice. PLoS One 6(7). https://doi.org/10.1371/journal.pone.0022442

Carey JB, Vrdoljak A, O'Mahony C, Hill AVS, Draper SJ, Moore AC (2014) Microneedle-mediated immunization of an adenovirus-based malaria vaccine enhances antigen-specific antibody immunity and reduces anti-vector responses compared to the intradermal route. Sci Rep 4:1–13. https://doi.org/10.1038/srep06154

Caucheteux SM, Mitchell JP, Ivory MO, Hirosue S, Hakobyan S, Dolton G, Ladell K, Miners K, Price DA, Kan-Mitchell J, Sewell AK, Nestle F, Moris A, Karoo RO, Birchall JC, Swartz MA, Hubbel JA, Blanchet FP, Piguet V (2016) Polypropylene Sulfide nanoparticle p24 vaccine promotes dendritic cell-mediated specific immune responses against HIV-1. J Investig Dermatol 136(6):1172–1181. https://doi.org/10.1016/j.jid.2016.01.033

Cha KJ, Kim T, Park SJ, Kim DS (2014) Simple and cost-effective fabrication of solid biodegradable polymer microneedle arrays with adjustable aspect ratio for transdermal drug delivery using acupuncture microneedles. J Micromech Microeng 24(11). https://doi.org/10.1088/0960-1317/24/11/115015

Champeau M, Jary D, Mortier L, Mordon S, Vignoud S (2020) A facile fabrication of dissolving microneedles containing 5-aminolevulinic acid. Int J Pharm 586(June):119554. https://doi.org/10.1016/j.ijpharm.2020.119554

Chen X, Kask AS, Crichton ML, McNeilly C, Yukiko S, Dong L, Marshak JO, Jarrahian C, Fernando GJP, Chen D, Koelle DM, Kendall MAF (2010) Improved DNA vaccination by skin-targeted delivery using dry-coated densely-packed microprojection arrays. J Control Release 148(3):327–333. https://doi.org/10.1016/j.jconrel.2010.09.001

Chen M-C, Lin Z-W, Ling M-H (2016) Near-Infrared Light-Activatable Microneedle System for Treating Superficial Tumors by Combination of Chemotherapy and Photothermal Therapy. ACS Nano 10(1):93–101. https://doi.org/10.1021/acsnano.5b05043

Chen F, Yan Q, Yu Y, Wu MX (2017a) BCG vaccine powder-laden and dissolvable microneedle arrays for lesion-free vaccination. J Control Release 255(2016):36–44. https://doi.org/10.1016/j.jconrel.2017.03.397

Chen Y, Chen BZ, Wang QL, Jin X, Guo XD (2017b) Fabrication of coated polymer microneedles for transdermal drug delivery. J Control Release 265(March):14–21. https://doi.org/10.1016/j.jconrel.2017.03.383

Chen M, Quan G, Sun Y, Yang D, Pan X, Wu C (2020a) Nanoparticles-encapsulated polymeric microneedles for transdermal drug delivery. J Control Release 325:163–175. https://doi.org/10.1016/j.jconrel.2020.06.039

Chen Y, Yang Y, Xian Y, Singh P, Feng J, Cui S, Carrier A, Oakes K, Luan T, Zhang X (2020b) Multifunctional graphene-oxide-reinforced dissolvable polymeric microneedles for transdermal drug delivery. ACS Appl Mater Interfaces 12(1):352–360. https://doi.org/10.1021/acsami.9b19518

Chen Z, Han B, Liao L, Hu X, Hu Q, Gao Y, Qiu Y (2020c) Enhanced transdermal delivery of polydatin via a combination of inclusion complexes and dissolving microneedles for treatment of acute gout arthritis. J Drug Deliv Sci Technol 55(October 2019):101487. https://doi.org/10.1016/j.jddst.2019.101487

Chew SWT, Zeng Y, Cui M, Chang H, Zheng M, Wei S, Zhao W, Xu C (2019) In situ generation of zinc oxide nanobushes on microneedles as antibacterial coating. SLAS Technol 24(2):181–187. https://doi.org/10.1177/2472630318812350

Chi J, Zhang X, Chen C, Shao C, Zhao Y, Wang Y (2020) Antibacterial and angiogenic chitosan microneedle array patch for promoting wound healing. Bioact Mater 5(2):253–259. https://doi.org/10.1016/j.bioactmat.2020.02.004

Choi HJ, Bondy BJ, Yoo DG, Compans RW, Kang SM, Prausnitz MR (2013) Stability of whole inactivated influenza virus vaccine during coating onto metal microneedles. J Control Release 166(2):159–171. https://doi.org/10.1016/j.jconrel.2012.12.002

Choi IJ, Kang A, Ahn MH, Jun H, Baek SK, Park JH, Na W, Choi SO (2018) Insertion-responsive microneedles for rapid intradermal delivery of canine influenza vaccine. J Control Release 286:460–466. https://doi.org/10.1016/j.jconrel.2018.08.017

Choi YH, Perez-Cuevas MB, Kodani M, Zhang X, Prausnitz MR, Kamili S, O'connor SM (2019) Feasibility of hepatitis B vaccination by microneedle patch: cellular and humoral immunity studies in rhesus macaques. J Infect Dis 220(12):1926–1934. https://doi.org/10.1093/infdis/jiz399

Christiansen D, Earnest-Silveira L, Grubor-Bauk B, Wijesundara DK, Boo I, Ramsland PA, Vincan E, Drummer HE, Gowans EJ, Torresi J (2019) Pre-clinical evaluation of a quadrivalent HCV VLP vaccine in pigs following microneedle delivery. Sci Rep 9(1):1–13. https://doi.org/10.1038/s41598-019-45461-z

Chun Y, Hee K, Edelhauser HF, Prausnitz MR (2015) Formulation to target delivery to the ciliary body and choroid via the suprachoroidal space of the eye using microneedles. Eur J Pharm Biopharm 95:398–406. https://doi.org/10.1016/j.ejpb.2015.05.020

Courtenay AJ, Mcalister E, Mccrudden MTC, Vora L, Steiner L, Levin G, Levy-nissenbaum E, Shterman N, Kearney M, Mccarthy HO, Donnelly RF (2020) Hydrogel-forming microneedle arrays as a therapeutic option for transdermal esketamine delivery. J Control Release 322(March):177–186. https://doi.org/10.1016/j.jconrel.2020.03.026

Davis SP, Martanto W, Allen MG, Prausnitz MR (2005) Hollow metal microneedles for insulin delivery to diabetic rats. IEEE Trans Biomed Eng 52(5):909–915. https://doi.org/10.1109/TBME.2005.845240

Dean CH, Alarcon JB, Waterston AM, Draper K, Early R, Guirakhoo F, Monath TP, Mikszta JA (2005) Cutaneous delivery of a live, attenuated chimeric flavivirus vaccine against Japanese encephalitis (ChimeriVax-JE) in non-human primates. Hum Vaccin 1(3):106–111. https://doi.org/10.4161/hv.1.3.1797

Demuth PC, Moon JJ, Suh H, Hammond PT, Irvine DJ (2012) Releasable layer-by-layer assembly of stabilized lipid Nanocapsules on microneedles for enhanced transcutaneous vaccine delivery, pp 8041–8051

Donnelly RF, Raj Singh TR, Woolfson AD (2010) Microneedle-based drug delivery systems: microfabrication, drug delivery, and safety. Drug Deliv 17(4):187–207. https://doi.org/10.3109/10717541003667798

Donnelly RF, Majithiya R, Singh TRR, Morrow DIJ, Garland MJ, Demir YK, Migalska K, Ryan E, Gillen D, Scott CJ, Woolfson AD (2011). Design, optimization and characterisation of polymeric microneedle arrays prepared by a novel laser-based micromoulding technique. Pharm Res 28(1):41–57. https://doi.org/10.1007/s11095-010-0169-8

Dsouza L, Ghate VM, Lewis SA (2020) Derma rollers in therapy: the transition from cosmetics to transdermal drug delivery. Biomed Microdevices 22(4):1–11. https://doi.org/10.1007/s10544-020-00530-3

Du G, Sun X (2020) Current advances in sustained release microneedles. Pharmaceut Front 02(01):e11–e22. https://doi.org/10.1055/s-0040-1701435

Edens C, Collins ML, Goodson JL, Rota PA, Prausnitz MR (2015) A microneedle patch containing measles vaccine is immunogenic in non human primates. Vaccine 176(3):139–148. https://doi.org/10.1016/j.vaccine.2015.02.074.A

Fernando GJP, Chen X, Prow TW, Crichton ML, Fairmaid EJ, Roberts MS, Frazer IH, Brown LE, Kendall MAF (2010) Potent immunity to low doses of influenza vaccine by probabilistic guided micro-targeted skin delivery in a mouse model. PLoS One 5(4). https://doi.org/10.1371/journal.pone.0010266

Frew PM, Paine MB, Rouphael N, Schamel J, Chung Y, Mulligan MJ, Prausnitz MR (2020) Acceptability of an inactivated influenza vaccine delivered by microneedle patch: results from a phase I clinical trial of safety, reactogenicity, and immunogenicity. Vaccine 38(45):7175–7181. https://doi.org/10.1016/j.vaccine.2020.07.064

Fu J-J, Li C-W, Liu Y, Chen M-Y, Zhang Q, Yu X-Y, Wu B, Li J-X, Du L-R, Dang Y-Y, Wu D, Wei M-Y, Lin Z-Q, Lei X-P (2020) The microneedles carrying cisplatin and IR820 to perform synergistic chemo-photodynamic therapy against breast cancer. J Nanobiotechnol 18(1):1–15. https://doi.org/10.1186/s12951-020-00697-0

Gala RP, Uz Zaman R, D'souza MJ, Zughaier SM (2018) Novel whole-cell inactivated Neisseria gonorrhoeae microparticles as vaccine formulation in microneedle-based transdermal immunization. Vaccine 6(3):1–17. https://doi.org/10.3390/vaccines6030060

Gill HS, Prausnitz MR (2007) Does needle size matter? J Diabetes Sci Technol 1(5):725–729. https://doi.org/10.1177/193229680700100517

Gill HS, Söderholm J, Prausnitz MR, Sällberg M (2010) Cutaneous vaccination using microneedles coated with hepatitis C DNA vaccine. Gene Ther 17(6):811–814. https://doi.org/10.1038/gt.2010.22

González-Vázquez P, Larrañeta E, McCrudden MTC, Jarrahian C, Rein-Weston A, Quintanar-Solares M, Zehrung D, McCarthy H, Courtenay AJ, Donnelly RF (2017) Transdermal delivery of gentamicin using dissolving microneedle arrays for potential treatment of neonatal sepsis. J Control Release 265(May):30–40. https://doi.org/10.1016/j.jconrel.2017.07.032

Guo T, Cheng N, Zhao J, Hou X, Zhang Y, Feng N (2019) Novel nanostructured lipid carriers-loaded dissolving microneedles for controlled local administration of aconitine. Int J Pharm 572(September):118741. https://doi.org/10.1016/j.ijpharm.2019.118741

Han M, Kim DK, Kang SH, Yoon HR, Kim BY, Lee SS, Kim KD, Lee HG (2009) Improvement in antigen-delivery using fabrication of a grooves-embedded microneedle array. Sensors Actuators B Chem 137(1):274–280. https://doi.org/10.1016/j.snb.2008.11.017

Henry S, McAllister DV, Allen MG, Prausnitz MR (1998) Microfabricated microneedles: a novel approach to transdermal drug delivery. J Pharm Sci 87(8):922–925. https://doi.org/10.1021/js980042+

Hiraishi Y, Nandakumar S, Choi SO, Lee JW, Kim YC, Posey JE, Sable SB, Prausnitz MR (2011) Bacillus Calmette- Guerin vaccination using a microneedle patch. Vaccine 29(14):2626–2636. https://doi.org/10.1016/j.vaccine.2011.01.042.Bacillus

Hirobe S, Azukizawa H, Hanafusa T, Matsuo K, Quan YS, Kamiyama F, Katayama I, Okada N, Nakagawa S (2015) Clinical study and stability assessment of a novel transcutaneous influenza vaccination using a dissolving microneedle patch. Biomaterials 57:50–58. https://doi.org/10.1016/j.biomaterials.2015.04.007

Huang H, Fu C (2007) Different fabrication methods of out-of-plane polymer hollow needle arrays and their variations. J Micromech Microeng 17:393–402. https://doi.org/10.1088/0960-1317/17/2/027

Hyun S, Hyun D, Sik C, Lee K, Hyoung J (2018) Depthwise-controlled scleral insertion of microneedles for drug delivery to the back of the eye. Eur J Pharm Biopharm 133(September):31–41. https://doi.org/10.1016/j.ejpb.2018.09.021

Ito Y, Kashiwara S, Fukushima K, Takada K (2011) Two-layered dissolving microneedles for percutaneous delivery of sumatriptan in rats. Drug Dev Ind Pharm 37(12):1387–1393. https://doi.org/10.3109/03639045.2011.576426

Jain AK, Lee CH, Gill HS (2016) 5-Aminolevulinic acid coated microneedles for photodynamic therapy of skin tumors. J Control Release 239:72–81. https://doi.org/10.1016/j.jconrel.2016.08.015

Jiang J, Gill HS, Ghate D, Mccarey BE, Patel SR, Edelhauser HF, Prausnitz MR (2007) Coated microneedles for drug delivery to the eye. Invest Ophthalmol Vis Sci 48(9):4038–4043. https://doi.org/10.1167/iovs.07-0066

Joyce JC, Carroll TD, Collins ML, Chen MH, Fritts L, Dutra JC, Rourke TL, Goodson JL, McChesney MB, Prausnitz MR, Rota PA (2018) A microneedle patch for measles and rubella vaccination is immunogenic and protective in infant rhesus macaques. J Infect Dis 218(1):124–132. https://doi.org/10.1093/infdis/jiy139

Kalluri H, Kolli CS, Banga AK (2011) Characterization of microchannels created by metal microneedles: formation and closure. AAPS J 13(3):473–481. https://doi.org/10.1208/s12248-011-9288-3

Kapoor Y, Milewski M, Dick L, Zhang J, Bothe JR, Gehrt M, Manser K, Nissley B, Petrescu I, Johnson P, Burton S, Moseman J, Hua V, Grunewald T, Tomai M, Smith R (2020) Coated microneedles for transdermal delivery of a potent pharmaceutical peptide, pp 1–10

Kim YC, Quan FS, Compans RW, Kang SM, Prausnitz MR (2010) Formulation and coating of microneedles with inactivated influenza virus to improve vaccine stability and immunogenicity. J Control Release 142(2):187–195. https://doi.org/10.1016/j.jconrel.2009.10.013

Kim JY, Han MR, Kim YH, Shin SW, Nam SY, Park JH (2016) Tip-loaded dissolving microneedles for transdermal delivery of donepezil hydrochloride for treatment of Alzheimer's disease. Eur J Pharm Biopharm 105:148–155. https://doi.org/10.1016/j.ejpb.2016.06.006

Kim MC, Kim KH, Lee JW, Lee YN, Choi HJ, Jung YJ, Kim YJ, Compans RW, Prausnitz MR, Kang SM (2019) Co-delivery of M2e virus-like particles with influenza split vaccine to the skin using microneedles enhances the efficacy of cross protection. Pharmaceutics 11(4):1–16. https://doi.org/10.3390/pharmaceutics11040188

Kim E, Erdos G, Huang S, Kenniston TW, Balmert SC, Carey CD, Raj VS, Epperly MW, Klimstra WB, Haagmans BL, Korkmaz E, Falo LD, Gambotto A (2020a) Microneedle array delivered recombinant coronavirus vaccines: immunogenicity and rapid translational development. EBioMedicine 55:102743. https://doi.org/10.1016/j.ebiom.2020.102743

Kim H, Theogarajan L, Pennathur S (2020b) A repeatable and reliable fabrication method for sharp, hollow silicon microneedles. In: 21st International Conference on Miniaturized Systems for Chemistry and Life Sciences, MicroTAS 2017, pp 344–345

Kim S, Yang H, Eum J, Ma Y, Fakhraei S, Jung H (2020c) Implantable powder-carrying microneedles for transdermal delivery of high- dose insulin with enhanced activity. Biomaterials 232(December 2019):119733. https://doi.org/10.1016/j.biomaterials.2019.119733

Kochhar JS, Zou S, Chan SY, Kang L (2012) Protein encapsulation in polymeric microneedles by photolithography. Int J Nanomedicine 7:3143–3154. https://doi.org/10.2147/IJN.S32000

Kolli CS (2015) Microneedles: bench to bedside. Ther Deliv 7(2):117–138

Koutsonanos DG, Vassilieva EV, Stavropoulou A, Zarnitsyn VG, Esser ES, Taherbhai MT, Prausnitz MR, Compans RW, Skountzou I (2012) Delivery of subunit influenza vaccine to skin with microneedles improves immunogenicity and long-lived protection. Sci Rep 2:1–10. https://doi.org/10.1038/srep00357

Lan X, Zhu W, Huang X, Yu Y, Xiao H, Jin L, Pu JJ, Xie X, She J, Lui VWY, Chen HJ, Su YX (2020) Microneedles loaded with anti-PD-1-cisplatin nanoparticles for synergistic cancer immuno-chemotherapy. Nanoscale 12(36):18885–18898. https://doi.org/10.1039/d0nr04213g

Lanza JS, Vucen S, Flynn O, Donadei A, Cojean S, Loiseau PM, Fernandes APSM, Frézard F, Moore AC (2020) A TLR9-adjuvanted vaccine formulated into dissolvable microneedle patches or cationic liposomes protects against leishmaniasis after skin or subcutaneous immunization. Int J Pharm 586. https://doi.org/10.1016/j.ijpharm.2020.119390

Lee JW, Choi S-O, Felner EI, Prausnitz MR (2011) Dissolving Microneedle Patch for Transdermal Delivery of Human Growth Hormone. Small 7(4):531–539. https://doi.org/10.1002/smll.201001091

Lee JW, Han MR, Park JH (2013) Polymer microneedles for transdermal drug delivery. J Drug Target 21(3):211–223. https://doi.org/10.3109/1061186X.2012.741136

Lee JW, Choi SO, Felner EI, Prausnitz MR (2014) Dissolving microneedle patch for transdermal delivery of human growth hormone. Small 7(4):531–539. https://doi.org/10.1002/smll.201001091.Dissolving

Lee HS, Ryu HR, Roh JY, Park JH (2017) Bleomycin-coated microneedles for treatment of warts. Pharm Res 34(1):101–112. https://doi.org/10.1007/s11095-016-2042-x

Li QY, Zhang JN, Chen BZ, Wang QL, Guo XD (2017) A solid polymer microneedle patch pretreatment enhances the permeation of drug molecules into the skin. RSC Advances 7(25):15408–15415. https://doi.org/10.1039/C6RA26759A

Li X, Xu Q, Zhang P, Zhao X, Wang Y (2019) Cutaneous microenvironment responsive microneedle patch for rapid gene release to treat subdermal tumor. J Control Release 314(October):72–80. https://doi.org/10.1016/j.jconrel.2019.10.016

Littauer EQ, Mills LK, Brock N, Esser ES, Romanyuk A, Pulit-Penaloza JA, Vassilieva EV, Beaver JT, Antao O, Krammer F, Compans RW, Prausnitz MR, Skountzou I (2018) Stable incorporation of GM-CSF into dissolvable microneedle patch improves skin vaccination against influenza. J Control Release 276(February):1–16. https://doi.org/10.1016/j.jconrel.2018.02.033

Liu Y, Ye L, Lin F, Gomaa Y, Flyer D, Carrion R, Patterson JL, Prausnitz MR, Smith G, Glenn G, Wu H, Compans RW, Yang C (2018) Intradermal immunization by Ebola virus GP subunit vaccines using microneedle patches protects mice against lethal EBOV challenge. Sci Rep 8(1):1–11. https://doi.org/10.1038/s41598-018-29135-w

Luo Z, Sun W, Fang J, Lee K, Li S, Gu Z, Dokmeci MR, Khademhosseini A (2018) Biodegradable gelatin methacryloyl microneedles for transdermal drug delivery. Adv Health Mater 1801054:1–9. https://doi.org/10.1002/adhm.201801054

Luzuriaga MA, Berry DR, Reagan JC, Smaldone RA, Gassensmith JJ (2018) Lab on a Chip for transdermal drug delivery †. 1223–1230. https://doi.org/10.1039/c8lc00098k

Ma Y, Gill HS (2015) Coating solid dispersions on microneedles via a molten dip coating method: development and in vitro evaluation for transdermal delivery of a water insoluble drug. 103(11):3621–3630. https://doi.org/10.1002/jps.24159.Coating

Ma Y, Tao W, Krebs SJ, Sutton WF, Haigwood NL, Gill HS (2014) Vaccine delivery to the oral cavity using coated microneedles induces systemic and mucosal immunity. Pharm Res 31(9):2393–2403. https://doi.org/10.1007/s11095-014-1335-1

Ma Y, Boese SE, Luo Z, Nitin N, Gill HS (2015) Drug coated microneedles for minimally-invasive treatment of oral carcinomas: development and in vitro evaluation. Biomed Microdevices 17(2). https://doi.org/10.1007/s10544-015-9944-y

Matsuo K, Yokota Y, Zhai Y, Quan Y, Kamiyama F, Mukai Y (2012) A low-invasive and effective transcutaneous immunization system using a novel dissolving microneedle array for soluble and particulate antigens. 161:10–17. https://doi.org/10.1016/j.jconrel.2012.01.033

Maurya A, Nanjappa SH, Honnavar S, Salwa M, Murthy SN (2018) Rapidly dissolving microneedle patches for transdermal iron replenishment therapy. J Pharm Sci 107(6):1642–1647. https://doi.org/10.1016/j.xphs.2018.02.011

Maurya A, Rangappa S, Bae J, Dhawan T, Srinivas S (2019) Evaluation of soluble fentanyl microneedles for loco-regional anti- nociceptive activity. Int J Pharm 564(April):485–491. https://doi.org/10.1016/j.ijpharm.2019.04.066

McAllister DV, Wang PM, Davis SP, Park JH, Canatella PJ, Allen MG, Prausnitz MR (2003) Microfabricated needles for transdermal delivery of macromolecules and nanoparticles: fabrication methods and transport studies. Proc Natl Acad Sci U S A 100(SUPPL. 2):13755–13760. https://doi.org/10.1073/pnas.2331316100

Migalska K, Morrow DIJ, Garland MJ, Thakur R, Woolfson AD, Donnelly RF (2011) Laser-engineered dissolving microneedle arrays for transdermal macromolecular drug delivery. Pharm Res 28(8):1919–1930. https://doi.org/10.1007/s11095-011-0419-4

Mikszta JA, Sullivan VJ, Dean C, Waterston AM, Alarcon JB, Dekker JP, Brittingham JM, Huang J, Hwang CR, Ferriter M, Jiang G, Mar K, Saikh KU, Stiles BG, Roy CJ, Ulrich RG, Harvey NG (2005) Protective immunization against inhalational anthrax: a comparison of minimally invasive delivery platforms. J Infect Dis 191(2):278–288. https://doi.org/10.1086/426865

Mikszta JA, Dekker JP, Harvey NG, Dean CH, Brittingham JM, Huang J, Sullivan VJ, Dyas B, Roy CJ, Ulrich RG (2006) Microneedle-based intradermal delivery of the anthrax recombinant protective antigen vaccine. Infect Immun 74(12):6806–6810. https://doi.org/10.1128/IAI.01210-06

Mir M, Permana AD, Ahmed N, Khan GM, Rehman, A. ur, & Donnelly, R. F. (2020) Enhancement in site-specific delivery of carvacrol for potential treatment of infected wounds using infection responsive nanoparticles loaded into dissolving microneedles: a proof of concept study. Eur J Pharm Biopharm 147(October 2019):57–68. https://doi.org/10.1016/j.ejpb.2019.12.008

Moon S, Wang Y, Edens C, Gentsch JR, Prausnitz MR, Jiang B (2013) Dose sparing and enhanced immunogenicity of inactivated rotavirus vaccine administered by skin vaccination using a microneedle patch. Vaccine 31(34):3396–3402. https://doi.org/10.1016/j.vaccine.2012.11.027

Morefield GL, Tammariello RF, Purcell BK, Worsham PL, Chapman J, Smith LA, Alarcon JB, Mikszta JA, Ulrich RG (2008) An alternative approach to combination vaccines: intradermal administration of isolated components for control of anthrax, botulism, plague and staphylococcal toxic shock. J Immune Based Therapies Vaccines 6:1–11. https://doi.org/10.1186/1476-8518-6-5

Moreira AF, Rodrigues CF, Jacinto TA, Miguel SP, Costa EC, Correia IJ (2020) Poly (vinyl alcohol)/chitosan layer-by-layer microneedles for cancer chemo-photothermal therapy. Int J Pharm 576(October 2019):118907. https://doi.org/10.1016/j.ijpharm.2019.118907

Moreno E, Schwartz J, Calvo A, Blanco L, Larrea E, Irache JM, Sanmartín C, Coulman SA, Soto M, Birchall JC, Espuelas S (2017) Skin vaccination using microneedles coated with a plasmid DNA cocktail encoding nucleosomal histones of Leishmania spp. Int J Pharm 533(1):236–244. https://doi.org/10.1016/j.ijpharm.2017.09.055

Muller DA, Depelsenaire ACI, Shannon AE, Watterson D, Corrie SR, Owens NS, Agyei-Yeboah C, Cheung STM, Zhang J, Fernando GJP, Kendall MAF, Young PR (2019) Efficient delivery of dengue virus subunit vaccines to the skin by microprojection arrays. Vaccine 7(4):1–13. https://doi.org/10.3390/vaccines7040189

Na YG, Kim M, Han M, Huh HW, Kim JS, Kim JC, Park JH, Lee HK, Cho CW (2020) Characterization of hepatitis B surface antigen loaded polylactic acid-based microneedle and its dermal safety profile. Pharmaceutics 12(6):1–14. https://doi.org/10.3390/pharmaceutics12060531

Naito S, Ito Y, Kiyohara T, Kataoka M, Ochiai M, Takada K (2012) Antigen-loaded dissolving microneedle array as a novel tool for percutaneous vaccination. Vaccine 30(6):1191–1197. https://doi.org/10.1016/j.vaccine.2011.11.111

Nguyen HX, Banga AK (2017) Fabrication, characterization and application of sugar microneedles for transdermal drug delivery. Ther Deliv 8(5):249–264. https://doi.org/10.4155/tde-2016-0096

Nguyen TT, Choi JAH, Kim JS, Park H, Yang E, Lee WJ, Baek SK, Song M, Park JH (2019) Skin immunization with third-generation hepatitis B surface antigen using microneedles. Vaccine 37(40):5954–5961. https://doi.org/10.1016/j.vaccine.2019.08.036

Park J, Allen MG, Prausnitz MR (2005) Biodegradable polymer microneedles: fabrication, mechanics and transdermal drug delivery. 104:51–66. https://doi.org/10.1016/j.jconrel.2005.02.002

Park SY, Lee HU, Lee YC, Kim GH, Park EC, Han SH, Lee JG, Choi S, Heo NS, Kim DL, Huh YS, Lee J (2014) Wound healing potential of antibacterial microneedles loaded with green tea extracts. Mater Sci Eng C 42:757–762. https://doi.org/10.1016/j.msec.2014.06.021

Pastor Y, Larrañeta E, Erhard Á, Quincooces G, Peñuelas I, Irache JM, Donnelly R, Gamazo C (2019) Dissolving microneedles for intradermal vaccination against shigellosis. Vaccine 7(4). https://doi.org/10.3390/vaccines7040159

Patel SR, Lin ASP, Edelhauser HF, Prausnitz MR (2011) Suprachoroidal drug delivery to the back of the eye using hollow microneedles. Pharm Res 28:166–176. https://doi.org/10.1007/s11095-010-0271-y

Pattani A, McKay PF, Garland MJ, Curran RM, Migalska K, Cassidy CM, Malcolm RK, Shattock RJ, McCarthy HO, Donnelly RF (2012) Microneedle mediated intradermal delivery of adjuvanted recombinant HIV-1 CN54gp140 effectively primes mucosal boost inoculations. J Control Release 162(3):529–537. https://doi.org/10.1016/j.jconrel.2012.07.039

Pearson FE, O'Mahony C, Moore AC, Hill AVS (2015) Induction of CD8+ T cell responses and protective efficacy following microneedle-mediated delivery of a live adenovirus-vectored malaria vaccine. Vaccine 33(28):3248–3255. https://doi.org/10.1016/j.vaccine.2015.03.039

Perez Cuevas MB, Kodani M, Choi Y, Joyce J, O'Connor SM, Kamili S, Prausnitz MR (2018) Hepatitis B vaccination using a dissolvable microneedle patch is immunogenic in mice and rhesus macaques. Bioeng Trans Med 3(3):186–196. https://doi.org/10.1002/btm2.10098

Permana AD, Mir M, Utomo E, Donnelly RF (2020) Bacterially sensitive nanoparticle-based dissolving microneedles of doxycycline for enhanced treatment of bacterial biofilm skin infection: a proof of concept study. Int J Pharm 2(April):100047. https://doi.org/10.1016/j.ijpx.2020.100047

Poirier D, Renaud F, Dewar V, Strodiot L, Wauters F, Janimak J, Shimada T, Nomura T, Kabata K, Kuruma K, Kusano T, Sakai M, Nagasaki H, Oyamada T (2017) Hepatitis B surface antigen incorporated in dissolvable microneedle array patch is antigenic and thermostable. Biomaterials 145:256–265. https://doi.org/10.1016/j.biomaterials.2017.08.038

Qiu Y, Guo L, Zhang S, Xu B, Gao Y, Hu Y, Hou J, Bai B, Shen H, Mao P (2016) DNA-based vaccination against hepatitis B virus using dissolving microneedle arrays adjuvanted by cationic liposomes and CpG ODN. Drug Deliv 23(7):2391–2398. https://doi.org/10.3109/10717544.2014.992497

Raphael AP, Prow TW, Crichton ML, Chen X, Fernando GJP, Kendall MAF (2010) Targeted, needle-free vaccinations in skin using multilayered, densely-packed dissolving microprojection arrays. Small 6(16):1785–1793. https://doi.org/10.1002/smll.201000326

Rodgers AM, McCrudden MTC, Vincente-Perez EM, Dubois AV, Ingram RJ, Larrañeta E, Kissenpfennig A, Donnelly RF (2018) Design and characterisation of a dissolving microneedle patch for intradermal vaccination with heat-inactivated bacteria: a proof of concept study. Int J Pharm 549(1–2):87–95. https://doi.org/10.1016/j.ijpharm.2018.07.049

Ronnander JP, Simon L, Koch A (2019) Transdermal delivery of sumatriptan succinate using iontophoresis and dissolving microneedles. J Pharm Sci 108(11):3649–3656. https://doi.org/10.1016/j.xphs.2019.07.020

Ronnander P, Simon L, Koch A (2020) Experimental and mathematical study of the transdermal delivery of sumatriptan succinate from polyvinylpyrrolidone-based microneedles. Eur J Pharm Biopharm 146(September 2019):32–40. https://doi.org/10.1016/j.ejpb.2019.11.007

Rouphael NG, Paine M, Mosley R, Henry S, McAllister DV, Kalluri H, Pewin W, Frew PM, Yu T, Thornburg NJ, Kabbani S, Lai L, Vassilieva EV, Skountzou I, Compans RW, Mulligan MJ, Prausnitz MR, Beck A, Edupuganti S et al (2017) The safety, immunogenicity, and acceptability of inactivated influenza vaccine delivered by microneedle patch (TIV-MNP 2015): a randomised, partly blinded, placebo-controlled, phase 1 trial. Lancet 390(10095):649–658. https://doi.org/10.1016/S0140-6736(17)30575-5

Roy G, Galigama RD, Thorat VS, Mallela LS, Roy S, Garg P, Venuganti VVK (2019) Amphotericin B containing microneedle ocular patch for effective treatment of fungal keratitis. In: International Journal of Pharmaceutics, vol 572. Elsevier B.V. https://doi.org/10.1016/j.ijpharm.2019.118808

Sabri AH, Cater Z, Gurnani P, Ogilvie J, Segal J, Scurr DJ, Marlow M (2020) Intradermal delivery of imiquimod using polymeric microneedles for basal cell carcinoma. Int J Pharm 589(August):119808. https://doi.org/10.1016/j.ijpharm.2020.119808

Saurer EM, Flessner RM, Sullivan SP, Prausnitz MR, Lynn DM (2010) Layer-by-layer assembly of DNA and protein containing films on microneedles for drug delivery to the skin. Biomacromolecules 11(11):3136–3143. https://doi.org/10.1021/bm1009443.Layer-by-Layer

Schepens B, Vos PJ, Saelens X, van der Maaden K (2019) Vaccination with influenza hemagglutinin-loaded ceramic nanoporous microneedle arrays induces protective immune responses. Eur J Pharm Biopharm 136(September 2018):259–266. https://doi.org/10.1016/j.ejpb.2019.02.002

Siddhapura K, Harde H, Jain S (2016) Immunostimulatory effect of tetanus toxoid loaded chitosan nanoparticles following microneedles assisted immunization. Nanomedicine: Nanotechnology Biology and Medicine 12(1):213–222. https://doi.org/10.1016/j.nano.2015.10.009

Srivastava PK, Thakkar HP (2020) Vinpocetine loaded ultradeformable liposomes as fast dissolving microneedle patch: tackling treatment challenges of dementia. Eur J Pharm Biopharm 156(August):176–190. https://doi.org/10.1016/j.ejpb.2020.09.006

Sullivan SP, Koutsonanos DG, Del Pilar Martin M, Lee JW, Zarnitsyn V, Choi SO, Murthy N, Compans RW, Skountzou I, Prausnitz MR (2010) Dissolving polymer microneedle patches for influenza vaccination. Nat Med 16(8):915–920. https://doi.org/10.1038/nm.2182

Turvey ME, Uppu DSSM, Mohamed Sharif AR, Bidet K, Alonso S, Ooi EE, Hammond PT (2019) Microneedle-based intradermal delivery of stabilized dengue virus. Bioeng Trans Med 4(2):1–11. https://doi.org/10.1002/btm2.10127

Uddin MJ, Scoutaris N, Economidou SN, Giraud C, Chowdhry BZ, Donnelly RF, Douroumis D (2020) 3D printed microneedles for anticancer therapy of skin tumours. Mater Sci Eng C 107:110248. https://doi.org/10.1016/j.msec.2019.110248

Ullah A, Khan H, Choi HJ, Kim GM (2019) Smart microneedles with porous polymer coatings for pH-responsive drug delivery. Polymers 11(11). https://doi.org/10.3390/polym11111834

Van Der Maaden K, Sekerdag E, Schipper P, Kersten G, Jiskoot W, Bouwstra J (2015) Layer-by-layer assembly of inactivated poliovirus and N-trimethyl chitosan on pH-sensitive microneedles for dermal vaccination. Langmuir 31(31):8654–8660. https://doi.org/10.1021/acs.langmuir.5b01262

Vassilieva EV, Wang S, Li S, Prausnitz MR, Compans RW (2017) Skin immunization by microneedle patch overcomes statin-induced suppression of immune responses to influenza vaccine. Sci Rep 7(1):1–9. https://doi.org/10.1038/s41598-017-18140-0

Wang T, Zhen Y, Ma X, Wei B, Li S, Wang N (2015) Mannosylated and lipid A-incorporating cationic liposomes constituting microneedle arrays as an effective oral mucosal HBV vaccine applicable in the controlled temperature chain. Colloids Surf B: Biointerfaces 126:520–530. https://doi.org/10.1016/j.colsurfb.2015.01.005

Wang Y, Vlasova A, Velasquez DE, Saif LJ, Kandasamy S, Kochba E, Levin Y, Jiang B (2016) Skin vaccination against rotavirus using microneedles: proof of concept in gnotobiotic piglets. PLoS One 11(11):1–15. https://doi.org/10.1371/journal.pone.0166038

Wang M, Hu L, Xu C (2017) Recent advances in the design of polymeric microneedles for transdermal drug delivery and biosensing. Lab Chip 17(8):1373–1387. https://doi.org/10.1039/C7LC00016B

Weldon WC, Zarnitsyn VG, Esser ES, Taherbhai MT, Koutsonanos DG, Vassilieva EV, Skountzou I, Prausnitz MR, Compans RW (2012) Effect of adjuvants on responses to skin immunization by microneedles coated with influenza subunit vaccine. PLoS One 7(7). https://doi.org/10.1371/journal.pone.0041501

Wermeling DP, Banks SL, Hudson DA, Gill HS, Gupta J, Prausnitz MR, Stinchcomb AL (2008) Microneedles permit transdermal delivery of a skin-impermeant medication to humans

Widera G, Johnson J, Kim L, Libiran L, Nyam K, Daddona PE, Cormier M (2006) Effect of delivery parameters on immunization to ovalbumin following intracutaneous administration by a coated microneedle array patch system. Vaccine 24(10):1653–1664. https://doi.org/10.1016/j.vaccine.2005.09.049

Xu J, Danehy R, Cai H, Ao Z, Pu M, Nusawardhana A, Rowe-Magnus D, Guo F (2019) Microneedle patch-mediated treatment of bacterial biofilms [research-article]. ACS Appl Mater Interfaces 11(16):14640–14646. https://doi.org/10.1021/acsami.9b02578

Yan Q, Liu H, Cheng Z, Xue Y, Cheng Z, Dai X, Shan W, Chen F (2017) Immunotherapeutic effect of BCG-polysaccharide nucleic acid powder on mycobacterium tuberculosis-infected mice using microneedle patches. Drug Deliv 24(1):1648–1653. https://doi.org/10.1080/10717544.2017.1391892

Yan Q, Cheng Z, Liu H, Shan W, Cheng Z, Dai X, Xue Y, Chen F (2018) Enhancement of Ag85B DNA vaccine immunogenicity against tuberculosis by dissolving microneedles in mice. Vaccine 36(30):4471–4476. https://doi.org/10.1016/j.vaccine.2018.06.025

Yang HW, Ye L, Guo XD, Yang C, Compans RW, Prausnitz MR (2017) Ebola vaccination using a DNA vaccine coated on PLGA-PLL/γPGA nanoparticles administered using a microneedle patch. Adv Healthc Mater 6(1):1–7. https://doi.org/10.1002/adhm.201600750

Yang P, Lu C, Qin W, Chen M, Quan G, Liu H, Wang L, Bai X, Pan X, Wu C (2020) Construction of a core-shell microneedle system to achieve targeted co-delivery of checkpoint inhibitors for melanoma immunotherapy. Acta Biomater 104:147–157. https://doi.org/10.1016/j.actbio.2019.12.037

Yi X, Wang C, Yu X, Yuan Z (2020) A novel bacterial biofilms eradication strategy based on the microneedles with antibacterial properties. Procedia CIRP 89:159–163. https://doi.org/10.1016/j.procir.2020.05.136

Yu J, Zhang Y, Ye Y, DiSanto R, Sun W, Ranson D, Ligler FS, Buse JB, Gu Z, Ho D (2015) Microneedle-array patches loaded with hypoxia-sensitive vesicles provide fast glucose-responsive insulin delivery. Proc Natl Acad Sci U S A 112(27):8260–8265. https://doi.org/10.1073/pnas.1505405112

Yu K, Yu X, Cao S, Wang Y, Zhai Y, Yang F, Yang X, Lu Y, Wu C, Xu Y (2020) Layered dissolving microneedles as a need based delivery system to simultaneously alleviate skin and joint lesions in psoriatic arthritis. Acta Pharm Sin B. https://doi.org/10.1016/j.apsb.2020.08.008

Zaric M, Lyubomska O, Poux C, Hanna ML, McCrudden MT, Malissen B, Ingram RJ, Power UF, Scott CJ, Donnelly RF, Kissenpfennig A (2015) Dissolving microneedle delivery of nanoparticle-encapsulated antigen elicits efficient cross-priming and th1 immune responses by murine langerhans cells. J Investig Dermatol 135(2):425–434. https://doi.org/10.1038/jid.2014.415

Zhang Y, Brown K, Siebenaler K, Determan A, Dohmeier D, Hansen K (2012) Development of lidocaine-coated microneedle product for rapid, safe, and prolonged local analgesic action. Pharm Res 29(1):170–177. https://doi.org/10.1007/s11095-011-0524-4

Zhang Y, Liu Q, Yu J, Yu S, Wang J, Qiang L, Gu Z (2017) Locally induced adipose tissue browning by microneedle patch for obesity treatment. ACS Nano. https://doi.org/10.1021/acsnano.7b04348

Zhang Y, Feng P, Yu J, Yang J, Zhao J, Wang J, Shen Q, Gu Z (2018) ROS-responsive microneedle patch for acne vulgaris treatment. Adv Ther 1(3):1800035. https://doi.org/10.1002/adtp.201800035

Zhang T, Sun B, Guo J, Wang M, Cui H, Mao H, Wang B, Yan F (2020) Active pharmaceutical ingredient poly(ionic liquid)-based microneedles for the treatment of skin acne infection. Acta Biomater 115:136–147. https://doi.org/10.1016/j.actbio.2020.08.023

Zhao X, Li X, Zhang P, Du J, Wang Y (2018) Tip-loaded fast-dissolving microneedle patches for photodynamic therapy of subcutaneous tumor. J Control Release 286(July):201–209. https://doi.org/10.1016/j.jconrel.2018.07.038

Zhen Y, Wang N, Gao Z, Ma X, Wei B, Deng Y, Wang T (2015) Multifunctional liposomes constituting microneedles induced robust systemic and mucosal immunoresponses against the loaded antigens via oral mucosal vaccination. Vaccine 33(35):4330–4340. https://doi.org/10.1016/j.vaccine.2015.03.081

Zhou Q, Wang F, Yang F, Wang Y, Zhang X, Sun S (2010) Augmented humoral and cellular immune response of hepatitis B virus DNA vaccine by micro-needle vaccination using Flt3L as an adjuvant. Vaccine 28(5):1357–1362. https://doi.org/10.1016/j.vaccine.2009.11.006

Zhu Z, Luo H, Lu W, Luan H, Wu Y, Luo J, Wang Y, Pi J, Lim CY, Wang H (2014) Rapidly dissolvable microneedle patches for transdermal delivery of exenatide. Pharm Res 31(12):3348–3360. https://doi.org/10.1007/s11095-014-1424-1

Zhu W, Pewin W, Wang C, Luo Y, Gonzalez GX, Mohan T, Prausnitz MR, Wang BZ (2017) A boosting skin vaccination with dissolving microneedle patch encapsulating M2e vaccine broadens the protective efficacy of conventional influenza vaccines. J Control Release 261(March):1–9. https://doi.org/10.1016/j.jconrel.2017.06.017

Zhu DD, Wang XL, Zhang XP, Ma JJ, Kong DL, Zhang MM, Guo XD, Wang C (2019) A dissolvable microneedle formulation of Bordetella pertussis subunit vaccine: translational development and immunological evaluation in mice. ACS Appl Bio Mater 2(11):5053–5061. https://doi.org/10.1021/acsabm.9b00730

Chapter 10
pH- and Ion-Sensitive Materials for Controlled Drug Delivery

Ankit Soni, Anuja Paprikar, Neeraj Kaushal, and Senshang Lin

10.1 Introduction

The conventional drug delivery approaches result in a peak in plasma drug concentrations, followed by a plateau and finally a decline in the curve of plasma drug level versus time. Consequently, these approaches of drug delivery may cause toxic plasma drug concentrations or ineffective plasma levels (Bawa et al. 2009). In addition, multiple-dose administration is required to maintain the effective drug level in the plasma during the course of therapeutic treatment. As a result, an interest in the design and development of dosage forms regulating drug release in a controlled manner from the dosage forms into the surrounding body fluids has increased in the past four decades (Jones 2004). Controlled release formulation approaches for controlling the rate of drug release were developed with aim of reduced toxicity and increase patient compliance and convenience. Controlled drug release systems have numerous advantages over conventional drug delivery systems. Mainly, plasma drug levels can be continuously maintained in a therapeutically desirable range; harmful side effects associated with systemic administration can be minimized; drugs having short biological half-lives may be protected from degradation.

Controlled drug delivery systems can be classified into various categories based on the mechanism controlling the release of the incorporated drug. One of the categories is diffusion-controlled systems, which can be further classified into reservoir (membrane device) and matrix (monolithic device) systems. Another category is chemically controlled systems, which are further classified into bio-erodible systems and pendant chain systems. Swelling and magnetically controlled systems are also the categories of controlled release dosage forms (Langer and Peppas 1981).

A. Soni · A. Paprikar · N. Kaushal · S. Lin (✉)
College of Pharmacy and Health Sciences, St. John's University, Queens, NY, USA
e-mail: linse@stjohns.edu

J.-C. Kim et al. (eds.), *Smart Nanomaterials in Biomedical Applications*,
Nanotechnology in the Life Sciences,
https://doi.org/10.1007/978-3-030-84262-8_10

The developments of all the categories of controlled drug delivery systems require the use of materials regulating the rate of drug release from the systems and subsequently provide a better therapeutic efficacy (Jones 2004). While therapeutically advantageous over conventional systems, controlled release systems remain insensitive to the changing metabolic states of the body. To regulate drug release profiles with changing physiological conditions, mechanisms capable of responding to these physiological variations must be provided. Ideally, a drug delivery system should respond to physiological requirements due to the changes in biological environment and subsequently adjust the drug-release profile accordingly (Bawa et al. 2009; Gupta et al. 2002).

For the developments of controlled drug release systems responding to the change of physiological conditions, stimuli-sensitive materials can be used. The stimuli-sensitive materials are able to respond sharply to small changes in physical or chemical conditions with relatively large phase or property changes. These materials can be sensitive to a number of factors that include pH, ionic strength, temperature, light, electrical field, biomaterials, solvent, chemical agents, and magnetic field (Kocak et al. 2017). In response to these factors, the stimuli-sensitive materials change in shape, surface characteristics, solubility, formation of an intricate molecular assembly, or a sol-to-gel transition (Ashok Kumar et al. 2007). The physical changes that occur are reversible, and the system is capable of returning to its initial state when the trigger is removed (Bawa et al. 2009). These materials are also termed as smart or intelligent, environmentally sensitive, and stimuli-sensitive polymers (Hoffman 2013; Jeong and Gutowska 2002).

Among all the recent advances in stimuli-**sensitive** materials, this chapter overviews the fields of pH- and ion-sensitive polymeric materials for controlled drug delivery. In addition, to achieve site-specific drug release, these materials have been widely employed in the controlled drug delivery systems due to their abilities to implement specific pH or ion gradients in the human body.

10.2 pH-Sensitive Materials for Controlled Drug Delivery

In response to a pH change of physiological environment, pH-sensitive materials can be used since they undergo structural and properties modifications due to the pH change. pH-**sensitive** materials have weak acidic or basic groups in their structures and are commonly used to define polyelectrolytes. Depending on the variation of the pH environment, they either accept or release protons, based on their functional group, and lead to physical transformation (Hafeli 2002). Over the past few years, research and applications of pH-sensitive materials have increased considerably. Especially, pH-sensitive materials offer the possibility of developing smart functional materials on demand. These materials have several potential applications in biomedical field, personal care, industrial processes, and water remediation (Ju et al. 2009; Kan et al. 2013; Y. Wang et al. 2016; Yu et al. 2017). Among all, the biomedical field is the most widely explored in the research of pH-sensitive materials to develop controlled drug delivery systems, gene carriers, and biosensors (Ankit Kumar et al. 2017; Liu et al. 2017; Mathew et al. 2017; Sousa et al. 2017).

10.2.1 Mechanism of Phase Transition of pH-Sensitive Materials

Under a particular pH environment, pH-sensitive materials transition themselves from coiled structure to expanded state. This behavior is advantageous to develop controlled release systems. When a drug-loaded controlled release systems is designed with these pH-sensitive materials, the materials will be in expanded state at the particular pH environment and subsequently regulate the release of drug in a controlled manner for prolonged period of therapeutic treatment.

There are various ionizable groups present in the pH-sensitive materials, which can be either acidic groups (i.e., carboxylic acids and sulfonic acids) or basic groups (tertiary amines). Similar to acidic or basic groups of monoacids or monobasic molecules, the acidic or basic groups of these polyelectrolytes can be ionized. However, because of the electrostatic effects exerted by other nearby ionized groups, complete ionization of these polyelectrolytes is more complicated. This differentiates the apparent constant of dissociation of polyelectrolytes from that of the related monoacid or monobasic molecules. By modifying the charges along the polymer backbone or electrolyte concentrations, the physical properties of pH-sensitive materials, such as solubility, chain conformation, configuration, and volume, can be modified. This modification can be regulated by electrostatic repulsion forces between adjacent molecules. Any factor that modifies electrostatic repulsion, such as pH, ionic strength, and type of counter-ions, affects the transformation between a tightly coiled structure and an expanded state polymer. This transformation from collapsed structure to highly expandable structure can be elucidated by modification in the osmotic pressure produced by mobile counter ions that neutralize the network charges (S. Dai et al. 2008).

For example, polyacid polymers have acidic groups (negative charge) in their structure that remain protonated and unionized at a lower pH environments, and thereby these polymers will remain collapsed. On the other hand, these polymers swell with increase in pH of the surrounding environment. The polybasic polymers have opposite behavior, since they transform from collapsed structure to highly expandable structure, when the pH of the surrounding environment reduces (Y. Dai et al. 2017). The pH at which phase transition occurs can be modulated by selection of ionization group with a pKa value that is desired for the pH range. Also, the phase transition can be modulated by the incorporation of hydrophobic functional groups into backbone of polymer (Na et al. 2004). The addition of a more hydrophobic moiety to the polymeric backbone leads to formation of more compact structure formation in the uncharged state and allows additional accused phase transition (Kocak et al. 2017).

10.2.2 Categories and Physicochemical Properties of pH-Sensitive Materials

Although pH-**sensitive** materials can be obtained from natural source, they can be synthesized using conventional or controlled radical polymerization techniques. Emulsion polymerization is among the most popular synthetic routes for preparing vinyl-based, pH-sensitive particulate systems, especially microgel systems (Gregory and Stenzel 2012)(Chuang et al. 2009). Broadly, the pH-sensitive materials can be categorized as polymers containing acidic groups and basic groups for the developments of controlled drug delivery systems and are described in brief below (Table 10.1).

10.2.2.1 Polymers Containing Acidic Groups

These polymers contain acidic functional groups in their chemical structures and are called polyacids or polyanions. The variations of pH in the external aqueous media decide the total number of negatively charged groups on the polymer chain

Table 10.1 List of commonly used pH-sensitive materials for controlled drug release characteristics

Category of polymer	Example of materials with their functional groups	References
Anionic	Carboxylic acid derivatives: Poly(acrylic acid) and poly(methacrylic acid), poly(methacrylic acid-co-methyl methacrylate), HPMC acetate succinate, and Carbopol grade polymers	Arya et al. (2017), Bazban-Shotorbani et al. (2017), Islam et al. (2004), Kang and Bae (2002), Soni (2014) and Yoshida et al. (2013)
	Sulfonic acid derivatives: Poly(2-acrylamido-2-methylpropane sulfonic acid), poly(methylpropane sulfonic acid), and poly(4-styrene sulfonic acid)	S. J. Kim et al. (2004), Liu et al. (2017) and Thakkar and Soni (2015)
	Sulfonamide groups: p-amino benzene-sulfonamide, sulfadimethoxine, sulfamethazine, sulfisomidine, and sulfapyridine	Kang and Bae (2002, 2003) and Liu et al. (2017)
Cationic	Polypyridine derivatives: Poly(2-vinyl pyridine) and poly(4- vinyl pyridine)	Korsmeyer et al. (1983) and Pinkrah et al. (2003)
	Imidazole derivatives: Poly(N- vinyl imidazole) and poly(4-vinyl imidazole)	Asayama et al. (2007) and Fodor et al. (2012)
	Other functional group derivatives: Poly(N,N-dialkyl aminoethyl methacrylates), chitosan, poly(l-lysine), and poly(ethylenimine), amino alkyl methacrylate copolymer, and poly(vinyl acetal diethylaminoacetate)	Gan et al. (2001), Soni (2015), Sousa et al. (2017), Velasco et al. (2011) and B. Wang et al. (2014)

which impart for hydrophilicity tuning in the media. The common example of polymers containing acidic groups is poly(carboxylic acids), such as poly(acrylic acid), poly(methacrylic acid), and poly(sulfonamide), which are derivatives of *p*-aminobenzene sulfonamide (Arya et al. 2017; Kang and Bae 2002). In addition, various types of polyacids can be classified by their functional groups.

10.2.2.1.1 Polymers Containing Sulfonic Acid and Sulfonamide Groups

Poly(2-acrylamido-2-methylpropane sulfonic acid) and poly(methylpropane sulfonic acid) are the most commonly used polymers containing sulfonic classes. These polymers constitute a class of effective polyelectrolyte hydrogels with a high degree of ionization. In their structure, they have pendant sulfonate groups and with pKa values in the range of 2 to 3. (S. J. Kim et al. 2004). These polymers represent a gradual transition over a wide-ranging pH interval, due to their high degree of ionization. However, this phase transition over wider pH range limits the applications of these polymers reacting to pH in a more narrow specific range.

To address this limitation of polymers containing sulfonic acid groups, a new class of pH-sensitive materials has been created. This class contains sulfonamide groups in the structure (S. Y. Park and Bae 1999). These polymers are p-amino benzenesulfonamide derivatives, which consist of sulfonamides group bound to aniline. The sulfonyl group from the amine nitrogen atom can function as an electron-withdrawing element, leading ionization and further attachment of hydrogen. The pKa of these polymers relies on the chemical composition of the substitution group such as sulfadimethoxine, sulfamethazine, sulfisomidine, and sulfapyridine. Also, these polymers have a small pH transition range between 0.2 and 0.3 units (Huh et al. 2012; Kang and Bae 2003).

10.2.2.1.2 Polymers Containing Carboxylic Acid Group

Poly(acrylic acid) is the most commonly studied polymer containing carboxylic acid attached to the polymer chain. At high pH values, available concentration of hydrogen ions is low. At low pH media, the COOH groups of the polymer begin to de-protonate and release their hydrogen ions, leading to an increase in negatively charged groups in the polymer chain. Hence, at high pH environment, the polymer converts from folded state to an expanded state. On the other hand, in low pH environment, the polymer does not swell. The mechanism of polymer conversion from folded state to an expanded state can be described by ionization of polymer in different pH environment based on the pKa value of polymer. For acidic polymer, the pH for initiating ionization of the polymer is determined by the value of the dissociation constant of the acid. The pKa values of the polymers depend on the monoacid, polymer structure, and molecular weight. For instance, poly(acrylic acid) having pKa value of 4.28 becomes predominantly uncharged at pH values below 4.28. However, at pH values below 4.28, the polymer has anionic charge and leads to its expansion (Bazban-Shotorbani et al. 2017; Gandhi et al. 2015; Kurkuri and Aminabhavi 2004).

10.2.2.2 Polymers Containing Basic Groups

Polymers containing basic functional groups are referred to as poly-bases or polycations. In response to external pH variation, these polymers can alter the number of positively charged groups by accepting protons below their pKa. The ionization-deionization transition of these polymers normally occurs from pH 7 to pH 11. Some of the usual monomers that can be used to synthesize pH-sensitive materials are the polycations containing either primary, secondary, or tertiary amine groups in the polymer chain and in particular, vinyl, acrylamide, meth-acrylamide, acrylates, and meth-acrylates comprising tertiary amine groups. The common examples of these polymers include poly(*N*,*N*-dialkyl aminoethyl methacrylates), chitosan, poly(l-lysine), and poly(ethylenimine) (Huh et al. 2012; Liu et al. 2017).

In comparison with the alkali-swellable carboxyl group as seen in polymer contains acid groups, these polyelectrolytes are acid-swellable groups. The polybasic groups are protonated under acidic conditions, which lead to an increase in the internal charge repulsions between adjacent protonated polybasic groups. This charge repulsion in the polymer groups contributes to an increase in the total dimensions and leads to swelling of polymer molecules. On the other hand, these groups become less ionized at higher pH values, which lead to decrease in the charge repulsion and increased polymer-polymer interactions. Hence, it leads to reduction in the polymer's overall hydrodynamic diameter and shrinkage at higher pH (Schmalz et al. 2010).

Also, nitrogen-containing groups, such as pyridine, pyrrolidine, morpholino, and imidazole belong to the group of cationic polymers. Out of all these, the pyridine and imidazole derivatives are the most commonly used pH-sensitive polycations having a lone pair of electrons available for proton bonding.

10.2.2.3 pH- Sensitive Natural Polymers

There are various synthetic polymers available that are advantageous in controlled drug delivery with their biodegradable nature. However, the natural polymers have better characteristics in terms of their good biocompatibility, abundance in nature, and ability to be readily modified by simple chemistry. These polymers are favorable for the development of protein-based molecules due to their nontoxic nature on administration, compatibility, and degradation behavior. They are widely used for small drug molecules as well for a better drug delivery system. In the past few decades, their applications in controlled drug delivery have been increased considerably. For instance, chitosan, dextran, hyaluronic acid, pullulan, and alginate are most common applied in controlled drug delivery systems (Alvarez-Lorenzo et al. 2013; Basu et al. 2015; Hoffman 2012).

10.2.3 Applications of pH-Sensitive Materials for Controlled Drug Delivery

In response to changes in the pH, these materials represent various potential applications due to their capacity to transition their structure and properties, such as volume, chain conformation, and solubility. Many formulation applications have been reported in past few decades and are available in the literature. Some of most attractive examples for application of pH-sensitive materials are discussed below.

As the human body has extreme physiological pH differences, pH-sensitive materials are a very promising medium for controlled drug delivery. The pH along the gastrointestinal tract is varying from pH 2 in the stomach, pH 7 in the colon, and up to pH 8.2 in the lower duodenum. Also, the pH varies in certain areas, such as certain tissues (and tumor zones) or subcellular compartments. Hence, based on the changes of pH, it is possible to formulate pH-sensitive drug delivery systems with the use of various formulation applications. These formulation applications include hydrogel, nanocarriers, solid dosage forms, and polymeric brushes.

10.2.3.1 Hydrogel

Hydrogels are the three-dimensional crosslinked networks of polymeric chains. Significant amount of water can be absorbed and hold within the interstitial spaces between the polymeric chains. Hydrogels are beneficial as potential materials for biomedical applications for controlled drug delivery owing to their softness, hydrophilicity, viscoelasticity, biodegradability, biocompatibility, and super-absorbency. The use of novel "smart polymer" involves the use of pH-sensitive materials to prepare the pH-sensitive hydrogels as a controlled drug delivery system. The swelling characteristic of pH-sensitive hydrogels is dependent on various factors including hydrophilicity, pH of the swelling medium, ionic charge, pK_a/pK_b values of ionizable groups, polymer concentration, and degree of ionization. Among all these factors, pH and the chemical nature of functional groups are the key factors contributing to the swelling characteristic of pH-sensitive hydrogels.

There are various reports available in the literatures that involve the applications of pH-sensitive hydrogels for controlled drug delivery. For instance, one of the applications of cationic hydrogels for antibiotic delivery in the stomach for treatment of ulcers has been prepared using chitosan and poly(ethylene imine) (Schmaljohann 2006). This hydrogel swells at low pH (acidic medium) environment owing to protonation of amino/imine groups. These protonated charged moieties on the polymer chains lead to repulsion between adjacent molecules resulted in the swelling of polymeric structure under acidic environment. In addition, this hydrogel system can be used as a carrier for an injectable drug delivery system. On the other hand, the anionic hydrogel containing carboxymethyl chitosan swells at alkaline pH environment and can be applied for controlled drug delivery. The degree of swelling is based on the ionization of the acidic functional groups presented on the polymer. It leads to ionization of the negatively charged groups on the polymer chains producing repulsion between adjacent molecules leading to swelling. This anionic hydrogel have been applied for controlled drug delivery at pH 7.4 environment (Du et al. 2015).

In addition to systemic applications, the pH-sensitive hydrogels have been employed for local drug delivery. The delivery of contraceptives and cervical cancer treatment agents in the vaginal canal can be accomplished using pH-sensitive hydrogel. In brief, the pH of vaginal cavity is slightly acidic (pH 3.8–4.5). With the use of polymer swelling at vaginal pH, a controlled drug delivery system can be developed. For instance, chitosan has high swelling characteristics in acidic environment. Furthermore, chitosan is cationic which promotes interaction of polymer with anionic mucosal surface of vaginal cavity. Hence, both muco-adhesive and controlled drug release features can be designed by using chitosan-based hydrogel. Furthermore, doxorubicin-loaded in situ hydrogel was developed using chitosan for the treatment of common cervical cancer as a controlled drug delivery system (Jalalvandi and Shavandi 2018; Rizwan et al. 2017).

10.2.3.2 Nanocarriers

Polymers containing carboxylic acid as an ionizable groups are the ideal candidates for fabricating pH-sensitive nanocarriers as controlled drug delivery systems. Some unique advantages, such as drug delivery, diagnostic, and image enhancement, are associated with nanocarriers. To improve therapeutic efficacy and minimize the undesirable side effects of the controlled drug delivery systems at specific sites in the body, the nanocarriers are generally considered (Felber et al. 2012). An example of nanocarriers can be pH-sensitive micellar systems, designed with elegant block copolymer system using poly(ethylene oxide) as the hydrophilic component and series of tertiary amine monomers as hydrophobic blocks (Ma et al. 2014). When the pH is higher than the pK_a of pH-sensitive micelle systems, the hydrophobic group remains hydrophobic. This effect leads to formation of self-assembly and quenching of fluorescence signal through photoinduced electron transfer. On the other hand, when pH is lower, the hydrophobic group converts to protonated making disassemble the micelles and increasing the fluorescence significantly. The dissembled micellar structure at lower pH (below 6) triggers the release of drug from pH-sensitive micelle systems into the endosomal or lysosomal cell compartments (Ma et al. 2014).

10.2.3.3 Polymer Brushes

Biomaterials can be distinguished according to supramolecular structures designed for medical device applications. Among the various biomaterials, polymer brushes and nanoparticle structures are the examples. Polymer brushes are monolayers of pH-sensitive material as a polymer connected to a solid support. Polymer brush surfaces have potential applications, such as low friction biological surfaces, organizers of nanoparticles, and stabilizers of nanoparticles for nuclear magnetic resonance contrast agents, mimetic and antifouling biological surfaces, and high dielectric polymer coatings (Kocak et al. 2017). pH-sensitive polymer brushes are composed of ionizable functional groups that can accept or donate protons in response to pH environmental changes. Polymeric brushes are generally composed of weakly acidic or basic groups with pK_a values about which the degree of ionization is significantly altered. Upon alteration to pH, it leads to alteration in the net charge of the functional group, which generates significant change in the hydrodynamic volume of the polymer chains to cause their rapid swelling. This rapid swelling results from the changes in the osmotic pressure exerted by surrounding counter ions. Moreover, the polymeric brushes are sensitive to effect of ionic strength. Moreover, pH-sensitive polymer brushes are typically also sensitive to changes in ionic strength, where screening of repulsive electrostatic interactions increases with increasing ionic strength resulting in the collapse of polymeric brushes (Chen et al. 2010).

Another interesting application of polymeric brushes is to deliver antimicrobial agent. The delivery system consists of a layer of poly(methyl methacrylate), covalently immobilized over a layer of antimicrobial peptides. The poly(methyl methacrylate) chains remain extended under physiological conditions and make the hierarchical surface biocompatible with mammalian cells (Yan et al. 2016). This work broadens the therapeutic applications of cationic antimicrobial peptides and provides the foundations for the production of smart materials with multi-functionalities to practical biomedical applications.

Apart from these examples, various pH-sensitive materials have been synthesized with or without modifications for controlled drug delivery. Some of them are listed in Table 10.2 along with the mechanism of the polymerization reaction, effect of pH on polymer structure, and advantages of the drug delivery system.

10.2.3.4 Ion-Sensitive Materials for Controlled Drug Delivery

Ion-sensitive materials are the polymers that contain ionizable groups in their structure which are responsible for their property of being **sensitive** to ionic strength. The variations in ionic strength of environment trigger physical changes in polymer solubility, swelling, polymer micelle size, and fluorescence quenching kinetics of electrolyte-bound chromophores. The effect of ionic strength in polymers is closely linked to the pH. Ionic strength variation leads to phase transition of acidic or basic polymers, but it would not affect the action of a non-ionizable polymer. However, at a given ionic concentration (i.e., sodium chloride) in an aqueous solution, a distinct transformation occurs in this polymer.

10.2.4 Mechanism of Phase Transition of Ion-Sensitive Materials

The mechanism of phase transition involving ionic activity can be described by two widely utilized approaches: effect of ionic strength and in situ gelation. Also, ion exchange resin is an approach where drug release can be controlled by effect of complexation of drug with materials.

Table 10.2 Formulation application of pH-**sensitive** materials for controlled drug release characteristics

Formulation approach	Materials	Mechanism of polymerization
Hydrogel	Polymer: Poly(L-lactide)-co-polyethylene glycol-co-poly(L-lactide) di-methacrylates along with L-lactide oligomers as a crosslinker and copolymerized with acrylic acid and N-isopropyl acrylamide Drug: Doxorubicin and tetracycline	Polyethylene glycol and L-lactide mixed and heated at 130 °C. after it cooled down to room temperature, the solid was dissolved in dichloromethane before pouring into excess diethyl ether. That leads to form precipitate comprising of polymer diols. After drying of the precipitate, resulting polymeric diols dissolved in dichloromethane, and triethylamine and methacrylic anhydride in dichloromethane were added, and reacted to give the end-capped polymer diols
	Polymer: Acrylamide and methyl acrylic acid with N-N′-methylene bisacrylamide Drug: Theophylline	Prepared using free radical polymerization method in presence of sodium bicarbonate as a blowing agent
	Polymer: Synthetic polymers, poly(2-(alkylamino) ethyl methacrylate) derivatives	Block copolymers (with pKa values between 6.3 and7.5) can be developed by reversible-deactivation radical polymerization. This polymerization leads to formation of pH-**sensitive** free-standing gels
	Polymer: Block copolymers consisting of a pH-sensitive block that is poly(2-(diisopropylamino) ethyl methacrylate) and a biocompatible zwitterionic poly(2-(methacryloyloxy) ethyl phosphorylcholine) block	Hydrogel films prepared by spin-coating of methanolic solutions. The basic building blocks formed by the physical entanglement of “flower-like” micelles of pH-sensitive unionized poly(2-(diisopropylamino) ethyl methacrylate) surrounded by biocompatible ionizable poly(2-(methacryloyloxy) ethyl phosphorylcholine blocks
	Polymer: Carboxymethyl cellulose/2-hydroxyethyl acrylate Drug: Naringenin	Naringenin-loaded hydrogel was prepared using carboxymethyl cellulose dissolved in distilled water at 75 °C and 400 rpm in a water bath. Nitrogen gas was then injected into the dissolved CMC solution to reduce the partial pressure of oxygen in the flask. Potassium persulfate was added to it as a radical initiator, and the mixture was left to react. Upon completion of the reaction, the compound was cooled sufficiently at room temperature
	Polymer: Chitosan-poly(ethylene oxide) hydrogels Drug: Metronidazole	Hydrogel matrix was synthesized by crosslinking chitosan and poly(ethylene oxide) in a blend to form semi-interpenetrating polymer network
Solid dosage form	Polymer: Matrix of hydroxypropyl methylcellulose (HPMC) and Carbopol® 934 Drug: Diclofenac sodium	HPMC and carboxy polymer were used as hydrophilic polymeric matrix formers with diclofenac sodium as a model drug. The diluent was lactose, starch, and magnesium stearate were used as a release modifier and lubricant, respectively.

Table 10.2 (continued)

Effect of pH on polymer structure	Advantage	References
The hydrogel has high swelling property at physiological pH, due to ionization and subsequent repulsion between the poly(acrylic acid) units at this pH 7.4	This hydrogel is "smart drug carriers" that allow controlled dual drug release for the anti-cancer drug doxorubicin and antibiotic tetracycline. The hydrogel found to be nontoxic based on cell culture study	Xu et al. (2018)
The hydrogel has minimum swelling in gastric pH 1.2, and maximum swelling and release of drugs in intestinal medium pH 6.8	Can be used as a versatile platform for targeted controlled release of gastric-sensitive therapeutics into the small intestine	Hibbins et al. (2017)
Insoluble at neutral pH but readily dissolved as cationic polyelectrolytes in acidic media due to protonation of the tertiary amine residues	Has potential as controlled-release delivery vehicles for drugs, pesticides, perfumes, dyes, and other active compounds	Bütün et al. (2001) and Castelletto et al. (2004)
Tri-block copolymer forms a free-standing transparent gel comprising a 3D network above pH 7. At higher gel pH, the poly(2-(diisopropylamino) ethyl methacrylate) chains become more deprotonated and hence more hydrophobic	Upon adjustment of pH over a narrow range (physiologically relevant pH 7–8), elasticity of the hydrogels could be conveniently tuned over a wide range	Bütün et al. (2001) and Yoshikawa et al. (2011)
The swelling ratio of this hydrogel increased at pH 7.5 and 8.5 with a stable gel network	The hydrogel may be useful for transdermal delivery system of naringenin to treat atopic dermatitis. Helpful for controlled delivery of naringenin via Fickian diffusion mechanisms from hydrogel matrix	S. H. Park et al. (2018)
Chitosan has pH-dependent swelling in acidic environment that leads to swollen hydrogel in stomach environment	Localized delivery of the loaded drug in the acidic environment of the gastric fluid	Patel and Amiji (1996)
Due to the chemical nature of Carbopol® 934, higher swelling of the polymer is seen at pH 5 to 9	Ideal for providing controlled drug release throughout the gastrointestinal track, since. Carbopol® 934 controls the drug release in intestinal media based on high swelling capacity. HPMC controls the drug release in both acidic and intestinal media	Bravo et al. (2004)

(continued)

Table 10.2 (continued)

Formulation approach	Materials	Mechanism of polymerization
	Polymer: HPMC 15 M and poly(methacrylic acid, ethyl acrylate) Drug: Diclofenac sodium	Matrix tablets with HPMC 15 M and poly(methacrylic acid, ethyl acrylate) were formulated using a fluidized bed using diclofenac as a model drug. Microgranules were prepared using drug and polymers, followed by the matrix tablets prepared by direct compression on a tablet press
	Polymer: Graft copolymer of poly(ethylene glycol) with poly(methacrylic acid)	Matrix was synthesized by copolymerizing poly(ethylene glycol) methacrylate macromonomers with methacrylic acid in the presence of a crosslinking agent, tetraethylene glycol dimethacrylate
	Polymer: Alginate-based microsphere	The alginate crosslinks with calcium ions to form temporary microspheres, which are stabilized by incubation with 0.2% poly-l-lysine
Nanocarrier	Polymer: Poly(methacrylic-grafted-ethylene glycol) in the presence of hydrophobic nanoparticles Drug: Doxorubicin	Poly(methacrylic-grafted-ethylene glycol) hydrogels matrix were dispersed with poly(methyl methacrylate) nanoparticles by first forming the nanoparticles core and then polymerizing the hydrogel in the presence of poly(methyl methacrylate) nanoparticles
	Polymer: Cationic pullulan and anionic dextran derivatives Drug: Piroxicam	Cationic derivative of pullulan was synthesized by grafting reaction and used together with dextran sulfate to form polysaccharide-based nanohydrogel crosslinked via electrostatic interactions using piroxicam as a drug. Due to the polycation-polyanion interactions nanohydrogel particles were formed instantly and spontaneously in aqueous medium
	Polymer: Poly(L-histidine-co-phenylalanine) nanogel Drug: Doxorubicin	The nanogel matrix of doxorubicin composed of a hydrophobic polymer core, and two layers of hydrophilic shell. The core was made of poly(L-histidine-co-phenylalanine). Polyethylene glycol forms the inner hydrophilic shell, whereas polyethylene glycol end is linked to polymer core, and bovine serum albumin forms a capsid-like outer shell. The core and inner shell were constructed by an oil-in-water emulsion method

Table 10.2 (continued)

Effect of pH on polymer structure	Advantage	References
Poly(methacrylic acid, ethyl acrylate) is capable of controlling the release of drug in stomach environment due to its pH-dependent swelling capability in lower pH	Poly(methacrylic acid, ethyl acrylate) polymer is effective in providing controlled drug release in acid environments. Whereas, HPMC K15M has a low sensitivity to pH and ionic strength and has better swelling characteristic in intestinal media. The drug release rate is controlled by HPMC K15M in intestinal media	Lamoudi et al. (2012)
Changing the pH from acidic to basic (pH > 7) resulted in a large increase in swelling.	These polymeric matrix forming polymer may be useful in controlled release applications due to their large, controllable swelling transitions	Peppas and Klier (1991)
Gelling properties of alginate microsphere is pH dependent. At low pH (gastric environment), alginate shrinks, and the encapsulated drug is not released. It has good swelling at intestinal pH environment	Alginate microspheres are better vehicle for oral delivery for protein antigens, and intestinal IgA antibody responses are induced by antigens encapsulated in alginate microspheres without any additional mucosal adjuvant	B. Kim (2002)
Poly(methacrylic-grafted-ethylene glycol) contains methacrylic acid with a pKa value around 4.8 that leads to increase in the degree of swelling as it travels to the colon environment (pH 7–8)	Provide local, direct delivery of doxorubicin to the colon for the treatment of colon cancer	Schoener et al. (2013)
The pH-dependent drug release observed at pH 4 for prolonged period of time	Targeted controlled drug delivery of piroxicam can be possible using nanogel approach. Nanoparticles were colloidally stable and their size and the surface charge could be controlled by the polycation/ polyanion ratio	Lachowicz et al. (2019)
pH-induced reversible swelling/ deswelling of the polymeric core is closely linked to the release rate of incorporated doxorubicin. The nanogels help in release of significant amount of drug at endosomal pH 6.4 to provide targeted intracellular delivery	No apparent cytotoxicity or systemic toxicity was observed in the mouse models for the polymer. The BSA shell is conjugated with folate ligands that leads to enhanced targeted delivery of nanogel for antitumor activity	Lee et al. (2008)

Polyelectrolytes are the hydrophilic polymers containing ionizable groups. After dispersion in water, polyelectrolytes dissociate and generate charged polymer chains (macroions) and small counter ions (Dobrynin and Rubinstein 2005). By varying ionic strength, highly charged polyelectrolytes can be shielded; thus, it leads to change in the charge density and adsorption behavior (Shiratori and Rubner 2000). The polymer coil can outspread to a high expansion in low ionic-strength environments. However, upon increase in the ionic strength, the repulsion between the charged polymer segments is increased by ions in the solution involving less-expanded polymer coil (Bolto and Gregory 2007).

The in situ gel is a solution dosage form and, after administration, forms a gel upon contacting with the mucosa or physiological environmental. The formation of gel is due to the viscosity of solution increased. This behavior is attributed to the presence of the smart materials in the dosage form. For instance, gellan gum, being one of the most promising in situ gelling materials, is an anionic, exocellular, deacetylated bacterial polysaccharide secreted by *Sphingomonas paucimobilis* (formerly known as pseudomonas elodea) with a tetrasaccharide repeating unit of 1β-l-rhamnose, 1β-d-glucuronic acid, and 2β-d-glucose (Shi-lei et al. 2009). The presence of cations is important for this gelation mechanism, which involves the formation of double-helical junction zones, followed by the aggregation of double-helical segments by complexation with cations and hydrogen bonding with water to form a three-dimensional network (Shi-lei et al. 2009; Salunke and Patil 2016). Divalent cations are much more effective at fostering gelation than monovalent ones. It has to be noted that the human nasal mucosa consists of sodium, potassium, and calcium ions. In response to presence of these cations in the nasal mucosa, a solution-gel phase transition can be expected (Shi-lei et al. 2009).

The use of ion-exchange resin involves complexation with drug in order to implement its application in controlled drug delivery. Ion-exchange resins are insoluble polymers comprised of a polystyrene backbone crosslinked with divinylbenzene and side chains of ion-active groups (Yoshida et al. 2013). Majority of ion-active groups in the ion-exchange resin include tertiary amine substitutes, quaternary ammonium, sulfonic acid, and carboxylic acid. These groups are useful in complexation based on electrostatic interaction between the drug and the resins which impacts the drug loading of the ion-exchange resins. The drug and ion exchange resin complex can be implemented for oral formulations, such as liquid suspensions, capsules, and beads, the drug-resin complexes (Guo et al. 2009b; Yoshida et al. 2013). Upon contact with a solution containing the corresponding counter ions to that of particular ion exchange resin, drug release from the resins is controlled by an ion exchange equilibrium reaction. Therefore, drug-resin binding can be preserved during storage by keeping the drug-resin complexes free of the resin counter ions in the liquid. In contrast, owing to the presence of ions in the saliva and gastrointestinal fluids, upon oral administration, their secretion drives the equilibrium forward and promotes the drug release (Guo et al. 2009b).

10.2.5 Applications of Ion-Sensitive Materials for Controlled Drug Delivery

The role of ion-**sensitive** materials has been widely utilized for controlled drug release applications. These include natural and synthetic polymers that are beneficial for not only controlled delivery but also targeted drug delivery. For example, gellan gum, a natural polymer, has been used for controlled drug release in ophthalmic applications. A controlled ocular delivery of brinzolamide from a liquid formulation containing gellan gum has been reported (Sun and Zhou 2018). This liquid formulation, upon contacting with tear fluid after the administration into eye, was converted into a gel form for controlling the drug release. In this study, not only the in vitro release profiles showed the controlled release of brinzolamide from the in situ gel, but also prolonged lowering of the intraocular pressure was observed after administration. Another study has shown application of gellan gum as an ion-sensitive gel forming material for in situ estradiol gel solution eye drops for the prevention of age-related cataracts (Kotreka et al. 2017). In brief, the solution eye drops resulted in an in situ phase change to gel state, upon mixing with simulated tear fluid, for controlled release of estradiol from the formed gel following non-Fickian release mechanism.

Carrageenan is another type of polymers with similar features to gellan gum, especially with regard to their gelling properties. Carrageenan is the generic name for a family of high molecular weight sulfated linear polysaccharides of d-galactose and 3, 6-anhydro-d-galactose (Bazban-Shotorbani et al. 2017). These polysaccharides can be traditionally split into six basic forms: kappa (κ)-, iota (τ)-, lambda (λ)-, mu (μ)-, nu (ν)-, and theta (θ)-carrageenan. Based on their physicochemical properties, κ- and τ-carrageenan can form gel easily, especially in the presence of monovalent and divalent ions (Li et al. 2014). Out of these forms, τ-carrageenan showed suitable rheological and controlled release characteristics for potential applications as vehicles for oral delivery of acetaminophen to dysphagic patients and children (Miyazaki et al. 2011). In another study, when comparing the topical ophthalmic formulation containing a radiotracer labeled gellan gum and κ-carrageenan with respect to an aqueous solution, it was observed that ophthalmic hydrogel formed with κ-carrageenan and gellan gum remained on the corneal surface for long periods (Fernández-Ferreiro et al. 2017).

Alginate has been studied widely as an attractive ion-**sensitive** polysaccharide, especially because of its biocompatibility, biodegradability, and reversible aqueous gelation chemistry with di- or trivalent cation. Alginate is a naturally occurring anionic polymer typically (Smidsrød and Skja 1990). Alginate-based ion-sensitive hydrogel can be fabricated using oil-in-water high internal phase emulsion technique. The hydrogel formed by use of methacrylate-modified alginate was covalently crosslinked with in the water phase (external phase) and the oil phase (internal phase). This highly porous and biocompatible hydrogels displayed a controllable ion-sensitive function and well-controlled morphology of the pore (Zhou et al. 2013).

For a better absorption characteristic of nicotine from buccal mucosa, ion exchange resin-based chewing gum has been applied (Guo et al. 2009a). To achieve better absorption, a carboxylic acid-based ion-exchange resin was used. The resin effectively suppressed the release of nicotine to just 5% in pure salt-free water; on the other hand, 70–90% of the drug release was observed in saliva within 30 min (Guo et al. 2009a). In another study, a sulfonic acid ion-exchange resin was used to achieve the continuous release of dextromethorphan from a liquid suspension. The drug-resin complexes were partly coated with ethylcellulose. The complexes (i.e., coated ethylcellulose-drug resin complexes and the uncoated complexes with an estimated ratio of 2:1) produced a constant concentration of plasma drugs for 12 h (Nadkar and Lokhande 2010). The coating of water-insoluble polymers on drug particles without resin does not have a driving force to facilitate the release of drugs and leads to incomplete drug release into gastrointestinal fluids. The drug-resin complexed drug delivery systems have the big benefit of the absolute release of the drug triggered by the ionic equilibrium. The drug release rate from drug-resin complexes, on the other hand, depends on the strength of drug-resin interactions. Interestingly, nicotine has sufficiently strong interaction with the resin to achieve controlled drug release. Nevertheless, pseudoephedrine and dextromethorphan have insufficient drug-resin interactions; hence, coating of the drug-resin complexes with water-insoluble polymers is essential for controlled drug release. For balancing drug release suppression and achieving maximum drug release, the combined use of drug-resin complexes and polymer coatings is very useful. As oral disintegrating tablets and pediatric liquid formulations become more common, ion-exchange resins in those formulations will certainly play a more important role in not only controlled drug release but also taste-masking.

To control the drug release via a polymer membrane, in addition to the charge, the swelling capacity of the polymer needs to be considered. Anion-responsive drug release can be obtained with the use of cationic polymers containing quaternary ammonium groups. For instance, Eudragit RL and RS copolymers [i.e., poly(ethyl acrylate-methyl methacrylate-trimethylammonium ethyl methacrylate chloride) copolymers] are practically insoluble in water. However, it can be hydrated and swell in water based on surrounding ionic environment. Therefore, depending on the swelling capacity of the polymer coating, polymers may be coated with beads and tablets containing drugs for oral formulations to control drug release via the polymer membrane. Good interaction in the media between the quaternary ammonium groups and counter ions causes less hydration or swelling and leads to slower release of the compound. It can be noted here that Eudragit RL- and RS-coated beads have been reported to exhibit buffer species-dependent drug release kinetics of diltiazem. In a study, based on the unique strength of interaction between the buffer anions and the quaternary ammonium groups of Eudragit RL and RS, the chloride ion in buffer media indicated highest order of drug release as compared to acetate and formate (Bodmeier et al. 1996). Additionally, Eudragit RS coating has been reported to aid in achieving a sigmoidal drug release profile. It consists of a bead system that had been prepared by coating Eudragit RS onto a drug core containing drug and organic acid. It was reported that the orders of hydration and drug release rates were highest for acetic acid, followed by succinic acid, glutamic acid, malic acid, tartaric acid, and citric acid. After oral administration of the bead system to dogs, the controlled drug release profiles were observed (Narisawa et al. 1997).

Another interesting approach includes use of polymers exhibiting lower critical solution temperature transitions for ion-sensitive drug delivery systems. Several polymers, such as poly(N-isopropylacrylamide), cellulose derivatives, poly(vinyl ether), and poly(N-vinyl caprolactam), exhibit lower critical solution temperature transitions in aqueous solutions (Crespy and Rossi 2007). At temperatures below lower critical solution temperature, these polymers are soluble in water owing to favorable hydration. However, at temperatures above lower critical solution temperature, they are insoluble to the entropy-driven dehydration and collapse of the hydrophobic polymers. The lower critical solution temperature of polymers can be reduced by use of highly water-soluble substances, such as electrolytes and sugars, which surround the polymers and remove water molecules from the polymers. This is referred to as the salting-out effect. To effectively control drug release, various oral formulations containing these polymers and salts have been reported. For instance, tablets containing terbutaline sulfate and Na_2SO_4 and coated with a poly(N-isopropylacrylamide) layer were prepared, aiming for colon drug delivery by generating long drug release lag time (Eeckman et al. 2002).

Although tablets comprising of ion-sensitive materials and ions is an interesting concept; the expected in vitro drug release performance may be tricky. However, with the use of FDA-approved polymers such as hydroxypropyl methylcellulose (HPMC) and polyethylene oxide, the in vitro performance is simpler. The dissolved salts, such as sodium carbonate, in the saliva, prevent the water-soluble polymers, for example, HPMC, from dissolving based on the salting-out effect. The drug release is suppressed due to the insolubilized part of the water-soluble polymers. However, in the gastrointestinal tract, most of the salts dissolve in the salting-out layer and further are released from the system. Although the salting-out effect gets minimized leading to immediate drug, this system generates long lag time in drug release as well. Studies have demonstrated that the major mechanism of the controlled drug release of this system is the salting-out effect (Tasaki et al. 2009). Ion-**sensitive** materials not only are limited to oral drug delivery system but also has found various applications in local drug delivery systems, such as ophthalmic, subcutaneous depot, and transdermal formulations. Integrated use of polymers, ions, and other materials may lead to improved sensitivity to slight changes in ion concentrations, enabling drug delivery systems at target sites to have greater accuracy in site-specific drug release.

10.3 Conclusions

As for controlled drug delivery, pH- and ion-sensitive materials have been utilized to overcome limitations associated with conventional dosage forms. By using these pH- and ion-sensitive materials, drug delivery systems can be developed to control not only the drug release but also its release triggered by the pH and ion strength environments of the delivery systems. Numerous pH- and ion-sensitive materials have been studied in past two decades with mechanism of their phase transition in response to variation in physiological environment. This phase transition characteristic of pH- and ion-sensitive materials has found various applications to provide local and/or systemic site-specific controlled drug delivery via oral, parenteral, ocular, and vaginal drug delivery by use of various types of dosage forms. However, limited research has been performed for the toxicity for their in vivo application. Further research is required to deeply investigating the mechanisms in the physiological setting by conducting in vivo experiments to further advance their therapeutic applications.

References

Alvarez-Lorenzo C, Blanco-Fernandez B, Puga AM, Concheiro A (2013) Crosslinked ionic polysaccharides for stimuli-sensitive drug delivery. Adv Drug Deliv Rev. https://doi.org/10.1016/j.addr.2013.04.016

Arya A, Majumdar DK, Pathak DP, Sharma AK, Ray AR (2017) Design and evaluation of acrylate polymeric carriers for fabrication of pH-sensitive microparticles. Drug Dev Ind Pharm. https://doi.org/10.1080/03639045.2016.1239629

Asayama S, Sekine T, Kawakami H, Nagaoka S (2007) Design of aminated poly(1-vinylimidazole) for a new pH-sensitive polycation to enhance cell-specific gene delivery. Bioconjug Chem. https://doi.org/10.1021/bc700205t

Basu A, Kunduru KR, Abtew E, Domb AJ (2015) Polysaccharide-based conjugates for biomedical applications. Bioconjug Chem. https://doi.org/10.1021/acs.bioconjchem.5b00242

Bawa P, Pillay V, Choonara YE, Du Toit LC (2009) Stimuli-responsive polymers and their applications in drug delivery. Biomed Mater. https://doi.org/10.1088/1748-6041/4/2/022001

Bazban-Shotorbani S, Hasani-Sadrabadi MM, Karkhaneh A, Serpooshan V, Jacob KI, Moshaverinia A, Mahmoudi M (2017) Revisiting structure-property relationship of pH-responsive polymers for drug delivery applications. J Control Release. https://doi.org/10.1016/j.jconrel.2017.02.021

Bodmeier R, Guo X, Sarabia RE, Skultety PF (1996) The influence of buffer species and strength on diltiazem HC1 release from beads coated with the aqueous cationic polymer dispersions, Eudragit RS, RL 30D. Pharm Res 13(1):52–56

Bolto B, Gregory J (2007) Organic polyelectrolytes in water treatment. Water Res 41(11):2301–2324

Bravo SA, Lamas MC, Salomon CJ (2004) Swellable matrices for the controlled-release of diclofenac sodium: formulation and in vitro studies. Pharm Dev Technol. https://doi.org/10.1081/PDT-120027420

Bütün V, Armes SP, Billingham NC (2001) Synthesis and aqueous solution properties of near-monodisperse tertiary amine methacrylate homopolymers and diblock copolymers. Polymer 42(14):5993–6008

Castelletto V, Hamley IW, Ma Y, Bories-Azeau X, Armes SP, Lewis AL (2004) Microstructure and physical properties of a pH-responsive gel based on a novel biocompatible ABA-type triblock copolymer. Langmuir 20(10):4306–4309

Chen T, Ferris R, Zhang J, Ducker R, Zauscher S (2010) Stimulus-responsive polymer brushes on surfaces: transduction mechanisms and applications. Progress Polymer Sci (Oxford). https://doi.org/10.1016/j.progpolymsci.2009.11.004

Chuang CY, Don TM, Chiu WY (2009) Synthesis of chitosan-based thermo- and ph-responsive porous nanoparticles by temperature-dependent self-assembly method and their application in drug release. J Polym Sci A Polym Chem. https://doi.org/10.1002/pola.23564

Crespy D, Rossi RM (2007) Temperature-responsive polymers with LCST in the physiological range and their applications in textiles. Polym Int 56(12):1461–1468

Dai S, Ravi P, Tam KC (2008) pH-responsive polymers: synthesis, properties and applications. Soft Matter. https://doi.org/10.1039/b714741d

Dai Y, Wang S, Shi W, Lang M (2017) pH-responsive carboxymethyl chitosan-derived micelles as apatinib carriers for effective anti-angiogenesis activity: preparation and in vitro evaluation. Carbohydr Polym. https://doi.org/10.1016/j.carbpol.2017.08.011

Dobrynin AV, Rubinstein M (2005) Theory of polyelectrolytes in solutions and at surfaces. Prog Polym Sci 30(11):1049–1118

Du H, Liu M, Yang X, Zhai G (2015) The design of pH-sensitive chitosan-based formulations for gastrointestinal delivery. In Drug Discovery Today. https://doi.org/10.1016/j.drudis.2015.03.002

Eeckman F, Moës AJ, Amighi K (2002) Evaluation of a new controlled-drug delivery concept based on the use of thermoresponsive polymers. Int J Pharm 241(1):113–125

Felber AE, Dufresne MH, Leroux JC (2012) PH-sensitive vesicles, polymeric micelles, and nanospheres prepared with polycarboxylates. Adv Drug Delivery Rev. https://doi.org/10.1016/j.addr.2011.09.006

Fernández-Ferreiro A, Silva-Rodríguez J, Otero-Espinar FJ, González-Barcia M, Lamas MJ, Ruibal A, Luaces-Rodríguez A, Vieites-Prado A, Lema I, Herranz M (2017) In vivo eye surface residence determination by high-resolution scintigraphy of a novel ion-sensitive hydrogel based on gellan gum and kappa-carrageenan. Eur J Pharm Biopharm 114:317–323

Fodor C, Bozi J, Blazsó M, Iván B (2012) Thermal behavior, stability, and decomposition mechanism of poly(N-vinylimidazole). Macromolecules. https://doi.org/10.1021/ma301712k

Gan LH, Roshan Deen G, Loh XJ, Gan YY (2001) New stimuli-responsive copolymers of N-acryloyl-N′-alkyl piperazine and methyl methacrylate and their hydrogels. Polymer. https://doi.org/10.1016/S0032-3861(00)00296-2

Gandhi A, Paul A, Sen SO, Sen KK (2015) Studies on thermoresponsive polymers: phase behaviour, drug delivery and biomedical applications. Asian J Pharm Sci. https://doi.org/10.1016/j.ajps.2014.08.010

Gregory A, Stenzel MH (2012) Complex polymer architectures via RAFT polymerization: from fundamental process to extending the scope using click chemistry and nature's building blocks. Progress Polymer Sci (Oxford). https://doi.org/10.1016/j.progpolymsci.2011.08.004

Guo X, Chang R-K, Hussain MA (2009a) Ion-exchange resins as drug delivery carriers. J Pharm Sci 98(11):3886–3902

Guo X, Chang RK, Hussain MA (2009b) Ion-exchange resins as drug delivery carriers. J Pharm Sci. https://doi.org/10.1002/jps.21706

Gupta P, Vermani K, Garg S (2002) Hydrogels: from controlled release to pH-responsive drug delivery. Drug Discov Today. https://doi.org/10.1016/S1359-6446(02)02255-9

Hafeli U (2002) BOOK REVIEW: controlled drug delivery: challenges and strategies, edited by Kinam Park. Ann Biomed Eng. https://doi.org/10.1114/1.137

Hibbins AR, Kumar P, Choonara YE, Kondiah PPD, Marimuthu T, du Toit LC, Pillay V (2017) Design of a versatile pH-responsive hydrogel for potential oral delivery of gastric-sensitive bioactives. Polymers 9(10). https://doi.org/10.3390/polym9100474

Hoffman AS (2012) Hydrogels for biomedical applications. Adv Drug Deliv Rev. https://doi.org/10.1016/j.addr.2012.09.010

Hoffman AS (2013) Stimuli-responsive polymers: biomedical applications and challenges for clinical translation. Adv Drug Deliv Rev. https://doi.org/10.1016/j.addr.2012.11.004

Huh KM, Kang HC, Lee YJ, Bae YH (2012) pH-sensitive polymers for drug delivery. Macromol Res. https://doi.org/10.1007/s13233-012-0059-5

Islam MT, Rodríguez-Hornedo N, Ciotti S, Ackermann C (2004) Rheological characterization of topical carbomer gels neutralized to different pH. Pharm Res. https://doi.org/10.1023/B:PHAM.0000033006.11619.07

Jalalvandi E, Shavandi A (2018) In situ-forming and pH-responsive hydrogel based on chitosan for vaginal delivery of therapeutic agents. J Mater Sci Mater Med. https://doi.org/10.1007/s10856-018-6166-x

Jeong B, Gutowska A (2002) Lessons from nature: stimuli-responsive polymers and their biomedical applications. Trends Biotechnol. https://doi.org/10.1016/S0167-7799(02)01962-5

Jones D (2004) Pharmaceutical applications of polymers for drug delivery. Polysaccharides Drug Delivery Pharm Appl

Ju XJ, Xie R, Yang L, Chu LY (2009) Biodegradable "intelligent" materials in response to physical stimuli for biomedical applications. Expert Opin Ther Pat. https://doi.org/10.1517/13543770902771282

Kan KHM, Li J, Wijesekera K, Cranston ED (2013) Polymer-grafted cellulose nanocrystals as pH-responsive reversible flocculants. Biomacromolecules. https://doi.org/10.1021/bm400752k

Kang SI, Bae YH (2002) pH-induced solubility transition of sulfonamide-based polymers. J Control Release. https://doi.org/10.1016/S0168-3659(02)00021-4

Kang SI, Bae YH (2003) A sulfonamide based glucose-responsive hydrogel with covalently immobilized glucose oxidase and catalase. J Control Release. https://doi.org/10.1016/S0168-3659(02)00409-1

Kim B (2002) Mucosal immune responses following oral immunization with rotavirus antigens encapsulated in alginate microspheres. J Control Release 85(1–3):191–202. https://doi.org/10.1016/S0168-3659(02)00280-8

Kim SJ, Lee CK, Kim SI (2004) Electrical/pH responsive properties of poly(2-acrylamido2-methylpropane sulfonic acid)/hyaluronic acid hydrogels. J Appl Polym Sci. https://doi.org/10.1002/app.20133

Kocak G, Tuncer C, Bütün V (2017) PH-responsive polymers. Polym Chem. https://doi.org/10.1039/c6py01872f

Korsmeyer RW, Gurny R, Doelker E, Buri P, Peppas NA (1983) Mechanisms of solute release from porous hydrophilic polymers. Int J Pharm. https://doi.org/10.1016/0378-5173(83)90064-9

Kotreka UK, Davis VL, Adeyeye MC (2017) Development of topical ophthalmic in situ gel-forming estradiol delivery system intended for the prevention of age-related cataracts. PLoS One 12(2):e0172306

Kumar A, Srivastava A, Galaev IY, Mattiasson B (2007) Smart polymers: physical forms and bioengineering applications. Progress Polymer Sci (Oxford). https://doi.org/10.1016/j.progpolymsci.2007.05.003

Kumar A, Montemagno C, Choi HJ (2017) Smart microparticles with a pH-responsive macropore for targeted Oral drug delivery. Sci Rep. https://doi.org/10.1038/s41598-017-03259-x

Kurkuri MD, Aminabhavi TM (2004) Poly(vinyl alcohol) and poly(acrylic acid) sequential interpenetrating network pH-sensitive microspheres for the delivery of diclofenac sodium to the intestine. J Control Release. https://doi.org/10.1016/j.jconrel.2003.12.025

Lachowicz D, Mielczarek P, Wirecka R, Berent K, Karewicz A, Szuwarzyński Michałand Zapotoczny S (2019) Nanohydrogels based on self-assembly of cationic pullulan and anionic dextran derivatives for efficient delivery of piroxicam. Pharmaceutics 11(12):1–16. https://doi.org/10.3390/pharmaceutics11120622

Lamoudi L, Chaumeil JC, Daoud K (2012) Development of gastro intestinal sustained release tablet formulation containing acryl-EZE and pH-dependent swelling HPMC K 15 M. Drug Dev Ind Pharm. https://doi.org/10.3109/03639045.2011.620966

Langer RS, Peppas NA (1981) Present and future applications of biomaterials in controlled drug delivery systems. Biomaterials. https://doi.org/10.1016/0142-9612(81)90059-4

Lee ES, Kim D, Youn YS, Oh KT, Bae YH (2008) A virus-mimetic Nanogel vehicle. Angew Chem Int Ed 47(13):2418–2421. https://doi.org/10.1002/anie.200704121

Li L, Ni R, Shao Y, Mao S (2014) Carrageenan and its applications in drug delivery. Carbohydr Polym 103:1–11

Liu L, Yao WD, Rao YF, Lu XY, Gao JQ (2017) pH-responsive carriers for oral drug delivery: challenges and opportunities of current platforms. Drug Deliv. https://doi.org/10.1080/10717544.2017.1279238

Ma X, Wang Y, Zhao T, Li Y, Su LC, Wang Z, Huang G, Sumer BD, Gao J (2014) Ultra-pH-sensitive nanoprobe library with broad pH tunability and fluorescence emissions. J Am Chem Soc. https://doi.org/10.1021/ja5053158

Mathew AP, Cho KH, Uthaman S, Cho CS, Park IK (2017) Stimuli-regulated smart polymeric systems for gene therapy. Polymers. https://doi.org/10.3390/polym9040152

Miyazaki S, Ishitani M, Takahashi A, Shimoyama T, Itoh K, Attwood D (2011) Carrageenan gels for oral sustained delivery of acetaminophen to dysphagic patients. Biol Pharm Bull 34(1):164–166

Na K, Lee KH, Bae YH (2004) pH-sensitivity and pH-dependent interior structural change of self-assembled hydrogel nanoparticles of pullulan acetate/oligo-sulfonamide conjugate. J Control Release. https://doi.org/10.1016/j.jconrel.2004.04.005

Nadkar S, Lokhande C (2010) Current trends in novel drug delivery-an OTC perspective. Pharma Times 42(4):17–23

Narisawa S, Nagata M, Hirakawa Y, Kobayashi M, Yoshino H (1997) An organic acid-induced sigmoidal release system for oral controlled-release preparations. III. Elucidation of the anomalous drug release behavior through osmotic pumping mechanism. Int J Pharm 148(1):85–91

Park SY, Bae YH (1999) Novel pH-sensitive polymers containing sulfonamide groups. Macromol Rapid Commun. https://doi.org/10.1002/(SICI)1521-3927(19990501)20:5<269::AID-MARC269>3.0.CO;2-3

Park SH, Shin HS, Park SN (2018) A novel pH-responsive hydrogel based on carboxymethyl cellulose/2-hydroxyethyl acrylate for transdermal delivery of naringenin. Carbohydr Polym 200:341–352. https://doi.org/10.1016/j.carbpol.2018.08.011

Patel VR, Amiji MM (1996) Preparation and characterization of freeze-dried chitosan-poly(ethylene oxide) hydrogels for site-specific antibiotic delivery in the stomach. Pharm Res 13(4):588–593. https://doi.org/10.1023/A:1016054306763

Peppas NA, Klier J (1991) Controlled release by using poly(methacrylic acid-g-ethylene glycol) hydrogels. J Control Release 16(1–2):203–214. https://doi.org/10.1016/0168-3659(91)90044-E

Pinkrah VT, Snowden MJ, Mitchell JC, Seidel J, Chowdhry BZ, Fern GR (2003) Physicochemical properties of poly(N-isopropylacrylamide-co-4-vinylpyridine) cationic polyelectrolyte colloidal microgels. Langmuir. https://doi.org/10.1021/la0262831

Rizwan M, Yahya R, Hassan A, Yar M, Azzahari AD, Selvanathan V, Sonsudin F, Abouloula CN (2017) pH sensitive hydrogels in drug delivery: brief history, properties, swelling, and release mechanism, material selection and applications. Polymers. https://doi.org/10.3390/polym9040137

Salunke SR, Patil SB (2016) Ion activated in situ gel of gellan gum containing salbutamol sulphate for nasal administration. Int J Biol Macromol. https://doi.org/10.1016/j.ijbiomac.2016.02.044

Schmaljohann D (2006) Thermo- and pH-responsive polymers in drug delivery. Adv Drug Delivery Rev. https://doi.org/10.1016/j.addr.2006.09.020

Schmalz A, Hanisch M, Schmalz H, Müller AHE (2010) Double stimuli-responsive behavior of linear and star-shaped poly(N,N-diethylaminoethyl methacrylate) in aqueous solution. Polymer. https://doi.org/10.1016/j.polymer.2009.11.023

Schoener CA, Hutson HN, Peppas NA (2013) pH-responsive hydrogels with dispersed hydrophobic nanoparticles for the oral delivery of chemotherapeutics. J Biomed Mater Res A 101A(8):2229–2236. https://doi.org/10.1002/jbm.a.34532

Shi-lei C, Ren X, Zhang Q, Chen E, Xu F, Chen J, Liu L-C, Jiang X (2009) In situ gel based on gellan gum as new carrier for nasal administration of mometasone furoate. Int J Pharm 365(1–2):109–115. https://doi.org/10.1016/j.ijpharm.2008.08.042

Shiratori SS, Rubner MF (2000) pH-dependent thickness behavior of sequentially adsorbed layers of weak polyelectrolytes. Macromolecules 33(11):4213–4219

Smidsrød O, Skja G (1990) Alginate as immobilization matrix for cells. Trends Biotechnol 8:71–78

Soni A (2014) Recent developments in ophthalmic drug delivery system. Lap Lambert Academic Publishing

Soni A (2015) Gastroretentive drug delivery systems: a comprehensive Review. Lap Lambert Academic Publishing

Sousa MP, Gonzalez de Torre I, Oliveira MB, Rodríguez-Cabello JC, Mano JF (2017) Biomimetic click assembled multilayer coatings exhibiting responsive properties. Mat Today Chem. https://doi.org/10.1016/j.mtchem.2017.04.001

Sun J, Zhou Z (2018) A novel ocular delivery of brinzolamide based on gellan gum: in vitro and in vivo evaluation. Drug Des Devel Ther 12:383

Tasaki H, Yoshida T, Maeda A, Katsuma M, Sako K (2009) Effects of physicochemical properties of salting-out layer components on drug release. Int J Pharm 376(1–2):13–21

Thakkar V, Soni A (2015) Vaginal Drug Delivery System. Lap Lambert Academic Publishing

Velasco D, Danoux CB, Redondo JA, Elvira C, San Román J, Wray PS, Kazarian SG (2011) PH-sensitive polymer hydrogels derived from morpholine to prevent the crystallization of ibuprofen. J Control Release. https://doi.org/10.1016/j.jconrel.2010.10.015

Wang B, Liu HJ, Jiang TT, Li QH, Chen Y (2014) Thermo-, and pH dual-responsive poly(N-vinylimidazole): preparation, characterization and its switchable catalytic activity. Polymer. https://doi.org/10.1016/j.polymer.2014.09.051

Wang Y, Shi Y, Xu M, Wu L, Jia X, Wei T, Zhang S, Guo X (2016) Smart flocculant with temperature and pH response derived from starch. RSC Adv. https://doi.org/10.1039/c6ra04060h

Xu L, Qiu L, Sheng Y, Sun Y, Deng L, Li X, Bradley M, Zhang R (2018) Biodegradable pH-responsive hydrogels for controlled dual-drug release. J Mater Chem B 6(3):510–517. https://doi.org/10.1039/C7TB01851G

Yan S, Shi H, Song L, Wang X, Liu L, Luan S, Yang Y, Yin J (2016) Nonleaching bacteria-responsive antibacterial surface based on a unique hierarchical architecture. ACS Appl Mater Interfaces. https://doi.org/10.1021/acsami.6b08436

Yoshida T, Lai TC, Kwon GS, Sako K (2013) PH-and ion-sensitive polymers for drug delivery. Expert Opin Drug Deliv. https://doi.org/10.1517/17425247.2013.821978

Yoshikawa HY, Rossetti FF, Kaufmann S, Kaindl T, Madsen J, Engel U, Lewis AL, Armes SP, Tanaka M (2011) Quantitative evaluation of mechanosensing of cells on dynamically tunable hydrogels. J Am Chem Soc 133(5):1367–1374

Yu J, Zhang Y, Kahkoska AR, Gu Z (2017) Bioresponsive transcutaneous patches. Curr Opin Biotechnol. https://doi.org/10.1016/j.copbio.2017.03.001

Zhou S, Bismarck A, Steinke JHG (2013) Ion-responsive alginate based macroporous injectable hydrogel scaffolds prepared by emulsion templating. J Mater Chem B 1(37):4736–4745

Chapter 11
Thermo-Responsive Polymers and Their Application as Smart Biomaterials

Jittima Amie Luckanagul, Khent Primo Alcantara, Bryan Paul I. Bulatao, Tin Wui Wong, Pornchai Rojsitthisak, and Pranee Rojsitthisak

11.1 Introduction

Polymers play an essential role in developing biomaterials (Wichterle and Lim 1960; Kwon et al. 1991). Their subunit and macromolecular structure varieties lead to intra- and intermolecular forces that determine the characters and functionalities of polymers. The control over the structural design and synthesis of polymers is crucial in developing materials for biomedical use. The continued efforts in research and development focus on the efficiency and proficiency in biomaterials fabrication and processing approach with recent success in clinical applications (Liu et al. 2008; Yang et al. 2008). There is a considerable number of research studies in developing biocompatible polymers in which artificial organs and drug delivery systems have been designed and applied clinically (Hoffman 2002). Utilizing polymers for

J. A. Luckanagul
Department of Pharmaceutics and Industrial Pharmacy, Faculty of Pharmaceutical Sciences, Chulalongkorn University, Bangkok, Thailand

Natural Products for Ageing and Chronic Diseases Research Unit, Faculty of Pharmaceutical Sciences, Chulalongkorn University, Bangkok, Thailand

K. P. Alcantara · B. P. I. Bulatao
Natural Products for Ageing and Chronic Diseases Research Unit, Faculty of Pharmaceutical Sciences, Chulalongkorn University, Bangkok, Thailand

Pharmaceutical Sciences and Technology Program, Faculty of Pharmaceutical Sciences, Chulalongkorn University, Bangkok, Thailand

T. W. Wong
Non-Destructive Biomedical and Pharmaceutical Research Centre, iPROMISE, Universiti Teknologi MARA, Puncak Alam, Selangor, Malaysia

J.-C. Kim et al. (eds.), *Smart Nanomaterials in Biomedical Applications*, Nanotechnology in the Life Sciences,
https://doi.org/10.1007/978-3-030-84262-8_11

such applications relies mainly on the self-organizing and biocompatible properties of these macromolecules (Khan et al. 2015). Polymeric biomaterials are generally categorized by how they are acquired, either by synthetic or natural origin. They can be further classified as biodegradable or non-biodegradable types. Biodegradable polymers, including synthetic and natural ones, have drawn great attention as biomedical materials due to their biocompatible nature that facilitates material-cells/tissue interaction (Khan et al. 2015). With the advances in materials science, more investigative research studies have been carried out to develop more efficient and clinically acceptable smart materials (Mura et al. 2013).

Smart materials refer to systems that have dynamic intrinsic properties allowing them to respond to slight environmental changes. Among the smart materials, stimuli-responsive polymers have been employed to develop drug delivery systems (Ward and Georgiou 2011; Kim and Matsunaga 2017; Shahriari et al. 2020; Sarwan et al. 2020). Many stimuli can trigger the transition of materials and support smart polymers' function, such as temperature, pH, ion concentration, electric/magnetic fields, and light (Schmaljohann 2006; Thornton et al. 2007; Bajpai et al. 2008). Thermo-responsive polymers are among the most investigated stimuli-responsive polymers for biomedical applications due to their reversible phase transition with temperature changes (Kim and Matsunaga 2017). Their physical properties can be modulated by internal physiological temperature or external heat triggers (Doberenz et al. 2020). Upon exposure to body temperature, the thermo-responsive polymer functions as a controlled release matrix for local or systemic delivery of therapeutic molecules or as a growth medium for tissue engineering cells (Sarwan et al. 2020). The thermo-responsive polymers have been developed in the form of solid, semi-solid, or liquid dosage forms such as liposomes, nanogels, hydrogel-coated metal nanoparticles, nanoparticle-hydrogel composites, polymeric nanoparticles, nanofibers, polymeric micelles, elastin-like peptide-drug conjugates, interpenetrating networks, micelles, crosslinked micelles, polymersomes, and films (Ward and Georgiou 2011; Ghosh Dastidar and Chakrabarti 2019).

Over the years, hybridization of thermo-responsive polymers with thermo- or non-thermo-responsive molecules or polymers has been carried out to improve their mechanical strength, adjust transition temperature, and improve their clinical applicability (Haladjova et al. 2014; Bordat et al. 2019; Sarwan et al. 2020). Fabrication

P. Rojsitthisak
Natural Products for Ageing and Chronic Diseases Research Unit, Faculty of Pharmaceutical Sciences, Chulalongkorn University, Bangkok, Thailand

Department of Food and Pharmaceutical Chemistry, Faculty of Pharmaceutical Sciences, Chulalongkorn University, Bangkok, Thailand

P. Rojsitthisak (✉)
Natural Products for Ageing and Chronic Diseases Research Unit, Faculty of Pharmaceutical Sciences, Chulalongkorn University, Bangkok, Thailand

Metallurgy and Materials Science Research Institute, Chulalongkorn University, Bangkok, Thailand
e-mail: pranee.l@chula.ac.th

of thermo-responsive polymers with stimuli-responsive materials to construct dual-responsive materials has also gained attention to enhance material characteristics and their performance in cell and animal models (Shahriari et al. 2020). In terms of drug and gene delivery, numerous attempts have been carried out to improve the temperature response at which controlled and targeted release of the therapeutic molecules are warranted to maximize safety and prevent suboptimal concentrations in the target site. However, the promising results of all the research studies undertaken on complex systems have yet to reach clinical trials and be applied clinically. In tissue engineering, significant efforts have been put into constructing thermo-responsive matrices with enhanced mechanical integrity, biodegradability, and similarity to the extracellular matrix of human tissues (O'Brien 2011). Numerous studies are evolved to construct complex tissue substitutes, and some have paved their way into clinical trials (Ward and Georgiou 2011; Haraguchi et al. 2012; Moschouris et al. 2016; Takahashi and Okano 2019; Doberenz et al. 2020). Optimizing the material and process parameters will eventually lead to the development of innovative and clinically acceptable thermo-responsive platforms. The reversible property and versatility of thermo-responsive polymers have made them attractive smart materials in the fields of drug delivery, gene delivery, and tissue engineering (Kim and Matsunaga 2017). An overview of the biomedical materials used to support the applications of thermo-responsive polymers is depicted in Fig. 11.1.

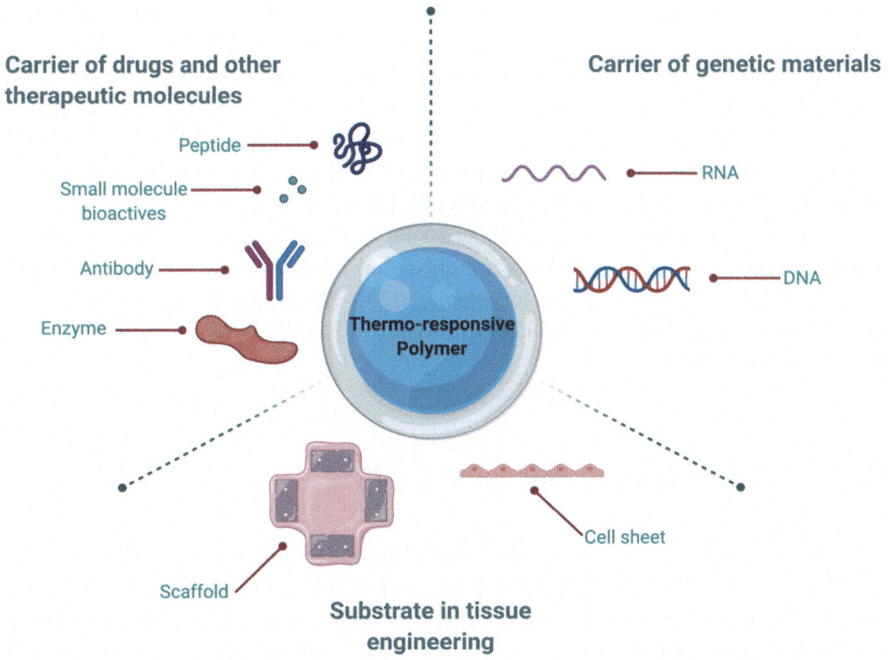

Fig. 11.1 Overview of the biomedical applications of thermo-responsive polymers

11.2 Mechanisms of Thermo-Responsive Polymers

Thermo-responsive polymers can perform reversible phase (or volume) transition with an alteration in the surrounding temperature (Hoffman 2013). Such property provides the polymers with the smartness to remotely take commands and cause desirable temperature-manipulative effects. The human body temperature ranges between 35 and 37 °C. The thermo-responsive polymers can have specific physical properties switchable in response to a change in the temperature. The alteration of the physical properties is brought about by hydrophilic/hydrophobic changes across its transition temperature (Kim and Matsunaga 2017). The external temperature change modulates the hydrophobic interaction capacity of the thermo-responsive polymer (Nagase and Okano 2016).

The thermal responsiveness of polymers is accompanied by linear and solubilized smart macromolecules pass from monophasic to the biphasic system giving rise to a reversible change of morphology and physical properties. Chain reorganization of smart polymeric networks results in the expansion of the network and vice versa as a function of temperature. Smart surfaces can have their surface hydrophilicity/hydrophobicity altered against a stimulus, and polymers swell or shrink in a solvent (Flory 1953). The smart responsiveness of polymers can be adopted in the development of minimally invasive injectable systems such as temperature-induced in situ gel of triblock copolymers containing amino-modified poly(ε-caprolactone) (PCN) with poly(ethylene glycol) (PEG). PCN-*b*-PEG-*b*-PCN is developed as sustained-release drug carrier (Guo et al. 2020); pulsatile drug delivery systems such as copolymer N-isopropylacrylamide-co-acrylamide (NiPAAm-co-AAm) gel is designed with its phase volume being reduced at 45 °C, releasing drug through raising temperature via the absorption of acoustic energy of focused ultrasound transducer applied on tissues (Ordeig et al. 2016; Ciancia et al. 2020), or new substrates for cell culture or tissue engineering where the in situ gel is used as a stem cell carrier to treat cartilage defect and locally immobilize bone marrow mesenchymal stem cells for regeneration of cartilage (Yang et al. 2020; Sponchioni et al. 2019).

The thermo-responsive polymers are classified into three categories: (1) shape-memory polymers (SMPs) (Strzelec et al. 2020), (2) liquid crystalline polymers (LCPs) (Risteen et al. 2018), and (3) stimuli-responsive polymers (SRP) (Ghaeini-Hesaroeiye et al. 2020).

11.2.1 Shape Memory Polymers (SMPs)

SMPs are constituted of a composition that responds to a thermal trigger within the temperature range of interest, with simultaneous polymer crosslinkage to introduce a permanent shape to the materials (Hager et al. 2015; Safranski 2017). Temporary and dormant morphology and properties of these materials can be induced under specific temperature and stress conditions. The initial configuration can be completely recovered through its elastic deformation under environmental command. SMPs are available as thermoplastic elastomers or thermosets as a function of the crosslinked nature (Safranski 2017). The shape memory has two phases: one is responsible for shape fixation, and the other accounts for shape reversibility, where the polymer molecules gain a higher level of mobility (Mazurek-Budzyńska et al. 2019). The memory capacity of SMPs is governed by the number molecular weight of each segment, melting temperature, or glass transition temperature (Zhou et al. 2012; Meng et al. 2016). Thermoplastic polymers with lower glass transition temperatures or molecular weight are presented with inferior shape fixation than thermoset polymers. This is attributed to the reduced crosslinking tendency of the former. Practically speaking, when the materials possess low mechanical strength, the accuracy of their recovery and memory is poor (Mu et al. 2018). The current scientific interest focuses on biocompatible thermally induced SMPs for the applications as surgical materials (Hager et al. 2015; Hamdy Makhlouf et al. 2020) and drug delivery systems (Erkeço lu et al. 2016). The thermoset SMPs are intrinsically stable with greater tensile strength than the thermoplastic variants (Xiao et al. 2015). They are widely used in applications under harsh stress.

11.2.2 Liquid Crystalline Polymers (LCPs)

Liquid crystal (LC) is a self-organizable material that is easily fabricated (Ge and Zhao 2020). One example of LCPs is ABA triblock-copolymers, where the A is a hard block (high T_g or crystalline material) grafted to the B, a soft block with liquid crystalline nature. The block copolymers can be synthesized by controlled polymerization methods such as atom transfer radical polymerization (ATRP) and ring-opening metathesis polymerization (ROMP) (Wei and Xiong 2018; Ndaya et al. 2019; Zenati and Thammalangsy 2018). Chemical and physical crosslinking can be applied on LCPs to create LC elastomers (LCEs) at the side chain (Kempe et al. 2004) and main chain (Supardi et al. 2015) of LCPs. Two main united properties required for LCE are an anisotropic property generated by the mesogenic units within the LCEs and elasticity of the polymer networks. The system can undergo self-rearrangement to an organized structure with certain directionality. The mobilization with the structural alignment is yet allowed with respect to the anisotropic organization. A phase transition causing contraction or expansion of LCEs can be reversibly induced by the heating-cooling of LCPs (Yusuf 2017). However, the LCPs do not always offer the possibility of a permanent shape memory effect. Heating LCPs above the T_g of the hard block, in some cases without shape-memory feature in the hard block, results in the loss of polymer orientation information (Fu et al. 2016; Martinez et al. 2021).

The application of thermo-responsive LCPs in the biomedical field raises an utmost important concern for their biocompatibility. Cellulose-based LCPs were developed through the patchy decoration of cellulose nanocrystals with a thermo-responsive polymer poly(*N*-isopropylacrylamide) (PNIPAM), resulting in a

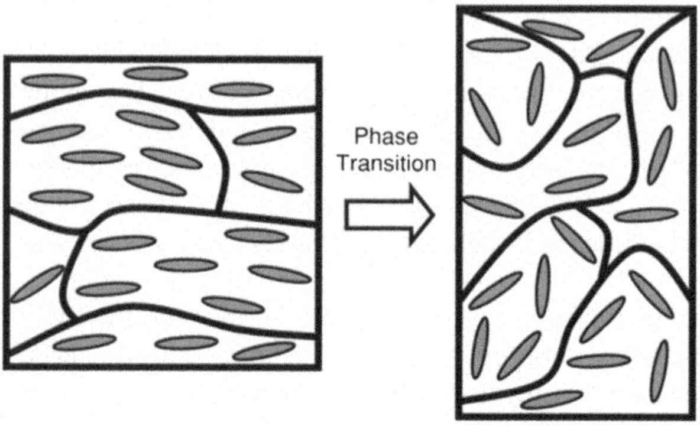

Fig. 11.2 In the LC phase (left), the polymer backbones enter an anisotropic environment with an extended chain conformation. In the isotropic phase (right), the polymer regains its coiled conformation at the phase transition. (Adopted from Ohm et al. 2010)

bio-derived, temperature-switchable liquid crystal (Risteen et al. 2018). LCPs have been implicated in micromechanical systems, such as valves in microfluidic systems and artificial muscles in robots (Yin et al. 2020).

Figure 11.2 represents the graphical model based on the shape variation of LCEs in the transition between anisotropic-isotropic conformations. The difference between the parallel and perpendicular arrangement of the radius of gyration (R_G) to the LC plane is observed when the polymer chains enter an anisotropic environment, departing from the isotropic conformation. In this case, a macroscopic shape change is developed with an isotropic chain conformation being adopted when such an elastomer loses its anisotropy after heating (Ohm et al. 2010).

11.2.3 Stimuli-Responsive Polymers (SRPs)

SRPs undergo a liquid-liquid phase transition in response to a variation in the surrounding temperatures. The temperature changes induce the separation of liquid into concentrated and diluted phases of polymers from a homogeneous solution.

11.2.3.1 Critical Solution Temperature

The thermo-responsive polymers may be characterized by lower and upper critical solution temperatures (Bordat et al. 2019; Kang et al. 2019). A polymer liquid turns into a solution above the upper critical solution temperature (UCST) or below the lower critical solution temperature (LCST). Hydrogen bonding and dipole-dipole interactions play a major role in solubilizing polymers in aqueous solutions through the polar parts of the polymers. Water molecules may form a cage around the non-polar parts of the polymers. Compared with the hydrogen bonds between water molecules, a much weaker interaction arises between the non-polar groups of the polymers and water in the aqueous solution of polymers. A dilute polymer liquid with a temperature above its lower critical solution temperature limit, known as the cloud point temperature, appears cloudy. The clouding effect is energetically favored due to the resultant gain in entropy by the water molecules, thereby giving rise to the "hydrophobic effect" and polymer aggregation (García-Peñas et al. 2019). At the critical temperature, the LCST polymer chain undergoes dehydration, while the water molecules gain entropy. The LCST transition is considered as an entropy-driven phenomenon (Zhang et al. 2017). UCST, on the other hand, is an enthalpically driven effect (Ruzette and Mayes 2001). The Gibbs energy of dissolution becomes positive at a temperature lower than the cloud point of a UCST polymer (Niskanen and Tenhu 2017). Typically, the UCST polymers exhibit strong supramolecular interactions in water. One example of UCST polymer is poly(*N*-acryloyl glycinamide) (PNAGA, Fig. 11.3e). This non-ionic UCST polymer can form intra- and intermolecular complexes between its own units' side groups via hydrogen bonds. Primary amide on glycinamide group of PNAGA can perform as a hydrogen donor coupled with carbonyl group as a hydrogen acceptor (Sun et al. 2017).

N-isopropylacrylamide (PNIPAM, Fig. 11.3a) displays a lower critical solution temperature at 32 °C (Futscher et al. 2017; Naziris et al. 2021). The polymer is readily solubilized in water at temperatures below 32 °C. It precipitates very quickly with the temperature rise above 32 °C. This volume phase transition phenomenon is induced by the change in the hydration state of polymers due to the exchanging and competing hydrogen bonding properties between the polymer and water molecules

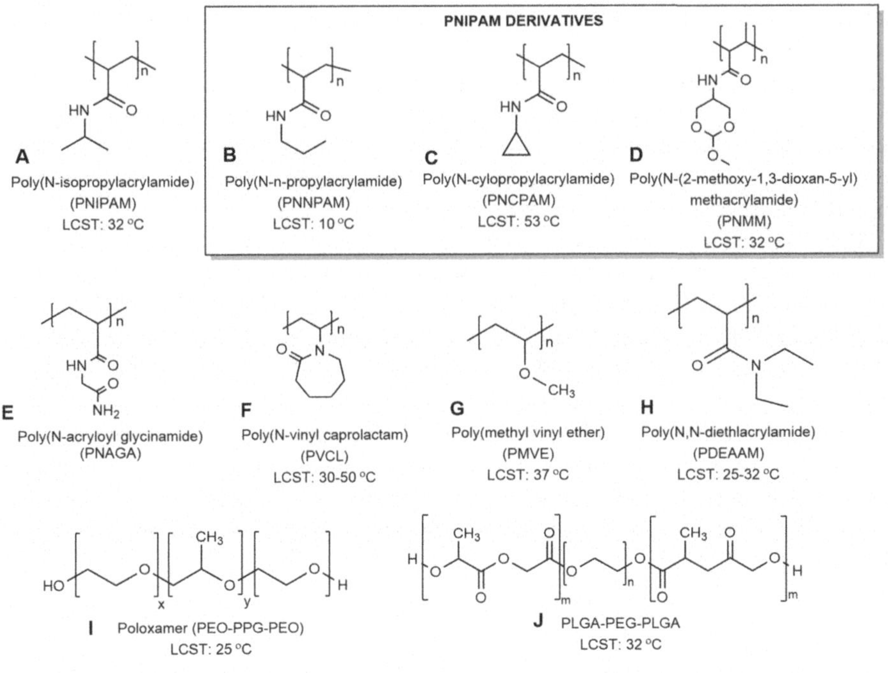

Fig. 11.3 Chemical structures of typical thermo-responsive synthetic (**a–h**) and hybrid triblock copolymers (**i** and **j**)

(Navarro et al. 2020; Podewitz et al. 2019). At lower temperatures, the intermolecular hydrogen bonds between the polymer and water molecules induce a chain conformation that hinders hydrophobic isopropyl groups of polymers from reaching the aqueous environment. At a temperature above the lower critical limit, the pre-existing bonds break with a loss in polymer-water interaction. The polymers aggregate and become water-insoluble. Conversely, a reduction in liquid temperature recovers polymer-water interaction and polymer solubility (Podewitz et al. 2019; Litowczenko et al. 2021).

11.2.3.2 Polymer Hydrophobicity

Polymer solubility, transition temperature, and gelation temperature of amphiphilic polymers in water can be altered by varying the hydrophobicity of the macromolecules. One recent study highlights that thermo-responsive polypeptoid acquires both LCST and UCST characteristics. The LCST or UCST behaviors of this polypeptoid can be regulated by the alkyl or ethylene glycol group modification. While hydrophilic ethylene glycol plays a role in the UCST phase transition, the hydrophobic residues are responsible for the LCST behavior (Liu and Sun 2020). Tahara et al. (2016) introduced a thermo-responsive building block based on cholesterol-bearing hydroxypropyl cellulose (Ch-HPC). They studied the effect of hydrophobic moieties consisting of cholesterol groups toward characteristics of self-assembled nanogel. The particle assembly of Ch-HPC with higher (i.e., 1.5 and 1.9) cholesterols per 100 glucose units into nanogels in deionized water at high temperature was influenced mainly by the hydrophobic interaction. A degree of cholesterols of 1.5 per 100 glucose units in the polymer was characterized by reduced polymer thermo-responsiveness. Interestingly, the Ch-HPC nanogels with high numbers of cholesterols as well as lower or no cholesterol-bearing nanogels were characterized by high thermo-responsiveness (Tahara et al. 2016).

Poly(D,L-lactic-co-glycolic acid)-b-poly(ethylene glycol)-b-poly(D,L-lactic-co-glycolic acid) (PLGA-PEG-PLGA) triblock copolymer (Fig. 11.3j) has been designed and processed into nanogel systems (Osorno et al. 2020). The PLGA is the hydrophobic block, while the PEG is the hydrophilic block. With an increase in PLGA molecular weight, the triblock copolymer solubility decreases. The polymer solubility can be further reduced by increasing the weight ratio of d,l-lactide, and glycolide, where the d,l-lactide is more hydrophobic than the glycolide. Polyethylene glycol is a waxy water-soluble excipient. The introduction of a shorter polyethylene glycol chain translates to lower gelation temperature where polymer aggregation is more prone to take place with fewer hydrophilic polyethylene glycol connections (McKenzie et al. 2015).

11.3 Sources of Thermo-Responsive Polymers

The sources of thermo-responsive polymers can be classified as natural, synthetic, and hybrid. The naturally occurring thermo-responsive polymers respond to a typical physiological stimulus without producing any immunological or toxic responses to humans. Nonetheless, these materials lack the mechanical strength required to develop drug delivery systems and biomedical devices (Matanović et al. 2014; Chatterjee et al. 2018). Synthetic thermo-responsive polymers are an alternative to natural counterparts. However, the biodegradability, cytotoxicity, and biocompatibility of synthetic polymers cannot be assured (Chatterjee et al. 2018). Hence, hybridization of polymeric materials has been exploited to create a biocompatible and nontoxic material with ideal mechanical properties.

11.3.1 Natural Thermo-Responsive Polymers

Natural thermo-responsive polymers and their derivatives have been extensively used in biomedical, food, and pharmaceutical industries with great potentials in tissue engineering and drug delivery system development (Cheng et al. 2010; Teotia et al. 2015; Bruschi et al. 2017; Chatterjee et al. 2018). In particular, the use of naturally occurring polysaccharides with thermo-responsive gelation properties is of interest because of their availability, versatility, biodegradability, and biocompatibility (Graham et al. 2019). These polysaccharides are harvested from biological resources such as plant and marine materials (Graham et al. 2019). Examples of natural thermo-responsive polymers are agarose (Miguel et al. 2014), carrageenan (Liu et al. 2016), and cellulose derivatives (Jain et al. 2013), which exhibits UCST, transitioning from solution to gel when cooled (Table 11.1).

Table 11.1 Selected naturally occurring thermo-responsive biopolymers and their properties

Natural polymer	Sol-gel transition temperature (°C)	Properties	References
Collagen	20–37	Has the highest biocompatibility, which makes it suitable as a tissue engineering substrate; has a small pore size and minimal fiber bundling	Holder et al. (2018)
Gelatin	< 30	Good mechanical properties that are ideal for pharmaceutical, cosmetics, and food application	Djabourov et al. (1988)
Agarose	< 35	Has better microcapsule quality and more stable in the membrane in vivo	Zhang et al. (2011)
Matrigel™	37	Excellent cellular compatibility (especially in chondrocytes and endothelial cells)	Li and Guan (2011)
Amylopectin	70–75	Undergo irreversible gelation at a temperature greater than 60 °C	Chen (1990)
Carrageenan	17–77	Has good stiffness and viscoelastic properties	Iijima et al. (2013)

11.3.1.1 Cellulose

Cellulose, commonly found in plant cell walls, is the most abundant naturally occurring polymer (Jain et al. 2013). It is a linear polysaccharide with repeating ß-D-anhydroglucopyranose units covalently bound through acetal functions between the equatorial group of the C4 carbon atom and the C1 carbon atom forming ß-1,4-glycosidic bond (Fig. 11.4) (Butardo and Sreenivasulu 2016). Cellulose is insoluble in water due to the limited access to C6-OH groups in its structure necessary for the intermolecular hydrogen bonds for solvation (Eo et al. 2016). Hence, several cellulose derivatives such as methylcellulose, carboxymethyl cellulose, hydroxyethylcellulose, hydroxypropyl methylcellulose, and cellulose acetate have been made soluble in an aqueous medium at low temperature (Benslimane et al. 2016)

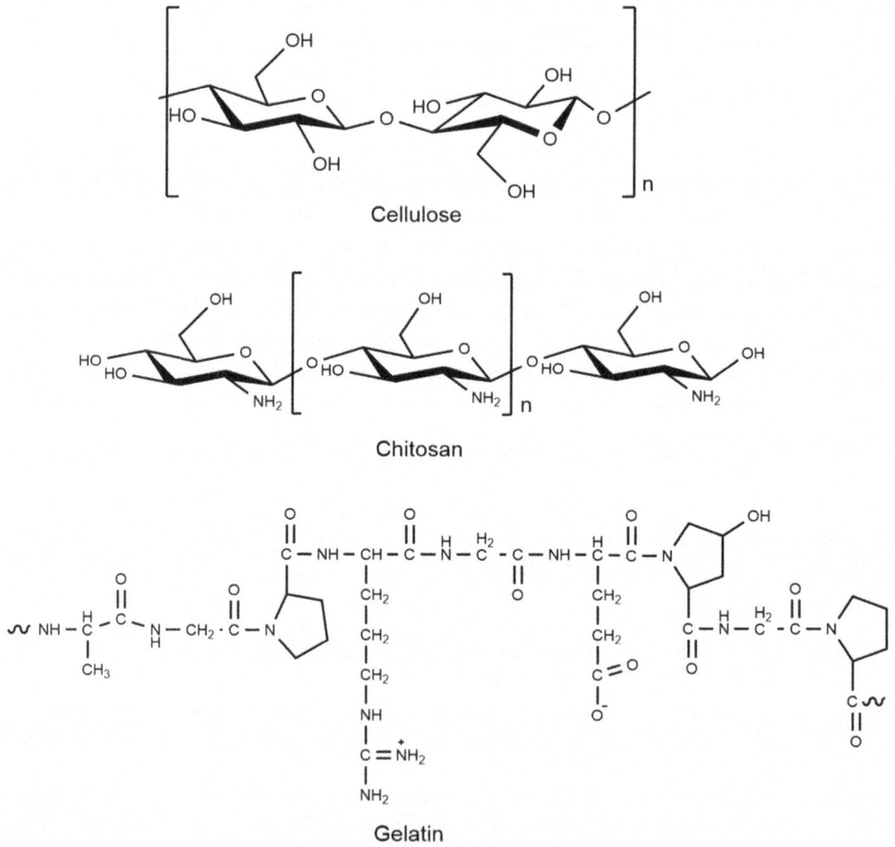

Fig. 11.4 Structures of commonly used naturally occurring polymers in biomedicine

Table 11.2 Common water-soluble thermo-responsive cellulose derivatives in pharmaceutical and biomedical application

Cellulose derivative	Sol-gel transition temperature (°C)	Characteristics	Application	References
Methylcellulose (MC)	75–90	Has good bioadhesive property and sustained drug release efficiency	Thickener, binder, emulsifying agent, and coating for different dosage forms	Kim et al. (2012)
Carboxymethyl cellulose (CMC)	60–80	Excellent film-forming ability	Film-forming agent	Benslimane et al. (2016)
Ethyl (hydroxyethyl) cellulose (EHEC)	~35	Good stabilizing property and biocompatibility	Smart material and drug delivery	Jain et al. (2013)
Hydroxypropyl methylcellulose (HPMC)	Starts at 55 (depending on the presence of methoxy residues)	High swelling ability and thermal gelation properties	Mucoadhesive and drug carrier (for controlled release)	Deshmukh et al. (2017) and Joshi (2011)

(Table 11.2). When subjected to elevated temperatures, strong hydrophobic interaction takes place between the macromolecular chains of the polymer, which results in the phase separation of the polymer from water, and the conformational changes in the polymer-promoting gelation (Bekkour et al. 2014; Benslimane et al. 2016).

11.3.1.2 Chitosan

Chitosan, a polysaccharide, is one of the most explored thermo-responsive polymers. It is the product of alkaline deacetylation of chitin, or poly (*N*-acetylglucosamine), isolated from the exoskeleton of crustaceans (Fig. 11.4) (Quiñones et al. 2018). It is biocompatible, mucoadhesive, biodegradable, nontoxic, and available in abundance and has low immunogenicity (Bowman and Leong 2006; Indulekha et al. 2016; Luckanagul et al. 2018; Quiñones et al. 2018). Chitosan receives widespread applications in tissue engineering, wound healing, and drug and gene delivery (Tran et al. 2011; Banerjee and Ganguly 2019; Ghaee et al. 2019). Chitosan (pKa = 6.5) can only be dissolved in acidic medium through the protonation of the free amino groups in its structure (Saravanan et al. 2019). The neutralization of chitosan solution to pH >6.5 at room temperature can rapidly induce the formation of a gel-like precipitate (Matanović et al. 2014). Chitosan can be physically or chemically linked to glycerophosphate (GP) polyol salt to modulate its thermo-responsiveness (Li and Guan 2011). In response to an average human body temperature of 37 °C, hydrogen bonds from between chitosan and GP composite exhibits an immense rise in the storage modulus (G') corresponding to sol-gel

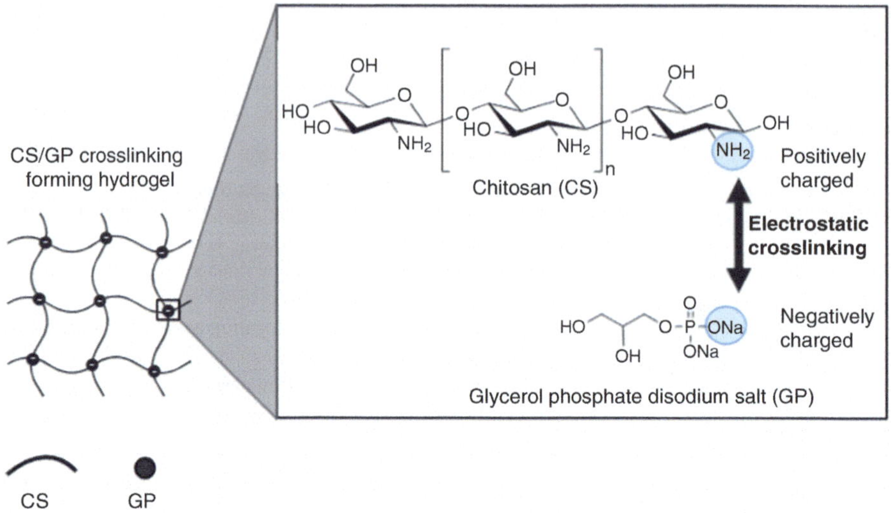

Fig. 11.5 The electrostatic crosslinking of positively charged chitosan (CS) to negatively charged glycerol phosphate disodium salt (GP), rendering its thermo-responsiveness

transition (Fig. 11.5). The molecular gelation mechanism of the chitosan/GP composite occurs in three steps: (1) the addition of basic GP increases the hydrogen bonding between the chitosan, leading to reduced electrostatic repulsion; (2) electrostatic attractions between the amino and phosphate groups of chitosan and GP, respectively; and (3) enhanced chitosan-chitosan hydrophobic interaction via glycerol affinity for water (Saravanan et al. 2019). The chitosan/GP composite has been adopted as a scaffold in tissue engineering and regenerative medicine due to its biocompatibility, remarkable mucoadhesive, antibacterial property, and minimal immune response (Zhou et al. 2015; Huang et al. 2016). It is also employed in smart-controlled release of drugs to avoid frequent dosing and secondary side effects in patients by slowly releasing the drug from the gel in response to changes in temperature (Ruel-Gariépy et al. 2004; Chen et al. 2013; Peng et al. 2013).

11.3.1.3 Gelatin

Gelatin is a natural peptide-based polymer with thermo-responsive properties. It is obtained via acidic, alkaline, or enzymatic hydrolysis of collagen (Rodríguez-Rodríguez et al. 2020). The gelatin polymeric chains are composed of Gly-Xaa-Yaa tripeptide units, with proline and hydroxyproline amino acids being commonly attached to Xaa and Yaa positions (Fig. 11.4) (Gornall and Terentjev 2007; Wang et al. 2017). Gelatin is advantageous from the perspectives of biodegradability, biocompatibility, easy modification, low cost, and ease of production (Song et al. 2018). Most commercially available gelatin is extracted from the pigskin (46%), the most abundant gelatin source (Duconseille et al. 2015). At a concentration lower than 2%w/v, the gelatin can transform from solution to gel at a temperature below 30 °C. During the sol-gel transition, the gelatin molecules come together and undergo conformational changes from a random coil to a triple helix. The intermolecular hydrogen bonds are formed between the macro fractions of gelatin chains (Xing et al. 2014). The molecular interaction and conformational changes can be easily disrupted when the gelatin is subjected to a temperature higher than 30 °C, thereby giving rise to gel-sol transformation. Pure gelatin is met with limited biomedical application due to its instability and poor mechanical and elastic properties, especially in the physiological environment (Xing et al. 2014). Covalent crosslinking of gelatin through amide bond formation via carbodiimide chemistry in removing divalent (Ca^{2+} and Fe^{2+}) present in gelatin structure paves the way to significantly enhance its stability and physical properties (Xing et al. 2014).

11.3.2 Synthetic Thermo-Responsive Polymers

Poly(N-isopropylacrylamide), also known as PNIPAM, is a thermo-responsive water-soluble polymer developed from acrylamide monomer (Yang et al. 2014; Luckanagul et al. 2018). It has a sol-gel transition temperature close to the regulated human body temperature of 32 °C (Heskins and Guillet 1968; Matanović et al. 2014; Kojima 2018). The thermo-responsive behavior of PNIPAM is mainly dominated by hydrogen bonding (Hirotsu 1993; Matanović et al. 2014). At temperatures below its lower critical solution temperature, the water molecules can solvate the amide functionalities of PNIPAM, rendering it with good water solubility (Pişkin 2004; Alexander et al. 2014). Above the critical temperature limit, the polymer becomes insoluble in water exhibiting a "coil to globule" transition due to stronger inter-polymer interaction between the hydrophobic propyl groups, while the hydrogen bonds between the polymer and water molecules turn weaker (Bischofberger et al. 2014; Bischofberger and Trappe 2015). The LCST of PNIPAM can be tuned by copolymerization with hydrophilic or hydrophobic monomers (Rzaev et al. 2007). Copolymerization of PNIPAM with a more lipophilic monomer will reduce its LCST and weaken polymer-water interaction (Feil et al. 1993; Kim et al. 2009; Matanović et al. 2014). Copolymerization of hydrophilic monomer like acrylamide (Fundueanu et al. 2009), N-(2-(dimethylamino)ethyl) methacrylamide (Wang et al. 2008), and PEG-methacrylamide (Trongsatitkul and Budhlall 2013) leads to increased hydrophilicity and LCST and a stronger affinity for water and gel formation (Feil et al. 1993).

In 2010, Azarbayjani et al. (2010) investigated the potential of thermosensitive PNIPAM for localization and accumulation of levothyroxine (T4) in the skin to mitigate adipose tissue deposition with minimal systemic absorption that can incur adverse drug reactions. PNIPAM was copolymerized with polyvinyl alcohol (PVA), a hydrophilic polymer with a high degree of swelling. Interestingly, formulation with PVA has a slower release rate than PNIPAM alone due to its slow degradation. PVA can be dissolved in water at a temperature above 90 °C. Literature suggests that at body temperature, PVA will disintegrate after 7 weeks; thus, a combination of PVA and PNIPAM results in a prolonged drug release of T4 in the skin (Azarbayjani et al. 2010).

Thermo-responsive PNIPAM has also been copolymerized with pH-responsive acrylic acid (AAc) to control caffeine release through multiple responsive properties of the nanogel (Abu Samah and Heard 2013). PNIPAM-co-AAC is thermo-responsive. The pH-responsive AAc has a carboxyl group ionized in response to a pH rise. A strong electrostatic repulsion is developed between the carboxylate anions within the network. The expanded network introduces water in flux. Coupling of gel microstructure changes due to thermo- and pH-responsiveness augments the drug release control. The same phenomenon was also concluded by (Singka et al. 2010) using butyl acrylate (BA) as the copolymer in transdermal delivery of methotrexate (MTX), causing a significant flux of MTX and its progressive release that is incomparable to the saturated solution of MTX. The typical synthetic thermo-responsive polymers (Fig. 11.3a–h) and hybrid triblock copolymers (Fig. 11.3i, j) with their chemical structures are depicted in Fig. 11.3. Among them, PNIPAM and its derivatives (Fig. 11.3b–d) are the most common synthetic polymers that are used as thermo-responsive polymeric materials in biomedicine. Other selected synthetic thermo-responsive polymers with their characteristics and biomedical applications are summarized in Table 11.3.

11.3.3 Hybrid Thermo-Responsive Polymers

The thermo-responsive property of a polymer can be modulated through a hybridization approach. The hybrid thermo-responsive polymers are typically constituted of three or more monomers in a single copolymer. Poloxamers (Pluronics) is a hybrid triblock copolymer of poly(ethylene oxide)-b-poly(propylene oxide)-b-poly(ethylene oxide) (PEO-PPO-PEO) (Fig. 11.3i) (Alexandridis and Alan Hatton 1995; Klouda and Mikos 2008; Batrakova and Kabanov 2008; Bruschi et al. 2017; Ban et al. 2017). Poloxamers have no actual lower critical solution temperature; however, the gelling temperature of this polymer can be manipulated by controlling its concentration, composition, and molecular weight (Klouda and Mikos 2008). Various gelation mechanisms of Pluronics have been proposed: (1) an increase in temperature initiates micellar aggregation and gelation (Rassing and Attwood 1982; Abou-Shamat et al. 2019), (2) an increase in temperature induces reorganization of water molecules at the core of PPO blocks and translates to the formation of non-polar surfaces favoring micellar aggregation and gelation (Vadnere et al. 1984), and (3) the dehydration of poly-propylene-oxide reduces the lubricative effect between the copolymeric micelles, thereby causing the micellar aggregation and gel formation (Rassing et al. 1984).

Table 11.3 Selected synthetic thermo-responsive polymers and their phase transition characteristics and biomedical applications

Synthetic polymer	Sol-gel transition temperature in aqueous solution (°C)	Characteristics and application
Poly(methyl vinyl ether) **(PMVE)**	37	PMVE is sensitive to nucleophiles like alcohol and amino acids, hence limiting its potential for further modification (Gandhi et al. 2015)
Poly(N-vinyl caprolactam) **(PVCL)**	30–50	Used for medical and biotechnological applications due to its good solubility in water and organic solvents. It also possesses good biocompatibility and absorption properties (Gandhi et al. 2015; Makhaeva et al. 1998)
Poly(pentapeptide) of elastin	28–30	Mostly used in targeted tumor strategy, since its thermo-responsiveness can be easily modulated by balancing its hydrophobic and hydrophilic residues (Gandhi et al. 2015)
Poly(N-ethyl oxazoline) **(PEtOx)**	~62	Currently explored as a drug delivery system due to their micellar aggregation above LCST (Gandhi et al. 2015)
Poly(acrylic acid-co-acrylamide)	~25	This polymer possesses UCST behavior due to the hydrogen bonding between the AAc and AAm units (Gandhi et al. 2015; Mohan et al. 2005)

The Pluronic F127 or the Poloxamer 407 is characterized by a % PEO content of 70% (Simões et al. 2015). The gelation of Pluronic F127 is mainly dependent on its concentration. It can transform from solution to clear gel at a room (25 °C) to body temperature (37 °C) with 20 to 30% of polymer content (Russo and Villa 2019). The hydrophilic PEO component attached on both sides of the Poloxamer unit renders its excellent water solubility; however, when subjected to high temperatures, the water-polymer bond interaction is weakened, resulting in the association of the hydrophobic PPO and sol-gel transition triggered (Dou et al. 2016). Further hybridization of Pluronic F127 with methylcellulose promotes thermo-gelation at body temperature at a polymer concentration as low as 12% and solution at room temperature (Rangabhatla et al. 2016).

Pluronic F127 has been hybridized with methacryloisobutyl-polyhedral oligosesquioxane (MA-POSS) via atomic transfer radical polymerization (ATRP) to modulate the thermal and mechanical properties of F127 (Dou et al. 2016). POSS is a hybrid organic-inorganic organosilicon with a rigid nano-block structure (Loh et al. 2007). The grafting of rigid POSS with a linear polymer chain of F127 can affect the segment mobility and the molecular interaction between the polymer backbone segments. As a result, the LCST value of the F127-POSS increases by 10 °C from 23.5 to 33.5 °C. This can be attributed to the addition of a hydrophobic rigid nano-block structure of the POSS that hinders PPO association with water; hence, higher sol-gel temperature is required (Dou et al. 2016). Generally, in the hybridization of Pluronics, the increased hydrophilic component in a polymer can reduce the hydrophobic interaction, which leads to low micellar aggregation (Park et al. 2008). The sol-gel transition can be difficult and require a higher transition temperature (Park et al. 2008; Dou et al. 2016). Table 11.4 summarizes various hybrids of Pluronics.

Table 11.4 Examples of the notable studies in tuning the gelation of thermo-responsive Poloxamer/Pluronics

Polymer	Sol-gel transition temperature (°C)	Advantages of hybridization of Pluronics	Application	References
Methylcellulose	37	Extends the in vitro release of Etidronate Na for more than 28 days; MC can also improve the gel strength of the Pluronics and its degradation.	Drug delivery for osteogenesis/bone remodeling	Rangabhatla et al. (2016)
Alginate and HA	37	The polymeric hydrogel provides good scaffolds that are essential for cellular growth	Scaffolds for cellular and tissue growth	Izhar Haider Abdi et al. (2012)
Alginate	36–40	F127 activation in alginate has a successful tuning in the gelation at a temperature that is close to normal physiological temperatures	Wound dressing, tissue engineering, injection, and dental treatments	Quah et al. (2018)
MA-POSS	33.5	The degradation and gelation temperatures were increased by 15 °C and 10 °C, respectively, due to the addition of hydrophobic rigid nano-block structure of the POSS that hinders PPO association with water	Biomaterials for food, cosmetics, and pharmaceuticals	Dou et al. (2016)
CMCS	32–37	Phase transition into gel can be achieved at normal physiological temperature; it also helps with the enhanced solvents and drug uptake through its connective pores	Local delivery of chemotherapeutics	Khan et al. (2018)

PNIPAM has been hybridized with poly(3-methacryloxypropyltrimethoxysil ane) into a triblock thermo-responsive copolymer consisting of PNIPAM-b-pMEMO-b-pPIPAM via reverse addition-fragmentation chain transfer polymerization approach (Wu et al. 2020a, b). Hybridization of PNIPAM causes gelation to occur at a higher temperature of ~35.3 °C at PNIPAM to pMEMO ratio of 28:1 compared to pure PNIPAM. The increase in the LCST is due to the hydrolysis of Si-O-CH_3 of pMEMO, which leads to stronger hydrogen bond interaction with water, thus enhancing its hydrophilicity (Yang et al. 2015; Wu et al. 2020a, b). It is noteworthy that the increase in the LCST of the polymer can be attributed to the increase in the pMEMO concentration (Wu et al. 2020a, b).

11.4 Current Application of Thermo-Responsive Polymers in the Biomedical Field

The thermo-responsive polymers have been widely investigated for a wide variety of applications. Table 11.5 summarizes the type of polymers and their uses for drug/gene delivery and tissue engineering.

11.4.1 Drug Delivery

Metal-based thermo-responsive polymeric nanoparticles have gained increased attention, notably in the field of cancer therapy. Kurdtabar et al. (2018) developed a thermo-responsive hydrogel system of doxorubicin (DOX) by coating gold nanorods (GNR) with PNIPAM grafted into carboxymethyl cellulose (CMC) as a backbone. Copolymerization of PNIPAM with CMC improved the swelling behavior and thermal stability of the hydrogel. The transition from swollen gel to flowable sol occurring at 43 °C and the repulsive effect of GNR to the polymer network producing larger pores have caused a rapid collapse of the polymer network, significantly improving the DOX release rate at 98%. Another study has DOX encapsulated using two oligo(ethylene glycol) monomers – 2-(2-methoxy)ethyl methacrylate (MEO_2MA) and oligo(ethylene glycol)methacrylate (OEGMA), with a tunable LCST between 26 and 90 °C by varying their ratio (Ferjaoui et al. 2019). They were covalently grafted on the surface of the iron oxide NPs (IONPs) to ensure the stability of the IONPs, entrapment success of DOX, and its release in response to body temperature. The DOX releases faster at 42 °C than at 37 °C. This is due to the expulsion of water molecules from the grafted polymer network upon reaching its LCST at 41 °C, leading to shrinkage and collapse of the polymer chains, thus breaking the hydrogen bonds between DOX and the ether-oxide moieties of the copolymer leading to the release of DOX.

Table 11.5 Summary of recent studies on the use of thermo-responsive polymers in drug/gene delivery and tissue engineering

Thermo-responsive polymer/s	Payload/main artifact	Preparation	Adjunct components and their use	Reversible behavior (swelling-deswelling)	Target disease/ potential application	Remarks/salient findings	References
Drug delivery							
PNIPAM	Doxorubicin	Polymer composite-coated gold nanorod	Carboxymethyl-cellulose serves as backbone polymer	2 cycles	Cancer	At 90 min, the release rate of DOX was reported as 98% at 43 °C compared to 33% at 37 °C; DOX release was also higher in the presence of GNR compared to the DOX-polymer composite alone due to the repulsive effect of GNR to the polymeric chains; biocompatible with L929 cell line (mouse fibroblast cells); potential for photothermal therapy through GNR by application of NIR laser irradiation	Kurdtabar et al. (2018)

Thermo-responsive polymer/s	Payload/main artifact	Preparation	Adjunct components and their use	Reversible behavior (swelling-deswelling)	Target disease/ potential application	Remarks/salient findings	References
MEO_2MA and OEGMA	Doxorubicin	Polymer composite-coated iron oxide	–	5 cycles	Ovarian cancer	Varying the ratio of MEO_2MA and OEGMA can adjust the LCST between 26 and 90 °C; maximum DOX release was observed at 42 °C; DOX-loaded polymer-coated magnetic NP was more cytotoxic toward SKOV-3 cell line (human ovarian cancer cells) than free DOX; potential for local and targeted heating of cancer cells upon application of an alternating current magnetic field	Ferjaoui et al. (2019)

(continued)

Table 11.5 (continued)

Thermo-responsive polymer/s	Payload/main artifact	Preparation	Adjunct components and their use	Reversible behavior (swelling-deswelling)	Target disease/potential application	Remarks/salient findings	References
PNVCL	Oxaliplatin	Polymer composite-coated graphene oxide	Poly(glycolic acid) serves as a pH-sensitive polymer	–	Breast cancer	Maximum release of Oxaliplatin was at 45 °C and pH 5.7 (endosomal pH of cancer cells); oxaliplatin-loaded GO-polymer composite was more cytotoxic against MCF-7 cell line (breast cancer cells) than free oxaliplatin; preparation can respond to both temperature and pH triggers	Kazempour et al. (2019)

Thermo-responsive polymer/s	Payload/main artifact	Preparation	Adjunct components and their use	Reversible behavior (swelling-deswelling)	Target disease/ potential application	Remarks/salient findings	References
PNVCL	Benzalkonium chloride and ciprofloxacin	Hydrogel film	Poly(2-hydroxyethyl methacrylate) is used to strengthen the mechanical stability of the film	–	Anti-biofilm in medical device-associated infections	Optimization of grafting method was successful using parameters such as swelling, drug release, and thermal behavior; grafted PP films produced a lower water contact angle to prevent adhesion of microbes and blood products; grafted PP films showed controlled release of benzalkonium chloride and ciprofloxacin for 8 h	González-Hernández et al. (2019)

(continued)

Table 11.5 (continued)

Thermo-responsive polymer/s	Payload/main artifact	Preparation	Adjunct components and their use	Reversible behavior (swelling-deswelling)	Target disease/potential application	Remarks/salient findings	References
Pentaerythritol, DL-lactic acid, and e-caprolactone	Bevacizumab and aflibercept	Hydrogel	–	–	Ocular inflammatory diseases (age-related macular degeneration)	Non-cytotoxic in vitro using human neuroblastoma (SH5Y) and rat fibroblast (R12) cell lines and in vivo using rats; sustained-release profile was demonstrated at 37 °C for 183 days showing 95% of bevacizumab and 25% of aflibercept; pathological and in vivo imaging studies on the intravitreally injected rats showed no adverse effects and abnormalities on the retinal layers and retinal function, respectively	Balachandra et al. (2019)

Thermo-responsive polymer/s	Payload/main artifact	Preparation	Adjunct components and their use	Reversible behavior (swelling-deswelling)	Target disease/ potential application	Remarks/salient findings	References
PLGA-PEG-PLGA	N′-dodecanoylisonicotino-hydrazide (DINH)	Liposome in hydrogel	–	3 cycles	Bone TB	Non-cytotoxic to mouse embryos osteoblast precursor cells (MC3T3-E1) cells; sustained-release of DINH; self-healing property of PLGA-PEG-PLGA was demonstrated for the first time; microdialysis and optical imaging provided the evidence of the sustained-release profile of DINH in the synovial joint of rabbits	Liu et al. (2019)

(continued)

Table 11.5 (continued)

Thermo-responsive polymer/s	Payload/main artifact	Preparation	Adjunct components and their use	Reversible behavior (swelling-deswelling)	Target disease/ potential application	Remarks/salient findings	References
PNIPAM	$G(IIKK)_3I\text{-}NH_2$	Polymer-based nanogel	I_3K serves as an amphiphilic peptide nanofibril added to improve drug loading	20 cycles	Injectable drug delivery and wound dressing	The formulation showed a promising application in sustained release of an antibacterial agent effective against *B. subtilis*, *P. aeruginosa*, *S. aureus*, and *E. coli*; nontoxic in 3 T3 fibroblast cells due to the morphological similarity of the polymer-based nanogel with the extracellular matrix of normal tissues	Cao et al. (2019)

Thermo-responsive polymer/s	Payload/main artifact	Preparation	Adjunct components and their use	Reversible behavior (swelling-deswelling)	Target disease/ potential application	Remarks/salient findings	References
Gene delivery							
PNIPAM	siRNA	Polymer/siRNA nanogel	AMPTMA was added to increase the LCST of pNIPAAm and confers positive charges to the hydrogel system	–	Cancer	Length of the PNIPAAM (n) or number of PAMPTMA blocks (m) determines the phase transition temperature and overall charge of the hydrogel; formulation containing 48:20 (n:m) provides the best properties such as a relatively low toxicity (65% of cell viability) and highly efficient gene silencing system, with 68% of GFP knockdown and only 33% of unspecific gene silencing using HT1080 cells (human fibrosarcoma cell line)	Cardoso et al. (2014)

(continued)

Table 11.5 (continued)

Thermo-responsive polymer/s	Payload/main artifact	Preparation	Adjunct components and their use	Reversible behavior (swelling-deswelling)	Target disease/ potential application	Remarks/salient findings	References
PNIPAM	siRNA	Polymeric liposome	DMAPAAm functions to increase the LCST of pNIPAAm and confers positive charges to the hydrogel system	–	Cancer	LCST of the copolymer is at 41.7 °C; the polymer-liposome complex indicated high transfection efficiency and low toxicity proving its potential as a siRNA carrier; the preparation also showed suppression of VEGF activity in HeLa cells at 42 °C, indicating its potential as an anticancer agent	Nagase et al. (2019)

Thermo-responsive polymer/s	Payload/main artifact	Preparation	Adjunct components and their use	Reversible behavior (swelling-deswelling)	Target disease/potential application	Remarks/salient findings	References
PNIPAM	pDNA	Au-modified polymer composite	Polydopamine (PDA) generates pores on the cell membrane for the loading of plasmids		Ex vivo cell-based therapies/tissue repair	Delivery of plasmids into the endothelial cells has a reported ~99% transfection efficiency; membrane-based delivery of plasmids into HeLa cells, mouse embryonic fibroblasts (mEFs), human umbilical vein endothelial cells (HUVECs), and mouse dendritic cells (mDCs) did not affect cell viability	Wu et al. (2019)

(continued)

Table 11.5 (continued)

Thermo-responsive polymer/s	Payload/main artifact	Preparation	Adjunct components and their use	Reversible behavior (swelling-deswelling)	Target disease/ potential application	Remarks/salient findings	References
Tissue engineering (scaffolds)							
Chitosan	Vancomycin	Injectable chitosan-based hydrogel	Glycerol phosphate disodium enhances the mechanical stability of the hydrogel	–	Osteomyelitis	Sustained release of vancomycin from the formulation showed a total release of 65% within 26 days; biocompatible and promotes osteoblast cell proliferation; vancomycin NP showed greater activity against *S. aureus* than free vancomycin; implantation of the vancomycin-loaded NP in the rabbit model showed anti-infective activity and promoted proliferation of osteoblast cells	Tao et al. (2020)

Thermo-responsive polymer/s	Payload/main artifact	Preparation	Adjunct components and their use	Reversible behavior (swelling-deswelling)	Target disease/ potential application	Remarks/salient findings	References
Chitosan	Human dental pulp stem and progenitor cells	Injectable cellulose-based scaffold	Nanocellulose provides the matrix for cell attachment; glycerol phosphate disodium enhances the mechanical strength of the hydrogel	–	Temporomandibular disorders	The scaffold system allowed the chondrogenic differentiation of embedded human dental pulp stem and progenitor cells (hDPSCs); biocompatible, injectable, mechanically stable	Talaat et al. (2020)
Hydroxypropyl chitin (HPCT)	Auricular chondrocytes	Injectable HPCT-based scaffolds	–	–	Cartilage defects	The hydrogel showed good gelation properties, biocompatibility, and mechanical strength when implanted in vivo, indicating a promising scaffold for cartilage regeneration	Xu et al. (2020)

(continued)

Table 11.5 (continued)

Thermo-responsive polymer/s	Payload/main artifact	Preparation	Adjunct components and their use	Reversible behavior (swelling-deswelling)	Target disease/ potential application	Remarks/salient findings	References
Poly(urea-urethane)/ collagen	Human bronchial fibroblasts	3D-TIPS printed elastomeric scaffold coated with collagen hydrogel	–	–	Tracheal defects	The hybrid scaffolds support the growth and differentiation of bronchial epithelial cells; promotes epithelial cell growth, epithelial ciliation, and epithelial barrier formation and mucin expression	Wu et al. (2020a, b)
PNIPAM	Human-induced pluripotent stem (hiPS) cells	Prior to culture in PNIPAM-coated dish, hiPS were treated with Wnt signaling molecules and labeled with superparamagnetic iron oxide	–	–	Ischemic cardiomyopathy (ICM)	The hiPS cells differentiated into cardiomyocytes (CM); a combination of omentum, a vascular-rich tissue, and cell sheet improved cardiac function and the long-term engraftment of hiPS-CM due to improved blood supply source for the cells in porcine ICM model; the cell sheets have potential for treating severe heart failure	Kawamura et al. (2017)

Thermo-responsive polymer/s	Payload/main artifact	Preparation	Adjunct components and their use	Reversible behavior (swelling-deswelling)	Target disease/ potential application	Remarks/salient findings	References
PNIPAM	Periodontal ligament (PDL)-derived cells from patients	PDL-derived cells were cultured on a PNIPAM-coated culture dish for 2 weeks at 37 °C	b-tricalcium phosphate granule was used as a biodegradable filler to improve bone repair of tooth	–	Periodontitis	The three-layered PDL-derived cell sheets that were fabricated and transplanted in 10 patients were found to be safe and efficacious in treating severe periodontitis; improved bone regeneration at 6 months after transplantation was observed with no serious adverse events after a mean follow-up of 55 months	Iwata et al. (2018)

(continued)

Table 11.5 (continued)

Thermo-responsive polymer/s	Payload/main artifact	Preparation	Adjunct components and their use	Reversible behavior (swelling-deswelling)	Target disease/ potential application	Remarks/salient findings	References
PNIPAM	Submandibular gland (SMG) cells	SMGs were cultured on PNIPAM-coated polystyrene dish for 8 days at 37 °C	–	–	Mucosal wounds and hyposalivation	Differentiated SMGs showed single layer cell sheets with a closely packed columnar epithelial pattern, tight junctions, and secretory granules, which are characteristics of salivary glands; a combination of single-cell layer sheets formed a double layer showing the glandular-like appearance, expressed AQP5 as an epithelial marker, and promoted saliva secretion and wound closure in a mouse model for 8 days	Nam et al. (2019)

Thermo-responsive polymer/s	Payload/main artifact	Preparation	Adjunct components and their use	Reversible behavior (swelling-deswelling)	Target disease/potential application	Remarks/salient findings	References
PNIPAM and glycidyl methacrylate (GMA)	Adipose mesenchymal stem cells (ASCs)	ASCs were cultured on NGMA-coated cell culture dishes for 14 days at 37 °C	–	–	Bilateral limbal stem cell deficiency	Differentiated ASCs showed positive expression of CK3/12 as a corneal epithelial marker; transplanting cell sheet in rabbit models of corneal injury showed clear and smooth cornea; histology reports revealed a sheet of cells aligned and integrated onto the injured corneal surface, 1-month post-transplantation	Venugopal et al. (2020)

(continued)

Kazempour et al. (2019) demonstrated the copolymerization of thermo-responsive poly(N-vinylcaprolactam (PNVCL) and pH-sensitive poly(glycolic acid) over the surface of graphene oxide (GO) that can provide various functional groups for successful attachment of polymers due to its hydrophilic surface and loading of hydrophobic drugs due to its hydrophobic nature. The anticancer drug oxaliplatin, loaded in the composite GO-hydrophilic polymers, showed a higher cytotoxic effect against MCF-7 breast cancer cells than the free drug. The grafting of poly(glycolic acid) with PNVCL modifies the LCST of PNVCL from 32–34 °C to 36–37 °C. Accordingly, the release extent of oxaliplatin was higher at 45 °C and pH 5.7 (endosomal pH of cancer cells) than at 27 °C and pH 7.4 (normal physiological pH). Similar to other polymers exhibiting LCST, PNVCL is soluble in water below its LCST but collapses into mesoglobules or precipitates (Siirilä et al. 2019) upon increasing the temperature, consequently increasing the diffusion kinetics of oxaliplatin. PNVCL has also been used to coat the surface of polypropylene (PP) films to overcome biofilm formation in medical devices (González-Hernández et al. 2019). The PP's surface was modified by radiation grafting of PNVCL and an acid/alkali stable poly(2-hydroxyethyl methacrylate) (PHEMA) to make it hydrophilic with benzalkonium chloride and ciprofloxacin as antibacterial agents. The grafted PP films produced a lower water contact angle to prevent the adhesion of microbes and blood products. It also showed a controlled-release profile of benzalkonium chloride and ciprofloxacin over 8 h. The grafting of PNVCL and PHEMA polymers on PP films makes them hydrophilic due to the polymer brushes, causing the temperature-sensitive swelling behavior of the system.

A novel thermo-responsive copolymer, consisting of the polyesters of glycolic, lactic, and caprolactone, demonstrates promising results for the ocular delivery of anti-vascular endothelial growth factor (VEGF) agents bevacizumab and aflibercept for age-related macular degeneration (Balachandra et al. 2019). It was found to be non-cytotoxic in vitro using human (SH5Y) and rat (R12) cell lines and in vivo. Sustained-release profiles of both drugs were demonstrated at 37 °C for 183 days and showed 95% release for bevacizumab and 25% for aflibercept. The drug-containing hydrogel is injected into the eye in the liquid state and transforms into a gel in response to body temperature. The gel state then facilitates sustained intraocular drug release, preventing frequent intravitreal injections. The release of drugs can be governed by two mechanisms: an initial diffusion-controlled phase where the loosely bound drug diffuses from the polymer gel and provides an initial burst release to achieve a therapeutic concentration of the drug followed by a degradation-controlled release phase as the polymer degrades, thus preventing the rapid depletion of the drug in the polymer matrix, allowing for sustained intraocular delivery of the drug.

Liu et al. (2019) developed a hybrid system composed of a liposome and a thermo-responsive block copolymer PLGA-PEG-PLGA for the intraarticular delivery of N′-dodecanoylisonicotinohydrazide (DINH), a derivative of isoniazid, for the localized treatment of bone tuberculosis. At low temperatures, the amphiphilic nature of PLGA-PEG-PLGA leads to its self-assembly into micelles in an aqueous solution. The micelles aggregate and form a non-flowable gel phase upon temperature increase due to the pronounced hydrophobic interactions of PLGA blocks and the partial dehydration of PEG moieties. Notably, this work demonstrates for the first time the self-healing property of the PLGA-PEG-PLGA copolymer system, with the biggest gel window obtained at 30% PLGA-PEG-PLGA and a sol-gel transition temperature of 32 °C. The DINH-loaded liposome in the hydrogel system was subjected to an applied force during medication administration. Increasing the applied force reduces the storage modulus (G′), showing that the polymer is structurally damaged. The self-healing property was demonstrated by the recovery of G′ after reducing the applied force due to the rapid reformation of the hydrophobic bonding among the PLGA fragments. The self-healing property provides a stronger polymer network to recover from the constant strain of synovial joints that could weaken the polymer crosslinks and release drugs prematurely. Combining the DINH-loaded liposome and PLGA-PEG-PLGA copolymer system leads to sustained release of the drug into the synovial fluid of rabbits, decreasing the frequency of intraarticular administration.

Cao et al. (2019) fabricated a thermo-responsive physically crosslinked hydrogel consisting of PNIPAM and I_3K, a peptide nanofibril that functions to improve the loading efficiency of an antimicrobial peptide $G(IIKK)_3I\text{-}NH_2$. The hydrogel system appeared transparent and flew freely at 25 °C while showing an opaque and non-flowable behavior at >33 °C, with 20 cycles of thermo-reversibility demonstrated. Since the LCST of PNIPAM is ~33 °C, gelation can be triggered by the body temperature indicating its use as an injectable hydrogel that can offer a controlled-release property. The gelation of PNIPAM was due to its coil to globule transformation, resulting in its collapse and physically driving the crosslinking of $G(IIKK)_3I\text{-}NH_2$ and I_3K nanofibrils. The sustained release of $G(IIKK)_3I\text{-}NH_2$ from the hydrogel system was attributed to its hydrophobic interactions with the peptide nanofibril I_3K and the exposed propyl groups of the collapsed PNIPAM network at 40 °C. Accordingly, the slow and linear release was an effective strategy against *B. subtilis*, *P. aeruginosa*, *S. aureus*, and *E. coli* growth.

11.4.2 Gene Delivery

The essential features of a gene delivery system include the ability of the copolymers to protect the genetic material, facilitate cell uptake, and promote dissociation of polymer and gene. Thermo-responsive polymers provide efficient delivery of genetic materials into cells in response to temperature and confer protection of their fragile nature (Fig. 11.6). The temperatures at which the complexation and incubation/transfection are undertaken are usually optimized to significantly improve transfection efficiency. Cardoso et al. (2014) developed a thermo-responsive block copolymer consisting of PNIPAM and PAMPTMA (3-acrylamidopropyl)trimethylammonium chloride) to deliver short interfering ribonucleic acids (siRNA) to silence a specific cancer-causing gene in a human fibrosarcoma cell line. The thermo-responsive behavior of PNIPAM and its effects on siRNA was demonstrated

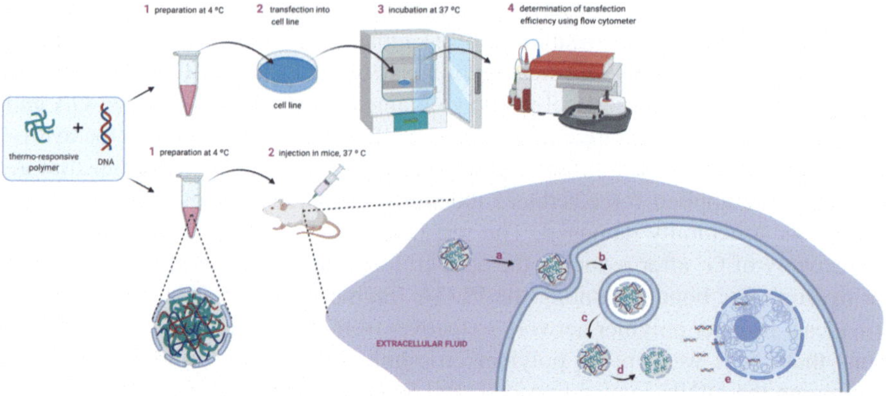

Fig. 11.6 Schematic presentation of the in vitro and in vivo delivery of genetic material into the nucleus of the cell (transfection process) through a thermo-responsive carrier. The basic steps involve: (**a**) cell uptake by endocytosis, (**b**) endosome/lysosome engulfment, (**c**) endosomal escape, (**d**) release of genetic material, (**e**) nuclear transport and gene expression

in several aspects. The copolymer system efficiently shielded the siRNA at 25 °C by about 70–87%. However, only a 30% protection level was shown at 37 °C. The decreased siRNA protection was due to the collapsed polymer network via increased hydrophobic interactions among the PNIPAM blocks, producing an aggregated structure and losing its capacity to protect the siRNAs. The decreased hydrodynamic diameter at 37 °C was a reflection of the collapsed network structure. The contraction of the polymers at 37 °C favors dissociation of the polymer, and siRNA was the reason for the exposure of the entrapped siRNA into the cytosol and ultimately silencing the target gene. More recently, Nagase et al. (2019) developed a thermo-responsive polymeric liposome for the transfection of siRNA using PNIPAM and N,N-dimethylaminopropyl acrylamide (DMAPAAM) copolymers for the suppression of luciferase and vascular endothelial growth factor (VEGF) expressions in cells. In contrast to the work of Cardoso et al. (2014), this study showed that an increase in temperature leads to a reduction of the aqueous layer of the liposome and subsequent aggregation of the liposomes due to reduction of the positive charges of the liposome upon shrinkage of the polymeric chain. At 42 °C, the hydrophobic chains of PNIPAM are exposed on the liposomal surface, increasing the liposome-cell interaction, producing a chain of events leading to high transfection efficiency. Wu et al. (2019) combined the thermo-responsive properties of PNIPAM and the photothermal effect of gold surface-modified polydopamine (PDA). This study features the roles of thermo-responsive polymers on efficient transfection and cell harvesting. This preparation takes advantage of the near-infrared laser irradiation response of Au-surface-modified PDA producing localized heat that creates fissures on the cell membrane allowing delivery of plasmids into the endothelial cells with a reported ~99% transfection efficiency. The thermo-responsive nature of PNIPAM enables the detachment of the adherent cells and harvesting viable cells from the films by simply reducing the temperature. This becomes possible by lowering the temperature (<LCST of PNIPAM), resulting in a stronger interaction between the amide groups and water molecules and forming extended coils of the polymer chains, leading to the detachment of the cells. The recovered cells may be recultured and applied in regenerative medicine. Overall, this work demonstrates a highly efficient transfection process and enables the harvesting of cells that may be further processed for tissue engineering applications.

11.4.3 Tissue Engineering

Tissue engineering may be classified as scaffold-based tissue engineering and cell sheet engineering. The examples that are given for each type include non-viral vectors that will specifically focus on thermo-responsive polymers. Some illustrative examples are depicted in Fig. 11.7.

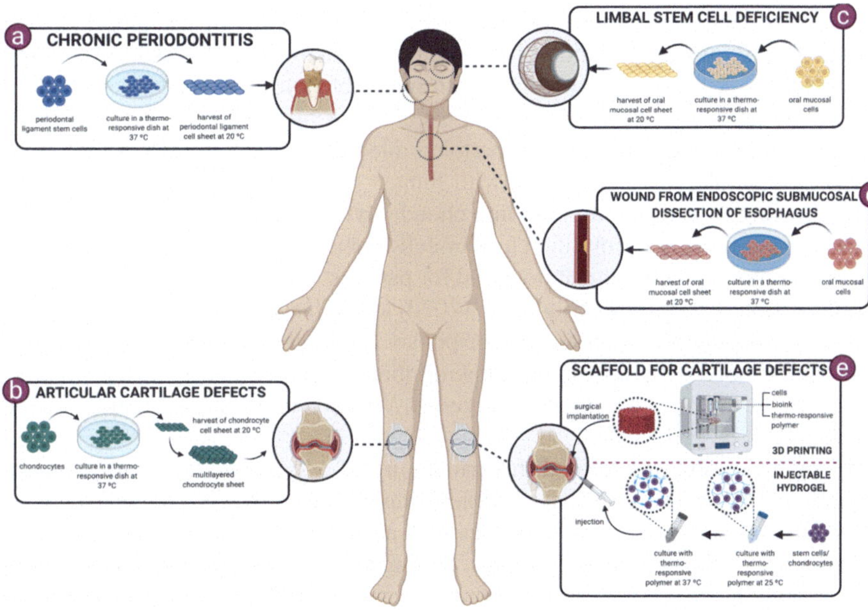

Fig. 11.7 Schematic presentation of the applications of thermo-responsive polymers in tissue engineering, (**a**–**d**) cell sheets for various disease conditions, and (**e**) formation of scaffolds through 3D printing and injection into the diseased area

11.4.3.1 Scaffold-Based Tissue Engineering

Injectable hydrogels that form gels in situ can serve as carriers of bioactive molecules and an alternative to invasive surgical implantation (Hogan and Mikos 2020). Tao et al. (2020) fabricated injectable chitosan (CS)/glycerophosphate (GP)-based thermosensitive hydrogel for the local delivery of vancomycin to treat *S. aureus* infection in osteomyelitis-induced rabbits and as a scaffold to accelerate bone repair. The injectable solution appeared translucent and slightly viscous at room temperature but turned into a semisolid gel upon contact with the target tissues. At room temperature, the hydroxyl groups of GP attach to the amine groups of CS, thus increasing the solubility, stability, and hydrophilicity of the CS polymeric chains. At temperatures above the LCST of CS, CS chains become dehydrated, resulting in increased interchain hydrophobic interactions and alteration of the distribution of hydrogen bonding, thus favoring more polymer to polymer than polymer to solvent interactions leading to the formation of rigid hydrogels. The gelation process increases the mechanical properties of the hydrogel through increased intermolecular interactions (Argüelles-Monal et al. 2017; Talaat et al. 2020). Further, the low viscosity of the CS-based hydrogel can fill the irregular bone surfaces. Results have shown osteoblast proliferation and anti-infective properties in the rabbit model, indicating a promising potential of the multifunctional nanogel system in treating osteomyelitis. A similar study utilizing an injectable hydrogel made from CS/GP but in combination with cellulose was exemplified by the work of Talaat et al. (2020), demonstrating the biocompatible, injectable, mechanically stable, and slowly degradable scaffold system with human dental pulp stem and progenitor cells, allowing their differentiation into chondrocytes in Sprague-Dawley rats. This work has successfully exemplified a stem-cell-based strategy in cartilage regeneration with a promising treatment potential for patients with temporomandibular joint disorders. Both the studies of Tao et al. (2020) and Talaat et al. (2020) have demonstrated the injectability of the thermo-responsive hydrogel systems as they display a liquid phase at room temperature, allowing them to be injected into the defective areas without the need for invasive surgeries. With 3D bioprinting, complex tissue structures can be constructed.

Zhang et al. (2020) explored the printability of a synthetic thermo-responsive polymer PNIPAM and the construction of multilayered skin grafts that may be applied to extensive surface skin injuries such as burns, which is an alternative platform for conventional wound dressings. During the printing process, the collecting plate containing the cell culture medium is heated to 37 °C, turning the loaded PNIPAM from a transparent color to cloudy white hydrogel. The fibrin was incorporated to enhance the mechanical strength of PNIPAM and retain the 3D architecture. The bioengineered skin constructs showed a multilayered structure composed of human umbilical vein endothelial cells (HUVECs) as subcutaneous capillaries, 3T3-J2 as the dermis, and cultured human keratinocyte cells (HaCaT) as epidermis to simulate the structural layout of skin tissues, finding its potential application in wound healing. Xu et al. (2020) explored the feasibility of thermosensitive hydroxypropyl chitin as an injectable hydrogel for cartilage regeneration. This study also explored the feasibility of an in vitro-constructed cartilage that will allow the growth and proliferation of chondrocytes. Lastly, the study evaluated the maturity of the subcutaneously implanted in vitro-constructed cartilage in nude mice. Below the LCST, the hydrophilic hydroxypropyl groups in the polymeric chain are responsible for keeping its solubility in an aqueous solution. The gelation process of hydroxypropyl chitin at physiological temperatures is attributed to the dehydration of the polymeric chains, thus exposing the hydrophobic N-acetyl groups on the chitin chains, expelling water, and forming the semisolid gel. The chondrocyte-hydrogel construct was established by encapsulating goat auricular chondrocytes into the hydrogel and implanted as an injectable hydrogel in nude mice. The results showed good gelation properties and biocompatibility and maintained its original shape when implanted in vivo, indicating a promising scaffold for cartilage regeneration.

Recently, Wu et al. (2020a, b) developed a 3D-printed human trachea-bronchial epithelium using thermo-responsive elastomer/collagen hybrid scaffolds and human bronchial epithelial cells. This construct is a promising tool for tracheal repair and reconstruction as it created a tissue construct that mimics the upper respiratory tract. The thermo-responsive display of this system exhibits a stiffness softening behavior due to a poly(urea-urethane) elastomeric scaffold. This behavior gives the system the dynamic elasticity required during tissue development. The stiffness softening at physiological temperatures allows the growth and differentiation of cells and vascularization of the tissues in vivo.

11.4.3.2 Cell Sheet Engineering

In principle, polymers can be used to grow adherent mammalian cells and produce monolayers in culture flasks and well plates, preserving the functional properties and integrity of the cells such as membrane surface and transport proteins, tight junctions, and the extracellular matrix that are otherwise damaged when conventional cell culture methods are performed. Essential features of the biomaterial components include biocompatibility and similarity with the human extracellular matrix to facilitate protein adsorption, thereby promoting cell adhesion. The latter process is a prerequisite for growth, proliferation, sometimes differentiation, and ultimately the survival of cells on the surface of biomaterials. Hydrophobic polymers are desirable over hydrophilic ones as the former would result in higher interfacial energy with water molecules, thus driving protein adsorption. Moreover, positively charged polymer surfaces can promote protein adsorption through electrostatic interaction. Based on these principles, thermo-responsive polymers can serve as substrates in harvesting cell monolayers with profound cell-to-cell junction integrity (Doberenz et al. 2020). Once the cells have proliferated, the temperature is reduced to 20 °C. Structural alterations in the polymers in response to a decrease in temperature convert them from a hydrophobic adsorptive to a hydrophilic repulsive state allowing protein desorption leading to cell detachment. The harvested intact cell sheets can then be utilized to adhere to diseased or injured tissues or organs (Takahashi and Okano 2019).

A promising study with in vivo demonstration is shown by the work of Venugopal et al. (2020) in engineering cell sheets of corneal epithelium from adipose mesenchymal stem cells using a synthetic thermo-responsive copolymer of PNIPAM (N) and glycidyl methacrylate (GMA). The thermo-responsive behavior is exhibited by the shift in the solvation state of the NGMA leading to changes in the inter- and intramolecular hydrogen bonding of the polymeric chains. These changes result in a detachment of an interconnected cell structure from the polymer-coated plate upon lowering the temperature to 20 °C. This work is essential to find alternative treatment strategies for limbal stem cell deficiency (LSCD), characterized by loss of limbal stem cells in the junction of cornea and sclera, which can result in blindness. The cell sheets were transplanted in corneal damage-induced rabbits. After a month, histology reports that sheets of cells aligned to the injured cornea, and it showed a clear and smooth cornea. This non-invasive transplantation is a promising approach in patients with bilateral LSCD conditions to restore vision and address the significant limitations of donor tissue availability and the high risk of immune rejection.

Several literatures have summarized the promising applications of cell sheet technology in regenerating damaged tissues. This evolved many studies constructing tissue substitutes for skin, salivary glands, esophageal ulcerations, and periodontal issues. Moreover, complex structures such as cardiac tissue, lung air leaks, pancreatic islet cells, kidneys, or hepatic tissues could be fabricated from multilayered cell sheets (Moschouris et al. 2016; Takahashi and Okano 2019; Doberenz et al. 2020). Currently, some cell sheets are undergoing clinical trials for the treatment of various conditions (Table 11.6).

11.4.4 Commercially Available Thermo-Responsive Products

Thermo-responsive polymers have typically been used in chemotherapy delivery wherein high drug concentration must be obtained at the target site to lessen systemic side effects and improve patient outcomes. ReGel™ is an example of a thermo-responsive biopolymer used to regulate the release of chemotherapy in the system. ReGel™ is a hybrid polymer that is composed of poly(d,l lactide-co-glycolide) (PLGA) and polyethylene glycol (PEG) with the basic structure of PLGA–PEG–PLGA (Zentner et al. 2001; Elstad and Fowers 2009). It is a water-soluble polymer with low viscosity at temperatures between 2 and 15 °C and becomes viscous (gel state) at normal body temperature (Elstad and Fowers 2009). ReGel™ has been used in the sustained delivery of paclitaxel, a chemotherapeutics that inhibits microtubule stabilization and angiogenesis and acts as a radiation sensitizer (Matthes et al. 2007; Elstad and Fowers 2009). OncoGel™ is the commercially available product of paclitaxel in ReGel™. This system was designed for the local delivery of paclitaxel to solid tumors without reaching the systemic circulation. In its clinical study, OncoGell™ can directly release paclitaxel in the tumor and its surrounded tissues for 6 weeks. Once injected at the tumor site, the OncoGel™ transitions into a gel, and the high tissue binding of paclitaxel continues, thus maintaining a high local drug concentration (Matthes et al. 2007).

Table 11.6 Summary of studies using cell sheet tissue engineering under clinical trial (clinicaltrials.gov)

Identifier	Disease/condition	Intervention/treatment	Status
NCT01694823	Articular cartilage defects	Cell sheet-autologous chondrocyte implantation	Phase 2
NCT02455648	Barrett esophagus	Endoscopic submucosal dissection	Not applicable
NCT03015779	Treating limbal stem cell deficiency disease	Autologous oral mucosal epithelial stem cell sheet	Phase 2
NCT02866019	Superficial esophageal cancer	CLS2702C (cell sheet) transplanted (applied) to the wound site after ESD using CLS2702D (transplantation device)	Phase 3
NCT01082822	Chronic periodontitis	Periodontal ligament stem cell implantation	Phase 2
NCT02149732	Limbal stem cell deficiency Stevens-Johnson syndrome Ocular cicatricial pemphigoid Chemical burn	Cultivated oral mucosal epithelial sheet transplantation	Not mentioned
NCT03949881	Bilateral limbal stem cell deficiency	Cultured autologous oral mucosa epithelial sheet	Phase 2
NCT02415218	Limbal stem cell deficiency	Transplantation of autologous oral mucosal epithelial sheets	Phase 2

Furthermore, OncoGel™ provides the advantage of enhancing the anti-neoplastic activity of paclitaxel through continuous exposure of the tumor. Clinical studies also suggest that the localized therapy provided by OncoGel™ can be advantageous in combination cancer therapy since it will not add up to the burden of possible systemic toxicities from different cancer treatment approaches. Clinical studies also proved that combining the OncoGel™ as chemotherapy and other cancer approaches such as radiation therapy can improve patient outcomes by targeting complementary cellular mechanisms of cancer. In rationale, paclitaxel can inhibit mitosis in the G2/M phase while subsequently acts as a radiosensitizer, making the cells more sensitive to radiation damage (Bhalla 2003). Oncogel™ can also provide physical targeting of the tumor via intralesional injection or the administration of the drug in tumor cavity after resection (0.6 to 36 mL of Oncogel™), with local and tolerated pain being managed by co-administering analgesics (Bagley et al. 2007; Vukelja et al. 2007; Duvall et al. 2009).

The safety profiling of Oncogel™ in normal tissue was done in rats, dogs, and pigs by injecting through various tissues such as the skin (via subcutaneous injection), central nervous system (via intracranial and spinal injections), and the pancreas (Zentner et al. 2001; Linghu et al. 2005; Bagley et al. 2007; Matthes et al. 2007; Gok et al. 2009). OncoGel™ has demonstrated an excellent local administration safety profile in nonclinical and clinical studies with no reported systemic toxicity (Elstad and Fowers 2009). Results of the nonclinical trial suggest that paclitaxel was localized in the tumor, with a negligible amount ($<0.2\%$) detected in the blood, tissues, and urine (Zentner et al. 2001). Other commercially available thermo-responsive products in biomedicine and pharmacotherapy are listed in Table 11.7.

Table 11.7 Commercially available thermo-responsive polymers in biomedicine and pharmacotherapy

Brand name	Thermo-responsive polymer	Target	Indication	Developer/ manufacturer	Related research	References
BST-CarGel®	Chito-san -β-glycerophosphate	Marrow access holes in the cartilage lesions	For articular cartilage repair due to degeneration, injury, and osteoarthritis	Bio Syntech Canada Inc., (2010); Pinamal (2010–2015); Smith & Nephew Inc. (2019)	*BST-CarGel*® can be effectively combined with bone marrow concentrate to produce a scaffold that is comparable to BST-*CarGel*® – blood combination (Snow et al. 2018) A retrospective cohort study suggests that patients treated with *BST-CarGel*® did not require any reoperations. *BST-CarGel*® also did not show any irritation, allergic reactions, or any signs of toxicity and is reported to reduce swelling and improve pain (Steinwachs et al. 2019)	Matanović et al. (2014), Rhee et al. (2018), Snow et al. (2018) and Steinwachs et al. (2019)
LeGOO®	Poloxamer 407	Infrainguinal arteries	For clamp less bypass thus preventing damage to vulnerable blood vessels and arteries	Bio Syntech	There is no proof of its comparative effect with the conventional "occlusion balloon" and other techniques to prevent vascular damage. *LeGOO*® has been used and found beneficial to larger arteries (e.g., Femoral arteries) (Shalhoub et al. 2013) *LeGOO*® provides safe, effective, and surgical ease than conventional vessel loops (Wimmer-Greinecker et al. 2011)	Matanović et al. (2014), Shalhoub et al. (2013) and Wimmer-Greinecker et al. (2011)

Brand name	Thermo-responsive polymer	Target	Indication	Developer/ manufacturer	Related research	References
Gantrez® AN	Poly (methyl vinyl ether-co-maleic anhydride) or PVM/MA	Nonspecific	Adjuvant for vaccines or as drug carrier	GAF Corporation	*Gantrez® AN* – PEG conjugates loaded with docetaxel (DTX) enhances its oral bioavailability to 56% (Ruiz-Gatón et al., 2019)	Arbós et al., (2002), Matanović et al. (2014), Moreno et al. (2014) and Ruiz-Gatón et al. (2019)
OncoGel™	Triblock polymer (PLGA-PEG-PLGA) – Paclitaxel	Local delivery of paclitaxel to solid tumors	As local anti-cancer therapy	BTG International Inc.	Provides targeted (Elstad and Fowers 2009) and continuous release to the tumor and surrounding tissues giving a depot effect of paclitaxel Matthes et al. (2007)	DeWitt et al. (2017), Elstad and Fowers (2009) and Matthes et al. (2007)

11.5 Conclusion

The field of biomedicine has been greatly evolving through time. Drug delivery systems have upgraded from conventional dosage forms to a more complex system of employing polymers in optimizing the therapeutic efficacy of drugs. Drug delivery technologies have focused on delivering drugs in a controlled or sustained-release manner to achieve high drug concentration at the target site. The utilization and commercialization of thermo-responsive biomedical products are becoming a breakthrough in medicine. Various natural, synthetic, and hybrid polymers that show thermo-responsive behavior are now being used in drug and gene delivery and tissue engineering due to their excellent adherent property, loading capacity, controlled-release property, and site-specific delivery capacity. Furthermore, advanced strategies for developing tunable thermo-responsive materials have been adopted. The hybridization of the polymer by controlling its hydrophobicity or copolymerizing with other smart materials (e.g., pH-responsive materials) is some of the approaches that can be used to modulate the sol-gel transition temperature and the rheological properties of the polymer. However, challenges that hinder its clinical translation still need to be addressed, like a deeper understanding of the toxicity of these materials, especially with their clearance and the possibility of accumulation in off-target organs (e.g., liver and kidney). Nonetheless, research opportunities in developing multi-responsive polymers can be explored to provide new flexible materials with expanded applicability in biomedicine design and development.

Acknowledgments This work was supported by the Chulalongkorn University's Graduate Scholarship Programme for ASEAN or Non-ASEAN countries (KPA and BPIB), the Chulalongkorn Academic Advancement into its Second Century (CUAASC) Project (JAL and Pornchai R.), the National Research University Project, Office of Higher Education Commission (NRU59-047-AM) (JAL and Pornchai R.), the National Research Council of Thailand (IRN FY2020 507/2563) (Pornchai R.), and the Thailand Science Research and Innovation (TSRI) Fund (CU_FRB640001_01_33_3) (Pornchai R. and Pranee R.).

References

Abdi SI, Choi JY, Lee JS, Lim HJ, Lee C, Kim J, Chung HY, Lim JO (2012) In vivo study of a blended hydrogel composed of pluronic F-127-alginate-hyaluronic acid for its cell injection application. Tissue Eng Regen Med 9:1–9

Abou-Shamat MA, Calvo-Castro J, Stair JL, Cook MT (2019) Modifying the properties of thermo-gelling poloxamer 407 solutions through covalent modification and the use of polymer additives. Macromol Chem Phys 220:1900173

Abu Samah NH, Heard CM (2013) Enhanced in vitro transdermal delivery of caffeine using a temperature- and pH-sensitive nanogel, poly(NIPAM-co-AAc). Int J Pharm 453:630–640

Alexander A, Ajazuddin KJ, Saraf S, Saraf S (2014) Polyethylene glycol (PEG)-Poly(N-isopropylacrylamide) (PNIPAAm) based thermosensitive injectable hydrogels for biomedical applications. Eur J Pharm Biopharm 88:575–585

Alexandridis P, Alan Hatton T (1995) Poly(ethylene oxide)poly(propylene oxide)poly(ethylene oxide) block copolymer surfactants in aqueous solutions and at interfaces: thermodynamics, structure, dynamics, and modeling. Colloid Surf A-Physicochem Eng Asp 96:1–46

Arbós P, Wirth M, Arangoa MA, Gabor F, Irache JM (2002) Gantrez® AN as a new polymer for the preparation of ligand-nanoparticle conjugates. J Control Release 83:321–330

Argüelles-Monal W, Recillas-Mota M, Fernández-Quiroz D (2017) Chitosan-based thermosensitive materials. In: Shalaby EA (ed) Biological activities and application of marine polysaccharides. InTech, Croatia, pp 279–302

Azarbayjani AF, Venugopal JR, Ramakrishna S, Lim PFC, Chan YW, Chan SY (2010) Smart polymeric nanofibers for topical delivery of levothyroxine. J Pharm Pharm Sci 13:400–410

Bagley CA, Bookland MJ, Pindrik JA, Ozmen T, Gokaslan ZL, Witham TF (2007) Local delivery of OncoGel delays paresis in rat metastatic spinal tumor model. J Neurosurg Spine 7:194–198

Bajpai AK, Shukla SK, Bhanu S, Kankane S (2008) Responsive polymers in controlled drug delivery. Prog Polym Sci 33(11):1088–1118

Balachandra A, Chan EC, Paul JP, Ng S, Chrysostomou V, Ngo S, Mayadunne R, van Wijngaarden P (2019) A biocompatible reverse thermoresponsive polymer for ocular drug delivery. Drug Deliv 26(1):343–353

Ban E, Park M, Jeong S, Kwon T, Kim EH, Jung K, Kim A (2017) Poloxamer-based thermoreversible gel for topical delivery of emodin: influence of P407 and P188 on solubility of emodin and its application in cellular activity screening. Molecules 22(2):246

Banerjee A, Ganguly S (2019) Alginate–chitosan composite hydrogel film with macrovoids in the inner layer for biomedical applications. J Appl Polym Sci 136(22):47599

Batrakova EV, Kabanov AV (2008) Pluronic block copolymers: evolution of drug delivery concept from inert nanocarriers to biological response modifiers. J Control Release 130:98–106

Bekkour K, Sun-Waterhouse D, Wadhwa SS (2014) Rheological properties and cloud point of aqueous carboxymethyl cellulose dispersions as modified by high or low methoxyl pectin. Food Res Int 66:247–256

Benslimane A, Bahlouli IM, Bekkour K, Hammiche D (2016) Thermal gelation properties of carboxymethyl cellulose and bentonite-carboxymethyl cellulose dispersions: rheological considerations. Appl Clay Sci 132–133:702–710

Bhalla KN (2003) Microtubule-targeted anticancer agents and apoptosis. Oncogene 22:9075–9086

Bischofberger I, Trappe V (2015) New aspects in the phase behaviour of poly-N-isopropyl acrylamide: systematic temperature dependent shrinking of PNiPAM assemblies well beyond the LCST. Sci Rep 5:15520

Bischofberger I, Calzolari DC, De Los RP, Jelezarov I, Trappe V (2014) Hydrophobic hydration of poly-N-isopropyl acrylamide: a matter of the mean energetic state of water. Sci Rep 4(1):1–7

Bordat A, Boissenot T, Nicolas J, Tsapis N (2019) Thermoresponsive polymer nanocarriers for biomedical applications. Adv Drug Deliv Rev 138:167–192

Bowman K, Leong KW (2006) Chitosan nanoparticles for oral drug and gene delivery. Int J Nanomedicine 1:117–128

Bruschi ML, Borghi-Pangoni FB, Junqueira MV, de Souza Ferreira SB (2017) Nanostructured therapeutic systems with bioadhesive and thermoresponsive properties. In: Ficai D, Grumezescu AM (eds) Nanostructures for novel therapy: synthesis, characterization and applications. Elsevier, pp 313–342

Butardo VM, Sreenivasulu N (2016) Tailoring grain storage reserves for a healthier rice diet and its comparative status with other cereals. In: Joen KW (ed) International review of cell and molecular biology. Elsevier, pp 31–70

Cao M, Wang Y, Hu X, Gong H, Li R, Cox H, Zhang J, Waigh TA, Xu H, Lu JR (2019) Reversible thermoresponsive peptide-PNIPAM hydrogels for controlled drug delivery. Biomacromolecules 20(9):3601–3610

Cardoso AM, Calejo MT, Morais CM, Cardoso AL, Cruz R, Zhu K, Pedroso de Lima MC, Jurado AS, Nyström B (2014) Application of thermoresponsive PNIPAAM-b-PAMPTMA diblock copolymers in siRNA delivery. Mol Pharm 11(3):819–827

Chatterjee S, Hui PC, Kan CW (2018) Thermoresponsive hydrogels and their biomedical applications: special insight into their applications in textile based transdermal therapy. Polymers (Basel) 10(5):480

Chen JF (1990) Effects of amylose and amylopectin on the functional properties of starch. Retrosp Theses Diss:1–72

Chen C, Wang L, Deng L, Hu R, Dong A (2013) Performance optimization of injectable chitosan hydrogel by combining physical and chemical triple crosslinking structure. J Biomed Mater Res – Part A 101(3):684–693

Cheng YH, Yang SH, Su WY, Chen YC, Yang KC, Cheng WTK, Wu SC, Lin FH (2010) Thermosensitive chitosan-gelatin-glycerol phosphate hydrogels as a cell carrier for nucleus pulposus regeneration: an in vitro study. Tissue Eng – Part A 16(2):695–703

Ciancia S, Cafarelli A, Zahoranova A, Menciassi A, Ricotti L (2020) Pulsatile drug delivery system triggered by acoustic radiation force. Front Bioeng Biotechnol 8:317

Deshmukh K, Basheer Ahamed M, Deshmukh RR, Khadheer Pasha SK, Bhagat PR, Chidambaram K (2017) Biopolymer composites with high dielectric performance: interface engineering. In: Sadasivuni KK, Ponnamma D, Kim J, Cabibihan JJ, AlMaadeed MA (eds) Biopolymer composites in electronics. Elsevier, pp 27–128

DeWitt JM, Murthy SK, Ardhanari R, DuVall GA, Wallner G, Litka P, Daugherty C, Fowers K (2017) EUS-guided paclitaxel injection as an adjunctive therapy to systemic chemotherapy and concurrent external beam radiation before surgery for localized or locoregional esophageal cancer: a multicenter prospective randomized trial. Gastrointest Endosc 86:140–149

Djabourov M, Leblond J, Papon P (1988) Gelation of aqueous gelatin solutions. I. Structural investigation. J Phys 49:319–332

Doberenz F, Zeng K, Willems C, Zhang K, Groth T (2020) Thermoresponsive polymers and their biomedical application in tissue engineering – a review. J Mater Chem B 8(4):607–628

Dou Q, Abdul Karim A, Loh X (2016) Modification of thermal and mechanical properties of PEG-PPG-PEG copolymer (F127) with MA-POSS. Polymers (Basel) 8:341

Duconseille A, Astruc T, Quintana N, Meersman F, Sante-Lhoutellier V (2015) Gelatin structure and composition linked to hard capsule dissolution: a review. Food Hydrocoll 43:360–376

DuVall GA, Tarabar D, Seidel RH, Elstad NL, Fowers KD (2009) Phase 2: a dose-escalation study of OncoGel (ReGel/paclitaxel), a controlled-release formulation of paclitaxel, as adjunctive local therapy to external-beam radiation in patients with inoperable esophageal cancer. Anti-Cancer Drugs 20(2):89–95

Elstad NL, Fowers KD (2009) OncoGel (ReGel/paclitaxel) – clinical applications for a novel paclitaxel delivery system. Adv Drug Deliv Rev 61:785–794

Eo MY, Fan H, Cho YJ, Kim SM, Lee SK (2016) Cellulose membrane as a biomaterial: from hydrolysis to depolymerization with electron beam. Biomater Res 20(1):1–3

Erkeçoglu S, Sezer A, Bucak S (2016) Smart delivery systems with shape memory and self-folding polymers. In: Sezer A (ed) Smart drug delivery system. InTech, Croatia, pp 1–30

Feil H, Bae YH, Feijen J, Kim SW (1993) Effect of comonomer hydrophilicity and ionization on the lower critical solution temperature of N-Isopropylacrylamide copolymers. Macromolecules 26:2496–2500

Ferjaoui Z, Jamal Al Dine E, Kulmukhamedova A, Bezdetnaya L, Soon Chang C, Schneider R, Mutelet F, Mertz D, Begin-Colin S, Quiles F, Gaffet E, Alem H (2019) Doxorubicin-loaded thermoresponsive superparamagnetic nanocarriers for controlled drug delivery and magnetic hyperthermia applications. ACS Appl Mater Interfaces 11(34):30610–30620

Flory PJ (1953) Principles of polymer chemistry. Cornell University Press, Ithaca

Fu S, Zhang H, Zhao Y (2016) Optically and thermally activated shape memory supramolecular liquid crystalline polymers. J Mater Chem C 4(22):4946–4953

Fundueanu G, Constantin M, Ascenzi P (2009) Poly(N-isopropylacrylamide-co-acrylamide) cross-linked thermoresponsive microspheres obtained from preformed polymers: influence of the physico-chemical characteristics of drugs on their release profiles. Acta Biomater 5:363–373

Futscher MH, Philipp M, Müller-Buschbaum P, Schulte A (2017) The role of backbone hydration of poly(N-isopropyl acrylamide) across the volume phase transition compared to its monomer. Sci Rep 7(1):17012

Gandhi A, Paul A, Sen SO, Sen KK (2015) Studies on thermoresponsive polymers: phase behaviour, drug delivery and biomedical applications. Asian J Pharm Sci 10(2):99–107

García-Peñas A, Biswas CS, Liang W, Wang Y, Yang P, Stadler FJ (2019) Effect of hydrophobic interactions on lower critical solution temperature for poly(N-isopropylacrylamide-co-dopamine Methacrylamide) copolymers. Polymers 11(6):991

Ge F, Zhao Y (2020) Microstructured actuation of liquid crystal polymer networks. Adv Funct Mater 30(2):1901890

Ghaee A, Bagheri-Khoulenjani S, Amir Afshar H, Bogheiri H (2019) Biomimetic nanocomposite scaffolds based on surface modified PCL-nanofibers containing curcumin embedded in chitosan/gelatin for skin regeneration. Compos Part B Eng 177:107339

Ghaeini-Hesaroeiye S, Razmi Bagtash H, Boddohi S, Vasheghani-Farahani E, Jabbari E (2020) Thermoresponsive nanogels based on different polymeric moieties for biomedical applications. Gels 6(3):20

Ghosh Dastidar D, Chakrabarti G (2019) Thermoresponsive drug delivery systems, characterization and application. In: Mohapatra S, Ranjan S, Dasgupta N, Mishra R, Thomas S (eds) Micro and nano technologies, applications of targeted nano drugs and delivery systems. Elsevier, pp 133–155

González-Hernández G, Pino-Ramos VH, Islas L, Alvarez-Lorenzo C, Concheiro A, Bucio E (2019) Radiation-grafting of N-vinylcaprolactam and 2-hydroxyethyl methacrylate onto polypropylene films to obtain a thermo-responsive drug delivery system. Radiat Phys Chem 157:6–14

Gornall JL, Terentjev EM (2007) Concentration-temperature superposition of helix folding rates in gelatin. Phys Rev Lett 99:028304

Graham S, Marina PF, Blencowe A (2019) Thermoresponsive polysaccharides and their thermoreversible physical hydrogel networks. Carbohydr Polym 207:143–159

Guo J, Feng Z, Liu X, Wang C, Huang P, Zhang J, Deng L, Wang W, Dong A (2020) An injectable thermosensitive hydrogel self-supported by nanoparticles of PEGylated amino-modified PCL for enhanced local tumor chemotherapy. Soft Matter 16(24):5750–5758

Gok B, McGirt MJ, Sciubba DM, Garces-Ambrossi G, Nelson C, Noggle J, Bydon A, Witham TF, Wolinsky JP, Gokaslan ZL (2009) Adjuvant treatment with locally delivered OncoGel delays the onset of paresis after surgical resection of experimental spinal column metastasis. Neurosurgery 65(1):193–19

Hager MD, Bode S, Weber C, Schubert US (2015) Shape memory polymers: past, present and future developments. Prog Polym Sci 49–50:3–33

Haladjova E, Toncheva-Moncheva N, Apostolova MD, Trzebicka B, Dworak A, Petrov P, Dimitrov I, Rangelov S, Tsvetanov CB (2014) Polymeric nanoparticle engineering: from temperature-responsive polymer mesoglobules to gene delivery systems. Biomacromolecules 15(12):4377–4395

Hamdy Makhlouf AS, Abu-Thabit NY, Ferretiz D (2020) Chapter 12 – shape-memory coatings, polymers, and alloys with self-healing functionality for medical and industrial applications. In: Makhlouf ASH, Abu-Thabit NY (eds) Advances in smart coatings and thin films for future industrial and biomedical engineering applications. Elsevier, pp. 335–358

Haraguchi Y, Shimizu T, Yamato M, Okano T (2012) Scaffold-free tissue engineering using cell sheet technology. RSC Adv 2(6):2184–2019

Heskins M, Guillet JE (1968) Solution properties of poly(N-isopropylacrylamide). J Macromol Sci Part A – Chem 2:1441–1455

Hirotsu S (1993) Coexistence of phases and the nature of first-order phase transition in poly-N-isopropylacrylamide gels. Adv Polym Sci 110:1–26

Hoffman AS (2002) Hydrogels for biomedical applications. Adv Drug Deliv Rev 54(1):3–12

Hoffman AS (2013) Stimuli-responsive polymers: biomedical applications and challenges for clinical translation. Adv Drug Deliv Rev 65(1):10–16

Hogan KJ, Mikos AG (2020) Biodegradable thermoresponsive polymers: applications in drug delivery and tissue engineering. Polymer 211:123063

Holder AJ, Badiei N, Hawkins K, Wright C, Williams PR, Curtis DJ (2018) Control of collagen gel mechanical properties through manipulation of gelation conditions near the sol-gel transition. Soft Matter 14:574–580

Huang CL, Bin CY, Lo YL, Lin YH (2016) Development of chitosan/β-glycerophosphate/glycerol hydrogel as a thermosensitive coupling agent. Carbohydr Polym 147:409–414

Iijima M, Takahashi M, Hatakeyama T, Hatakeyama H (2013) Detailed investigation of gel-sol transition temperature of κ-carrageenan studied by DSC, TMA and FBM. J Therm Anal Calorim 114:895–901

Indulekha S, Arunkumar P, Bahadur D, Srivastava R (2016) Thermoresponsive polymeric gel as an on-demand transdermal drug delivery system for pain management. Mater Sci Eng C 62:113–122

Iwata T, Yamato M, Washio K, Yoshida T, Tsumanuma Y, Yamada A, Onizuka S, Izumi Y, Ando T, Okano T, Ishikawa I (2018) Periodontal regeneration with autologous periodontal ligament-derived cell sheets – a safety and efficacy study in ten patients. Regen Ther 9:38–44

Jain S, Sandhu S, Malvi R, Gupta B (2013) Cellulose derivatives as thermoresponsive polymer: an overview. J Appl Pharm Sci 3:139–144

Joshi SC (2011) Sol-gel behavior of hydroxypropyl methylcellulose (HPMC) in ionic media including drug release. Materials (Basel) 4:1861–1905

Kang H, Suich DE, Davies JF, Wilson AD, Urban JJ, Kostecki R (2019) Molecular insight into the lower critical solution temperature transition of aqueous alkyl phosphonium benzene sulfonates. Commun Chem 2(1):51

Kawamura M, Miyagawa S, Fukushima S, Saito A, Miki K, Funakoshi S, Yoshida Y, Yamanaka S, Shimizu T, Okano T, Daimon T, Toda K, Sawa Y (2017) Enhanced therapeutic effects of human iPS cell derived-cardiomyocyte by combined cell-sheets with omental flap technique in porcine ischemic cardiomyopathy model. Sci Rep 7(1):8824

Kazempour M, Edjlali L, Akbarzadeh A, Davaran S, Farid SS (2019) Synthesis and characterization of dual pH-and thermo-responsive graphene-based nanocarrier for effective anticancer drug delivery. J Drug Deliv Sci Technol 54(999):101158

Kempe MD, Scruggs NR, Verduzco R, Lal J, Kornfield JA (2004) Self-assembled liquid-crystalline gels designed from the bottom up. Nat Mater 3(3):177–182

Khan F, Tanaka M, Ahmad SR (2015) Fabrication of polymeric biomaterials: a strategy for tissue engineering and medical devices. J Mater Chem B 3(42):8224–8249

Khan S, Akhtar N, Minhas MU (2018) Fabrication, rheological analysis, and in vitro characterization of in situ chemically cross-linkable thermogels as controlled and prolonged drug depot for localized and systemic delivery. Polym Adv Technol 30:pat.4514

Kim YJ, Matsunaga YT (2017) Thermo-responsive polymers and their application as smart biomaterials. J Mater Chem B 5(23):4307–4321

Kim S, Kim JH, Jeon O, Kwon IC, Park K (2009) Engineered polymers for advanced drug delivery. Eur J Pharm Biopharm 71:420–430

Kim JK, Won YW, Lim KS, Kim YH (2012) Low-molecular-weight methylcellulose-based thermo-reversible gel/pluronic micelle combination system for local and sustained docetaxel delivery. Pharm Res 29:525–534

Klouda L, Mikos AG (2008) Thermoresponsive hydrogels in biomedical applications. Eur J Pharm Biopharm 68:34–45

Kojima H (2018) Studies on the phase transition of hydrogels and aqueous solutions of thermosensitive polymers. Polym J 50:411–418

Kurdtabar M, Baghestani G, Bardajee GR (2018) Development of a novel thermo-responsive hydrogel-coated gold nanorods as a drug delivery system. Gold Bull 52(1):9–17

Kwon IC, Bae YH, Kim SW (1991) Electrically erodible polymer gel for controlled release of drugs. Nature 354(6351):291–293

Li Z, Guan J (2011) Thermosensitive hydrogels for drug delivery. Expert Opin Drug Deliv 8:991–1007

Linghu E, Matthes K, Mino-Kenudson M, Brugge WR (2005) Feasibility of endoscopic ultrasound-guided OncoGel (ReGel/Paclitaxel) injection into the pancreas in pigs. Endoscopy 37:1140–1142

Litowczenko J, Gapiński J, Markiewicz R, Woźniak A, Wychowaniec JK, Peplińska B, Jurga S, Patkowski A (2021) Synthesis, characterization and in vitro cytotoxicity studies of poly-N-isopropyl acrylamide gel nanoparticles and films. Mater Sci Eng C 118:111507

Liu D, Sun J (2020) Thermoresponsive Polypeptoids. Polymers (Basel) 12(12):2973

Liu W, Griffith M, Li F (2008) Alginate microsphere-collagen composite hydrogel for ocular drug delivery and implantation. J Mater Sci Mater Med 19(11):3365–3371

Liu S, Huang S, Li L (2016) Thermoreversible gelation and viscoelasticity of κ-carrageenan hydrogels. J Rheol (N Y N Y) 60:203–214

Liu P, Guo B, Wang S, Ding J, Zhou W (2019) A thermo-responsive and self-healing liposome-in-hydrogel system as an antitubercular drug carrier for localized bone tuberculosis therapy. Int J Pharm 558:101–109

Loh XJ, Goh SH, Li J (2007) Hydrolytic degradation and protein release studies of thermogelling polyurethane copolymers consisting of poly[(R)-3-hydroxybutyrate], poly(ethylene glycol), and poly(propylene glycol). Biomaterials 28:4113–4123

Luckanagul JA, Pitakchatwong C, Ratnatilaka Na Bhuket P et al (2018) Chitosan-based polymer hybrids for thermo-responsive nanogel delivery of curcumin. Carbohydr Polym 181:1119–1127

Makhaeva EE, Tenhu H, Khokhlov AR (1998) Conformational changes of poly(vinylcaprolactam) macromolecules and their complexes with ionic surfactants in aqueous solution. Macromolecules 31:6112–6118

Martinez AM, Cox LM, Killgore JP, Bongiardina NJ, Riley RD, Bowman CN (2021) Permanent and reversibly programmable shapes in liquid crystal elastomer microparticles capable of shape switching. Soft Matter 17:467–474

Matanović MR, Kristl J, Grabnar PA (2014) Thermoresponsive polymers: insights into decisive hydrogel characteristics, mechanisms of gelation, and promising biomedical applications. Int J Pharm 472:262–275

Matthes K, Mino-Kenudson M, Sahani DV, Holalkere N, Fowers KD, Rathi R, Brugge WR (2007) EUS-guided injection of paclitaxel (OncoGel) provides therapeutic drug concentrations in the porcine pancreas (with video). Gastrointest Endosc 65(3):448–453

Mazurek-Budzyńska M, Razzaq MY, Behl M, Lendlein A (2019) Shape-memory polymers. In: Jafar Mazumder MA, Sheardown H, Al-Ahmed A (eds) Functional polymers. Springer, Cham, pp 605–663

McKenzie M, Betts D, Suh A, Bui K, Kim LD, Cho H (2015) Hydrogel-based drug delivery systems for poorly water-soluble drugs. Molecules 20(11):20397–20408

Meng Y, Jiang J, Anthamatten M (2016) Body temperature triggered shape-memory polymers with high elastic energy storage capacity. J Polym Sci B Polym Phys 54. https://doi.org/10.1002/polb.23990

Miguel SP, Ribeiro MP, Brancal H et al (2014) Thermoresponsive chitosan-agarose hydrogel for skin regeneration. Carbohydr Polym 111:366–373

Mohan YM, Murthy PSK, Sreeramulu J, Raju KM (2005) Swelling behavior of semi-interpenetrating polymer network hydrogels composed of poly(vinyl alcohol) and poly(acrylamide-co-sodium methacrylate). J Appl Polym Sci 98:302–314

Moreno E, Schwartz J, Larrañeta E, Nguewa PA, Sanmartín C, Agüeros M, Irache JM, Espuelas S (2014) Thermosensitive hydrogels of poly(methyl vinyl ether-co-maleic anhydride) – Pluronic® F127 copolymers for controlled protein release. Int J Pharm 459:1–9

Moschouris K, Firoozi N, Kang Y (2016) The application of cell sheet engineering in the vascularization of tissue regeneration. Regen Med 11(6):559–570

Mu T, Liu L, Lan X, Liu Y, Leng J (2018) Shape memory polymer and its composite: function and application. In: Beaumont PWR, Zweben CH (eds) Comprehensive composite materials II. Elsevier, Oxford, pp 454–486
Mura S, Nicolas J, Couvreur P (2013) Stimuli-responsive nanocarriers for drug delivery. Nat Mater 12(11):991–1003
Nagase K, Okano T (2016) Thermoresponsive-polymer-based materials for temperature-modulated bioanalysis and bioseparations. J Mater Chem B 4(39):6381–6397
Nagase K, Hasegawa M, Ayano E, Maitani Y, Kanazawa H (2019) Effect of polymer phase transition behavior on temperature-responsive polymer-modified liposomes for siRNA transfection. Int J Mol Sci 20(2):430
Nam K, Kim K, Dean SM, Brown CT, Davis RS, Okano T, Baker OJ (2019) Using cell sheets to regenerate mouse submandibular glands. NPJ Regen Med 4:16
Navarro L, Theune LE, Calderón M (2020) Effect of crosslinking density on thermoresponsive nanogels: a study on the size control and the kinetics release of biomacromolecules. Eur Polym J 124:109478
Naziris N, Skandalis A, Mavromoustakos T, Pispas S, Demetzos C (2021) Association of the thermodynamics with the functionality of thermoresponsive chimeric nanosystems. In: Mavromoustakos T, Tzakos AG, Durdagi S (eds) Supramolecules in drug discovery and drug delivery: methods and protocols. Springer US, New York, pp 221–233
Ndaya D, Bosire R, Kasi RM (2019) Cholesteric–azobenzene liquid crystalline copolymers: design, structure and thermally responsive optical properties. Polym Chem 10(28):3868–3878
Niskanen J, Tenhu H (2017) How to manipulate the upper critical solution temperature (UCST)? Polym Chem 8(1):220–232
O'Brien FJ (2011) Biomaterials & scaffolds for tissue engineering. Mater Today 14(3):88–95
Ohm C, Brehmer M, Zentel R (2010) Liquid crystalline elastomers as actuators and sensors. Adv Mater 22(31):3366–3387
Ordeig O, Chin SY, Kim S, Chitnis PV, Sia SK (2016) An implantable compound-releasing capsule triggered on demand by ultrasound. Sci Rep 6(1):22803
Osorno LL, Maldonado DE, Whitener RJ, Brandley AN, Yiantsos A, Medina JDR, Byrne ME (2020) Amphiphilic PLGA-PEG-PLGA triblock copolymer nanogels varying in gelation temperature and modulus for the extended and controlled release of hyaluronic acid. J Appl Polym Sci 137(25):48678
Park CW, Lee SJ, Kim D, Lee DS, Kim SC (2008) Micelle formation and sol–gel transition behavior of comb-like amphiphilic poly((PLGA-b-PEG)MA) copolymers. J Polym Sci Part A Polym Chem 46:1954–1963
Peng Y, Li J, Li J, Fei Y, Dong J, Pan W (2013) Optimization of thermosensitive chitosan hydrogels for the sustained delivery of venlafaxine hydrochloride. Int J Pharm 441(1–2):482–490
Pişkin E (2004) Molecularly designed water soluble, intelligent, nanosize polymeric carriers. Int J Pharm 277(1–2):105–118
Podewitz M, Wang Y, Quoika PK, Loeffler JR, Schauperl M, Liedl KR (2019) Coil–globule transition thermodynamics of poly(N-isopropylacrylamide). J Phys Chem B 123(41):8838–8847
Quah SP, Smith AJ, Preston AN, Laughlin ST, Bhatia SR (2018) Large-area alginate/PEO-PPO-PEO hydrogels with thermoreversible rheology at physiological temperatures. Polymer (Guildf) 135:171–177
Quiñones JP, Peniche H, Peniche C (2018) Chitosan based self-assembled nanoparticles in drug delivery. Polymers (Basel) 10:235
Rangabhatla ASL, Tantishaiyakul V, Oungbho K, Boonrat O (2016) Fabrication of pluronic and methylcellulose for etidronate delivery and their application for osteogenesis. Int J Pharm 499:110–118
Rassing J, Attwood D (1982) Ultrasonic velocity and light-scattering studies on the polyoxyethylene-polyoxypropylene copolymer Pluronic F127 in aqueous solution. Int J Pharm 13:47–55

Rassing J, Mckenna WP, Bandyopadhyay S, Eyring EM (1984) Ultrasonic and 13C-NMR studies on gel formation in aqueous solutions of the ABA block polymer Pluronic F 127. J Mol Liq 27:165–178

Rhee C, Amar E, Glazebrook M, Coday C, Wong IH (2018) Safety profile and short-term outcomes of BST-CarGel as an adjunct to microfracture for the treatment of chondral lesions of the hip. Orthop J Sport Med 6:2325967118789871

Risteen B, Delepierre G, Srinivasarao M, Weder C, Russo P, Reichmanis E, Zoppe J (2018) Thermoresponsive liquid crystals: thermally switchable liquid crystals based on cellulose nanocrystals with patchy polymer grafts (Small 46/2018). Small 14(46). https://doi.org/10.1002/smll.201870218

Rodríguez-Rodríguez R, Espinosa-Andrews H, Velasquillo-Martínez C, García-Carvajal ZY (2020) Composite hydrogels based on gelatin, chitosan and polyvinyl alcohol to biomedical applications: a review. Int J Polym Mater Polym Biomater 69:1–20

Ruel-Gariépy E, Shive M, Bichara A, Berrada M, Le Garrec D, Chenite A, Leroux JC (2004) A thermosensitive chitosan-based hydrogel for the local delivery of paclitaxel. Eur J Pharm Biopharm 57:53–63

Ruiz-Gatón L, Espuelas S, Huarte J, Larrañeta E, Martin-Arbella N, Irache JM (2019) Nanoparticles from Gantrez® AN-poly(ethylene glycol) conjugates as carriers for oral delivery of docetaxel. Int J Pharm 571:118699

Russo E, Villa C (2019) Poloxamer hydrogels for biomedical applications. Pharmaceutics 11:671

Ruzette A-VG, Mayes AM (2001) A simple free energy model for weakly interacting polymer blends. Macromolecules 34(6):1894–1907

Rzaev ZMO, Dinçer S, Pişkin E (2007) Functional copolymers of N-isopropylacrylamide for bioengineering applications. Prog Polym Sci 32:534–595

Safranski DL (2017) Introduction to shape-memory polymers. In: Safranski DL, Griffis JC (eds) Shape-memory polymer device design. William Andrew Publishing, pp 1–22

Saravanan S, Vimalraj S, Thanikaivelan P, Banudevi S, Manivasagam G (2019) A review on injectable chitosan/beta glycerophosphate hydrogels for bone tissue regeneration. Int J Biol Macromol 121:38–54

Sarwan T, Kumar P, Choonara YE, Pillay V (2020) Hybrid thermo-responsive polymer systems and their biomedical applications. Front Mater 7. https://doi.org/10.3389/fmats.2020.00073

Schmaljohann D (2006) Thermo- and pH-responsive polymers in drug delivery. Adv Drug Deliv Rev 58(15):1655–1670

Shahriari M, Torchilin VP, Taghdisi SM, Abnous K, Ramezani M, Alibolandi M (2020) "Smart" self-assembled structures: toward intelligent dual responsive drug delivery systems. Biomater Sci 8(21):5787–5803

Shalhoub J, Hinchliffe RJ, Powell JT (2013) The world of legoo assessed: a short systematic and critical review. Eur J Vasc Endovasc Surg 45:44–45

Siirilä J, Karesoja M, Pulkkinen P, Malho J-M, Tenhu H (2019) Soft poly(N-vinylcaprolactam) nanogels surface-decorated with AuNPs. Response to temperature, light, and RF-field. Eur Polym J 115:59–69

Simões SM, Figueiras AR, Veiga F, Concheiro A, Alvarez-Lorenzo C (2015) Polymeric micelles for oral drug administration enabling locoregional and systemic treatments. Expert Opin Drug Deliv 12:297–318

Singka GSL, Samah NA, Zulfakar MH, Yurdasiper A, Heard CM (2010) Enhanced topical delivery and anti-inflammatory activity of methotrexate from an activated nanogel. Eur J Pharm Biopharm 76:275–281

Snow M, Williams R, Pagkalos J, Grover L (2018) An in vitro study to determine the feasibility of combining bone marrow concentrate with BST-CarGel as a treatment for cartilage repair. Cartilage 12(2):226–236

Song R, Murphy M, Li C, Ting K, Soo C, Zheng Z (2018) Current development of biodegradable polymeric materials for biomedical applications. Drug Des Devel Ther 12:3117–3145

Sponchioni M, Capasso Palmiero U, Moscatelli D (2019) Thermo-responsive polymers: applications of smart materials in drug delivery and tissue engineering. Mater Sci Eng C 102:589–605

Steinwachs M, Cavalcanti N, Reddy SM, Werner C, Tschopp D, Choudur HN (2019) Arthroscopic and open treatment of cartilage lesions with BST-CARGEL scaffold and microfracture: a cohort study of consecutive patients. Knee 26(1):174–184
Strzelec K, Sienkiewicz N, Szmechtyk T (2020) Classification of shape-memory polymers, polymer blends, and composites. In: Parameswaranpillai J, Siengchin S, George JJ, Jose S (eds) Shape memory polymers, blends and composites: advances and applications. Springer Singapore, Singapore, pp 21–52
Sun W, An Z, Wu P (2017) UCST or LCST? Composition-dependent thermoresponsive behavior of Poly(N-acryloylglycinamide-co-diacetone acrylamide). Macromolecules 50(5):2175–2182
Supardi, Yusuf Y, Harsoyo S (2015) Characterization of main-chain liquid crystal elastomers by using differential scanning calorimetry (DSC) method. Adv Mater Res 1123:69–72
Tahara Y, Sakiyama M, Takeda S, Nishimura T, Mukai S-A, Sawada S-I, Sasaki Y, Akiyoshi K (2016) Self-assembled nanogels of cholesterol-bearing hydroxypropyl cellulose: a thermoresponsive building block for nanogel tectonic materials. Langmuir 32(47):12283–12289
Takahashi H, Okano T (2019) Thermally-triggered fabrication of cell sheets for tissue engineering and regenerative medicine. Adv Drug Deliv Rev 138:276–292
Talaat W, Aryal Ac S, Al Kawas S, Samsudin ABR, Kandile NG, Harding DRK, Ghoneim MM, Zeiada W, Jagal J, Aboelnaga A, Haider M (2020) Nanoscale thermosensitive hydrogel scaffolds promote the chondrogenic differentiation of dental pulp stem and progenitor cells: a minimally invasive approach for cartilage regeneration. Int J Nanomedicine 15:7775–7789
Tao J, Zhang Y, Shen A, Yang Y, Diao L, Wang L, Cai D, Hu Y (2020) Injectable chitosan-based thermosensitive hydrogel/nanoparticle-loaded system for local delivery of vancomycin in the treatment of osteomyelitis. Int J Nanomedicine 15:5855–5871
Teotia AK, Sami H, Kumar A (2015) Thermo-responsive polymers: structure and design of smart materials. Switch responsive surfaces. Mater Biomed Appl:3–43
Thornton PD, Mart RJ, Ulijn RV (2007) Enzyme-responsive polymer hydrogel particles for controlled release. Adv Mater 19(9):1252–1256
Tran DL, Pham GD, Nguyen XP, Nhuyen NT, Tran V, Mai TTT, Ba TC (2011) Some biomedical applications of chitosan-based hybrid nanomaterials. Adv Nat Sci Nanosci Nanotechnol 2(4):045004
Trongsatitkul T, Budhlall BM (2013) Microgels or microcapsules? Role of morphology on the release kinetics of thermoresponsive PNIPAm-co-PEGMa hydrogels. Polym Chem 4:1502–1516
Vadnere M, Amidon G, Lindenbaum S, Haslam JL (1984) Thermodynamic studies on the gel-sol transition of some pluronic polyols. Int J Pharm 22:207–218
Venugopal B, Shenoy SJ, Mohan S, Anil Kumar PR, Kumary TV (2020) Bioengineered corneal epithelial cell sheet from mesenchymal stem cells-a functional alternative to limbal stem cells for ocular surface reconstruction. J Biomed Mater Res B Appl Biomater 108(3):1033–1045
Vukelja SJ, Anthony SP, Arseneau JC, Berman BS, Cunningham CC, Nemunaitis JJ, Fowers KD (2007) Phase 1 study of escalating-dose OncoGel® (ReGel®/paclitaxel) depot injection, a controlled-release formulation of paclitaxel, for local management of superficial solid tumor lesions. Anti-Cancer Drugs 18:283–289
Wang B, Xu XD, Wang ZC, Cheng SX, Zhang XZ, Zhuo RX (2008) Synthesis and properties of pH and temperature sensitive P(NIPAAm-co-DMAEMA) hydrogels. Colloids Surf B: Biointerfaces 64(1):34–41
Wang X, Ao Q, Tian X et al (2017) Polymers gelatin-based hydrogels for organ 3D bioprinting. Polymers (Basel) 9(9):401
Ward MA, Georgiou TK (2011) Thermoresponsive polymers for biomedical applications. Polymers 3(3):1215–1242
Wei W, Xiong H (2018) Orientation and morphology control of the liquid crystalline block copolymer thin film by liquid crystalline solvent. Langmuir 34(50):15455–15461
Wichterle O, Lim D (1960) Hydrophilic gels for biological use. Nature 185(4706):117–118

Wimmer-Greinecker G, Bouchot O, Verhoye JP, Perrault LP, Börgermann J, Diegeler A, Van Garsse L, Rastan AJ (2011) Randomized clinical trial comparing a thermosensitive polymer (LeGoo) with conventional vessel loops for temporary coronary artery occlusion during off-pump coronary artery bypass surgery. Ann Thorac Surg 92:2177–2183

Wu J, Zheng Y, Jiang S, Qu Y, Wei T, Zhan W, Wang L, Yu Q, Chen H (2019) Two-in-one platform for high-efficiency intracellular delivery and cell harvest: when a photothermal agent meets a thermoresponsive polymer. ACS Appl Mater Interfaces 11(13):12357–12366

Wu L, Magaz A, Huo S, Darbyshire A, Loizidou M, Emberton M, Birchall M, Song W (2020a) Human airway-like multilayered tissue on 3D-TIPS printed thermoresponsive elastomer/collagen hybrid scaffolds. Acta Biomater 113:177–195

Wu J, Zhai W, Gao X, Zhang R, Yu Y (2020b) Preparation and self-assembly of thermosensitive triblock copolymers with N-isopropylacrylamide and 3-methacryloxypropyltrimethoxysilane as monomers. Polym Bull:1–16. https://doi.org/10.1007/s00289-020-03131-5

Xiao X, Kong D, Qiu X, Zhang W, Liu Y, Zhang S, Zhang F-h HY, Jinsong L (2015) Shape memory polymers with high and low temperature resistant properties. Sci Rep 5:14137

Xing Q, Yates K, Vogt C, Qian Z, Frost MC, Zhao F (2014) Increasing mechanical strength of gelatin hydrogels by divalent metal ion removal. Sci Rep 4:4706

Xu Y, Xu Y, Bi B, Hou M, Yao L, Du Q, He A, Liu Y, Miao C, Liang X, Jiang X, Zhou G, Cao Y (2020) A moldable thermosensitive hydroxypropyl chitin hydrogel for 3D cartilage regeneration in vitro and in vivo. Acta Biomater 108:87–96

Yang F, Wang Y, Zhang Z, Hsu B, Jabs EW, Elisseeff JH (2008) The study of abnormal bone development in the Apert syndrome Fgfr2+/S252W mouse using a 3D hydrogel culture model. Bone 43(1):55–63

Yang HW, Lee AW, Huang CH, Chen JK (2014) Characterization of poly(N-isopropylacrylamide)-nucleobase supramolecular complexes featuring bio-multiple hydrogen bonds. Soft Matter 10:8330–8340

Yang W, Zhu L, Chen Y (2015) One-step fabrication of 3-methacryloxypropyltrimethoxysilane modified silica and investigation of fluorinated polyacrylate/silica nanocomposite films. RSC Adv 5:58973–58979

Yang L, Fan X, Zhang J, Ju J (2020) Preparation and characterization of thermoresponsive Poly(N-Isopropylacrylamide) for cell culture applications. Polymers (Basel) 12(2):389

Yin L, Han L, Ge F, Tong X, Zhang W, Soldera A, Zhao Y (2020) A novel side-chain liquid crystal elastomer exhibiting anomalous reversible shape change. Angew Chem Int Ed 59(35):15129–15134

Yusuf Y (2017) Influence of cross-linker concentration on physical properties of main-chain liquid crystalline elastomers. Mater Res 20:1541–1547

Zenati A, Thammalangsy S (2018) Characteristics and self-assembly behaviors of photochromic triblock azo-copolymers based on hard-soft-hard segments synthesized via RAFT process. J Polym Sci Part A: Polym Chem 56(15):1617–1629

Zentner GM, Rathi R, Shih C, McRea JC, Seo MH, Oh H, Rhee BG, Mestecky J, Moldoveanu Z, Morgan M, Weitman S (2001) Biodegradable block copolymers for delivery of proteins and water-insoluble drugs. J Control Release 72:203–215

Zhang Y, Yu W, Lv G, Zhu J, Wang W, Ma X, Liu X (2011) The artificial organ: cell encapsulation. In: Comprehensive biotechnology, 2nd edn. Elsevier Inc, pp 99–114

Zhang Q, Weber C, Schubert US, Hoogenboom R (2017) Thermoresponsive polymers with lower critical solution temperature: from fundamental aspects and measuring techniques to recommended turbidimetry conditions. Mater Horiz 4(2):109–116

Zhang J, Yun S, Karami A, Jing B, Zannettino A, Du Y, Zhang H (2020) 3D printing of a thermosensitive hydrogel for skin tissue engineering: a proof of concept study. Bioprinting 19:e00089

Zhou B, Liu YJ, Lan X, Leng JS, Yoon SH (2012) A glass transition model for shape memory polymer and its composite. Int J Mod Phys B 23(06n07):1248–1253

Zhou HY, Jiang LJ, Cao PP, Li JB, Chen XG (2015) Glycerophosphate-based chitosan thermosensitive hydrogels and their biomedical applications. Carbohydr Polym 117:524–536

Chapter 12
Polymeric Micelles for Drug Delivery

Anuja Paprikar, Ankit Soni, Neeraj Kaushal, and Senshang Lin

Abbreviation

CMC	critical micelle concentration
MePEG	methoxy-polyethylene glycol
PEG	polyethylene glycol
PICXM	polyion complex micelles
PM	polymeric micelles
QbD	quality by design

12.1 Introduction

The advanced and targeted drug delivery market has been forecasted to grow at a 10.4% compound annual growth rate to \$319 billion by 2021 from \$168 billion in 2016 (Valcourt et al. 2018). In recent years, polymeric micelles (PM) have been on the radar as novel colloidal delivery systems that can comply with the requirements of an ideal and versatile drug carrier. The concept of using PM as drug carriers was first proposed in 1984 (Bader et al. 1984). Since then, they have found applications in delivering variety of drugs ranging from small molecules to large molecules including DNA, viruses, as well as lipoproteins. Especially, PM have been extensively investigated to facilitate the delivery of hydrophobic drugs. They have proven to be better as compared to other carrier systems in terms of their targeting applications in disease conditions such as cancer as discussed later in this chapter.

PM are constructed from synthetic block copolymers or graft copolymers to form macromolecular. The polymers have been widely characterized since they govern not only loading of the drug into PM but also the release of the drug at the site of action. In contrast with surfactant micelles systems, employing PM as a

A. Paprikar · A. Soni · N. Kaushal · S. Lin (✉)
College of Pharmacy and Health Sciences, St. John's University, Queens, NY, USA
e-mail: linse@stjohns.edu

J.-C. Kim et al. (eds.), *Smart Nanomaterials in Biomedical Applications*,
Nanotechnology in the Life Sciences,
https://doi.org/10.1007/978-3-030-84262-8_12

delivery system has several advantages attributed to their enhanced stability, low critical micelle concentration (CMC), lower degree of dissociation, controlled drug release, and ability to be retained for longer time systemically. This chapter aims at providing an overview of the formation of PM, polymers involved, in vitro–in vivo characterization, formulation, and clinical applications of PM.

12.2 Polymeric Micelles

12.2.1 Formation Mechanism of Polymeric Micelles

Amphiphilic agents, such as surface-active agents, tend to form association or amphiphilic colloids. At low concentrations in a liquid medium, these amphiphilic agents exist separately as sub-colloidal particles. The increase in concentration leads to aggregation over a narrow concentration range. These aggregates containing 50 or more monomers are called micelles (Martin 1993). Like amphiphilic agents, amphiphilic polymers consisting of hydrophobic and hydrophilic blocks orient themselves during PM formation to minimize the free energy of the system. Upon addition of water, the hydrophobic blocks break down to form aggregates leading to the formation of a hydrophobic cavity and a hydrophilic shell and increasing the entropy of the system (Nagarajan 1996).

It has been reported that micelle formation involves two types of forces. The attractive force is responsible for the association of molecules to form micelles, whereas the repulsive force inhibits unlimited growth of the micelles (Astafieva et al. 1993). However, the reduction in free energy of system produced by the breakdown of hydrophobic blocks from the aqueous phase remains the main driving force for the formation of PM. On the other hand, the hydrophilic blocks sustain interaction with water molecules at the surface of the particle which in turn aids in stabilizing PM. In case of amphiphilic copolymers, at very low concentrations (i.e., below CMC), single chains of polymers exist. However, when the concentration of amphiphilic copolymer in the solvent reaches a critical value (i.e., CMC), polymers start to associate to form micelles, therefore reducing the free energy of the system (Jones and Leroux 1999).

12.2.2 Types of Polymeric Micelles

12.2.2.1 Block Copolymeric Micelles

Block copolymer micelles are one of the most promising particulate carriers utilized in drug delivery. The inner core consists of hydrophobic block, while hydrophilic block forms the outer shell. Block copolymers are produced by one homopolymer (repeated throughout the structure), which are engineered so that one or more desired properties can be rendered to the copolymer. For example, block copolymers of poly(ethylene oxide) as hydrophilic segments and drug-conjugated poly(aspartic acid) as hydrophobic blocks have been reported to chemically entrap drug by drug-polymer interaction. On the other hand, Pluronic block copolymer micelles conjugated with insulin or antibody effectively delivered haloperidol, which was physically entrapped inside PM (Cammas-Marion et al. 1999).

12.2.2.2 Polyion Complex Micelles

Polyion complex micelles (PICM) are formed by mixing two oppositely charged block copolymers in an aqueous solution and were first reported by Harada and Kataoka in 1995 (Harada and Kataoka 1995). These formed micelles are driven by electrostatic interactions along with a narrow size distribution. PICM are known as promising carriers for charged molecules like siRNA, DNA, and proteins. On the contrary, in case of proteins with charges, neutral, hydrophobic surfaces, RNA, and DNA are highly charged which affect the binding process to the polymer. The core of these PICM hosts these charged compounds which allows modulation of their inherent stability, solubility, and reactivity properties (Chen and Stenzel 2018). The methods, such as supercharging of proteins or the addition of more charged polymers, have been applied to stabilize PICM. Although there are many challenges involved in drug delivery by PICM, some applications have found to be promising. The PEGylated albumin PICM has shown tremendous potential for the delivery of anticancer agents. Also, PICM can be a pH-responsive carrier for colon-specific drug delivery as well as intracellular drug delivery (Chen et al. 2015; Luo et al. 2009).

12.2.2.3 Polymers Used to Formulate Polymeric Micelles

The polymers used to formulate PM mainly involve the utilization of amphiphilic diblock copolymers, in addition to graft and triblock copolymers. The properties of the three copolymer types, which form the core of PM, are beneficial for prolonging the drug circulation time, controlling the drug-release profile, and/or having the ability to add targeting ligands.

The amphiphilic diblock copolymers consist of a hydrophobic core, which may assume certain hydrophilicity with the use of one or more polymers like polyethylene glycol (PEG). This provides certain hydrophilicity (i.e., stealth properties to PM) assisting their escape from reticuloendothelial system and subsequently increasing their half-life in systemic circulation. For instance, the hydrophobic polystyrene core and a mixed shell comprised of poly(methacrylic acid) (PMA) and poly(ethylene oxide) (PEO) blocks formed hybrid water soluble micelles. Such modified systems were able to enhance the stability and the drug release behavior of these micellar systems (Štěpánek et al. 2001). In another study, methotrexate release and micelle stability were reported to be influenced by the degree of hydrophobicity of the block copolymer which was modulated by drug conjugation (Yu Li and Kwon

2000). Although polyethers, such as PEG and PEO, have been extensively reported to be used as the outer hydrophilic shell, other shell polymers like poly(trimethylene carbonate), poly(acryloylmorpholine), and poly(vinylpyrrolidone) have been used as well (V P Torchilin et al. 2001; Z. Zhang et al. 2006).

In the case of Pluronics, by changing the lengths of chain segments, a remarkable alterability of CMC and partition coefficient values can be attained. CMC governs the thermodynamic stability of PM against the dilution in body fluids, while partition coefficient gives an idea of the amount of drug which can be incorporated into the micelle (Kozlov et al. 2000). Furthermore, Pluronic copolymers have an exceptional characteristic to hinder P-glycoprotein-mediated efflux which is correlated with multiple drug resistance. Since P-glycoprotein mainly associated with sustained ATP levels, the utilization of Pluronic escalated membrane fluidity and changed intracellular ATP levels in endothelial and multiple drug resistance cells (Batrakova et al. 2001).

The graft-type copolymer can be derived by random derivatization of poly(l-amino acids) (PAAs) amino or carboxyl groups. PAAs are non-toxic, economical, and biocompatible. Poly(ε-caprolactone), poly(glycolide), and poly(D,L-lactide) polyesters are frequently used to synthesize nanoparticles and PM. Polyesters like poly(lactic-co-glycolic acid) have been widely stated for the targeted delivery of not only small hydrophilic molecules but also plasmids and oligonucleotides (Brannon-Peppas et al. 2007).

Alternatively, PM may be classified based on their applications such as oral pH-sensitive, mucoadhesive, and P-glycoprotein inhibition. As discussed in this section, it can be noted that careful evaluation of the properties of the polymer is important during the formulation development stage.

12.3 Development and Characterization of Polymeric Micelles

A lot of literature has discussed the specifics of the formulation development of PM in the past. However, very few have compiled the application of quality by design (QbD)-design of experiments (DoE) approach for the development and characterization of polymeric micelle formulations. This section aims to not only highlight the application of QbD-DoE principles for the development of PM but also discuss their characterization.

12.3.1 Formulation Optimization of Polymeric Micelles Using Design of Experiments

QbD is a systematic industrial approach, based on quality risk management, for developing pharmaceutical products. QbD originates with predefined objectives and focus on product and process understanding as well as process control. In addition, the introduction of QbD elements to define and manage quality is promoted by various regulatory authorities and marketing authorization submissions. The core elements of the QbD approach are the identification of quality target product profile, the selection of critical quality attributes critical process parameters, risk evaluation, and experimental design. Also, the creation of a design space with a proper control plan, and ultimately the management of lifecycle of pharmaceutical product are core QbD elements. Although the critical quality attributes are linked to the product's efficiency, the protection and efficacy profiles, the critical material attributes, and the critical process parameters are linked to the production method chosen. By rating critical quality attributes and critical process parameters in terms of their degree of effect on the targeted product quality, a true risk-and-knowledge-based quality management system can be developed (Csóka et al. 2018).

The application of QbD approach has been extensively discussed in the literature for micro-particulate systems, such as levodopa-loaded microspheres and lamotrigine-loaded nanoparticles as dry powder nasal formulations (Bartos et al. 2018; Gieszinger et al. 2017). Recently, a more systematic application of QbD principles for nose-to-brain delivery of meloxicam-loaded PM has been published (Sipos et al. 2020). The development of knowledge space was feasible for this system with the application of Ishikawa diagram. This cause-and-effect diagram covered not only the material and product characteristics but also the production method parameters and therapeutic expectations. Furthermore, based on this and the severity scores, risk assessment was performed to indicate that particle size, dissolution time, encapsulation efficiency of the formulation, as well as the excipients involved, drug solubility, and permeability could have the highest impacts among critical quality attributes on the quality and efficacy of meloxicam-loaded PM. On the other hand, it was observed that the highest influence on critical process parameters was determined by the composition of the formulation. Using the Box-Behnken experimental design, the effect of three formulation excipients (independent variables) was studied on particle size and polydispersity index (dependent variables).

Another study highlighted the importance of a QbD approach to evaluate the material attributes and processing parameters, ultimately, affecting the formulation critical quality attributes of PM. In this study, a full factorial design ($3 \times 3 \times 4$) investigation was performed (Gupta et al. 2020). Furthermore, a Doehlert matrix design was applied to evaluate the effect of four formulation variables on the response variables of paclitaxel-loaded PM (Wei et al. 2009). Also, a new micellar delivery system was designed to investigate in vitro reversal of resistant ovarian tumor cells, based on Pluronic P105 (nonionic triblock copolymer) and paclitaxel. The central composite design-response surface methodology was employed for optimization of these paclitaxel-loaded PM (Y. Z. Wang et al. 2008). As discussed in the above examples, it can be noted that the use of QbD-DoE principles has become an integral part of formulation development of polymeric micelle systems.

12.3.2 Characterization of Polymeric Micelles

The following section aims at giving general background of the characterization of PM including in vitro as well as in vivo studies.

12.3.2.1 In Vitro Characterization of the Polymeric Micelles

12.3.2.1.1 Particle Size, Size Distribution, and Shape

The PM particle size in a range of 10–100 nm diameters is applicable in drug targeting through the bloodstream. This is because carriers in this size range cannot only efficiently avoid clearance at the kidney but also capture the carriers at reticuloendothelial systems. This enables accumulation of the carrier particles into target organ tissues via selective extravasation from the bloodstream (Ishida et al. 1999). As compared to nanoparticles and liposomes, PM exhibit smaller particle size (lesser than 50 nm). In addition, the size can be better controlled for PM unlike the latter, since it primarily depends on the chemical structure of polymers and is not related directly to the preparation process (Yokoyama 2014).

The commonly used techniques for particle size measurements are dynamic light scattering and laser diffraction. In principle, the particle movement causes fluctuation of the intensity of the scattered light, which is measured by the dynamic light scattering. In addition to the high speed of analysis and ease of sample preparation, this method can detect particles in a size range from few nanometers to up to 3 microns. In addition to size, the polydispersity index can also be determined. Practically, blue color is indicative of a mono-dispersion, which is considered stable owing to less aggregation (white color). In contrast to dynamic light scattering, laser diffraction aids in providing information about the particle shape. PM have been reported to have various shapes, and these include, but not limited to, star shapes, flower-like, spherical, worm-shaped, vesicles, toroids, and helices. However, spherical PM have been observed frequently (Amin et al. 2017). The size and shape can be determined by scanning electron microscopy, transmission electron microscopy techniques, when visualizing with respect to blank PM, scanning electron microscopy and transmission electron microscopy images have revealed that PM get swollen due to incorporation of the drug (Ahmad et al. 2014). Atomic force microscopy is used for the direct visualization of block copolymer micelles in the dried or liquid state. This method can differentiate single, aggregate, and fused particles, which is useful in multimodal size distribution unlike dynamic light scattering.

CMC is a key factor for the stability of PM. The different methods employed to determine CMC in the aqueous dispersion of PM CMC include surface tension, conductivity, osmotic pressure, X-ray scattering, and differential scanning calorimetry. In addition, techniques, such as dynamic light scattering, gel permeation chromatography, and fluorescence, are used. Although dynamic light scattering is a powerful technique, it can predict the onset of micellization only CMC occurring in the concentration range that is sensitive to this technique. For example, the sensitivity of this technique is not within range to determine CMC for block copolymers in water. In case of gel permeation chromatography, the polymer may get adsorbed on the column limiting the determination of CMC of PM. Pyrene fluorescence is a promising option for the determination of CMC of PM. Fluorescence spectra of pyrene are highly sensitive to minor changes in solution and polarity of the probe microenvironment (G. Kwon et al. 1993). It has to be noted that although the apparent pyrene concentration was unchanged upon increasing polymer concentration, its fluorescence intensity increased after reaching the CMC of polymer. Following the micellization, pyrene could accumulate at the micellar core due to partitioning phenomenon. Therefore, the CMC of polymer can be easily measured from the graph of fluorescence intensity and concentration of polymer (Kabanov et al. 1995).

12.3.2.1.2 Association

PM are found to be effective in drug targeting because of their ability to evade the clearance from kidney owing to their relatively larger size. On the other hand, eventually, they disassemble into small single polymer chains being removed by kidney out from the body. Thus, the ability to exist in dual statuses of relatively larger and smaller sizes has increased the targeting efficiency as well as reduced the related toxicity.

12.3.2.1.3 Drug Loading and Release

PM are known for their large volume of the hydrophobic inner core encapsulating with hydrophobic drug. The hydrophobic interactions between drug and polymeric core impact the drug loading capacity. It has been reported that the strong hydrophobic interactions have led to high drug loading. Also, factors like hydrogen bonding and steric hindrance have an effect on drug loading. Hydrogen bonding can occur easily since the presence of the functional groups, such as carboxyl, hydroxyl, and amide, in both the drug molecules and polymers are commonly found. The popular drug molecule doxorubicin is believed to have high incorporation owing to the π-π interaction-rich anthracycline ring in its chemical structure (Johnson et al. 2013). On the other hand, nucleic acids are favorably incorporated because of the presence of a negative charge at every monomer unit unlike proteins. Consequently, nucleic acids can be incorporated easily than proteins.

The drug loading for PM may not be efficient upon simply equilibrating the drug and PM in water. Hence, the loading of insoluble drugs can be improved by placing them into PM by either chemical conjugation or physical entrapment (i.e., via dialysis or emulsification). In the chemical conjugation method, the incorporation of drug in the core of PM involves the formation of a covalent bond between the specific chemical group of drug and hydrophobic core of PM. The chemical conjugation has been reported to have impact on drug release rate, wherein with an increase in the drug's conjugation resulted in decreased drug release (Yu Li and Kwon 2000). The physical method is more feasible than the chemical method for drug incorporation, since no conjugation chemistry is involved. Physical incorporation of drugs is usually performed by dialysis and/or oil-in-water emulsion process. In dialysis, a selective solvent for the hydrophilic part of the polymer is selected which brings the drug and polymer together. On the other hand, using this selective solvent, the hydrophobic portion of the polymer associates with the drug to integrate it via micellar core formation. In oil-in-water emulsion, an aqueous copolymer solution is prepared and the drug solution using a water-insoluble volatile solvent is added to physically entrap the drug inside the oil droplets. As the solvent evaporates, the micelle drug conjugate is formed. The dialysis method is advantageous over the oil-in-water method, since potentially toxic solvents can be avoided. In contrast, it has been reported that during the incorporation of doxorubicin in PEO–PBLA micelles, the emulsification method was more efficient than dialysis (G. S. Kwon et al. 1995).

The amount of drug added directly affects the incorporation efficiency as well as the loading capacity. The excess amount of drug gets precipitated after maximum capacity has been reached (Yokoyama et al. 1998b). The aggregation number of PM may influence the drug loading efficiency. In other words, PM with a high aggregation number help to solubilize the drug in the inner micelle core (Hagan et al. 1996). Interestingly, the polymeric core can be engineered to develop the drug interaction with the micelle core which helps to control the drug release from PM. The two major pathways for drugs to get released from micellar core are dissociation and diffusion. Upon the dissociation of the micelles into monomers, the drug molecules separate from the monomers, eventually breaking the drug-polymer bond within the micelles. Thereafter, the drug molecules escape via diffusional from the delivery system.

12.3.2.1.4 Cytotoxicity Studies

Although drug carriers, such as PM, are considered low in toxicity, in the case of anticancer drugs, toxicity of such carriers may be hidden due to the strong effects of drugs. However, in the case of milder anticancer drugs, it is important to study the toxicity of carriers. The in vitro cytotoxicity of paclitaxel-loaded P123/F127 mixed micelles was investigated and compared with that of the free drug and Taxol injection using human SPC-A1 and A-549 cells (S. C. Lee et al. 2007). In addition to the verification of toxicity, the effect of method of preparation can also be evaluated using cytotoxicity studies. For instance, the cytotoxic effect of the drug, upon physical or chemical entrapment, was evaluated using P388D1 mouse leukemia cells. It was observed that the physically entrapped adriamycin expressed cytotoxic activity, but not the case of the chemically conjugated adriamycin (Yokoyama et al. 1998a).

12.3.2.2 In Vivo Studies

There is a plethora of literature on in vivo studies for PM that focus on delivering a variety of drugs ranging from siRNAs to small molecules. In a study, stable PM were formulated to consist of the siRNA incorporated into the cRGD peptide, which increased gene silencing ability, enhanced cell uptake, and led to broader in vitro distribution. Furthermore, following intravenous injection into mice, PM improved accumulation in both tumor mass and tumor-associated blood vessels. Additionally, stable and targeted PM depressed the growth of subcutaneous HeLa tumor models indicating that the proposed siRNA-based therapy is suitable for cancer treatments (Christie et al. 2012). On the other hand, a study reported that formulated itraconazole PM showed similar outcome between PM and commercial product using cyclodextrin as a solubilizing agent. Also, they were well tolerated in rats and dogs at dose levels corresponding to clinical doses (Yi et al. 2007). In addition, some studies focus on the in vivo fate of the polymeric system. For instance, the block copolymer was investigated following the intravenous administration of three different doses of the copolymer to female Balb/C mice. After analysis of the plasma samples, it was revealed that even after 24 h of administration, the copolymer stayed assembled as intact PM significantly. In other words, this investigation reveals that the hydrophobic and semi-crystalline nature of the PCL core conveys a high degree of kinetic stability to this micelle system (J. Liu et al. 2007).

As per the reported studies, even though PM may improve the solubilization of poorly water-soluble drugs, absorption is not observed to improve. This can be attributed to the low availability of readily absorbable form of drug in the gastrointestinal tract. Some studies have taken a step forward to understand the behavior of PM in fasted conditions. The pH-responsive fenofibrate micelles were administered by oral gavage to male Sprague-Dawley rats which showed enhanced bioavailability as compared to the commercial formulation (Sant et al. 2005).

12.4 Formulation Applications of Polymeric Micelles

PM are widely considered convenient nanocarriers for various types of applications and have several beneficial properties, such as biocompatibility, durability properties, good in vitro and in vivo stabilities, the ability to effectively solubilize poorly water-soluble drugs, the enhancement in the release profile of drugs, and the capacity to accumulate the drug in target region. Therefore, the formulation applications of PM are to enhance these fundamental drug performance parameters that include applications for solubilization of poorly water-soluble drugs, for the routes of administration, for the disease state, and for drug targeting to specific cell and/or organ.

12.4.1 Application for Solubilization of Poorly Water-Soluble Drugs

It is currently estimated that about 90% of new chemical entities produced by drug discovery have low aqueous solubility and are classified as Class II or Class IV in the Biopharmaceutical Classification System (Movassaghian et al. 2015; C. Y. Wu and Benet 2005). Formulation of these chemical entities poses a significant challenge for the pharmaceutical industry. In addition, although these entities have potential pharmacodynamic properties, they do not enter the market owing to low bioavailability (due to poor aqueous solubility) (Fournier et al. 2004). Micellar solubilization is a dominant alternative for dissolving poorly water-soluble drugs in aqueous environments.

Conventional solubilizing agents utilized to resolve solubility issues are either ineffective or harmful and contribute to barriers associated with drug use, such as embolisms due to drug aggregation (Gelderblom et al. 2001). Recent research using PM has focused on the effects of PM on solubilization enhancement, dissolution rate, and enhancement of aqueous solubility. By application of PM, the aqueous solubility of many drugs can be increased as 10- to 8400-fold. PM provide a hydrophobic reservoir where drugs can partition by chemical conjugation, physical entrapment, or ionic interactions, depending on the nature of the drug and physicochemical properties of the polymer (Chiappetta et al. 2011; Savić et al. 2006). Paclitaxel micellar formulation is one of the most popular examples. The use of diblock copolymers containing one block of methoxy-polyethylene glycol (MePEG) and one block of poly(D,L-lactide), which dissolved in aqueous media to form PM with a hydrophobic polyester core and a water-soluble MePEG shell. By incorporation of paclitaxel in PM resulted in a 5000-fold increase of the solubilized paclitaxel levels in an aqueous vehicle (Burt et al. 1999). Also, the results of a range of in vitro and in vivo biocompatibility/toxicology tests in rats showed the poly(D, L-lactide)-MePEG micelles were biocompatible and non-toxic. Another example is for the corticosteroids; the aqueous solubility of dihydrotestosterone was enhanced 300-fold upon the incorporation into core of polycaprolactone-b-polyethylene glycol micelles (Bromberg 2008).

12.4.2 Application for the Route of Administration

A few examples of drug-loaded PM are given in Table 12.1, which is based on the route of administration along with their description, advantages, and disadvantages. As detailed in Table 12.1, the intravenous route has found a lot of applications for the use of PM for especially anticancer drug targeting. The oral route is the preferred route of drug administration owing to the advantages as most convenient, especially for chronic therapy, and painless self-medication from the patient's viewpoint. Along with that, many variables such as intrinsic properties (poor aqueous

Table 12.1 Applications of polymeric micelles based on the route of administration

Route of administration	Type of excipients used	CMC	Description	Advantages	Disadvantages	References
Intra-venous	Polymeric micelle: Polyethylene glycol-SS-poly(2,4,6-trimethoxybenzylidene-pentaerythritol carbonate) (PEG-SS-PTMBPEC) copolymer Drug: Doxorubicin	2.0×10^{-5} to 0.2 mg/mL	pH-sensitive micelles promote drug release in endosomes at pH 5 and along with redox-responsive potential intended to disassemble the micelle structure and release drugs into the cytosol, which contains 100- to 1000-fold higher concentration of reducing glutathione tripeptide	Targeted intracellular anticancer drug release of doxorubicin (DOX) Higher anti-tumor activity based on MTT assay Biodegradable	May lead to cytotoxicity at higher concentration of micelles	Chen et al. (2013)
	Polymeric micelle: Amphiphilic spiropyran-modified hyperbranched polyphosphate HPHEEP-SP Model drug: Coumarin 102	2.5×10^{-4} to 0.5 mg/mL	Spiropyran is photochromic molecule that can undergo an isomerization between hydrophobic spiropyran and hydrophilic merocyanine. Under UV light irradiation, the colorless and nonpolar spiropyran molecule with a closed-form isomerizes convert into the colored and polar merocyanine molecule with an open form	Light-responsive micelles with biocompatibility, biodegradability Highly promising as smart carriers for controlled delivery of hydrophobic molecules for biomedical applications	Requires UV radiations for the modification of drug release	Chen et al. (2012)
	Folate-labeled Pluronic P105 micelles Drug: Doxorubicin	Not available	Application of low frequency ultrasound as an external stimulus on the amount of drug release from folate-labeled micelles. The drug release increases as the power intensity of ultrasound increases	Active targeting to cancer cells with the use of folate molecule on the micelles	Requires external stimuli to trigger the drug release	Husseini et al. (2013)
	Poly(ketal adipate)-co-polyethylene glycol block copolymers (PKA-PEG) Model drug: Camptothecin	2×10^{-4} to 1×10^{-2} mg/mL	pH-dependent drug release property of the polymeric micelle due to backbone of PKA-PEG copolymers has ketal linkages, which can be cleaved rapidly in acidic environments. This mechanism trigger drug releases selectively at tumor tissues or endosomes and lysosomes of cells	Allows drug to release in the tumor cell that has a slightly acidic pH environment	Camptothecin-loaded PKA-PEG micelles showed higher cytotoxicity than free camptothecin	Lee et al. (2013)
	Monomethyl poly(ethylene glycol)-poly (ε-caprolactone)-poly (trimethylene carbonate) Drug: Curcumin	Not available	In vitro anticancer studies confirmed apoptosis induction and cellular uptake on CT26 cells. Also, in vivo on mice shows effective suppression of the tumor growth of subcutaneous CT26 colon	Improved solubility and stability of curcumin can be achieved. Enhanced cellular uptake of curcumin can be obtained with use of PM for anticancer activity	Not available	Yang et al. (2015)

Table 12.1 (continued)

Route of administration	Type of excipients used	CMC	Description	Advantages	Disadvantages	References
Oral	N-octyl-N-arginine chitosan (OACS) Drug: Insulin	Not available	In vitro degradation experiment showed that the OACS has a relative protective effect for insulin from proteolysis (pepsin and trypsin). Also, hypoglycemic effect in diabetic rats were enhanced with the micelles	Using arginine PM for the enhancement of oral insulin delivery with reduced degradation in the GIT	Long-term stability for the micelles containing peptide is a limiting factor	Zhang et al. (2013)
	Monomethoxy polyethylene glycol-b-poly(lactic acid) and poly(lactic acid) Drug: Itraconazole	Not available	The polymeric micelle of itraconazole increased the solubility up to 15 mg/mL in aqueous media and provided stable solutions at a wide range of concentrations and pH Pharmacokinetic study in dog with single oral administration reveals higher bioavailability compared to itraconazole cyclodextrin complex	Improve solubility and oral bioavailability of poorly water-soluble drug	Acute toxicity of polymeric following i.v. bolus injection to rats. At a dose of 45 mg/kg in male rats, 2 animals died	Yi et al. (2007)
	Monomethyl ether poly(ethyleneglycol)750-poly(caprolactone-co-trimethylene carbonate) Model drugs: Carbamazepine (BCS class II) or chlorothiazide (BCS class IV)	0.001 to 0.05%	Drug-loaded micelles were absorbed by both passive diffusion and fluid-phase endocytosis in order to cross a Caco-2 model of the intestinal barrier	Increase in solubilization of BCS class II and improve both solubility and absorption for BCS IV drug	Drug-loaded micelles cannot inhibit the intestinal P-glycoprotein efflux pumps below or above its CMC	Mathot et al. (2007)

Route of administration	Type of excipients used	CMC	Description	Advantages	Disadvantages	References
Ophthalmic	Poly(L-lactide)-b-poly(methacrylic acid-co-3-acrylamidophenylboronic acid) block Drug: Cyclosporin A	0.035 to 0.073 mg/mL	Drug-loaded mucoadhesive micelles have the potential to significantly improve the bioavailability of topically applied ophthalmic drugs based on improved contact time. Improved mucoadhesion was compared to a commonly known mucoadhesive chitosan and has demonstrated ability to improve delivery of cyclosporin A	Reduce dosage frequency of administration and unintentional systemic exposure	Not available	Prosperi-Porta et al. (2016)
	Polyvinyl caprolactam–polyvinyl acetate–polyethylene glycol graft copolymer Drug: Myricetin	0.076 mg/mL	Drug-loaded PM greatly improved their in vitro cellular uptake and antioxidant activity. Along with that, improved in vivo (in rabbit) corneal permeation as well as anti-inflammatory activity	Improved aqueous solubility, stability, corneal permeation of poorly soluble drug	Not available	Sun et al. (2019)
	Poly(ethylene oxide)-poly (propylene oxide) Drug: Lornoxicam	Not available	Improve solubility from 0.0318 to more than 2.34 mg/mL of lornoxicam Histopathological and confocal laser studies revealed the non-irritant nature and good corneal penetrating (in vivo study in rabbit) power of the proposed nanoformulation	Simple and cost-effective techniques with improved physical stability and solubility of the drug	Not available	Salama and Shamma (2015)

solubility and stability) and physiological conditions (transit time, surface area, enzymatic activity, and regional pH) must be considered which may affect the oral bioavailability. Moreover, ophthalmic drug delivery is extremely challenging due to poor solubility, permeation, and reduced retention time. To overcome these obstacles, aqueous formulations of drug-loaded PM can be used (Salama and Shamma 2015).

12.4.3 Application for the Disease State

PM owing to their ability to load a variety of drugs, target the drug to site of action, alter pharmacokinetics of the loaded drug and reduce nonspecific cytotoxicity have received attention for their usage in different disease states.

12.4.3.1 Cancer

In terms of formulation design, amphiphilic copolymers that spontaneously form self-assembled micellar structure are attractive for their application to tumors. These micellar structures are nano-sized, colloidal particles with a hydrophobic core and a hydrophilic corona (Aliabadi and Lavasanifar 2006; V P Torchilin 2006a). The hydrophobic core aids in the solubilization of hydrophobic drug molecules, and hydrophilic segment forms the corona that protect the loaded drug within PM by preventing interaction with blood components. This may lead to increased circulation of the delivery system, ultimately resulting in enhanced accumulation within the tumor cells.

Pluronics, also known as Poloxamers, can undergo self-assembly to form micelles. They are amphiphilic, nonionic block copolymer, consisting of a central poly(propylene oxide) block, which represents the hydrophobic core surrounded by two hydrophilic chains of poly(ethylene oxides) for the formation of hydrophilic corona (Rowe et al. 2009). Interestingly, this micellar system has been explored for oral delivery of paclitaxel (Yoncheva et al. 2012). Paclitaxel, owing to its low aqueous solubility and activity by P-glycoprotein efflux transporters in the intestine, impacts its oral administration negatively. The oral administration of paclitaxel-loaded PM in mice showed elevated area under the curve of paclitaxel (similar to intravenous administration of Taxol®), longer (i.e., 9-times) mean residence time, and a high (i.e., two-fold) volume of distribution as compared to that of intravenous administration of Taxol®, indicating improved oral absorption of paclitaxel by PM. A similar micelle-like structure of Poloxamer-methotrexate conjugate was developed and evaluated for methotrexate delivery (Ren et al. 2015). Methotrexate was physically entrapped and chemically conjugated (i.e., ester linkage) to PM, which led to increase in drug loading. The conjugates could self-assemble into the micelle-like structures in aqueous phase, and methotrexate was in the inner-core of PM. The resulting methotrexate-loaded PM showed a biphasic release of methotrexate, representing a rapid burst release (up to 6–7 h), followed by a sustained release of methotrexate. Based on the pharmacokinetics study, the mean residence time was extended from methotrexate-loaded PM, indicating a delayed elimination of methotrexate from the systemic circulation. Consequently, the area under the curve of methotrexate-loaded PM was greater than the free drug, indicating an enhanced bioavailability.

Poly(lactic) acid (PLA) represents another synthetic biodegradable polymer and is weakly hydrophilic. Copolymerization with polyethylene glycol (PEG) has greatly enhanced its hydrophilicity, stealth properties, and rate of degradation (Bazile et al. 1995; Hu and Liu 1993). The diblock copolymer comprising of PEG-PLA undergoes self-assembly in water to form PM (Letchford and Burt 2007). Genexol-PM® utilizing PEG-PLA as a micelle-forming polymer has been approved clinically for paclitaxel. Genexol-PM® is a solid-state formulation and is reconstituted with sterile water before intravenous infusion. It has been shown that Genexol-PM® does not increase the circulation time of paclitaxel, which is consistent with rapid release of paclitaxel from paclitaxel-loaded PM (S. C. Kim et al. 2001). However, Genexol-PM® was found to be less toxic than its commercially available counterpart Taxol®. Lower toxicity as exhibited by Genexol-PM® permits dose escalation of paclitaxel in preclinical studies. Consequently, Genexol-PM® was found to be much more effective than Taxol® in ovarian and breast murine tumor models, suggesting its use in human clinical trials. PEG-PLA micelles have shown the ability for multi-drug solubilization. In other words, they can take up 2 or 3 anticancer drugs within the same micelles (Shin et al. 2009). Simultaneously dissolving paclitaxel, 17-allylamino-17-demethoxygeldanamycin, and rapamycin within PEG-PLA micelles have been reported (Shin et al. 2011). These systems were found to be 40 nm in diameter which were found to be stable for 24 h at 25 °C. The half-life for in vitro drug release from these systems was 1–15 h under sink conditions. The correlation between the half-life values with log $P_{o/w}$ values is implicating a diffusion-controlled mechanism for drug release.

12.4.3.2 Fungal Infection

Systemic fungal infections pose a universal health challenge due to the significant rise in mortality in immunocompromised patients (Ravikumar et al. 2015). Broad-spectrum antifungal agents are a crucial therapeutic treatment option. However, their usage is limited due to toxicity associated with the treatment regimen (Halperin et al. 2016). To circumvent this challenge, the development of PM to reduce the toxicity induced by the antifungal agent (i.e., amphotericin B) has been evaluated (Y. Wang et al. 2016). Amphotericin B is a poorly soluble and toxic antifungal drug, which was encapsulated by conjugation into PM self-assembled from phenylboronic acid-functionalized polycarbonate/PEG (PEG-PBC). This micellar formulation had a narrow particle size distribution with a close to neutral zeta potential. Furthermore, UV–Vis absorption analysis indicated that amphotericin B-loaded PM significantly reduced amphotericin B aggregation due to the interactions between amphotericin B and the micellar core, while the marketed formulation (Fungizone®) exhibited no effect. The amphotericin B-loaded PM exhibited a sustained drug release and showed comparable antifungal activity to free amphotericin B and Fungizone®. Histological examination showed that amphotericin B-loaded PM led to a significantly lower number of apoptotic cells in the kidneys compared to Fungizone®, indicating reduced nephrotoxicity of PM formulations.

12.4.3.3 Diabetes-Associated Corneal Disease

α-Lipoic acid is an antioxidant for the prevention and treatment of ophthalmic complications. A micelle-based formulation (i.e., Soluplus®) enhancing the solubility, stability, and corneal permeability of α-Lipoic acid has been developed and compared to a conventional surfactant formulation (Alvarez-Rivera et al. 2016). Interestingly, Soluplus® prepared with 1- or 2-mM block copolymer concentration exhibited in situ gelling capability at 35 °C and could transform into gels, which consequently increased the corneal residence time. Therefore, Soluplus® was suggested to be a better dosage from with improved therapeutic response as compared to currently available commercial product.

In another study, a block polymer composed of cationic chitosan and methoxypolyethylene glycol-poly(ε-caprolactone), which could self-assemble into cationic micelles to solubilize hydrophobic drugs such as diclofenac was evaluated (Shi et al. 2015). A similar in vitro release profiles for diclofenac from commercial eye drops and the micellar system was obtained. A biphasic release pattern was observed for achieving therapeutic drug levels quickly and maintaining them for prolonged period of time. In vitro corneal penetration studies demonstrated enhanced penetration of diclofenac from the micellar system relative to the commercial eye drop containing diclofenac. A passive diffusion as the primary mechanism of diclofenac transport into the cornea from both formulation was observed. Corneal permeation was further confirmed under in vivo conditions using Nile red as a fluorescent agent. The Nile red micellar system resulted in higher fluorescence intensity in the corneal endothelium than the aqueous solution of the fluorescent agent 60 min post-instillation. In addition, the diclofenac micellar system demonstrated higher concentrations of diclofenac in the aqueous humor of male New Zealand albino rabbits in comparison to commercial diclofenac containing eye drops (2.3-fold higher area under the curve) indicating greater drug retention and improved drug bioavailability.

12.4.4 Application for Drug Targeting to Specific Cell and/or Organ

The role of PM as a drug delivery system remains vast with debate circumventing the distinction between therapeutic efficacies of passively versus actively targeting potential of this delivery system. It is widely accepted that methodologies (e.g., surface modification) result in homing a high concentration of drug at the target site. The intracellular trafficking mechanism of these carriers to dictate the overall therapeutic outcome is seldom described. The purpose of this section is to provide the reader with a condensed discussion on different mechanisms by which PM targets the drug to the cell and/or organ of interest.

Before discussing the drug targeting mechanism of drug-loaded PM, an overview of the desired characteristics of this system in circulation is presented. After their administration, drug-loaded PM must pass through several physiological and biological barriers before reaching the target site. Briefly, these carriers must be able (1) to remain intact in the circulation, (2) to escape from the reticuloendothelial system and do not get captured by the mononuclear phagocyte system, (3) to cross the vasculature as well as penetrate the tissue interstitial fluid, and finally, (4) to interact with the target site (tissue and/or receptor) exclusively (Barar and Omidi 2013; Omidi and Barar 2014). Since drug targeting to tumor cells is widely explored by drug delivery scientists, the drug targeting efficacy of PM to tumor cells has been discussed.

12.4.4.1 Passive Targeting

Passive targeting involves the accumulation of PM within the tumor cells mainly via the enhanced permeation and retention effect (Maeda et al. 2000). PM owing to their unique physicochemical properties (i.e., surface properties, particle size) not only leak into the tumor cells but also get retained within due to the poor lymphatic drainage. However, this unique feature does not apply to small drug molecules (molecular weight < 40 KDa) which may have short circulation time and fast washout from the tumor cells. Thus, the encapsulation of small molecule drugs within PM is advisable. Such encapsulation also alters the pharmacokinetics (i.e., prolonged systemic circulation) of the drugs and provides some tumor selectivity. This type of tumor-targeting is termed as "passive" which solely relies on the carrier and the tumor microenvironment characteristics (i.e., particle size and enhanced permeation and retention effect) (Fang et al. 2011; Maeda 2001).

Additionally, the stability of PM in the circulation is crucial for passive targeting (Vladimir P Torchilin and Trubetskoy 1995). There is a consensus that the stability of PM can be enhanced by hydrophilic polyethylene glycol coatings. One key consideration not been reported in the literature is the selection criteria of polyethylene glycol. Based on the molecular weight, their different grades of polyethylene glycol are available commercially. A literature review indicates that a surface polyethylene glycol chain with molecular weight of ≥2000 may be required to achieve increased mononuclear phagocyte system avoidance characteristics. This minimum molecular weight requirement can be linked to the loss in the flexibility of shorter chains of low molecular weight polyethylene glycol. Also, it has been shown that as molecular weight increases beyond 2000 Da, the blood circulation half-life of the PEGylated (i.e., overcoated with polyethylene glycol) carriers is also increased, which may be partly due to the increased chain flexibility of higher molecular weight polymers

(Gref et al. 1994). Furthermore, surface chain density and conformation are also critical factors for achieving improved stealth characteristics (Owens III and Peppas 2006). Therefore, the grades and concentration of overcoating with polyethylene glycol may need optimization.

The size of PM also plays a crucial role in the enhanced permeation and retention effect-mediated accumulation. PM, which are below the cutoff limit for most tumors, ensure them remaining in circulation for longer intervals and gradually enter the tumor vasculature. As a result, they are not up taken by the mononuclear phagocyte system.

12.4.4.2 Active Targeting

Tumor cells exhibit an expression of specific proteins (i.e., receptors or antigens) and are normally not expressed on the surface of normal cells. Active targeting utilizes this feature to selectively target anticancer drugs to the tumor cells (Nie et al. 2007). The surface of PM can be fabricated for active targeting with targeting ligands having strong specificity toward receptors or antigens expressed on cancer cells (V P Torchilin 2006b). Usually, the targeting ligands can be attached to the hydrophilic blocks of micelles, in such orientation they avoid steric hindrance when binding to their target receptors (Vladimir P Torchilin and Trubetskoy 1995).

A wide collection of ligands has been reported in literature for active targeting of PM to tumor cells. Commonly used targeting ligands can be broadly classified as antibodies, peptides and proteins, sugar moieties, and small molecules (Chung et al. 2014; Jin et al. 2011; Sawant et al. 2014; C. Yang et al. 2013). Few important considerations need to take into account when selecting a ligand. Firstly, a clear understanding of the type of cancer cells being targeted. Secondly, the type of receptors expressed in those cells needs to be identified. Thirdly, limitation of the ligand, for example, protein- and peptide-overcoated PM may have low stability, and these ligands may be prone to enzymatic degradation by the peptidases before reaching the target tissue. Additionally, the larger the molecular weight of the ligand (i.e., antibodies) the lower is its density on the surface. However, such a challenge can be overcome by reducing the particle size and/or conjugating the ligand on the surface of PM (K. Chen and Conti 2010). An example of each class and their usage as targeting moiety has been briefly described herein.

Antibodies by far have been the most common targeting ligands for PM because of the diversity of their target receptors and selectivity of their interaction with them. To overcome the challenge of low surface density, conjugation of antibody fragment with PM for selectively delivering platinum drugs to pancreatic tumors has been reported. By such modifying efforts, more than a fivefold increase in cellular binding and rapid cellular internalization as compared to non-targeted PM were observed in in vitro cell culture model. In in vivo study, conjugated PM significantly suppressed the growth of pancreatic tumor xenografts for a prolonged period (i.e., > 40 days), in comparison to non-targeted PM and free drug (controls) (Ahn et al. 2015).

Proteins and peptides have been utilized extensively as targeting ligands for PM. Specifically, transferrin receptors (overexpressed in many cancers) have been used for the development of transferrin-targeted polymeric micellar system (Singh 1999). Transferrin-decorated PM were synthesized to deliver paclitaxel across the brain for glioma therapy. Due to the significant expression of transferrin on brain capillary endothelial cells and glioma cells, these PM could cross the blood-brain barrier and target the glioma cells. Additionally, their being rapidly be uptaken by the glioma cells had been reported from an in vitro model study. In in vivo study, this system presented a better anti-glioma effect with a prolonged median survival of nude mice bearing glioma than the unmodified PM which served as control (P. Sun et al. 2020).

Hyaluronic acid is a polysaccharide that can actively target the cell surface receptors. It binds to cluster determinant 44 (better known as CD44) receptors overexpressed on the surface of cancer cells and is also one of the main components of the extracellular matrix along with collagen (Aruffo et al. 1990). In addition to acting as a targeting moiety, hyaluronic acid can be used simultaneously as a hydrophilic backbone polymer (Choi et al. 2011). In a study, a hyaluronic acid-coated redox-sensitive micellar system was prepared for tumor-specific intracellular rapid delivery of gambogic acid. The resulting cationic PM had hyaluronic acid as a coat on the surface for CD44-mediated active targeting together with protection from cation-associated defects. The hyaluronic receptor-mediated cellular uptake and burst drug release intracellularly have been further confirmed from the results of flow cytometry and confocal microscopy studies. Consequently, the coated PM displayed the highest apoptosis induction and cytotoxicity over the nonselective hyaluronic acid un-coated controls against A549 NSCLC model both in vitro and in vivo studies. Furthermore, reduced systemic cytotoxicity was observed in mice treated with coated PM as compared to un-coated cationic PM and free drug solution (Xu et al. 2019).

12.4.4.3 Multifunctional Targeting

PM provide a platform allowing the formulator for the integration of multiple components in single micelle, which makes the delivery system complex, however, enhances its functionality. For example, polyethylene glycol and hyaluronic acid-coated PM for enhanced stability in the circulation and selectively targeting the CD44 receptors in cancer cells. The resulting PM combine several distinct functions in a single carrier having each component functioning distinctively and in flawless harmony with other components and is termed "multifunctional" micelles (Jhaveri and Torchilin 2014). Additionally, tracer substances can also be incorporated within PM permitting real-time imaging of PM and their existence within cells. Therefore, an ideal multi-functional PM may simultaneously circulate in the body for prolonged periods for passive and/or active targeting of the loaded therapeutics. Although targeted therapeutic treatments have been established, tumor cells may still escape the targeted treatments leading to acquired resistance and the failure of targeted therapeutic treatments (Blanco et al. 2009). Therefore, a multi-functionality methodology for targeting cancer seems crucial. Recently, a lot of the focus of current research has been in the direction of the development of multifunctional PM for improving the efficacy, increasing the safety, and promoting novel chemotherapeutics.

Multifunctional PM have been investigated extensively for the delivery of therapeutics, a selected few of which are discussed here. In one study, nano-supramolecular PM with a spherical core-shell structure was prepared. To prepare such PM [i.e., folate-polyethylene glycol-poly(aspartate hydrazone adriamycin)], self-assembling amphiphilic block copolymers were designed and synthesized by mounting a molecular promoter to enhance intracellular transport (i.e., folate) at the end of the shell-forming polyethylene glycol chain and conjugating with the anticancer drug (i.e., adriamycin). The tetrazolium dye method using human pharyngeal cancer cells showed that folate-bound PM significantly improved cell growth inhibitory activity. Subsequent flow cytometric analysis revealed that cellular uptake of folate-bound PM significantly increased (Bae et al. 2005).

MRI-responsive PM were developed in a novel approach, wherein hybrid PM were fabricated from Pluronic F127 and a peptide-amphiphile in which the magnetic nanoparticles were embedded in the shell (Yongyong Li et al. 2013). The shell-embedded magnetic nanoparticles significantly improved stability and retarded the release of doxorubicin from the hybrid PM. Another novel approach, targeting-clickable and tumor-cleavable multi-block polyurethane nanomicelle containing antitumor drugs formulation were reported (Song et al. 2013). In this study, polyurethane was synthesized from biodegradable poly(ε-caprolactone) and l-lysine ethyl ester diisocyanate, then extended by a new designed l-cystine-derivatized chain extender bearing a redox-responsive disulfide bond and clickable alkynyl groups, and finally terminated by a detachable methoxy polyethylene glycol with a highly pH-sensitive benzoic-imine linkage. The fabricated polymers exhibited astonishing self-assembly characteristics and stimuli-responsiveness, good cytocompatibility, as well as high loading capacity for doxorubicin. Furthermore, folic acid was used as a model targeting ligand was conjugated to the polyurethane micelles via an efficient click reaction. The conjugation of folic acid showed an enhanced cellular uptake and improved drug efficacy toward folic acid receptor-positive HeLa cancer cells.

12.5 Clinical Applications of Polymeric Micelles

There are various reports available in the literature for PM with improved solubility and stability. However, still many of drug-loaded PM are in clinical trials due to various hurdles. These include capability of scalability of the process that remains a challenge. Also, it demands a more systematic approach at the initial stages of research to accelerate and simplify the transfer from basic research to clinical application. To minimize these hurdles, there is a need to understand biodistribution, pharmacokinetics, pharmacodynamics, accelerated systemic clearance (i.e., immunogenic response), and in vivo degradation profiles (Deshmukh et al. 2017).

One of the critical aspects of using PM is to improve drug targeting and minimize the systemic side effect of anticancer drugs. Under the code name NC-6004 (Nanoplatin™), PM composed of PEG and complex of poly(glutamic acid) were loaded with cisplatin. Compared to free drugs, the NC-6004 displayed lower cytotoxicity when studied in various cancer cell lines based on the in vitro studies. Based on phase I clinical trials in patients suffering from colorectal carcinoma, the NC-6004 demonstrated superior antitumor activity as compared to that of the free drug. The NC-6004 showed effective results in metastatic pancreatic cancer patients to determine its appropriate dose and evaluate its safety, tolerability, and efficacy in conjunction with gemcitabine under phase I/II clinical studies initiated in Taiwan and Singapore. The formulation is currently under phase III clinical trial (Plummer et al. 2011).

To achieve better efficacy of the anticancer drug (i.e., oxaliplatin), NC-4016 is PM developed to reduce neurotoxicity, resolve drug resistance, and increase circulation time. NC-4016 consist of PEG-b-poly(L-glutamic acid) copolymer, stabilized by incorporating dichloro(1,2-diaminocyclohexane) platinum(II) (DACHPt). DACHPt, an active metabolite of oxaliplatin, has reduced toxicity which enhances the delivery of oxaliplatin to solid tumors. During the preclinical studies, NC-4016 was found to be more stable in physiological conditions as compared to oxaliplatin alone. Also, it not only extended blood circulation time of the drug significantly after bolus injection but also achieved a greater than 1000-fold increase in plasma concentration (Gong et al. 2012; Takahashi et al. 2013).

Doxorubicin is an effective anticancer drug, and the doxorubicin-loaded PM named NK911 is composed of doxorubicin conjugated to PEG-poly(aspartate). In this formulation, physically entrapped doxorubicin is conjugated with both hydrophobic poly(α,β-aspartic acid) core. The NK911 completed phase I clinical trial at the National Cancer Center Hospital, Tokyo, Japan. There was a minimum toxicity dose observed, and the recommended dose for the phase II study was determined to be 67 and 50 mg/m^2, respectively, based on phase I experimental findings. The NK911is currently under phase II clinical study (Gong et al. 2012; Matsumura and Kataoka 2009).

The NC-6300 micelle formulation was fabricated by conjugation of epirubicin, which is 4′-epimer of doxorubicin, and is less toxic as compared to doxorubicin. The epirubicin was conjugated with the core component hydrophobic poly(α,β-aspartic acid) substituted, partially, with hydrophobic benzyl groups to stabilize the formed epirubicin micelles. The NC-6300 shows similar anticancer activity to that of NK911, with significantly reduced the cardiac toxicity. The phase I clinical trial of NC-6300 demonstrated its safety, tolerability, and recommended dosage to determine the efficacy in patients with either advanced or metastatic solid tumors (Takahashi et al. 2013).

Multidrug resistance is the major obstacle in the development of an effective clinical cancer therapy. The nonionic surfactant PEO-PPO-PEO, commonly known as Pluronic, is the only copolymer that has demonstrated a reduction in the drug resistance via inhibition of cancer cell P-glycoprotein as reported in the literature (Valle et al. 2011). This Pluronic-based micelle formulation of doxorubicin SP1049C is in a clinical trial. This formulation SP1049C has a blend of Pluronic L61 and Pluronic F127. It can be noted that Pluronic F127 contributes to stability without any alteration in the cytotoxicity, whereas Pluronic L61 inhibits the P-glycoprotein efflux mechanism and modulates the activity of the doxorubicin in multidrug-resistant cells. The biodistribution profile showed an increased accumulation of micelles in brain tumors, suggesting its promise in solid tumors of the brain. Phase I clinical trials were conducted and demonstrated linear pharmacokinetic behavior. Based on the results of phase I trials, phase II clinical trials were initiated in patients with advanced adenocarcinoma of the esophagus and gastroesophageal junctions.

Paclitaxel has been widely studied to improve its solubility, stability, and targeted delivery for the treatment of cancer. There are two products already in the clinical trials for paclitaxel that includes Genexol®-PM and NK105 aiming to improve solubility and used for breast and lung cancer. Genexol®-PM is copolymer micelles containing mPEG-PDLLA. As compared to the currently available marketed formulation (i.e., Taxol®), Genexol®-PM has 25% more drug solubilizing capacity and has higher maximum tolerated doses and median lethal dose. Also, it is effective in reduced P-glycoprotein efflux, higher tumor accumulation of drugs, and decreased myelosuppression. Genexol®-PM is currently in Phase II trials in the US and already approved in Bulgaria, Hungary, and South Korea (Cabral and Kataoka 2014; Gong et al. 2012; Lu and Park 2013).

The NK105 micelle formulation was developed to solubilize and deliver physically entrapped paclitaxel to solid tumors. It consists of PEG-poly(aspartic acid) copolymer modified with 4-phenyl-1-butanol to enhance its hydrophobicity. The micelle formulation can be readily dissolved in 5% glucose solution for intravenous administration. The hydrophilic surface, due to the presence of PEG and smaller droplet size, makes increased blood circulation time. Hence, as compared to free drug, the micelle formulation can achieve maximum concentration of paclitaxel in tumor site when tested in HT-29 colon cancer cell lines (Lu and Park 2013; Ri et al. 2018; Saito et al. 2010).

12.6 Conclusions

As nanocarriers, PM have many advantages, such as effective drug-targeting ability, longer circulation time in the body, easy production, and relatively low toxicity, so PM have aroused widespread interest in research. In particular, owing to the feasibility of developing copolymeric structures, it is possible to match the same formulation requirement. Also, the exceptional solution behavior along with established characterization methodologies attracts the researchers to work upon and further explore these PM. Although there are some hurdles, many PM formulations have shown steady progression through clinical trials. To that end, PM unlike other nanocarriers are already ahead in terms of proven clinical success owing to their comparatively simple design and easy development. Several multifunctional micellar systems are under investigation currently, and as new technologies are evolving, further research will be accelerated on the applications of PM in the future.

References

Ahmad Z, Shah A, Siddiq M, Kraatz H-B (2014) Polymeric micelles as drug delivery vehicles. RSC Adv 4(33):17028–17038

Ahn J, Miura Y, Yamada N, Chida T, Liu X, Kim A, Sato R, Tsumura R, Koga Y, Yasunaga M, Nishiyama N, Matsumura Y, Cabral H, Kataoka K (2015) Antibody fragment-conjugated polymeric micelles incorporating platinum drugs for targeted therapy of pancreatic cancer. Biomaterials 39:23–30. https://doi.org/10.1016/j.biomaterials.2014.10.069

Aliabadi HM, Lavasanifar A (2006) Polymeric micelles for drug delivery. Expert Opin Drug Deliv 3(1):139–162. https://doi.org/10.1517/17425247.3.1.139

Alvarez-Rivera F, Fernández-Villanueva D, Concheiro A, Alvarez-Lorenzo C (2016) $α$-lipoic acid in Soluplus ®polymeric Nanomicelles for ocular treatment of diabetes-associated corneal diseases. J Pharm Sci 105(9):2855–2863. https://doi.org/10.1016/j.xphs.2016.03.006

Amin MCIM, Butt AM, Amjad MW, Kesharwani P (2017) Polymeric micelles for drug targeting and delivery. In: Nanotechnology-based approaches for targeting and delivery of drugs and genes. Elsevier, pp 167–202

Aruffo A, Stamenkovic I, Melnick M, Underhill CB, Seed B (1990) CD44 is the principal cell surface receptor for hyaluronate. Cell 61(7):1303–1313. https://doi.org/10.1016/0092-8674(90)90694-A

Astafieva I, Zhong XF, Eisenberg A (1993) Critical Micellization phenomena in block polyelectrolyte solutions. Macromolecules. https://doi.org/10.1021/ma00078a034

Bader H, Ringsdorf H, Schmidt B (1984) Watersoluble polymers in medicine. Die Angewandte Makromolekulare Chemie: Appl Macromolecular Chem Phys 123(1):457–485

Bae Y, Jang W-D, Nishiyama N, Fukushima S, Kataoka K (2005) Multifunctional polymeric micelles with folate-mediated cancer cell targeting and pH-triggered drug releasing properties for active intracellular drug delivery. Mol BioSyst 1(3):242. https://doi.org/10.1039/b500266d

Barar J, Omidi Y (2013) Targeted gene therapy of Cancer: second amendment toward holistic therapy. Bioimpacts 3(2):49–51. https://doi.org/10.5681/bi.2013.014

Bartos C, Pallagi E, Szabó-Révész P, Ambrus R, Katona G, Kiss T, Rahimi M, Csóka I (2018) Formulation of levodopa containing dry powder for nasal delivery applying the quality-by-design approach. Eur J Pharm Sci 123:475–483

Batrakova EV, Li S, Vinogradov SV, Alakhov VY, Miller DW, Kabanov AV (2001) Mechanism of pluronic effect on P-glycoprotein efflux system in blood-brain barrier: contributions of energy depletion and membrane fluidization. J Pharmacol Exp Ther 299(2):483–493

Bazile D, Prud'homme C, Bassoullet M, Marlard M, Spenlehauer G, Veillard M (1995) Stealth Me. PEG-PLA nanoparticles avoid uptake by the mononuclear phagocytes system. J Pharm Sci 84(4):493–498. https://doi.org/10.1002/jps.2600840420

Blanco E, Kessinger CW, Sumer BD, Gao J (2009) Multifunctional micellar nanomedicine for Cancer therapy. Exp Biol Med 234(2):123–131. https://doi.org/10.3181/0808-MR-250

Brannon-Peppas L, Ghosn B, Roy K, Cornetta K (2007) Encapsulation of nucleic acids and opportunities for cancer treatment. Pharm Res 24(4):618–627

Bromberg L (2008) Polymeric micelles in oral chemotherapy. J Control Release. https://doi.org/10.1016/j.jconrel.2008.01.018

Burt HM, Zhang X, Toleikis P, Embree L, Hunter WL (1999) Development of copolymers of poly(D,L-lactide) and methoxypolyethylene glycol as micellar carriers of paclitaxel. Colloids Surf B: Biointerfaces. https://doi.org/10.1016/S0927-7765(99)00067-3

Cabral H, Kataoka K (2014) Progress of drug-loaded polymeric micelles into clinical studies. J Control Release. https://doi.org/10.1016/j.jconrel.2014.06.042

Cammas-Marion S, Okano T, Kataoka K (1999) Functional and site-specific macromolecular micelles as high potential drug carriers. Colloids Surf B: Biointerfaces. https://doi.org/10.1016/S0927-7765(99)00071-5

Chen K, Conti PS (2010) Target-specific delivery of peptide-based probes for PET imaging. Adv Drug Deliv Rev 62(11):1005–1022. https://doi.org/10.1016/j.addr.2010.09.004

Chen F, Stenzel MH (2018) Polyion complex micelles for protein delivery. Aust J Chem 71(10):768–780

Chen CJ, Jin Q, Liu GY, Li DD, Wang JL, Ji J (2012) Reversibly light-responsive micelles constructed via a simple modification of hyperbranched polymers with chromophores. Polymer. https://doi.org/10.1016/j.polymer.2012.06.024

Chen W, Zhong P, Meng F, Cheng R, Deng C, Feijen J, Zhong Z (2013) Redox and pH-responsive degradable micelles for dually activated intracellular anticancer drug release. J Control Release. https://doi.org/10.1016/j.jconrel.2013.01.001

Chen J, Ding J, Zhang Y, Xiao C, Zhuang X, Chen X (2015) Polyion complex micelles with gradient pH-sensitivity for adjustable intracellular drug delivery. Polym Chem 6(3):397–405

Chiappetta DA, Hocht C, Taira C, Sosnik A (2011) Oral pharmacokinetics of the anti-HIV efavirenz encapsulated within polymeric micelles. Biomaterials. https://doi.org/10.1016/j.biomaterials.2010.11.082

Choi KY, Yoon HY, Kim J-H, Bae SM, Park R-W, Kang YM, Kim I-S, Kwon IC, Choi K, Jeong SY, Kim K, Park JH (2011) Smart Nanocarrier based on PEGylated hyaluronic acid for Cancer therapy. ACS Nano 5(11):8591–8599. https://doi.org/10.1021/nn202070n

Christie RJ, Matsumoto Y, Miyata K, Nomoto T, Fukushima S, Osada K, Halnaut J, Pittella F, Kim HJ, Nishiyama N (2012) Targeted polymeric micelles for siRNA treatment of experimental cancer by intravenous injection. ACS Nano 6(6):5174–5189

Chung EJ, Cheng Y, Morshed R, Nord K, Han Y, Wegscheid ML, Auffinger B, Wainwright DA, Lesniak MS, Tirrell MV (2014) Fibrin-binding, peptide amphiphile micelles for targeting glioblastoma. Biomaterials 35(4):1249–1256. https://doi.org/10.1016/j.biomaterials.2013.10.064

Csóka I, Pallagi E, Paál TL (2018) Extension of quality-by-design concept to the early development phase of pharmaceutical RandD processes. Drug Discov Today 23(7):1340–1343

Deshmukh AS, Chauhan PN, Noolvi MN, Chaturvedi K, Ganguly K, Shukla SS, Nadagouda MN, Aminabhavi TM (2017) Polymeric micelles: basic research to clinical practice. Int J Pharm. https://doi.org/10.1016/j.ijpharm.2017.09.005

Fang J, Nakamura H, Maeda H (2011) The EPR effect: unique features of tumor blood vessels for drug delivery, factors involved, and limitations and augmentation of the effect. Adv Drug Deliv Rev 63(3):136–151. https://doi.org/10.1016/j.addr.2010.04.009

Fournier E, Dufresne MH, Smith DC, Ranger M, Leroux JC (2004) A novel one-step drug-loading procedure for water-soluble amphiphilic nanocarriers. Pharm Res. https://doi.org/10.1023/B:PHAM.0000029284.40637.69

Gelderblom H, Verweij J, Nooter K, Sparreboom A (2001) Cremophor EL: the drawbacks and advantages of vehicle selection for drug formulation. Eur J Cancer. https://doi.org/10.1016/S0959-8049(01)00171-X

Gieszinger P, Csóka I, Pallagi E, Katona G, Jójárt-Laczkovich O, Szabó-Révész P, Ambrus R (2017) Preliminary study of nanonized lamotrigine containing products for nasal powder formulation. Drug Des Devel Ther 11:2453

Gref R, Minamitake Y, Peracchia MT, Trubetskoy V, Torchilin V, Langer R (1994). Biodegradable longcirculating polymeric nanospheres. Science 263:1600–1603

Gong J, Chen M, Zheng Y, Wang S, Wang Y (2012) Polymeric micelles drug delivery system in oncology. J Control Release. https://doi.org/10.1016/j.jconrel.2011.12.012

Gupta A, Costa AP, Xu X, Lee S-L, Cruz CN, Bao Q, Burgess DJ (2020) Formulation and characterization of curcumin loaded polymeric micelles produced via continuous processing. Int J Pharm 119340

Hagan SA, Coombes AGA, Garnett MC, Dunn SE, Davies MC, Illum L, Davis SS, Harding SE, Purkiss S, Gellert PR (1996) Polylactide– poly (ethylene glycol) copolymers as drug delivery systems. 1. Characterization of water dispersible micelle-forming systems. Langmuir 12(9):2153–2161

Halperin A, Shadkchan Y, Pisarevsky E, Szpilman AM, Sandovsky H, Osherov N, Benhar I (2016) Novel water-soluble amphotericin B-PEG conjugates with low toxicity and potent in vivo efficacy. J Med Chem 59(3):1197–1206

Harada A, Kataoka K (1995) Formation of polyion complex micelles in an aqueous milieu from a pair of oppositely-charged block copolymers with poly (ethylene glycol) segments. Macromolecules 28(15):5294–5299

Hu DS-G, Liu H-J (1993) Effect of soft segment on degradation kinetics in polyethylene glycol/ poly (L-lactide) block copolymers. Polym Bull 30(6):669–676

Husseini GA, Velluto D, Kherbeck L, Pitt WG, Hubbell JA, Christensen DA (2013) Investigating the acoustic release of doxorubicin from targeted micelles. Colloids Surf B: Biointerfaces. https://doi.org/10.1016/j.colsurfb.2012.05.025

Ishida O, Maruyama K, Sasaki K, Iwatsuru M (1999) Size-dependent extravasation and interstitial localization of polyethyleneglycol liposomes in solid tumor-bearing mice. Int J Pharm 190(1):49–56

Jhaveri AM, Torchilin VP (2014) Multifunctional polymeric micelles for delivery of drugs and siRNA. Front Pharmacol 5. https://doi.org/10.3389/fphar.2014.00077

Jin C, Yang W, Bai L, Wang J, Dou K (2011) Preparation and characterization of targeted DOX–PLGA–PEG micelles decorated with bivalent fragment HAb18 F(ab′) 2 for treatment of hepatocellular carcinoma. J Control Release 152:e14–e15. https://doi.org/10.1016/j.jconrel.2011.08.093

Johnson, R. P., Jeong, Y. Il, John, J. V, Chung, C.-W., Kang, D. H., Selvaraj, M., Suh, H., and Kim, I. (2013). Dual stimuli-responsive poly (N-isopropylacrylamide)-b-poly (L-histidine) chimeric materials for the controlled delivery of doxorubicin into liver carcinoma. Biomacromolecules, 14(5), 1434–1443

Jones MC, Leroux JC (1999) Polymeric micelles - a new generation of colloidal drug carriers. Eur J Pharm Biopharm. https://doi.org/10.1016/S0939-6411(99)00039-9

Kabanov AV, Nazarova IR, Astafieva IV, Batrakova EV, Alakhov VY, Yaroslavov AA, Kabanov VA (1995) Micelle formation and solubilization of fluorescent probes in poly (oxyethylene-b-oxypropylene-b-oxyethylene) solutions. Macromolecules 28(7):2303–2314

Kim SC, Kim DW, Shim YH, Bang JS, Oh HS, Kim SW, Seo MH (2001) In vivo evaluation of polymeric micellar paclitaxel formulation: toxicity and efficacy. J Control Release 72(1–3):191–202

Kozlov MY, Melik-Nubarov NS, Batrakova EV, Kabanov AV (2000) Relationship between pluronic block copolymer structure, critical micellization concentration and partitioning coefficients of low molecular mass solutes. Macromolecules 33(9):3305–3313

Kwon G, Naito M, Yokoyama M, Okano T, Sakurai Y, Kataoka K (1993) Micelles based on AB block copolymers of poly (ethylene oxide) and poly (. Beta.-benzyl L-aspartate). Langmuir 9(4):945–949

Kwon GS, Naito M, Yokoyama M, Okano T, Sakurai Y, Kataoka K (1995) Physical entrapment of adriamycin in AB block copolymer micelles. Pharm Res 12(2):192–195

Lee SC, Huh KM, Lee J, Cho YW, Galinsky RE, Park K (2007) Hydrotropic polymeric micelles for enhanced paclitaxel solubility: in vitro and in vivo characterization. Biomacromolecules 8(1):202–208

Lee I, Park M, Kim Y, Hwang O, Khang G, Lee D (2013) Ketal containing amphiphilic block copolymer micelles as pH-sensitive drug carriers. Int J Pharm. https://doi.org/10.1016/j.ijpharm.2013.03.017

Letchford K, Burt H (2007) A review of the formation and classification of amphiphilic block copolymer nanoparticulate structures: micelles, nanospheres, nanocapsules and polymersomes. Eur J Pharm Biopharm 65(3):259–269

Li Y, Kwon GS (2000) Methotrexate esters of poly (ethylene oxide)-block-poly (2-hydroxyethyl-L-aspartamide). Part I: effects of the level of methotrexate conjugation on the stability of micelles and on drug release. Pharm Res 17(5):607–611

Li Y, Ma J, Zhu H, Gao X, Dong H, Shi D (2013) Green synthetic, multifunctional hybrid micelles with Shell embedded magnetic nanoparticles for Theranostic applications. ACS Appl Mater Interfaces 5(15):7227–7235. https://doi.org/10.1021/am401573b

Liu J, Zeng F, Allen C (2007) In vivo fate of unimers and micelles of a poly (ethylene glycol)-block-poly (caprolactone) copolymer in mice following intravenous administration. Eur J Pharm Biopharm 65(3):309–319

Lu Y, Park K (2013) Polymeric micelles and alternative nanonized delivery vehicles for poorly soluble drugs. Int J Pharm. https://doi.org/10.1016/j.ijpharm.2012.08.042

Luo Y, Yao X, Yuan J, Ding T, Gao Q (2009) Preparation and drug controlled-release of polyion complex micelles as drug delivery systems. Colloids Surf B: Biointerfaces 68(2):218–224

Maeda H (2001) The enhanced permeability and retention (EPR) effect in tumor vasculature: the key role of tumor-selective macromolecular drug targeting. Adv Enzym Regul 41:189–207. https://doi.org/10.1016/s0065-2571(00)00013-3

Maeda H, Wu J, Sawa T, Matsumura Y, Hori K (2000) Tumor vascular permeability and the EPR effect in macromolecular therapeutics: a review. J Control Release 65(1–2):271–284. https://doi.org/10.1016/S0168-3659(99)00248-5

Martin A (1993) Physical pharmacy: physical chemical principles in the pharmaceutical sciences, 6th edn. BI Waverly. Pvt Ltd

Mathot F, des Rieux A, Ariën A, Schneider YJ, Brewster M, Préat V (2007) Transport mechanisms of mmePEG750P(CL-co-TMC) polymeric micelles across the intestinal barrier. J Control Release. https://doi.org/10.1016/j.jconrel.2007.09.001

Matsumura Y, Kataoka K (2009) Preclinical and clinical studies of anticancer agent-incorporating polymer micelles. Cancer Sci. https://doi.org/10.1111/j.1349-7006.2009.01103.x

Movassaghian S, Merkel OM, Torchilin VP (2015) Applications of polymer micelles for imaging and drug delivery. Wiley Interdiscip Rev Nanomed Nanobiotechnol. https://doi.org/10.1002/wnan.1332

Nagarajan R (1996) Solubilization of hydrophobic substances by block copolymer micelles in Aqeous solutions. Solvents Self-Organ Polymers. https://doi.org/10.1007/978-94-009-0333-3_8

Nie S, Xing Y, Kim GJ, Simons JW (2007) Nanotechnology applications in Cancer. Annu Rev Biomed Eng 9(1):257–288. https://doi.org/10.1146/annurev.bioeng.9.060906.152025

Omidi Y, Barar J (2014) Targeting tumor microenvironment: crossing tumor interstitial fluid by multifunctional nanomedicines. Bioimpacts 4(2):55–67. https://doi.org/10.5681/bi.2014.021

Owens DE III, Peppas NA (2006) Opsonization, biodistribution, and pharmacokinetics of polymeric nanoparticles. Int J Pharm 307(1):93–102

Plummer R, Wilson RH, Calvert H, Boddy AV, Griffin M, Sludden J, Tilby MJ, Eatock M, Pearson DG, Ottley CJ, Matsumura Y, Kataoka K, Nishiya T (2011) A phase i clinical study of cisplatin-incorporated polymeric micelles (NC-6004) in patients with solid tumours. Br J Cancer. https://doi.org/10.1038/bjc.2011.6

Prosperi-Porta G, Kedzior S, Muirhead B, Sheardown H (2016) Phenylboronic-acid-based polymeric micelles for Mucoadhesive anterior segment ocular drug delivery. Biomacromolecules. https://doi.org/10.1021/acs.biomac.6b00054

Ravikumar S, Win MS, Chai LYA (2015) Optimizing outcomes in immunocompromised hosts: understanding the role of immunotherapy in invasive fungal diseases. Front Microbiol 6:1322

Ren J, Fang Z, Yao L, Dahmani FZ, Yin L, Zhou J, Yao J (2015) A micelle-like structure of poloxamer–methotrexate conjugates as nanocarrier for methotrexate delivery. Int J Pharm 487(1–2):177–186

Ri M, Suzuki K, Iida S, Hatake K, Chou T, Taniwaki M, Watanabe N, Tsukamoto T (2018) A phase I/II study for dose-finding, and to investigate the safety, pharmacokinetics and preliminary efficacy of NK012, an SN-38-incorporating macromolecular polymeric micelle, in patients with multiple myeloma. Intern Med. https://doi.org/10.2169/internalmedicine.9567-17

Rowe RC, Sheskey P, Quinn M (2009) Handbook of pharmaceutical excipients. Libros Digitales-Pharmaceutical Press

Saito Y, Yasunaga M, ichiro Kuroda, J., Koga, Y., and Matsumura, Y. (2010) Antitumour activity of NK012, SN-38-incorporating polymeric micelles, in hypovascular orthotopic pancreatic tumour. Eur J Cancer. https://doi.org/10.1016/j.ejca.2009.11.014

Salama AH, Shamma RN (2015) Tri/tetra-block co-polymeric nanocarriers as a potential ocular delivery system of lornoxicam: in-vitro characterization, and in-vivo estimation of corneal permeation. Int J Pharm. https://doi.org/10.1016/j.ijpharm.2015.07.010

Sant VP, Smith D, Leroux J-C (2005) Enhancement of oral bioavailability of poorly water-soluble drugs by poly (ethylene glycol)-block-poly (alkyl acrylate-co-methacrylic acid) self-assemblies. J Control Release 104(2):289–300

Savić R, Eisenberg A, Maysinger D (2006) Block copolymer micelles as delivery vehicles of hydrophobic drugs: micelle-cell interactions. J Drug Target. https://doi.org/10.1080/10611860600874538

Sawant RR, Jhaveri AM, Koshkaryev A, Zhu L, Qureshi F, Torchilin VP (2014) Targeted transferrin-modified polymeric micelles: enhanced efficacy in vitro and in vivo in ovarian carcinoma. Mol Pharm 11(2):375–381. https://doi.org/10.1021/mp300633f

Shi S, Zhang Z, Luo Z, Yu J, Liang R, Li X, Chen H (2015) Chitosan grafted methoxy poly (ethylene glycol)-poly ($ε$-caprolactone) nanosuspension for ocular delivery of hydrophobic diclofenac. Sci Rep 5:11337

Shin H-C, Alani AWG, Rao DA, Rockich NC, Kwon GS (2009) Multi-drug loaded polymeric micelles for simultaneous delivery of poorly soluble anticancer drugs. J Control Release 140(3):294–300

Shin H-C, Alani AWG, Cho H, Bae Y, Kolesar JM, Kwon GS (2011) A 3-in-1 polymeric micelle nanocontainer for poorly water-soluble drugs. Mol Pharm 8(4):1257–1265

Singh M (1999) Transferrin as a targeting ligand for liposomes and anticancer drugs. Curr Pharm Des 5(6):443–451

Sipos B, Szabó-Révész P, Csóka I, Pallagi E, Dobó DG, Bélteky P, Kónya Z, Deák Á, Janovák L, Katona G (2020) Quality by design based formulation study of meloxicam-loaded polymeric micelles for intranasal administration. Pharmaceutics 12(8):697

Song N, Ding M, Pan Z, Li J, Zhou L, Tan H, Fu Q (2013) Construction of targeting-clickable and tumor-cleavable polyurethane Nanomicelles for multifunctional intracellular drug delivery. Biomacromolecules 14(12):4407–4419. https://doi.org/10.1021/bm401342t

Štěpánek M, Podhájecká K, Tesařová E, Procházka K, Tuzar Z, Brown W (2001) Hybrid polymeric micelles with hydrophobic cores and mixed polyelectrolyte/nonelectrolyte shells in aqueous media. 1. Preparation and basic characterization. Langmuir 17(14):4240–4244

Sun F, Zheng Z, Lan J, Li X, Li M, Song K, Wu X (2019) New micelle myricetin formulation for ocular delivery: improved stability, solubility, and ocular anti-inflammatory treatment. Drug Deliv. https://doi.org/10.1080/10717544.2019.1622608

Sun P, Xiao Y, Di Q, Ma W, Ma X, Wang Q, Chen W (2020) Transferrin receptor-targeted PEG-PLA polymeric micelles for chemotherapy against glioblastoma Multiforme. Int J Nanomedicine 15:6673–6687. https://doi.org/10.2147/IJN.S257459

Takahashi A, Yamamoto Y, Yasunaga M, Koga Y, Kuroda JI, Takigahira M, Harada M, Saito H, Hayashi T, Kato Y, Kinoshita T, Ohkohchi N, Hyodo I, Matsumura Y (2013) NC-6300, an epirubicin-incorporating micelle, extends the antitumor effect and reduces the cardiotoxicity of epirubicin. Cancer Sci. https://doi.org/10.1111/cas.12153

Torchilin VP (2006a) Micellar Nanocarriers: pharmaceutical perspectives. Pharm Res 24(1):1–16. https://doi.org/10.1007/s11095-006-9132-0

Torchilin VP (2006b) Multifunctional nanocarriers☆. Adv Drug Deliv Rev 58(14):1532–1555. https://doi.org/10.1016/j.addr.2006.09.009

Torchilin VP, Trubetskoy VS (1995) Which polymers can make nanoparticulate drug carriers long-circulating? Adv Drug Deliv Rev 16(2–3):141–155. https://doi.org/10.1016/0169-409X(95)00022-Y

Torchilin VP, Levchenko TS, Lukyanov AN, Khaw BA, Klibanov AL, Rammohan R, Samokhin GP, Whiteman KR (2001) p-Nitrophenylcarbonyl-PEG-PE-liposomes: fast and simple attachment of specific ligands, including monoclonal antibodies, to distal ends of PEG chains via p-nitrophenylcarbonyl groups. Biochimica et Biophysica Acta (BBA)-Biomembranes 1511(2):397–411

Valcourt DM, Harris J, Riley RS, Dang M, Wang J, Day ES (2018) Advances in targeted nanotherapeutics: from bioconjugation to biomimicry. Nano Res 11(10):4999–5016

Valle JW, Armstrong A, Newman C, Alakhov V, Pietrzynski G, Brewer J, Campbell S, Corrie P, Rowinsky EK, Ranson M (2011) A phase 2 study of SP1049C, doxorubicin in P-glycoprotein-targeting pluronics, in patients with advanced adenocarcinoma of the esophagus and gastro-esophageal junction. Investig New Drugs. https://doi.org/10.1007/s10637-010-9399-1

Wang YZ, Fang XL, Li YJ, Zhang ZW, Han LM, Sha XY (2008) Preparation, characterization of paclitaxel-loaded Pluronic P105 polymeric micelles and in vitro reversal of multidrug resistant tumor. Yao Xue Xue Bao= Acta Pharmaceutica Sinica 43(6):640–646

Wang Y, Ke X, Voo ZX, Yap SSL, Yang C, Gao S, Liu S, Venkataraman S, Obuobi SAO, Khara JS, Yang YY, Ee PLR (2016) Biodegradable functional polycarbonate micelles for controlled release of amphotericin B. Acta Biomater 46:211–220. https://doi.org/10.1016/j.actbio.2016.09.036

Wei Z, Hao J, Yuan S, Li Y, Juan W, Sha X, Fang X (2009) Paclitaxel-loaded Pluronic P123/F127 mixed polymeric micelles: formulation, optimization and in vitro characterization. Int J Pharm 376(1–2):176–185

Wu CY, Benet LZ (2005) Predicting drug disposition via application of BCS: transport/absorption/elimination interplay and development of a biopharmaceutics drug disposition classification system. Pharm Res. https://doi.org/10.1007/s11095-004-9004-4

Xu W, Wang H, Dong L, Zhang P, Mu Y, Cui X, Zhou J, Huo M, Yin T (2019) Hyaluronic acid-decorated redox-sensitive chitosan micelles for tumor-specific intracellular delivery of gambogic acid. Int J Nanomedicine 14:4649–4666. https://doi.org/10.2147/IJN.S201110

Yang C, Zhao H, Yuan H, Yu R, Lan M (2013) Preparation and characterization of thermosensitive and folate functionalized Pluronic micelles. J Nanosci Nanotechnol 13(10):6553–6559. https://doi.org/10.1166/jnn.2013.7520

Yang X, Li Z, Wang N, Li L, Song L, He T, Sun L, Wang Z, Wu Q, Luo N, Yi C, Gong C (2015) Curcumin-encapsulated polymeric micelles suppress the development of colon cancer in vitro and in vivo. Sci Rep. https://doi.org/10.1038/srep10322

Yi Y, Yoon HJ, Kim BO, Shim M, Kim SO, Hwang SJ, Seo MH (2007) A mixed polymeric micellar formulation of itraconazole: characteristics, toxicity and pharmacokinetics. J Control Release. https://doi.org/10.1016/j.jconrel.2006.10.001

Yokoyama M (2014) Polymeric micelles as drug carriers: their lights and shadows. J Drug Target 22(7):576–583

Yokoyama M, Fukushima S, Uehara R, Okamoto K, Kataoka K, Sakurai Y, Okano T (1998a) Characterization of physical entrapment and chemical conjugation of adriamycin in polymeric micelles and their design for in vivo delivery to a solid tumor. J Control Release 50(1–3):79–92

Yokoyama M, Satoh A, Sakurai Y, Okano T, Matsumura Y, Kakizoe T, Kataoka K (1998b) Incorporation of water-insoluble anticancer drug into polymeric micelles and control of their particle size. J Control Release 55(2–3):219–229

Yoncheva K, Calleja P, Agüeros M, Petrov P, Miladinova I, Tsvetanov C, Irache JM (2012) Stabilized micelles as delivery vehicles for paclitaxel. Int J Pharm 436(1–2):258–264. https://doi.org/10.1016/j.ijpharm.2012.06.030

Zhang Z, Grijpma DW, Feijen J (2006) Thermo-sensitive transition of monomethoxy poly (ethylene glycol)-block-poly (trimethylene carbonate) films to micellar-like nanoparticles. J Control Release 112(1):57–63

Zhang ZH, Abbad S, Pan RR, Waddad AY, Hou LL, Lv HX, Zhou JP (2013) N-octyl-N-arginine chitosan micelles as an oral delivery system of insulin. J Biomed Nanotechnol. https://doi.org/10.1166/jbn.2013.1572

Chapter 13
Applications of Dendrimers in Drug Delivery Systems

Raja Abhilash Punagoti, Mallikarjun Vasam, and Rita Mourya

13.1 Introduction

Dendrimers or macromolecules with a multi-branched tree-like structure are a relatively unique class of polymers, which attract researchers to explore their properties and architecture. Several drugs had limitations like poor aqueous solubility and undesirable half-life. Therefore, it is necessary to design a delivery system, which can deliver a drug efficiently. For an instance, poor water solubility in formulating a liquid dosage form has a great impact in formulation R&D. The importance of this issue is underlined by the fact that 10–30% of commercial drugs and 60–70% of drugs under preclinical stage have solubility problems, which in turn possess bioavailability constrains (Kovács et al. 2009). Conventional drug delivery systems are useful to some extent in overcoming these shortcomings, but they have failed to prove their effectiveness in delivering many drugs. One can surpass these issues by carefully selecting the excipients or carriers or the pharmaceutical formulation, improving the dissolution profile of the drug (Santos et al. 2020). Owing to their uniform nanosize (1–100 nm), water solubility, modifiable surface functionality, and available internal cavities made dendrimers as suitable drug carrier.

R. A. Punagoti (✉)
Department of Pharmaceutical Chemistry, College of Medicine and Health Sciences, University of Gondar, Gondar, Ethiopia
e-mail: Raja.Abhilash@uog.edu.et

M. Vasam
Chaitanya (Deemed to be) University- Pharmacy, Warangal, Telangana, India

R. Mourya
Department of Pharmaceutical Chemistry, School of Pharmacy, SAM Global University, Bhopal, India

J.-C. Kim et al. (eds.), *Smart Nanomaterials in Biomedical Applications*, Nanotechnology in the Life Sciences,
https://doi.org/10.1007/978-3-030-84262-8_13

Dendrimers are new class of carriers which have been recognized as promising nanostructures, and their applicability as drug carrier has been explored, due to their distinctive physicochemical and structural properties (Roy et al. 2015). The term "dendrimer" arises from two Greek word: "dendron" meaning tree and "meros" meaning part, which was coined by Vögtle and his colleagues in the late 1970s, and ever since numerous studies have been reported. In the past three decades, its presence as pharmaceutical excipient has been dramatically increased, due to its unique physicochemical properties and distinct and divergent applications in pharmaceutical development.

Moreover, these cascade molecules consist of internal cavities that enable the encapsulation of a wide range of drugs, genes, and specific targeted agents and or otherwise onto the surface through branching or covalent bonding. Additionally, dendrimers can be used to minimize the drug toxicity and for the improvement of drug efficacy. All these diversified and unique physicochemical and biological properties offer a wide range of potential applications using dendrimers as nanomedicine agents. Dendrimer consists of three basic architectural components. They are (1) a core unit or focal point from which all the branching originates, (2) branching dendrons that incorporate a branching point, and (3) surface ligands or terminal groups that form the chain ends. Figure 13.1 illustrates the basic structure of dendrimers.

The number of branches originating from the center can be regarded as subsequent "generations." The morphology of the dendrimers is going to affect solubility, degradability, and biological activity. One of the fascinating attributes of dendrimers is that as the number of generation is proportional to the dendrimer size and is responsible for the rigidity in terminal groups, which in turn not only regulates release rates from the dendrimer interior but also allows for multiple ligand attachment sites and increases the probability of an affinity interaction. Further, the terminal groups are vital because they can be hydrophobic or hydrophilic, anionic or cationic, to accommodate a variety of molecules (Lee et al. 2005, 2006). By regulating dendrimer synthesis, it is possible to precisely manipulate both their molecular

Fig. 13.1 Basic dendrimer components

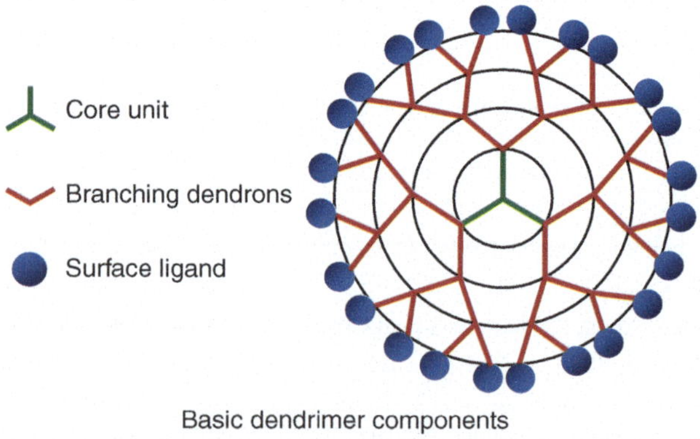

weight and chemical composition, thereby allowing predictable tuning of their biocompatibility and pharmacokinetics, thereby offering a plethora of biomedical applications.

13.2 Structure and physicochemical Properties of Dendrimers

It is essential to study the physicochemical properties of dendrimers, for better understanding of their solubility capability. They comprised of layers between each cascade point popularly known as "generations." The morphology of dendrimers consists of inner core, circularly attached branches (generations) not only equipped with various functional groups at external surface, that can be easily modified but also offers encapsulation of guest drugs (Tomalia et al. 1985; Dufès et al. 2005a, b). Applications of dendrimers typically involve conjugating other chemical groups onto their surface that can function as detecting agents, affinity ligands, targeting components, radioligands, imaging agents, or pharmaceutically active compounds (Otto and de Villiers 2018). As targeted drug delivery systems, dendrimers could change the current scenario in cancer research, diagnosis, and therapeutics (Kyriazis and Papaioannidou 2010). Figures 13.2 and 13.3 illustrates schematic representation of dendrimer components and dendrimers as multifunctional nanoplatforms.

Due to their proven ability as prodrugs or drug carriers, they can be produced either by encapsulating drugs in the core or coupled via electrostatic or covalent bonds to their surface. There are many chemical classes of dendrimers; among all, poly(propylene imine) (PPI) and poly(propylene amine) (POPAM) dendrimers, poly(amidoamine) (PAMAM) dendrimers, and pegylated dendrimers (Papageorgiou and Papaioannidou 2010) are popular. Each type has their significant uses and applications.

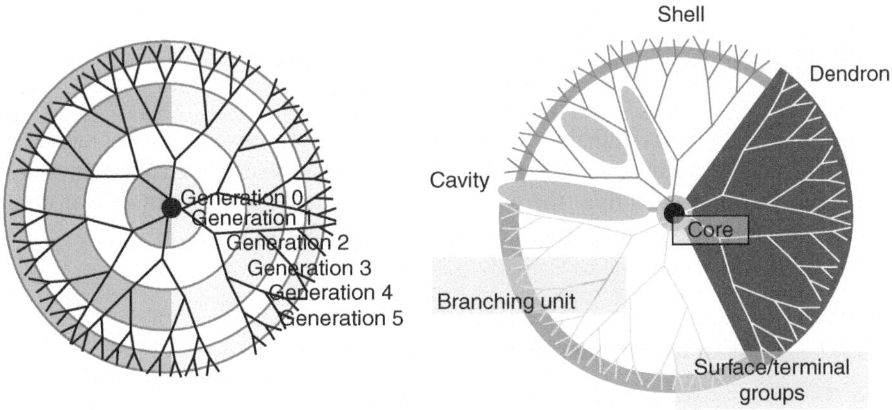

Fig. 13.2 Schematic representation of dendrimer components. (Adopted from Dufès et al. 2005a, b)

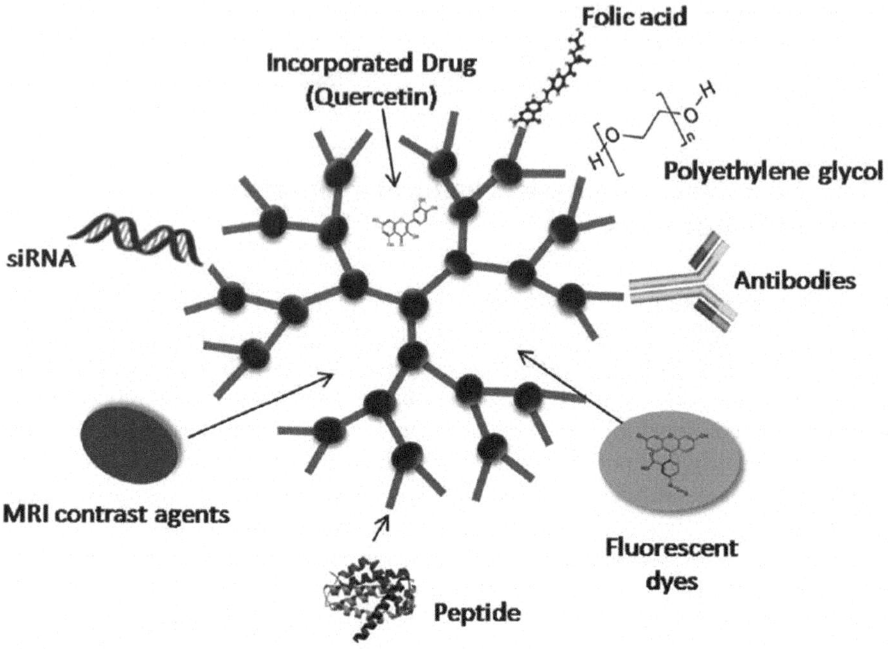

Fig. 13.3 Dendrimers as multifunctional nanoplatforms

13.2.1 Types of Dendrimers

Based on their chemical composition, fabrication strategy, and characteristics, dendrimers can be of different types, furnished in Table 13.1.

13.3 Fabrication Methods

Theoretically, synthesis of dendrimers is easy and straightforward, but in practice it is tedious and time consuming and should be prepared under controlled environment. For our understanding, there are two approaches for the synthesis of dendrimers, namely, classical approach and accelerated approach.

13.3.1 Classical Approach

Based on the target end applications, two synthetic methods, divergent and convergent methods, are popular. Each method has its own merits and demerits.

Table 13.1 Classification of dendrimers based on composition

S. No.	Type	Composition	Applications	References
1	**PAMAM**	Layered architecture of spheroidal or ellipsoidal in shape, obtained from ethylene diamine or ammonia. Usually prepared by the divergent method	Suitable delivery vehicles of various drugs including antiviral, antitubercular, anticancer, etc.	Yiyun et al. (2007) and Vinicius et al. (2018)
2	**PAMAMOS**	Hydrophilic, nucleophilic polyamidoamine (PAMAM) interiors and hydrophobic organosilicon (OS) exteriors	Useful precursors for the preparation of honeycomb-like networks with nanoscopic PAMAM and OS domains	Abbasi et al. (2014)
3	**PPI**	Core structure contains diamino butane (DAB) along with primary amines as terminal groups and tertiary propylene amines as inner core	Pharmaceutical excipient to enhance drug solubility, diagnostic agents, fabrication of electrochemical biosensors and sensors	Chaudhary et al. (2016).
4	**Tecto**	Blend of two or more dendrimers; each performs a specific function; one act as diagnostic agent and other perform drug delivery	Multidrug delivery	Boas et al. (2006)
5	**Frechet**	Poly(benzyl ether) as poly branched interior with carboxylic group as surface functional groups	Improves drug solubility of poorly soluble drugs both in aqueous and organic solvents	Sherje et al. (2018)
6	**Amphiphilic**	Comprises of two different sites for electron transfer reactions	Drug encapsulation, micellar drug delivery systems, and preparing nanovesicles	Park et al. (2016).
7	**Hybrid**	Fused skeletons of dendritic and linear polymers	Pharmaceutical excipient and in targeted drug delivery	Sherje et al. (2018).
8	**Chiral**	Chiral core consists of enantioselective branching	Chiral hosts for enantiomeric resolutions and as chiral catalysts for asymmetric synthesis	Quintana et al. (2017)
9	**Peptide**	On polylysine skeleton, amino acids are of branching or core	Biological applications, such as in vaccine, gene delivery, and diagnostic research	Luo et al. (2012)

13.3.1.1 Divergent Method

This is proposed by Tomalia, in which the dendrimer grows from core to periphery in a sequential step-by-step process (Tomalia et al. 1985). The core molecule reacts with the monomer molecule having two dormant and one reactive group (AB_n

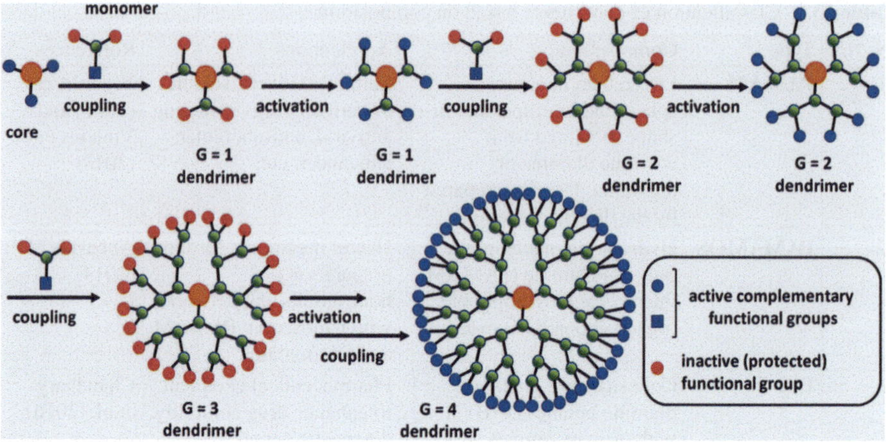

Fig. 13.4 Synthesis of dendrimers according to the divergent method. (Adopted from Sowińska and Lipkowska 2014)

monomer). Sequentially, there is a coupling of the monomers to a core facilitated by chemical bond formation between the A-functional group of the monomer, and one of the previously activated B-functional groups of the substrate. These two steps can be repeated multiple times to get various generations (tiers) of dendrimers with each generation having arms from previous generations (Boas et al. 2006). Figure 13.4 depicts the divergent approach.

Divergent method is advantageous in terms of simplicity, high yield attributed to the commercial scale preparations, but it also suffers from the demerits such as accumulation of impurities and "de Gennes dense packing" (De Gennes and Hervet 1983). In the synthesis of higher generation dendrimers by this method, results in dendrimer products with defects such as missing arms and dimers/trimers. Due to these factors, asymmetric and imperfect dendrimers are synthesized. However, by controlling various factors, one can fabricate large volume dendrimers. Most popular dendrimers such as PAMAM and PPI dendrimers are known to be prepared by this method.

13.3.1.2 Convergent Method

Hawker and FrÈchet (1990) developed this strategy, where dendrimer grows starting from end groups and progresses inward. The synthesis of dendrons is conducted by using the conventional AB_2 monomer, which has reactive B-functionalities and the deactivated/protected A-functionality. Small molecules come together and reaction proceeds inward, eventually the dendrons become attached to the core.

The shortcomings with the divergent method can be overcome by this method, as at first, the dendrons are prepared, purified, and then coupled with core to complete the dendrimeric architecture (Bosman et al. 1999). However, more synthetic steps, low yield, and steric hindrance are the demerits of this approach. In order to

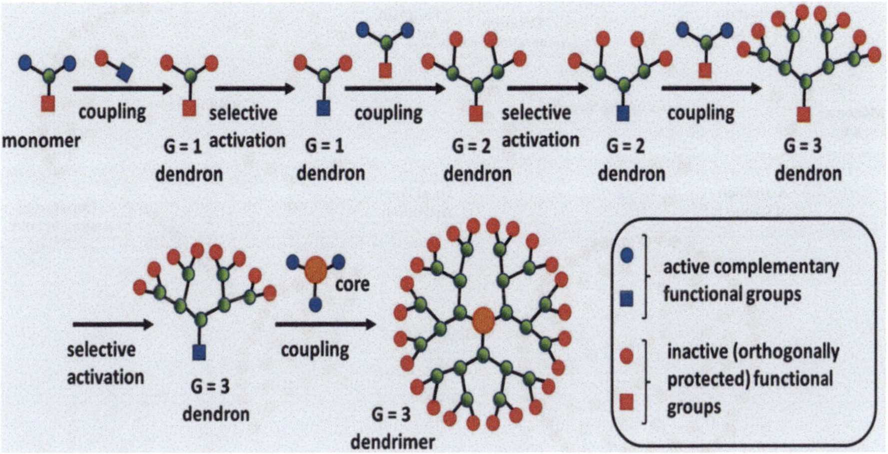

Fig. 13.5 Synthesis of dendrimers according to the convergent method. (Adopted from Sowińska and Lipkowska 2014)

overcome these shortcomings, accelerated approaches are gaining importance. Synthesis of dendrimers according to the convergent method is illustrated in Fig. 13.5.

13.3.2 Accelerated Approach

13.3.2.1 Double Exponential and Mixed Growth

In this method, two monomers from both convergent and divergent growth are reacted together to give an orthogonally protected trimer and can be synthesized further to obtain higher generation dendrimer molecule. The benefit of this method is less reaction steps to fabricate suitable polymer (Kesharwani et al. 2016; Majoral and Caminade 1999; Mathur et al. 2015). Synthesis of dendrimers by the double exponential growth technique is shown in Fig. 13.6.

13.3.2.2 Hypercores and Branched Monomers Growth

This approach deals with the prior assembled oligomeric species which can be linked together to give dendrimers in fewer steps or higher yields in a circular, with more branches. In branched monomers growth, core is allowed to with two or more moles of solvent containing at least two protecting branching sites, followed by removal of the protecting groups. The resultant free activated sites lead to the first generation dendrimers (Parata and Felder-Flescha 2016). Synthesis of dendrimers by the hypercore approach is diagrammatically shown in Fig. 13.7.

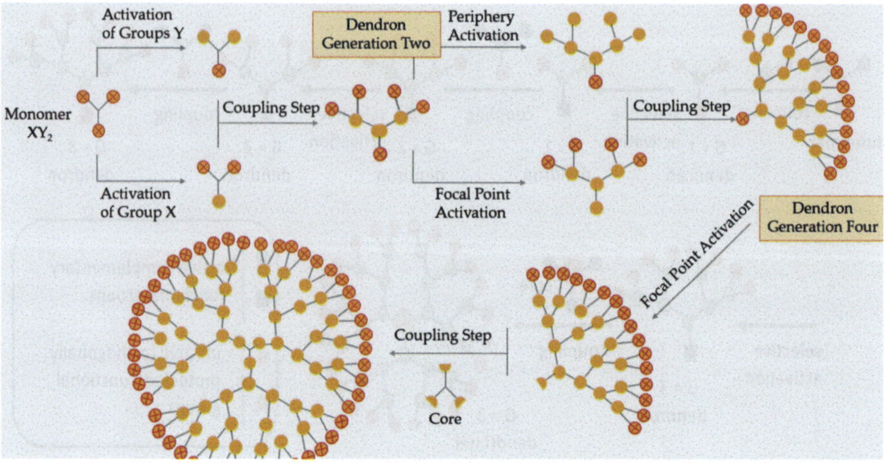

Fig. 13.6 Synthesis of dendrimers by the double exponential growth technique. (Adopted from Santos et al. 2020)

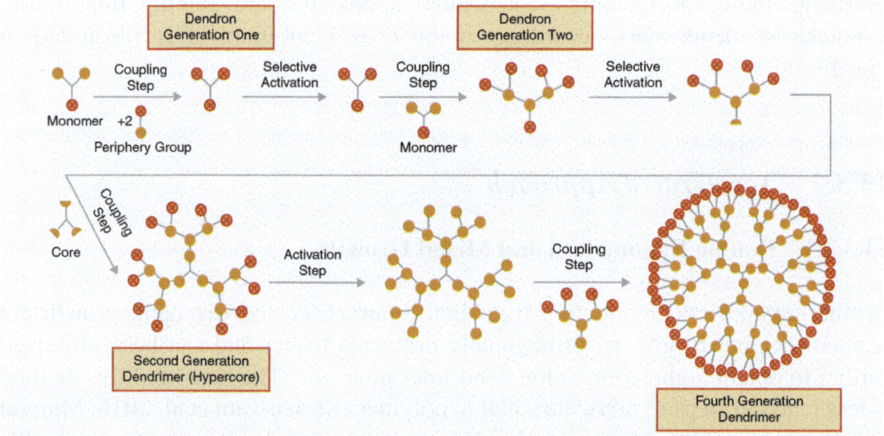

Fig. 13.7 Synthesis of dendrimers by the hypercore approach. (Adopted from Santos et al. 2020)

13.4 Applications

The advent of dendrimers opened a plethora of opportunities for novel drug delivery and pharmaceutical applications. Due to the versatility and affordability of dendrimers, their use is dramatically increased in the recent past. These drug delivery scaffolds have been already evaluated in various routes of drug administration systems and yielded promising results. In this chapter, we review vast and vital applications of dendrimers as a typical drug-delivery agent in various routes of administration (Cheng et al. 2008). Owing to its structural flexibility and surface modification, dendrimers have found to be efficient in various routes of drug delivery such as parental, oral, transdermal, and ocular drug delivery systems. Among various classes of dendrimers, PAMAM and PPI were evaluated as promising alternatives in drug delivery.

13.4.1 Parenteral Drug Delivery

It is obvious that the parenteral route of administration is most popular because of its rapid onset of drug action. However, most of the drugs, for instance, anti-neoplastic agents, that possess poor or weak solubility raise concern in the development of formulations (Gebbia and Puozzo 2005). Moreover, traditional parenteral dosage forms have biocompatibility and biotoxicity issues such as hemolysis and phlebitis that need to be addressed (Markman 2003). Thanks to many researchers who developed dendrimers that can be conveniently administered through intravenous, intratumoral, and intraperitoneal routes by overcoming aforementioned shortcomings. Research findings indicated that intravenous and intratumoral administration of anticancer drugs such cisplatin proved to be reliable with minimal side effects.

Malik et al. (1999) proposed a novel approach for cancer chemotherapy by fabricating G3.5PAMAM dendrimer conjugate of cisplatin for IV route, which yielded equal action of cisplatin alone but 3- to 15-fold less toxic. Bhadra et al. (2003) and Zhuo et al. (1999), individually demonstrated that 5-fluorouracil encapsulated in PEGmodified G4PAMAM dendrimer via IV route has shown high drug loading capability, less hemolytic activity with prolonged drug delivery. Chauhan et al. (2004) improved the solubility of indomethacin, an anti-inflammatory drug, by developing G4PAMAM dendrimer by electrostatic interaction, and when administered intravenously to albino rats, it indicated enhanced drug concentration at the inflamed site and 2.29 times superior targeting efficiency than free drug. An anti-inflammatory drug flurbiprofen as G4PAMAM dendrimer possessed five times greater distribution and produced controlled, prolonged, and site-specific activity (Asthana et al. 2005). In similar fashion, folic acid-coated PAMAM conjugate of methotrexate in IV route not only enhanced its antitumor activity but also decreased its toxicity (Kukowska-Latallo et al. 2005).

Apart from the above mentioned therapeutic applications, many researches assured dendrimers as potential and safe alternative to traditional dosage form. Bhadra et al. (2005) used an antimalarial drug, primaquine phosphate, to prevent relapse of malaria, as a guest molecule encapsulated in galactose-coated PPI, when administered intravenously, demonstrated safe, prolonged release rate (5–6 days) with remarkably negligible adverse reactions. Intravenous and intratumoral injection of a fourth-generation starburst dendrimer that was boronated and linked to epidermal growth factor, assessed for its ability in Boron neutron capture therapy (used to treat many gliomas), produced satisfactory results (Yang et al. 2002). Similarly, various other authors (Barth et al. 1994, 2004; Wu et al. 2004; Yang et al. 1997) have evaluated efficiency of boronated dendrimers as potential option as boron neutron captures therapy for treating brain tumors.

Additionally, dendrimers were promising carrier (vector) in gene transfection to treat a variety of genetic disorders. Mainly cationic dendrimers were preferred as nonviral carriers to establish desired functions. Both DNA and RNA can be conveniently condensed and thereby prevent enzymatic degradation. Both PPI and PAMAM dendrimers have high-density tertiary amine groups as their terminals, responsible for "proton effect," which in turn favors in endosomal escape (Vruti et al. 2020).

In the recent past, plenty of research has been conducted on dendrimers as gene delivery agents. G9 PAMAM dendrimer-DNA complex was investigated for pulmonary gene transfection, (Kukowska-Latallo et al. 2004) and was successful in gene transfer to lungs. With the repeated IV administration of G9 PAMAM dendrimer-DNA complex, it was observed to have prolonged transgene expression. In another study carried out by Kihara et al. in 2003, PAMAM-cyclodextrin conjugate that hosts DNA with varying degrees of substitution (D.S) were fabricated and inferred that the dendrimer DNA complex with D.S 2.4 has demonstrated as suitable non-vector in both in vivo and in vitro. Wada et al. (2005) evaluated mannose-PAMAM-cyclodextrin conjugated with DNA by electrostatic interaction in in vivo and in vitro which resulted in successful kidney gene transfection. Their study showed the complex with D.S 3.3 as a promising non-vector. In another interesting research by Mamede et al. (2004), In-111-labeled-oligo with G4 dendrimers established optimum uptake by the kidney and the liver. It was observed that PPI dendrimers were popular liver targeted gene expression (Dufès et al. 2005a, b; Schatzlein et al. 2005).

13.4.2 Oral Drug Delivery

Oral route of administration is the most preferred route of administration among other routes due to its specific advantages, such as noninvasive administration, patient compliance, and economic and high stability. After the oral administration, the fate of the drug is governed by various factors, for instance, due to the high surface area (300–400 m^2) offered by microvilli present on the intestinal epithelium, responsible for absorption of the drug and then distributed in systemic circulation (Cheng et al. 2008). These lymphoid areas, Peyer's patches, are associated with M cells and are vital for drug delivery as it offers high transcytotic capacity. Despite its merits, it also suffers from certain demerits which include poor solubility, first pass metabolism, instability in gastric environment, drug penetration via P-glycoprotein (P-gp) mucus membrane, and difficulty in formulation. Numerous researches have been carried out to address this problem by adopting crystallization, micellar formation, solid dispersions and nanoparticles, liposomes and dendrimers, etc. Particularly, the novel class of delivery systems, such as dendrimers, gained attention in the recent past and was acknowledged by various researchers as a promising one as drug is carried in oral drug delivery (Luciana et al. 2018).

To cite some of the works, D'Emanuele et al. (2004) investigated the oral bioavailability of dendrimer-propranolol prodrug. In their study, they fabricated G3 PAMAM loaded with the drug and concluded that this conjugate allows the bypassing of the efflux transporter and thereby increasing the bioavailability. In similar fashion, Tripathi et al. (2002) improved the bioavailability of 5-fluorouracil in the form of dendrimer grafts than conventional dosage form. They also observed absorption through the lymphatic route. Additionally, according to Man et al. (2006), the use of G5 PAMAM dendrimers as drug carrier for ketoprofen (anti-inflammatory drug) proved to increase oral bioavailability and possess sustained release behavior. These results indicated the dendrimers used as novel orally administrated formulations.

13.4.3 Ocular Drug Delivery

Due to the complexity of anatomy and physiology of the eye, ocular delivery approach faces many difficulties. There are two main constrains that affect bioavailability, namely, static and dynamic barriers. While static barrier raises from various segments of the eye such as cornea, sclera, retina, and blood-retinal barriers and dynamic barrier from choroidal and conjunctival blood flow, lymphatic clearance, and tear dilution, has to be considered during formulation (Sikandar et al. 2011). The biggest challenge associated with ocular drug delivery system is rapid and extensive elimination of conventional eye drops from the eye which favors extensive loss of drug; as a result, only a fraction of the drug penetrates the corneal layer and reached internal tissue of the eye. Moreover, conventional formulations have low bioavailability and limited residence time at the site.

Many formulations have been developed to overcome the abovementioned drawbacks, especially novel delivery systems such as niosomes, liposomes, dendrimers, etc., with their unique characteristics (Shimpi et al. 2005). Dendrimers being nanosize possess distinct advantages as ophthalmic drug carriers, such as decreased dosing frequency, prolonged residence time, and enhanced bioavailability. However, blurred vision is the associated demerit of these formulations, due to the diminished secretion of lachrymal fluid (Loftsson and Stefansson 1997).

Vandamme et al. in 2005 reported the use of PAMAM dendrimers (generations 1.5 and 2–3.5 and 4) loaded with pilocarpine nitrate and tropicamide to have shown prolonged miotic activity and mydriatic activity, respectively. Dendrimers containing –COOH and –OH groups at their terminal ends exhibited longer residence time and is dependent on molecular weight and size. Another interesting application of PPI dendrimers in ophthalmic therapy was studied by Duan and Sheardown (2006) and Duan et al. (2007), where the scientists used G2 PPI, and cell adhesion peptide-modified PPI dendrimers consists of collagen was evaluated for their efficiency as a corneal tissue-engineering scaffold, yielded fruitful results, and therefore assured the efficiency of artificial collagen. All these research findings referred dendrimers as promising and reliable agent in ocular drug delivery.

13.4.4 Transdermal Drug Delivery (TDDD)

Transdermal delivery aimed for the treatment of localized tissues below the skin or on surface of the skin in a controlled way. Due to their versatility, TDDDS is gaining attention in the recent past, which offers benefits for the end users by ensuring patient compliance, pain-free administration, and minimal or almost no side effects. Regardless of its merits, TDDDS is challenging and complex, as skin acts as potential barrier and thereby influence dermato-pharmacokinetics. Very few classes of drugs exhibit the ability to permeate across the stratum corneum in optimum concentration to elicit pharmacological activity (Dave and Venkata 2017). In order to facilitate drug transdermal absorption, various strategies have been investigated, reported, and patented. Advancements in skin permeation-enhancement technologies have led to renewed interest in transdermal drug delivery which include but not limited to iontophoresis, electroporation, ultrasound, micro-needles to open up the skin, and more recently the use of transdermal nanocarriers. These various classes of TDDD systems are now commercially available in the market and are found to be useful in the treatment of various ailments and diseases. The cascading structure of dendrimers with uniform shape, size, and molecular weights increases the permeability through skin. Moreover, dendrimers also act like solubility enhancers, favor the penetration of lipophilic drugs, and, to be honest, limit the application for administration of hydrophilic drugs.

Reliable literatures have been advocated that various drugs can be conveniently administered topically in the form of dendrimers. Tamsulosin hydrochloride upon simple encapsulation to fabricate G3 PAMAM has shown enhanced transdermal delivery efficacy. The scientists concluded that when the dendrimer-drug complex was applied on snake skin, it fulfilled the penetration amount of tamsulosin hydrochloride required to conduct clinical trial (Wang et al. 2003a, b). The same group of authors used G3 PAMAM dendrimer with $-NH_3$ groups and G2.5 PAMAM dendrimer with –CooH groups to demonstrate the drug permeation enhancement effect (Wang et al. 2003a, b).

A class of anti-inflammatory drugs such as indomethacin in three different types of formulations with varying concentrations was evaluated by Chauhan et al. (2003) and concluded that when applied to the shaved abdominal skin of Wistar rats, the G4-NH2 dendrimer at 0.2% w/v concentration has shown high permeation enhancement effect than that of pure drug suspension. Furthermore, ketoprofen and diflunisal, as PAMAM dendrimers applied on excised rat skins, yielded promising results by reflecting good permeation ability and greater bioavailability when compared to the pure drug suspension without dendrimers (Cheng et al. 2006). These results assured that PAMAM dendrimers can effectively favors skin permeability of NSAIDs and continue to generate interest among researchers.

13.4.5 Dendrimers in Market and Under Clinical Studies

Almost 40 dendrimer-based entities are currently in clinical and preclinical stages of development, and dozens are commercially available as different treatment regimen. PAMAM, PPI, PLL, Priostar®, Astramol®, etc. are popular and known for industrial applications. Starpharma being pioneer in the domain of dendrimers released its first commercial dendrimer-based product, VivaGel®, for treatment and prevention of bacterial vaginosis and sexually transmitted diseases (Roots Analysis Private Ltd 2016). This is marketed under the brand names Betafem® BV Gel (UK), Betadine BV™ (Europe), Betadine™ BV Gel (Asia), and Fleurstat BV gel (Australia and New Zealand). Moreover, other products, SuperFect®, Stratus® CS, PrioFect™, Alert Ticket™, are viable in the market (Starpharma n.d.). Several other dendrimers currently in late-stage clinical studies include NKTR-181, NKTR-102 (Nektar Therapeutics), CRLX-101 (Cerulean Pharma), and Opaxio™ (CTI BioPharma). The success of these molecules in clinical trial studies will determine the fate of the dendrimers and can be anticipated as block buster for the pharmaceutical industry.

13.5 Conclusions

Dendrimers with diversified properties, such as high degree of branching, multivalency, size uniformity, and surface functionality flexibility, were rendered to be a promising tool in drug delivery systems. These cascade molecules can be fabricated by two approaches, namely, divergent growth method and convergent growth method. Recently, cutting edge chemical technologies such as double exponential growth method, click chemistry, and Lego chemistry yielded perfect dendrimers. The association of biodegradable properties with dendrimer can remarkably uptrend its applicability.

Moreover, these cascade molecules consist of internal cavities enables the encapsulation of a wide range of drugs, genes, and specific targeted agents and/or otherwise onto the surface through branching or covalent bonding. Additionally, dendrimers can be used for minimizing the drug toxicity and for the improvement of the drug efficacy. Many studies already did show evidence that dendrimers are ideal drug carrier in various drug delivery systems. However, more in vivo studies need to be performed to establish their biocompatibility, efficacy, and long-term toxicity.

The evolving era of dendrimers assures remarkable promise to the pharmaceuticals.Among the diversified applications, drug delivery holds a major opportunity, especially with PAMAM and PPI dendrimers. The dendrimer market is yet to unveil its potential and definitely will revolutionize the biomedical industry. Dendrimers are being exploited for numerous therapeutic areas. With success of the already marketed drugs, the next generations of dendrimers are poised to witness an accelerated growth as pharmaceuticals, diagnostic agents, and excipients.

References

Abbasi E, Aval SF, Akbarzadeh A, Milani M, Nasrabadi HT (2014) Dendrimers: synthesis, applications, and properties. Nanoscale Res Lett 9:1–10
Asthana A, Chauhan AS, Diwan PV, Jain NK (2005) Poly(amidoamine) (PAMAM) dendritic nanostructures for controlled site-specific delivery of acidic anti-inflammatory active ingredient. AAPS PharmSciTech 6:E536–E542
Barth RF, Adams DM, Soloway AH, Alam F, Darby MV (1994) Boronated starburst dendrimer monoclonal-antibody immunoconjugates – evaluation as a potential delivery system for neutron-capture therapy. Bioconjug Chem 5:58–66
Barth RF, Wu G, Yang WL, Binns PJ, Riley KJ, Patel H, Coderre JA, Tjarks W, Bandyopadhyaya AK, Thirumamagal BTS, Ciesielski MJ, Fenstermaker RA (2004) Neutron capture therapy of epidermal growth factor (plus) gliomas using boronated cetuximab (IMC-C225) as a delivery agent. Appl Radiat Isot 61:899–903
Bhadra D, Bhadra S, Jain S, Jain N (2003) A PEGylated dendritic nanoparticulate carrier of fluorouracil. Int J Pharm 257(1):111–124
Bhadra D, Yadav AK, Bhadra S, Jain NK (2005) Glycodendrimeric nanoparticulate carriers of primaquine phosphate for liver targeting. Int J Pharm 295:221–233
Boas U, Christensen JB, Heegaard PMH (2006) Dendrimers: design, synthesis and chemical properties. In: Dendrimers in medicine and biotechnology. New molecular tools. The Royal Society of Chemistry, Cambridge, pp 1–27
Bosman AW, Janssen HM, Meijer EW (1999) About dendrimers: structure, physical properties, and applications. Chem Rev 99(7):1665–1688
Chaudhary S, Gothwal A, Khan I, Srivastava S, Malik R, Gupta U (2016) Polypropyleneimine and polyamidoamine dendrimer mediated enhanced solubilization of bortezomib: comparison and evaluation of mechanistic aspects by thermodynamics and molecular simulations. Mater Sci Eng 72:611–619
Chauhan AS, Sridevi S, Chalasani KB, Jain AK, Jain SK, Jain NK, Diwan PV (2003) Dendrimer mediated transdermal delivery: enhanced bioavailability of indomethacin. J Control Release 90:335–343
Chauhan AS, Jain NK, Diwan PV, Khopade AJ (2004) Solubility enhancement of indomethacin with poly(amidoamine) dendrimers and targeting to inflammatory regions of arthritic rats. J Drug Target 12:575–583
Cheng YY, Man N, Xu TW (2006) Transdermal delivery of nonsteroidal anti-inflammatory drugs mediated by polyamidoamine (PAMAM) dendrimers. J Pharm Sci 96:595–602
Cheng Y, Xu Z, Ma M, Xu T (2008) Dendrimers as drug carriers: applications in different routes of drug administration. J Pharm Sci 97(1):123–143
D'Emanuele A, Jevprasesphant R, Penny J, Attwood D (2004) The use of a dendrimer-propranolol prodrug to bypass efflux transporters and enhance oral bioavailability. J Control Release 95:447–453
Dave K, Venkata VV (2017) Dendritic polymers for dermal drug delivery. Ther Deliv 8:1077–1096
De Gennes PG, Hervet H (1983) Statistics of "starburst" polymers. J Phys Lett 44:351–360
Duan X, Sheardown H (2006) Dendrimer crosslinked collagen as a corneal tissue engineering scaffold: mechanical properties and corneal epithelial cell interactions. Biomaterials 27:4608–4617
Duan X, McLaughlin C, Griffith M, Sheardown H (2007) Biofunctionalization of collagen for improved biological response: scaffolds for corneal tissue engineering. Biomaterials 28:78–88
Dufès C, Keith WN, Bilsland A, Proutski I, Uchegbu IF, Schatzlein AG (2005a) Synthetic anticancer gene medicine exploits intrinsic antitumor activity of cationic vector to cure established tumors. Cancer Res 65:8079–8084
Dufès C, Uchegbu IF, Schätzlein AG (2005b) Dendrimers in gene delivery. Adv Drug Deliv Rev 57:2177–2202
Gebbia V, Puozzo C (2005) Oral versus intravenous vinorelbine: clinical safety profile. Expert Opin Drug Saf 4:915–928
Hawker CJ, FrÈchet JMJ (1990) Preparation of polymers with controlled molecular architecture. A new convergent approach to dendritic macromolecules. J Am Chem Soc 112:7638–7647

Kesharwani S, Jaiswal PK, Kesharwani R, Kumar V, Patel DK (2016) Dendrimer: a novel approach for drug delivery. J Pharm Sci Innov 5:54–62

Kihara F, Arima H, Tsutsumi T, Hirayama F, Uekama K (2003) In vitro and in vivo gene transfer by an optimized alpha-cyclodextrin conjugate with polyamidoamine dendrimer. Bioconjug Chem 14:342–350

Kovács K, Orosz T, Stampf G, Antal I, Klebovich I, Ludányi K (2009) Difficulties encountered during formulation of a parenteral dosage form containing a poorly soluble drug. Acta Pharm Hung 79(1):35–44

Kukowska-Latallo JF, Chen C, Raczka E, Qunintana A, Rymaszewski M, Baker JR (2004) Intravascular and endobronchial DNA delivery to murine lung tissue using a novel, nonviral vector. Hum Gene Ther 11:1385–1395

Kukowska-Latallo JF, Candido KA, Cao ZY, Nigavekar SS, Majoros IJ, Thomas TP, Balogh LP, Khan MK, Baker JR (2005) Nanoparticle targeting of anticancer drug improves therapeutic response in animal model of human epithelial cancer. Cancer Res 65:5317–5324

Kyriazis G, Papaioannidou P (2010) Dendrimers as drug delivery systems. Front Pharmacol. Conference Abstract: 8th Southeast European Congress on Xenobiotic Metabolism and Toxicity – XEMET 2010. https://doi.org/10.3389/conf.fphar.2010.60.00155

Lee CC, MacKay JA, Fréchet JM, Szoka FC (2005) Designing dendrimers for biological applications. Nat Biotechnol 23(12):1517–1526

Lee CC, Gillies ER, Fox ME, Guillaudeu SJ, Fréchet JM, Dy EE, Szoka FC (2006) A single dose of doxorubicin-functionalized bow-tie dendrimer cures mice bearing C-26 colon carcinomas. Proc Natl Acad Sci 103(45):16649–16654

Loftsson T, Stefansson E (1997) Effect of cyclodextrins on topical drug delivery to the eye. Drug Dev Ind Pharm 23:473–481

Luciana AN, Ricardo GA, Daniele MLO, Lucas RA, Luciana MH, Classius FS, Eliana BS, Patrícia S (2018) Applications of nanocomposite materials in the delivery of anticancer drugs. In: Inamuddin AM, Ali M (eds) Applications of nanocomposite materials in drug delivery. Woodhead Publishing, Cambridge, MA, pp 339–352

Luo K, Li C, Li L, She W, Wang G, Gu Z (2012) Arginine functionalized peptide dendrimers as potential gene delivery vehicles. Biomaterials 33(19):4917–4927

Majoral JP, Caminade AM (1999) Dendrimers containing heteroatoms (si, p, B, ge, or bi). Chem Rev 99:845–880

Malik N, Evagorou EG, Duncan R (1999) Dendrimer- platinate: a novel approach to cancer chemotherapy. Anti-Cancer Drugs 10:767–776

Mamede M, Saga T, Ishimori T, Higashi T, Sato N, Kobayashi H, Brechbiel MW, Konishi J (2004) Hepatocyte targeting of 111In-labeled oligo-DNA with avidin or avidin-dendrimer complex. J Control Release 95:133–141

Man N, Cheng YY, Xu TW, Ding Y, Wang XM, Li ZW, Chen ZC, Huang GY, Shi YY, Wen LP (2006) Dendrimers as potential drug carriers. Part II. Prolonged delivery of ketoprofen by in vitro and in vivo studies. Eur J Med Chem 41:670–674

Markman M (2003) Intraperitoneal antineoplastic drug delivery: rationale and results. Lancet Oncol 4:277–283

Mathur V, Satrawala Y, Rajput MS (2015) Dendrimers: a review. Inventi Impact NDDS 1

Otto DP, de Villiers MM (2018) Poly (amidoamine) dendrimers as a pharmaceutical excipient. Are we there yet? J Pharm Sci 107:75–83

Papageorgiou A, Papaioannidou P (2010) Dendrimers: chemical structure and properties. Front Pharmacol. Conference Abstract: 8th Southeast European Congress on Xenobiotic Metabolism and Toxicity – XEMET 2010. https://doi.org/10.3389/conf.fphar.2010.60.00175

Parata A, Felder-Flescha D (2016) General introduction on dendrimers, classical versus accelerated syntheses and characterizations. In: Dendrimers in nanomedicine. Jenny Stanford Publishing, New York, pp 1–22

Park EJ, Kim J, Kim K, Park J (2016) Glucosamine-conjugated anionic poly (amidoamine) dendrimers inhibit Interleukin-8 production by helicobacter pylori in gastric epithelial cells. Bull Kor Chem Soc 37:596–599

Quintana S, Garcı'a MA, Marina ML, Go'mez R, de la Mata FJ, Ortega PJ (2017) Synthesis of chiral carbosilane dendrimers with l-cysteine and N-acetyl-l-cysteine on their surface and their application as chiral selectors for enantiomer separation by capillary electrophoresis. Tetrahedron Asymmetry 28(12):1797–1802

Roots Analysis Private Ltd (2016) Dendrimers and polymer drug conjugates market, 2016–2026

Roy U, Rodríguez J, Barber P, Das Neves J, Sarmento B, Nair M (2015) The potential of HIV-1 nanotherapeutics: from in vitro studies to clinical trials. Nanomedicine 10:3597–3609

Santos A, Veiga F, Figueiras A (2020) Dendrimers as pharmaceutical excipients: synthesis, properties, toxicity and biomedical applications. Materials 13:65

Schatzlein AG, Zinselmeyer BH, Elouzi A, Dufes C, Chim YTA, Roberts CJ, Davies MC, Munro A, Gray AI, Uchegbu IF (2005) Preferential liver gene expression with polypropyleniminedendrimers. J Control Release 101:247–258

Sherje AP, Jadhav M, Dravyakar BR, Kadam D (2018) Dendrimers: A versatile nanocarrier for drug delivery and targeting. Int J Pharm 548:707–720

Shimpi S, Chauhan B, Shimpi P (2005) Cyclodextrins: application in different routes of drug administration. Acta Pharma 55:139–156

Sikandar K, Sharma PK, Visht S (2011) Ocular drug delivery system: an overview. Int J Pharm Sci Res 9:1168–1175

Sowińska M, Lipkowska Z (2014) Advances in the chemistry of dendrimers. New J Chem 38:2168–2203

Starpharma (n.d.) VivaGel® availability. https://www.starpharma.com/vivagel/vivagel_availability

Tomalia DA, Baker H, Dewald J, Hall M, Kallos G, Martin S, Roeck J, Ryder J, Smith P (1985) A new class of polymers: starburst-dendritic macromolecules. Polym J 17:117–132

Tripathi PK, Khopade AJ, Nagaich S, Shrivastava S, Jain S, Jain NK (2002) Dendrimer grafts for delivery of 5-fluorouracil. Pharmazie 57:261–264

Vandamme TF, Brobeck L (2005) Poly (amidoamine) dendrimers as ophthalmic vehicles for ocular delivery of pilocarpine nitrate and tropicamide. J Control Release 102:23–38

Vinicius R, Ara D, Santos S, Ferreira EI, Giarolla J (2018) New advances in general biomedical applications of PAMAM dendrimer. Molecules 23:2849

Vruti P, Chitra R, Debleena P, Pooja B, Kuldeep R, Youngren-Ortiz SR, Rakesh KT (2020) Chapter 8 – dendrimers as novel drug-delivery system and its applications. In: Rakesh KT (ed) Advances in pharmaceutical product development and research, drug delivery systems. Academic Press, pp 333–392

Wada K, Arima H, Tsutsumi T, Chihara Y, Hattori K, Hirayama F, Uekama K (2005) Improvement of gene delivery mediated by mannosylated dendrimer/ [alpha]-cyclodextrin conjugates. J Control Release 104:397–413

Wang ZX, Itoh Y, Hosaka Y, Kobayashi I, Nakano Y, Maeda I, Umeda F, Yamakawa J, Kawase M, Yag K (2003a) Novel transdermal drug delivery system with polyhydroxyalkanoate and starburst polyamidoamine dendrimer. J Biosci Bioeng 95:541–543

Wang ZX, Yoshiaki I, Yoshifumi H (2003b) Mechanism of enhancement effect of dendrimer on transdermal drug permeation through polyhydroxyalkanoate matrix. J Biosci Bioeng 96:537–540

Wu G, Barth RF, Yang WL, Chatterjee M, Tjarks W, Ciesielski MJ, Fenstermaker RA (2004) Sitespecific conjugation of boron-containing dendrimers to anti-EGF receptor monoclonal antibody cetuximab (IMC-C225) and its evaluation as a potential delivery agent for neutron capture therapy. Bioconjug Chem 15:185–194

Yang WL, Barth RF, Adams DM, Soloway AH (1997) Intratumoral delivery of boronated epidermal growth factor for neutron capture therapy of brain tumors. Cancer Res 57:4333–4339

Yang WL, Barth RF, Adams DM, Ciesielski MJ, Fenstermaker RA, Shukla S, Tjarks W, Caligiuri MA (2002) Convection-enhanced delivery of boronated epidermal growth factor for molecular targeting of EGF receptor-positive gliomas. Cancer Res 62:6552–6558

Yiyun C, Na M, Tongwen X, Rongqiang F, Xueyuan W, Xang W, Longping W (2007) Transdermal delivery of nonsteroidal anti-inflammatory drugs mediated by Polyamidoamine (PAMAM) dendrimers. J Pharm Sci 96:595–602

Zhuo RX, Du B, Lu ZR (1999) In vitro release of 5-fluorouracil with cyclic core dendritic polymer. J Control Release 57(3):249–257

Chapter 14
Use of Solid Lipid Nanoparticles to Improve the Oral Bioavailability of Poorly Water-Soluble Drugs

Neeraj Kaushal, Anuja Paprikar, Ankit Soni, and Senshang Lin

14.1 Introduction

With advancement of high-throughput screening tools employed during initial phase of drug discovery, a large number of lipophilic substances have been under investigation for product development. This can be attributed to the dependence of lipophilic substances on kinetic or non-equilibrium solubility, rather than equilibrium solubility (Jain et al. 2015). The poor water solubility of these lipophilic substances substantially lowers their dissolution rate. A low dissolution rate may result in low oral bioavailability. This is essential in oral drug administration where complete dissolution from the drug product is desired during their intestinal transit to facilitate the drug absorption. Based on the aqueous solubility and membrane permeability of a substance, Biopharmaceutical Classification System (BCS) classifies drug substances into four classes. These four classes are defined in terms of high and low aqueous solubility as well as high and low membrane permeability (Amidon et al. 1995; Yu et al. 2002). Specifically, BCS Class II (i.e., low solubility and high permeability) and IV (i.e., low solubility and low permeability) drugs impede the product development due to their low dissolution rate. In conjunction with the BCS, the Biopharmaceutics Drug Disposition Classification System predicts drug deposition by an interplay between the transport, absorption, and elimination; majority of the BCS Class II drugs have poor aqueous solubility and an extensive first-pass metabolism in humans resulting in poor oral bioavailability (Ku 2008; Wu and Benet 2005). Therefore, the development of a delivery system, which can bypass the first-pass metabolism and showcase an enhanced dissolution, is needed to improve the therapeutic outcomes of poorly water-soluble drugs.

N. Kaushal · A. Paprikar · A. Soni · S. Lin (✉)
College of Pharmacy and Health Sciences, St. John's University, Queens, NY, USA
e-mail: linse@stjohns.edu

J.-C. Kim et al. (eds.), *Smart Nanomaterials in Biomedical Applications*,
Nanotechnology in the Life Sciences,
https://doi.org/10.1007/978-3-030-84262-8_14

In addition to the drug physicochemical characteristics, the drug delivery technologies may impact dosage form related factors by affecting their oral bioavailability. Substantial advancements have been produced in these drug delivery technologies over the last few decades by employing the use of lipid (natural or synthetic) excipients as a strategy in drug product development for enhancing the bioavailability of orally administered poorly water-soluble drugs. This takes into consideration the role of food intake in improving the bioavailability of many orally administered poorly water-soluble drugs, wherein the enhanced drug absorption is often related to the lipid composition of the food. Although the chemistry and the intake amount of such dietary lipids are in stark contrast to what would be acceptable in a drug product, the concept of fed-state bioavailability can be adapted as a potential approach for designing formulations for such drugs.

Lipid-based drug delivery systems have gained substantial popularity as an approach for drug delivery for incorporating the poorly water-soluble drugs in a solubilized form for oral administration. They have gained attention for exhibiting oral bioavailability enhancement of several poorly water-soluble drugs (Chudasama et al. 2015; Gupta et al. 2015; Weng et al. 2014), by eliminating the rate-limiting dissolution step, which plays a significant role in the poor oral bioavailability of most poorly water-soluble drugs (O'Driscoll and Griffin 2008; Porter et al. 2007). Self-emulsifying systems, liposomes, and microemulsions have been explored as lipid-based drug delivery systems for their potential for the oral bioavailability enhancement (Jain et al. 2017). However, lipid-based drug delivery systems offer limited solubilizing capacity for high-melting drugs and require a high amount of emulsifier resulting in toxicity issues as well as their in vivo stability is influenced by physiological parameters such as digestive enzymes and pH (Cole et al. 2008). These challenges can be overcome by the SLNs which can subsequently aid in enhancement of the bioavailability of orally administered poorly water-soluble drugs. This chapter highlights the role of SLNs as a delivery system. The general preparation and characterization techniques of drug-loaded SLNs as well as the in vitro and in vivo techniques to understand the absorption mechanism pathways of drug-loaded SLNs were presented. Finally, an attempt was made to present some of the current research investigations to resolve the current issues of low bioavailability of orally administered poorly water-soluble drugs.

14.2 SLNs to Improve the Oral Bioavailability of Poorly Water-Soluble Drugs

SLNs are lipid nanocarriers introduced in the early 1990s. Characteristically, they comprise lipids, which are intact or solid at physiological and room temperature. These lipids form a solid lipid core which is stabilized by an emulsifier at the interface. Comprising of lipid constituents and having a solid nature, SLNs have the dual advantage of both the lipid emulsion systems and polymeric nanoparticle systems.

The drug-loaded SLNs have been successfully used for oral bioavailability enhancement of poorly water-soluble drugs. For example, oral bioavailability studies of cyclosporin A-loaded SLNs and the nanocrystals were performed and compared with the marketed formulation as a reference (i.e., Neoral®) in young pigs. The mean plasma profile generated after oral administration of cyclosporin A-loaded SLNs had similarity to that of the reference but was devoid of the initial peak concentration of more than 1000 ng/ml presented in the plasma profile generated by the reference, while the nanocrystals of cyclosporin A generated a mean plasma profile with very low concentrations of cyclosporin A. This finding suggests that the lipid constituent of drug-loaded SLN system may play a crucial role in promoting the drug absorption and recommends that SLNs have the potential to be considered as an alternative delivery system for cyclosporin A to elicit the desired response. Moreover, amphotericin B-loaded SLNs were evaluated and found to be stable in the simulated gastric and intestinal fluids (Chaudhari et al. 2016). In vivo studies using rats as an animal model revealed that the oral bioavailability of amphotericin B-loaded SLNs was 1.05-fold higher as compared to that of the marketed formulation (i.e., Fungizone®).

It must be noted that two case studies presented above using cyclosporin A and amphotericin B represent the BCS Class II and IV drugs, respectively. The commonality between the physicochemical characteristics of these drugs is their poorly water solubility. However, they are different in terms of their permeability characteristics (i.e., cyclosporin A is more permeable than amphotericin B). The findings from both case studies suggest that in addition to enhancement of oral bioavailability of the poorly water-soluble drugs, SLNs can also facilitate the oral absorption of permeation-limited compounds. The fundamentals and absorption mechanisms responsible for bioavailability enhancement of orally administered poorly water-soluble drug molecules by SLNs are discussed below.

14.2.1 Fundamentals of Improving Oral Bioavailability by SLNs

14.2.1.1 Dissolution/Solubilization

Owing to the presence of lipids, orally administered drug-loaded SLNs stimulate the various events (i.e., the gallbladder contractions, pancreatic and biliary secretions) within the gastrointestinal tract. These along with simultaneous gastric shear movements facilitate the formation of a crude emulsion which facilitates the solubilization of the co-administered lipophilic drug. Furthermore, the precipitation of the lipophilic drug is impeded when micelles and mixed micelles are formed (Nanjwade 2011).

14.2.1.2 Stimulation of Intestinal Lymphatic Transport

The intestinal lymphatic route can transport the lipophilic drugs (i.e., LogP > 5) and make them available into the systemic circulation (Nordskog et al. 2001). The main advantage of the intestinal lymphatic transport of the drugs is a reduction in the required dose and its potential side effects. Since the drugs are absorbed via the intestinal lymphatic system, they would be safeguarded against the hepatic first-pass metabolism, in contrast to drugs entering the systemic circulation directly via portal blood (Beg et al. 2011; Porter and Charman 2001).

In general, the process of digestion of lipids (predominantly triglycerides) initiates in the stomach, where the gastric acid lipase facilitates the breakdown of the long chain of lipids (Abrams et al. 1988; Cohen et al. 1971; Hamosh et al. 1981). Once the content reaches the pyloric antrum (which separates the stomach and the duodenum), gastric chyme is released and combined with the peristaltic movements, which facilitates the emulsification of triglycerides as they empty into the duodenum (Pantaleo et al. 1994; Tso 1994). The emulsification of triglycerides results in the formation of corresponding monoglycerides and fatty acids, which get absorbed and enter the enterocyte (absorptive cells of the intestines) (Shiau 1990; Simmonds 1972). Following their absorption into the enterocyte, migration of these digested products (i.e., monoglycerides and fatty acids) across the enterocyte is dependent on the lipid chain length, wherein, the short- and/or medium-chain lipids (i.e., carbon-chain length ≤C12) passively diffuse across the enterocyte. Conversely, long-chain lipids (>C14) generally migrate to the endoplasmic reticulum where they get re-acylated and assembled into lipoproteins before secretion into the mesenteric lymph (Chaikoff et al. 1951; Kiyasu et al. 1952) (Fig. 14.1).

Conversely, for lipophilic drugs, following its absorption and transiting across the enterocytes, two potential scenarios can occur (Fig. 14.1). First, drugs may enter the systemic circulation passively, and, second, they may enter the systemic circulation via the lymphatic capillaries. The former approach facilitates the drug entry into the portal circulation and is the most common mechanism of drug absorption, because of the high flow rate (i.e., 500-fold) of the portal blood compared to that of the intestinal lymph (Trevaskis et al. 2008). Upon entry of lipophilic drugs into the enterocyte, lymphatic access occurs via association with lipid absorption and lipoprotein assembly during the diffusion across the enterocytes (Trevaskis et al. 2008; Yáñez et al. 2011). The drug-lipoprotein complex then gains access to the lymphatics. It should be noted that the intestinal lymphatic transport of these drugs is only substantial when administered with the source of lipid, as lipids promote the formation of lipoproteins (Khoo et al. 2003).

The above scenarios best explain the role of co-administered lipids on the bioavailability of the orally administered poorly water-soluble drugs. Similarly, SLNs can be used as a lipid carrier to facilitate the lymphatic absorption of the drug, ultimately increasing oral bioavailability. To this end, evidence exists of the uptake and transport of the intact drug-loaded SLNs in the lymph when drug-loaded SLNs at particle size about 80 nm were administered intraduodenally to rats (Bargoni et al. 1998). This particle size enables the transport of intact particles through the

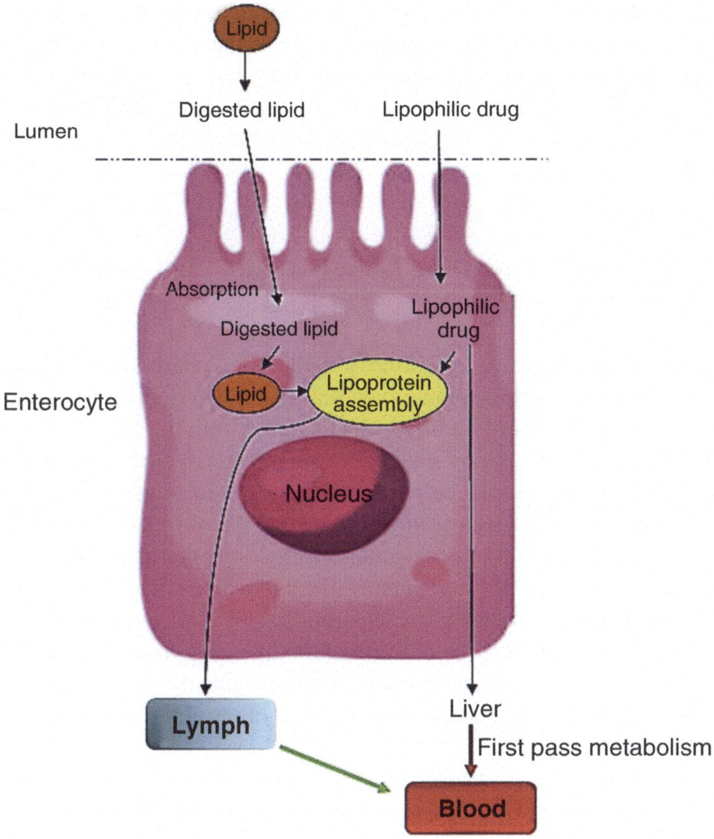

Fig. 14.1 Schematic representation of the pathway of lipid and lipophilic drug access to the gut-associated lymphatic following oral administration. (Adapted from Trevaskis et al. 2015)

intestinal cell layer into the vascular or lymphatic system. The quantity of gut uptake of SLNs and their transport to other organs seem strongly dependent on their particle size, hydrophobicity, surface charge, bio-adhesion to the gut, and the lipid matrix (Müller et al. 2000).

14.2.2 Mechanism of Drug Absorption Enhancement by SLNs

Following the oral administration of drug-loaded SLNs, the released free drug along with the intact drug-loaded SLNs would be available for absorption. Depending upon the physicochemical properties, the free form of the drug could be absorbed passively, via active transport, or via facilitated passive diffusion across the epithelial cell barrier. On the other hand, the intact drug-loaded SLNs could be absorbed via an active absorption mechanism. To this end, after the oral administration of camptothecin-loaded SLNs, two peaks (C_{max} values) in the plasma concentration-time curves of camptothecin were obtained. The values of t^1_{max} and t^2_{max} (i.e., the time to reach the maximum concentration-peak 1 and 2) were 30 min and 6 h, respectively, indicating the absorption of released camptothecin from camptothecin-loaded SLNs and the intact camptothecin-loaded SLNs, respectively (Yang et al. 1999).

14.2.2.1 Absorption of Free Drug Released from Drug-Loaded SLNs via Gastrointestinal Tract

Generally, orally administered free drugs get absorbed via the passive paracellular and transcellular routes and become bioavailable in the systemic circulation via the portal vein. For poorly water-soluble drugs loaded within the SLNs, their oral bioavailability is enhanced in various ways. For example, unstable poorly water-soluble drugs loaded into the SLNs may be protected during their transit from the gastrointestinal environment to the absorption site. Once at the absorption site, these drug-loaded SLNs may facilitate the drug release in a controlled/sustained manner and subsequent absorption, thereby enhancing the bioavailability. Conversely, poorly water-soluble drugs stable in the gastrointestinal tract are released from drug-loaded SLNs, which do not allow the precipitation of the free drug from the resulting supersaturated solution in gastrointestinal fluids. This prevents the precipitation of poorly water-soluble drugs, hence resulting in higher oral bioavailability (Nanjwade 2011).

14.2.2.2 Passive Absorption of Intact Drug-Loaded SLNs via Blood Capillary

Although the majority of orally administered drugs can be absorbed and gain access to the systemic circulation via portal blood capillary, a report suggests that intact drug-loaded SLNs can be absorbed by passive absorption in the gastrointestinal tract. For example, for quercetin-loaded SLNs with particle size of 155.3 nm, it was observed that the absorption from the stomach was 6.2% (Li et al. 2009). Absorption from the stomach indicates the absorption of intact drug-loaded SLNs can occur through passive transport too. Additionally, passive diffusion across the bilayer lipid membranes of intact drug-loaded SLN suspension occurred across the intestine (Li et al. 2009).

14.2.2.3 Passive Absorption of Intact Drug-Loaded SLNs via Lymph Capillary

Generally, absorbed substances via the gastrointestinal tract become available in the systemic circulation via portal blood capillary. However, extremely lipophilic substances can gain access to the systemic circulation via the lymph capillary (Makwana et al. 2015). The bioavailability of poorly water-soluble drugs could be enhanced by the stimulation of the intestinal lymphatic transport pathways as shown in Fig. 14.1. Furthermore, drug-loaded SLNs may facilitate the lymphatic transport of lipophilic substances by stimulating the production of chylomicrons from the enterocytes. The intact drug-loaded SLNs can then be transported to the lymph by association with chylomicrons via chylomicron paracellular transport (O'Driscoll and Griffin 2008). Other proposed mechanisms are transcellular transport through the lymphatic capillary or junctional entry through the tip of villus (Dixon 2010). Owing to the nature of the lipids employed for their preparation, intact drug-loaded SLNs have shown to be transported to the lymph and blood capillaries (Bargoni et al. 1998). The size of these intact drug-loaded SLNs was determined to be 143 nm in the lymph. Furthermore, it has been reported that lymph capillary had a higher concentration of intact drug-loaded SLNs in comparison to blood capillary. And increased accumulation with the increase in administered dose has also been reported (Bargoni et al. 1998). Additionally, when methotrexate-loaded SLNs were orally administered, methotrexate was detected in the lymph indicating the lymphatic uptake mechanism along with the transcellular mechanism (Paliwal et al. 2009a). Thus, drug-loaded lipid nanocarriers composed of suitable lipids can be used as a carrier to facilitate the lymphatic uptake consequently increasing the oral bioavailability of drugs (Chakraborty et al. 2009). In addition, transport via the intestinal lymphatic has various advantages, such as a reduction in the first-pass metabolism, delivery of high concentrations of the drug into the lymphatic system and subsequently to the systemic circulation, and thereby reduction of required dose and any associated side effects (Beg et al. 2011; Porter and Charman 2001). Further, it has been observed that using chylomicron flow blocking approach, efavirenz significantly bypassed the portal system due to chylomicron uptake mechanism. About 45% reduction in the liver accumulation of efavirenz enhancement of oral bioavailability was attributed to the use of lipid nanocarriers (Makwana et al. 2015).

14.2.2.4 Active Absorption of Intact Drug-Loaded SLNs Through Intestinal Epithelium

The intact drug-loaded SLNs can be absorbed through the intestinal epithelium by endocytosis. And an interplay of macropinocytosis pathway and clathrin- and caveolae-related routes has been reported to mediate the endocytosis (Chai et al. 2016). Following their internalization within the cell, the role of lysosomes in the degradation of drug-loaded SLNs has also been reported (Chai et al. 2016), wherein the drug-loaded SLNs taken up by lysosomes are degraded and the residuum may be either transported back to the apical or translocates across the basolateral membrane.

14.2.2.5 Active Absorption of Intact Drug-Loaded SLNs via Peyer's Patches

Active absorption of intact drug-loaded SLNs can occur through Peyer's patches in the ileum region of the small intestine (Ali Khan et al. 2013). The mechanism of transport across Peyer's patches may occur via the combination of endocytosis, phagocytosis, pinocytosis, and micropinocytosis (Jung et al. 2010). The transport initiates by drug-loaded SLNs adhering to the surface of the M cells and forming vesicles, followed by M cell-facilitated vesicular transfer of particulates into the underlying lymphocytes (Reddy and Murthy 2002). Since M cells lack lysosomes, there is a reduced degradation of the drug-loaded SLNs during their transit; as a result, intact drug-loaded SLNs get transported across the lymphocytes (Reddy and Murthy 2002). However, the uptake of particles by M cells is particle size-dependent, wherein smaller particles (preferentially <200 nm) get uptaken more (Bargoni et al. 1998). It has been reported that absorption of quercetin-loaded SLNs in the intestinal segments of the ileum and colon was more than in the duodenum and jejunum. This can be attributed to the abundance of Peyer's patches and M cells in the ileum and the colon region than in other intestinal segments (Li et al. 2009). In another study, higher hydrophobicity of nanoparticles exhibited relatively higher accumulation in Peyer's patches (Bachhav et al. 2017). Also, neutral and negatively charged SLNs exhibited higher uptake than positively charged SLNs (Bagby et al. 2012). And surface modification by attaching specific ligands like lectin further enhances the M cell uptake (Harde et al. 2011).

14.3 Fabrication of Drug-Loaded SLNs

14.3.1 *Lipid Components and Selection Criteria*

Lipid nanocarriers such as SLNs are composed of mainly solid lipids and/or liquid lipids for their preparation (Table 14.1). The lipids employed in SLN formulations are regulatorily accepted, well-tolerated, and amenable for large-scale production. However, to improve drug incorporation and release properties, solid lipids are mixed with spatially incompatible liquid lipids (Müller et al. 2002). Further, lipids are classified as wax-based carriers or glyceride-based carriers. However, SLNs prepared from these two carriers may vary with respect to drug encapsulation efficacy, particle size and size distribution, storage, and crystal packing. For example, glyceride-based SLNs have shown good encapsulation of the drug, but poor physical stability, whereas wax-based SLNs possessed good physical stability but insufficient drug encapsulation in the solidified state. These differences were contributed by the different crystal packing of the lipids. The reduction in the orderliness of the crystal structure increases upon addition of drug inclusion which is evident from the case of glyceryl monostearate and glyceryl behenate SLNs. For instance, it was

Table 14.1 Physicochemical characteristics of lipids for the fabrication of drug-loaded SLNs

Lipids			Chemical composition	Effect of lipid on SLN characteristics	Storage stability	References
Solid lipids	Partial glycerides	Glyceryl monostearate (Imwitor 900, Geleol)	Mixture of mono- and diacylglycerols	Decreased degree of lipid phase crystallinity and increased loading capacity	Less stable as compared to wax-based	Ebrahimi et al. (2015)
		Glyceryl behenate (Compritol 888 ATO)	Diacylglycerols of behenic acid and with different quantities of mono- and triacylglycerols	Imperfections in crystal lattice lead to higher entrapment efficiency	Less stable as compared to wax-based	Shukla et al. (2018)
		Glyceryl palmitostearate	Mixture of mono-, di-, and tri-acylglycerols of palmitic and stearic fatty acids	Helps in sustained-release action	Less stable as compared to wax-based	Shukla et al. (2018)
	Triglycerides	Trimyristin	Triglyceride of myristic acid	Trimyristin SLNs have shown to have larger lipophilic drug incorporation in liquid state than in solid	Degrade faster and show a change in degradation behavior with increasing storage time	Bunjes et al. (1996) and Olbrich et al. (2002)
		Tripalmitin	Triglyceride of palmitic acid	Tripalmitin SLNs have larger particle size owing to long carbon chain and high molecular weight	Polymorphic change observed on storage	Helgason et al. (2008) and Öztürk et al. (2019)
		Tristearin	Triglyceride of stearic acid	Improved hydrolytic stability of drug in SLN (mechanism unknown)	Forms SLN at room temperature	Chirio et al. (2009)
	Wax	Cetyl palmitate	Ester derived from hexadecanoic acid and hexadecanol	Highly ordered crystals lead to drug expulsion from carrier	More stable as compared to partial glyceride-based	Jenning and Gohla (2001)
	Fatty acid	Stearic acid	Fatty acid (C-18)	Encapsulation efficiency and loading capacity were highest as compared to palmitic acid and tetradecanoic acid	SLNs prepared with stearic acid and liquid lipid-like oleic acid can increase stability up to 6 months at various temperatures	Kelidari et al. (2017)

(continued)

Table 14.1 (continued)

Lipids			Chemical composition	Effect of lipid on SLN characteristics	Storage stability	References
Liquid lipids	Hydrocarbons	Squalene	Unsaturated hydrocarbon	Loaded drug was easily released. Higher flux observed since squalene increased imperfection of the solid lipid matrix	SLNs consisting of squalene indicated smaller particle size and indicated better stability with respect to charge	Fang et al. (2008)
	Vitamin	Vitamin E	Tocotrienol	Inclusion of vitamin E in SLNs improved UVB absorption of the synthesized new compound	Vitamin E can also reinforce stability of the new compound and play a role in retarding the progress of life senescence	Song and Liu (2005)
	Triglyceride	Miglyol 812	55% triglycerides of C8 and 45% triglycerides of C10 fatty acids	Improved drug load levels were encapsulated by lipid particles supplemented with oily constituents	Laser diffraction measurements revealed similar sizes and distribution for SLNs with or without oil after day 1	Jenning and Gohla (2001)
	Ether	Transcutol HP	Highly purified diethylene glycol monoethyl ether	The solubility of drug was found to be highest in Transcutol HP leading to higher drug loading	After a period of 90 days, the particle size, PDI, and %EE were not significantly changed compared with the initial freeze-dried nanoparticles	Soni et al. (2020)
	Mixtures	Labrafil	Consists of mono-, di-, and triglycerides and PEG-6 (MW 300) mono- and diesters of oleic (C18:1) acid	SLN formulations modified with Labrafil®(HLB 4) showed significantly higher encapsulation efficiency than Labrasol® (HLB 14) owing to higher lipophilicity	Labrafil® with HLB value of 4 increases its miscibility in Tristearin®(solid lipid) and thus led to stabilization of the polymorphic form of Tristearin® in the desired α or β′ forms	Abdel-Salam et al. (2016)
		Lipophile WL 1349	Mixture of medium chain triglycerides, mainly from caprylic (C8) and capric (C10) acids	LWL was able to complex and showed highest pDNA binding capacity as compared to other lipids	All formulations are stable after 3-month storage at room temperature	Oner et al. (2020)

observed that during storage, SLN destabilization of the crystal matrix occurred with glyceryl behenate leading to gelation. By the incorporation of inhibitors, the transition of lipid modification into crystalline form can be avoided (Freitas and Müller 1999). The general criteria for lipid selection for the formulation of SLNs have been reported elsewhere (Shah et al. 2019b).

In addition to the reported criteria, the selection of lipids is also dependent on the chemical stability of other adjuvants, such as emulsifiers, stabilizers, and/or cryoprotectants. The choice of lipids affects the formation and stabilization of SLNs. High recrystallization index of lipid leads to increased chances of particle size growth during storage. In other words, the recrystallization index of SLNs is maintained below that of the bulk material, since it increases with an increase in storage time (Freitas and Müller 1999). Furthermore, an increase in lipid content (i.e., >5–10%) may impact the particle size and polydispersity of the size distribution of the resulting SLNs. This can be attributed to the increase in viscosity of the formulation, which impedes the homogenization efficiency resulting in larger particle size and polydispersity (Müller et al. 2000). However, this challenge can be circumvented by using higher proportions of emulsifier and co-emulsifier system. Other properties like the hydrophilicity, shape, surface area of the lipid crystals, and source of lipids may also affect the formation of SLNs (Paliwal et al. 2009a). Specifically, these factors influence the zeta potential and crystallization and eventually the stability of SLNs. SLNs formulated using triglycerides were reported to be more stable than with mono- and diglycerides. In general, lipid stability is an important prerequisite for the consideration of the introduction of SLNs into the pharmaceutical market (Radomska-Soukharev 2007). There are several theoretical approaches to screen lipids for a respective drug to be formulated within SLNs. The use of solubility parameter, wherein lipid with a solubility parameter value similar to the drug results in a higher probability of drug entrapment in the lipid melt during the preparation of SLNs. Consequently, higher drug loading can be achieved (Das et al. 2012). Although the difference between the solubility parameter of the drug and the lipid (i.e., $\Delta\delta T$ value) should be small, if components are to be miscible, it may be difficult to establish a threshold for the $\Delta\delta T$ value. Generally, the limits for $\Delta\delta T$ which indicate components are likely to be miscible, if $\Delta\delta T < 7.0\ MPa^{1/2}$, while $\Delta\delta T > 10.0\ MPa^{1/2}$, suggest the likelihood of components being immiscible has been proposed in the literature (Greenhalgh et al. 1999). However, more stringent limits have also been proposed (Forster et al. 2001) for predicting miscibility (i.e., $\Delta\delta T < 2.0\ MPa^{1/2}$), while immiscibility is anticipated for systems when $\Delta\delta T > 10.0\ MPa^{1/2}$. Such inconsistency in the threshold of $\Delta\delta T$ suggests a need for systemic screening of various lipids (Shah et al. 2014).

As discussed above, the solubility of drugs in lipid melt plays an important role in entrapment efficiency. A study has reported that corticosteroid solubility in the lipids used in SLN formation was dependent on the monoglyceride content (i.e., polarity of the lipid) (Jensen et al. 2010). Further, SLNs prepared using lipids of less ordered crystal lattices (e.g., glyceryl monostearate and glyceryl behenate) facilitate better drug loading as compared to those prepared using highly ordered crystal packing lipids (e.g., beeswax and cetyl palmitate). However, their long-term stabilities are quite different. For instance, beeswax-based SLNs provided excellent physical long-term stability as well as particle size distribution (Jenning and Gohla 2000). It was also found that higher lipid chain length yielded higher plasma concentration of the nitrendipine when formulated in SLNs. In this case, not only the higher lipid chain length but also the incorporation of charge modifiers (e.g., stearyl amine) enhanced bioavailability (Manjunath et al. 2011). It was reported that the lipid matrix affects drug release and stability. Also, higher uptake from Peyer's patches and longer circulation resulted in an enhanced bioavailability of drug-loaded SLNs formulated with longer lipid chains (Harde et al. 2011).

14.3.2 Emulsifiers

Emulsifiers are essential for stabilizing the SLNs and also aid in increasing the entrapment efficiency of drugs. In general, the emulsifier should be compatible with other components of the SLNs and aid in producing desired particle size at a minimum concentration. At a lower concentration, emulsifiers may facilitate particle agglomeration due to insufficient coverage of the surface of the particle. On the other hand, a higher concentration may decrease the drug entrapment, may cause burst release of the loaded drug, and can lead to toxicity. Hence, an optimum concentration of emulsifier should be a part of preformulation studies. Specifically, hydrophilic-lipophilic balance (HLB) of the emulsifier should be considered while screening an ideal emulsifier for a particular lipid matrix (Harde et al. 2011).

It has been observed that the emulsifier-mediated charged SLNs may also affect the stability and drug release from drug-loaded SLNs (Manjunath and Venkateswarlu 2006). In the case when the emulsifier is used as a coating, for example, adsorption of poloxamer 188 and 407 on the surface of SLNs inhibits their uptake from (both small and large) intestine, leading to a reduction in adhesion to gut-associated lymphoid tissue owing to the coating (Florence et al. 1995). Further, it is observed that the particle size of SLNs stabilized with nonionic emulsifiers is generally larger than those obtained with ionic emulsifiers. A study has reported that the combination of nonionic emulsifiers with ionic emulsifiers like lecithin led to an increase in the particle size of the SLNs (Bunjes et al. 2003). Also, the combination of two or more emulsifying agent's forms mixed emulsifier films at the interface that not only covers the surface efficiently but also produces adequate viscosity which increases the stability. Among four nonionic emulsifiers (Cremophor EL, Tween 20, Tween 80, and Pluronic F68), SLNs prepared using Cremophor EL demonstrated the lowest size, lower polydispersity index, and higher drug encapsulation efficiency (Shah et al. 2019a). It has also been proposed that an HLB value of 12–16 is desirable for an emulsifier, to achieve a lower particle size of SLNs. Any emulsifier with HLB out of this specification, like Pluronic F68, may yield in bigger particle size as well as a higher polydispersity index of the formulated SLNs (Das et al. 2012). However, the choice of selection of suitable emulsifiers is substantially influenced by the lipids, and there is no universal emulsifier; therefore, it would be recommended that screening of emulsifiers should be performed (Manjunath et al. 2005). In addition, although the emulsifier distributes at the o/w interface while in a cooled state, it aids in avoiding the agglomeration of SLNs. Also, the emulsifier plays a crucial role in reducing the crystallization of lipid and hence directly influencing the stability of the formed SLNs (Awad et al. 2008).

14.3.3 Fabrication Methods

There are various methods available for the fabrication of drug-loaded SLNs. The selection of fabrication methods mainly depends upon the physicochemical properties of the drug and the lipid being used. Briefly, drug characteristics (i.e., partition coefficient, thermal sensitivity, and solubility of drugs in an aqueous solution containing emulsifier) and lipid characteristics (i.e., melting point of lipid and solubility of the drug in lipid) at temperature during the preparation affect the choice of method. In addition to characteristics of the drug and the lipid, desired particle size range and combination of the lipid and the emulsifier may alter the choice of preparation method and its process variables. A condensed discussion on the commonly used methods, along with their application and advantages and disadvantages, is presented in Table 14.2. For a detailed description of the methods employed for the fabrication of drug-loaded SLNs, interested readers are referred to work reported elsewhere (Battaglia et al. 2014).

14.4 Evaluations of Drug-Loaded SLNs

Several techniques are available for the evaluation of drug-loaded SLNs and may be broadly classified into physicochemical, in vitro drug release and absorption, ex vivo, and preclinical evaluations of these drug-loaded SLNs. Particle size is the most prominent property of drug-loaded SLNs. However, other parameters, such as drug loading (either adsorbed on the surface and/or entrapped within the lipid matrix) and lipid crystallinity, may largely influence the drug release and degradation of the lipid matrix, whereas surface characteristics, such as surface charge, may significantly influence the interaction with the biological environment, ultimately impacting the body distribution. In addition, drug-lipid interaction collectively aids in providing pertinent information on the stability of drug-loaded SLNs. This information can then be extrapolated to understand the in vivo stability as well as pharmacokinetic profile of the drug-loaded SLNs. It can be said that collectively evaluating the data generated from these different techniques can better assist the formulation scientists in choosing suitable excipients. Therefore, a careful selection and validation of appropriate techniques are imperative to translate and correlate the obtained in vitro data to in vivo performance.

Table 14.2 Methods employed for the fabrication of drug-loaded SLNs

Methods	Type of drug/excipients used	Description	Application	Advantages	Disadvantages	References
Hot homogenization	Lipid: Cetyl palmitate (Cutina® CP) Surfactants: Alkyl poly-glycosides (APG) surfactants such as Plantacare® 810 UP, Glucopon® 425 NH/H, Glucopon® 600 CS UP, and Plantacare® 1200 UP	SLN dispersions composed of 10% (w/w) solid lipid and 1% (w/w) surfactant. The solid lipid melted at 75 °C; hydrophilic surfactants and water were heated to the same temperature and added to the melted lipid. The mixture dispersed with a high-shear mixer for 30 s at 8000 rpm and subjected to high-pressure homogenization applying a pressure of 500 bar at 75 ° C to get SLNs	Improved skin permeation potential for dermal application SLNs prepared with APGs revealed reduced skin irritation potential	In comparison to SLN stabilized with classical surfactants, e.g., APG-stabilized SLNs possess a smaller size and improved physical stability and contain less surfactant	Increase in the number of the homogenization cycles may result in agglomeration of SLNs	Keck et al. (2014)
	Drug: Retinol Lipid: Compritol 888 ATO® Surfactant: Sodium cocoamphoacetate	The lipid melted at 85 °C, and the model drug retinol was added to it. The hot lipid phase was dispersed in a surfactant solution, and a premix can be obtained using an ultra Turrax. The coarse premix was passed through the high-pressure homogenizer to obtain SLNs	To develop SLNs, produced in different batch sizes using high-pressure homogenization with medium-scale production To determine batch-to-batch variability in terms of distribution and physical storage stability	Can be used to develop scale-up batch size range (i.e., from 40 ml to 50 liters) with less batch to batch variability in particle size (i.e., less than10 nm in mean particle diameter) To get excellent reproducibility of SLNs produced	The cooling rate identified as a critical production parameter. Temperature fluctuations can lead to particle aggregation	Jenning et al. (2002)

Methods	Type of drug/excipients used	Description	Application	Advantages	Disadvantages	References
Cold homogenization	Drug: Lysozyme Lipid: Witepsol E 85, Softisan 142, and cetyl alcohol Surfactant: Poloxamer 182	Lysozyme and Poloxamer 182 above its critical micellar concentration dissolved in distilled water at room temperature, and this solution added dropwise to 20 g of melted lipids to obtain SLNs	SLNs can be developed as antigen carriers for vaccine delivery	Reliable, can be used for heat-sensitive drugs	The incorporation of polypeptides into hydrophobic media, together with the presence of surfactants, would probably damage peptide molecules	Almeida et al. (1997)
	Drug: Vinorelbine bitartrate Lipid: Glyceryl monostearate, lecithin E80 and oleic acid Surfactant: Poloxamer 188	Lipid phase (i.e., glyceryl monostearate) melted at 60 °C; vinorelbine bitartrate, lecithin, and oleic acid dissolved in 1 mL ethanol, and the organic solution then added into melted lipid phase dropwise at 60 °C. the mixture was cooled by pouring the mixture in liquid nitrogen to remove ethanol. Resultant solid dispersion milled to form microparticles. The microparticles were then suspended in 40 ml aqueous phase containing 1% Poloxamer 188 and 20% sugar and homogenized at 15,000 rpm for 0.5 h by a homogenizer to form SLNs	Vinorelbine bitartrate has antitumor activity for the advanced breast and non-small cell lung cancer Toxicity associated with vinorelbine bitartrate can be reduced with use of SLNs	Vinorelbine bitartrate is easy to be degraded on condition of high temperature. Hence drug-loaded SLNs can be prepared by cold homogenization technique Also, the cellular cytotoxicity tests demonstrated lower cellular cytotoxicity by the entrapment of drug in lipid matrix	The drug loading was observed to be 6.6% due to hydrophilic nature of the drug	You et al. (2007)

(continued)

Table 14.2 (continued)

Methods	Type of drug/excipients used	Description	Application	Advantages	Disadvantages	References
Ultrasound	Drug: Risperidone Lipid: Imwitor® 900 K surfactant: Sodium deoxycholate	The solid lipid heated 5–10 °C above its melting point and then added to a mixture of surfactants and water to the lipid at the same temperature. This leads to form a pre-emulsion obtained under stirring at 8000 rpm for 5 min. A sonication probe placed in this pre-emulsion, by means of an ultrasonic processor with amplitude of 70% applied for 20 minutes, leads to formation of nanoparticles	By using SLNs, solubility of risperidone (BCS class II drug) can be improved, and that leads to improved dissolution rate and oral absorption	SLNs prepared with ultrasound method produce smaller size particles compared to the SLNs prepared with hot homogenization method	Compared to ultrasound method, the dispersions prepared by high-pressure homogenization have a lower degree of crystallinity, indicating the presence of more unstable polymorphic forms with long-term storage	Silva et al. (2011)
Warm microemulsion	Lipid: Tripalmitin Surfactant: Soybean lecithin, taurocholate sodium salt	First step is the preparation of the defined microstructure of warm microemulsions of lecithin/water/tripalmitin/1-butanol/taurocholate sodium salt at 70 °C. SLNs prepared by dropping warm microemulsions maintained at 70 ± 0.1 °C to aqueous phases at 4 ± 0.1 °C, under continuous stirring at 7000 rpm for 20 min	Well-defined droplet structure contained in the microemulsion, and therefore, no energy is required to achieve submicron particle sizes	Uniform particle size and size distribution can be obtained	Not applicable for heat-sensitive drug	Fadda et al. (2013)

Methods	Type of drug/excipients used	Description	Application	Advantages	Disadvantages	References
Microemulsion extrusion	Lipid: Stearic acid Surfactant: Pluronic F68 Cosurfactant: N- [1-(2,3-dioleoyloxy) propyl]-N,N,N-trimethyl-ammonium chloride (DOTAP)	Solution of Pluronic F68 (surfactant) and DOTAP (co-surfactant) pre-heated above the melting temperature of the solid lipid used added to the melted lipid followed by homogenization using vigorous stirring. The prepared hot emulsion loaded into a syringe and extruded several times at a temperature above the melting temperature of the lipid, using a heating block, and the lipid nanoparticles transferred to an ice bath and stored at 4 °C	Suitable for small-scale preparations of SLNs prepared with cationic lipids for gene delivery or radio-labeled lipids for nanoparticle tracking/diagnosis	Does not require the use of organic solvents and allows large-scale production	The equipment is expensive with energy-intensive operating conditions. Also, huge amounts of surfactants/co-surfactants during the preparation	De Jesus et al. (2013)
Solvent evaporation	Drug: Cholesteryl acetate, Lipid: Soy phosphatidylcholine Surfactant: Lecithin, bile salt Cosurfactant: Tween 80	Cholesteryl acetate was dissolved in cyclohexane containing lecithin. The organic solution was emulsified in an aqueous solution containing a cosurfactant. A stable o/w emulsion resulted by treatment in a colloid mill for 2 min at room temperature followed by high-pressure homogenization. The solvent was evaporated from the emulsion and cholesteryl acetate precipitated as emulsion droplets	The method can manufacture spherical particles down to a diameter of 25 nm using a mixture of bile salts and lecithin	They are biodegradable and composed of biocompatible emulsifiers and oils	Ostwald ripening and flocculation may occur leading to increased particle size on long-term storage	Sjöström and Bergenståhl (1992)

(continued)

Table 14.2 (continued)

Methods	Type of drug/excipients used	Description	Application	Advantages	Disadvantages	References
Solvent diffusion	Lipid: Glyceryl monostearate Surfactant: Tween 80, soya lecithin Solvent: Benzyl alcohol	Benzyl alcohol and water mutually saturated at 47 ± 2 °C for 10 min in order to ensure initial thermodynamic equilibrium of both liquids. Lipid dissolved in 1000 ml of water-saturated solvent. Emulsification obtained by addition of lipid phase with 9.0 ml of solvent-saturated aqueous solution containing 100 mg emulsifier at same temperature using an ultra Turrax at 12,000 rpm for 1 min. The SLNs can be obtained by precipitation by quickly adding water into the initial emulsion to extract the solvent into the continuous phase	Use of physiological lipids leads to decreases in the danger of acute and chronic toxicity	Does not require heat	Use of organic solvent during the preparation	Trotta et al. (2003)
	Drug: Clobetasol propionate Lipid: Monostearin Stabilizer: Polyvinyl alcohol Solvent: Ethanol and acetone	Lipid and drug were dissolved completely in a mixture of acetone and ethanol in water bath at 50 °C. the resultant organic solution poured into acidic aqueous phase containing 1% polyvinyl alcohol under mechanical agitation with 400 rpm at room temperature for 5 min. The entire dispersed system was then centrifuged (4000 rpm for 10 minutes) and resuspended in distilled water. The resultant dispersion was dried by lyophilization	Prolonged drug release (up to 4 days) can be obtained, with this novel solvent diffusion method in aqueous system	Improved stability in terms of nearly neutral charge of the prepared SLNs under acidic environment Minimum burst release during drug release study	Use of organic solvent for the preparation	Hu et al. (2002)

Methods	Type of drug/excipients used	Description	Application	Advantages	Disadvantages	References
Membrane contractor	Drug: Vitamin E Lipid: Gelucire 44/14 Surfactant: Tween 20	The lipid phase is pressed, at a temperature above the melting point of the lipid, through the membrane pores. That leads to the formation of small droplets. The aqueous phase circulates inside the membrane module and sweeps away the droplets forming at the pore outlets. SLNs are formed by the following cooling of the preparation to room temperature	This method can be used for preparation of polymeric nanoparticles as well by methods involving a polymerization of dispersed monomers or a dispersion of preformed polymers	Ease of use, the control of the particle size by an appropriate choice of process parameters, its scaling-up abilities	Stability of vitamin E-loaded SLNs represents degradation of the drug when stored at daylight environment	Charcosset et al. (2005)
Supercritical fluid	Drug: Praziquantel Lipid: Cetyl palmitate Surfactant: Tween 20 Cosurfactant: Soy lecithin Solvent: Dichloromethane	Oil phase, composed of solid lipid, drug, surfactant, and co-surfactant, solubilized in dichloromethane using magnetic stirring for 30 minutes. The prepared solution then introduced into an extraction column from the top at a constant flow rate of supercritical CO_2 at a constant pressure of ~80 bar and 35 °C to produce SLNs. The equipment allows the production of size batches from 100 mL to 5000 ml. The batch size of our PZQ-SLNs was 100 mL	Prepare SLNs, within few minutes along with scale-up batch with uniform particle size distribution. This method allows the scale-up batches (i.e., from 100 ml to 5000 ml)	Produces SLNs with a narrow size distribution, less residual organic solvent, and with high encapsulation efficiency for different types of drugs	Use of organic solvent to dissolve lipid phase	Andrade et al. (2019)

(continued)

14.4.1 Physicochemical Evaluations

An adequate characterization of drug-loaded SLNs is necessary to ensure the quality of the product. The parameters discussed below have a direct effect on the stability and drug release profile of the drug-loaded SLNs aiding in predicting their fate in a biological system.

14.4.1.1 Particle Size and Zeta Potential

The physical stability of drug-loaded SLNs can be directly correlated with the particle size and zeta potential. The commonly used techniques for measurements of particle size are dynamic light scattering and laser diffraction. In principle, the particle movement causes fluctuation of the intensity of the scattered light, which is measured by the dynamic light scattering. The drug-loaded SLNs intended for oral administration should have a size between 10 and 200 nm to be taken up by the gut-associated lymphatic system (Swartz 2001). Hence, during the formulation of drug-loaded SLNs, it is critical to control the particle size as the lymphatic uptake becomes more selective and slower as the particle size increases. In contrast to dynamic light scattering, laser diffraction aids in providing information about the particle shape.

In addition to the colloidal stability of drug-loaded SLNs, zeta potential measurements can provide information regarding their shelf life and interaction of this delivery system with the biological membrane. As a rule of thumb, under the given conditions, high magnitude (i.e., ±30 mV) of zeta potential is expected to stabilize the colloidal dispersion of drug-loaded SLN via electrostatic repulsion (Mehnert and Mäder 2012). In other words, measuring the zeta potential in the dispersion medium such as water is a measure of the thickness of the diffuse layer. The lower the zeta potential, the thinner the diffusion layer and the less stable is the colloidal dispersion of drug-loaded SLNs (Obeidat et al. 2010). The formulations with zeta potential near zero under storage conditions may also be stabilized upon storage by creating a physical barrier against aggregation with the hydrophilic polymer (i.e., polyethylene glycol) coating.

Another approach called acoustic spectroscopy for the determination of size and surface charge of drug-loaded SLNs can be employed. This technique measures the attenuation of sound waves. Upon application of the acoustic energy, a charge is introduced to the drug-loaded SLNs because of the movement and generation of the oscillating electric field. The generated electric field is utilized to describe the surface charge information (Mishra et al. 2018). By use of photoacoustic spectroscopy, no chemical reactions were revealed in between drugs and the lipids during the characterization of hesperidin-loaded SLNs (Ferrari et al. 2019).

Mammalian biological membranes comprise phospholipids that possess exclusively zwitter- and anionic character (Klausen et al. 2016). The main part of the phospholipids carries a net neutral charge (van Meer and de Kroon 2011), while multiple negatively charged membrane components are present which are important for special functions (i.e., signaling pathways) of the cells (Magee et al. 2002). Charged-based distribution of drug-loaded SLNs employing in vitro cell culture model has been explored (Chirio et al. 2014). In that study, the surface of fabricated paclitaxel-loaded SLNs was functionalized with stearylamine or glycol chitosan, which imparted a net positive charge (i.e., +8–20 mV) on the paclitaxel-loaded SLNs. Thereafter, the cytotoxicity of modified paclitaxel-loaded SLNs was evaluated. However, no difference in cytotoxicity between the neutral and positively charged paclitaxel-loaded SLNs was observed. This could be attributed to the fact that cytotoxicity is a function of the drug rather than the charge of drug-loaded delivery systems. Conversely, in another study chitosan- or hydroxypropyl trimethyl ammonium chloride chitosan (HACC)-modified docetaxel-loaded SLNs bearing a net positive charge were evaluated both in in vitro cell culture model (i.e., Caco-2 cells) and in vivo (i.e., oral administration to rats) (Shi et al. 2016). In both cases, chitosan-loaded SLNs with surface modified by HACC and bearing a mean surface charge of +23.10 mV exhibited superior efficacy as compared to both chitosan-loaded SLNs (8.83 mV) and blank SLNs (−20.24 mV). A superior efficacy in cell culture studies, in this case, was correlated to the transport of the positively charged chitosan-loaded SLNs across the monolayer, which is the function of the delivery system. These results indicate that zeta potential may play a crucial role in the transport of the chitosan-loaded SLNs across biological membranes.

14.4.1.2 Surface and Shape Characteristics

The particle surface and shape of drug-loaded SLNs impact features such as a loading capacity and controlled drug release characteristics. For instance, due to smaller lipid-water interfaces and longer diffusion pathways, a spherical-shaped particle can provide high loading capacity (Saupe et al. 2006). The electron microscopy techniques, such as scanning electron microscopy (SEM), give three-dimensional images of the particles as well as surface morphology. Transmission electron microscopy (TEM), which provides two-dimensional projections of the sample, provides information about the size and shape of nanoparticles as well as internal structure. Although both microscopy tools provide information about the surface and shape characteristics, it is solely based on the type of information desired from the fabricated delivery system. For example, SEM provides information on the surface composition of drug-loaded SLNs and may be useful in contamination detection, while TEM provides additional information on the inner structure of the drug-loaded SLNs.

14.4.1.3 Degree of Crystallinity and Lipid Modification

The degree of crystallinity of drug-loaded SLNs can be assessed by determining the presence or absence of geometric scattering of radiation from the crystal planes in a solid. Differential scanning calorimetry, a thermo-analytical technique, can be used to reveal the nature and the speciation of crystallinity within the drug-loaded SLNs by measuring the glass and melting point temperature (K Manjunath et al. 2011). Also, it is a fast and accurate method for determining the degree of crystallinity of lipids based on the enthalpy of the lipid. However, due to the small size of the drug-loaded SLNs and the presence of emulsifiers, lipid modifications may not be easily captured. Therefore, powder X-ray diffractometry may be employed simultaneously for the description of crystalline materials and to analyze the crystal structure of drug-loaded SLNs (Jenning et al. 2000).

14.4.1.4 Drug Loading and Entrapment Efficiency

The amount of drug loaded in SLNs may influence the drug release. Hence, it is important to quantify the amount of drug loaded per unit weight of SLNs. To calculate the drug loading and entrapment efficiency, drug content of drug-loaded SLNs has to be determined. Various techniques such as centrifugation, filtration, or gel permeation chromatography can be employed for the determination of drug content. Indirect determination of the drug content is performed by determining the drug content in the supernatant after centrifugation filtration/ultracentrifugation of drug-loaded SLNs suspension. Alternatively, sediment may be dissolved in an appropriate solvent followed by analysis using standard analytical techniques (i.e., high-performance liquid chromatography, spectrometry, and liquid scintillation counting). In gel permeation chromatography, specific gels are used for the removal of the free drug from drug-loaded SLN preparations. Alternatively, the drug content can also be determined directly in drug-loaded SLNs after extracting with a suitable solvent (Huang et al. 2008; Yadav et al. 2013).

14.4.2 *In Vitro Drug Release and Absorption Evaluations*

14.4.2.1 Drug Release

The release profile and mechanism of the drug from drug-loaded SLNs depend on the lipid and its composition, which govern drug release by either degradation, erosion, or diffusion. The drug is either dispersed into the matrix or on the surface drug-loaded SLNs causing dual release (i.e., initial immediate release followed by sustained release). It is postulated that the drug on the surface gets dispersed from the surface of drug-loaded SLNs to show an immediate release effect. Thereafter, the matrix drug-loaded SLNs may degrade and/or erode depending on the lipid composition contributing to the sustained release effect. The drug release from drug-loaded SLNs is affected by certain factors, such as particle size, temperature, and drug properties. With smaller particle size, the surface area of the same quantity of drug-loaded SLNs is larger which leads to higher drug release as compared to the larger particles. Furthermore, it has been reported that higher temperatures may cause burst drug release from drug-loaded SLNs (Parhi and Suresh 2010). Also, drug properties such as their solubility in lipid and drug-lipid interaction can influence the drug release from drug-loaded SLNs.

Various methods have been employed to study in vitro drug release, such as dialysis (diffusion or reverse dialysis) and side-by-side diffusion cells with an artificial membrane. The dialysis method is the most common method used, wherein hydrated dialysis tubing is used. The drug-loaded SLNs are placed in a dialysis sac. This dialysis sac is dialyzed against a suitable dissolution medium. The aliquots are withdrawn from the dissolution medium and analyzed for the drug content using a suitable analytical method. On the other hand, in the reverse dialysis technique, several small dialysis sacs filled with a small volume (i.e., 1 ml) of dissolution medium are placed in drug-loaded SLN dispersion. The drug-loaded SLNs are then directly placed into the dissolution medium. At predetermined time intervals, upon withdrawing the sacs, the contents of the sacs are analyzed for drug content (Wang et al. 2002). Although frequently used, the major limitation of the dialysis method is the barrier properties which results in an inherent delay of the free drug availability in the dissolution medium, which may underestimate the actual drug release from the drug-loaded SLNs (Modi and Anderson 2013). Therefore, there is a need to screen the dialysis membrane before its usage for drug release studies. In the side-by-side diffusion cell method, an artificial membrane is sandwiched between two compartments (i.e., the donor and the receptor compartment). The drug-loaded SLNs dispersion is placed in the donor compartment of the diffusion cell. Upon dialyzing the dispersion against a suitable dissolution medium, the samples are withdrawn from the dissolution medium from the receptor compartment at specific intervals ensuring maintenance of sink condition. The aliquoted samples are then analyzed for drug content using a suitable analytical method.

14.4.2.2 Drug Absorption

The accessibility of high-throughput screening approaches for evaluation of the absorption mechanism is needed, especially for drug-loaded SLNs (Porter and Charman 2001). Dissolution rate of drugs from drug-loaded SLNs in the gastrointestinal fluids, drug permeability through the lipid bilayer of the epithelial cell lining, and the intestinal membranes are considered the most important factors. Although several methods have been investigated, there are no adequate in vitro tools for evaluating absorption of the released drug and/or the uptake of intact drug-loaded lipid nanocarriers through the lymphatic system, available currently (Corti et al. 2006a, b). Among others, the method based on the use of human excised Caco-2 cell monolayer (a human colorectal adenocarcinoma cell line) has been demonstrated to be a particularly valuable tool for estimating absorption of intact drug-loaded SLNs (Shah et al. 2015).

14.4.2.2.1 Caco-2 Cell Model

The US Food and Drug Administration endorses the use of Caco-2 cell line of human origin as a model for intestinal cells for absorption studies. Once confluent, Caco-2 cell monolayer has been a remarkable in vitro tool owing to its similarity to intestinal epithelium cells. Upon exposure to lipid-based formulations, Caco-2 cells would produce and secrete the chylomicrons and have been widely used as an alternative to the animal models for the investigation and prediction of particle transport across the small intestinal epithelium (Gaumet et al. 2009). It has been reported that biodegradable nanoparticles were able to stimulate the uptake and transport of intact nanoparticles suggesting the involvement of active transport (Desai et al. 1996). The research has been reported for lipidic excipients using Caco-2 cell line to determine the effect of lipid-based excipients as formulation variable for the lymphatic delivery of therapeutic agents (Caliph et al. 2000). It has been also reported that clathrin-mediated endocytosis is involved in the uptake/transport of drug-loaded SLNs across Caco-2 cell lines (Shah et al. 2015). Therefore, the Caco-2 cell culture model could be a beneficial tool during the preliminary screening procedure for evaluating the lymphatic absorption for lipid-based formulations (Shah et al. 2015).

14.4.2.2.2 Chylomicrons Model

The association of drugs/lipids with chylomicrons is a complex process. Chylomicrons isolated from human plasma represent a classical in vitro model for evaluating the lymphatic transport of intact drug-loaded lipid comprising nanocarriers (Karpe et al. 1996). However, characterization and quantification of the harvested chylomicrons should be performed in terms of density measurements (refractometry), vesicle size (dynamic light scattering), and zeta potential before its use (Shahnaz et al. 2011). For evaluating the uptake and understanding the association of lipid nanocarriers with chylomicrons, isolated harvested chylomicron emulsion should be treated with test formulations upon incubation at 37 °C for a predetermined time. At the end of the incubation period, chylomicrons should be separated by density gradient ultracentrifugation technique for subsequent sample analysis (Gershkovich and Hoffman 2005). This model has been validated, and the results have shown that the SLNs with a different type of lipids significantly affect the affinity to chylomicrons and hence the uptake and absorption. Therefore, the chylomicron model could be a useful screening/optimizing tool for oral bioavailability determination before in vivo experiments.

14.4.3 Ex Vivo Evaluations

The ex vivo evaluations across the gut help in understanding the permeation potential of drug-loaded SLNs. The tissue from the animal (i.e., part of the intestine) is appropriately washed and cut into segments and mounted on the side-by-side diffusion cells of a specific surface area. In short, the parts of the intestine, such as the duodenum, jejunum, ileum, and colon, are ligated on both sides and subsequently used for the permeability studies. In some other cases, only rat jejunum or duodenum has been reported to determine permeability across the gut as demonstrated for the passage of enalaprilat SLNs (Yadav et al. 2013).

14.4.4 Preclinical Evaluations

Although rats have been widely employed to study the bioavailability of drug-loaded SLNs, rabbit and pig animal models have been used as well. Traditionally, rats have been used as a suitable animal model to conduct pharmacokinetic studies. For example, after an oral administration of 10 mg/kg of vinpocetine in different formulations, it was found that the relative bioavailability of vinpocetine in vinpocetine-loaded SLNs was significantly higher as compared to that of the vinpocetine solution (Luo et al. 2006). In another study, curcumin-loaded PEGylated SLNs were evaluated in rats. These drug-loaded SLNs rapidly permeated the epithelium due to the neutral surface charge of the micelles, resulting in a greater than 12-fold increase in bioavailability as compared to curcumin solution (Ban et al. 2020). Further, New Zealand white rabbits are also employed in bioavailability studies of drug-loaded SLNs. For instance, carvedilol when formulated in suspension had a negative effect on the bioavailability studied in rabbits. But when formulated as carvedilol-loaded SLNs, the bioavailability of carvedilol enhanced by more than twofold. The incorporation of carvedilol in carvedilol-loaded SLNs not only enhanced its solubility but also enhanced its tissue permeability due to the small nano-sized nature of carvedilol-loaded SLNs (El-Say and Hosny 2018). In another study, healthy pigs have been employed to study the bioavailability of drug-loaded SLNs. It has been proved that using SLNs was suitable as a drug carrier for oral administration of cyclosporine A (Müller et al. 2006). Additionally, the in vivo data along with in vitro can be used to establish an in vitro/in vivo correlation (IVIVC) for furosemide-loaded SLNs. With the application of GastroPlus®, various pharmacokinetic parameters were evaluated, and IVIVC was established with IVIVCPlus™ (Ali et al. 2017).

14.5 Recent Pharmaceutical Development of Drug-Loaded SLNs

SLNs have exhibited the potential to improve the oral bioavailability of many poorly water-soluble drugs. BCS Class II and IV drugs impede product development because of their low solubility and poor oral bioavailability. SLNs, thus, offer a highly versatile platform for oral administration to improve gastrointestinal absorption and oral bioavailability of drugs. Some of these reported attempts are discussed below.

14.5.1 Simvastatin

Simvastatin, an HMG-CoA reductase inhibitor, is a BCS Class II drug with poor oral bioavailability (i.e., < 5%). In a study, simvastatin was successfully loaded on SLNs (Padhye and Nagarsenker 2013). The anticholesterolemic effects of different simvastatin-loaded SLN formulations after oral administration were compared by poloxamer-induced hyperlipidemia in Sprague Dawley rat as an animal model. The increase in total cholesterol from the basal level was chosen as a parameter for evaluation of the efficacy of simvastatin-loaded SLN formulation in comparison to the control. The increase in total cholesterol values after the treatment was significantly ($P < 0.05$) higher in the control group as compared to groups treated with simvastatin-loaded SLNs. Since statins are also known to increase the high-density lipoprotein cholesterol, the high-density lipoprotein cholesterol levels for all groups (i.e., controls and test) were evaluated. A significant increase in high-density lipoprotein cholesterol was observed in groups treated with simvastatin-loaded SLN formulation when compared with the control group. Based on this, the superior efficacy of simvastatin-loaded SLN formulations could be attributed to the increased bioavailability of simvastatin, as a result of loading it in simvastatin-loaded SLNs. However, the major caveat of this study was the substantial variability (i.e., higher standard deviation) in results which was acknowledged by the authors.

14.5.2 Nisoldipine

Nisoldipine, a calcium channel blocker, is also a BCS Class II drug and consequently has low oral bioavailability (i.e., about 5%). In a study, nisoldipine-loaded SLNs were formulated and evaluated (Dudhipala et al. 2018). Pharmacokinetic studies were conducted after oral administration in rats, and results were compared with various controls. The serum concentration versus time profiles following single-dose administration of nisoldipine-loaded SLNs and respective controls were parametrized based on the non-compartmental analysis. From the results, the maximum plasma concentration and time to reach the peak drug concentration for nisoldipine-loaded SLNs formulation were found to be significantly higher ($P < 0.001$). A higher time to reach the peak drug concentration for nisoldipine-loaded SLNs indicates a delayed absorption of the nisoldipine-loaded SLNs. Furthermore, the mean residence time of nisoldipine-loaded SLNs was almost double as compared to the free drug formulation. Furthermore, a 2.24-fold enhancement in bioavailability was observed of nisoldipine from nisoldipine-loaded SLN formulations when compared to the control (i.e., free drug formulation).

14.5.3 Docetaxel

Docetaxel is a potent anticancer drug and a BCS Class IV drug. In an interesting study, various surface-modified [i.e., with Tween 80 or D-alpha-tocopheryl poly(ethylene glycol 1000) succinate as an emulsifier] drug-loaded SLNs were prepared and evaluated in terms of their feasibility as oral delivery systems for docetaxel in comparison with a marketed docetaxel formulation (Taxotere®) (Cho et al. 2014). Specifically, Tween 80-emulsified SLNs showed enhanced intestinal absorption. Compared with Tween 80-emulsified SLNs, the intestinal absorption of docetaxel in rats was further improved in TPGS 1000-emulsified SLNs, which was attributed to better inhibition of drug efflux by TPGS 1000. To evaluate the intestinal lymphatic uptake of docetaxel from Taxotere® and docetaxel-loaded SLNs, the docetaxel content was recovered from the mesenteric lymph node after intraduodenal administration of docetaxel formulations. The docetaxel content from both surface-modified SLNs was significantly higher than that of Taxotere®, which was attributed to tristearin-mediated lymphatic uptake of docetaxel-loaded SLNs. Taken together, it is worth noting that these surface-modified docetaxel-loaded SLNs may serve as efficient oral delivery systems for docetaxel.

14.5.4 Methotrexate

Methotrexate is an effective anticancer drug and belongs to BCS Class IV. To overcome the limitation of misleading data due to lymphatic drainage from other areas of the body obtained from cannulation of the thoracic duct, mesenteric duct cannulation was performed to confirm the lymphatic uptake of orally administered methotrexate-loaded SLNs (Paliwal et al. 2009b). The selected formulations were studied for pharmacokinetics. Periodic lymphatic concentration following oral administration of respective formulations was also determined by mesenteric duct cannulation and sample collection. Results revealed that Compritol 888 ATO-based methotrexate MTX-loaded SLNs could noticeably improve the oral bioavailability of methotrexate, which was attributed to lipid digestion of SLNs and co-absorption through the lymphatic transport route. Similarly, the lymphatic methotrexate concentration profile revealed enhanced lymphatic absorption of the Compritol-based SLNs. This was attributed to the co-absorption of methotrexate and carrier lipid into chylomicrons and subsequent transportation and, finally, release of the methotrexate at the albuminal site of the endothelium during reassembling of derivatized lipid. Based on these results, it was concluded the lymphatic uptake and well-defined transcellular mechanism of lipid processing played important roles in the increased oral bioavailability of methotrexate and thus may account for a delay in time to attain peak plasma concentration.

14.5.5 Apomorphine

Apomorphine is a mixed dopamine D1/D2 agonist used for the treatment of Parkinson's disease and has a poor oral bioavailability of only 1.7% (Subramony 2006). In a study, apomorphine-loaded SLNs were formulated for improving the oral bioavailability and subsequent brain distribution of the drug (Tsai et al. 2011). The pharmacokinetic study revealed that apomorphine-loaded SLN formulations exhibited a higher bioavailability (i.e., 13-fold) than the free form of the drug in solution. Furthermore, in the same study, the anti-Parkinsonian activity of apomorphine was evaluated with 6-hydroxydopamine-induced lesions in rats (i.e., a model of studying Parkinson's disease). And the contralateral rotation behavior was examined after oral administration of apomorphine-loaded SLN formulation as well as controls. An increase in the total number of rotations from 20 up to 115 from apomorphine-loaded SLN formulations was observed. These experimental results indicated that apomorphine-loaded SLNs may offer a promising strategy for apomorphine delivery via the oral route.

14.5.6 Sumatriptan

Sumatriptan is the first drug approved by the US Food and Drug Administration for the treatment of acute migraine. However, sumatriptan undergoes incomplete absorption and extensive first-pass metabolism, leading to an oral bioavailability of 15% (Jagdale and Pawar 2014). In a study, chitosan-coated sumatriptan-loaded SLNs were formulated (Hansraj et al. 2015). The biodistribution study exhibited brain accumulation of 5.28 μg/g. And the AUC_{brain}/AUC_{plasma} ratio after 2 h of oral administration of sumatriptan-loaded SLNs was found to be 4.5, indicating a brain targeting efficiency of sumatriptan-loaded SLNs rather than confinement in the plasma. Within the same study, the mouse model of photophobia (i.e., behavioral study) demonstrated that the time spent in the lit chamber of a light/dark box was significantly increased by sumatriptan-loaded SLNs when compared to the free drug. Based on these results, it could be elucidated that sumatriptan-loaded SLN formulations successfully crossed the blood-brain barrier. Thus, exhibiting that, sumatriptan-loaded SLNs with a chitosan overcoat can be successfully targeted to the brain via oral delivery for the management of migraine. However, it may be worthwhile to note that brain/plasma ratio, as reported in this study, may lead to erroneous interpretation about the brain targeting potential of the drug-loaded SLNs, since the brain tissue was isolated and drug content was analyzed at the end of the pharmacokinetic study (i.e., after 2 h). Reporting it as AUC_{brain} may negatively impact data interpretation, as single point data collection of tissue does not take into consideration clearance from the tissue. Therefore, caution needs to be employed during such data interpretation.

14.6 Conclusions

SLNs are lipid nanocarriers employed for drug delivery of poorly water-soluble drugs. Despite the advantages of oral administration of drug-loaded SLNs, several challenges about their formulation remain to be resolved for better application in the future. For example, the coalescence and unpredictable gelation tendency due to the presence of lipids lead to stability issues of the drug-loaded SLN formulations. On the other hand, the burst release of the surface adsorbed drug may cause toxicity. Conversely, the very slow drug release can lead to sub-therapeutic levels required for treating the disease. Therefore, optimization of drug-loaded SLN formulations based on the property of the drug and lipid is necessary. Furthermore, there is a need for more preclinical studies to better understand the role of SLNs as a carrier for poorly water-soluble drugs. Furthermore, from a commercial perspective, at least a fivefold improved oral bioavailability needs to be achieved to justify the use of SLNs over conventional approaches. Therefore, there is a need for a comprehensive understanding of the full potential of SLNs to aids its transition from the laboratory bench to large-scale commercialization.

References

Abdel-Salam FS, Elkheshen SA, Mahmoud AA, Ammar HO (2016) Diflucortolone valerate loaded solid lipid nanoparticles as a semisolid topical delivery system. Bull Fac Pharm Cairo Univ 54(1):1–7

Abrams CK, Hamosh M, Lee TC, Ansher AF, Collen MJ, Lewis JH, Benjamin SB, Hamosh P (1988) Gastric lipase: localization in the human stomach. Gastroenterology 95(6):1460–1464

Ali Khan A, Mudassir J, Mohtar N, Darwis Y (2013) Advanced drug delivery to the lymphatic system: lipid-based nanoformulations. Int J Nanomedicine 8:2733–2744. https://doi.org/10.2147/IJN.S41521

Ali H, Verma PRP, Dubey SK, Venkatesan J, Seo Y, Kim S-K, Singh SK (2017) In vitro–in vivo and pharmacokinetic evaluation of solid lipid nanoparticles of furosemide using Gastroplus™. RSC Adv 7(53):33314–33326

Almeida AJ, Runge S, Müller RH (1997) Peptide-loaded solid lipid nanoparticles (SLN): influence of production parameters. Int J Pharm. https://doi.org/10.1016/S0378-5173(97)04885-0

Amidon GL, Lennernäs H, Shah VP, Crison JR (1995) A theoretical basis for a biopharmaceutic drug classification: the correlation of in vitro drug product dissolution and in vivo bioavailability. Pharm Res 12(3):413–420. https://doi.org/10.1023/A:1016212804288

Andrade LN, Oliveira DML, Chaud MV, Alves TFR, Nery M, da Silva CF, Gonsalves JKC, Nunes RS, Corrêa CB, Amaral RG, Sanchez-Lopez E, Souto EB, Severino P (2019) Praziquantel-solid lipid nanoparticles produced by supercritical carbon dioxide extraction: physicochemical characterization, release profile, and cytotoxicity. Molecules. https://doi.org/10.3390/molecules24213881

Awad TS, Helgason T, Kristbergsson K, Weiss J, Decker EA, McClements DJ (2008) Temperature scanning ultrasonic velocity study of complex thermal transformations in solid lipid nanoparticles. Langmuir 24(22):12779–12784

Bachhav SS, Dighe VD, Kotak D, Devarajan PV (2017) Rifampicin Lipid-Polymer hybrid nanoparticles (LIPOMER) for enhanced Peyer's patch uptake. Int J Pharm 532(1):612–622. https://doi.org/10.1016/j.ijpharm.2017.09.040

Bagby TR, Duan S, Cai S, Yang Q, Thati S, Berkland C, Aires DJ, Laird Forrest M (2012) Lymphatic trafficking kinetics and near-infrared imaging using star polymer architectures with controlled anionic character. Eur J Pharm Sci 47(1):287–294. https://doi.org/10.1016/j.ejps.2012.04.016

Ban C, Jo M, Park YH, Kim JH, Han JY, Lee KW, Kweon D-H, Choi YJ (2020) Enhancing the oral bioavailability of curcumin using solid lipid nanoparticles. Food Chem 302:125328

Bargoni A, Cavalli R, Caputo O, Fundarò A, Gasco MR, Zara GP (1998) Solid lipid nanoparticles in lymph and plasma after duodenal administration to rats. Pharm Res 15(5):745–750. https://doi.org/10.1023/a:1011975120776

Battaglia L, Gallarate M, Panciani PP, Ugazio E, Sapino S, Peira E, Chirio D (2014) Techniques for the preparation of solid lipid nano and microparticles. Appl Nanotechnol Drug Deliv. https://doi.org/10.5772/58405

Beg S, Swain S, Rizwan M, Irfanuddin M, Malini DS (2011) Bioavailability enhancement strategies: basics, formulation approaches and regulatory considerations. Curr Drug Deliv 8(6):691–702. https://doi.org/10.2174/156720111797635504

Bunjes H, Westesen K, Koch MHJ (1996) Crystallization tendency and polymorphic transitions in triglyceride nanoparticles. Int J Pharm 129(1–2):159–173

Bunjes H, Koch MHJ, Westesen K (2003) Influence of emulsifiers on the crystallization of solid lipid nanoparticles. J Pharm Sci 92(7):1509–1520

Caliph SM, Charman WN, Porter CJ (2000) Effect of short-, medium-, and long-chain fatty acid-based vehicles on the absolute oral bioavailability and intestinal lymphatic transport of halofantrine and assessment of mass balance in lymph-cannulated and non-cannulated rats. J Pharm Sci 89(8):1073–1084. https://doi.org/10.1002/1520-6017(200008)89:8<1073::aid-jps12>3.0.co;2-v

Chai G-H, Xu Y, Chen S-Q, Cheng B, Hu F-Q, You J, Du Y-Z, Yuan H (2016) Transport mechanisms of solid lipid nanoparticles across Caco-2 cell monolayers and their related cytotoxicology. ACS Appl Mater Interfaces 8(9):5929–5940. https://doi.org/10.1021/acsami.6b00821

Chaikoff IL, Bloom B, Stevens BP, Reinhardt WO, Dauben WG (1951) Pentadecanoic acid-5-C14: its absorption and lymphatic transport. J Biol Chem 190:431–435

Chakraborty S, Shukla D, Mishra B, Singh S (2009) Lipid – an emerging platform for oral delivery of drugs with poor bioavailability. Eur J Pharm Biopharm 73(1):1–15. https://doi.org/10.1016/j.ejpb.2009.06.001

Charcosset C, El-Harati A, Fessi H (2005) Preparation of solid lipid nanoparticles using a membrane contactor. J Control Release. https://doi.org/10.1016/j.jconrel.2005.07.023

Chaudhari MB, Desai PP, Patel PA, Patravale VB (2016) Solid lipid nanoparticles of amphotericin B (AmbiOnp): in vitro and in vivo assessment towards safe and effective oral treatment module. Drug Deliv Transl Res 6(4):354–364. https://doi.org/10.1007/s13346-015-0267-6

Chirio D, Gallarate M, Trotta M, Carlotti ME, Gaudino EC, Cravotto G (2009) Influence of α-and γ-cyclodextrin lipophilic derivatives on curcumin-loaded SLN. J Incl Phenom Macrocycl Chem 65(3–4):391

Chirio D, Gallarate M, Peira E, Battaglia L, Muntoni E, Riganti C, Biasibetti E, Capucchio MT, Valazza A, Panciani P, Lanotte M, Annovazzi L, Caldera V, Mellai M, Filice G, Corona S, Schiffer D (2014) Positive-charged solid lipid nanoparticles as paclitaxel drug delivery system in glioblastoma treatment. Eur J Pharm Biopharm 88(3):746–758. https://doi.org/10.1016/j.ejpb.2014.10.017

Cho H-J, Park JW, Yoon I-S, Kim D-D (2014) Surface-modified solid lipid nanoparticles for oral delivery of docetaxel: enhanced intestinal absorption and lymphatic uptake. Int J Nanomedicine 9:495–504. https://doi.org/10.2147/IJN.S56648

Chudasama A, Patel V, Nivsarkar M, Vasu K, Shishoo C (2015) Role of lipid-based excipients and their composition on the bioavailability of antiretroviral self-emulsifying formulations. Drug Deliv 22(4):531–540. https://doi.org/10.3109/10717544.2014.891270

Cohen M, Morgan RGH, Hofmann AF (1971) Lipolytic activity of human gastric and duodenal juice against medium and long chain triglycerides. Gastroenterology 60(1):1–15

Cole ET, Cadé D, Benameur H (2008) Challenges and opportunities in the encapsulation of liquid and semi-solid formulations into capsules for oral administration. Adv Drug Deliv Rev 60(6):747–756. https://doi.org/10.1016/j.addr.2007.09.009

Corti G, Maestrelli F, Cirri M, Furlanetto S, Mura P (2006a) Development and evaluation of an in vitro method for prediction of human drug absorption I. assessment of artificial membrane composition. Eur J Pharm Sci 27(4):346–353. https://doi.org/10.1016/j.ejps.2005.11.004

Corti G, Maestrelli F, Cirri M, Zerrouk N, Mura P (2006b) Development and evaluation of an in vitro method for prediction of human drug absorption II. Demonstration of the method suitability. Eur J Pharm Sci 27(4):354–362. https://doi.org/10.1016/j.ejps.2005.11.005

Das S, Ng WK, Tan RBH (2012) Are nanostructured lipid carriers (NLCs) better than solid lipid nanoparticles (SLNs): development, characterizations and comparative evaluations of clotrimazole-loaded SLNs and NLCs? Eur J Pharm Sci 47(1):139–151

De Jesus MB, Radaic A, Zuhorn IS, De Paula E (2013) Microemulsion extrusion technique: a new method to produce lipid nanoparticles. J Nanopart Res. https://doi.org/10.1007/s11051-013-1960-3

Desai MP, Labhasetwar V, Amidon GL, Levy RJ (1996) Gastrointestinal uptake of biodegradable microparticles: effect of particle size. Pharm Res 13(12):1838–1845. https://doi.org/10.1023/a:1016085108889

Dixon JB (2010) Mechanisms of chylomicron uptake into lacteals. Ann N Y Acad Sci 1207(Suppl):E52–E57. https://doi.org/10.1111/j.1749-6632.2010.05716.x

Dudhipala N, Janga KY, Gorre T (2018) Comparative study of nisoldipine-loaded nanostructured lipid carriers and solid lipid nanoparticles for oral delivery: preparation, characterization, permeation and pharmacokinetic evaluation. Artif Cells Nanomed Biotechnol 46(sup2):616–625. https://doi.org/10.1080/21691401.2018.1465068

Ebrahimi HA, Javadzadeh Y, Hamidi M, Jalali MB (2015) Repaglinide-loaded solid lipid nanoparticles: effect of using different surfactants/stabilizers on physicochemical properties of nanoparticles. DARU J Pharm Sci 23(1):46

El-Say KM, Hosny KM (2018) Optimization of carvedilol solid lipid nanoparticles: an approach to control the release and enhance the oral bioavailability on rabbits. PLoS One 13(8):e0203405

Fadda P, Monduzzi M, Caboi F, Piras S, Lazzari P (2013) Solid lipid nanoparticle preparation by a warm microemulsion based process: influence of microemulsion microstructure. Int J Pharm. https://doi.org/10.1016/j.ijpharm.2013.02.027

Fang J-Y, Fang C-L, Liu C-H, Su Y-H (2008) Lipid nanoparticles as vehicles for topical psoralen delivery: solid lipid nanoparticles (SLN) versus nanostructured lipid carriers (NLC). Eur J Pharm Biopharm 70(2):633–640

Ferrari PC, Correia MK, Somer A, Ribeiro MA, Astrath NGC, Sato F, Novatski A (2019) Hesperidin-loaded solid lipid nanoparticles: development and physicochemical properties evaluation. J Nanosci Nanotechnol 19(8):4747–4757

Florence AT, Hillery AM, Hussain N, Jani PU (1995) Nanoparticles as carriers for oral peptide absorption: studies on particle uptake and fate. J Control Release 36(1–2):39–46

Forster A, Hempenstall J, Tucker I, Rades T (2001) Selection of excipients for melt extrusion with two poorly water-soluble drugs by solubility parameter calculation and thermal analysis. Int J Pharm 226(1–2):147–161. https://doi.org/10.1016/S0378-5173(01)00801-8

Freitas C, Müller RH (1999) Correlation between long-term stability of solid lipid nanoparticles (SLN™) and crystallinity of the lipid phase. Eur J Pharm Biopharm 47(2):125–132

Gaumet M, Gurny R, Delie F (2009) Localization and quantification of biodegradable particles in an intestinal cell model: the influence of particle size. Eur J Pharm Sci 36(4–5):465–473. https://doi.org/10.1016/j.ejps.2008.11.015

Gershkovich P, Hoffman A (2005) Uptake of lipophilic drugs by plasma derived isolated chylomicrons: linear correlation with intestinal lymphatic bioavailability. Eur J Pharm Sci 26(5):394–404. https://doi.org/10.1016/j.ejps.2005.07.011

Greenhalgh DJ, Williams AC, Timmins P, York P (1999) Solubility parameters as predictors of miscibility in solid dispersions. J Pharm Sci 88(11):1182–1190. https://doi.org/10.1021/js9900856

Gupta B, Poudel BK, Tran TH, Pradhan R, Cho H-J, Jeong J-H, Shin BS, Choi H-G, Yong CS, Kim JO (2015) Modulation of pharmacokinetic and cytotoxicity profile of imatinib base by employing optimized nanostructured lipid carriers. Pharm Res 32(9):2912–2927. https://doi.org/10.1007/s11095-015-1673-7

Hamosh M, Scanlon JW, Ganot D, Likel M, Scanlon KB, Hamosh P (1981) Fat digestion in the newborn: characterization of lipase in gastric aspirates of premature and term infants. J Clin Invest 67(3):838–846

Hansraj GP, Singh SK, Kumar P (2015) Sumatriptan succinate loaded chitosan solid lipid nanoparticles for enhanced anti-migraine potential. Int J Biol Macromol 81:467–476

Harde H, Das M, Jain S (2011) Solid lipid nanoparticles: an oral bioavailability enhancer vehicle. Expert Opin Drug Deliv 8(11):1407–1424. https://doi.org/10.1517/17425247.2011.604311

Helgason T, Awad TS, Kristbergsson K, McClements DJ, Weiss J (2008) Influence of polymorphic transformations on gelation of tripalmitin solid lipid nanoparticle suspensions. J Am Oil Chem Soc 85(6):501–511

Hu FQ, Yuan H, Zhang HH, Fang M (2002) Preparation of solid lipid nanoparticles with clobetasol propionate by a novel solvent diffusion method in aqueous system and physicochemical characterization. Int J Pharm. https://doi.org/10.1016/S0378-5173(02)00081-9

Huang G, Zhang N, Bi X, Dou M (2008) Solid lipid nanoparticles of temozolomide: potential reduction of cardial and nephric toxicity. Int J Pharm 355(1–2):314–320

Jagdale SC, Pawar CR (2014) Application of design of experiment for polyox and xanthan gum coated floating pulsatile delivery of sumatriptan succinate in migraine treatment. Biomed Res Int 2014

Jain S, Patel N, Lin S (2015) Solubility and dissolution enhancement strategies: current understanding and recent trends. Drug Dev Ind Pharm 41(6):875–887. https://doi.org/10.3109/03639045.2014.971027

Jain S, Patel N, Shah MK, Khatri P, Vora N (2017) Recent advances in lipid-based vesicles and particulate carriers for topical and transdermal application. J Pharm Sci 106(2):423–445. https://doi.org/10.1016/j.xphs.2016.10.001

Jenning V, Gohla S (2000) Comparison of wax and glyceride solid lipid nanoparticles (SLN®). Int J Pharm 196(2):219–222

Jenning V, Gohla SH (2001) Encapsulation of retinoids in solid lipid nanoparticles (SLN). J Microencapsul 18(2):149–158

Jenning V, Thünemann AF, Gohla SH (2000) Characterisation of a novel solid lipid nanoparticle carrier system based on binary mixtures of liquid and solid lipids. Int J Pharm 199(2):167–177

Jenning V, Lippacher A, Gohla SH (2002) Medium scale production of solid lipid nanoparticles (SLN) by high pressure homogenization. J Microencapsul 19(1):1–10

Jensen LB, Magnussson E, Gunnarsson L, Vermehren C, Nielsen HM, Petersson K (2010) Corticosteroid solubility and lipid polarity control release from solid lipid nanoparticles. Int J Pharm 390(1):53–60

Jung C, Hugot J-P, Barreau F (2010) Peyer's patches: the immune sensors of the intestine. Int J Inflamm 2010:823710. https://doi.org/10.4061/2010/823710

Karpe F, Hamsten A, Uffelman K, Steiner G (1996) Apolipoprotein B-48. Methods Enzymol 263:95–104. https://doi.org/10.1016/s0076-6879(96)63007-9

Keck CM, Kovačević A, Müller RH, Savić S, Vuleta G, Milić J (2014) Formulation of solid lipid nanoparticles (SLN): the value of different alkyl polyglucoside surfactants. Int J Pharm. https://doi.org/10.1016/j.ijpharm.2014.08.008

Kelidari HR, Saeedi M, Akbari J, Morteza-Semnani K, Valizadeh H, Maniruzzaman M, Farmoudeh A, Nokhodchi A (2017) Development and optimisation of spironolactone nanoparticles for enhanced dissolution rates and stability. AAPS PharmSciTech 18(5):1469–1474

Khoo SM, Shackleford DM, Porter CJH, Edwards GA, Charman WN (2003) Intestinal lymphatic transport of halofantrine occurs after oral administration of a unit-dose lipid-based formulation to fasted dogs. Pharm Res 20(9):1460–1465. https://doi.org/10.1023/A:1025718513246

Kiyasu JY, Bloom B, Chaikoff IL (1952) The portal transport of absorbed fatty acids. J Biol Chem 199(1):415–419

Klausen LH, Fuhs T, Dong M (2016) Mapping surface charge density of lipid bilayers by quantitative surface conductivity microscopy. Nat Commun 7(1):12447. https://doi.org/10.1038/ncomms12447

Ku MS (2008) Use of the biopharmaceutical classification system in early drug development. AAPS J 10(1):208–212. https://doi.org/10.1208/s12248-008-9020-0

Li H, Zhao X, Ma Y, Zhai G, Li L, Lou H (2009) Enhancement of gastrointestinal absorption of quercetin by solid lipid nanoparticles. J Control Release 133(3):238–244. https://doi.org/10.1016/j.jconrel.2008.10.002

Luo Y, Chen D, Ren L, Zhao X, Qin J (2006) Solid lipid nanoparticles for enhancing vinpocetine's oral bioavailability. J Control Release 114(1):53–59

Magee T, Pirinen N, Adler J, Pagakis SN, Parmryd I (2002) Lipid rafts: cell surface platforms for T cell signaling. Biol Res 35(2):127–131. https://doi.org/10.4067/s0716-97602002000200003

Makwana V, Jain R, Patel K, Nivsarkar M, Joshi A (2015) Solid lipid nanoparticles (SLN) of Efavirenz as lymph targeting drug delivery system: elucidation of mechanism of uptake using chylomicron flow blocking approach. Int J Pharm 495(1):439–446. https://doi.org/10.1016/j.ijpharm.2015.09.014

Manjunath K, Venkateswarlu V (2006) Pharmacokinetics, tissue distribution and bioavailability of nitrendipine solid lipid nanoparticles after intravenous and intraduodenal administration. J Drug Target 14(9):632–645

Manjunath K, Reddy JS, Venkateswarlu V (2005) Solid lipid nanoparticles as drug delivery systems. Methods Find Exp Clin Pharmacol 27(2):127–144

Manjunath K, Venkateswarlu V, Hussain A (2011) Preparation and characterization of nitrendipine solid lipid nanoparticles. Die Pharmazie 66(3):178–186

Mehnert W, Mäder K (2012) Solid lipid nanoparticles: production, characterization and applications. Adv Drug Deliv Rev 64:83–101

Mishra V, Bansal KK, Verma A, Yadav N, Thakur S, Sudhakar K, Rosenholm JM (2018) Solid lipid nanoparticles: emerging colloidal nano drug delivery systems. Pharmaceutics 10(4):191

Modi S, Anderson BD (2013) Determination of drug release kinetics from nanoparticles: overcoming pitfalls of the dynamic dialysis method. Mol Pharm 10(8):3076–3089. https://doi.org/10.1021/mp400154a

Müller RH, Mäder K, Gohla S (2000) Solid lipid nanoparticles (SLN) for controlled drug delivery – a review of the state of the art. Eur J Pharm Biopharm 50(1):161–177. https://doi.org/10.1016/s0939-6411(00)00087-4

Müller RH, Radtke M, Wissing SA (2002) Nanostructured lipid matrices for improved microencapsulation of drugs. Int J Pharm 242(1–2):121–128

Müller RH, Runge S, Ravelli V, Mehnert W, Thünemann AF, Souto EB (2006) Oral bioavailability of cyclosporine: solid lipid nanoparticles (SLN) versus drug nanocrystals. Int J Pharm 317(1):82–89. https://doi.org/10.1016/j.ijpharm.2006.02.045

Nanjwade BK (2011) Functions of lipids for enhancement of oral bioavailability of poorly water-soluble drugs. Sci Pharm 79(4):705–727. https://doi.org/10.3797/scipharm.1105-09

Nordskog BK, Phan CT, Nutting DF, Tso P (2001) An examination of the factors affecting intestinal lymphatic transport of dietary lipids. Adv Drug Deliv Rev 50(1–2):21–44. https://doi.org/10.1016/s0169-409x(01)00147-8

O'Driscoll CM, Griffin BT (2008) Biopharmaceutical challenges associated with drugs with low aqueous solubility – the potential impact of lipid-based formulations. Adv Drug Deliv Rev 60(6):617–624. https://doi.org/10.1016/j.addr.2007.10.012

Obeidat WM, Schwabe K, Müller RH, Keck CM (2010) Preservation of nanostructured lipid carriers (NLC). Eur J Pharm Biopharm 76(1):56–67

Olbrich C, Kayser O, Müller RH (2002) Lipase degradation of Dynasan 114 and 116 solid lipid nanoparticles (SLN) – effect of surfactants, storage time and crystallinity. Int J Pharm 237(1–2):119–128

Oner E, Kotmakci M, Kantarci AG (2020) A promising approach to develop nanostructured lipid carriers from solid lipid nanoparticles: preparation, characterization, cytotoxicity and nucleic acid binding ability. Pharm Dev Technol:1–13

Öztürk AA, Aygül A, Şenel B (2019) Influence of glyceryl behenate, tripalmitin and stearic acid on the properties of clarithromycin incorporated solid lipid nanoparticles (SLNs): formulation, characterization, antibacterial activity and cytotoxicity. J Drug Deliv Sci Technol 54:101240

Padhye SG, Nagarsenker MS (2013) Simvastatin solid lipid nanoparticles for oral delivery: formulation development and in vivo evaluation. Indian J Pharm Sci 75(5):591–598

Paliwal R, Rai S, Vaidya B, Khatri K, Goyal AK, Mishra N, Mehta A, Vyas SP (2009a) Effect of lipid core material on characteristics of solid lipid nanoparticles designed for oral lymphatic delivery. Nanomed Nanotechnol Biol Med 5(2):184–191. https://doi.org/10.1016/j.nano.2008.08.003

Paliwal R, Rai S, Vaidya B, Khatri K, Goyal AK, Mishra N, Mehta A, Vyas SP (2009b) Effect of lipid core material on characteristics of solid lipid nanoparticles designed for oral lymphatic delivery. Nanomedicine 5(2):184–191. https://doi.org/10.1016/j.nano.2008.08.003

Pantaleo G, Graziosi C, Demarest JF, Cohen OJ, Vaccarezza M, Gantt K, Muro-Cacho C, Fauci AS (1994) Role of lymphoid organs in the pathogenesis of human immunodeficiency virus (HIV) infection. Immunol Rev 140:105–130

Parhi R, Suresh P (2010) Production of solid lipid nanoparticles-drug loading and release mechanism. J Chem Pharm Res 2(1):211–227

Porter CJ, Charman WN (2001) In vitro assessment of oral lipid based formulations. Adv Drug Deliv Rev 50(Suppl 1):S127–S147. https://doi.org/10.1016/s0169-409x(01)00182-x

Porter CJH, Trevaskis NL, Charman WN (2007) Lipids and lipid-based formulations: optimizing the oral delivery of lipophilic drugs. Nat Rev Drug Discov 6(3):231–248. https://doi.org/10.1038/nrd2197

Radomska-Soukharev A (2007) Stability of lipid excipients in solid lipid nanoparticles. Adv Drug Deliv Rev 59(6):411–418

Reddy LHV, Murthy RSR (2002) Lymphatic transport of orally administered drugs. Indian J Exp Biol 40(10):1097–1109

Saupe A, Gordon KC, Rades T (2006) Structural investigations on nanoemulsions, solid lipid nanoparticles and nanostructured lipid carriers by cryo-field emission scanning electron microscopy and Raman spectroscopy. Int J Pharm 314(1):56–62

Shah MK, Madan P, Lin S (2014) Preparation, in vitro evaluation and statistical optimization of carvedilol-loaded solid lipid nanoparticles for lymphatic absorption via oral administration. Pharm Dev Technol 19(4):475–485. https://doi.org/10.3109/10837450.2013.795169

Shah MK, Madan P, Lin S (2015) Elucidation of intestinal absorption mechanism of carvedilol-loaded solid lipid nanoparticles using Caco-2 cell line as an in-vitro model. Pharm Dev Technol 20(7):877–885. https://doi.org/10.3109/10837450.2014.938857

Shah MK, Khatri P, Vora N, Patel NK, Jain S, Lin S (2019a) Chapter 5 – lipid nanocarriers: preparation, characterization and absorption mechanism and applications to improve oral bioavailability of poorly water-soluble drugs (A. M. B. T.-B. A. of N. Grumezescu (ed), pp 117–147). William Andrew Publishing. https://doi.org/10.1016/B978-0-12-816506-5.00003-6

Shah MK, Khatri P, Vora N, Patel NK, Jain S, Lin S (2019b) Lipid nanocarriers: preparation, characterization and absorption mechanism and applications to improve oral bioavailability of poorly water-soluble drugs. In: Biomedical applications of nanoparticles. Elsevier, pp 117–147

Shahnaz G, Hartl M, Barthelmes J, Leithner K, Sarti F, Hintzen F, Rahmat D, Salvenmoser W, Bernkop-Schnürch A (2011) Uptake of phenothiazines by the harvested chylomicrons ex vivo model: influence of self-nanoemulsifying formulation design. Eur J Pharm Biopharm 79(1):171–180. https://doi.org/10.1016/j.ejpb.2011.01.025

Shi L-L, Xie H, Lu J, Cao Y, Liu J-Y, Zhang X-X, Zhang H, Cui J-H, Cao Q-R (2016) Positively charged surface-modified solid lipid nanoparticles promote the intestinal transport of docetaxel through multifunctional mechanisms in rats. Mol Pharm 13(8):2667–2676. https://doi.org/10.1021/acs.molpharmaceut.6b00226

Shiau YF (1990) Mechanism of intestinal fatty acid uptake in the rat: the role of an acidic microclimate. J Physiol 421(1):463–474

Shukla T, Upmanyu N, Prakash Pandey S, Gosh D (2018) Chapter 1 – lipid nanocarriers (A. M. B. T.-L. N. for D. T. Grumezescu (ed), pp 1–47). William Andrew Publishing. https://doi.org/10.1016/B978-0-12-813687-4.00001-3

Silva AC, González-Mira E, García ML, Egea MA, Fonseca J, Silva R, Santos D, Souto EB, Ferreira D (2011) Preparation, characterization and biocompatibility studies on risperidone-loaded solid lipid nanoparticles (SLN): high pressure homogenization versus ultrasound. Colloids Surf B: Biointerfaces. https://doi.org/10.1016/j.colsurfb.2011.03.035

Simmonds WJ (1972) The role of micellar solubilization in lipid absorption. Aust J Exp Biol Med Sci 50(4):403–421

Sjöström B, Bergenståhl B (1992) Preparation of submicron drug particles in lecithin-stabilized o/w emulsions I. Model studies of the precipitation of cholesteryl acetate. Int J Pharm. https://doi.org/10.1016/0378-5173(92)90303-J

Song C, Liu S (2005) A new healthy sunscreen system for human: solid lipid nanoparticles as carrier for 3, 4, 5-trimethoxybenzoylchitin and the improvement by adding Vitamin E. Int J Biol Macromol 36(1–2):116–119

Soni NK, Sonali LJ, Singh A, Mangla B, Neupane YR, Kohli K (2020) Nanostructured lipid carrier potentiated oral delivery of raloxifene for breast cancer treatment. Nanotechnology 31(47):475101

Subramony JA (2006) Apomorphine in dopaminergic therapy. Mol Pharm 3(4):380–385

Swartz MA (2001) The physiology of the lymphatic system. Adv Drug Deliv Rev 50(1–2):3–20

Trevaskis NL, Charman WN, Porter CJH (2008) Lipid-based delivery systems and intestinal lymphatic drug transport: a mechanistic update. Adv Drug Deliv Rev 60(6):702–716. https://doi.org/10.1016/j.addr.2007.09.007

Trevaskis NL, Kaminskas LM, Porter CJH (2015) From sewer to saviour – targeting the lymphatic system to promote drug exposure and activity. Nat Rev Drug Discov 14(11):781–803. https://doi.org/10.1038/nrd4608

Trotta M, Debernardi F, Caputo O (2003) Preparation of solid lipid nanoparticles by a solvent emulsification-diffusion technique. Int J Pharm. https://doi.org/10.1016/S0378-5173(03)00135-2

Tsai M-J, Huang Y-B, Wu P-C, Fu Y-S, Kao Y-R, Fang J-Y, Tsai Y-H (2011) Oral apomorphine delivery from solid lipid nanoparticles with different monostearate emulsifiers: pharmacokinetic and behavioral evaluations. J Pharm Sci 100(2):547–557. https://doi.org/10.1002/jps.22285

Tso P (1994) Intestinal lipid absorption. Physiol Gastrointestinal Tract 2:1867–1908

van Meer G, de Kroon AIPM (2011) Lipid map of the mammalian cell. J Cell Sci 124(1):5–8. https://doi.org/10.1242/jcs.071233

Wang J-X, Sun X, Zhang Z-R (2002) Enhanced brain targeting by synthesis of 3′, 5′-dioctanoyl-5-fluoro-2′-deoxyuridine and incorporation into solid lipid nanoparticles. Eur J Pharm Biopharm 54(3):285–290

Weng T, Qi J, Lu Y, Wang K, Tian Z, Hu K, Yin Z, Wu W (2014) The role of lipid-based nano delivery systems on oral bioavailability enhancement of fenofibrate, a BCS II drug: comparison with fast-release formulations. J Nanobiotechnol 12:39. https://doi.org/10.1186/s12951-014-0039-3

Wu C-Y, Benet LZ (2005) Predicting drug disposition via application of BCS: transport/absorption/elimination interplay and development of a biopharmaceutics drug disposition classification system. Pharm Res 22(1):11–23. https://doi.org/10.1007/s11095-004-9004-4

Yadav N, Khatak S, Sara UVS (2013) Solid lipid nanoparticles-a review. Int J Appl Pharm 5(2):8–18

Yáñez JA, Wang SWJ, Knemeyer IW, Wirth MA, Alton KB (2011) Intestinal lymphatic transport for drug delivery. Adv Drug Deliv Rev 63(10–11):923–942. https://doi.org/10.1016/j.addr.2011.05.019

Yang S, Zhu J, Lu Y, Liang B, Yang C (1999) Body distribution of camptothecin solid lipid nanoparticles after oral administration. Pharm Res 16(5):751–757. https://doi.org/10.1023/A:1018888927852

You J, Wan F, de Cui F, Sun Y, Du YZ, qiang Hu, F. (2007) Preparation and characteristic of vinorelbine bitartrate-loaded solid lipid nanoparticles. Int J Pharm. https://doi.org/10.1016/j.ijpharm.2007.07.003

Yu LX, Amidon GL, Polli JE, Zhao H, Mehta MU, Conner DP, Shah VP, Lesko LJ, Chen M-L, Lee VHL, Hussain AS (2002) Biopharmaceutics classification system: the scientific basis for biowaiver extensions. Pharm Res 19(7):921–925. https://doi.org/10.1023/a:1016473601633

Chapter 15
Liposomes for Drug Delivery: Progress and Problems

Ramachandran Chelliah, Imran Khan, Eric Banan-Mwine Daliri, L. Tamizhini, K. S. Pravitha, Mahamuda Begum, Kandasamy Saravanakumar, Myeong-Hyeon Wang, and Deog Hwan Oh

15.1 Introduction

The design and production of novel agents with impending action against a widespread variety of drug targets in vitro had resulted in recent developments in biomedical science and biotechnological applications. Some of these experimental medications, however, fail in the clinic to live up to their promise. For example, while there are several anti-malignance agents that are toxic to the tumor cells, the absence of an in vivo selective anti-malignance effect prevents their clinical use. A

R. Chelliah (✉) · E. B.-M. Daliri · D. H. Oh
Department of Food Science and Biotechnology, College of Agriculture and Life Sciences, Kangwon National University, Chuncheon, Gangwon-do, South Korea

I. Khan
Department of Food Science and Technology, The University of Haripur, Haripur, Khyber Pakhtunkh, Pakistan

L. Tamizhini
Department of Biotechnology, Bhupat and Jyoti Mehta School of Biosciences, Indian Institute of Technology (IIT) Madras and Initiative for Biological Systems Engineering (IBSE), Chennai, Tamil Nadu, India

K. S. Pravitha
Department of Bioinformatics, School of Biosciences, Sri Krishna Arts and Science College, Coimbatore, Tamil Nadu, India

M. Begum
PG and Research Department of Biotechnology, Marudhar Kesari Jain College for Women, Vaniyambadi, Tamil Nadu, India

K. Saravanakumar · M.-H. Wang
Department of Medical Biotechnology, College of Biomedical Sciences, Kangwon National University, Chuncheon, Gangwon-do, Korea

J.-C. Kim et al. (eds.), *Smart Nanomaterials in Biomedical Applications*, Nanotechnology in the Life Sciences,
https://doi.org/10.1007/978-3-030-84262-8_15

low therapeutic index (TI) is one of the key drawbacks of anti-neoplastic medicines, i.e., the dosage needed to achieve an anti-carcinogenic outcome is toxic on normal cells. One of their failures was attaining therapeutic concentrations at the drug target site (solid tumor's); (ii) non-specific cell toxicity to vital regular cells such as kidney, gastrointestinal tract, and cardiac tissue; and (iii) difficulties correlated with drug formulation, such as low aqueous solubility in pharmacologically appropriate medium, resulting in the use of the drug, for example. Therefore, effective delivery systems are required which act as function of formulation aid but also adjust the distribution of drugs to reach the target site with higher concentration. Liposomes, typically 0.05–5.0/~m in diameter, are micro-particulate or colloidal carriers that impulsively develop, when certain lipids are hydrated in aqueous media (Mansoori et al. 2012).

Liposomes are constructed based on relatively biocompatible and biodegradable substance and consist of an aqueous volume of natural and/or synthetic lipids trapped by one or more bilayers (Fig. 15.1). Liposomes, is a bilayer, which compressed of water molecule and the phospholipid, based on the type of liposome, different drugs with different lipid layer may be encapsulated (Fig. 15.1). As a reservoir of numerous chemotherapeutic agents such as antimicrobial agent, liposomal amphotericin B indicated higher efficacy than the antibiotic-based amphotericin B deoxycholate (Daraee et al. 2016). More effective methods for monitoring the shelf-life and reactivity of liposomes after systematic dosage are now possible because of recent advances in liposome technology (Misra et al. 2009). At least two types of liposomes can be currently applied to build on the basis to intermingle with tissues or blood-based compounds. By integrating water polar-dependent long-chain

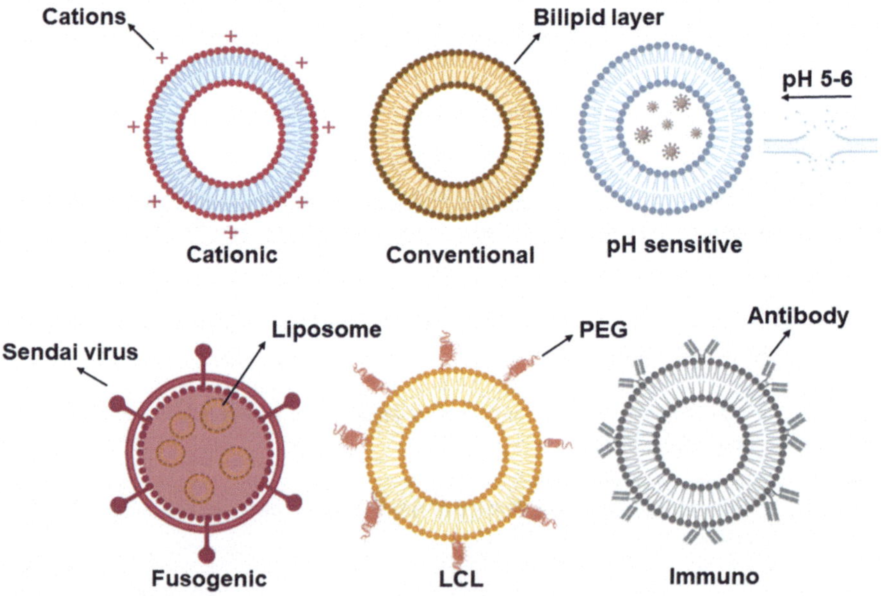

Fig. 15.1 Classification of liposomes based on composition and system of drug distribution

polymers into the bilayer that can form a coated layer on the liposome surface and prevent the penetration and adsorption of opsonin, sterically stabilized liposomes can be formulated. Reduction of the "marking" of opsonins contributes to slower reticuloendothelial cell absorption of these long-circulating liposomes (LCL).

Liposomes are microscopic, circular, concentric bilayer vesicles such as natural or synthetic phospholipid structures that can fully and widely use aqueous volume to trap drug release. Alec D. Bangham, a British hematologist at the Babraham Institute, gave the liposome concept for the first time in 1961 (Zhou and Zhao 2015). The electron microscope was applied to identify the structure after lysosome as liposome. The framework seen by the microscope appeared to be similar to the plasma lemma and was the first proof of the cell membrane bilayer lipid structure. Therefore, liposomes were found to be structurally equivalent to naturally occurring cells and found to be epitome drug carriers. Liposomal goods take a long time to catch up with the market because the traditional method of preparation makes it impossible to scale up. The major structural components of the liposome are phospholipids and cholesterol. Phospholipids are known as amphipathic molecules, due to their hydrophobic and hydrophilic tail; they have an association for aqueous and polar molecules. The head section is polar and consists of 2 chains of fatty acids containing 10–24 carbon atoms and 0–6 double bonds per chain (Sipai et al. 2012).

15.2 Liposome Formation and Classification

Liposomes play a significant role in targeted therapy, dermatology, gene transfer, and modulation of the immune system. Thus, it is important to understand the formation and classification of liposomes for the liposomal drug delivery system.

15.2.1 Liposome Formation

The fundamental portion of the liposome is built by phosphorylated lipids. The polar-dependent part is principally phosphoric corrosive bound to a water-dissolvable particle, while the non-polar part comprises two unsaturated fat chains with 10–24 carbon molecules and 0–6 double bonds in each chain (Vyas and Khar 2002). Due to the amphipathic property of phospholipids, liposomes act as a targeted organization for drug transport.

The liposomes are formed by dissolving phospholipids into an aqueous medium. The phospholipids form a lamellar sheet by arranging phosphoric acid confrontations to the aqueous environment, resulting in a spherical sac-like structure known as liposomes. The hydrophilic head is associated with an aqueous channel attached with conserving fatty acids. The phospholipids are capable of constructing closed concentric bilayered vesicles by means of amphiphilic traits when dissolving in water together with applying energy like warming, sonication, and homogenization

(Kalepu et al. 2013). At the point when phosphorylated lipids were hydrated, alongside the contribution of energy-like sonication and homogenization and so forth, it is the polar/non-polar associations between lipid-lipid and lipid-water atoms that lead to the development of bilayered vesicles to accomplish a thermodynamic harmony in the watery stage (Lasic et al. 2001). The purposes behind bilayered arrangement include:

- The troublesome connections made among the polar and non-polar stages can be limited by collapsing into shut concentric vesicles.
- Enormous bilayered vesicle development advances the decrease of substantial free energy distinction between the polar and non-polar environments.
- Higher shelf-life sustainability to supramolecular self-amassed assembly can be accomplished by framing into vesicles.

15.2.2 Organization of Liposomes

Different types of liposomes were accounted in literature. They are grouped depending on their size, number of bilayers, structure, and technique for arrangement. In view of the size and shape of bilayers, liposomes are delegated multi-lamellar vesicles (MLV), large unilamellar vesicles (LUV), and small unilamellar vesicles (SUV). In light of arrangement, they are named conventional liposomes (CL), pH-sensitive liposomes, cationic liposomes, long circulating liposomes (LCL), and immunoliposomes. For the reference of the strategy for arrangement, they are delegated reverse phase evaporation vesicles (REV), French press vesicles (FPV), and ether injection vesicles (EIV). In this specific circumstance, the order dependent on size and number of bilayers is examined beneath.

15.2.3 Classification Based on Dimension

Liposome dimensions can vary from shortest (0.025/~m) to huge (2.5/~m) vesicles. Furthermore, liposomes possess single or different bilayer films. Based on the type of lipid layer and vesicle diameter, the shelf life of the liposomes was determined. In addition the dimension of the bilayers affects the degree of the drug absorption in the liposomes. Based on their size and number of bilayers, liposomes can likewise be differentiated into one of three classes: (I) multi-lamellar vesicles (MLV); (ii) large unilamellar vesicles (LUV); and (iii) small unilamellar vesicles (SUV). The detailed description is explained below.

15.2.4 Multi-lamellar Vesicles (MLV) with Several Lamellar Phase Lipid Bilayers

MLV have a dimension more prominent than 0.1 μm, besides comprising at least dual bilayers. Their technique for readiness is straightforward, which incorporates slight film hydrated strategy or hydration of lipids in abundance of natural dissolvable. It is considered as precisely steady on extended capacity. Because of the huge size, they are cleared quickly by the reticuloendothelial framework (RES) cells and consequently can be helpful for focusing on the organs of RES (Sharma and Sharma 1997). MLV have a modest caught volume, i.e., measure of fluid capacity to lipid proportion. The medication ensnarement into the vesicles can be improved by slower pace of hydration and delicate blending (Pando et al. 2013). Liposome are risen up out of self-shaping lipid bi-layer upon hydration (Barenholz et al. 1977). Ensuing lyophilization and rehydration in the wake of blending in with the watery stage (containing the medication) can yield MLV with 40% epitome productivity (Ohsawa et al. 1984; Kirby and Gregoriadis 1984).

15.2.5 Large Unilamellar Vesicles (LUV)

This course of liposomes comprises a solitary dual layer and has a dimension more prominent than ≤0.2 μm (Vemuri and Rhodes 1995). Reverse-phase evaporation (RPE) was used for LUV preparation with 60–65% drug entrapment efficiency. In RPE, a buffer in the organic phase causes the formation of a water-in-oil emulsion of the phospholipid. Preferred position of LUV is that fewer measure of lipid is needed for exemplifying enormous amount of medication. Like MLV, they are quickly cleared by RES cells, because of their larger dimension (Pick 1981). LUV can be set up by different strategies like ether infusion, cleanser dialysis, and opposite stage dissipation procedures. Aside from these techniques, freeze-thawing of liposomes, lack of hydration/rehydration of SUV, and moderate expanding of the lipids in non-electrolyte arrangement can likewise be utilized to plan LUV (Kasahara and Hinkle. 1977; Shew and Deamer 1985; Reeves and Dowben 1969).

15.2.6 Small Unilamellar Vesicles (SUV)

SUV are more modest in size (under 0.1 μm dimension) when contrasted with MLV and LUV and have a solitary dual layer. They have a minor ensnared watery capacity to lipid proportion and been described as having extended course of shelf-life. SUV can be set up by utilizing dissolvable infusion strategy (ethanol or ether infusion strategies) (Deamer and Bangham 1976) or on the other hand by decreasing the dimension of MLV or LUV utilizing sonication or expulsion measure under a dormant air like nitrogen or argon. The sonication can be accomplished utilizing whichever a shower or test type sonicator. SUV can likewise be accomplished under high tension. These SUV are helpless to accumulation and combination at inferior or unimportant charge (Hamilton et al. 1980).

15.2.7 Classification Based on Composition

Liposomes are made out of common and additionally manufactured lipids and may likewise contain other dual-layer elements (Sharma and Sharma 1997). The nature and degree of liposome-cell interactions thusly decide the method of intracellular delivery of medications. Along these lines, the transcendent component behind intracellular distribution of medications by liposomes may for the most part rely upon their piece.

Liposomes can be ordered as far as piece and component of intracellular conveyance into five sorts as (i) traditional liposomes (CL); (ii) pH-dependent liposomes; (iii) cationic-based liposomes; (iv) immune-based liposomes; and (v) long-mingling liposomes (LCL) as shown in Fig. 15.1 with detailed description and composition of liposomes explained.

15.2.7.1 Conventional Liposomes

Ordinary liposomes, the original of liposomes, were a group of pumiceous structure or a scoriaceous structure dependent on lipid bilayers encompassing fluid compartments. These particles are commonly made out of just phospholipids (neutral or negatively charged) and additionally cholesterol without alteration (Tsermentseli et al. 2018). With its hydrophobic lipid bilayer and the hydrophilic fluid space, different kinds of medication mixes could be joined likewise. Off-the-rack particles are regularly provided as lyophilized or pre-liposome powder structure (Navarro et al. 2011; Cattel et al. 2003). Uniquely crafted liposomal plan can be sanely planned with high adaptability to build drug conveyance productivity, by being truly adaptable frameworks through changing their physical and chemical properties, for example, dimension-based lipid synthesis, surface charge and number, and ease of the phospholipid bilayer. Contrasted with free medications, traditional liposomal plans upgrade drug delivery by decreasing the harmfulness of mixes, improving medication biodistribution, and altering pharmacokinetics (Lila and Ishida 2017).

15.2.7.2 Cationic Liposomes

Cationic liposomes are structures that are made of positively charged lipids and are progressively being investigated for use in quality treatment because of their great cooperation with adversely charged DNA and cell layers (Simoes and Filipe 2005). After connecting with contrarily charged DNA, cationic liposomes structure bunches of collected vesicles. At a basic thickness, the DNA is consolidated and gets epitomized inside a lipid bilayer, in spite of the fact that it is conceivable that the liposomes tie along the outside of the DNA, holding its shape. They are additionally ready to connect with adversely charged cell layers more promptly than traditional liposomes. Combination between cationic vesicles and cell surfaces can convey the DNA legitimately over the plasma layer. This cycle sidesteps the endosomal-lysosomal course which prompts corruption of anionic liposome (Vinothini and Rajan 2019).

15.2.7.3 Immunoliposomes

Immunoliposomes are liposomes which have immunizer appended to the layer surface. The functionalization of immunoliposomes has arisen as a promising system for focused delivery to and take-up by cells overexpressing the antigens to these antibodies, with a subsequent decrease in results (Way 2013). Antibodies or different ligands can be connected to liposomes either previously or after their readiness. Restricting is accomplished either covalently or non-covalently. For covalent connection, it must be mulled over that the immunizer particle contains utilitarian compound gatherings. (Manjappa et al. 2011; Zhang et al. 2003). Regardless, this gathering happens rarely in neutralizer atoms and must be either produced by the decrease of disulfide gatherings or through proper thiolation specialists. Thiolated antibodies containing sulfhydryl gatherings will at that point respond with antibodies containing an artificially receptive atom, similar to a lipid-containing maleimide, shaping a thioether linkage. The lipid can be combined preceding its joining into the liposome's bilayer, and a response is done between the ligand and the anchor followed by blending the subsequent ligand in with different constituents of the liposomes (Mendelsohn and Baselga 2003; Guin et al. 2009). On the other hand, the anchor might be recently fused in the liposomal layer, and afterward a coupling response is completed on the outside of vehicles. Furthermore, antibodies can be non-covalently connected to liposomes, despite the fact that this system is more uncommon. The response among biotin and neutravidin or streptavidin is especially valuable. All the while, the ligand is bound to the outside of liposomes through hydrophobic anchor with functional groups (Clayton et al. 2009, Josimar et al. 2017).

Table 15.1 Liposome classification and mode of drug delivery are represented

Type	Composition	Attributes
Cationic liposome	Electronegative and/or neutrally charged phospholipids (DOTAP/DOPE, DOPE) and cholesterol	Conjugate with endosome membranes or cell Distribution of macromolecules like DNA and RNA Extreme dose becomes toxic Transformation efficiency reduces with time
pH-sensitive liposome	1,2-Dioleoyl-sn-glycero-3--phosphoethanolamine (DOPE) or phosphatidylethanolamine (PE) with cholesteryl hemisuccinate (CHEMS) or oleic acid (OA)	Coated-pit endocytosis, biodistribution Conjugate with endosome membranes and discharge the constituents into cytoplasm Intracellular distribution of weak bases
Long-circulating liposome (LCL)	Neutral lipids and cholesterol under high temperature and 5–10% PEG-DSPE	Long half-life Inferior antibody opsonization Deliquescent surface membrane Dose-dependent pharmacokinetics
Immunoliposome (IL)	LCL or CL with mAbs or recognition sequences	Clathrin-dependent endocytosis Target cell-specific binding
Fusogenic liposome (RSVE)	Reconstituted Sendai virus envelope	Endocytosis-independent Discharge their encapsulated constituents directly to the cytoplasm using fusion mechanism of Sendai virus

15.2.7.4 Long-Circulating Liposomes

Liposomes have been examined for use as medication transporters, and numerous past examinations have shown upgraded adequacy of typified drugs and the decrease of the results of medications so ensnared (Table 15.1). By and large liposomal drugs are controlled through the circulatory system. The steadiness in the circulation system, leeway, and biodistribution are reliant on the organization, size, and charge of the liposomes. Inflexible, little size (100–200 nm) liposomes will in general be held in the blood without debasement. Since the ordinary liposomes are caught in the reticuloendothelial framework (RES), RES focusing by methods for liposomes is effectively accomplished. This propensity of liposomes, in any case, is the most genuine impediment when their objective isn't the RES. Numerous endeavors have been made to dodge the RES catching and to delay the course season of liposomes with mono-sialoganglioside GM1, polyethylene glycol, glucuronide subsidiaries, etc. At the point when the objectives are tumor tissues, these RES keeping away from long-circulating liposomes inactively gather in such tissues because of extravasation through the defective vasculature in the tumor tissues. Accordingly, long-flowing liposomes are helpful instruments, particularly for tumor imaging and treatment (Oku and Namba 1994).

15.3 Margins, Which Impact on In Vivo Performance of Liposomes

Liposome technology has great potentials as drug delivery system and revolutionized the current drug delivery system. Liposome has unique characteristics such as biocompatibility, biodegradability, good encapsulation efficiency for both hydrophobic and hydrophilic drugs, etc. As discussed above, liposomes can be prepared using a number of techniques with different characteristics. In addition, during liposomal formulation optimization, several physicochemical parameters are modified for the purpose to get a desirable biodistribution, target delivery, and cellular uptake of drugs. These parameters are discussed below.

15.3.1 Bilayer Fluidity

The liposomal bilayer membrane fluidity, when it is prepared from a one type of lipid, relies on the phase transition temperature (Tc) of lipid (Rahman et al. 2018). Also the fluidity depends on the lipid relative position compared to ambient temperature (Frezard 1999). In the latter case, the lipid molecules have higher freedom of moment (Roger 1990). The liposomes with lower than 37 °C Tc behave like fluid and are susceptible to leakage of the carrier drug contents at physiological temperature. However, liposomes with greater than 37 °C Tc are well-ordered, are more rigid, and are less likely to leak at physiological temperature (Sharma et al. 2018). The Tc of liposome influences the interaction with cells, and those with low Tc lipids have increased uptake by reticuloendothelial system as compared to those with higher Tc. Moreover, modification in liposome formulation by addition of fat in the dual layer can change the lipid bilayer membrane flexibility at a temperature higher. With these modifications, liposomes have shown greater stability (Chrai et al. 2001). By increasing cholesterol level in the liposomes, bilayer made of unsaturated phospholipids exhibited reduction in the fluidity and elasticity (Kotouček et al. 2020).

15.3.2 Surface Charge

The surface charge on the liposomal vesicles has greater influence on the cell-lipid interaction. Changing the nature and density of charge on lipid composition changes the charge on the surface of liposomes. Small unilamellar vesicle liposomes start aggregation if no surface charge is present. Lack of surface charge causes reduction in liposomal stability and interaction with cell (Chibowski and Szcześ 2016). Surface charge is commonly determined in terms of zeta potential. Adsorption of ions and enzymes to the liposome surface leads to changing the zeta potential value or even its sign (positive or negative), stability, and hydrolysis reaction (Ikonen et al. 2010). The uptake of liposome by different cell types depends on the lipid composition (Düzgüneş and Nir 1999). Moreover, liposome with higher electrostatic surface charge may enhance the interaction between lipid and cells. The lipid-cell interaction magnitude is highly influenced by negatively charged density and improves the intracellular uptake of liposomes by target cells (Kelly et al. 2011). However, liposome with positive charge has enhanced passive topical delivery and can be a good strategy to improve topical ophthalmic treatments (Johnston et al. 2007).

15.3.3 Three-Dimensional Surface-Based Hydration

Surface coating controls recognition and binding behavior of particle surface to plasma components before cellular uptake. Hydrophilic surface coating of liposomes is less disposed to opsonization, and therefore their uptake by reticuloendothelial system is reduced (Senior 1992). This reduced uptake can be attributed to the hydrophilic coating on liposome which limits the interaction with blood components and cells (Sharma et al. 2018). The liposomes with hydrophilic surface coating are more stable in the biological circulation and exhibit extended shelf-life, for example, GMi and phosphatidylinositol (Allen 1994), PEG (Klibanov et al. 1990), poly(acrylamide), poly(vinyl pyrrolidone) (Torchilin et al. 1994), or poly(methyl or ethyl oxazoline) as compared to liposomes with hydrophobic surface coatings.

15.3.4 Liposome Size

Another important parameter which governs the fate of liposome in the biological circulation is liposomal vesicle size, as it determines the portion cleared by reticuloendothelial system (Daraee et al. 2016). The rate of liposomal uptake is directly related to its size. Increasing the diameter of liposome vesicle will increase the uptake by reticuloendothelial system (Dexi Liu et al. 1992; Sun et al. 2017). Liposome vesicles with size greater than 0.1 μm are rapidly taken up by reticuloendothelial system as compared to the liposomes with size smaller than 0.1 μm (Harashima et al. 1994; Ong et al. 2016). Various strategies have been proposed in order to extend the circulation time such as sonication, extrusion, and homogenization (Ong et al. 2016). Due to higher permeability of tumor capillaries, fluids along with nano-sized liposome can reach to tumor and accumulate the drug (Sharma et al. 2018).

15.3.5 Method of Preparation

The stability and in vivo circulation of liposome are also dependent on the preparation techniques. There are a number of methods used to prepare liposomal formulations. The method of choice depends on many factors including circulation time and interaction with plasma components (Bozzuto and Molinari. 2015). There are many important variants of liposome vesicle such as size and size distribution, trapped volume and osmotic activity, number of lamellae, compositional heterogeneity dimension, and dissemination of fat constituents in the bilayer. These properties are controlled by the method of preparation of the vesicles. For example, liposomal doxorubicin formulation was prepared to be medium-sized oligolamellar liposomes, negatively surface charged, in which the drug was passively entrapped by the lipid hydration technique (Hofheinz, et al. 2005).

15.3.6 Formulation Aid

Effective mechanism of the liposome system in vivo by controlling its physicochemical characteristics is vital to apprehend shelf-life and drug treatment therapy in the biological system (Rotman et al. 2015). In addition, Hu et al. (2017) reported based on the in vivo rat model, which reveled the impact of two PEG liposomal-methotrexate formulations on hydrogenated soy phosphatidylcholine.

15.4 Applications of Liposomes in Drug Delivery

Earlier, drugs were offered to any person circulating in the entire body, while drug delivery was only needed at one specific body site and not throughout the body. This was the primary explanation for the consideration of drug targeting. Drug targeting distributes medications to particular body locations, thereby improving drug efficacy and lowering the level of toxic effects. Liposomes may be handled with antibodies, sugar residues, Apo proteins, or hormones as a drug targeting agent for the compounds. Liposomes are now being developed to produce various immunogens and have been used to treat diseases like cancer a number of times (Lofthouse and Kemp 2002). Because of lack of adequate tumor-to-normal tissue drug ratios, chemotherapeutics was risky before, but cancer treatment can now be achieved with the aid of liposomes (Andresen et al. 2005). Liposomes are not appropriate for ingestion due to breakdown in the gastrointestinal tract. For different skin conditions, liposomes may be used. They may decrease the amounts of medications, thereby reducing the side effects. They improve transdermal delivery for many trapped drugs. A variety of antiretroviral nucleotide derivatives have entered the market to tackle the disease AIDS. They have a number of uses both in cosmetics and skin care formulations. The task of antiviral drugs is to combat HIV replication and, through viral DNA synthesis, to interrupt reverse transcriptase. By embedding the drug with the aid of liposomes, excess use and drug toxicity may be minimized. Preferable drug absorption contributes to a higher liposomal product value (Oussoren et al. 1999).

Liposomes are able to trap genetic code and distribute it. In gene therapy, chelating liposomes are most fitting. The biggest challenge to be solved is that the embedded genetic information should try to function and reduce its DNA and RNA susceptibility to nucleases and liposome formation at the same time. As they have extended shelf life, they can fuse to identify the target cell (Chonn and Cullis 1998). It enables plasmid condensation of DNA and contributes to highly organized DNA structure and defends toward deterioration. When DNA trapping in UV fused and when the target cell was encountered, diseases such as cystic fibrosis, infectious agents, and some tumors could be feel energized (Chen et al. 2000). Liposomes were marked with viral envelope protein so that they could identify and bind to the target cell membrane. It is easy to take up genetic material encapsulated inside liposomes and thus ensures high synaptic transmission to the cell on fusion.

15.4.1 Liposomes in Human Therapy

In human medicine, the use of liposomes is limited to systemic fungal infections and cancer therapy, given the positive and promising results obtained by using liposomes as drug carriers in multiple diseased animal models. Nevertheless, vaccines based on liposomes play a vital role in pharmaceutical marketing.

Liposomes being amphiphilic molecules are novel drug delivery systems (N.D.D.S.) having the properties to entrap drugs; studies have been successfully done on animal models (Gregoriadis and Perrie 2010). They aim to administer the medication at the time of the body's requirement during the treatment and control the drug target. Liposomes have widely helped optimize the properties of drug treatment loading and control of drug release rate, surpassing the liposome authorization pace and intracellular delivery of medications. Apart from liposomes, other drug delivery systems include niosomes, transfersomes, and pharmacosomes. The liposomes-based therapeutic index has amplified therapeutic efficacy in humans compared to the already existing non-liposomal formulation methods (Yadav et al. 2017). Various steps involved in drug delivery by liposomes are shown in Fig. 15.2.

The therapeutic applications of liposomes are categorized as:

1. Formulation aid
2. Intracellular drug delivery
3. Sustained release drug delivery
4. Gene therapy
5. Site-avoidance delivery

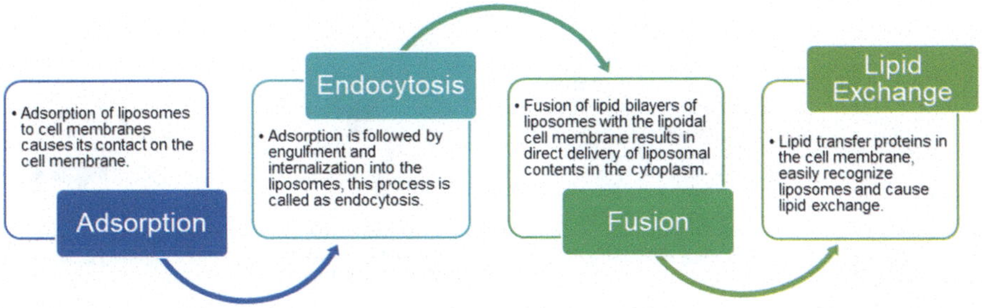

Fig. 15.2 Steps involved in drug delivery by liposomes

15.4.2 Formulation Aid

Drug formulation plays an important role in the efficiency of drugs for their targeted action and the probability of adverse reactions. The available hydrophobic medications prescribed to patients are considered toxic due to the solubilizer in which they were formulated such as surfactants and natural cosolvents for fundamental audits, which generally cause poisonousness. However, liposomes are made of phospholipid, which involves standard biological mechanism in the human body. They are considered as biocompatible and biodegradable atom; therefore, they have significant nontoxic and non-immunogenic properties (Sharma and Sharma 1997). Additionally, lipid-phosphide combination is used as solubilizer excipients to prepare better-tolerated formulations of several lipophilic drugs (Sen and Satpathy 2014). Studies reported that, liposomes are utilized to convey paclitaxel and different structures into the ethanol/cremophor drugs where the drug was systemically being delivered and increased the therapeutic index of the drug as studied in ovarian cancer models (Sharma and Sharma 1997).

15.4.3 Intracellular Drug Delivery

Medications for intracellular receptors have challenges to cross the cytomembrane to demonstrate the pharmacological activity. Liposome based drugs increase the specificity, and towards drug target mechanism in the cells by pinocytosis are generally dynamic and have satisfactory medication contrasts contrasted with extracellular fluid (Heath and Brown 1989). Therefore, liposomal delivery for pinocytosis has made it efficient rather than normal because liposomes carry comparatively high concentrations of drugs (Loira-Pastoriza et al. 2014). The drug delivery across the cytosol using drug liposome encapsulation has shown a noticeable increase in its potential action against tumor cell lines, which is regularly ineffective for free drugs (Ramkrishna and Singh 2014) as shown in Figure 15.3.

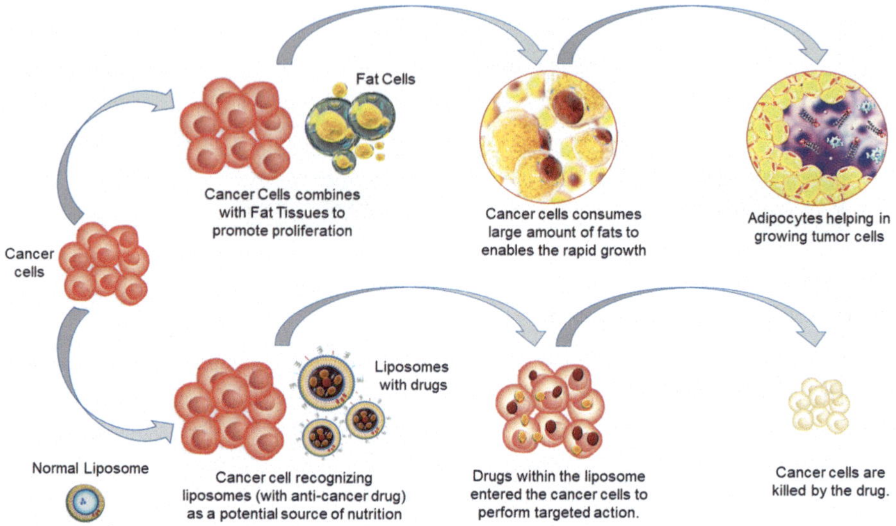

Fig. 15.3 Liposome drug delivery (targeted) can kill cancer cells

15.4.4 Sustained Release Drug Delivery

The therapeutic efficacy mainly support the delivery frameworks, which were needed to accomplish the grouping of medication managed inside the remedy, and it's frequently essential on medication conveyance frameworks of drugs that need plasma concentrations (Yadav et al. 2017). Sustained drug delivery index was also required to maintain the concentration of drugs delivered within the therapeutically effective range of medication. Due to the repetitive intake, it may influence outcomes into unfortunate poisonousness and deserted proficiency. Therefore, to weaken these fluctuations, a novel medication conveyance framework has been built up that incorporates liposomes and is efficiently being in use (Kalepu et al. 2013).

15.4.5 Gene Therapy

Liposomes are utilized broadly inside the systematic sciences besides concerning medication and quality conveyance. Various foundational infections are brought due to the absence of proteins or components resulting from absent or blemished qualities. Numerous diseases are caused by defective genes, and to resolve these diseases, several attempts have been made in recent years. As of late, a few endeavors are made to restore natural wonder by conveyance of the important exogenous DNA. This can be achieved by restoring regular gene expression by delivering relevant genes to cells. To stabilize gene expression, cationic liposome formulations are considered as a potential human gene delivery system. Some of the commonly used cationic liposomes are Lipofectin, Lipofectamine, transfectAce, Cytofectin, etc. (Sharma and Sharma 1997). For instance, Allovectin-7 is a gene transfer product under medical trials as an immune-based drug treatment therapy for metastatic melanoma (Nabel 1995). Plasmid-based liposome complexes have numerous advantages over drug delivery (Crystal et al. 1995), such as the following: (i) these complexes of non-immunogenic due to absence of proteins; (ii) Lipid-based complexes can be applied for transfection on large-dimension nuclear material; (iii) Plasmid-based liposome complexes.

15.4.6 Site-Avoidance Delivery

The drug conveyance to the targeted site plays a significant role in medications, which helps in reducing their acquaintance to other normal tissues. Drugs used in treating complex diseases like cancer are considered highly toxic to the normal tissue due to their narrow therapeutic index (Kalepu et al. 2013). This toxic impact of these drugs can be minimized by using a targeted drug delivery system rather than usual delivery, which will be profoundly harmful to normal tissues. Studies have shown that it can be achieved by targeted distribution of the drug by liposome encapsulating, reducing the toxicity at a considerable range (Szoka and Papahadjopoulos 1980). Liposomes are engrossed inadequately by tissues like the heart, kidney, and other sensitive organs, which are significant destinations for harmful results of an assortment of antineoplastic medications, and liposome detailing may improve the T.I. by adjusting the bio-conveyance of medicine a long way from drugs affecting typical tissues. For instance, the drug (H.C.Q.), which was earlier shown useful for COVID-19 patients, was later found to have an adverse impact in terms of cardiac toxicity (Stevenson et al. 2020). Therefore, liposome reformulation of these drugs could give better results in addition to reduced toxicity without any change in its therapeutic efficacy.

15.4.7 Site-Specific Targeting

The role of site-specific targeting Liposomes was initially proposed by Paul Ehrlich, for conveyance of significant portion of medication to the objective site and subsequently declined the presentation to typical tissues (Ehrlich 1906). There are two different ways of site-explicit focusing, for example, dynamic and detached objective of medications:

(i) Active targeting
(ii) Passive targeting
(iii) Intraperitoneal administration
(iv) Immunological adjuvants in vaccines

15.4.7.1 Active Targeting

Dynamic focusing of liposome wrap medications might be consummate by coupling explicit antibodies to vesicles (immunoliposomes). Immunoliposomes containing diphthera poison (D.T.) have created assurance against the vague poisonousness of D.T. during malignancy chemotherapy (Vingerhoeds et al. 1996). Long-coursing immunoliposomes might show to perceive and tie with higher particularity to zero in on cells following foundational organization (Gregoriadis 1995). It's been demonstrated that long-circling immunoliposomes (L.C.I.) increment restorative adequacy of cover doxorubicin during a murine cariogenic model (Ahmad et al.1993). The impact on dimension of biodistribution of L.C.I. has shown essentially lower aggregation in RES contrasted with CL with or without AM. Dynamic focusing on utilizing immunoliposomes has a few focal points over that of immunizer drug forms (Gregoriadis 1995): (a) immunoliposomes can convey a more significant number of medication atoms contrasted with straightforward ways; (b) immune-based liposomes can encompass medication with broadly changing physical and chemical properties; and (c) medications can even arrive at their intra-tissue-dependent objective by dissemination after delivery from immune-based liposomes related to the targeted tissue. Accordingly, dissimilar to antibodies-drug forms, sometimes immunoliposomes may not go through receptor interceded endocytosis to convey their substance intracellularly (Torchilin 1996).

15.4.7.2 Passive Targeting

Traditional liposome (CL) details of medication and immune-based stimulators have been significantly utilized for focusing on the cells of RES and show ground-breaking improvement inside the T.I. of the medicines (evaluated by Alving 1991). In clinical preliminaries, the foundational organization of CL containing muramyl peptide subsidiaries caused improvement inside the tumoricidal properties of monocytes in patients with repetitive osteosarcoma (Killion and Fidler 1994). Liposomes have likewise been acclimated to improve the antigenicity of specific atoms for spic and span antibody details. Besides, CLs have also been utilized to focus on immunosuppressive medications to lymphatic tissues, such as the spleen (Binder et al. 1994).

15.4.7.3 Intraperitoneal Administration

The intraperitoneal organizations have antineoplastic specialists that cover into the intraperitoneal whole and are restoratively helpful for malignancy advancement in the macrodimension to the peritoneal movement (Dedrick et al. 1978; Markman et al. 2004). In any case, the latitude of liposomes from the peritoneal depression is enough dawdling than that of free medicines. Consequently, substantial medication fixations should be possible in the closeness of the objective site for drawn-out timeframes with the utilization of liposome formulations. Further reformulation drugs were examined of medication doxorubicin (Forssen and Tokes. 1983).

15.4.7.4 Immunological Adjuvants in Vaccines

Liposomes were first utilized as immunological adjuvants to improve the insusceptible reaction to embodied diphtheria pathogen (Allison and Gregoriadis 1974). Therefore, since then, liposomes are applied as nontoxic complex with microbes, tumor, and other antigens (Kersten and Crommelin 1995; Gregoriadis and Panagiotidi 1989). Furthermore, counteracting agent interceded, focusing on liposomal to antigen-introducing cells may likewise improve immunostimulatory impacts (Alving 1991). The effect of physicochemical properties of the liposomes, for example, charge thickness, film ease, and epitope thickness, on the safe reaction of the antigen has been broadly examined (Kersten and Crommelin 1995), and it is observed that stage progress temperature (Tc) of the lipids additionally appears to influence immunogenicity, for instance, the immunogenicity of haptens was higher in liposomes made out of lipids with a high Tc than in those with a low Tc (Yasuda et al 1977).

15.5 Limitations of Liposome Technology

15.5.1 Based on Stability

One of the great lipid-based problems is stabilization. Therefore all physicochemical volatiles have a shorter shelf life. It is chemically unstable, or both may be the cause, because of ester bond hydrolysis or unsaturated oxidation of acyl chains. Therefore all physicochemical volatiles have a shorter shelf life. It is chemically unstable, because of ester bond hydrolysis or unsaturated oxidation of acyl chains leads to the instability of drug diffusion and intracellular fusion. Product is retained under stable dry conditions to improve shelf life and lyophilization system reliability. Liposome products are marketed in drug testing as a lyophilized powder. Liposome products are marketed in drug testing as a lyophilized powder such as, AmBisomeTM (Sharma and Sharma 1997). Multilayered liposomes of chitosan and sodium alginate resist ecological stress, such as pH and processing (liu et al. 2016). Unlike the complete specification of liposome drugs preceding, stabilization connectors such as pH-sensitive copolymers of the bola type can be incorporated into liposome drugs (Hao et al. 2017).

15.5.2 Insufficient Sterilization

One of the key problems facing liposomes is finding an efficient method for disinfecting liposomes. Liposome filtration by 0.22 μm of sterilized membranes is a method of euthanizing liposomes, but it can't be used with larger vesicles (>0.2 μm). Sterilization cannot be accomplished by irradiation or other chemical agents, since it leaves harmful contaminants in the waste. The autoclave method also sterilizes liposomes, which contributes to phospholipid deterioration and material loss (Zuidam et al. 1995). Besides the above sterilization procedures, gamma-irradiation sterilization, saturated steam sterilization, dry heat sanitization, ethylene oxide contraceptives, aseptic manufacturing, and dense gas methods are used for liposome sterilization. Yet they have their limits in sterilizing liposomes. It was thought that combining filtration, aseptic processing, and dense gas techniques to make liposomes may now solve a lot of problems, but it could take time.

15.5.3 Drawbacks in Encapsulation Efficiency

Research is also underway to improve encapsulation efficacy. Active loading also improves encapsulation efficiency. Hydrophobic drugs like paclitaxel, whose zeta potential is less than 3% mole, should not be used. This is due to the lower lipid bilayer affinity (Straubinger and Balasubramanian 2005). Today's micro-encapsulation vesicle technique is used to boost drug encapsulation efficacy (Nii and Ishii 2005).

15.5.4 Drawbacks in Gene Therapy

In viral vectors, transfecting ability is higher than in liposomes. Again, due to long-term particle size, DNA-lipid compounds are unstable. Thus the toxicity of chelating lipids restricts the DNA-lipid complex's guided dose to liposomes lacking in vivo targeting. Plasmid-liposome complexes were found to be more efficient in delivering genetic information via supervision (Gao and Huang 1996).

15.5.5 Lysosomal Degradation

Cells at the active site can't always consume the drug. If the resistant liposome which enters the cells cannot release the trapped material, it will inevitably be ruined by the lysosomes (Gregoriadis 1995).

15.5.6 Gene Therapy

In the case of virulence factors, the ability to transfect is higher than in liposomes. Again, DNA-lipid molecules are unstable because of their particle size over a long period of time. The potency of chelating lipids thereby restricts the regulated dose of lipid-DNA complex to liposomes which lack in vivo targeting. Plasmid-liposome combinations have been found to be more efficient in supplying genetic data by supervision (Gao and Huang 1995).

15.6 Future Perspectives Toward Liposome Formulations in Market

Day by day, the application of liposomes is growing significantly. The active specific knowledge of the liposome makes it unique and aids in drug delivery. It was first used for the delivery of medications, but now it is used in gene therapy, diagnostic devices, the food processing, etc. The main barriers of liposome research to improve liposome application are stability and encapsulation performance.

In the food sector, liposomes were applied for encapsulation applications. In the type of micro- or nano-encapsulation, sugar, protein, minerals, vitamins, and antioxidants are encapsulated. The food industry uses liposomes to encapsulate acids, bases, flavors, and nutrition substances (Gibbs et al. 1999). Cheese and food emulsions such as spreads, margarine, and mayonnaise are now made using liposomes. Regulated release of food content can be accomplished by liposome use. Sensitive food components can be encapsulated by the use of liposomes. By encapsulation, the shelf life and durability of sensitive food components can be increased.

References

Allen T (1994) The use of glycolipids and hydrophilic polymers in avoiding rapid uptake of liposomes by the mononuclear phagocyte system. Adv Drug Deliv Rev 13(3):285–309

Allison AC, Gregoriadis G (1974) Liposomes as immunological adjuvants. Nature 252.5480:252–25.

Andresen TL, Jensen SS, Jørgensen K (2005) Advanced strategies in liposomal cancer therapy: problems and prospects of active and tumor specific drug release. Prog Lipid Res 44(1):68–97

Barenholz Y, Gibbes D, Litman BJ, Gall J, Thompson TE, Carlson RD (1977) A simple method for the preparation of homogeneous phospholipid vesicles. Biochemistry 16:2806–2810

Bozzuto G, Molinari A (2015) Liposomes as nanomedical devices. Int J Nanomedicine 10:975–999. https://doi.org/10.2147/IJN.S68861

Binder J, Mishina E, Jusko WJ, Kupiec-Weglinski JW (1994) Prolongation of cardiac allograft survival in rats by liposome-encapsulated methylprednisolone. Transplantation 58(5):633–635

Crystal RG (1995) Transfer of genes to humans: early lessons and obstacles to success. Science 270(5235):404–41.

Cattel L, Ceruti M, Dosio F (2003) From conventional to stealth liposomes a new frontier in cancer chemotherapy[J]. Tumori J 89(3):237–249
Chen EYJ, Bartlett MC, Clarke DM (2000) Cystic fibrosis transmembrane conductance regulator has an altered structure when its maturation is inhibited. Biochemistry 39(13):3797–3803
Chibowski E, Szcześ A (2016) Zeta potential and surface charge of DPPC and DOPC liposomes in the presence of PLC enzyme. Adsorption 22(4-6):755–765
Chonn A, Cullis PR (1998) Recent advances in liposome technologies and their applications for systemic gene delivery. Adv Drug Deliv Rev 30(1-3):73–83
Chrai SS, Murari R, Ahmad I (2001) Liposomes: A review part I: Manufacturing issues. Biochem Pharmacol 14(11):10–14
Clayton R et al (2009) Sustained and specific vivo inhibition of HIV-1 replication by a protease inhibitor encapsulated in gp-120 targeted liposomes. Antivir Res 84:142–149
Daraee H, Etemadi A, Kouhi M, Alimirzalu S, Akbarzadeh A (2016) Application of liposomes in medicine and drug delivery. Artif Cells Nanomed Biotechnol 44(1):381–391
Deamer DW, Bangham AD (1976) Large volume liposomes by an ether vaporization method. Biochim Biophys Acta 443:629–634
Dedrick RL, Myers CE, Bungay PM, DeVita VT (1978) Pharmacokinetic rationale for peritoneal drug administration. Cancer Treat Rep 62:1–13
Ehrlich E (1906) Die Tatsachen des Gewohnheitsrechts.
Frezard F (1999) Liposomes: from biophysics to the design of peptide vaccines. Braz J Med Biol Res 32(2):181–189
Gao X, Huang L (1996) Potentiation of cationic liposome-mediated gene delivery by polycations. Biochemistry 35(3):1027–1036
Gregoriadis G (1995) Engineering liposomes for drug delivery: progress and problems. Trends Biotechnol 13(12):527–537
Guin S, Yao HP, Wang MH (2009) RON receptor tyrosine kinase as a target for delivery of chemo-drugs by antibody directed pathway for cancer cell cytotoxicity. Mol Pharm 7:386–397
Gao XZ, Huang LB (1995) Cationic liposome-mediated gene transfer. Gene therapy 2(10):710–722
Gibbs BF, Kermasha S, Alli I, Mulligan CN (1999) Encapsulation in the food industry: a review. Int J Food Sci Nutr 50(3):213–224
Gregoriadis G, Panagiotidi C (1989) Immunoadjuvant action of liposomes: comparison with other adjuvants. Immunol Lett 20(3):237–240
Gregoriadis G. (1995) Engineering liposomes for drug delivery: progress and problems. Trends in biotechnology 13(12):527–537
Gregoriadis G, Perrie, Y. (2010) Liposomes. eLS
Hamilton RL, Goerke J, Guo L (1980) Unilamellar liposomes made with the French pressure cell: A simple preparative and semi-quantitative technique. J Lipid Res 21:981–992
Hao J, Guo B, Yu S, Zhang W, Zhang D, Wang J, Wang Y (2017) Encapsulation of the flavonoid quercetin with chitosan-coated nano-liposomes. LWT - Food Sci Technol 85:37–44
Harashima H, Sakata K, Funato K, Kiwada H (1994) Enhanced hepatic uptake of liposomes through complement activation depending on the size of liposomes. Pharm Res 11(3):402–406
Heath TD, Brown CS (1989) Liposome dependent delivery of N-(phosphonacetyl)-L-aspartic acid to cells in vitro. Journal of Liposome Research 1(3):303–317
Hofheinz R-D, Gnad-Vogt SU, Beyer U, Hochhaus A (2005) Liposomal encapsulated anti-cancer drugs. Anti-Cancer Drugs 16(7):691–707
Hu Y, Rip J, Gaillard PJ, de Lange EC, Hammarlund-Udenaes M (2017) The Impact of liposomal formulations on the release and brain delivery of methotrexate: an in vivo microdialysis study. J Pharm Sci 106(9):2606–2613
Ikonen M, Murtomäki L, Kontturi K (2010) Microcalorimetric and zeta potential study on binding of drugs on liposomes. Colloids Surf B Biointerfaces 78(2):275–282
Johnston MJ, Semple SC, Klimuk SK, Ansell S, Maurer N, Cullis PR (2007) Characterization of the drug retention and pharmacokinetic properties of liposomal nanoparticles containing dihydrosphingomyelin. Biochimica et Biophysica Acta (BBA)-Biomembranes 1768(5):1121–1127

Josimar OE et al (2017) Immunoliposomes: a review on functionalization strategies and targets for drug delivery. Colloids Surf B Biointerfaces. https://doi.org/10.1016/j.colsurfb.2017.07.085

Kalepu S, KT S, Betha S et al (2013) Liposomal drug delivery system – a comprehensive review. Int J Drug Dev Res 5:62–75

Kasahara M, Hinkle PC (1977) Reconstitution and purification of the o-glucose transporter from human erythrocytes. J Biol Chem 252:7384–7390

Kelly C, Jefferies C, Cryan S-A (2011) Targeted liposomal drug delivery to monocytes and macrophages. J Drug Deliv 2011:727241–727241. https://doi.org/10.1155/2011/727241

Kirby CJ, Gregoriadis G (1984) A simple procedure for preparing liposomes capable of high encapsulation efficiency under mild conditions. In: Liposome technology, vol 1. CRC Press, Boca Raton, pp 19–27

Klibanov AL, Maruyama K, Torchilin VP, Huang L (1990) Amphipathic polyethyleneglycols effectively prolong the circulation time of liposomes. FEBS Lett 268(1):235–237

Kotouček J, Hubatka F, Mašek J, Kulich P, Velínská K, Bezděková J, Stráská J (2020) preparation of nanoliposomes by microfluidic mixing in herring-bone channel and the role of membrane fluidity in liposomes formation. Sci Rep 10(1):1–11

Kersten GF, Crommelin DJ (1995) Liposomes and ISCOMS as vaccine formulations. Biochim Biophys Acta 1241(2):117–138

Loira-Pastoriza C, Todoroff J, Vanbever R (2014) Delivery strategies for sustained drug release in the lungs. Adv Drug Deliv Rev 75:81–91

Lasic DD, Joannic R, Keller BC, Frederik PM, Auvray L (2001) Spontaneous vesiculation. Adv Colloid Interf Sci 89-90:337–349

Lila ASA, Ishida T (2017) Liposomal delivery systems: design optimization and current applications[J]. Biol Pharm Bull 40(1):1–10

Liu D, Mori A, Huang L (1992) Role of liposome size and RES blockade in controlling biodistribution and tumor uptake of GM1-containing liposomes. Biochimica et Biophysica Acta (BBA)-Biomembranes 1104(1):95–101

Liu W, Liu W, Ye A, Peng S, Wei F, Liu C, Han J (2016) Environmental stress stability of microencapsules based on liposomes decorated with chitosan and sodium alginate. Food Chem 196:396–404

Lofthouse S, Kemp J (2002) Manipulating the immune response; applications in livestock breeding. J Reprod Immunol 57(1-2):239–253

Manjappa AS, Chaudhari KR, Venkataraju MP, Dantuluri P, Nanda B, Sidda C et al (2011) Antibody derivatization and conjugation strategies: application in preparation of stealth immunoliposome to target chemotherapeutics to tumor. J Control Release 150(1):2–22

Mansoori MA, Agrawal S, Jawade S, Khan MI (2012) A review on liposome. Int J Adv Res Pharm Bio-sci 2(4):453–464

Mendelsohn J, Baselga J (2003) Status of epidermal growth factor receptor antagonists in the biology and treatment of cancer. J Clin Oncol 21:2787–2799

Misra A, Jinturkar K, Patel D, Lalani J, Chougule M (2009) Recent advances in liposomal dry powder formulations: preparation and evaluation. Expert Opin Drug Deliv 6(1):71–89

Markman M, Gordon AN, McGuire WP, Muggia FM (2004). Liposomal anthracycline treatment for ovarian cancer. In Seminars in oncology (Vol. 31, pp. 91-105). WB Saunders

Nabel EG (1995) Gene therapy for cardiovascular disease. Circulation 91(2):541–548

Navarro G, Cabral P, Cabrera M et al (2011) 99mTc-labeling and biological evaluation of conventional liposomes [J]. Alasbimn J 13:51

Nii T, Ishii F (2005) Encapsulation efficiency of water-soluble and insoluble drugs in liposomes prepared by the microencapsulation vesicle method. Int J Pharm 298(1):198–205

Ohsawa T, Miura H, Harada K (1984) A novel method for preparing liposome with a high capacity to encapsulate proteinous drugs: freeze-drying method. Chem Pharm Bull 32:2442–2445

Oku N, Namba Y (1994) Long-circulating liposomes. Crit Rev Ther Drug Carrier Syst 11(4):231–270

Ong SGM, Chitneni M, Lee KS, Ming LC, Yuen KH (2016) Evaluation of extrusion technique for nanosizing liposomes. Pharmaceutics 8(4):36

Oussoren C, Magnani M, Fraternale A, Casabianca A, Chiarantini L, Ingebrigsten R et al (1999) Liposomes as carriers of the antiretroviral agent dideoxycytidine-5′-triphosphate. Int J Pharm 180(2):261–270

Pando D, Gutiérrez G, Coca J, Pazos C (2013) Preparation and characterization of niosomes containing resveratrol. J Food Eng 117(2):227–234

Pick UI (1981) Liposomes with a large trapping capacity prepared by freezing and thawing of sonicated phospholipid mixtures. Arch Biochem Biophys 212:186–194

Rahman A, Uahengo V, Likius D (2018) Mini review on emerging methods of preparation of liposome and its application as Liposome drug delivery systems. Open J Pharmacol Pharmacother 3(1):005–021

Reeves JP, Dowben RM (1969) Formation and properties of thin-walled phospholipid vesicles. J Cell Physiol 734:49–60

Roger R (1990) Liposomes: a practical approach. IRL Press, Oxford/New York/Tokyo

Rotman M, Welling MM, Bunschoten A, de Backer ME, Rip J, Nabuurs RJ et al (2015) Enhanced glutathione PEGylated liposomal brain delivery of an anti-amyloid single domain antibody fragment in a mouse model for Alzheimer's disease. J Control Release 203:40–50

Ramkrishna D, Singh MR (2014) Population balance modeling: current status and future prospects. Annu Rev Chem Biomol Eng 5:123–146

Sen R, Satpathy S (2014) Liposomes as drug delivery system: A brief review

Stevenson AJ, Vanwalleghem G, Stewart TA, Condon ND, Lloyd-Lewis B, Marino N, ... Davis FM (2020) Multiscale imaging of basal cell dynamics in the functionally mature mammary gland. Proc Natl Acad Sci 117(43):26822–26832

Szoka Jr F, Papahadjopoulos D (1980) Comparative properties and methods of preparation of lipid vesicles (liposomes). Annu Rev Biophys Bioeng 9(1):467–508

Senior JH (1992) How do hydrophilic surfaces determine liposome fate in vivo? J Liposome Res 2(3):307–319

Sharma A, Sharma US (1997) Liposomes in drug delivery: progress and limitations. Int J Pharm 154(2):123–140

Sharma D, Ali AAE, Trivedi LR (2018) An updated review on: liposomes as drug delivery system. PharmaTutor 6(2):50–62

Shew RL, Deamer D (1985) A novel method for encapsulation of macromolecules in liposomes. Biochim Biophys Acta 816:1–8

Simoes S, Filipe A (2005) Cationic liposomes for gene delivery. Expert Opin Drug Deliv 2(2):237–254. https://doi.org/10.1517/17425247.2.2.237. PMID 16296751. S2CID 32349478

Sipai ABM, Vandana Y, Mamatha Y, Prasanth VV (2012) Liposomes: an overview. J Pharm Sci Innov 1(1):13–21

Straubinger RM, Balasubramanian SV (2005) Preparation and characterization of taxane-containing liposomes. In: Methods in enzymology, vol 391. Academic Press, pp 97–117

Sun X, Yan X, Jacobson O, Sun W, Wang Z, Tong X et al (2017) Improved tumor uptake by optimizing liposome based RES blockade strategy. Theranostics 7(2):319–328. https://doi.org/10.7150/thno.18078

Torchilin VP (1996) Liposomes as delivery agents for medical imaging. Mol Med Today 2(6):242–249

Torchilin VP, Shtilman MI, Trubetskoy VS, Whiteman K, Milstein AM (1994) Amphiphilic vinyl polymers effectively prolong liposome circulation time in vivo. Biochimica et Biophysica Acta (BBA)-Biomembranes 1195(1):181–184

Tsermentseli SK, Kontogiannopoulos KN, Papageorgiou VP et al (2018) Comparative study of PEGylated and conventional liposomes as carriers for shikonin [J]. Fluids 3(2):36

Vemuri S, Rhodes CT (1995) Preparation and characterization of liposomes as therapeutic delivery systems: a review. Pharm Acta Helv 70:95–111

Vinothini K, Rajan M (2019) Mechanism for the nano-based drug delivery system. In: Characterization and biology of nanomaterials for drug delivery. Elsevier, pp 219–263

Vingerhoeds MH, Haisma HJ, Belliot SO, Smit RH, Crommelin DJ, Storm G (1996) Immunoliposomes as enzyme-carriers (immuno-enzymosomes) for antibody-directed enzyme prodrug therapy (ADEPT): optimization of prodrug activating capacity. Pharm Res 13(4):604–610

Vyas SP, Khar RK (eds) (2002) Targeted and controlled drug delivery: Novel carrier systems. CBS publishers, pp 173–248

Way CJ (2013) Folate receptor-targeting liposomes for the delivery of antisense molecules to cancer cells

Yadav LR, Lingaraju K, Manjunath K, Raghu GK, Kumar KS, Nagaraju G (2017) Synergistic effect of MgO nanoparticles for electrochemical sensing, photocatalytic-dye degradation and antibacterial activity. Materials Research Express 4(2):025028

Zhang Y, Boado RJ, Pardridge WM (2003) In vivo knockdown of gene expression in brain cancer with intravenous RNAi in adult rats. J Gene Med 5(12):1039–1045

Zhou X, Zhao G (2015) Global liposome research in the period of 1995–2014: a bibliometric analysis. Scientometrics 105(1):231–248

Zuidam NJ, Gouw HME, Barenholz Y, Crommelin DJ (1995) Physical (in) stability of liposomes upon chemical hydrolysis: the role of lysophospholipids and fatty acids. Biochimica et Biophysica Acta (BBA)-Biomembranes 1240(1):101–110

Chapter 16
Niosomes: A Smart Drug Carrier Synthesis, Properties and Applications

Madhusudhan Alle, Noufel Samed, and Jin-Chul Kim

16.1 Introduction

Even though liposomes can encapsulate the various types of drugs and are capable of sustained release, they are also associated with several disadvantages like high formulation cost, brief shelf life, low stability of their aqueous suspensions at different pH levels and toxicity. These factors led to diverting the focus on developing vesicular carriers that were more biocompatible and could nullify the abovementioned shortcomings. Hence, niosomes came into being. They were first formulated by the cosmetic company L'Oréal, and the first commercial product containing niosomes was released in 1987 (Singh and Sharma 2016).

Niosomes possess most of the physical and structural properties of liposomes having unilamellar, bilamellar or multilamellar structure enclosing an aqueous core. Their ability to self-assemble easily in an aqueous medium during preparation is because of the high interfacial tension between water and the hydrocarbon portion of the surfactant involved. In addition, the presence of strong steric or ionic repulsion between the head groups of the surfactant molecule makes sure that these groups always stay in contact with water. These two strong contradicting forces impart strong structural strength to niosomes (Uchegbu and Florence 1995). The formation of stable niosomes is also governed by membrane-stabilizing agents,

M. Alle
Institute of Forest Science, Kangwon National University, Chuncheon, Republic of Korea

Department of Biomedical Science & Institute of Bioscience and Biotechnology, Kangwon National University, Chuncheon, Republic of Korea

N. Samed · J.-C. Kim (✉)
Department of Biomedical Science & Institute of Bioscience and Biotechnology, Kangwon National University, Chuncheon, Republic of Korea
e-mail: jinkim@kangwon.ac.kr

J.-C. Kim et al. (eds.), *Smart Nanomaterials in Biomedical Applications*, Nanotechnology in the Life Sciences,
https://doi.org/10.1007/978-3-030-84262-8_16

primarily cholesterol (Abdelkader et al. 2014). The property of high-interfacial activity also gives the niosomes the ability to simultaneously load many hydrophilic and hydrophobic drugs, for example, doxorubicin and curcumin (anticancer drugs) (Mahale et al. 2012) . Moreover, the presence of nonionic surfactants enhances the fluidity of biological membrane at the delivery site, hence making niosomes an ideal drug delivery vehicle for a variety of drugs via the transdermal route. It has also been observed that the stability of peptide-based drugs is more when encapsulated in niosomes. However, niosomes are also associated with disadvantages like drug leakage and its hydrolysation in aqueous suspensions, and they also exhibit more irritability than liposomes (Moghassemi and Hadjizadeh 2014; Singh and Sharma 2016).

This chapter aims to provide a comprehensive account of niosomes, including their structural properties, synthesis, characterization techniques used and administration routes and an overview of their applications.

16.2 Advantages of Niosomes

When a comparative study is done, niosomes have significant advantages over liposomes which can be summarized as follows:

1. The risk of toxicity posed by niosomes is way less when compared to liposomes.
2. Niosomes are more chemically stable and offer a longer shelf life than liposomes.
3. They are capable of being degraded within any biological system and pose no threat of severe immune response by the body in which they have been administered.
4. Because of the high fluidity exhibited by niosomes, they can be administered effectively and safely to the body via many routes like oral, transdermal, pulmonary as well as ocular with better patient compliance.
5. The presence of high-interfacial activity in niosomes gives them the ability to perform multidrug delivery mainly as a combination.
6. The use of nonionic surfactants for synthesizing niosomes gives the advantage of higher biocompatibility than liposomes (Rajera et al. 2011; Chandu et al. 2012; Moghassemi and Hadjizadeh 2014; Verma and Utreja 2019).

A variety of substances ranging from simple and complex proteins like bovine serum albumin (BSA), α-interferon, immunosuppressants like cyclosporine A, hormones like vasopressin, insulin and luteinizing releasing hormone and antigens for influenza virus have been encapsulated effectively and delivered in a targeted and sustained mode using niosomes. Moreover, as niosomes can provide high drug concentration and bioavailability, many pharmaceutical drugs also have been encapsulated in niosomes for various applications like anti-leishmanial, anti-inflammatory, anti-tubercular and hormonal and anticancer therapy (Kumar and Rajeshwarrao 2011; Mahale et al. 2012).

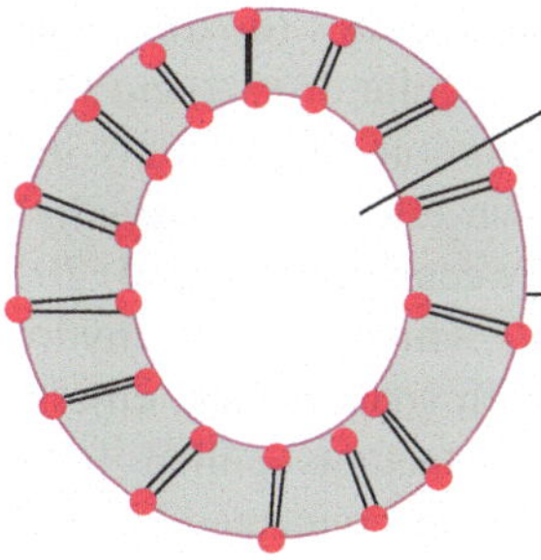

Fig. 16.1 Structure of a niosome. (Adapted from Mahale et al. 2012)

16.3 Structure of Niosomes

Niosomes have vesicular or micellar structure, generally with a hydrophilic core and hydrophobic outer shell as shown in Fig. 16.1. The formation of micellar structures mainly is by the self-assembly of the nonionic surfactants, which can be ascribed mainly to the hydrophobic and hydrophilic interactions at the interfacial level at the time of aggregation. The hydrophobic interactions take place at the hydrocarbon tail portion of the nonionic surfactant molecule. The hydrophilic interactions are between the hydrophilic heads and the aqueous medium. They tend to remain in constant contact with the aqueous medium in which the niosomes are prepared. Hence, these two opposing forces tend to decrease (hydrophobic) the interfacial area per surfactant molecule, while the hydrophilic force tends to increase the interfacial area (Abdelkader et al. 2014). Normally, hydrophilic drugs get encapsulated in the aqueous core, while hydrophobic drugs get entrapped at the hydrophobic shell (Abdelkader et al. 2014; Moghassemi and Hadjizadeh 2014).

However, the morphology of niosomes depends not only on the nonionic surfactant involved but also on other factors like stabilizing agents like cholesterol and other formulation parameters, which will be discussed in the following sections.

16.4 Factors Affecting the Structure and Properties of Niosomes

16.4.1 Nature of the Nonionic Surfactant Involved

Nonionic surfactants have a polar (hydrophilic) head and non-polar (hydrophobic) tail. Since they are amphiphilic and do not carry a net charge, they exhibit lesser toxicity, lesser haemolytic activity, more stability and lesser irritation when compared to other class of surfactants. They also show more potent inhibition towards P-proteins which enhance their drug absorption capacity. They also maintain an almost stable physiological pH in the solution. The inhibition of nonionic surfactants towards the multidrug resistant p-glycoprotein allows adsorption and thereby targeting of a variety of cancer drugs (doxorubicin, curcumin, morusin), steroids (hydrocortisone) and cardiovascular drugs (beta-blockers) (Agarwal et al. 2018).

Some of the significant nonionic surfactants include polyglycerol alkylethers, glucosyl dialkyl ethers, crown ethers, polyoxyethylene ethers and esters, such as series of Brijs, Spans and Tweens (Shilpa et al. 2011). A new class of surfactants known as Gemini surfactants which have two hydrophobic chains and two hydrophilic head groups linked with spacers have emerged as ideal candidates for niosomes. Another class of surfactants called bolaamphiphiles contains bipolar amphiphiles with two polar heads connected by one or two long hydrophobic spacers that are also being used for niosomal formulation (Pardakhty et al. 2007; Mahale et al. 2012).

The chain length and the size of the hydrophilic head group play a crucial role in the entrapment efficiency of the niosomes. Long stearyl chains increase entrapment efficiency, while shorter lauryl chains decrease entrapment efficiency. Longer alkyl group chain with extended hydrophilic head group favours entrapment of hydrophilic drugs, while shorter alkyl chain length with long hydrophilic head group is better suited for entrapment of hydrophobic drugs (Nasseri 2005). Some of the commonly used nonionic surfactants for formulating niosomes have been shown in Table 16.1 (Raymond et al. 2006; Kumar and Rajeshwarrao 2011; Ag Seleci et al. 2016; Gharbavi et al. 2018).

The other significant properties of the nonionic surfactants which determine the nature of niosomes formed are:

(a) Hydrophilic-lipophilic balance (HLB) value
(b) Critical packing parameter

(a) Hydrophilic-lipophilic balance (HLB) value

Table 16.1 Brief description of some of the commonly used nonionic surfactants used for niosome formulation

Name of the nonionic surfactant	Characteristics
Sorbitan fatty acid esters or Span	The gel transition temperature increases as the carbon chain length of the molecule increases Commonly used in making aqueous cosmetic products for enhancing oil solubilization
Alkyl ethers and alkyl glyceryl ethers polyoxyethylene 4 lauryl ether (Brij 30)	Having a HLB value of 9.7 and phase transition temperature below 10, they form LUV when combined with cholesterol Oxidation occurs upon interaction with oxidizable medications and results in discoloration
Polyoxyethylene cetyl ether (Brij 58)	Having a high HLB value of 15.7, it forms inverse vesicles
Polyoxyethylene stearyl ethers (Brij 72 and Brij 76)	As HLB value of Brij 72 is lesser than that of Brij 76 (4.9 and 12.4, respectively), Brij 72 has a higher entrapment efficiency than Brij 76
Tween or polyoxyethylene fatty acid esters	Derivatives of fatty acid-esterified ethoxylated sorbitans Tween 20, 40, 60 and 80 usually used for niosome formulation

The HLB value is a number is a ratio of the hydrophilic and lipophilic groups in a surfactant molecule. Its value ranges from 0 to 20 for nonionic surfactant molecules. Generally, as a thumb rule, surfactants with values greater than 10 have a greater affinity for water (hydrophilic), while those with a value less than 10 have an affinity towards oil (lipophilic). An HLB number ranging from 4 to 8 is highly favourable for vesicle formation (Uchegbu and Florence 1995). Attributing to their higher aqueous solubility, hydrophilic surfactants cannot form free hydrated units (vesicles), but instead, they aggregate and undergo coalescence forming lamellar-like structures (Jousma et al. 1988, 1989). On the other hand, surfactants with a higher HLB value ranging between 14 and 17 do not usually form niosomes (Shahiwala and Misra 2002). When it comes to surfactants with an HLB value of around 10, cholesterol plays a critical role in forming niosomes (Tavano et al. 2011). Table 16.2 summarizes the effect of HLB value on niosomal formation (Shahiwala and Misra 2002).

(b) Critical packing parameter

Another important factor that determines niosomal formation is the critical packing parameter (CPP) of the monomer. CPP value primarily determines the type of aggregates formed. Figure 16.2 shows the influence of CPP on the morphology of niosomes formed. It is given by the expression, CPP = $V/(a_0 \times l_c)$, where V corresponds to the molecule tail volume, a_0 corresponds to the surface area per molecule at the hydrocarbon-water interface and l_c corresponds to the critical span of a molecular chain in fluid when in aggregation. For better understanding, the diagrammatic representation of above-mentioned parameters is shown in Fig. 16.2.

However, the values of CPP and HLB alone cannot be reliable in foretelling the shape of the micelles formed. One more factor known as AC is the effective area per lipid chain, i.e. the cross-sectional of the molecule tail. If the value of A_c is above 0.43 nm^2, then a micellar structure is formed; if it is slightly below, then bilayer vesicles are formed; and if the value is way less than 0.43 nm^2, then multilamellar structures are formed (Israelachvili et al. 1976; Tavano et al. 2011). A point to be noted is that during niosomal preparation, addition of a water-insoluble nonionic surfactant (as a co-surfactant) to water-soluble nonionic surfactant can result in formation of large-sized niosomes with lesser entrapment efficiency. This can be

Table 16.2 Impact of HLB value on niosomal formulation

HLB value	Impact on formulation
Values lesser than 6	Cholesterol addition necessary to increase stability
>6	Cholesterol addition required in formation of bilayer vesicle
1.7–8.6	Reduction in entrapment efficiency
8.6	Enhancement in entrapment efficiency
14–16	Niosomes not produced

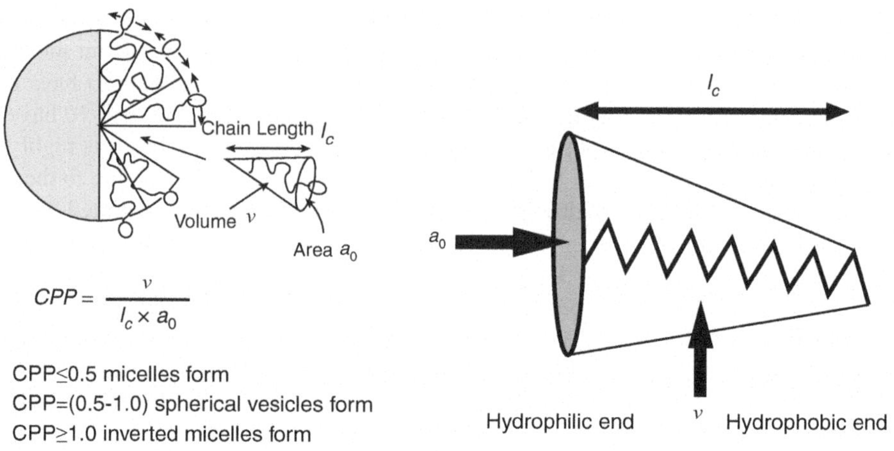

Fig. 16.2 Diagrammatic representation of critical packing parameter (CPP). (Adapted from Kumar and Rajeshwarrao 2011)

attributed to the competitive interaction between the hydrophobic co-surfactant and cholesterol during bilayer structure formation, resulting in forming a weak membrane, and hence drug leakage occurs (Yeo et al. 2019).

16.4.2 Addition of Cholesterol

Cholesterol is an inherent part of biological membranes that play a key role in the membrane properties like aggregation, fusion processes, ion permeability and elasticity and the size and shape of the vesicles. The addition of cholesterol during niosomal formulation mainly enhances the fluidity and rigidity of the niosomes (Liu et al. 2007). As shown in Fig. 16.3, cholesterol strengthens the structural integrity of the niosome by hydrogen bond formation with the surfactant involved (span 60) (Moghassemi and Hadjizadeh 2014). The addition of cholesterol is also dependent on the HLB value of the surfactant involved. When the surfactant has an HLB value of above 10, a minimum addition of cholesterol is necessary to compensate larger head groups (Bandyopadhyay and Johnson 2007). Cholesterol also improves the cohesion among the areas of non-polar portions of the vesicular bilayer formed (Manconi et al. 2006). Its action also sees the stabilizing effect of cholesterol on niosomes of inhibiting the destabilizing effects of serum and plasma components, decreasing the vesicle permeability (ROGERSON et al. 1988). Cholesterol also imparts better stability to the niosomes by improving the gel to the liquid transition temperature of the vesicles formed. Moreover, cholesterol addition helps more hydrophobic surfactants in vesicle formation by inhibiting aggregate formation (Kumar and Rajeshwarrao 2011).

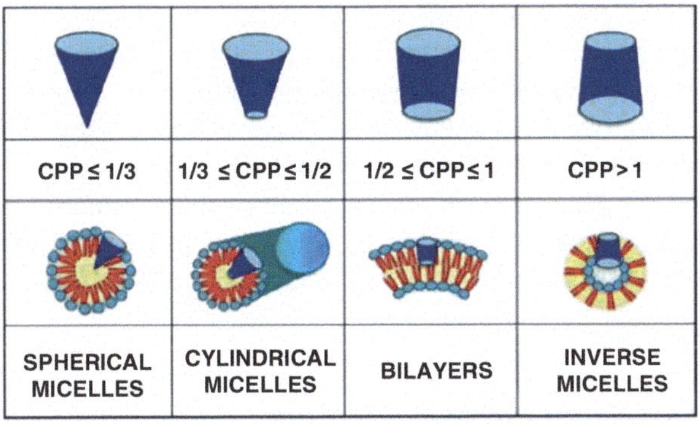

Fig. 16.3 Morphology of niosomes formed according to CPP value. (Adapted from Marianecci et al. 2014)

Generally on a quantitative basis, the maximum amount of the surfactant and lipids like cholesterol used is in the range of 1 to 2.5% (w/w). Changing the surfactant/cholesterol ratio can drastically affect the drug entrapment efficiency, the viscosity and, upon hydration, the assembly of the niosomes formed (Uchegbu and Vyas 1998; Kumar and Rajeshwarrao 2011). Studies have shown that the best entrapment efficiency is for those niosomes having 60% of cholesterol (Carafa et al. 2002; Moghassemi et al. 2015, 2017). Cholesterol also plays a key role in affecting or regulating the niosomal release profile of the entrapped drug.

16.4.3 Addition of Charge-Inducing Agents and Phase Transition Temperature

Charge-inducing agents are added for the purpose to impart a net charge on the surface of the niosomes for attaining bilayer stabilization and enhance mobility in an electrophoretic sense, as observed in the case of erythrocytes. However, adding such agents beyond a specific amount will inhibit niosomal formation (Waddad et al. 2013). Particle aggregation is also prevented due to electrostatic repulsion between like-charged vesicles. Charge-inducing agents also improve drug encapsulation efficiency in some cases and improve chances of skin permeability and chances of hybrid niosomal complex formation (Oh et al. 2006; Cametti 2008). A few examples of charge-inducing agents include dicetyl phosphate (negative charge inducing), phosphatidic acid (negative charge) and stearylamine (positive charge) (Uchegbu and Vyas 1998; Rajera et al. 2011; Abdelkader et al. 2014; Moghassemi and Hadjizadeh 2014).

Another essential factor that decides the entrapment efficiency of the surfactant involved during formulation will be the phase transition temperature of the surfactant. Phase or gel transition temperature is termed as the temperature that is required to change the inherent closely packed state of the surfactant to a more randomly oriented state with increased fluidity. An example where gel transition temperature plays a deciding factor is in the surfactant span 60. Span 60 has a trend of having higher entrapment efficiency because of its high gel transition temperature (Kumar and Rajeshwarrao 2011; Key et al. 2018). Studies conducted for combined delivery of hydrophobic and hydrophilic anticancer drugs (curcumin and doxorubicin, respectively) have shown that Tween 60 has a better entrapment efficiency than span 60 Tween 60 a longer hydrophobic and hydrophilic chains of than span 60 (Naderinezhad et al. 2017). Some of the commonly used nonionic surfactants for formulating niosomes have been shown in Table 16.1 (Manosroi et al. 2003; Kumar and Rajeshwarrao 2011; Ag Seleci et al. 2016; Gharbavi et al. 2018).

16.4.4 Hydration Medium

The type of hydration medium used to form niosomes is primal. Phosphate buffer solution (PBS) of various pH levels has been used as a hydration medium. The pH of the buffer medium is decided based on the solubility of the drug that has to be encapsulated. An increase in entrapment efficiency occurs when the hydration medium is acidic (pH=5.5), and it decreases under alkaline conditions (pH=6.8) (Kumar and Rajeshwarrao 2011). It has been observed that entrapment efficiency increases when hydration time is prolonged, but at the same time, the entrapment efficiency decreases when the volume of the hydration medium is increased (Kumar and Rajeshwarrao 2011; Mahale et al. 2012).

The temperature of the hydration medium should be maintained at a level above the gel to the liquid phase transition of the surfactant involved. The temperature determines the possibility of the aggregation or assembly of the surfactant into vesicles and the shape of the vesicles formed (Azmin et al. 1985). The duration of hydration time also influences entrapment efficiency and vesicular size, with shorter durations producing large-sized vesicles with lesser entrapment efficiency. It has been recommended that the optimal time for hydration will be 60 mins, with a volume of hydration medium being 5 ml (Yeo et al. 2019). Vesicles of smaller size are produced when hydration time is prolonged. The morphology or structural properties occur whenever a change is induced in the hydration temperature; this can also affect vesicle formation (Uchegbu and Vyas 1998; Rajera et al. 2011). In research conducted by Maryam, it was found that upon increasing the temperature of the hydration medium (from 25 °C to 55 °C) and hydration time (from 10 to 25 mins), the population of large-sized multilamellar vesicles increased (Homaei 2016). In another study, earlier formed polyhedral vesicles were transformed into spherical vesicles when the temperature was increased from 25 °C to 48 °C, and upon reverse cooling from 55 °C to 35 °C, polyhedral vesicles retained. During the transition stage, spherical vesicles of smaller size in a clustered state are formed at 49 °C (Arunothayanun et al. 2000).

16.4.5 Nature of the Drug to be Encapsulated

The degree of encapsulation of the concerned drug depends on its various properties like molecular weight and structural properties, whether hydrophilic or lipophilic (Mozafari 2007). Its subsequent interaction with the surfactant involved can affect vesicle size and entrapment efficiency. As far as entrapment of hydrophilic drugs is concerned, the efficiency is normally within 10–20% (Peltonen et al. 2002; Ferreira et al. 2004). But it has been observed if a charge interaction occurs between the surfactant and the drug, there is an enhancement in the entrapment efficiency. This was seen in a study conducted on transdermal delivery of a hydrophilic drug gallidermin in which interaction that occurred between the positive and negative charges of the drug and niosomes, respectively, brought up the entrapment efficiency up to 45% (Manosroi et al. 2010).

When it comes to the entrapment efficiency of hydrophobic drugs, an inverse relationship exists, i.e. entrapment efficiency decreases as the concentration of the drug increases as that would interfere with the vesicle formation (Marianecci et al. 2010). The process of drug entrapment ceases once the bilayer structural saturation occurs.

16.4.6 Method of Niosomal Preparation

The method of niosomal preparation can influence the vesicle size and drug entrapment efficiency. A study conducted concerning prednisolone-loaded ethoniosomes (ethanol-based niosomes) showed that the vesicles prepared by thin-film hydration method have better entrapment efficiency than those prepared by the ethanol injection method. But niosomes produced by the ethanol injection method were smaller in size (Gaafar et al. 2014). The niosomes formulated by the transmembrane pH method have high drug entrapment efficiency (Parthasarathi et al. 1994; Kazi et al. 2010; Rajera et al. 2011).

16.5 Methods of Preparation of Niosomes

16.5.1 Thin-Film Hydration Technique

Also called the "Hand shaking method," this is the most commonly used preparation method. As shown in Fig. 16.4, initially, the involved nonionic surfactant, hydrophobic drug and cholesterol taken in appropriate quantities are dissolved in a suitable organic solvent and introduced in a round bottom flask. The organic solvent is then evaporated preferably in a rotary evaporator leading to the formation of a thin layer or film, which, when wetted by a suitable hydration medium like

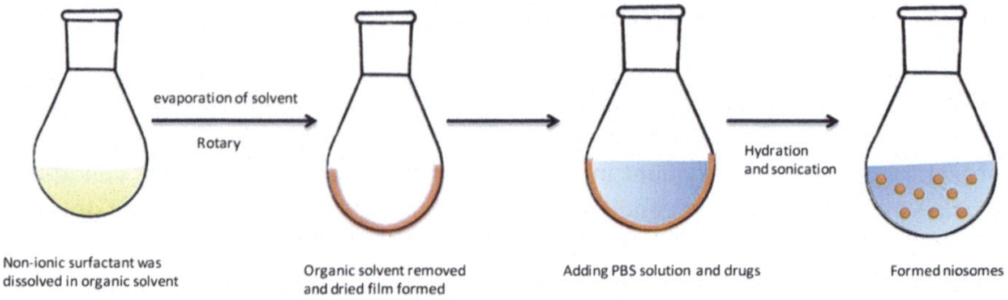

Fig. 16.4 Thin-film hydration technique or "Hand shaking method". (Adapted from Ge et al. 2019)

phosphate buffer (in which hydrophilic drug is dissolved), results in the formation of multilamellar vesicles (Rajera et al. 2011; Gandhi et al. 2012; Moghassemi and Hadjizadeh 2014; Ag Seleci et al. 2016). It can be seen that a large variety of hydrophilic and hydrophobic drugs can be encapsulated in niosomes using this technique, for example, paclitaxel (Bayindir and Yuksel 2010), zidovudine (Ruckmani and Sankar 2010), green tea extract (Isnan and Jufri 2017), lornoxicam (Bini et al. 2012) and gallidermin (Manosroi et al. 2003).

16.5.2 Ether Injection Method

An organic solvent like diethyl ether is used to mix the surfactant and cholesterol together, and then the mixture gets introduced into an aqueous solution containing the drug. It has to be made sure the temperature of the aqueous medium should be maintained at a temperature above 60 °C (Moghassemi and Hadjizadeh 2014). The solvent is evaporated, which leads to the formation of unilamellar vesicles of diameter ranging from 50 to 1000 μm (Fig. 16.5) (Uchegbu and Vyas 1998; Verma et al. 2010). This method has been used to prepare stavudine-loaded niosomes (Shreedevi et al. 2016).

16.5.3 Reverse-Phase Evaporation Method

After mixing the surfactant and cholesterol in an organic solvent, an aqueous solution of the preferred hydrophilic drug is added to the organic mixture, leading to the formation of a two-phase system (Fig. 16.6). After homogenization, the organic phase is then removed in negative pressure mode. This method forms large unilamellar vesicles (LUVs) and has been used to prepare niosomal formulations containing bovine serum albumin (Moghassemi et al. 2017), ellagic acid (Junyaprasert et al. 2012) and isoniazid (Singh et al. 2011).

Fig. 16.5 Ether injection method. (Adapted from Chen et al. 2019)

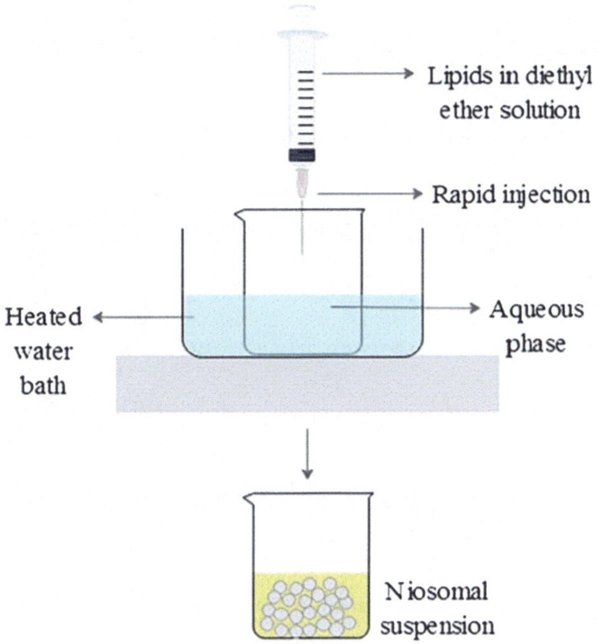

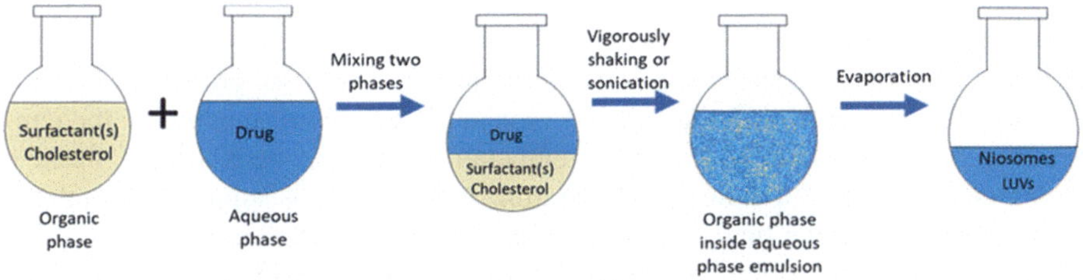

Fig. 16.6 Schematic diagram of reverse phase evaporation method. (Adapted from Durak et al. 2020)

16.5.4 Transmembrane pH Gradient

This method involves dissolving equal proportions of the surfactant and cholesterol in an organic solvent like chloroform, followed by removing the organic solvent under reduced pressure. A thin film like lipid layer gets formed, hydrated by vortexing using an acidic solution (usually citric acid). Once the mixture is freeze-thawed, an aqueous solution of the desired drug is added and mixed well by vortexing. The pH of the final mixture is then adjusted to the desired value using disodium hydrogen solution (Fig. 16.7). Remote drug loading can happen using this method (Kazi et al. 2010; Kumar and Rajeshwarrao 2011; Durak et al. 2020).

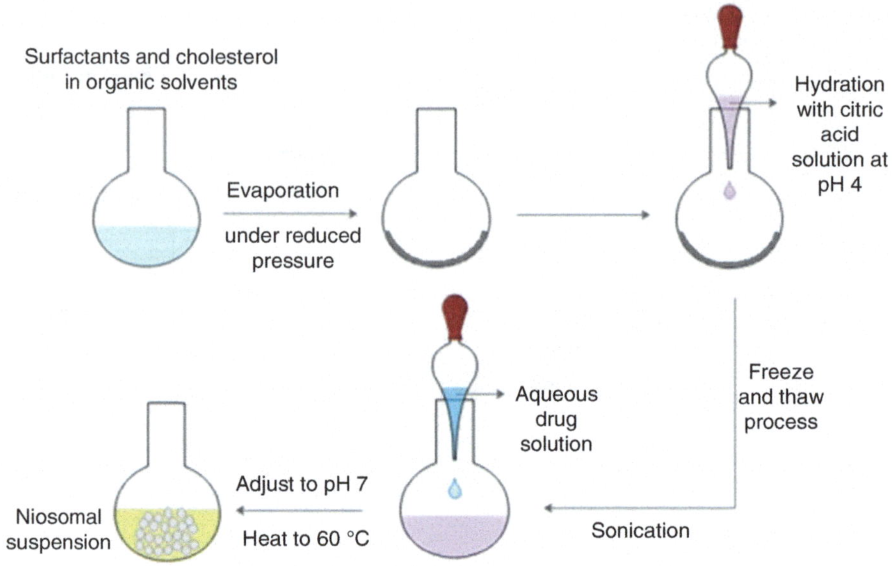

Fig. 16.7 Schematic diagram of transmembrane pH gradient. (Adapted from Chen et al. 2019)

16.5.5 The "Bubble" Method

This method is devoid of using an organic solvent. A layered system comprising of a flask having three necks is used. In the flask, proper mixing of the surfactant, cholesterol and phosphate buffer is ensured. A water bath is used to maintain the temperature of the flask at the desired temperature. At one neck of the flask, a thermometer is placed to note the temperature. Nitrogen gas is supplied through the second neck, while the final neck has an attachment for reflux cooling. After dispersing the mixture at a temperature of 70 °C and homogenizing for 15 s, nitrogen gas is supplied (Fig. 16.8) (Moghassemi and Hadjizadeh 2014). Large unilamellar vesicles formed by this method are required to undergo a suitable size reduction (Kazi et al. 2010; Mahale et al. 2012).

16.5.6 Microfluidization Method

In this method, a mixture of the desired drug and the surfactant upon dissolving in a reservoir is pumped under pressure to an interaction chamber maintained at low temperatures (using ice packs) because the resulting interaction produces heat. The solution is passed through a cooling loop for further cooling (Fig. 16.9). Comparatively, much uniformly smaller-sized niosomes get produced by this method (Sorgi and Huang 1996; Kazi et al. 2010; Kumar and Rajeshwarrao 2011).

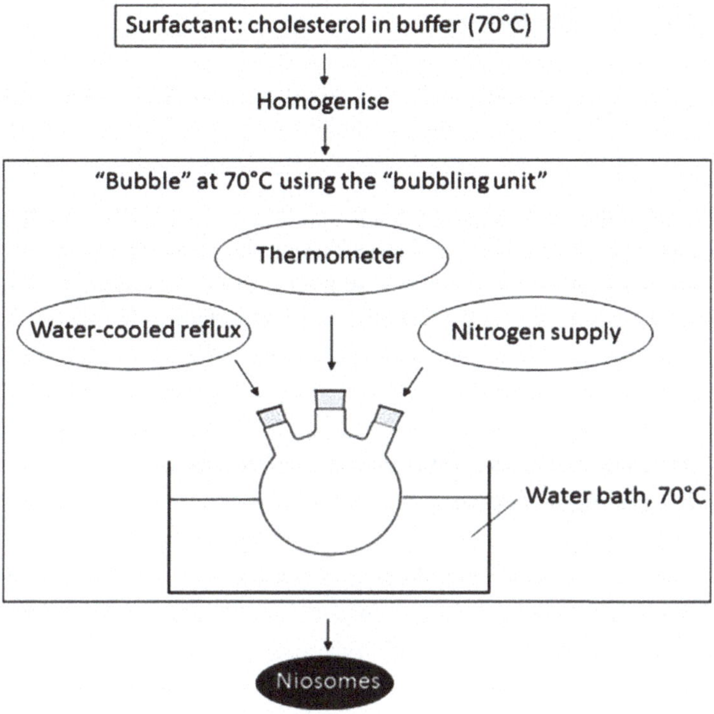

Fig. 16.8 Schematic diagram of the "Bubble" method. (Adapted from Yeo et al. 2017)

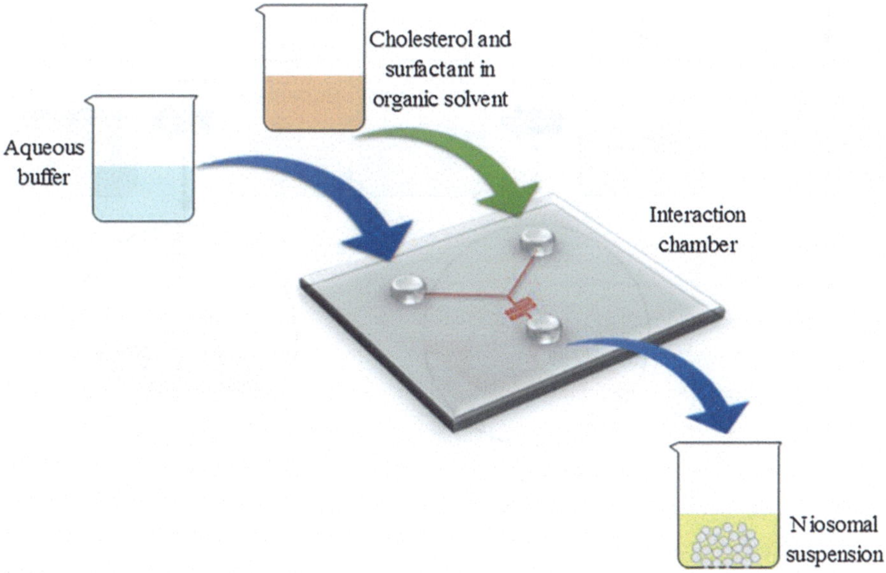

Fig. 16.9 Schematic diagram of microfluidization method. (Adapted from Chen et al. 2019)

16.5.7 Heating Method

In this method, the surfactant (and other additives) and cholesterol are hydrated separately with buffer at room temperature under an inert nitrogen atmosphere for about 1 h. The buffer solution of cholesterol is heated to 120 °C for an hour to ensure proper cholesterol dissolving. After cooling the solution to about 60 °C, a buffer solution of the surfactant and other ingredients are added while the solution is continuously stirred, leading to the formation of niosomes. The niosomes are first kept at room temperature for about half an hour, followed by storage at 4–5 °C in nitrogen atmosphere (Fig. 16.10) (Moghassemi and Hadjizadeh 2014; Ag Seleci et al. 2016).

16.5.8 Multi-membrane Extrusion Technique or Single-Pass Technique

This method involves passing a lipid-containing drug from a porous device through a nozzle. Passing through the nozzle allows forming uniformly sized niosomes with about 50–500 nm (Gharbavi et al. 2018).

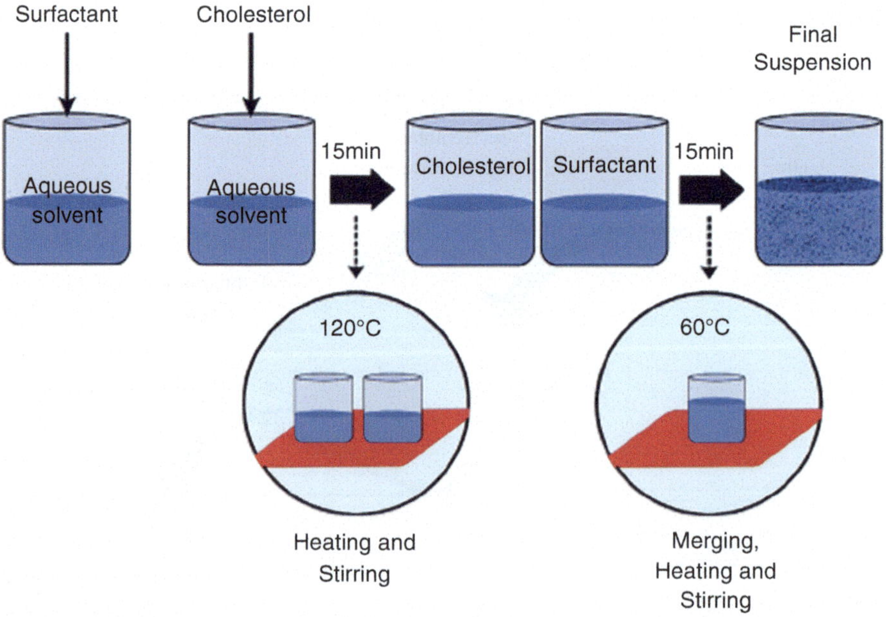

Fig. 16.10 Schematic diagram of a heating method. (Adapted from Moghassemi and Hadjizadeh 2014)

16.5.9 Dehydration-Rehydration Method

In this method, niosomes are first prepared by thin-film hydration method followed by overnight freeze-drying using liquid nitrogen, which results in the formation of powdered niosomes. The powered niosomes are then hydrated using phosphate buffer saline of pH 7.4 at 60 °C. Niosomes containing the drug naltrexone for ocular delivery have been formulated using this method (Abdelkader et al. 2012).

16.5.10 Freeze and Thaw Method

This method is somewhat similar to the method mentioned above, but the niosomes prepared by thin-film hydration undergo freezing cycles. The niosomes are first frozen using liquid nitrogen for a minute, followed by 1-min thawing in a water bath at 60 °C. This cycle is done five times. This alternate freeze method and thaw helps reduce the size of the niosomes formed, but the entrapment efficiency of the niosomes is compromised. This method produces multilamellar vesicles (Moghassemi and Hadjizadeh 2014; Bartelds et al. 2018).

16.5.11 Sonication Method

This method is just probe sonication of a mixture of the surfactant and cholesterol and a buffer solution of the drug at 60 °C, which produces multilamellar vesicles. Unilamellar vesicles can be further obtained using ultrasonication. This method was used to formulate niosomes containing the drug cefdinir (Bansal et al. 2013).

16.5.12 The Handjani-Vila Method

In this method, a mix of the surfactant and cholesterol is added to an aqueous solution of the desired drug followed by agitation or ultracentrifugation for the purpose of homogenization. Throughout the process, the temperature has to be maintained (Palani 2010).

16.5.13 Supercritical Carbon Dioxide Fluid

In this method, the supercritical carbon dioxide apparatus used to prepare niosomes is a view cell with two glass windows on both sides equipped with a high-pressure pump for carbon dioxide gas feeding. An appropriate amount of the surfactant mixed with cholesterol and the drug dissolved in Dulbecco's phosphate buffer solution is added to the view cell. The view cell is maintained at high pressure (about 200 bar) and temperature (about 60 °C), and carbon dioxide is introduced into the view cell. The mixture undergoes magnetic stirring for about 30 min, and once equilibrium is achieved, the pressure is released, and niosome dispersions are obtained. Throughout the process, the temperature has to be maintained constant (Manosroi et al. 2008).

16.5.14 Microfluidic Hydrodynamic Focusing

In this method, two miscible liquids undergo mixing in microchannels in a rapid but controlled manner. This method is very effective in producing niosomes with better size distribution than other regularly used methods. The parameters like surfactant chemical structure, the material used for microchannel fabrication and microfluid mixing conditions affect the assembly of the niosomes. When the flow rate ratio is increased, there is a subsequent decrease in the diffusive mixing time, hence producing small-sized niosomes. However, when a microchannel of a wider size is used, diffusive mixing time increases, thereby producing large-sized niosomes (Lo et al. 2010).

16.6 Type of Niosomes

Broadly, niosomes formed are of three types, as shown in Table 16.3 and Fig. 16.11.

Besides this general classification, there are some special niosomes formed as discussed below which was shown in Table 16.4.

16.6.1 Proniosomes

These are niosomes in the dehydrated state which, upon hydration in the aqueous medium, form dispersions. Since being in the dry form, the chances of the niosomes to aggregate, fuse or cake are less than average niosomes, and hence the problems

Table 16.3 Major types of niosomes

Type of niosomes	Characteristics
Small unilamellar vesicles (SUVs)	10–100 nm in size
Large unilamellar vesicles (LUVs)	100–3000 nm in size
Multilamellar vesicles (MLVs)	More than 1000 nm in size

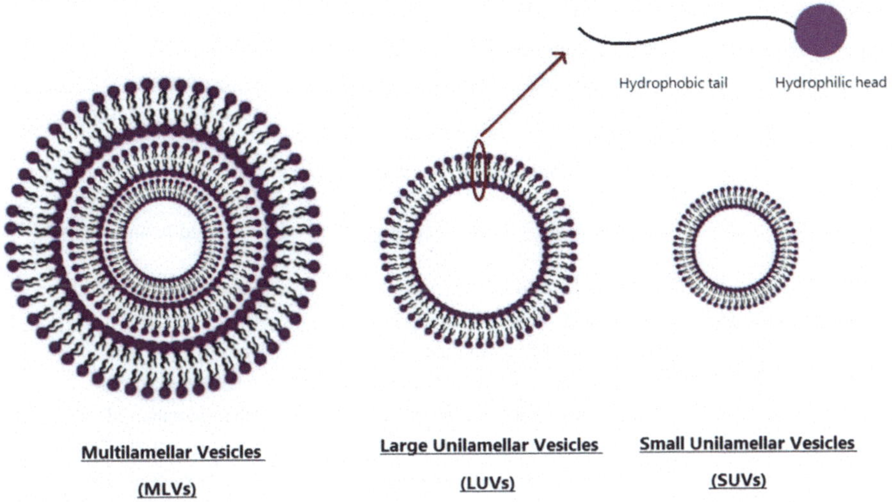

Fig. 16.11 Major types of niosomes formed

of transportation and distribution are minimized (Rajera et al. 2011; Gharbavi et al. 2018). Proniosomes, once hydrated, can be used for transdermal delivery of drugs. Their percutaneous absorption capacity is much better than currently used semisolid preparations (Touitou et al. 2000; Touitou and Godin 2007).

16.6.2 PEGylated Niosomes

These are niosomes that have undergone modification with polyethylene glycol (PEG), and they can evade the phagocytosis by mononuclear cells, thereby increasing the circulation time of the drug-encapsulated niosomes in the body (Moghassemi and Hadjizadeh 2014). When monostearate was attached to the hydrophilic tail of PEG, it could be incorporated better into the hydrophobic lipid vesicular core (He et al. 2017). Moreover, it has been observed that PEGylation helps in improving the solubility and bioavailability of certain drugs, for example, gambogenic acid. In this case, span 60 was the surfactant used along with cholesterol and diacetyl phosphate as additives, and niosomes were prepared by ether injection method followed by modification with PEG (Lin et al. 2013). Dual drug delivery (paclitaxel and curcumin) also is possible with PEGylated niosomes (Alemi et al. 2018).

Table 16.4 Types of niosomes and the large variety of drugs they can encapsulate

Types of niosomes	Drugs encapsulated	References
Proniosomes	Celecoxib (anti-inflammatory drug)	Nasr (2010)
	Vinpocetine (treating cerebrovascular disorders)	El-Laithy et al. (2011)
	Diphenyl dimethyl bicarboxylate (antiviral drug)	Aburahma and Abdelbary (2012)
	Isradipine (antihypertensive drug)	Veerareddy and Bobbala (2013)
	Nateglinide (anti-diabetic drug)	Sahoo et al. (2014)
PEGylated niosomes	Gambogenic acid (anticancer)	Lin et al. (2013)
	Paeonol (neurological disorders)	He et al. (2017)
	Paclitaxel and curcumin (anticancer drugs)	Alemi et al. (2018)
	Carboplatin (anti-tumour drug)	Davarpanah et al. (2018)
Elastic niosomes	Papain (scar treatment on skins)	Manosroi et al. (2013a)
	Calcitonin (treating osteoporosis)	Manosroi et al. (2013b)
Discomes	Naltrexone (to treat drug addiction)	Abdelkader et al. (2012)
Bola-surfactant niosomes	Fluorouracil (to treat skin cancer)	Muzzalupo et al. (2007)
	Ammonium glycyrrhizinate (anti-inflammatory)	Paolino et al. (2007)
Aspasomes	Zidovudine (anti-HIV)	Gopinath et al. (2004)
	Magnesium-ascorbyl phosphate (treating melasma)	Aboul-Einien et al. (2020)
Surfactant ethosomes	Paeonol (neurological disorders)	Ma et al. (2018)
	Testosterone propionate (hormone treatment)	Meng et al. (2013)
	Evodiamine (anti-inflammatory)	Lin et al. (2020)
	Raloxifene (to treat osteoporosis)	Mahmood et al. (2018)
	Curcumin (anticancer)	(Li et al. (2020)

16.6.3 Elastic Niosomes

These do come under the class of niosomes, but a nonionic surfactant is used along with ethyl alcohol and water. So we can say that it's a hybrid version of ethosomes. They are very suitable for transdermal drug delivery because of their ability to quickly move through the tight junction pores at the stratum corneum (Kumar and Rajeshwarrao 2011). They can be loaded with drugs independent of molecular weight and have a longer duration of action when compared to conventional niosomes. An example of a drug loaded in elastic niosomes is papain used for treating scars. The component of the drug carrier was Tween 61 along with cholesterol and methanol which are taken in the ratio 1:1 (Manosroi et al. 2013a).

16.6.4 Discomes

They are, as the name suggests, large disc-shaped niosomes (11–60 μm in size) formed after spherical niosomes undergo a morphological transformation when kept incubated in Solulan 24 of varied proportions at 74 °C and shaken for 1 h (Uchegbu et al. 1992; Abdelkader et al. 2014). They are preferably used for drug delivery via the ocular route (Abdelkader et al. 2014; Gharbavi et al. 2018), and the drug naltrexone has been delivered using discomes (Abdelkader et al. 2012).

16.6.5 Bola-Surfactant Niosomes

Bola-surfactant or bola-amphiphiles, unlike single-headed amphiphiles, have two hydrophilic groups attached to both ends of a long hydrophobic carbon chain and can aggregate or undergo self-assembly to form vesicles under certain conditions. Niosome formation has been reported to be formed when the bola-surfactant omega-hexadecyl-bis-(1-aza-18-crown-6), span 80 and cholesterol were taken in the molar ratio (2:3:1).

16.6.6 Aspasomes

Aspasomes are lamellar made from a mix of ascorbyl palmitate in water. Cholesterol and diacetyl phosphate are added for enhancing rigidity and improving stabilization, respectively. Apasomes are suitable for drug delivery via the transdermal route because of their excellent permeation properties and hence used in various cosmeceutical and pharmaceutical preparations. Moreover, they possess excellent antioxidative properties and hence are used in treatments involving reactive oxygen species (Rajera et al. 2011; Gandhi et al. 2012; Gharbavi et al. 2018; Aboul-Einien et al. 2020).

16.7 Characterization of Niosomes

16.7.1 Morphology, Size, Size Distribution and Other Structural Properties of Niosomes

Microscopic techniques like SEM (scanning electron microscopy) and TEM (transmission electron microscopy) are usually used to examine and the size and shape of niosomes formed (Moghassemi and Hadjizadeh 2014; Agarwal et al. 2018). AFM (atomic force microscopy), X-ray spectroscopy and NMR can determine the number of lamellae present in the niosomal structure (Madhav and Saini 2011; Rajera et al. 2011; Marianecci et al. 2014). X-ray scattering technique can be used for examining the thickness of the bilayer structure with the help of energy-dispersive X-ray diffraction (Liu and Guo 2007a, b; Domenici et al. 2009; Di Marzio et al. 2011). Fluorescence probe as a function of temperature can be used for determining the membrane rigidity of niosomes (Singh et al. 2004; Rajera et al. 2011).

To have a proper understanding of the niosomal membrane packing structure, fluorescence polarization is used, which is also used to determine the microviscosity of the membrane (Manosroi et al. 2008; Marianecci et al. 2014). Confocal laser scanning microscopy (CLSM) can be used to understand the location of entrapped entities (dyes). Usually, the hydrophilic substance gets embedded in the core of the niosomes and the hydrophobic substance in the outer shell of the niosomes as shown in Fig. 16.12. In Fig. 16.12, the dyes used were Nile red (hydrophobic) and FITC-dextran (hydrophilic). Nile red gets deposited at the niosome hydrophobic shell (Fig. 16.12a), while FITC-dextran gets encapsulated in the aqueous core (Fig. 16.12b). From Fig. 16.12c we can visualize the ability of niosomes to load both hydrophobic and hydrophilic entities simultaneously.

Average particle size is usually determined by the dynamic light scattering (DLS) method, but the actual particle size of the vesicle on an individual basis can be determined by TEM (Ma et al. 2018). Determining the overall or net charge on niosomes is essential to evaluate the stability of niosomes. Zeta potential is calculated to determine the stability of niosomes and is done using zeta potential analyser, DLS instrument, Mastersizer, pH-sensitive fluorophores, microelectrophoresis and high-performance capillary electrophoresis (Shilpa et al. 2011). A niosomal system, having a negative value of zeta potential within the range of −41.7 to −58.4 mV, is considered stable with enough electrostatic repulsion between the particles (Bayindir and Yuksel 2010).

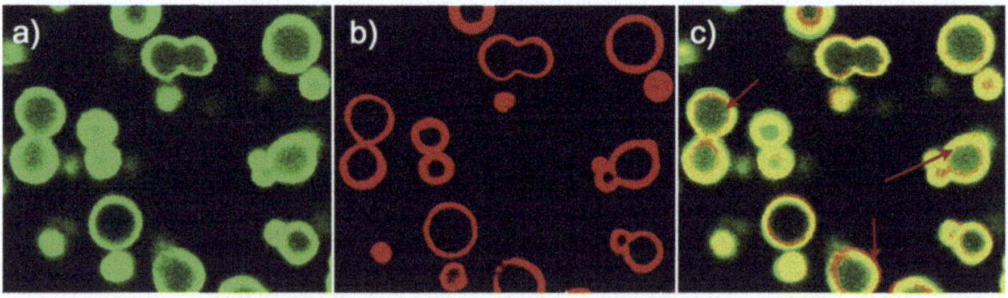

Fig. 16.12 CLSM image of niosomes showing the distribution of the Nile red and FITC-dextran. (**a**) Green channel (FITC-dextran), (**b**) red channel (Nile red) and (**c**) overlapped image of red and green channels. Scale bar = 5 μm. (Adapted from Sharma et al. 2015)

16.7.2 Entrapment Efficiency (EE)

It is usually calculated by subtracting the unloaded drug from the total amount of drug added (Kumar and Rajeshwarrao 2011). The quantity of unloaded drug is determined by filtration, chromatographic techniques like HPLC and gel chromatography and centrifugation, while the concentration of loaded drugs can be determined by dissolving the niosomes in a solvent like 0.1% TritonX-100 or 50% n-propanol and determining the assay of the solution using any specific method (Vyas et al. 2005; Manvi et al. 2012; Rinaldi et al. 2018).

The equation determines the entrapment efficiency (usually mentioned in percentage):

% Entrapment Efficiency = (Quantity of drug-loaded in the niosome/ Total quantity of drug taken) X 100

16.7.3 Drug Release Studies (In Vitro and In Vivo)

The dialysis membrane method is the most preferred to examine the in vitro drug release from niosomes. The niosomal suspension, after being put in a dialysis bag, is placed in a container having the dissolution media (usually buffer at the desired value of pH). The whole assembly is then placed on a magnetic stirrer, and the temperature is maintained at 37 °C (to simulate normal human body temperature) (Fig. 16.13). A sample is drawn at specific time intervals from the receptor, and the concentration of the drug is determined usually by spectroscopic methods like UV-Vis spectroscopy (Manosroi et al. 2010; Kumar and Rajeshwarrao 2011; Rajera et al. 2011).

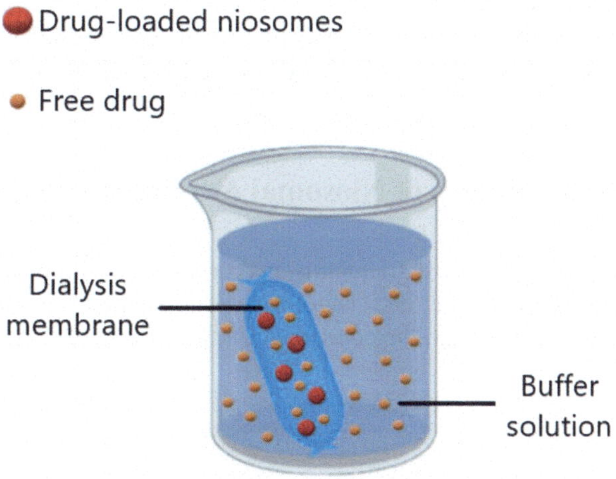

Fig. 16.13 The dialysis method, a prevalent method used to study the in vitro drug release from niosomes. (Adapted from Durak et al. 2020)

In another in vitro method, about 15 mg of the niosomal preparation is dissolved in 15 ml of phosphate buffer of pH 4.5 and 7.4. The pH of 4.5 and 7.4 was chosen to mimic the intracellular environment and blood plasma pH conditions. The samples are taken in Eppendorf tubes and continuously kept under rotation for a specific period of days. The tubes are removed at specific time intervals and centrifuged at high speeds of about 15000 rpm, and the resultant supernatant is analysed for determining the drug content by spectroscopic methods (Agarwal et al. 2018).

The in vivo study depends on the chosen route of drug administration or delivery, concentration of the drug taken and the presence or circulation time of the drug inside the body tissues, namely, the liver, spleen, lung and bone marrow (Verma et al. 2010; Moghassemi and Hadjizadeh 2014). Usually animal models like laboratory strain or species of rats, mice and guinea pigs are used to study the distribution of the drugs in the previously mentioned tissues. The distribution pattern is studied by sacrificing the drug administered animals, and its various tissues (as mentioned before) are removed, are washed with buffer and underwent homogenization and centrifugation. The drug concentration is analysed from the supernatant (Rajera et al. 2011).

16.7.4 Stability Studies

To check on the degree of drug leakage from the niosomal suspension, the stability studies are conducted at various temperatures like refrigeration temperature (4 °C), room temperature and elevated temperature (45 °C) (Kopermsub et al. 2011; Moghassemi and Hadjizadeh 2014). The usual parameters determined periodically during stability studies include entrapment efficiency, size and shape of the niosomes. The determination of these parameters is usually conducted after exposing the niosomal suspension to various conditions of humidity and irradiation with light of desired frequency range. In one study, the effect on niosomal stability under different body enzymes like trypsin, chymotrypsin and pepsin was studied, and it was discovered that the niosomes protect the encapsulated drug (paclitaxel) from these gastrointestinal enzymes (Bayindir and Yuksel 2010).

16.8 Routes of Niosomal Administration

Some of the most frequently used routes of administration for niosomes are discussed in this section. The route selection depends on the drug properties, the site of drug delivery and the concerned disease (Fig. 16.14).

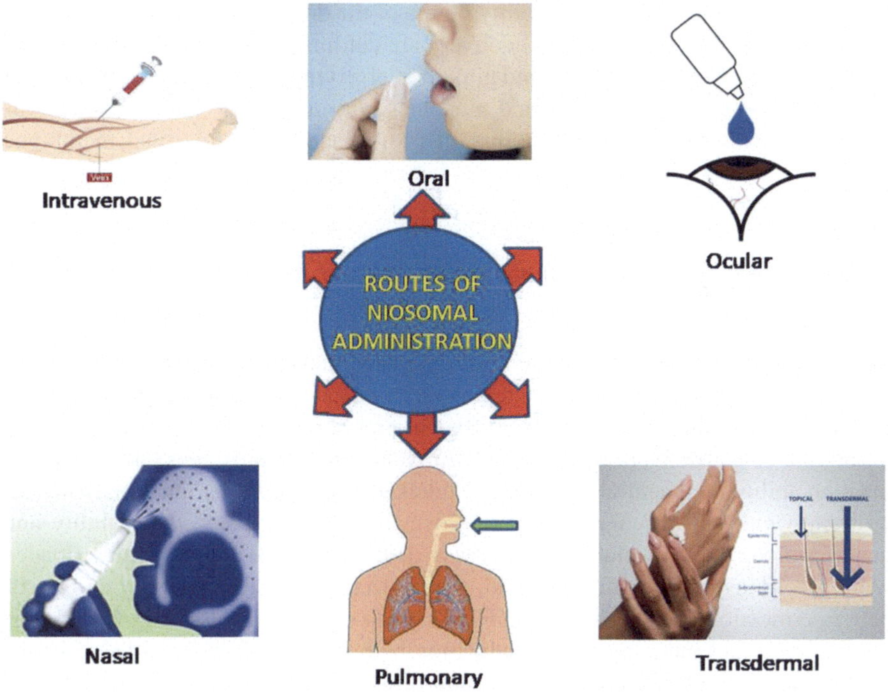

Fig. 16.14 An overview of the different routes of niosomal administration

16.8.1 Intravenous

The main advantage of the intravenous route is that the niosomes have a direct entry into systemic circulation, and niosomes ensure that the stability of the drug is maintained and its duration of being present in the blood is prolonged. To ensure that the prepared niosomes evade the monophagocytic system, the niosomes can undergo PEGylation in the synthesis stage. PEGylation also enhances the stability and bioavailability of niosomes. Morin hydrate and paeonol are two medications that were administered via an intravenous method include morin hydrate and paeonol (Waddad et al. 2013; He et al. 2017).

16.8.2 Oral

Generally, the oral route is the most preferred non-invasive route for any drug administration. A setback of oral administration is the destruction of the drug by the highly acidic environment and enzymes of the gastrointestinal system and, hence, a decrease in the drug's bioavailability (Bayindir and Yuksel 2010). However, it has been observed that niosomes can protect the drug and get adsorbed into the gastric mucosa, increasing oral bioavailability. Some of the drugs with the proven increase in oral bioavailability by niosomal delivery are cefdinir (Bansal et al. 2013), paclitaxel (Bayindir and Yuksel 2010) and ginkgo biloba (Onochie et al. 2013). In another study, by coating the insulin-loaded niosomes with trimethyl chitosan, there was enhancement in the permeation ability of insulin across the intestinal membrane (Moghassemi et al. 2015).

16.8.3 Ocular

When administered via the ocular route, the low bioavailability of a drug can be attributed to a loss in the precorneal portion because of tear formation and a short period of stay in the conjunctival sac (Biswas and Majee 2017). A study found that during the ocular delivery of the drug naltrexone, anionic niosomes had better permeation ability than neutral niosomes (Abdelkader et al. 2012). Functionalizing the niosomes with specific agents will improve their corneal permeation ability and hence prolong their availability. This was observed in coating niosomes (loaded with tacrolimus) with hyaluronic acid, which decreased its lipophilicity and hence better permeation (Zeng et al. 2016). Hence using suitable agents for surface charge modification and inducing functionalization help in improving the chances of the niosomes to be better candidates for ocular drug delivery.

16.8.4 Nasal

Nasal administration is a suitable route for drug delivery if the objective is to escape the gastric and hepatic metabolism, thereby improving the efficacy of delivering drugs to the brain. However, nasal administration has some shortcomings, including short residence time in the nasal cavity because if airflow obstruction occurs, the niosomes can be eliminated by the mucociliary presence. Moreover, the susceptible nature of the nasal mucosa can drastically reduce the drug bioavailability and permeation. The bioavailability of the drug diltiazem has increased when loaded in niosomes and faced fewer chances of elimination by the nasal mucosa (Ammar et al. 2017).

16.8.5 Pulmonary

The pulmonary route for drug administration shows better permeation, targeting, sustained or prolonged drug delivery and, hence, better therapeutically. Ciprofloxacin-loaded niosomes exhibited a lesser minimal inhibitory concentration than the free drug. In the study involving MTT assay on human carcinoma lung cell line (A549), the niosomal suspension showed higher cytotoxicity compared to the free drug form of ciprofloxacin, and the nebulization was also higher. These results can show that niosomes can be considered for effective pulmonary drug delivery (Moazeni et al. 2010).

16.8.6 Transdermal

The transdermal route is always preferred if the delivery of the drug is superficial and localized. The most significant benefit of this route is that the concerned drug does not undergo pre-systemic metabolism resulting in better bioavailability and better patient compliance. The primary challenging part for transdermal drug delivery is crossing the barrier made by the stratum corneum (SC). One of the primary reasons is that SC is tightly connected, limiting the drug permeation (Alexander et al. 2012). It has been observed that vesicular systems are ideal vehicles for transdermal drug delivery because of their malleable nature giving them better penetration ability through the SC and hence better bioavailability and lesser side effects. The sustained or prolonged release of drug is also ensured when loaded into niosomes.

Usually, ethosomes are preferred over niosomes for transdermal drug delivery. The predominant presence of ethanol in high concentration in ethosomes imparts them the ability to modify the highly dense alignment of the SC's lipid bilayers, thereby ensuring deeper penetration. However, niosomal gel loaded with the drug lopinavir showed better permeation and more site deposition than ethosomal gel loaded with the same drug (Patel et al. 2012). Moreover, the use of proniosomes helps in improving chances of the drug delivery. This was seen in the case of the anti-fungal drug fluconazole. The proniosomal formulation had a better drug deposition and more localization when compared to the available marketed formulation of the drug (Sandeep et al. 2014). The use of penetration enhancers can help in achieving sustained drug delivery. A study was conducted on the effect of chemical enhancers on the penetrating ability of ellagic acid. The potential of a penetration enhancer of two chemicals, namely, dimethyl sulphoxide (DMSO) and N-methyl-2 pyrrolidone, was examined. It was found that niosomes with DMP are better suited for dermal delivery, while DMSO niosomes were ideally better for epidermal drug delivery of ellagic acid (Junyaprasert et al. 2013).

16.9 Applications of Niosomes

Niosomes have been found to be versatile drug delivery carrier. As mentioned before, they are more stable and less cytotoxic than liposomes, and loading multiple drugs onto a single carrier and stimuli-responsive drug release have been possible with niosomes. Niosomes have been found to be effective in delivering drugs via

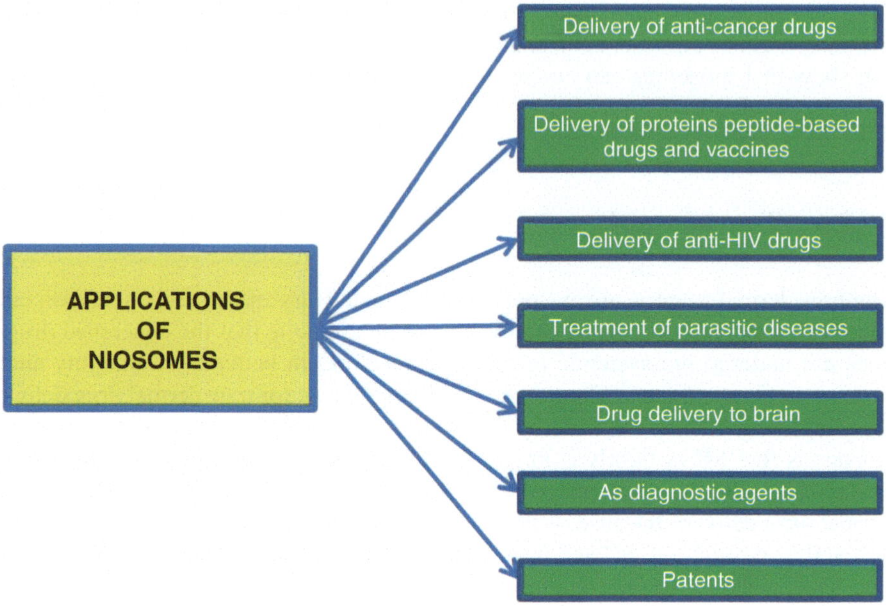

Fig. 16.15 Applications of niosomes

different routes, and some of the applications in recent years are being discussed in this section (Fig. 16.15).

16.9.1 Delivery of Anticancer Drugs

One of the vast and established applications of niosomes is their ability to deliver anticancer drugs. Normally for anticancer targeted drug delivery, targeting on the tumour is of two types:

- Passive, in which the drug carrier (in this case niosomes) gets deposited on the tumour cells on account of tumour cells having unique properties than normal cells (e.g. the enhanced permeability of tumour cell membranes allows better drug deposition in the tumour cells) (Attia et al. 2019).
- Active, in which drug targeting is achieved because of the abnormal presence of certain receptors on tumour cells. Active targeting involves the attachment of niosomes with specific ligands. For attaching ligands, either PEG or cholesterol conjugation of niosomes is done (Kim et al. 2012; Oswald et al. 2017; Gharbavi et al. 2018).

Sustained and more stable delivery of gambogenic acid for anticancer activity was possible with PEGylation of niosomes (Lin et al. 2013). The bioavailability of paclitaxel when orally delivered increased by loading them in niosomes. The simultaneous multidrug delivery for anticancer activity is also possible; this has been seen in the case of niosomes loaded with doxorubicin (hydrophilic) and curcumin (niosomes). This strategy increased the bioavailability of curcumin, and release happened for 7 days with effective cytotoxic activity on HeLa cell lines, while the release of doxorubicin was only for 2 days (Sharma et al. 2015) which shows further research has to be conducted on enhancing niosomal uptake and release of hydrophilic drugs. Another example of dual-drug delivery for anticancer activity is combined delivery of paclitaxel and curcumin and showed enhanced antitumour activity (Alemi et al. 2018). A stimuli-responsive drug release was seen in the case of morusin-loaded niosomes. Here the release of the drug from the niosomes was pH-sensitive with the higher release of the drug in an acidic environment (pH=4.5) than in a neutral or alkaline environment (pH=7.4) (Agarwal et al. 2018).

16.9.2 Delivery of Proteins- and Peptide-Based Drugs and Vaccines

The oral delivery of protein- and peptide-based drugs has been challenging because of the drug destruction by the highly acidic environment and enzymatic activity in the gastrointestinal tract. Nevertheless, studies have shown that they are better protected by niosomal loading of such drugs and hence better bioavailability (Kumar and Rajeshwarrao 2011). To enhance the bioavailability and permeation of insulin at the intestinal mucosa, a study was conducted by loading insulin in niosomes and then coating them with trimethyl chitosan. It was seen that the uptake of such coated niosomes was four times more than insulin alone in Caco-2 cell monolayer (Moghassemi et al. 2015). Another study involving oral delivery of peptides was an investigation done by H. Yoshida (Yoshida et al. 1992). The peptide/protein 9-desglycinamide 8-arginine vasopressin (DGAVP) was encapsulated in stable niosomes made of polyoxyethylene alkylethers ($C_{18}EO_3$). The in vitro release studies conducted on the rat intestinal loop model showed that the release of DGAVP from the niosomes was much higher than DGAVP solution alone. This study shows the role of the surfactant used in formulating the niosomes as penetration enhancers in the intestinal tract.

Bilosomes are niosomes formulated by including bile salt in the vesicle bilayers during the preparation of niosomes. These bilosomes have been found to be effective for oral delivery of vaccines because of their ability to protect antigens from being degraded by the proteolytic enzymes of the gastrointestinal tract (Wilkhu et al. 2013). Hence, niosomes can be considered for non-invasive administration of vaccines which will promote better patient compliance. Niosomes or proniosomes are highly permeable to oxygen and hence can be used as a haemoglobin carrier, and niosomes can be used to treat anaemia by making the presence of haemoglobin more in the blood (Radha et al. 2013).

16.9.3 Delivery of anti-HIV Drugs

Lopinavir is an anti-HIV drug that acts as a protease inhibitor. However, its high lipophilicity, vulnerability to P-glycoprotein efflux transporters and cytochrome sensitivity make the systemic bioavailability of the drug poor, and hence oral administration is least preferred. Transdermal delivery of the drug was possible by niosomal gel, and its properties were compared with that of an ethosomal gel. Even though in the comparative study, ethosomal gel had a better drug deposition, the niosomal gel showed better and deeper penetration, and the drug release profile was more sustained (Patel et al. 2012). Another example of increasing the bioavailability of an anti-HIV drug administered orally by niosomal delivery is the case of the anti-HIV drug zidovudine (Ruckmani and Sankar 2010).

The dual-drug delivery ability of niosomes has been found to an ideal candidate for targeted delivery of anti-HIV drugs. In a study, mannosylated niosomes were used as a dual-drug delivery system loaded with gold nanoparticles (GNPs) and the anti-retroviral medication efavirenz (Malik et al. 2018). Efaverinz being lipophilic got loaded in the hydrophobic vesicle membrane, while GNPs got encapsulated in the aqueous core of the niosomes. The niosomes were functionalized with a mannate appendage to enhance their targeting ability via receptor-based drug delivery. HIV host cells have receptors for the oligosaccharide mannan, and these functionalized niosomes can interact better with the target and improve the bioavailability of the drug in question. GNPs can effectively inhibit the entry of HIV into cells. Moreover, the GNPs also played an essential role in developing this niosomal formulation into a thermoresponsive gel, assuring non-invasive drug administration. This study shows one more example of how surface modification of niosomes helps in improving its targeting ability.

16.9.4 Treatment of Parasitic Diseases

Niosomes can also be used to treat a parasitic disease like leishmaniasis. Leishmaniasis mainly affects the liver and spleen, and loading the antimonials (drugs for treating leishmaniasis) in niosomes can improve the drug uptake by the liver and hence cause lesser side effects on other organs. Selenium is an important element necessary in our body for cancer prevention and anti-inflammatory and anti-oxidant effects and also has predatory action against *Leishmania major* and *Leishmania tropica* and *Leishmania infantum*. Vesicular delivery is one of the best approaches to deliver selenium, providing sufficient bioavailability. A dual-loaded niosomal formulation of selenium and glucantime was synthesized, which showed very effective lethality in vitro activity against leishmaniasis (Mostafavi et al. 2019).

Schistosomiasis is a disease caused by schistosomes which are parasitic flatworms. Praziquantel is the drug recommended by WHO for the early treatment of schistosomiasis. However, its low bioavailability when administered orally can affect its efficacy and bioavailability. In a study, niosomes were loaded with the drug to overcome these shortcomings (Zoghroban et al. 2019). The niosomes were further modified using Peceol to improve their absorption. The in vivo studies were conducted on rats, and the formulation was administered orally. The use of niosomal formulation resulted in death of 50% of the adult parasites in the rats, while the drug, when administered alone, only showed 10% death. Hence, encapsulation of the drug into niosomes enhanced its absorption and anti-parasitical activity as well.

16.9.5 Delivery of Drugs to the Brain

Functionalization of niosomes by surface modification can help in enhancing their delivery to the brain to treat neurological disorders like Alzheimer's disease. The drug pentamidine was able to cross the BBB when they were loaded in niosomes modified with chitosan-glutamate coating (Rinaldi et al. 2018). The niosomes were delivered intranasally which helped the drug in bypassing the hepatic metabolism The specific targeting of the oral alkylating agent temozolomide to treat glioblastoma was possible by surface modification of niosomes with the peptide choloro-toxin (Mamelak and Jacoby 2007).

During the onset of neurological disorders, the blood-brain barrier (BBB) deteriorates, and the membrane loses its highly selective permeability property. The presence of specific receptors also becomes more at the BBB, making receptor-mediated drug delivery possible. The process of PEGylation of niosomes helps in the association of ligands with niosomes. One way is to attach the concerned ligand to PEG chains at their distal end if it's PEGylated niosomes. The other way is direct incorporation of a conjugate made of cholesterol, PEG and the ligand into niosomes during the formulation or synthesis step (Kim et al. 2012; Oswald et al. 2017). Some examples that show ligand attachment to niosomes have been beneficial: the delivery of the neuropeptide vasoactive intestinal peptide (VIP) to the brain using niosomes containing glucose (Dufes et al. 2004).

16.9.6 Niosomes as Diagnostic Agents

The ability of niosomes to load a hydrophilic and hydrophobic compound simultaneously has been beneficial in developing niosomes as a diagnostic imaging agent. The dyes or contrast agents can be incorporated either into the vesicle layer or in the aqueous compartment of niosomes or on the surface of niosomes by conjugation (Masotti 2013). The conjugation of gadobenate dimeglumine-loaded niosomes with n-palmitoylglucosamine (NPG) and PEG 4400 helped improve the targeting and imaging capability of gadolinium associated with MR imaging. This study again shows the importance of PEG surface modification of niosomes, improving their functionality (Luciani et al. 2004).

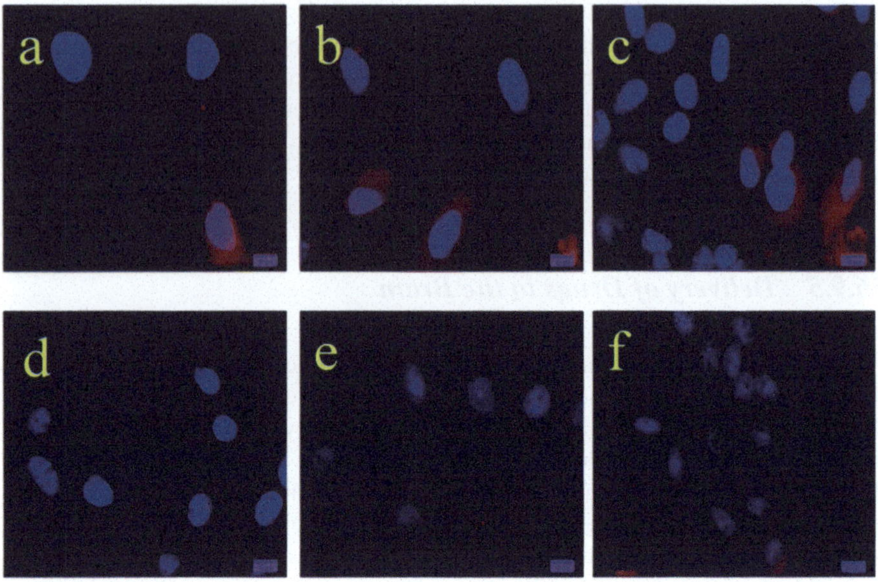

Fig. 16.16 Fluorescence imaging of niosomes loaded with protoporphyrin IX. (Adapted from Barlas et al. 2016)

For a multifunctional approach, theranostic nanoparticles are garnering significant interest in recent times, which can also act as imaging agents to visualize active cellular uptake of these nanoparticles. With this aim in mind, in one study, gadolinium nanoparticles and protoporphyrin IX were simultaneously loaded in niosomes (Barlas et al. 2016). The multifunctionality role of niosomes is the combined "radio therapy and photodynamic therapy" along with MRI and fluorescence imaging by gadolinium and protoporphyrin IX, respectively. As far as the anticancer effect of the agents was considered, they showed the excellent therapeutic effect on HeLa cells (human cervical cancer cell lines) and A549 cells (human alveolar type II). The imaging capabilities of the niosomes are shown in Fig. 16.16 (fluorescence imaging due to the presence of protoporphyrin IX) and Fig. 16.17 (MR imaging with the assistance of gadolinium nanoparticles).

Similarly, in another study, with one more aim of improving the targeting ability of niosomes, folic acid was attached to niosomes in which GNPs and the photosensitizer protoporphyrin IX were encapsulated. The tagging of folic acid helped in receptor-mediated delivery to the cancer cell lines used in this study: HeLa cells (human cervical cancer cell lines) and A549 cells (human alveolar type II). The GNPs and protoporphyrin IX were simultaneously loaded to provide both radio therapy and photodynamic therapy, respectively. GNPs, due to surface plasmon resonance (SPR), allowed live imaging and tracking of the niosomal entry into the cancer cells (Demir et al. 2018). So what we see here is that by surface modification of niosomes and loading them with multifunctional agents, we can make niosomal

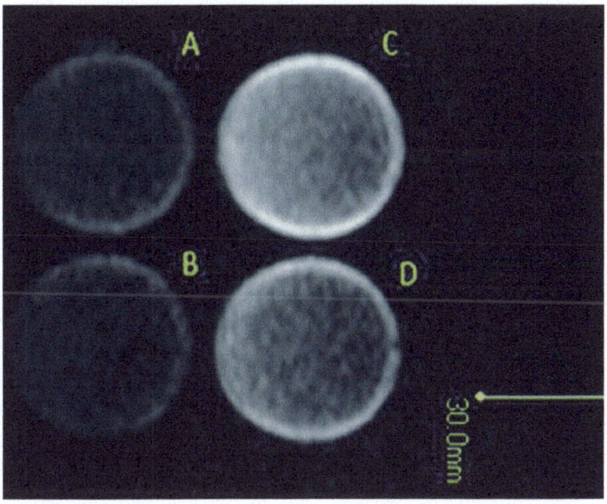

Fig. 16.17 MR imaging of niosomes loaded with gadolinium nanoparticles. (Adapted from Barlas et al. 2016)

assemblies just like assembling "building blocks" which can perform a variety of functions.

16.9.7 Patents on Niosomes

The release of patented and marketed formulations of niosomes has occurred, showing that the global acceptance of niosomes for drug delivery is also expanding. The bioavailability of doxorubicin has significantly increased with the help of niosomal formulations. A patented niosomal gel loaded with doxorubicin is used to treat cancers associated with the skin (Ramanunny et al. 2020). Another example of a patent is a dental gel developed for transdermal drug delivery for treating periodontal diseases (Samoylenko 2016). In another patent, the targeting ability of anticancer drug chlorotoxin increased when incorporated into a niosomal formulation, and thermal responsive drug release was possible when the niosomes were incorporated into the chitosan hydrogel network.

16.10 Conclusions

The main attractive feature of niosomes is their ability to encapsulate both hydrophilic and hydrophobic compounds simultaneously. Their ease of synthesis, lesser toxicity and cheaper cost than liposomes make them a better candidate for vesicular drug delivery. Hence, niosomes have been used to encapsulate many drugs and enzymes, peptides, vaccines, genes and imaging agents. The surface modification of niosomes with particular entities helps them in protecting the encapsulated drug from the highly acidic environment of the gastrointestinal tract and improves the drug absorption in the intestinal mucosa. This showcases the potential of niosomes being developed as a primal source for oral administration (the most preferred and non-invasive mode of drug administration) for many drugs. Since most niosomal formulations are in suspension, they can be easily ingested, which can be a great relief for those who have a phobia of swallowing tablets. The loading of niosomes with agents like metallic nanoparticles (gold and gadolinium) and photosensitizers (protoporphyrin IX) has paved the path of developing multifunctional vesicular assemblies for cancer treatment and imaging.

However, the increase in entrapment efficiency of hydrophilic drugs in niosomes remains a challenge, and surface modification by coating with suitable agents might be a solution. The method of ligand attachment on niosomes to improve their targeting ability remains unexplored to a great extent. The surge in the number of patents and marketed niosomal formulations proves that niosomes are getting much attention globally and are on the brink of being recognized as a well-established and universal pharmaceutical formulation for the non-invasive administration of drugs.

Acknowledgement This research was supported by Basic Science Research Program through the National Research Foundation of Korea (NRF) funded by the Ministry of Education (No. 2018R1A6A1A03025582). This work was supported by the Technology development Program (S2911350) funded by the Ministry of SMEs and Startups(MSS, Korea)

References

Abdelkader H, Wu Z, Al-Kassas R, Alany RG (2012) Niosomes and discomes for ocular delivery of naltrexone hydrochloride: morphological, rheological, spreading properties and photo-protective effects. Int J Pharm 433:142–148

Abdelkader H, Alani AWG, Alany RG (2014) Recent advances in non-ionic surfactant vesicles (niosomes): Self-assembly, fabrication, characterization, drug delivery applications and limitations. Drug Deliv 21:87–100. https://doi.org/10.3109/10717544.2013.838077

Aboul-Einien MH, Kandil SM, Abdou EM et al (2020) Ascorbic acid derivative-loaded modified aspasomes: formulation, in vitro, ex vivo and clinical evaluation for melasma treatment. J Liposome Res 30:54–67

Aburahma MH, Abdelbary GA (2012) Novel diphenyl dimethyl bicarboxylate provesicular powders with enhanced hepatocurative activity: preparation, optimization, in vitro/in vivo evaluation. Int J Pharm 422:139–150

Ag Seleci D, Seleci M, Walter J-G et al (2016) Niosomes as Nanoparticular Drug Carriers: Fundamentals and Recent Applications. J Nanomater 2016:7372306. https://doi.org/10.1155/2016/7372306

Agarwal S, Mohamed MS, Raveendran S et al (2018) Formulation, characterization and evaluation of morusin loaded niosomes for potentiation of anticancer therapy. RSC Adv 8:32621–32636. https://doi.org/10.1039/c8ra06362a

Alemi A, Reza JZ, Haghiralsadat F et al (2018) Paclitaxel and curcumin coadministration in novel cationic PEGylated niosomal formulations exhibit enhanced synergistic antitumor efficacy. J Nanobiotechnol 16:1–20

Alexander A, Dwivedi S, Ajazuddin et al (2012) Approaches for breaking the barriers of drug permeation through transdermal drug delivery. J Control Release 164:26–40. https://doi.org/10.1016/j.jconrel.2012.09.017

Ammar HO, Haider M, Ibrahim M, El Hoffy NM (2017) In vitro and in vivo investigation for optimization of niosomal ability for sustainment and bioavailability enhancement of diltiazem after nasal administration. Drug Deliv 24:414–421

Arunothayanun P, Bernard M-S, Craig DQM et al (2000) The effect of processing variables on the physical characteristics of non-ionic surfactant vesicles (niosomes) formed from a hexadecyl diglycerol ether. Int J Pharm 201:7–14

Aranya, Manosroi Paveena, Wongtrakul Jiradej, Manosroi Hideki, Sakai Fumio, Sugawara Makoto, Yuasa Masahiko, Abe (2003) Characterization of vesicles prepared with various non-ionic surfactants mixed with cholesterol. Colloids and Surfaces B: Biointerfaces 30(1-2) 129-138 10.1016/S0927-7765(03)00080-8

Attia MF, Anton N, Wallyn J et al (2019) An overview of active and passive targeting strategies to improve the nanocarriers efficiency to tumour sites. J Pharm Pharmacol 71:1185–1198. https://doi.org/10.1111/jphp.13098

Azmin MN, Florence AT, Handjani-Vila RM et al (1985) The effect of non-ionic surfactant vesicle (niosome) entrapment on the absorption and distribution of methotrexate in mice. J Pharm Pharmacol 37:237–242. https://doi.org/10.1111/j.2042-7158.1985.tb05051.x

Bandyopadhyay P, Johnson M (2007) Fatty alcohols or fatty acids as niosomal hybrid carrier: Effect on vesicle size, encapsulation efficiency and in vitro dye release. Colloids Surf B Biointerfaces 58:68–71. https://doi.org/10.1016/j.colsurfb.2007.01.014

Bansal S, Aggarwal G, Chandel P, Harikumar SL (2013) Design and development of cefdinir niosomes for oral delivery. J Pharm Bioallied Sci 5:318

Barlas FB, Demir B, Guler E et al (2016) Multimodal theranostic assemblies: double encapsulation of protoporphyrine-IX/Gd3+ inniosomes. RSC Adv 6:30217–30225. https://doi.org/10.1039/c5ra26737d

Bartelds R, Nematollahi MH, Pols T et al (2018) Niosomes, an alternative for liposomal delivery. PLoS One 13:e0194179

Bayindir ZS, Yuksel N (2010) Characterization of niosomes prepared with various nonionic surfactants for paclitaxel oral delivery. J Pharm Sci 99:2049–2060. https://doi.org/10.1002/jps.21944

Bini KB, Akhilesh D, Prabhakara P, Kamath JV (2012) Development and characterization of non-ionic surfactant vesicles (niosomes) for oral delivery of lornoxicam. Int J Drug Dev Res 4:147–154

Biswas GR, Majee SB (2017) Niosomes in ocular drug delivery. Eur J Pharm Med Res 4:813–819

Cametti C (2008) Polyion-induced aggregation of oppositely charged liposomes and charged colloidal particles: the many facets of complex formation in low-density colloidal systems. Chem Phys Lipids 155:63–73

Carafa M, Santucci E, Lucania G (2002) Lidocaine-loaded non-ionic surfactant vesicles: characterization and in vitro permeation studies. Int J Pharm 231:21–32. https://doi.org/10.1016/S0378-5173(01)00828-6

Chandu VP, Arunachalam A, Jeganath S et al (2012) International journal of novel trends in pharmaceutical sciences niosomes : a novel drug delivery system. Int J Nov Trends Pharm Sci 2:25–31

Chen S, Hanning S, Falconer J et al (2019) Recent advances in non-ionic surfactant vesicles (niosomes): Fabrication, characterization, pharmaceutical and cosmetic applications. Eur J Pharm Biopharm 144:18–39. https://doi.org/10.1016/j.ejpb.2019.08.015

Davarpanah F, Khalili Yazdi A, Barani M et al (2018) Magnetic delivery of antitumor carboplatin by using PEGylated-Niosomes. Daru 26:57–64. https://doi.org/10.1007/s40199-018-0215-3

Demir B, Barlas FB, Gumus ZP et al (2018) Theranostic niosomes as a promising tool for combined therapy and diagnosis: "all-in-one" approach. ACS Appl Nano Mater 1:2827–2835. https://doi.org/10.1021/acsanm.8b00468

Di Marzio L, Marianecci C, Petrone M et al (2011) Novel pH-sensitive non-ionic surfactant vesicles: comparison between Tween 21 and Tween 20. Colloids Surf B Biointerfaces 82:18–24. https://doi.org/10.1016/j.colsurfb.2010.08.004

Domenici F, Panichelli D, Castellano AC (2009) Alamethicin-lipid interaction studied by energy dispersive X-ray diffraction. Colloids Surf B Biointerfaces 69:216–220. https://doi.org/10.1016/j.colsurfb.2008.11.029

Dufes C, Gaillard F, Uchegbu IF et al (2004) Glucose-targeted niosomes deliver vasoactive intestinal peptide (VIP) to the brain. Int J Pharm 285:77–85. https://doi.org/10.1016/j.ijpharm.2004.07.020

Durak S, Rad ME, Yetisgin AA et al (2020) Niosomal drug delivery systems for ocular disease—recent advances and future prospects. Nanomaterials 10:1–29. https://doi.org/10.3390/nano10061191

El-Laithy HM, Shoukry O, Mahran LG (2011) Novel sugar esters proniosomes for transdermal delivery of vinpocetine: preclinical and clinical studies. Eur J Pharm Biopharm 77:43–55

Ferreira LS, Ramaldes GA, Nunan EA, Ferreira LAM (2004) In vitro skin permeation and retention of paromomycin from liposomes for topical treatment of the cutaneous leishmaniasis. Drug Dev Ind Pharm 30:289–296

Gaafar PME, Abdallah OY, Farid RM, Abdelkader H (2014) Preparation, characterization and evaluation of novel elastic nano-sized niosomes (ethoniosomes) for ocular delivery of prednisolone. J Liposome Res 24:204–215

Gandhi A, Sen SO, Paul A (2012) Abstract : introduction : structural components of niosomes 2:339–353

Ge X, Wei M, He S, Yuan WE (2019) Advances of non-ionic surfactant vesicles (niosomes) and their application in drug delivery. Pharmaceutics 11. https://doi.org/10.3390/pharmaceutics11020055

Gharbavi M, Amani J, Kheiri-Manjili H et al (2018) Niosome: a promising nanocarrier for natural drug delivery through blood-brain barrier. Adv Pharmacol Sci 2018:6847971. https://doi.org/10.1155/2018/6847971

D, Gopinath D, Ravi B.R, Rao S.S, Apte D, Renuka D, Rambhau (2004) Ascorbyl palmitate vesicles (Aspasomes): formation characterization and applications. International Journal of Pharmaceutics 271(1-2) 95-113 10.1016/j.ijpharm.2003.10.032

He R-X, Ye X, Li R et al (2017) PEGylated niosomes-mediated drug delivery systems for paeonol: preparation, pharmacokinetics studies and synergistic anti-tumor effects with 5-FU. J Liposome Res 27:161–170. https://doi.org/10.1080/08982104.2016.1191021

Homaei M (2016) Preparation and characterization of giant niosomes

Isnan AP, Jufri M (2017) Formulation of niosomal gel containing green tea extract (Camellia sinensis L. Kuntze) using thin-layer hydration. Int J Appl Pharm 9:38–43. https://doi.org/10.22159/ijap.2017.v9s1.23_28

Israelachvili JN, Mitchell DJ, Ninham BW (1976) Theory of self-assembly of hydrocarbon amphiphiles into micelles and bilayers. J Chem Soc Faraday Trans 2 72:1525. https://doi.org/10.1039/f29767201525

Jousma H, Joosten JGH, Junginger HE (1988) Mesophases in mixtures of water and polyoxyethylene surfactant: Variations of repeat spacing with temperature and composition. Colloid Polym Sci 266:640–651. https://doi.org/10.1007/BF01411505

Jousma H, Joosten JGH, Gooris GS, Junginger HE (1989) Changes of mesophase structure of BRIJ 96/water mixtures on addition of liquid paraffin. Colloid Polym Sci 267:353–364. https://doi.org/10.1007/BF01413630

Junyaprasert VB, Singhsa P, Suksiriworapong J, Chantasart D (2012) Physicochemical properties and skin permeation of Span 60/Tween 60 niosomes of ellagic acid. Int J Pharm 423:303–311

Junyaprasert VB, Singhsa P, Jintapattanakit A (2013) Influence of chemical penetration enhancers on skin permeability of ellagic acid-loaded niosomes. Asian J Pharm Sci 8:110–117. https://doi.org/10.1016/j.ajps.2013.07.014

Kazi KM, Mandal AS, Biswas N et al (2010) Niosome: a future of targeted drug delivery systems. J Adv Pharm Technol Res 1:374

Key A, Sch T, Pharm AJ, et al (2018) Scholars Academic Journal of Pharmacy (SAJP) Ethosomes: The Novel Drug Delivery Carriers. https://doi.org/10.21276/sajp.2018.7.6.9

Kim TH, Jo YG, Jiang HH et al (2012) PEG-transferrin conjugated TRAIL (TNF-related apoptosis-inducing ligand) for therapeutic tumor targeting. J Control Release 162:422–428. https://doi.org/10.1016/j.jconrel.2012.07.021

Kopermsub P, Mayen V, Warin C (2011) Potential use of niosomes for encapsulation of nisin and EDTA and their antibacterial activity enhancement. Food Res Int 44:605–612

Kumar GP, Rajeshwarrao P (2011) Nonionic surfactant vesicular systems for effective drug delivery—an overview. Acta Pharm Sin B 1:208–219. https://doi.org/10.1016/j.apsb.2011.09.002

Li Y, Xu F, Li X et al (2020) Development of curcumin-loaded composite phospholipid ethosomes for enhanced skin permeability and vesicle stability. Int J Pharm:119936. https://doi.org/10.1016/j.ijpharm.2020.119936

Lin T, Fang Q, Peng D et al (2013) PEGylated non-ionic surfactant vesicles as drug delivery systems for Gambogenic acid. Drug Deliv 20:277–284

Lin H, Lin L, Choi Y, Michniak-Kohn B (2020) Development and in-vitro evaluation of co-loaded berberine chloride and evodiamine ethosomes for treatment of melanoma. Int J Pharm 581:119278. https://doi.org/10.1016/j.ijpharm.2020.119278

Liu T, Guo R (2007a) Investigation of PEG 6000/Tween 80/Span 80/H 2 O niosome microstructure. Colloid Polym Sci 285:711–713

Liu T, Guo R (2007b) Structure and transformation of the niosome prepared from PEG 6000/Tween 80/Span 80/H2O lamellar liquid crystal. Colloids Surf A Physicochem Eng Asp 295:130–134. https://doi.org/10.1016/j.colsurfa.2006.08.041

Liu T, Guo R, Hua W, Qiu J (2007) Structure behaviors of hemoglobin in PEG 6000/Tween 80/Span 80/H2O niosome system. Colloids Surf A Physicochem Eng Asp 293:255–261. https://doi.org/10.1016/j.colsurfa.2006.07.053

Lo CT, Jahn A, Locascio LE, Vreeland WN (2010) Controlled self-assembly of monodisperse niosomes by microfluidic hydrodynamic focusing. Langmuir 26:8559–8566. https://doi.org/10.1021/la904616s

Luciani A, Olivier JC, Clement O et al (2004) Glucose-receptor mr imaging of tumors: study in mice with pegylated paramagnetic niosomes. Radiology 231:135–142. https://doi.org/10.1148/radiol.2311021559

Ma H, Guo D, Fan Y et al (2018) Paeonol-loaded ethosomes as transdermal delivery carriers: design, preparation and evaluation. Molecules 23:1756

Madhav NVS, Saini A (2011) Niosomes: a novel drug delivery system. Int J Res Pharm Chem 1:498–511

Mahale NB, Thakkar PD, Mali RG et al (2012) Niosomes: novel sustained release nonionic stable vesicular systems –an overview. Adv Colloid Interface Sci 183–184:46–54. https://doi.org/10.1016/j.cis.2012.08.002

Mahmood S, Mandal UK, Chatterjee B (2018) Transdermal delivery of raloxifene HCl via ethosomal system: Formulation, advanced characterizations and pharmacokinetic evaluation. Int J Pharm 542:36–46. https://doi.org/10.1016/j.ijpharm.2018.02.044

Malik T, Chauhan G, Rath G et al (2018) Efaverinz and nano-gold-loaded mannosylated niosomes: a host cell-targeted topical HIV-1 prophylaxis via thermogel system. Artif Cells Nanomed Biotechnol 46:79–90. https://doi.org/10.1080/21691401.2017.1414054

Mamelak AN, Jacoby DB (2007) Targeted delivery of antitumoral therapy to glioma and other malignancies with synthetic chlorotoxin (TM-601). Expert Opin Drug Deliv 4:175–186. https://doi.org/10.1517/17425247.4.2.175

Manconi M, Vila AO, Sinico C et al (2006) Theoretical and experimental evaluation of decypolyglucoside vesicles as potential drug delivery systems. J Drug Deliv Sci Technol 16:141–146. https://doi.org/10.1016/s1773-2247(06)50021-8

Manosroi A, Chutoprapat R, Abe M, Manosroi J (2008) Characteristics of niosomes prepared by supercritical carbon dioxide (scCO2) fluid. Int J Pharm 352:248–255. https://doi.org/10.1016/j.ijpharm.2007.10.013

Manosroi A, Khanrin P, Lohcharoenkal W et al (2010) Transdermal absorption enhancement through rat skin of gallidermin loaded in niosomes. Int J Pharm 392:304–310

Manosroi A, Chankhampan C, Manosroi W, Manosroi J (2013a) Transdermal absorption enhancement of papain loaded in elastic niosomes incorporated in gel for scar treatment. Eur J Pharm Sci 48:474–483

Manosroi A, Chankhampan C, Ofoghi H et al (2013b) Low cytotoxic elastic niosomes loaded with salmon calcitonin on human skin fibroblasts. Hum Exp Toxicol 32:31–44. https://doi.org/10.1177/0960327112454892

Manvi SR, Gupta VRM, Srikanth K, Devanna N (2012) Formulation and evaluation of candesartan niosomal suspension. Res J Pharm Technol 5:1570–1572

Marianecci C, Paolino D, Celia C et al (2010) Non-ionic surfactant vesicles in pulmonary glucocorticoid delivery: characterization and interaction with human lung fibroblasts. J Control Release 147:127–135

Marianecci C, Di Marzio L, Rinaldi F et al (2014) Niosomes from 80s to present: The state of the art. Adv Colloid Interface Sci 205:187–206. https://doi.org/10.1016/j.cis.2013.11.018

Masotti A (2013) Niosomes as candidate bioconjugates for imaging and pH-sensitive drug delivery nanocarriers for rare pediatric tumors. J Drug Deliv Sci Technol 23:22–24. https://doi.org/10.1016/S1773-2247(13)50003-7

Meng S, Chen Z, Yang L et al (2013) Enhanced transdermal bioavailability of testosterone propionate via surfactant-modified ethosomes. Int J Nanomedicine 8:3051–3060. https://doi.org/10.2147/IJN.S46748

Moazeni E, Gilani K, Sotoudegan F et al (2010) Formulation and in vitro evaluation of ciprofloxacin containing niosomes for pulmonary delivery. J Microencapsul 27:618–627. https://doi.org/10.3109/02652048.2010.506579

Moghassemi S, Hadjizadeh A (2014) Nano-niosomes as nanoscale drug delivery systems: an illustrated review. J Control Release 185:22–36

Moghassemi S, Parnian E, Hakamivala A et al (2015) Uptake and transport of insulin across intestinal membrane model using trimethyl chitosan coated insulin niosomes. Mater Sci Eng C 46:333–340

Moghassemi S, Hadjizadeh A, Omidfar K (2017) Formulation and characterization of bovine serum albumin-loaded niosome. Aaps Pharmscitech 18:27–33

Mostafavi M, Khazaeli P, Sharifi I et al (2019) A novel niosomal combination of selenium coupled with glucantime against leishmania tropica. Korean J Parasitol 57:1–8. https://doi.org/10.3347/kjp.2019.57.1.1

Mozafari MR (2007) Nanomaterials and nanosystems for biomedical applications. Springer

Naderinezhad S, Amoabediny G, Haghiralsadat F (2017) Co-delivery of hydrophilic and hydrophobic anticancer drugs using biocompatible pH-sensitive lipid-based nano-carriers for multidrug-resistant cancers. RSC Adv 7:30008. https://doi.org/10.1039/c7ra01736g

Nasr M (2010) In vitro and in vivo evaluation of proniosomes containing celecoxib for oral administration. Aaps Pharmscitech 11:85–89

Nasseri B (2005) Effect of cholesterol and temperature on the elastic properties of niosomal membranes. Int J Pharm 300:95–101. https://doi.org/10.1016/j.ijpharm.2005.05.009

Oh Y-K, Kim MY, Shin J-Y et al (2006) Skin permeation of retinol in Tween 20-based deformable liposomes: in-vitro evaluation in human skin and keratinocyte models. J Pharm Pharmacol 58:161–166. https://doi.org/10.1211/jpp.58.2.0002

Onochie ITO, Nwakile CD, Umeyor CE et al (2013) Formulation and evaluation of niosomes of benzyl penicillin. J Appl Pharm Sci 3:66

Oswald M, Geissler S, Goepferich A (2017) Targeting the Central Nervous System (CNS): a review of rabies virus-targeting strategies. Mol Pharm 14:2177–2196. https://doi.org/10.1021/acs.molpharmaceut.7b00158

G., Parthasarathi N., Udupa P., Umadevi G., Pillai (2008) (1994) Niosome Encapsulated of Vincristine Sulfate: Improved Anticancer Activity with Reduced Toxicity in Mice. Journal of Drug Targeting 2(2) 173-182 10.3109/10611869409015907

Palani S (2010) Niosomes in targeted drug delivery: some recent advances. Int J Pharm Sci Res 1:1–8

Paolino D, Muzzalupo R, Ricciardi A et al (2007) In vitro and in vivo evaluation of Bola-surfactant containing niosomes for transdermal delivery. Biomed Microdevices 9:421–433. https://doi.org/10.1007/s10544-007-9046-6

Pardakhty A, Varshosaz J, Rouholamini A (2007) In vitro study of polyoxyethylene alkyl ether niosomes for delivery of insulin. Int J Pharm 328:130–141. https://doi.org/10.1016/j.ijpharm.2006.08.002

Patel KK, Kumar P, Thakkar HP (2012) Formulation of niosomal gel for enhanced transdermal lopinavir delivery and its comparative evaluation with ethosomal gel. AAPS PharmSciTech 13:1502–1510. https://doi.org/10.1208/s12249-012-9871-7

Peltonen L, Koistinen P, Karjalainen M et al (2002) The effect of cosolvents on the formulation of nanoparticles from low-molecular-weight poly (I) lactide. Aaps Pharmscitech 3:52

Radha GV, Rani TS, Sarvani B (2013) A review on proniosomal drug delivery system for targeted drug action. J Basic Clin Pharm 4:42–48. https://doi.org/10.4103/0976-0105.113609

Rajera R, Nagpal K, Singh SK, Mishra DN (2011) Niosomes: a controlled and novel drug delivery system. Biol Pharm Bull 34:945–953

Ramanunny AK, Wadhwa S, Gulati M et al (2020) Nanocarriers for treatment of dermatological diseases: principle, perspective and practices. Eur J Pharmacol 173691. https://doi.org/10.1016/j.ejphar.2020.173691

Raymond CR, Paul JS, Siân CO (2006) Handbook of pharmaceutical excipients, 5th edn. Pharmaceutical Press, London

Rinaldi F, Hanieh PN, Chan LKN et al (2018) Chitosan glutamate-coated niosomes: a proposal for nose-to-brain delivery. Pharmaceutics 10:38

Rita, Muzzalupo Fiore Pasquale, Nicoletta Sonia, Trombino Roberta, Cassano Francesca, Iemma Nevio, Picci (2007) A new crown ether as vesicular carrier for 5-fluoruracil: Synthesis characterization and drug delivery evaluation. Colloids and Surfaces B: Biointerfaces 58(2) 197-202 10.1016/j.colsurfb.2007.03.010

Rogerson A, Cummings J, Willmott N, Florence AT (1988) The distribution of doxorubicin in mice following administration in niosomes. J Pharm Pharmacol 40:337–342. https://doi.org/10.1111/j.2042-7158.1988.tb05263.x

Ruckmani K, Sankar V (2010) Formulation and optimization of zidovudine niosomes. AAPS PharmSciTech 11:1119–1127. https://doi.org/10.1208/s12249-010-9480-2

Sahoo RK, Biswas N, Guha A, Kuotsu K (2014) Maltodextrin based proniosomes of nateglinide: bioavailability assessment. Int J Biol Macromol 69:430–434

Samoylenko VA (2016) The results of clinical testing of photodynamic therapy in the complex treatment of gingivitis, complicating orthodontic treatment with bracket systems. Medicni Perspekt (Medical Perspect 21). https://doi.org/10.26641/2307-0404.2016.2.72164

Sandeep G, Vasavi Reddy D, Devireddy SR (2014) Formulation and evaluation of fluconazole proniosomal gel for topical administration. J Appl Pharm Sci 4:98–104. https://doi.org/10.7324/JAPS.2014.40717

Shahiwala A, Misra A (2002) Studies in topical application of niosomally entrapped nimesulide. J Pharm Pharm Sci a Publ Can Soc Pharm Sci Soc Can des Sci Pharm 5:220–225

Sharma V, Anandhakumar S, Sasidharan M (2015) Self-degrading niosomes for encapsulation of hydrophilic and hydrophobic drugs: an efficient carrier for cancer multi-drug delivery. Mater Sci Eng C 56:393–400. https://doi.org/10.1016/j.msec.2015.06.049

Shilpa S, Srinivasan BP, Chauhan M (2011) Niosomes as vesicular carriers for delivery of proteins and biologicals. Int J Drug Deliv 3:14–24. https://doi.org/10.5138/ijdd.2010.0975.0215.03050

Shreedevi HM, Nesalin JAJ, Mani TT (2016) Development and evaluation of stavudine niosome by ether injection method. Int J Pharma Sci Res 7:38–46

Singh TG, Sharma N (2016) Nanobiomaterials in cosmetics: current status and future prospects. Elsevier Inc.

Singh A, Dally WJ, Gupta AK, Towles B (2004) Adaptive channel queue routing on K-Ary n-cubes. In: Proceedings of the sixteenth annual ACM symposium on parallelism in algorithms and architectures. Association for Computing Machinery, New York, pp 11–19

Singh G, Dwivedi H, Saraf SK, Saraf SA (2011) Niosomal delivery of isoniazid-development and characterization. Trop J Pharm Res 10
Sorgi FL, Huang L (1996) Large scale production of DC-Chol cationic liposomes by microfluidization. Int J Pharm 144:131–139
Tavano L, Alfano P, Muzzalupo R, De Cindio B (2011) Niosomes vs microemulsions: new carriers for topical delivery of Capsaicin. Colloids Surf B Biointerfaces 87:333–339. https://doi.org/10.1016/j.colsurfb.2011.05.041
Touitou E, Godin B (2007) Ethosomes for skin delivery. J Drug Deliv Sci Technol 17:303–308
Touitou E, Dayan N, Bergelson L et al (2000) Ethosomes – novel vesicular carriers for enhanced delivery: characterization and skin penetration properties. J Control Release 65:403–418. https://doi.org/10.1016/S0168-3659(99)00222-9
Uchegbu IF, Florence AT (1995) Non-ionic surfactant vesicles (niosomes): physical and pharmaceutical chemistry. Adv Colloid Interface Sci 58:1–55. https://doi.org/10.1016/0001-8686(95)00242-I
Uchegbu IF, Vyas SP (1998) Non-ionic surfactant based vesicles (niosomes) in drug delivery. Int J Pharm 172:33–70. https://doi.org/10.1016/S0378-5173(98)00169-0
Uchegbu IF, Bouwstra JA, Florence AT (1992) Large disk-shaped structures (discomes) in nonionic surfactant vesicle to micelle transitions. J Phys Chem 96:10548–10553
Veerareddy PR, Bobbala SKR (2013) Enhanced oral bioavailability of isradipine via proniosomal systems. Drug Dev Ind Pharm 39:909–917
Verma S, Utreja P (2019) Vesicular nanocarrier based treatment of skin fungal infections: Potential and emerging trends in nanoscale pharmacotherapy. Asian J Pharm Sci 14:117–129
Verma S, Singh SK, Syan N et al (2010) Nanoparticle vesicular systems: a versatile tool for drug delivery. J Chem Pharm Res 2:496–509
Vyas SP, Singh RP, Jain S et al (2005) Non-ionic surfactant based vesicles (niosomes) for noninvasive topical genetic immunization against hepatitis B. Int J Pharm 296:80–86
Waddad AY, Abbad S, Yu F et al (2013) Formulation, characterization and pharmacokinetics of Morin hydrate niosomes prepared from various non-ionic surfactants. Int J Pharm 456:446–458. https://doi.org/10.1016/j.ijpharm.2013.08.040
Wilkhu JS, McNeil SE, Anderson DE, Perrie Y (2013) Characterization and optimization of bilosomes for oral vaccine delivery. J Drug Target 21:291–299. https://doi.org/10.3109/1061186X.2012.747528
Yeo PL, Lim CL, Chye SM et al (2017) Niosomes: a review of their structure, properties, methods of preparation, and medical applications. Asian Biomed 11:301–313. https://doi.org/10.1515/abm-2018-0002
Yeo LK, Chaw CS, Elkordy AA (2019) The effects of hydration parameters and co-surfactants on methylene blue-loaded niosomes prepared by the thin film hydration method. Pharmaceuticals 12:46
Yoshida H, Lehr CM, Kok W et al (1992) Niosomes for oral delivery of peptide drugs. J Control Release 21:145–153. https://doi.org/10.1016/0168-3659(92)90016-K
Zeng W, Li Q, Wan T et al (2016) Hyaluronic acid-coated niosomes facilitate tacrolimus ocular delivery: Mucoadhesion, precorneal retention, aqueous humor pharmacokinetics, and transcorneal permeability. Colloids Surf B Biointerfaces 141:28–35. https://doi.org/10.1016/j.colsurfb.2016.01.014
Zoghroban HS, El-Kowrany SI, Aboul Asaad IA et al (2019) Niosomes for enhanced activity of praziquantel against schistosoma mansoni: in vivo and in vitro evaluation. Parasitol Res 118:219–234. https://doi.org/10.1007/s00436-018-6132-z

Chapter 17
Ethosome: A Potential Tool for Drug Delivery Through the Skin

Nimmathota Madhavi, Beeravelli Sudhakar, and K. V. N. Suresh Reddy

Abbreviations

GIT	gastrointestinal tract
PG	propylene glycol
SC	stratum corneum
TDDS	transdermal drug delivery system
Tm	phase transition temperature

17.1 Introduction

The transdermal drug delivery system (TDDS) is a potential delivery strategy to overcome the limitations of the conventional drug administration (Barry 1983). The significant advantages of TDDS are to avoid the gastrointestinal tract (GIT) problems and hepatic metabolism, provide the steady-state plasma drug concentration, reduce dosage frequency by controlled release tendency, and improve patient compliance (El Maghraby and Williams 2009).

Skin is the largest organ of the human body, with a total weight of more than 3 kg and a total area of 1.5–1.9 m^2. The skin regulates the influx of toxins and efflux of minerals. The human skin consists of three layers named the epidermis, dermis, and

N. Madhavi
Department of Pharmaceutics, CMR College of Pharmacy, Hyderabad, Telangana, India

B. Sudhakar
Department of Formulation Research and Development, Airis Pharma Pvt Ltd, Hyderabad, Telangana, India

K. V. N. Suresh Reddy (✉)
Department of Chemistry, GITAM Institute of Science, GITAM (Deemed to be University), Visakhapatnam, Andhra Pradesh, India

J.-C. Kim et al. (eds.), *Smart Nanomaterials in Biomedical Applications*, Nanotechnology in the Life Sciences,
https://doi.org/10.1007/978-3-030-84262-8_17

hypodermis. The epidermis is further divided into five anatomical layers, with the stratum corneum (SC) as the outermost layer (first layer of the skin) of the epidermis, exposing to the external environment. The SC is having 10–15 μm in thickness, and its structural order is like brick and mortar. The SC is made up of mostly keratin and lipids. The bricks, also called corneocytes, are made by cross-linked keratin fibers, which have a thickness in the range of 0.2–0.4 μm and about 40 μm wide (Benson and Watkinson 2012). These corneocytes are well organized by corneodesmosomes, which give structural stability to the SC (Martins et al. 2013). In SC, proteins and dried lipids have a significant role in protecting the lower layers of the skin and act as a barrier for bacteria and toxins. The mortar and whole of the stratum corneum are slightly acidic due to cellular processes, helping to prevent bacterial growth. The SC acts a rate-limiting role for permeating the molecules across the skin (Scheuplein 1976). The skin is widely recognized for its useful barrier properties compared to other biological membranes due to SC's impermeable nature, and it makes the skin a minor port for the entry for drugs (Scheuplein and Blank 1971). For many years there was speculation that active transport mechanisms occurring within skin were responsible for its unique barrier properties. Through extensive studies, it was conclusively proven that SC is the critical barrier to permeation, and it is at least three to five times less permeable than the dermis, and that permeability of the entire epidermis was indistinguishable from that of the SC alone (Elias and Menon 1991). The structure of the skin is shown in Fig. 17.1a.

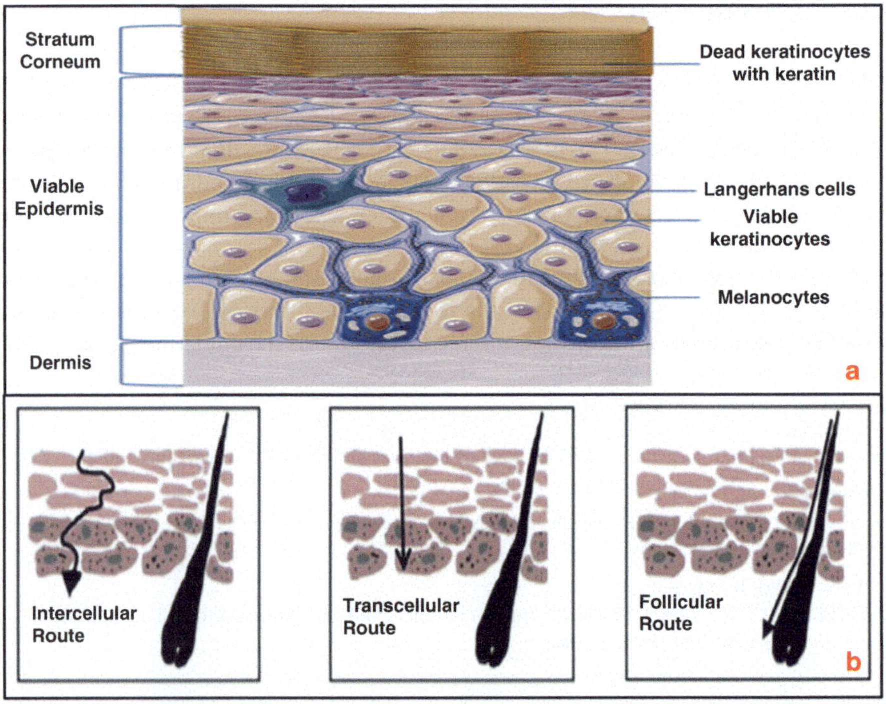

Fig. 17.1 (**a**) Schematic structure of skin barriers. (Adapted from Krishnan and Mitragotri. 2020). (**b**) Transport of drugs through intercellular, transcellular, and transappendageal route. (Adapted from Carter et al. 2019)

17.1.1 Transport of Drugs Through the Skin

Transdermal drug delivery bypasses the gastrointestinal tract (GIT) and overcomes the problems associated with oral administration, such as GIT irritation and first-pass metabolism by the liver. The permeation of drugs through the skin follows percutaneous absorption and shunt diffusion (through sweat glands and hair follicles). Diffusion through the intact SC was dominant in the steady-state stage of percutaneous absorption. Hence, SC is the primary barrier for drugs that must permeate across the skin (Scheuplein 1967). The skin permeation occurs by Fickian diffusion of the penetrating substances, with the concentration gradient across the entire skin being an essential factor (Andrews et al. 2013).

17.1.2 Routes of Drug Penetration Through the Skin

Three pathways, intercellular, transcellular, and transappendageal routes, were identified, through which the drug is administered, as mentioned in Fig. 17.1b. The intercellular route allows the drug through the gaps between the corneocytes, whereas the transcellular route contemplates crossing the corneocytes and intervening lipids. Most of the drug molecules permeate the skin through intercellular micro-routes. Hence, many researchers develop strategies to disrupt or bypass the elegant molecular structure of the SC on the excellent passage of drug molecules through the intercellular route (Scheuplein 1967; Andrews et al. 2013; Barry 2001). In the transappendageal route, the drug molecules penetrate neither through the hair follicles nor through the sebaceous glands. This kind of molecular transport is also termed as shunt pathway, and 0.1% of the fractional appendageal area is available to transport drugs. Thus, the transappendageal route has a limited role in the delivery of molecules through transdermal delivery (Ita 2014; Barry 2001).

17.1.3 Factors Affecting Permeation Across the Skin

The diffusion process controls the permeation of active substances across the skin. Permeation of various active substances such as hydrophilic and lipophilic substances followed Fick's first law. Several factors affect the permeation profiles of active substances through the skin, such as solubility, charge, partition coefficient (Log P), molecular weight (compounds with molecular weight more than 500 Daltons cannot permeate the skin), and concentration gradient. The permeation efficiency mainly depends on the drug solubility and partition coefficient (Naik et al. 2000; Karadzovska et al. 2013).

17.1.4 Model Drug Properties for Transdermal Delivery

An ideal drug for transdermal delivery should have the following properties:

1. The drug molecule should have a low molecular weight (not more than 500 Da).
2. It should have an affinity for lipophilic and hydrophilic phases.
3. It should have a short half-life.
4. It should be non-irritating and non-sensitizing.
5. The partition coefficient should be more than >1.

17.1.5 History of Medication Through the Skin

For many decades, various dosage forms have been used for local indications by the skin. Scopolamine transdermal film (innovator TRANSDERM SCOP®) is developed by GlaxoSmithKline Consumer Healthcare Ltd. and is used for motion sickness, approved by USFDA in the United States in 1982. Various conventional dosage forms available such as semisolid dosage forms (ointments, pastes, creams, and gels) are the most common. However, foams, sprays, medicated powders, solutions, lotions, and medicated adhesive plasters are also available (Barry 1983). They may be either rubbed on or sprayed on to the skin. These formulations are applied in the treatment of various skin disorders. These dosage forms fail to improve the drug's systemic availability over a period due to low permeation across the skin. Various novel drug delivery systems are available to overcome these drawbacks, i.e., active and passive tools, including iontophoresis, sonophoresis, electrophoresis, penetration enhancers, and vesicular systems (Barry 1988). Among them, the vesicular systems are one of the potential tools for effective transdermal drug delivery.

During the past few decades, there was a broad interest in exploring novel transdermal drug delivery strategies to improve the systemic availability of drugs, minimize the toxic manifestations of the drugs, and improve patient compliance. The development of novel drug delivery systems is associated with specific requirements, suitability to a patient, ensuring that the drug is properly targeted to its site of action, and designing an appropriate dose of the drug. For a few decades, a considerable research investigation has been done to enhance drug permeation through SC using lipid bilayer vesicles. Liposomes are one of the first vesicular carriers investigated in this field. Conventional liposomes are phospholipid-based water-filled colloidal spherical-shaped particles consisting of one or more concentrated bilayers. The classic liposomes have limited applications in improving the drug penetration into deeper layers of the skin (accumulate the drug molecules in the SC) and systemic circulation (limited delivery to hydrophilic deeper layers) (Bangham et al. 1965). Dermal delivery of triamcinolone acetonide liposomal lotion and liposomal gels increased drug concentrations in the epidermis and dermis, with lower systemic (minimum flux) levels than conventional formulations of the same drug concentration (Mezei and Gulasekharam 1982). The conventional liposomes have little or no value as carriers for transdermal drug delivery because they do not deeply penetrate the skin but rather remain confined to the upper layer of the SC. Thus, conventional liposomes failed to improve the steady-state plasma drug concentration in systemic circulation and therapeutic index. To overcome these disadvantages, conventional liposomes, alternatively, deformable liposomes or elastic liposomes or transfersomes, and novel ethosomes were developed (Cevc and Blume. 1992).

17.2 Ethosomes

Elka Touitou first developed ethosomes in 1997 (Touitou et al. 1997). They are soft and malleable liposomes embedded with high alcoholic content (up to 45%v/v) with the particle size available from 30 nm to 200 nm. The composition of ethosomes includes phospholipid, ethanol, and water, with or without cholesterol. The higher concentration of ethanol leads to the vesicle bilayer self-degradation by interpenetration and leads to smaller size entities' formation than conventional liposomes. Ethosomes, because of their ethanolic content, impart fluidity and penetrate easily through the SC. The combined effect of phospholipid and high ethanol concentration in vesicular formulations leads to in-depth distribution and penetration into the skin lipid bilayers (Godin and Touitou 2004). Ethosomes are biologically accustomed and biodegradable. The smaller size of ethosomes had a large surface area to ensure close contact of particles with the SC for effective drug delivery, thus improving drug permeation. The presence of high ethanolic concentration confers a negative charge to the ethosomes, which avoids aggregation of vesicles due to electrostatic repulsion (Verma and Pathak 2012).

17.2.1 Advantages of Ethosomes

- These are versatile carrier-based drug delivery systems and could be used to encapsulate hydrophilic and lipophilic molecules.
- Ethosomes can alter the solubility properties of active ingredients, which leads to improved encapsulation and loading efficiency.
- Ethosomes have better stability compared to liposomes (Touitou et al.1997; Godin and Touitou 2004).

17.2.2 Composition of Ethosomes

Generally, ethosomes contain phospholipids, short-chain alcohols (C2-C4), water, and polyols. Preferably ethosomes contain phospholipids, ethanol, double distilled water (DDW), and propylene glycol (PG) (Godin and Touitou 2004; Touitou 1998). The composition and role of various excipients are shown in Table 17.1.

Table 17.1 Composition and role of various excipients

S. No	Composition	Role	Examples
1	Phospholipids	Ethosomal vesicle former	Phosphatidylcholine, Phospholipon 90H, lipoid E80, hydrogenated soy phosphatidylcholine (HSPC)
2	Sterols	Rigidifies the ethosome vesicle	Cholesterol
3	Short chain alcohols (C2–C4)	Softening of vesicles, fluidizing agent, and preservative	Ethanol and isopropyl alcohol
4	Polyols or glycol	Acts as a permeation enhancer and stabilizing agent	Propylene glycol (PG) Transcutol®
5	Water	Dispersion system	Double distilled water, phosphate buffers
6	α-Tocopherol	Antioxidant	Tocopheryl acetate, tocopherol acetate, or vitamin E acetate
7	Organic solvent	Bilayer film forming agent	Chloroform or methanol

17.2.3 Types of Ethosomes

Based on composition, ethosomes may be classified into three types, such as classical ethosomes, binary ethosomes, and transethosomes. Types of ethosomes are shown in Fig. 17.2a.

17.2.3.1 Classical Ethosomes

These are prepared by modifying classical liposomes, composed of phospholipids, ethanol (45%v/v), and water. The classical ethosomes were reported to be superior to classical liposomes for transdermal drug delivery due to their soft malleability, smaller size, and higher entrapment efficiency. Moreover, classical ethosomes showed better skin permeation and stability profiles compared to classical liposomes (Li et al. 2012).

17.2.3.2 Binary Ethosomes

These ethosomes were developed by adding another type of alcohol to the classical ethosomes. The most commonly used alcohols are propylene glycol (PG) and isopropyl alcohol (IPA) (Zhang et al. 2012).

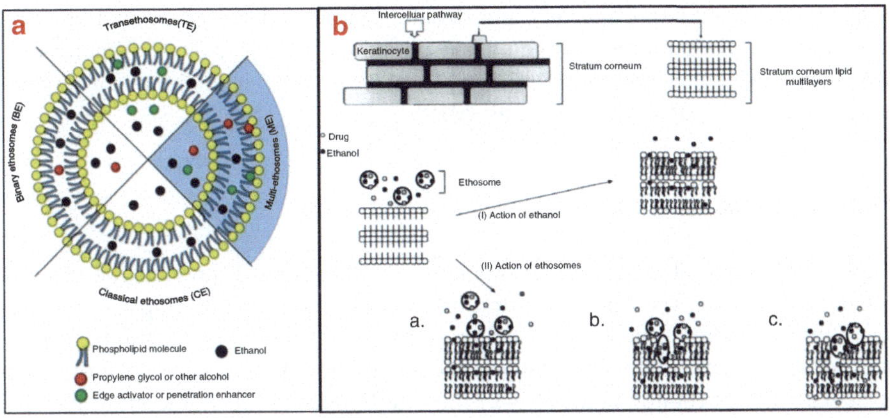

Fig. 17.2 (**a**) Schematic representations of the different types of ethosomal systems (adapted and reproduced from Zhang et al. 2020 under Creative Commons Attribution License). (**b**) Drug release mechanism of ethosomes. (Adapted from Touitou et al. 2000)

17.2.3.3 Transethosomes

These are the new generation of ethosomes containing penetration enhancers or an edge activator (surfactant). These novel vesicles were developed to combine the advantages of classical ethosomes and deformable liposomes (transfersomes) to produce transethosomes (Song et al. 2012).

17.2.4 Composition and Their Effect on Ethosomes

Composition and type of the development method influence the ethosomal profile.

17.2.4.1 Ethanol

Ethanol acts as a permeation enhancer and has an excellent role in transdermal delivery. The typical ethanol concentration is up to 45% *v/v*. Due to ethanol's presence, ethosomal vesicles are packed loosely compared to liposomes and possess a high degree of fluidity and lower transition temperature (Rakesh and Anoop 2012; Ahad et al. 2013).

17.2.4.2 Phospholipids

There are three types of phospholipids available such as natural, synthetic, and semisynthetic, respectively. The selection of phospholipid and its concentration is very important, which has an impact on the particle size, surface charge, entrapment efficiency, and loading efficiency and decides stability of the ethosomal system. Generally, the concentration range of phospholipids is 0.5%–5%(w/v). Increasing the phospholipid concentration will increase the vesicular size slightly or moderately (Zhou et al. 2010; Madhavi et al. 2019).

17.2.4.3 Cholesterol

Cholesterol is one of the critical components in ethosomes. It predicts the in vitro behavior of ethosomes. The concentration of cholesterol directly impacts the size distribution, loading efficiency, vesicle rigidity, permeability, and stability. It prevents drug or internal aqueous core leakage and reduces vesicular permeability and vesicular fusion (Touitou et al. 2014). Cholesterol does not form the bilayer on its own. During the formation of ethosomal bilayer vesicles, the cholesterol is used to fill the phospholipid bilayers' gaps and controls the drug leakage from the vesicles (Touitou et al. 1997; Madhavi et al. 2019).

17.2.4.4 Isopropyl Alcohol (IPA)

IPA has a vital role in the development of binary ethosomes. Presences of IPA in binary ethosomes have shown a significant difference from the conventional ethosomes. IPA had a pronounced effect on entrapment efficiency and in vitro percentage drug release profiles of the ethosomal vesicular systems; hence, binary ethosomal vesicles produce better skin permeation than classical ethosomes (Dave et al. 2010).

17.2.4.5 Propylene Glycol (PG)

It is widely used as a penetration enhancer, affecting the particle size, entrapment efficiency, drug permeation capability, and stability of the ethosomes. When PG concentration increases, it reduces particle size compared to the ethosomes without PG in their composition (Wu et al. 2012). Presence of both ethanol and PG in ethosomes provides better solubility of drugs and enhances the entrapment efficiency, which further improves the drug distribution throughout the vesicle (Akhtar et al. 2012).

17.2.4.6 α-Tocopherol

Generally, the phospholipids are the primary excipient to develop the ethosomal dispersion. It is a big challenge for researchers to develop stable ethosomes due to the hydrogenation of phospholipids. In an aqueous medium, phospholipids are more prone to oxidation or hydrolysis reactions, impacting stability. Addition of antioxidants like α-tocopherol to ethosomes increases stability. Hence, it is essential to keep its level to a minimum in a given composition (Rao 1984; RRC New 1990).

17.2.5 Preparation of Ethosomes

Ethosomes can be prepared by different methods as described below.

17.2.5.1 Cold Method

It is the simplest and most widely used method for the development of ethosomal systems. In this method, the organic phase is obtained by dissolving the phospholipids and cholesterol in 45%v/v ethanol in a well-closed container at room temperature (25 °C to 30 °C) by vigorous stirring with a mixer. In a separate vessel, purified water was heated to 30 °C. Based on drug solubility, the drug may be taken either in ethanol or in purified water. Once both the aqueous and ethanolic lipid mixtures reached 30 °C, the purified water is added to the ethanolic lipid solution dropwise under stirring in the range of 600 to 700 RPM and continued stirring till milky suspension is obtained. The obtained ethosomes are kept at room temperature and then stored in the refrigerator (2–8 °C) (Touitou 1996; Touitou et al. 1997).

17.2.5.2 Ethanol Injection Method

In this method, phospholipid and cholesterol are dissolved in 45% (*v/v*) ethanol in a suitable container to control ethanol's evaporation and mixed with purified water. The drug may be taken either in ethanol or in purified water, based on drug solubility. The obtained ethosomes were then homogenized with an ultrasonic probe for the required cycle at a suitable temperature. Then ethosomal suspension is filtered using a suitable size polycarbonate filter. The filtered ethosomes are transferred into a suitable container, and stored at 2–8 °C refrigerator (Liu et al. 2007).

17.2.5.3 Hot Method

In this method, the phospholipid is dispersed in water/buffer and then placed on a water bath at 40 °C until a colloidal suspension is formed. In another vessel, the ethanolic solution is heated to 40 °C. Phospholipid solution was added to the ethanolic solution drop by drop under continuous mixing using a mechanical or magnetic stirrer (Touitou 1996; Ashoniya and Meenakshi 2011).

17.2.6 Evaluation Parameters in the Development of Ethosomes

Various physicochemical parameters to be evaluated for ethosomes are mentioned in Table 17.2.

Generally, permeation of ethosomes through intact SC includes intercellular and transcellular routes. At body temperature, the SC lipid multilayers are densely packed and highly conformationally ordered. When ethosomes are applied to the skin, ethanol interacts with the polar head group region of dried phospholipids in SC, called the intercalation effect. The intercalation effect reduces the phase transition temperature (Tm) of the SC. The synergetic action of ethanol and ethosomal vesicles altered the membrane properties of SC and enhances fluidity. During this stage, the soft, malleable nature of ethosomal vesicles could easily penetrate the skin's deep layers. The penetrated ethosomal vesicle releases the drug in deep layers

Table 17.2 Various characterization parameters of ethosomes

Characterization methods	Analytical methods/instrumentation
Physical characterization	
Vesicle size and size distribution Vesicle shape, surface morphology	Dynamic light scattering, transmission electron microscope (TEM), scanning electron microscope (SEM)
Surface charge and surface pH	Zeta potential measurements using Zetasizer and pH probes
Lamellarity and lipid and ethanol interaction	Small angle X-ray scattering, P-NMR, freeze-fracture electron microscopy
Phase behavior and bilayer fluidity	Freeze-fracture electron microscopy, DSC
Percent encapsulation	Mini column centrifugation, gel exclusion, ion-exchange chromatography, protamine aggregation, radiolabelling
Drug release	Franz diffusion cell with artificial or biological membrane
Vesicle fusion measurement	FRET assay, probe dilution assay
Turbidity	Nephalometer
Vesicle skin interaction	Confocal laser scanning microscopy, fluorescence microscopy, TEM, eosin-hematoxylin staining, FTIR
Chemical characterization	
Phospholipid concentration	Bartlett assay/Stewart assay
Cholesterol concentration	Cholesterol oxidase assay and HPLC
Phospholipid peroxidation	UV absorbance, TBA (for endoperoxidase), iodometric (for hydroperoxidase) and GLC
Phospholipid hydrolysis	HPLC and TLC and fatty acid concentration
Cholesterol auto-oxidation	HPLC and TLC
Antioxidant degradation Ethanol quantification	HPLC and TLC Enzymatic diagnostic kit (sigma), oxidation of alcohol to acetaldehyde

of the skin, and the released drug absorption could result from the fusion of ethosomes with skin lipids. The drug release mechanism of ethosomes is shown in Fig. 17.2b. In conclusion, ethanol has a significant role in vesicle fluidity and the possibility of the interaction of ethosomes with SC, which may lead to superior delivery of drugs (Touitou et al. 1997; Godin and Touitou 2004; Touitou 1998).

17.3 Ethosomes as Drug Delivery Systems

Ethosomal dosage forms contain high alcohol concentration and are administered through either transdermal or topical delivery. Few novel ethosomal formulations, ethosomal gels, transdermal patches, and creams are described below.

17.3.1 Ethosomal Gels

Generally, ethosomes are available in the form of gels. Ethosomal gel has a superior skin permeation rate compared to ethosomal ointments and creams. Carbopols and hydroxypropyl methylcellulose polymers are commonly used as gel forming agents because of their compatibility and bioadhesive nature. Gels produce enough viscosity (pseudoplastic flow) and quickly spread onto the skin, thus enhancing the permeation property of the ethosomal gels.

Etodolac-loaded ethosomal and liposomal gels were investigated against marketed gel for transdermal delivery. In this study, they reported that ETO-EG showed maximum flux (0.0613 mg/cm^2/hr) and permeability coefficient 7.6×10^{-3} cm/hr. as compared to ETO-LG, marketed gel, and 45%*v/v* ethanolic ETO-Solution. The permeation enhancement ratio was found to be more for ETO-EG (1.47) comparative to 45%*v/v* ethanolic ETO-Solution (0.64) and ETO-LG (1.22), respectively. The flux was improved from the ETO-LG due to the synergetic action of the ethanol and ethosomal vesicle on SC lipid structure. The anti-inflammatory activity was tested on rats by carrageenan paw edema model. The percentage reduction in edema (%RIE) of test products showed 46.97 ± 1.23%, 70.02 ± 1.98%, and 38.67 ± 0.38% for ETO-LG, ETO-EG, and PROXYM®, respectively. The %RIE of 45% v/v ethanolic ETO-Solution was observed only 44.18 ± 0.17% up to 4 h. After 4 h the activity was decreased. The vesicular gels showed better %RIE compared to other test products. After 4 h the anti-inflammatory activity kept decreasing up to 8 hr. (Madhavi et al. 2018; Madhavi et al. 2019).

17.3.2 Ethosomal Patches

Ethosomal patch preparation is difficult than gels due to the requirement of molds. Various polymers are used to prepare patches such as polyvinylpyrrolidone/vinyl acetate, acrylic resin, and hydroxypropyl methylcellulose E15. Triethyl citrate was added to the formulation as a plasticizer. Ethosomal patches of various drugs are found in the literature. Touitou et al. (2000) have investigated testosterone ethosomes by transdermal delivery. Testosterone ethosomes showed superior results compared to marketed AndroGel® and hydroethanolic solutions due to synergistic mechanisms between ethanol, vesicles, and skin lipids. In ethosomes, ethanol alters the structural organization of the SC lipids and enhances the membrane fluidity. Hence the soft, malleable ethosomes can penetrate the SC. The release of drugs in deep layers of the skin and its transdermal absorption could result from the fusion of ethosomes with skin lipids.

In another study, the delivery pattern of ligustrazine-loaded ethosomal patch is compared to intragastric ligustrazine and conventional patch. The pharmacokinetic studies of ligustrazine ethosomal patch on rat show maximum AUC (168.38 μg·ml^{-1}.h) compared to intragastric ligustrazine (80.42 μg·ml^{-1}) and conventional patch (79.31 μg·ml^{-1}·h), respectively. The AUC value of ligustrazine ethosomal patch was 2.09 times greater than that of intragastric ligustrazine and 2.12 times greater than that of conventional transdermal film, respectively. These results indicated that ligustrazine ethosomal patches facilitate more permeation by altering the barrier properties of SC, which improves the drug absorption and bioavailability compared to other test preparations (Liu et al. 2011).

In the comparative study of novel anti-malarial transdermal nano-system (ethosomal anti-malarial cataplasm) derivative of artesunate (synthesized) with febrifugine (natural product), artesunate permeation was significantly increased at 8 h, which was 1.57 times as much as that of conventional cataplasm. Ethosomal cataplasm could make a large quantity of anti-malarial drug penetrate through the skin; then the remaining drug in the ethosomal cataplasm could be steadily released. These characteristics of ethosomal cataplasm are favorable for anti-malarial drugs to kill *Plasmodium*. The ethosomal cataplasm showed a superior result than artesunate and febrifugine conventional cataplasm (Shen et al. 2015).

17.3.3 Ethosomal Creams

Curcuma longa (*C. longa*) extract-loaded novel vesicular systems are investigated against the *Curcuma longa* extract for their photo-protective effect by assessment of skin hydration and sebum content using Cutometer and Sebumeter, respectively. The test product efficacy was in the order *C. longa* extract-loaded transfersomal creams > *C. longa* extract-loaded ethosomal creams > *C. longa* extract-loaded liposomal creams > *C. longa* extract-loaded creams > empty transfersome-loaded cream > empty ethosome-loaded cream > empty liposome-loaded cream > base cream (Kaur and Saraf et al. 2011).

From another study, *C. longa* extract ethosomal cream is investigated as a photoprotective agent or an anti-wrinkle agent (Gunjan and Swarnlata et al. 2014). *C. longa* ethosomes deposit the curcumin into the deeper layers of the skin, which leads to its antioxidant effect. The encapsulation of *C. longa* was significantly improved with 58%. The Cutometer results showed improvement in overall elasticity, biological elasticity, recovery of deformed skin, firmness, and fatigue reduction, correlated with the cream's anti-wrinkle properties. These favorable effects might have been due to the synergistic antioxidant, anti-inflammatory, and protective properties of *C. longa* and hydrant, moisturizing, and lipid components of ethosomes and cream.

In another study, the effect of azelaic acid ethosomal cream against the *Propionibacterium acnes* was studied. Azelaic acid is an anti-acne drug and has been shown to decrease the population of *P. acnes*. Azelaic acid penetrates through the SC into the cytoplasm by passing through thick peptidoglycan of *Propionibacterium acnes* (*P. acnes*). Thus, it is essential to improve the penetration of azelaic acid using ethosome. The %EE of azelaic acid ethosomes was found around 94% by film hydration method using 35%*v/v* of ethanol concentration. Antibacterial activity to *P. acnes* showed that azelaic acid ethosome cream was given better activity than marketed cream (Zelface® cream) due to the presence of ethanol and phospholipids in composition. Ethanol has the ability to change the physical properties of bacterial environments and bacterial interactions with their environment. Ethanol modifies the lipid composition in bacteria, so bacterial growth can be inhibited. In addition, ethanol also interferes with integrity of bacterial cell walls, thus making it easy for the ethosome to enter into bacterial cytoplasmic membrane. Teichoic acid present inside the cell walls of Gram-positive bacteria is hydrophilic in nature and dissolve in ethanol, and damage the bacterial cell wall structure. (Apriani et al. 2019).

17.3.4 Applications of Ethosomes for Various Skin Disorders

The ethosomes of cryptotanshinone (CPT) are investigated for transdermal delivery. CPT is a diterpene quinone extracted from the Chinese herb *Salvia miltiorrhiza*. CPT is useful for the clinical treatment of acne vulgaris. CPT ethosomes were developed using two different concentrations of soya phosphatidylcholine (SPC). The higher concentration of SPC (0.4 mg) in ethosomes showed mean vesicle size below 100 nm and had potential for CPT delivery into the skin, whereas the lower concentration of SPC (0.2 mg) showed larger vesicle size. The %EE of CPT ethosomes is reported higher than that of liposomes due to the greater amount of the drug in the ethosomal core. The transdermal permeation of CPT ethosomal gel through excised pig skin showed higher transdermal flux (19.36 ± 4.16 ng/h/cm^2) than conventional gels (7.913 ± 1.929 ng/h/cm^2). This is due to ethanol's presence in ethosomes that can interact with lipid molecules often and reduces the phase transition temperature of SC lipids, thereby enhancing their fluidity, and finally lipid multilayer density was decreased. The skin irritation of the CPT ethosomal gel was also studied on rabbits as per OECD guidelines. The result indicated that CPT itself causes no skin irritation, whereas CPT ethosomal gel can be considered to cause slight irritation (negligible). An in vivo study proved that CPT ethosomal gel had a better anti-acne effect than a conventional gel with only slight skin irritation (Yuet al. 2016).

Cetirizine liposomes and ethosomes were investigated for skin permeation and deposition studies via topical route for 8 hrs. The results found that ethosomal dispersion showed a higher cumulative permeation rate than other test products (ethosomal gel, liposomal dispersion, liposomal gel, conventional cream, and cetirizine aqueous solution). The result is due to flexible membrane packing of the ethosomal system over the conventional liposomal system, which helps in better penetration and permeation. Two ways could achieve the better transdermal absorption of drug: fusion of ethosomes with skin lipids and the other is the synergistic mechanism between ethanol, vesicles, and skin lipids, which promotes maximal permeation from the ethosomal vesicles (Goindi et al. 2014).

Anthralin ethosomes and liposomes were investigated for their effectiveness in treating psoriasis based on PASI (Psoriasis Area and Severity Index). The ex vivo cumulative amount of drug that permeated through rat abdominal skin for ethosomal gel was found to be higher than that of liposomal gel and drug hydroalcoholic solution, respectively. The presence of ethanol in ethosomes leads to more permeation across the rat abdominal skin. Ethanol can increase the vesicles' flexibility and fluidity and allow them to deform and penetrate through skin pores that are smaller than their size. Various factors influenced drug permeation from ethosomes, such as the lipid-softening effect of ethanol and ethosome lipids with the SC lipids (Dina et al. 2020).

17.3.5 Ethosomes for Microbial Skin Infection

Ethosomes had a significant role as an antibiotic in the treatment of skin infections by transdermal delivery. Erythromycin ethosomes were investigated against *Staphylococcus aureus*-induced deep dermal infections of the mice. No subdermal healing was found in infected animals, when treated with topical hydroethanolic erythromycin solution. The observation of animals was done for 10 days. At the end of the 10th day of treatment, the ethosomal erythromycin preparation-treated animals totally recovered from the infection and re-grew hair on their backs. When the dermal application of erythromycin is compared with systemic administration, in six mice, one mouse showed slight dermatonecroses. Hence, the obtained result has purely demonstrated the suitability of ethosomal erythromycin for dermal application. The cause for this curative effect is soft vesicle production in ethosomes because of ethanol (Godin et al. 2005).

Another study demonstrated that combination drug therapy effectively reduced acne. This includes ethosomal gel of clindamycin and salicylic acid (CLSA). The effectiveness of gel was tested on human beings of both sexes suffering from mild to moderate acne vulgaris. The gel is applied twice a day on the acne facial lesions, and it was observed for 8 weeks. When compared with previous treatment history data, it was shown very less lesion count. Tolerability of the gel can be evaluated by erythema, burning, peeling, and pruritis. After completion of the treatment period (8 weeks), a considerable improvement of the acne condition was observed. No single case of severe or moderate adverse skin reactions was reported with CLSA treatment (Touitou et al. 2008).

For the treatment of dermatophytosis, griseofulvin ethosomes and liposomes were investigated by transdermal delivery. Ethosome dispersion showed higher drug entrapment efficiency (72.94 ± 0.80%) as compared to liposome dispersion. The presence of ethanol in ethosomes showed superior results compared to liposomes. Ethanol improves the griseofulvin solubility, leading to improved drug encapsulation, whereas liposomes showed lower drug entrapment efficiency due to their rigid bilayer structure. The evaluation of the fluorescent microscope study of Rhodamine 123-loaded ethosomes and liposomes also reported in this work. It was found that ethosomes penetrated and deposited the Rhodamine 123 into the skin. In contrast, the liposomes showed low penetration due to the slow rate of drug diffusion and vesicles' rigidity. Hence the ethosomal penetration mechanism is suitable to target superficial fungal infections of the skin. In pharmacodynamics, liposome gel studies exhibited a slower cure rate than ethosome gel due to the rigid or less deformable nature of liposomes which leads to low penetration and diffusion of drug in the skin. The enhanced properties of griseofulvin ethosomes may provide better dermatopharmacokinetic and pharmacodynamics properties against *Microsporum* dermatophytosis as compared to liposomes (Aggarwal and Goindi 2013).

17.3.6 Ethosomes as Carriers in Peptides and Protein Delivery

Currently, most of the protein and peptide drugs are administered by parenteral routes such as subcutaneous (SC), intramuscular (IM), or intravenous (IV) injections. The intravenous injection produces hemolysis. Subcutaneous administration is easy; it is rarely painful and achieves steady-state drug levels in the blood. For example, insulin is an oligomeric protein used in the treatment of diabetes mellitus, having a molecular weight (MW) of 6000 Dalton, and is given by a subcutaneous route. Most researchers have focused their research on non-passive delivery of insulin through the skin using physical methods like iontophoresis, phonophoresis, sonophoresis, etc. (Sage et al. 1993; Tachibana 1992).

Thymosin β-4 (Tβ-4) is a macromolecular protein drug used for wound repair, but it has low membrane permeability and unstable physicochemical characteristics. Thymosin β-4 (Tβ-4)-loaded ethosomes were investigated for wound repair through transdermal delivery. The Tβ-4-loaded ethosomes were prepared by ethanol infusion method using orthogonal design(Fu et al. 2019). In invitro studies, they reported that the ethosomal gel had released the drug 1.67 times higher than the T-β4 gel within 5 hrs. Based on the in vivo pharmacodynamic study, they reported that the healing time for T-β4 ethosomal gel was significantly shorter than that of test products T-β4 gels and blank gel group, respectively.

17.3.7 Ethosomes for Anti-inflammatory and Anti-arthritis

Ammonium glycyrrhizinate ethosomes were investigated for topical application for the treatment of inflammation-based skin diseases. The composition of ethosomes showed impact on the particle size of ammonium glycyrrhizinate-loaded ethosomes and blank ethosomes. The mean size of ethosomes decreased with increasing ethanol concentrations, whereas the ethosomal mean vesicle size was increased with decreasing phospholipid concentration. Ammonium glycyrrhizinate ethosomes showed a maximum in vitro and in vivo percutaneous permeation by sustained release manner. Hence, a prolongation of ethosome therapeutic activity can be achieved (Paolino et al. 2005).

Cannabinoid (CBD) ethosomes were investigated to treat rheumatic arthritis. CBD is a highly lipid-soluble neutral compound due to the polar phenolic group's presence and interaction with one or more corresponding polar groups at the bilayer interface, which is involved in membrane perturbation. On the other hand, there is a formation of a eutectic mixture between CBD and phospholipid (PL)-90, enhancing the transdermal permeation through SC. When CBD interacted with PL-90, it may decrease the phase transition of PL-90. This phenomenon significantly increases the permeation of CBD and suppresses the inflammation (Lodzki et al. 2003).

In another study, curcumin-loaded propylene glycol liposomes (PGL), ethosomes, and traditional liposomes were investigated to determine their efficacy by transdermal delivery. Anti-inflammatory activity of test products is investigated on male Sprague Dawley rats using carrageenan-induced paw edema method. The result was that rapid evaporation of ethanol content results in anti-inflammatory activity, even though curcumin ethosomes were precipitated and then aggregated on the skin surface. Among all test products, curcumin ethosomes produce prolonged inhibition of edema on rats effectively. Apart from curcumin ethosomes, curcumin PGL also enhances skin penetration due to PG's presence in their composition, which will enhance the flexibility of lecithin bilayer and maintain liposome stability (Zhao et al. 2013).

17.3.8 Ethosomes on Antitumor Activity

Ethosomes were investigated for cancer treatment by dermal delivery. The dermal delivery of 5-fluorouracil (5-FU) was developed using Tumorep DS (tumor repressive delivery system) with n-decyl methyl sulfoxide (n-decylMSO; as a differentiation agent). Tumorep DS ethosomes and liposomes were developed with the same ethanol concentration (30%w/w) and phospholipids (5%w/w). After analyzing the in vivo skin tumor activity on nude female mice (animal model), it was concluded that 5-FU Tumorep DS's antitumor effect is significantly superior to the drug solution and a commercial product (Efudex®). They studied the dermal delivery of 5-FU from Tumorep DS and ethosomes, both of which contained the same amount of ethanol and phospholipids, and found that the ethosomes delivered more drug (7.98 %) compared to Tumorep DS (3.39 %). . Also, in Tumorep DS composition, replacement of 1% w/w n-decylMSO with 10% DMSO resulted in lower skin accumulation (2.3 times, i.e., 109.23 ± 12.35 μg) followed by higher permeation (2.9 times, i.e., 38.0 ± 8.86 μg) of the drug across the skin. The Tumorep DS gel antitumor activity was compared with Efudex®. The Tumorep DS showed drug accumulation of 10.17% and Efudex® showed 3.75%, respectively, but both formulations contain 5% w/w drug. Based on the result, it could be concluded that Tumorep DS efficiently enhanced the delivery of the anticancer drug into the deep layers of the skin and thus can improve topical skin cancer therapy (Ainbinder and Touitou 2011).

The mitoxantrone (MTO) ethosomal gel was developed for melanoma therapy. MTO can increase calreticulin (CRT) expression on the cell surface to induce the anticancer immune response effectively. The immunohistochemical study demonstrated that the anticancer mechanism of MTO improves CRT translocation from the intracellular spaces to the membranes. Furthermore, the membrane CRT is recognized by immune cells, and the subsequent immune activation leads to killing melanoma cells. The CRT expression improvement induced by the MTO ethosomal gel indicated a very effective anti-melanoma transdermal formulation. The cell biological study further revealed a similar result via flow cytometry, indicating a high level of cell uptake of MTO from the ethosomes. The tumor inhibitory rate of the MTO ethosomal gel was found to be 68.44% (Xiang et al. 2015).

17.3.9 Ethosomes for Antifungal Drugs

Terbinafine hydrochloride (TH) ethosomes, binary ethosomes, and transfersomes were investigated to compare the skin permeation of TH using mouse skin. Among all the test products, binary ethosomes showed a more skin permeation rate than transfersomes and ethosomes, respectively. The confocal laser scanning microscopy (CLSM) experiments, penetration depth, and fluorescence intensity of Rhodamine B confirm that binary ethosomes were much more significant than those from ethosomes and transfersomes (Zhang et al. 2012).

17.3.10 Marketed Ethosomal Products

Elka Touitou developed an ethosomal cream of acyclovir (Supra-vir) product by associating with Trima pharmaceutical company. The Supra-vir is indicated for herpes simplex virus infections of the skin and mucous membrane. Some other ethosomal marketed products are also available such as Nanominox, Cellutight EF, Skin Genuity, Decorin cream, etc. (Mbah et al. 2014; Sudhakar et al. 2012).

17.4 Conclusion

Ethosomes are novel tools for effective transdermal delivery to achieve steady-state plasma drug concentration across the skin on safe mode. Their composition, physicochemical profiles, and mechanism of action were shown superior compared to other vesicular systems. The soft and malleable nature of the ethosomes and ethanol and phospholipids' synergetic action promote the skin's permeation rate. Ethosomes are proved as efficient systems for transdermal delivery of various categories of pharmaceutically active molecules. Because of their composition, ethosomes are soft vesicles; such developments may further expand the field of transdermal applications. However, it required more extensive research studies to establish the ethosomal system. The entire chapter reveals that ethosomes are the potential dosage forms for dermal and transdermal delivery of therapeutic and cosmetic agents.

References

Aggarwal N, Goindi S (2013) Dermato pharmacokinetic and pharmacodynamic evaluation of ethosomes of griseofulvin designed for dermal delivery. J Nanopart Res 15:1983

Ahad A, Aqil M, Kohli K, Sultana Y, Mujeeb M (2013) Enhanced transdermal delivery of an antihypertensive agent via nanoethosomes statistical optimization, characterization and pharmacokinetic assessment. Int J Pharm 443(1–2):26–38

Ainbinder D, Touitou E (2011) A new approach for skin tumor treatment: from delivery system characterization to in vivo evaluation. Drug Deliv Transl Res 1:53–65

Andrews SN, Jeong E, Prausnitz MR (2013) Transdermal delivery of molecules is limited by full epidermis, not just stratum corneum. Pharm Res 30:1099–1109

Apriani EF, Rosana Y, Iskandarsyah I (2019) Formulation, characterization, and in vitro testing of azelaic acid ethosome-based cream against Propionibacterium acnes for the treatment of acne. J Adv Pharm Technol Res 10:75–80

Ashoniya S, Meenakshi C (2011) Ethosomes as vesicular carrier for enhanced transdermal delivery of ketoconazole formulation and evaluation. J Pharm Cosmetol 1(3):1–14

Bangham AD, Standish MM, Watkins JC (1965) Diffusion of univalent ions across the lamellae of swollen phospholipids. J Mol Biol 13:238–252

Barry BW (1983) Dermatological formulations: percutaneous absorption. Marcel Dekker, New York and Basel

Barry BW (1988) Action of skin penetration enhancers-the lipid protein partitioning theory. Int J Cosmet Sci 10(6):281–293

Barry BW (2001) Novel mechanisms and devices to enable successful transdermal drug delivery. Eur J Pharm Sci 14:101–114

Benson HAE, Watkinson AC (2012) Topical and transdermal drug delivery. Wiley, Hoboken

Carter P, Narasimhan B, Wang Q (2019) Biocompatible nanoparticles and vesicular systems in transdermal drug delivery for various skin diseases. Int J Pharm 555:49–62

Cevc G, Blume G (1992) Lipid vesicles penetrate into intact skin owing to the transdermal osmotic gradients and hydration force. Biochim Biophys Acta 1104(1):226–232

Dave V, Kumar D, Lewis S, Paliwal S (2010) Ethosome for enhanced transdermal drug delivery of aceclofenac. Int J Drug Deliv 2(1):81–92

Dina F, Youssef EMK, Soliman GM (2020) Liposomal and ethosomal gels for the topical delivery of Anthralin preparation, comparative evaluation and clinical assessment in psoriatic patients. Pharmaceutics 12:446
Elias PM, Menon GK (1991) Structural and lipid biochemical correlates of the epidermal permeability barrier. Adv Lipid Res 24:1–26
El Maghraby GM, Williams AC (2009) Vesicular systems for delivering conventional small organic molecules and larger macromolecules to and through human skin. Expert Opin Drug Deliv 6:149–163
Fu X, Shi Y, Wang H, Zhao X, Sun Q, Huang Y, Qi T, Lin G (2019) Ethosomal gel for improving transdermal delivery of Thymosin β-4. Int J Nanomedicine 14:9275–9284
Godin B, Touitou E (2004) Mechanism of bacitracin permeation enhancement through the skin and cellular membranes from an ethosomal carrier. J Control Release 94:365–379
Godin B, Touitou E, Rubinstein E, Athamna A, Athamna M (2005) A new approach for treatment of deep skin infections by an ethosomal antibiotic preparation an in vivo study. J Antimicrob Chemother 55:989–994
Goindi S, Dhatt B, Kaur A (2014) Ethosomes based topical delivery system of antihistaminic drug for treatment of skin allergies. J Microencapsul 31(7):716–724
Gunjan J, Swarnlata S (2014) Topical delivery of Curcuma longa extract loaded nanosized ethosomes to combat facial wrinkles. J Pharm Drug Deliv Res 3(1):100–118
Ita KB (2014) Transdermal drug delivery: progress and challenges. J Drug Del Sci Technol 24(3):245–250
Karadzovska D, Brooks JD, Monteiro Riviere NA, Riviere JE (2013) Predicting skin permeability from complex vehicles. Adv Drug Deliv Rev 65(2):265–277
Kaur CD, Saraf S (2011) Topical vesicular formulations of Curcuma longa extract on recuperating the ultraviolet radiation-damaged skin. J Cosmet Dermatol 10(4):260–265
Krishnan V, Mitragotri S (2020) Nanoparticles for topical drug delivery: potential for skin cancer treatment. Adv Drug Deliv Rev 153:87–108
Li G, Fan Y, Fan C (2012) Tacrolimus-loaded ethosomes: physicochemical characterization and in vivo evaluation. Eur J Pharm Biopharm 82(1):49–57
Liu X, Liu H, Liu J (2011) Preparation of a ligustrazine ethosome patch and its evaluation in vitro and in vivo. Int J Nanomedicine 6:241–247
Lodzki M, Godin B, Rakou L, Mechoulam R, Touitou E (2003) Cannabidiol transdermal delivery and anti-inflammatory effect in a murine model. J Control Release 93:377–387
Madhavi N, Sudhakar B, Reddy KVNS, Ratna JV (2018) Pharmacokinetic and pharmacodynamic studies of etodolac loaded vesicular gels on rats by transdermal delivery. Daru 26(1):43–56
Madhavi N, Sudhakar B, Reddy KVNS, Ratna JV (2019) Design by optimization and comparative evaluation of vesicular gels of etodolac for transdermal delivery. Drug Dev Ind Pharm 45(4):611–628
Mbah CC, Builders PF, Attama AA (2014) Nanovesicular carriers as alternative drug delivery systems: ethosomes in focus. Expert Opin Drug Deliv 11(1):45–59
Martins M, Azoia NG, Ribeiro A, Shimanovich U, Silva C, Cavaco Paulo A (2013) In vitro and computational studies of transdermal perfusion of nano formulations containing a large molecular weight protein. Colloid Surface B 108:271–278
Mezei M, Gulasekharam V (1982) Liposomes, a selective drug delivery system for the topical route of administration: gel dosage form. J Pharm Pharmacol 34:473–474
Naik A, Kalia YN, Guy RH (2000) Transdermal drug delivery: overcoming the skin's barrier function. Pharm Sci Technol 3:318–326
Paolino D, Lucania G, Mardente D, Alhaique F, Fresta M (2005) Ethosomes for skin delivery of ammonium glycyrrhizinate: in vitro percutaneous permeation through human skin and in vivo anti-inflammatory activity on human volunteers. J Control Release 106:99–110
Rakesh R, Anoop KR (2012) Formulation and optimization of nano-sized ethosomes for enhanced transdermal delivery of cromolyn sodium. J Pharm Bioallied Sci 4(4):333–340

Rao LS (1984) Preparation of liposomes on the industrial scale problems and perspectives. In: Gregoriadis G (ed) Liposome technology, vol 1. CRC Press, Florida
RRC New (1990) Liposomes: a practical approach. Oxford University Press, New York
Scheuplein RJ (1976) Permeability of the skin: a review of major concepts and some new developments. J Invest Dermatol 67(5):672–676
Scheuplein RJ, Blank IH (1971) Permeability of the skin. Physiol Rev 51(4):702–747
Scheuplein RJ (1967) Mechanism of percutaneous absorption II. Transient diffusion and the relative importance of various routes of skin penetration. J Invest Dermatol 48:79–88
Sage BH, Hoke RA, McFarland AC, Kowalcyk K (1993) The importance of skin pH in iontophoresis of peptides, prediction percutaneous penetration. STS Publishing Cardiff, pp 410–418
Shen S, Liu SZ, Zhang YS (2015) Compound antimalarial ethosomal cataplasm preparation evaluation, and mechanism of penetration enhancement. Int J Nanomedicine 10:4239–4253
Song CK, Balakrishnan P, Shim CK, Chung SJ, Chong S, Kim DD (2012) A novel vesicular carrier, transethosome, for enhanced skin delivery of voriconazole characterization and in vitro/in vivo evaluation. Colloids Surf B Biointerfaces 92:299–304
Sudhakar CK, Nitish U, Sanjay J, Charyulu RN (2012) Ethosomes as non-invasive loom for transdermal drug delivery. Nanomed Drug Deliv Apple Academic Press, pp 1–16
Tachibana K (1992) Transdermal delivery of insulin to alloxan diabetic rabbits by ultrasound exposure. Pharm Res 9:952–954
Touitou E, Alkabes M, Dayan N, Eliaz M (1997) Ethosomes novel vesicular carriers for enhanced skin delivery. Pharm Res 14(S305)
Touitou E (1998) Composition for applying active substances to or through the skin. U.S. Patent 5:638–716
Touitou E (1996) Compositions for applying active substances to or through the skin. United States patent US 5540934 a
Touitou E, Dayan N, Bergelson L, Godin B, Eliaz M (2000) Ethosomes novel vesicular carriers for enhanced delivery: characterization and skin penetration properties. J Control Release 65(3):403–441
Touitou B, Godin M, Shumilov N, Bishouty D, Ainbinder R, Shouval AI, Leibovici V (2008) Efficacy and tolerability of clindamycin phosphate and salicylic acid gel in the treatment of mild to moderate acne vulgaris. J Eur Acad Dermatol Venerol 22(5):629–631
Verma P, Pathak K (2012) Nanosized ethanolic vesicles loaded with econazole nitrate for the treatment of deep fungal infections through topical gel formulation. Nanomedicine 8(4):489–496
Wu X, Guy RH (2009) Applications of nanoparticles in topical drug delivery and in cosmetics. J Drug Deliv Sci Technol 19(6):371–384
Xiang Y, Lina D, Yu L, Guiying F, Jin Y (2015) Improved anti-melanoma effect of a transdermal mitoxantrone ethosome gel. Biomed Pharmacother 73:6–11
Yu Z, Lv H, Han G, Ma K (2016) Ethosomes loaded with Cryptotanshinone for acne treatment through topical gel formulation. PLoS One 11(7):e0159967
Zhang JP, Wei YH, Zhou Y, Li YQ, Wu XA (2012) Ethosomes, binary ethosomes and transfersomes of terbinafine hydrochloride a comparative study. Arch Pharm Res 35(1):109–117
Zhang L, Li X, Zhu S, Zhang T, Maimaiti A, Ding M, Shi (2020) Dermal targeting delivery of terbinafine hydrochloride using novel multi-ethosomes: a new approach to fungal infection treatment. Coatings 10(4):304
Zhao YZ, Lu CT, Zhang Y (2013) Selection of high efficient transdermal lipid vesicle for curcumin skin delivery. Int J Pharm 454(1):302–309
Zhou Y, Wei Y, Liu H, Zhang G, Wu X (2010) Preparation and in vitro evaluation of ethosomal total alkaloids of Sophora alopecuroides loaded by a transmembrane pH-gradient method. AAPS Pharm Sci Tech 11(3):1350–1358
Nida, Akhtar Kamla, Pathak (2012) Cavamax W7 Composite Ethosomal Gel of Clotrimazole for Improved Topical Delivery: Development and Comparison with Ethosomal Gel. AAPS PharmSciTech 13(1) 344-355 10.1208/s12249-012-9754-y

Chapter 18
Polysaccharides of Natural Gums-Based Biomedical Devices for Drug Delivery Application

Kasula Nagaraja, Kummara Madhusudana Rao, Kummari S. V. Krishna Rao, Khateef Riazunnisa, and K. V. N. Suresh Reddy

18.1 Introduction

Plant gums are one of the most accepted natural remedies for the protection of human health. These are more sustainable, eco-friendly, non-toxic and safer for production of biomaterials; it is due to the above properties that they have great potential application in various fields. Gums are readily dissolved in water and other aqueous media; it results in the formation of mucilage and slimy masses. Natural gums are pathological products, whereas mucilages are physiological products. Examples of natural gums are tragacanth gum, gum kondagogu, gum karaya, tamarind gums and guar gum. While mucilage is found in various parts of plants, such as epidermal cell walls, roots, seed coats, barks and middle lamella, plant-based natural gums contain hydrophilic carbohydrate polysaccharides. Most of the natural gums are high molecular weight polymers, and they are composed of monosaccharide units linked in glycoside bonds (Pandey and Khuller 2004; Chamarthy and Pinal 2008).

K. Nagaraja · K. S. V. Krishna Rao (✉)
Polymer Biomaterial Design and Synthesis Laboratory, Department of Chemistry, Yogi Vemana University, Kadapa, Andhra Pradesh, India
e-mail: ksvkr@yogivemanauniversity.ac.in

K. M. Rao
School of Chemical Engineering, Yeungnam University, Gyeongsan, Gyeongbuk, South Korea

K. Riazunnisa
Biotechnology and Bioinformatics, Yogi Vemana University, Kadapa, Andhra Pradesh, India

K. V. N. Suresh Reddy
Department of Chemistry, GITAM Institute of Technology, GITAM (Deemed to be University), Visakhapatnam, Andhra Pradesh, India

J.-C. Kim et al. (eds.), *Smart Nanomaterials in Biomedical Applications*, Nanotechnology in the Life Sciences,
https://doi.org/10.1007/978-3-030-84262-8_18

In recent years, plant gums are playing incredible role in various applications such as emulsifying agents, binders, disintegrates, thickeners, paints, paper making, protective colloids, film coating agents, buccal films, microsphere, beads, viscous liquid formulations, nanoparticle synthesis, bioadhesive gelling agents and drug delivery (Fig. 18.1 and Table 18.1). Newly enhanced excipients continue to be developed in drug delivery systems. Natural gum based polymeric material properties are highly suitable as a pharmaceutical excipient, needs a lot of attention in pharmaceutical and drug delivery researches. The gum polymeric materials in formulation and development of novel dosage forms are important, because they are considered safe, biodegradable and biocompatible. However, natural resource (gum/polysaccharide) is widely accepted and used commonly in the food and pharmaceutical industry (Madhusudhan et al. 2019; Alonso-Sande et al. 2009).

The natural polymers are generally engaged in floating drug delivery systems to target the drug to specific gastrointestinal region in the stomach. Natural gums gained a lot of importance during the last two decades in the area of drug delivery. Plant gum derivatives are a group of polymers, widely used in pharmaceutical applications and biomedical field of drug delivery system. The drug delivery systems' main advantages are attaining optimal drug concentration, enhancement of labile drugs' activity and usually time prolonged due to protection against environment and diminishing side effects due to the high initial blood concentration. Natural carbohydrate polymers are advantageous over the synthetic polymers; generally natural polymers are economically low-cost, non-toxic, freely available,

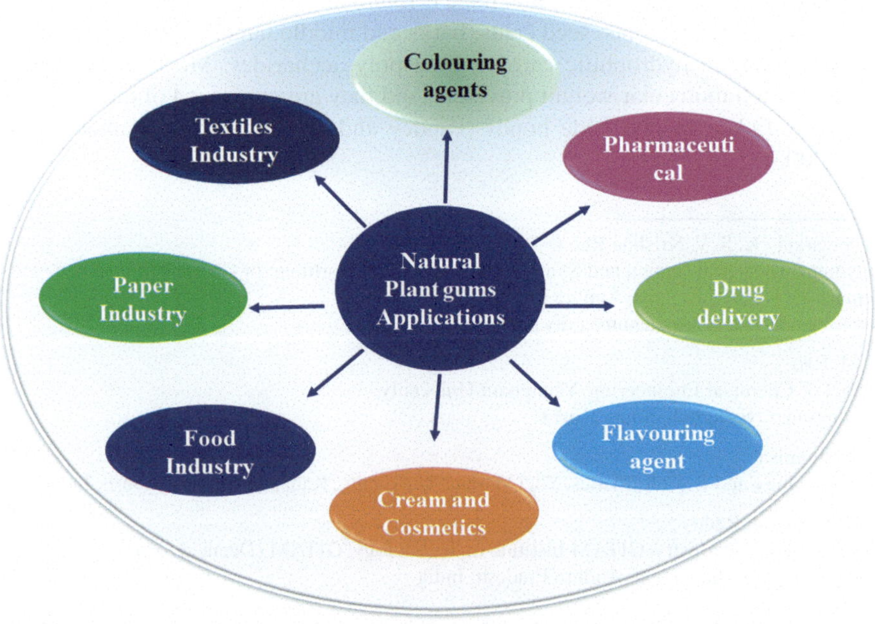

Fig. 18.1 Various potential applications of plant polysaccharide-based gums

Table 18.1 Natural gums' botanical name, family name and applications

S.No	Common name	Botanical name	Family	Applications	References
1	Tamarind gum	*Tamarindus indica*	Leguminosae (Fabaceae)	Stabilizing agent, binder, thickener, suspending agent, emulsifying agent, gelling agent and food industry and drug delivery	Rana V et al. (2011), Mishra and Malhotra (2011), and Nayak et al. (2015)
2	Tragacanth gum	*Astragalus gummifer*	Leguminosae	Cosmetic, emulsifier, textiles, pharmaceutical leather, food industry, thickener, stabilizer, moisture-retaining agent, binding agent, freezing agent, gelling agent, adhesive agent, drug delivery, wound healing, tissue engineering and soil fertility	Rao et al. (2018) and Hemmati and Ghaemy (2016)
3	Strychnos potatorum gum	*Strychnos potatorum*	Loganiaceae	Congo red dye Nanoparticles, antibacterial activity and medicinal properties	Yadav KN et al. (2014), Katherin Steffy et al. (2017), and Srikanth Kagithoju et al. (2015)
4	Cordia myxa	*Assyrian plum*	Boraginaceae	Emulsifier, binder, sustained releases of matrix making tablet, microcapsules, nanoparticles, beeswax and plasticizers	Vidyasagar et al. (2010), Munish Ahuja et al. (2013), and Abdul Haq et al. (2014)
5	Gum kondagogu	*Cochlospermum gossypium*	Bixaceae	Nanoparticles, binding agent, pharmaceutical, biosorbent, ocular delivery and plant culture media, mucoadhesive microcapsules and metal ions, emulsifying and gelling agent	Vinod et al. (2011) and Saravanan et al. (2012)
6	Guar gum	*Cyamopsis tetragonoloba*	Leguminosae	Drug delivery, wound healing, tissue engineering, paper industry, food, textiles, cosmetics, removal of dye and agriculture	Sathya Seeli and Prabaharan (2016) Ming Duan et al. (2019), and Thombare et al. (2018)
8	Konjac glucomannan	*Amorphophallus konjac*	Araceae	Pharmaceutical, biotechnology, controlled drug delivery; wound healing films blend membrane, coating material, cosmetics, emulsifier and packaging applications	Ian Ratcliffe et al. (2005), Xie et al. (2018), and Zhang et al. (2005)
9	Hibiscus esculentus	*Hibiscus esculentus*	Malvaceae	Pharmaceutical, emulsifier, binder, sustained release, retarding, release modifier, mucoadhesive and drug delivery	Ghor et al. (2014) and Raghu et al. (2019)

(continued)

Table 18.1 (continued)

S.No	Common name	Botanical name	Family	Applications	References
10	Mimosa scabrella		Leguminous Brazilian	Paints, food, cosmetics, pharmaceuticals, paper products and controlled drug delivery	Ganter et al. (1997) and Noleto et al. (2009)
11	Gum ghatti	*Anogeissus latifolia*	Combretaceae	Pharmaceutical, food, wax emulsifier, thickening and paper binding properties and gelling agent and metal ions	Rani et al. (2012), Ji Kang et al. (2011), and Kashma Sharma et al. (2013)
12	Locust bean gum	*Ceratonia siliqua*	Leguminosae	Gelling and emulsifying agent and non-teratogenic and drug delivery	Dionísio and Grenha (2012), Luis Braz et al. (2018), and Shukla and Tiwari (2012)
13	Aloe mucilage	*Aloe* species	Liliaceae	Cosmetic, food industry, creams, lotions, soaps, shampoos, facial cleansers and wound healing	Boudreau and Beland (2016), Ahlawat et al. (2011), and Jani et al. (2009)
14	Xanthan gum	*Xanthomonas campestris*	*Xanthomonadaceae*	Cosmetic, food, texture, paints, broadly available, biocompatible, colours, printing pastes, thickener, stabilizing agent, emulsifying and pharmaceutical, drug delivery, tissue engineering and removing metal ions	Kumar A et al. (2017) and Makhado et al. (2018)
15	Fenugreek	*Trigonella foenum-graecum*	Leguminous	Food industry, gelling agent, binding properties and controlled release	Ali Nokhodchi et al. (2008), Nayak et al. (2015), and Srinivasan and Mishra (2008)
16	Cashew gum	*Anacardium occidentale*	Anacardiaceae	Binding, suspending agent and controlled delivery	Ganesh et al. (2010) and Silva et al. (2018)
17	Carrageenan	*Chondrus crispus*	Rhodophyceae	Gelling agent, thinking agent, stabilize toothpaste preparation, pharmaceutical, tablet binder and food industry	Rastogi et al. (2018) and Takashi Masadome et al. (1997)
18	Almond gum	*Prunus amygdalus*	Rosaceae	Binding capacity, stabilizer, emulsifier, food industry and antimicrobial activity, methylene blue dye removal, nanoparticles and supercapacitors	Bouazizfatma et al. (2017), Theophil Anand et al. (2019), and Jaison et al. (2020)
19	Sterculia foetida	*Sterculia foetida*	Sterculiaceae	Mucoadhesive buccal films, controlled release and fix oils and time texture	Dixit et al. (2012) and Namdeo et al. (2008)

S.No	Common name	Botanical name	Family	Applications	References
20	Leucaena gum	*Leucaena leucocephala*	Leguminosae	Firewood, sustainable habitat development, organic fertilizer and paper industry, removal of Congo red aqueous solution, pharmaceutical and food industry	Vignesh et al. (2020) and Mittal et al. (2016)
21	Gum karaya	*Sterculia urens*	Sterculia urens	Pharmaceutical, suspending agent, thickening agent, stabilizing, dental adhesive, sustain release tablets and emulsifying agent and paper, food and textiles industries	(2008) and Setia et al. (2010)
22	Moringa oleifera	*Moringa oleifera*	Moringaceae	Gelling agent, stabilizer, suspending agent, surfactant, film forming and binder, release retardant in tablet formulations and disintegrate	Padayachee and Baijnath (2020) and Aderinola et al. (2020)
23	Khaya gum	*Khaya grandifoliola*	Meliaceae	Suspending agent, binding agent and drug delivery	Adetunji Adeniran et al. (2019) and Bonsu et al. (2016)
24	Tara gum	*Caesalpinia spinosa*	Leguminosae (Fabaceae)	Stabilizer, thickening agent and controlled release tablets	Fagiol et al. (2019) and Fernandes and Garcia-Rojas (2020)
25	Gum damar	*Shorea wiesneri*	Dipterocarpaceae	Stabilizer, emulsifier, production of colours, paints, inks and aromatic emulsions of food and cosmetic industries	Chick and Cassella (2011) and Morkhade and Joshi (2007)
26	Grewia gum	*Grewia mollis*	Tiliaceae	Binder, film-forming agent and release of retardant	Nep EI et al. (2016) and Ogaji and Hoag (2011)
27	Cassia tora	*Cassia tora*	Caesalpiniaceae	Suspending agent and binder tablets	Sharma et al. (2003) and Singh Y et al. (2018)
28	Gum copal	*Bursera bipinnata*	Burseraceae	Binding, stabilizer and emulsifying agent, paints and production of colours, sustaining the drug delivery and wastewater treatment	Umekar and Yeole (2008) and Kaur et al. (2018)
29	Bhara gum	*Terminalia bellirica* Roxb	Combretaceae	Sustained microencapsulated drug delivery	Shankar et al. (2008a, b)
30	Gellan gum	*Pseudomonas elodea*	Pseudomonas elodea	Pharmaceutical, disintegrating agent, food industry, sustainable drug delivery, ocular drug delivery and drug delivery applications	Balasubramaniam et al. (2003), Antony and Sanghavi (1997), and Agnihotri et al. (2006)

(continued)

Table 18.1 (continued)

S.No	Common name	Botanical name	Family	Applications	References
31	Neem gum	*Azadirachta indica*	Meliaceae	Textiles, binder, papers and pharmaceuticals	Malviya et al. (2019) and Velusamy et al. (2015)
32	Albizia gum	*Albizia zygia*	Leguminosae	Suspension food industry, binding agent, emulsifier and suspending agent	Singh et al. (2018) and Femi-Oyewo et al. (2004)
33	Okra gum	*Abelmoschus esculentus*	Malvaceae	Pharmaceutical, emulsion stabilizer, thickener and tablet binder, disintegrating agent, target colon drug delivery, industrial textile wastewater treatment	Kumar et al. (2009) and de Carvalho et al. (2011)
34	Ocimum basilicum gum	*Ocimum basilicum*	Lamiaceae	Emulsion, thickening, drug delivery, bioabsorption, tableting agent and removal of metal ions	Shamsnejati et al. (2015), Levankumar et al. (2009), and Zeynali et al. (2019)
35	Descurainia sophia seed gum	–	Brassicaceae	Thickener, plasticizer, emulsifier, stabilizer, edible coating, gelling agent, food and pharmaceutical industries	Soukoulis et al. (2018), Rezan Rezaeian et al. (2016), and Kian Aghaabbasi et al. (2014)
36	Honey locust gum	*Gleditsia triacanthos*	Leguminosae	Tablets matrix	Chuan-Rui Zhang et al. (2015)
37	Mangifera indica (mango gum)	*Mangifera indica*	Anacardiaceae	Food packaging, tablet binder, tablet retardant, disintegrate, sustained release and pharmaceutical	Rambabu et al. (2019), Danalache et al. (2016), and Sharma et al. (2016)
38	Hakea gum	*Hakea acicularis*	Proteaceae	Sustained release and mucoadhesive buccal tablets	Alur HH et al. (2001) and Alur et al. (1999)
39	Moi gum	*Lannea coromandelica*	Anacardiaceae	Controlled release matrix tablets	Eswaramma and Murthy (2019)
40	Balangu gum	*Lallemantia royleana*	Lamiaceae	Food formulations	Razavi SM et al. (2008)
41	Bael gum	*Aegle marmelos*	Rutaceae	Gelling agent, water-proofing material, adhesives and controlled drug release, bone tissue engineering and films	Balasubramanian et al. (2016), Dev et al. (2018), and Jindal et al. (2013)

S.No	Common name	Botanical name	Family	Applications	References
42	*Terminalia* gum	*Terminalia randii*	Combretaceae	Binding agent	Meka et al. (2012)
43	Jackfruit gum	*Artocarpus heterophyllus*	Moraceae	Timber industries, food industry and binder to pigment printing	Kabir et al et al. (2018) and Baslingappa Swami et al. (2012)
44	Gulmohar mucilage	*Delonix regia* gum	Fabaceae	Food industry, textiles, medicinal gelling agent, thickening, emulsifying, binding and formation of the films and textiles, suspending agent, wound dressing and drug delivery	Cheirmadurai et al. (2016), Okoye et al. (2014), and Nidhi Suhane et al. (2016)
45	Taro gum	*Colocasia esculenta*	Araceae	Thickener, binder, stabilizer and emulsifier, cosmetic, pharmaceutical and food industries and drug delivery	Andrade et al. (2020) and Ghumman et al. (2019)
46	Gum acacia	*Acacia seyal*	Fabaceae	Stabilizing agent, emulsifying, thickening, binding properties and gum food formulations, water-emulsions, wound dressing, nanoparticles, antimicrobial studies, cosmetic products (lotion and creams)	Jangra and Pothuraju (2020), Mohamed et al. (2020), and Patel and Goyal (2015)
47	Agar	*Gelidium amansii*	Rhodophyceae	Food packing, antibacterial wound dressing, cartilage regeneration, culture media of microorganism and films	Akkaya et al. (2020), Mostafavi and Zaeim (2020), and Shukla et al. (2012)
48	Cactus mucilage	*Opuntia ficus-indica*	Cactaceae	Cosmetic, food industry, coating films, heavy metal removal and pharmaceuticals	Zegbe et al. (2012) and Fox et al. (2012)
49	Cedrela gum	*Cedrela odorata*	Meliaceae	Chicken vaccinated infectious bursal diseases, veterinary vaccine delivery and binder and mucoadhesive component of ibuprofen tablet formulations	Odeniyi et al. (2013) and Emikpe et al. (2016)
50	Galbanum gum	*Umbelliferous*	Apiaceae	Cosmetics, textiles and various glues and antimicrobial activity	Najafabadi et al. (2017), Adhami et al. (2014), and Jalali et al. (2013)
51	Malva nut gum	-	Malvaceae	Microporous carbon for supercapacitors and pasting and textual properties and reducing high glucose levels by Caxo-2cell	Srichamroen (2018) and Phimolsiripol et al. (2011)

Table 18.1 (continued)

S.No	Common name	Botanical name	Family	Applications	References
52	Mucuna	*Mucuna flagellipes*	Fabaceae	Cosmetics, domestics and pharmaceuticals, emulsifying agent, stabilizer, thickening agent, suspending agent and electrical treatment of the biomaterial	Okwu and Okoro (2006) and Agba et al. (2019)
53	Sesbania gum	*Sesbania grandiflora*	Leguminosae	Gelling agent, paper making, beneficiation, thickening agent, mining industry, oil drilling food, cosmetic, pharmaceutical applications and painting, petroleum, wastewater treatment, warp sizing sustainable drug delivery	Lan et al. (2014) and Pal et al. (2018)
54	Olibanum gum			Cosmetic, pharmaceutical, ceramic, food industry, paints and textile industries, nanoparticles and drug delivery	Kora and Sashidhar (2015) and Kora and Rastogi (2016)
55	Salmalia malabarica gum		Malvaceae (Bombacaceae)	Expectorant, stimulant, anti-inflammatory, antiangiogenic, astringent, antioxidant, anti-asthma, anticancer and anti-hypertensive, anti-ulcerogenic effects and green synthesis, preparation of nanoparticles, catalytic reduction of used methylene blue and Congo red dyes	Rani and Khullar (2004), Saleem et al. (2003), Jain and Verma (2012), Krishna et al. (2016), and Reddy et al. (2015)

biodegradable, eco-friendly and biocompatible compared to synthetic polymers (Avachat et al. 2011; Nayyar et al. 2015).

The most important classification of carbohydrate polysaccharides is tree exudates with history of 5000 years. Gums are enormous quantities of different plant, marine, microbial and animal sources. The plant gums are very common and exhibit different structural properties commonly found in various families such as Combretaceae, Leguminosae, Sterculiaceae, Bixaceae, Rutaceae, Lamiaceae, Moringaceae, Meliaceae, Tiliaceae, Caesalpiniaceae, Anacardiaceae, Malvaceae, Liliaceae, Boraginaceae and Rosaceae. The available carbohydrate polysaccharides are classified as follows.

18.1.1 Gum Definition

"Gum" is a "plant derived exudate, seed mucilage and stem bark carbohydrate polymer". Gums are mainly grouped into three categories, namely, natural gum, synthetic gum and modified gums. Natural gums are produced in response to wounding (exudate gums) and extracted from seed legumes. Natural gums are found in a large number of families (Ahmad et al. 2019).

The term "gum" is defined as a group of the naturally occurring carbohydrate polysaccharides owning to their ability to form either viscous liquids or gummy mass.

18.1.2 Mucilage

The term mucilage is used to define the greasy aqueous solution dispersions by plants, microbials and animals, which mainly consist of water-soluble polysaccharides with starches and modified starches (BeMiller 2008).

18.1.3 Resins

Resins are amorphous mixture of essential oils, oxygenated products of terpene and carboxylic acids found as exudations from the trunk of a variety of trees (Dell and McComb 1979).

18.1.4 Classification of Natural Gums

Gums are mainly produced by different plant gums, animals, seaweeds and microbial sources for a number of structural purposes and metabolic functions (Fig. 18.2). Gums are classified into various types (Sarojini et al. 2010; Goswami and Naik 2014).

18.1.4.1 According to Charge

- *Anionic natural gums*
- **Natural:** Tragacanth gum, gum kondagogu, *Cordia myxa*, alginic acid, xanthan gum, gum arabic, gum kataya, pectin, *Hibiscus esculentus*, cashew gum, gum copal, gellan gum, okra gum and almond gum
- **Semi-natural:** Chitin, carboxymethyl cellulose gum
- *Cationic natural gums*
- **Natural:** Chitosan
- **Semi-natural:** Cationic guar gum
- *Non-ionic natural gums*
- **Natural:** Tamarind gum, *Strychnos potatorum* gum, guar gum, *Cassia tora*, tara gum, gum acacia and malva nut gum
- **Semi-natural:** Cellulose ethers (methyl cellulose, nitrocellulose, hydroxyl ethyl cellulose)
- *Amphoteric natural gums*
- **Semi-natural gums:** N-hydroxyl-di-carboxy ethyl chitosan, carboxy methyl cellulose
- *Hydrophobic natural gums*
- **Semi-natural gums:** Polyquaternium, cetyl hydroxy ethyl cellulose

18.1.4.2 According to Source

- *Plant origin gums*
- **Tree shrubs and exudates:** Gum kondagogu, gum karaya, gum ghatti, arabic gum, tragacanth gum, khaya gum, cashew gum, neem gum, almond gum, *Albizia* gum, moi gum, hakea gum, *Mangifera indica*, terminalia gum, bhara gum, *Leucaena leucocephala* gum, shatavari gum, *Lepidium sativum*, *Olibanum* gum and *Sterculia foetida*
- **Seeds gum:** *Strychnos potatorum*, tamarind gum, *Cordia myxa* (Assyrian plum), *Abelmoschus* gum, tara gum, guar gum, locust bean gum, bael gum, malva nut gum, sesbania gum and *Cassia tora* gum
- *Algal (seaweeds)/marine origin gums:* Carrageenan, agar, alginic acid and *Laminaria*

- *Animal origin gums:* Chitosan, chitin, hyaluronic acid and chondroitin sulphate
- *Microbial origin gums:* Dextrin, xanthan gum, pullulan, curdlan, Zanflo, Emulsan, Baker's yeast glycan, schizophyllan, lentinan and scleroglucan
- *Tuber and root gums:* Potato starch and grewia gum
- *Semi-synthetic gums*
- **Starch derivative gums:** Starch acetate, heta starch and phosphate starch
- **Cellulose derivative gums:** Hydroxy ethyl cellulose, carboxy methyl cellulose, hydroxy propyl methyl cellulose, methyl cellulose and microcrystalline cellulose

18.1.4.3 According to Shape

- **Linear gums:** Algins, pectin, amylose and cellulose
- **Branched gums:** Xanthan, amylopectin, xylan, gum arabic, galactomannans, gum kondagogu, *Strychnos potatorum*, tamarind gum, gum karaya and tragacanth gum

18.1.4.4 According to Monomeric Units of Chemical Structure

- **Homoglycans:** Arabinans, cellulose and amylose
- **Di-heteroglycans:** Carrageenans, galactomannan and algins
- **Tri-heteroglycans:** Xanthan and gellan gum
- **Tetra-heteroglycans:** Gum arabic and psyllium gum
- **Penta-heteroglycans:** Tragacanth gum, tamarind gum, gum karaya, *Strychnos potatorum* gum and gum kondagogu

18.1.5 Advantages of Natural Gums

The number of advantages of plant-derived natural gum based on materials has been scored, such as eco-friendliness, being economical low-cost, bio-compatibility, non-toxicity, biodegradability, local availability, better patient tolerance as well as public acceptance, edible sources and uncontrolled rate of hydration (Goswami and Naik 2014; Reddy and Manjunath 2013; Kulkarni Vishakha et al. 2012).

(i) **Eco-friendliness:** Gums and mucilage available from various sources are easily collected in different seasons.
(ii) **Low-cost:** Natural gums are always cheaper as they are available from natural sources. The production cost is also low compared to synthetic materials.
(iii) **Bio-compatibility and non-toxicity:** All these plant gums are carbohydrates chemically consisting of repeating monosaccharide units and non-toxic.

(iv) **Biodegradability:** Naturally available biodegradable polysaccharides are produced by living organisms. They are renewable sources, and they do not have adverse impact on human and environmental health.
(v) **Local availability:** Developing countries and governments promote the production of plants like tragacanth gum, guar gum and tamarind gum due to the wide applications in different industries.
(vi) **Better patient tolerance as well as public acceptance:** There are less side and adverse effects on natural materials compared to synthetic materials (Povidone and PMMA).
(vii) **Edible source:** Most natural gums and mucilage are obtained from the edible sources.

18.1.6 Disadvantages of Natural Gums

Disadvantages of natural gums are batch-to-batch variation, microbial contamination, reduced viscosity on storage and uncontrolled rate of hydration.

(i) **Reduced viscosity to storage:** The gums in contact with water increase in the viscosity of formulations due to complex nature of gums, it has been found that after storage is reduced.
(ii) **Batch-to-batch variation:** Synthetic manufacturing is a controlled procedure with fixed amounts of ingredients and production of gums; the mucilage depends on environmental and seasonal productions.
(iii) **Uncontrolled rate of hydration:** Due to variation in the collection times as well as difference in area, species and climate conditions of natural gums, percentage of chemical constituents in a given material differs. There is a need to develop suitable monographs on available mucilage and gums.
(iv) **Microbial contamination:** Natural gums are, structurally, carbohydrate, and during the production, they are exposed to an external environment; hence, there is chance of microbial contamination. This can be prevented by proper handling and use of preservatives.

18.2 Plant Gums and Drug Delivery

This chapter comprises various plant gums and their origin, plausible chemical structures (Table 18.2) and potential applications. However, we emphasized the results of drug delivery.

18.2.1 Tamarind Gum

Tamarind gum (TM) is a plant-extracted polysaccharide from seed kernel of *Tamarindus indica*, and endosperm of the plant belongs to Leguminosae family and is widely distributed in Southeast Asia. It is hydrophilic in nature, and the polysaccharide backbone consists of (1→4)-β-D-glucan with α-D-xylopyranose linked-β-D-galactopyranosyl(1→2)-α-D-xylopyranose linked (1→6) to glucose residues (Sougata et al. 2013; Viyoch et al. 2005). Owing to the high viscosity, mucoadhesive nature, biocompatibility, non-carcinogenesis and stability even in acidic pH media, TM can be used in various applications such as the pharmaceutical, food industry and cosmetic science (Rana et al. 2013). It is also widely used for various properties in a stabilizing agent, emulsifier, thickener, binder, gelling agent, suspending agent and food industry. However, the main disadvantage of TM in drug delivery is its fast microbial degradation. To enhance better drug delivery properties of TM and its derivatives, it is blended/copolymerized with gelatine for in vitro drug delivery using ciprofloxacin as model drug. The hydrogels developed by TM have exhibited pH-dependent drug release via diffusion (Mishra and Malhotra 2011; Nayak et al. 2015).

18.2.2 Tragacanth Gum

Tragacanth gum (TG) is an exudate from the bark of *Astragalus gummifer* belonging to the family Leguminosae. It is a natural hydrophilic polymer with eco-friendliness and good physiological properties. Hence it has been widely used for drug delivery, tissue engineering and biomedical applications. Due the presence of -COOH

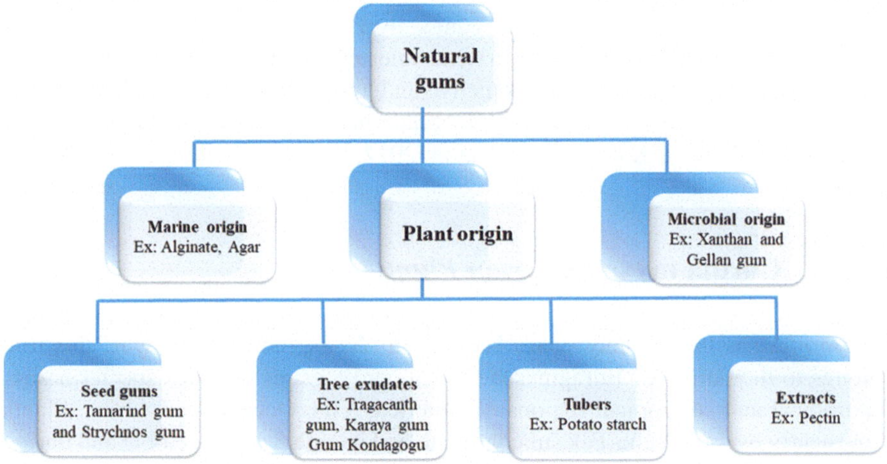

Fig. 18.2 Natural gum classification based on origin

functional groups in the backbone of (1→4)-D-galacturonic acid, it exhibits good pH sensitivity; this property helps the TG materials for biomedical applications. Tragacanth gum resembles pectin, but mainly differs from side chains and bonded to acidic groups (Saruchi et al., 2018; Eastwood et al. 1984; Singh et al. 2016). TG is a biodegradable, biocompatible, non-mutagenic, non-carcinogenic and non-toxic polysaccharide. Hence, TG is used for various applications such as cosmetic, leather, pharmaceutical, textiles, food additive, emulsifier, thickener, stabilizer, moisture-retaining agent, binding agent, freezing agent, gelling agent and adhesive agent. Tragacanth gum contains two major fractions: (i) tragacanthin (water-soluble) of minor amount and (ii) bassorin which is swellable but not water soluble. Tragacanth gum-based polysaccharide polymeric hydrogels are biodegradable and widely used for soil fertility, removal of acid dyes, metal recovery, wound healing, tissue engineering and drug delivery system (Rao et al. 2018; Hemmati and Ghaemy 2016).

18.2.3 Strychnos potatorum *Gum*

The *Strychnos potatorum* polysaccharide (SPP) extracted from the seeds of endosperm belongs to the family Loganiaceae and is found in India, Burma and Sri Lanka. The local name of this plant is most popular in India known as Nirmali or Kataka. The *Strychnos potatorum Lin.* seeds contain galactan and galactomannan mixture, which exhibit a coagulation activity. The SPP has β-(1→4)-linked D-galactopyranose residue as a backbone with side chain mannopyranosyl. The polysaccharide is composed of 1:7 ratio of galactan and galactomannan (Adinolfi et al. 1994; Sethurajan et al. 2011). The gums are widely used for household water purification in many Asian countries and water treatment plants. The SPP seeds are used in Ayurveda medicine (Indian traditional medicine). The polysaccharide possesses broad range of properties including antidiarrheal, diuretic, hepatoprotective, antioxidant, antimicrobial, antiulcer genic, liver and kidney. The decoction of *Strychnos potatorum* leaves is used to treat watering, aching eyes and cough activities. The SPP was used as hepatoprotective and antioxidant and in binding of cadmium and adsorption of Congo red dye for water purification (Saleh Saif et al. 2012; Yadav KN et al. 2014; Katherin Steffy et al. 2017; Srikanth Kagithoju et al. 2015).

18.2.4 Cordia myxa *(Assyrian Plum)*

Cordia myxa is locally known as “Lasura” in one of Indian languages; this tree belongs to the family of Boraginaceae. Gum cordia genus includes species across the tropical and subtropical regions of America, Asia, Oceania and Africa. Plants grow nearly over the India-Pak subcontinent, and its ripe fruits are edible and unripe fruits are used to prepare pickle. *Cordia myxa* gum is a hydrophilic polymer due to the presence of (1→6)-linked D-glucopyranosyl and (1→2)-linked L-arabinofuranosyl residues on the backbone (Samavati and Skandari 2014; Keshani-Dokht et al. 2018). It is used as emulsifier, stabilizer, thickener, surface coating for edible products, binder, sustained release of matrix making tablet, bead, microcapsules, nanoparticles, ophthalmic drug delivery, effect on beeswax and plasticizers. Due to high nutritional values, it has been used for the treatment of astringent, urinary infections, leprosy and lung diseases (Vidyasagar et al. 2010; Munish Ahuja et al. 2013; Abdul Haq et al. 2014).

18.2.5 Gum Kondagogu

The gum kondagogu is a naturally available polysaccharide and is an exudate from the bark of *Cochlospermum gossypium* belonging to the family Bixaceae (Kumar and Ahuja 2012,). It is widely available in the forest area of Andhra Pradesh and Telangana states in India. Gum kondagogu comes under the rhamnogalacturonans, which consist of "(1→2)-β-D-Gal p, (1→6)-β-D-Gal p, (1→4)-β-Glc pA, 4-OMe-α-D-Glc Pa, (1→2)- α-L-Rha and (1→4)- α-D-Gal pA". Recently, gum kondagogu has been studied for its morphological, physico-chemical, structural and rheological properties (Kora and Sashidhar 2010; Vinod et al. 2008); hence, it was used in textile, pharmaceutical industry as a biosorbent and ocular delivery device. In addition, it has been used for preparation of plant culture media, mucoadhesive microcapsules and metal nanoparticle, as antimicrobial and for catalytic activity (Reddy et al. 2015; Vinod et al. 2011; Saravanan et al. 2012).

18.2.6 Guar Gum

Guar gum is a powder of the endosperm of the seeds of *Cyamopsis tetragonoloba* Linn. In India, locally it is called as guaran, clusterbean, Calcutta lucern, cyamopsis gum, guarina, glucotard and guyarem. It is a non-ionic galactomannan that consists of β-(1→4) linked-D-mannopyranoses, but α-D-galactose (1→6) is connected to every alternate mannose (Sathya Seeli and Prabaharan 2016). Due to its non-toxicity, degradation at microbial environment in intestinal fluids, biocompatibility and biodegradability, it is used in various applications such pharmaceuticals, drug delivery, wound healing, tissue engineering, paper industry, food, textiles, cosmetics and also removal of dyes and moisture sensitive hydrogel for agriculture applications (Sathya Seeli and Prabaharan 2016; Ming Duan et al. 2019; Thombare et al. 2018).

18.2.7 Konjac Glucomannan

The konjac glucomannan (KG) is a derivative of tuber *Amorphophallus konjac* plant, and it is native from the southern part of China. KG is containing higher molecular weight 1000 kDa polysaccharide consisting of mannose and glucose sugars. It is a hydrophilic and non-ionic glucomannan consisting of linear macromolecule of (1→4)-linked β-D-mannose and β-D-glucose. In the aqueous environment, it exhibits high elasticity as well as large amount of wound exudate, due to the presence of hydroxyl and acetyl functional groups on KG backbone (Vipul Dave and McCarthy 1997; Yan Hu et al. 2019; Mu et al. 2018). KG has been cultivated in Japan and used in production of food jellies and noodles. It is mainly conferring solubility, high viscosity, swelling and specific biological activities. KG has been used for food science like jelly, cake and meat products, textiles, paper manufacturing (Ian Ratcliffe et al. 2005), pharmaceutical, controlled drug delivery, wound healing, films, coating material, cosmetics, emulsifier, biodegradable resins and packaging applications (Xie et al. 2018; Zhang et al. 2005).

18.2.8 Hibiscus esculentus

Hibiscus esculentus gum is extracted from the *Abelmoschus esculentu90s Linn*, and it comes under the family of Malvaceae. The chemical composition of the gum is L-rhamnose, D-galactose and L-galacturonic acid along with glucose, arabinose, mannose and xylose side chains (Gbenga and Zulikha 2013; Jani and Shah 2008). *Hibiscus* is widely distributed into tropical and subtropical regions in 250 species (3). It is widely used in various applications such as pharmaceutical, emulsifier, binder, sustained release, retarding, release modifier, mucoadhesive gel for nasal delivery and controlled drug delivery. Hibiscus gum-based tablets are developed with ibuprofen for colon-specific drug delivery (Ghori et al. 2014; Raghu et al. 2019).

18.2.9 Mimosa scabrella *(Bracatinga)*

The *Mimosa scabrella* is commonly known as bracatinga; it belongs to the family Leguminous native from the Brazilian tree of the Mimosaceae. The chemical structure consists of (1→4)-linked D-mannopyranosyl backbone and D-galactopyranosyl residue in mannose. The native species of *Mimosa scabrella* is having commercial values among producers in Brazilian subtropical areas (Ughini et al. 2004; Bonnet et al. 2002). Its polysaccharide exibits high viscosity in aqueous media and has more industrial applications such as paints, food, cosmetics, pharmaceuticals, paper products and oil drilling muds. Also, it has desired product properties such as texture, stability and controlled drug device. It has been used for various kind of

Table 18.2 Plausible chemical structures of various polysaccharides present in the plant gums

S.NO	Polysaccharide name	Chemical constituents	Chemical structure
1	Tamarind gum	(1→4)-β-D-glucan backbone with substituted side chains -α-D-xylopyranose linked-β-D-galactopyranosyl (1→2)-α-D-xylopyranose linked (1→6) to glucose residues	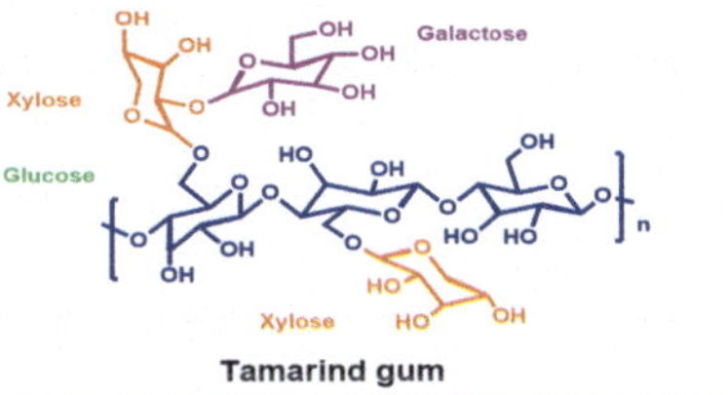
2	Tragacanth gum	Tragacanth gum consists of (1→4)-D-galacturonic acid as a main chain backbone	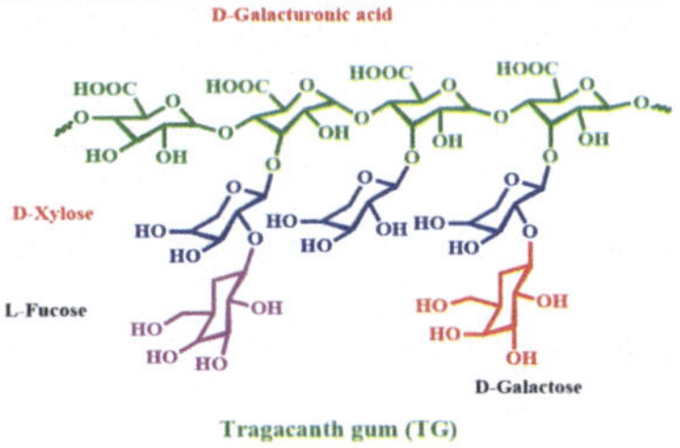
3	Strychnos potatorum gum	Strychnos potatorum polysaccharide backbone consists of β-(1→4)-linked D-galactopyranose residue and that side chain mannopyranosyl	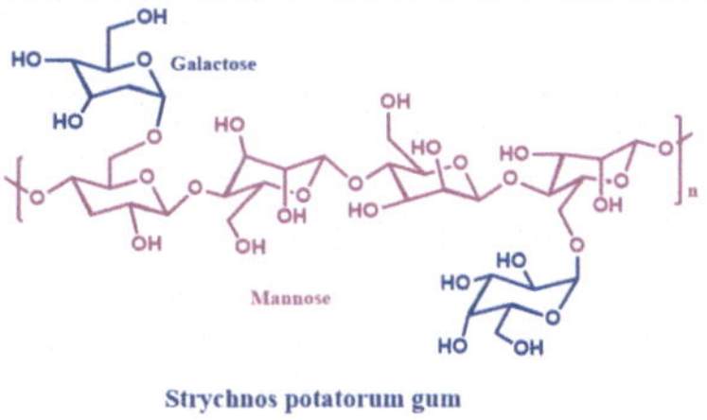
4	Kondagogu gum	Gum kondagogu contains chemically glycoside linkage of (1→2)-β-D-Gal p, (1→6)-β-D-Gal p, (1→4)-β-Glc pA, 4-OMe-α-D-Glc Pa, (1→2)- α-L-Rha and (1→4)-α-D-Gal pA and side units backbone structure group and substituted rhamnogalacturonan	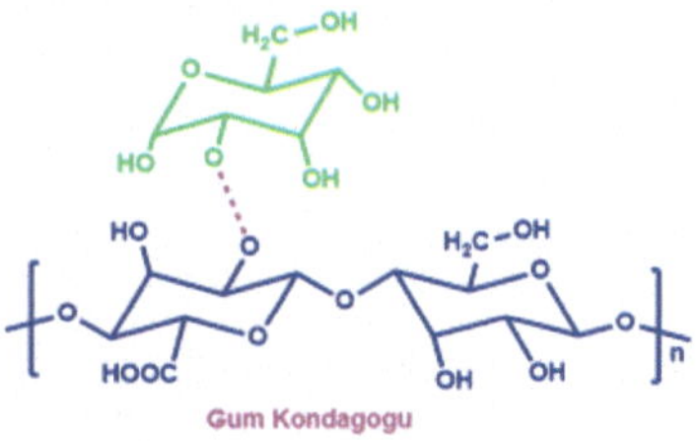
5	Guar gum		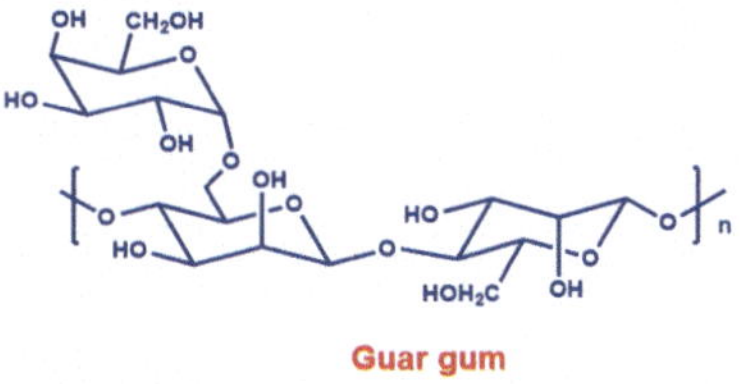
6	Konjac glucomannan	KG – (1→4)-linked β-D mannose and β-D glucose	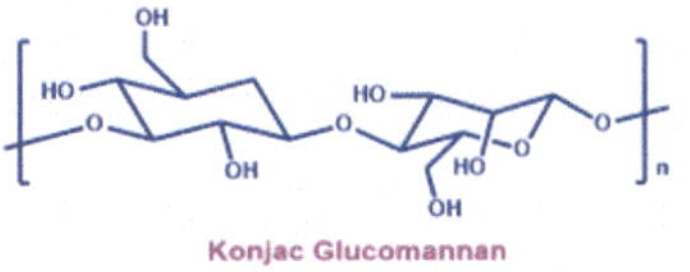

(continued)

Table 18.2 (continued)

S.NO	Polysaccharide name	Chemical constituents	Chemical structure
7	Mimosa scabrella (bracatinga)	Structure consisting of (1→4) linked D-mannopyranosyl backbone and D-galactopyranosyl residue in mannose	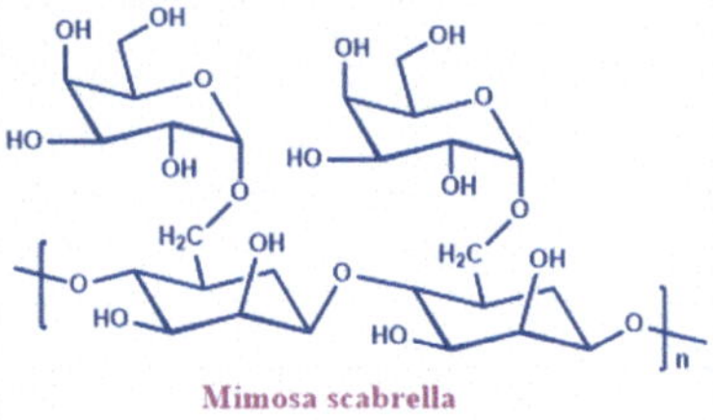
8	Locust bean gum	Chemical consisting of linear chain -(1→4) -linked -β-D-mannopyranosyl unit with (1→6)-linked α-D-galactopyranosyl residue chain	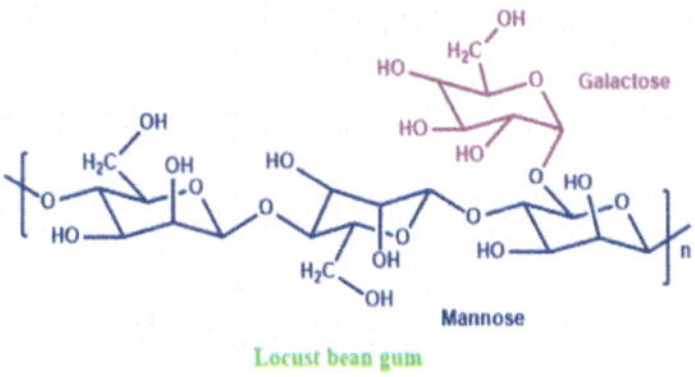
9	Aloe mucilage	Different ways (1→2, 1→3, 1→4, 1→5 and 1→6) in a α- or β-configuration) due to the presence of branched side-chain units	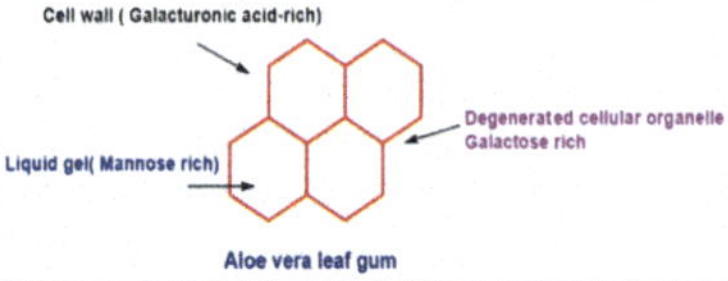
10	Xanthan gum	Chemical composition of primary chain consits of β-D-(1→4) glucose backbone and glucuronic acid residue linked (1→4) mannose terminal unit, (1→2) connect second mannose unit backbone of the structure	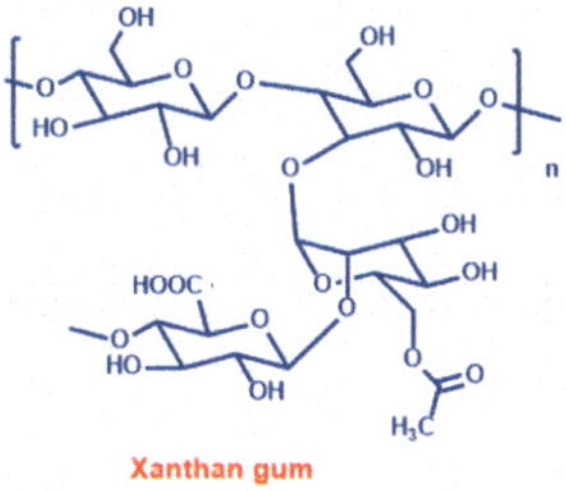
11	Fenugreek mucilage	Chemical constituents of –(1→4)-β-D-mannan backbone with D-galactosyl subunits are attached to –(1→6)-glycosidic linkage	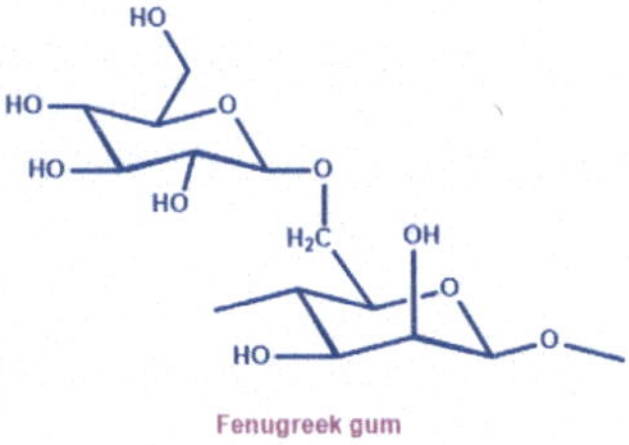
12	Cashew gum	Galactomannan consisting of (1→3)-linked-β–D-galactopyranosyl units interspersed with β-(1→ 6) linkages	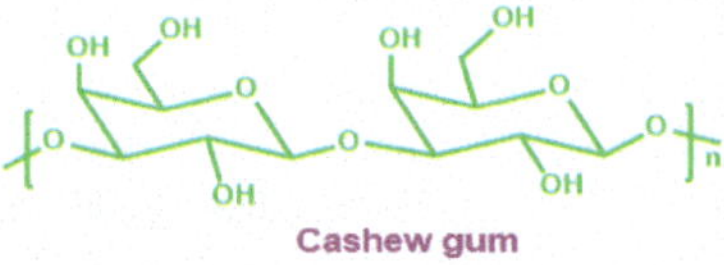
13	Carrageenan	Chemical structure consisting of altering – (1→3)-linked β-D-galactopyranose and (1→4)-linked α-D-galactopyranose joined in glycoside linkage	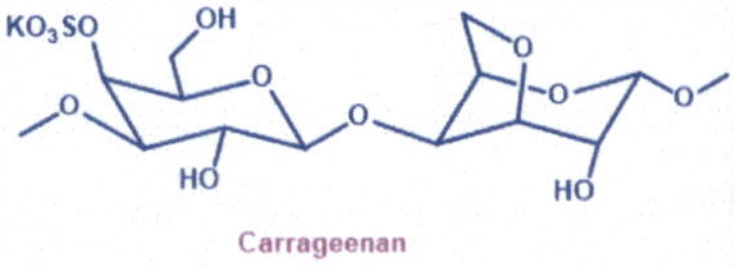

(continued)

Table 18.2 (continued)

S.NO	Polysaccharide name	Chemical constituents	Chemical structure
14	Karaya gum	Chemical consisting of rhamnogalacturonan main chain α-(1→4)-linked D- galacturonic acid and α-(1→2)- linked L-rhamnosyl residues and side chain units (1→6)-linked β-D-galactose in galacturonic acid	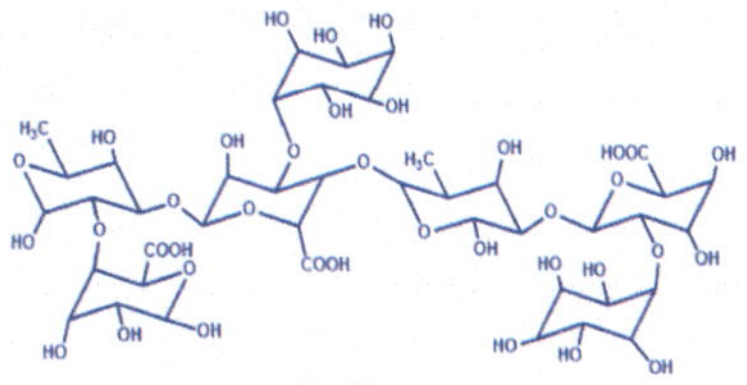 Gum Karaya
15	Moringa oleifera gum	Arabinogalactan unit consists of (1→6), (1→3) and (1, 3, 6)-linked Galp units	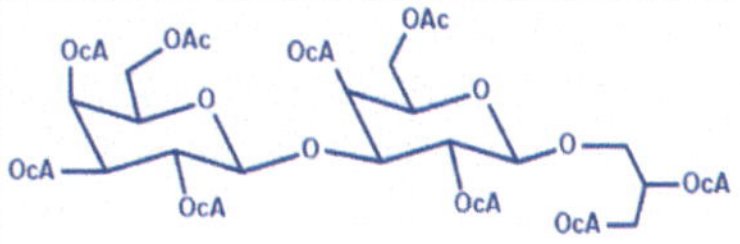 Moringa oleifera gum
16	Tara gum	Linear polysaccharide chain (1→4)-β-D-mannopyranose units with α-D-galactopyranose units attached by (1→6)- linkage	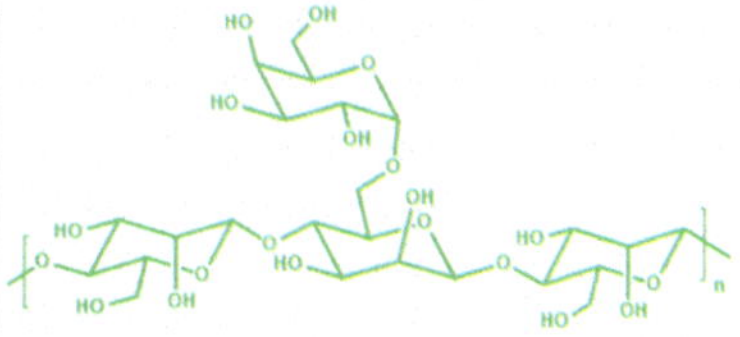 Tara gum
17	Cassia tora mucilage	Linear polysaccharide chain -(1→4)-linked-β-D-mannopyranose and –(1→6)-α-D-galactopyranose and sugars contain side chain units	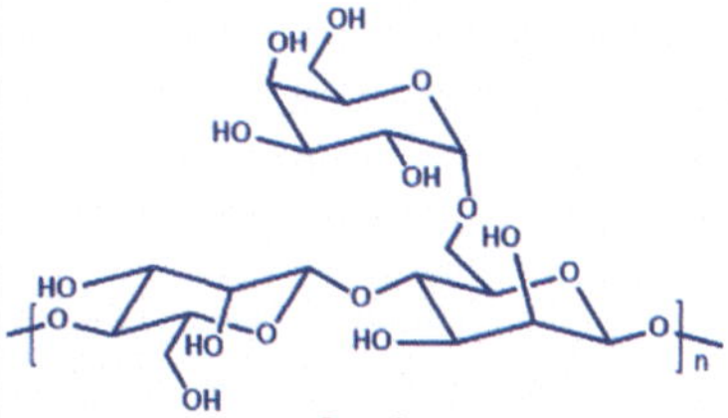 Cassia tora gum
18	Gellan gum	Gellan gum chemical consisting of glucose, rhamnose, glucuronic acid	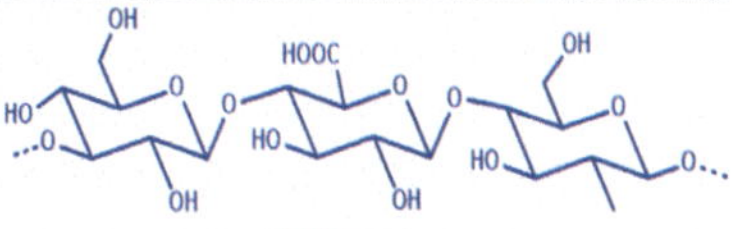 Gellan gum
19	Abelmoschus gum or okra gum	Chemical consisting of different molar ratios L-galactonic acid, D-galactose, L-rhamnose, mannose, glucose, xylose and arabinose	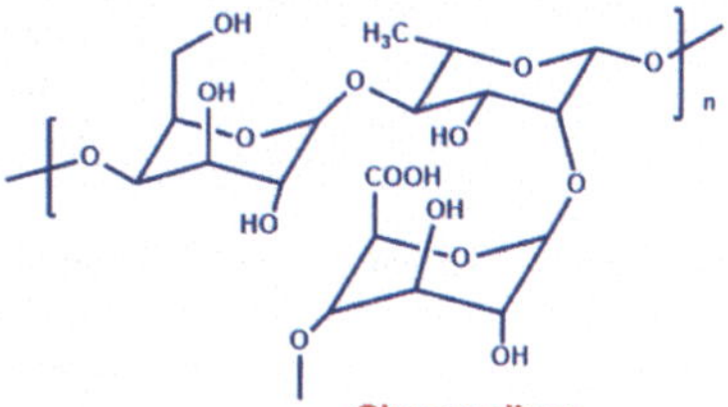 Okra mucilage

(continued)

Table 18.2 (continued)

S.NO	Polysaccharide name	Chemical constituents	Chemical structure
20	Ocimum basilicum gum	D-mannose and D-glucose repeated units of chain β-(1→4)-linked glycosidic linkage of acetyl groups are attached	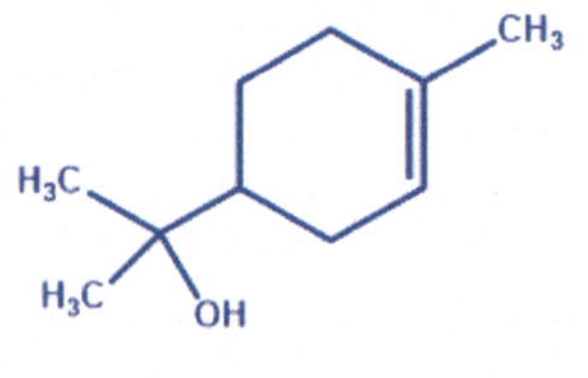
21	Mangifera indica (mango)	Chemical structure established a 2-C-β-D-glucopyranosyl-1, 3, 6, 7-tetrahydroxyxanthone	
22	Delonix regia gum (gulmohar mucilage)	Delonix regia gum chemical consisting of α-D-mannose-(1→4)-linkage and α-D-galactose-(1→6)-branches	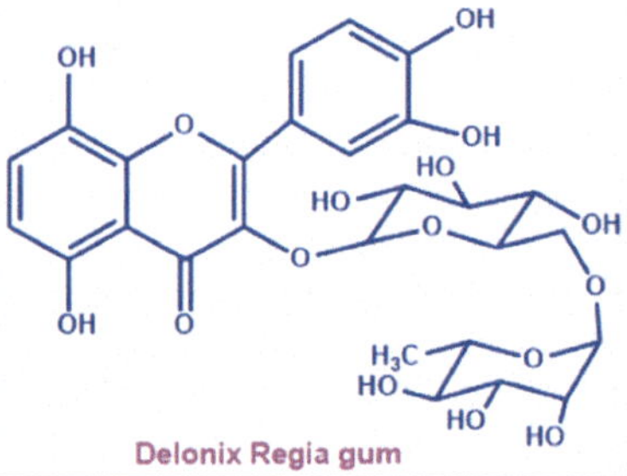
23	Gum Acacia	Chemical consisting of –(1→3)-linked β-D-galactopyranosyl units with side chains β-L-rhamnopyranose, α-L-arabinofuranosyl, β-D-di glucuronic acid and β-D-digalactopyranose chain	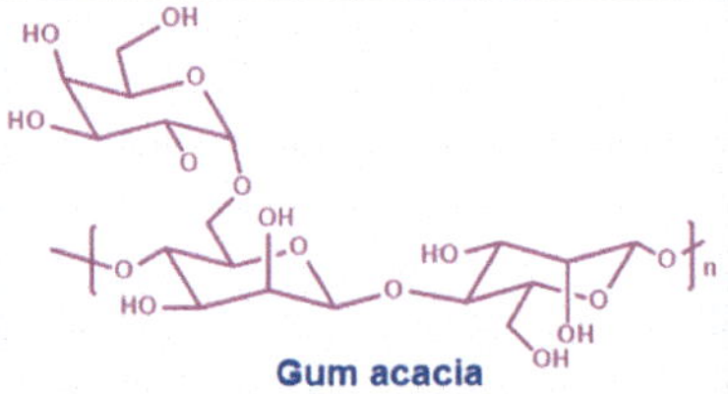
24	Agar	Chemical consisting of D-galactose and 3-6 anhydro-L-galactose linked by α-(1→3) and β-(1→6) glycoside bonds	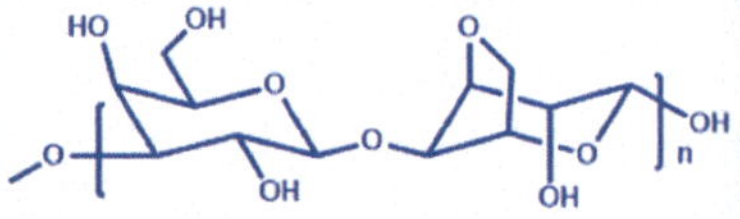
25	Sesbania gum	Chemical structure consisting of mannose connect β-(1→4)-glycosidic bond and galactose linked –α-(1→6)-glycoside side chain unit	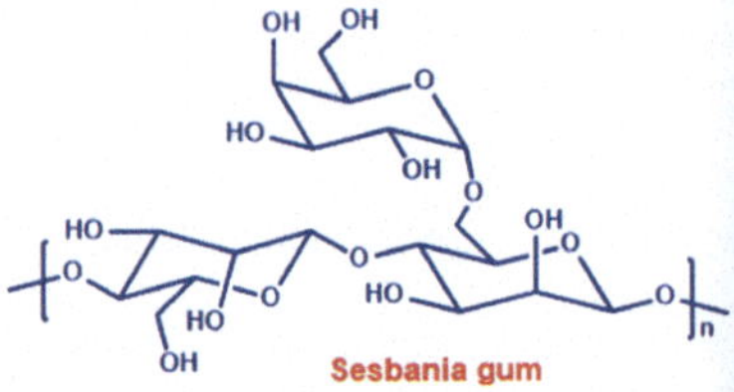

applications, such as controlled drug delivery, in vivo antiherpetic and antirotaviral activity, cytotoxicity on HeLa cells and sustained release matrix (Ganter et al. 1997; Noleto et al. 2009).

18.2.10 Gum Ghatti

The gum ghatti is extracted from the exudate of *Anogeissus latifolia* tree; it belongs to the family of Combretaceae and tree found in dry areas. Gum ghatti is a anionic polysaccharide; its chemical structure constitutes the (1→6)-linked D-galactopyranose units altering 4-O-substituted and 2-O-substituted α-D-mannopyranose units with side chain containing arabinose unit (Deshmukh AS et al. 2012; Mittal et al. 2015). It is not completely soluble in cold water, but solubility increased by heating. It is mainly used in pharmaceutical and food industries as wax emulsifier and thickening and additive agent (Rani et al. 2012). Also, it is non-toxic, highly viscous, pH-stable, a gelling agent, biocompatible and biodegradable. It is successfully used for the treatment of various neurological and autoimmune diseases, diabetes and mutagenicity and removal of dyes and metal ions from aqueous media (Ji Kang et al. 2011; Kashma Sharma et al. 2013).

Locust bean gum (LBG) is extracted from the endosperm of carob tree seed of *Ceratonia siliqua*, and it comes under Leguminosae family. It is widely available in Asia, North Africa and South America. The chemical constituents of LBG polysaccharide have (1→4)-linked-β-D-mannopyranosyl unit along with (1→6)-linked α-D-galactopyranosyl residue chain (Maiti et al. 2010; Santanu Kaity et al. 2013; Barak and Mudgil 2014). LBG is used in various applications such as controlled drug delivery, specific colon drug delivery, bioactivity, ocular drug delivery, topical drug delivery and oral drug delivery (63–68); due to its potential properties, i.e. biodegradable, high water absorption isotherm, biocompatible, non-mutagenic, gelling and emulsifying agent and non-teratogenic (Dionísio and Grenha 2012; Braz et al. 2018; Shukla and Tiwari 2012).

18.2.11 Aloe vera *Mucilage*

Aloe vera mucilage (AVM)-based polysaccharide is isolated from the *Aloe barbadensis*, which belongs to the family of Liliaceae. AVM is locally used for topical treatment of burn skin and wounds (Nirmal Pugh et al. 2001; Channe gowda et al. 1987). The chemical constituents of AVM are arabinorhamnogalactan, galactan, arabinan, galactogalacturan, galactoglucoarabinomannan, glucogalactomannan and glucuronic acid. These monosaccharides are linked as (1→2, 1→3, 1→4, 1→5 and 1→6) in a α- or β-configuration. Commercially, AVM (leaf pulp) has worldwide demand due to its significance in various products like food, health drinks, beverages, cosmetic, toiletry, creams, lotions, soaps, shampoos and facial cleansers and wound healing (Boudreau and Beland 2016; Ahlawat et al, 2011; Jani et al. 2009).

18.2.12 Xanthan Gum

Xanthan gum (XG) is produced by the *Xanthomonas campestris* gram-negative) bacterium under the fermentation of glucose and sucrose at aerobic conditions. This bacterium belongs to the family of *Xanthomonadaceae.* Usually, it consists of high molecular weight anionic polysaccharide with the chain β-D-(1→4) glucose backbone and glucuronic acid residue linked (1→4) mannose terminal unit, (1→2) connect second mannose unit backbone of the structure (Akito Kawano et al. 2019; Wang B et al. 2016; Gemma Leonea et al. 2018). The xanthan gum is mainly used for drug delivery, tissue engineering, nanoparticle and anticancer activity and antimicrobial and controlled agrochemical release and as adsorbent for removal of dyes and metal ions from aqueous solution (Kumar a et al. 2016; Madhusudhan et al. 2020; Makhado et al. 2018).

18.2.13 Fenugreek Mucilage

Fenugreek is generally known as "*Trigonella foenum-graceum*"; it belongs to the Leguminous family (Mishra et al. 2006). These seeds contain high degree of mucilage on the seed surface. These seeds are not only used for the food preservation but also used as traditional medicine in India, to treat dysentery, colic flatulence, diarrhoea, chronic cough, dropsy, rickets, gout and diabetes. Fenugreek mucilage has better physical property, i.e. swelling in aqueous media. Hence it is used as matrix for controlled release of propranolol hydrochloride and Methocel K4M (Ali Nokhodchi et al. 2008; Nayak et al. 2015; Srinivasan and Mishra 2008).

18.2.14 Cashew Gum

Cashew gum is an exudate of stem bark of *Anacardium occidentale* Linn.; it belongs to the *Anacardiaceae*. The chemical composition of cashew gum includes the galactose, arabinose, rhamnose, glucose and glucuronic acid along with some sugar moieties. However, the chemical constituents of gum are branched galactomannan consisting of (1→3)-linked -β-D-galactopyranosyl units along with β-(1→6) linkages (da Silva et al. 2007; Lima et al. 2018). Due to potential physico-chemical properties of gum, it is used as tablet binder for controlled release of metronidazole, as well as drug delivery system for controlled release of diclofenac sodium (Ganesh et al. 2010; Silva et al. 2018).

18.2.15 Carrageenan Gum

Carrageenan, a polysaccharide, contains high molecular weight sulphated polymer extracted from the *Chondrus crispus* (red seaweeds; mainly produced by cell wall materials) which belongs to the Rhodophyceae family (Rastogi et al. 2010). Carrageenan is also known as especially crispus, *Eucheuma* spp., *Gigartina stellata* and *Iridaea* spp. The chemical composition of carrageenan polysaccharide is (1→3)-linked β-D-galactopyranose and (1→4)-linked α-D-galactopyranose joined in glycoside linkage. The k-type of carrageenan form the brittle gel, lambda-type carrageenan forms the viscous solution but not gelling, and L-type of carrageenan produces the elastic gel (Takashi Masadome et al. 1997). It is used for various applications such as pharmaceutical, tablet binder, anticoagulant and antithrombotic activity, antiviral activity, antitumour, immunodulatory activity and food industry applications (Raja et al. 2016; Yadav et al. 2019).

18.2.16 Almond Gum

Almond gum is an anionic polysaccharide extracted from the exudate of *Prunus amygdalus*, which belongs to the Rosaceae family. Its chemical composition includes the galactose, xylose, arabinose and glucuronic acid along with trace amount of rhamnose, glucose and mannose; hence the suggested structure contains the high molecular weight arabinogalactan chain (Hussain et al. 2017a; Bouaziz et al. 2015). Gum is not soluble in aqueous solution at low temperature but soluble in hot conditions, i.e. around at 60 °C. Polysaccharide contains good physico-chemical properties, such as antioxidant, binder, biodegradability, biocompatibility, non-toxicity, metal uptake capacity, water binding capacity, bioactivity, fat binding capacity, stabilizer, emulsifier, food industry and antimicrobial activity (Bouazizfatma et al. 2017). Recently, it has been used for removal of malachite green and methylene blue dyes from aqueous solution. Also, it is used for green synthesis of metal and metal oxide nanoparticles, viz. Ag and ZnO, and for supercapacitor, antimicrobial, anticancerous, antioxidant and anti-inflammatory activities. In addition, it is used for various food- and non-food-related applications (Theophil Anand et al. 2019; Jaison et al. 2020).

18.2.17 Sterculia foetida

Sterculia foetida gum is obtained from dried gummy exudates of tropical plant stem bark which belongs to the family Sterculiaceae, which is also called "Java olive", Bastard tree and hazel sterculia, and it is known in India as "jangali badam" (Hindi and Bengali) and gurapu-badam (Tamil). It is the species of Thailand, commonly known as sam-rong seed. It is a partially acetylated polymer which contains D-galactose, D-galacturonic acid and L-rhamnose (Prajapati et al. 2010; Kale et al. 2011). It is mainly used in different kinds of applications, pharmaceutical investigated polymer hydrophilic network for controlled release preparation, analysis of fix oils, time texture probing and mucoadhesive buccal films (Dixit et al. 2015; Namdeo et al. 2008).

18.2.18 Leucaena leucocephala

The plant *Leucaena leucocephala* belongs to the family Fabaceae and subfamily Mimosoideae. It is a persistent thornless tropical plant with height around up to 7.5–12 m and also native from tropical Southern Mexico and Northern Central America. It is also known as petai cina, lamtoro and popinac in Indonesia, but native plant is commonly found in colony in tropical countries. It is also known as Kubabul/Sababul in Andhra Pradesh and Gujarat states in India (Kristianto et al. 2019; Awang et al. 2019). The chemical composition of seed endosperm polysaccharide chain contains (1→4)-D-mannose units and side chain substituted single galactose unit. The gum is used as binding agent in granules, suspending agent, emulsifier, tablet and disintegrating agent and organic fertilizer. In addition, it has different properties like anti-inflammatory, anticancer, antiviral, anti-thrombotic, anticoagulant, antibiotic and immune stimulant (Vignesh et al. 2020; Mittal et al. 2016).

18.2.19 Karaya Gum

Gum karaya is generally called as Indian tragacanth gum or sterculia gum; it is an exudate obtained from the *Sterculia urens* tree which belongs to the Malvaceae family, native of India and Burma. It consists of high molecular weight of acetylated polysaccharide galactose, rhamnose and galacturonic acid with small amount of glucuronic acid. The chemical composition is rhamnogalacturonan main chain α-(1→4)-linked D-galacturonic acid and α-(1→2)-linked L-rhamnosyl residues and side chain units (1→6)-linked β-D-galactose in galacturonic acid (Alange et al. 2017; Marvelys L et al. 2006). It is widely used as a good emulsifying agent, acid stabilizer, high viscosity agent, suspending agent, inherent nature of antimicrobial activity, and binder. It has been used in the treatment of ulcers, diarrhoea, irritable bowel syndrome and chronic colonic disease and in reducing cholesterol. It has been used in various industrial applications, such as production of nanoparticles, catalytic activity, pharmaceutical, dental adhesive (Venkatesham et al. 2014, Vinod Vellora et al. 2015), tablets, paper, food and textiles (Baljit Singh et al. 2008, Setia et al. 2010).

18.2.20 Moringa oleifera *Gum*

Moringa oleifera gum is an exudate of a stem of *Moringa oleifera* tree; it belongs to the family Moringaceae. It is widely available in Afghanistan, Bangladesh, India and Pakistan (Sarwar et al. 2020; Anwa et al. 2005). The chemical constituents of gum are (1→6)-, (1→3)- and (1, 3, 6)-linked Galp units. Due to its potential physico-chemical properties (Moyo et al. 2011), it has been used for the preparation of tablet for controlled release of calcium sulphate dihydrate and propranolol hydrochloride (Padayachee and Baijnath 2020; Aderinola et al. 2020).

18.2.21 Khaya Gum

Khaya gum is obtained from an incised trunk of *Khaya grandifoliola* tree; it belongs to the family Meliaceae. The chemical constituents of gum are D-galactose, L-rhamnose, D-galacturonic acid and 4-O-6 methyl-D-glucuronic acid (Kuevi et al. 2019; Hussain et al. 2017b). Due to its potential physico-chemical properties, it is blended with hydroxypropylmethylcellulose and *Albizia* gum individually, for controlled release of paracetamol and indomethacin. Khaya gum showed better drug delivery profiles of paracetamol when compared to the *Acacia sieberiana* and *Acacia* senegal gums (Adedokun et al. 2017; Adetunji Adeniran et al. 2019; Bonsu et al. 2016).

18.2.22 Tara Gum

Tara gum is obtained from the endosperm of *Caesalpinia spinosa* plant, and it belongs to the Leguminosae or Fabaceae family (Chi et al. 2018). The chemical constituents of gum are (1→4)-β-D-mannopyranose attached α-D-galactopyranose with (1→6) linkage along with some sugar side chains. It has the ability to form highly viscous solution at low concentration (~1% w/v); hence, it is used for development of controlled release formulations with metformin hydrochloride and ciprofloxacin hydrochloride (Yong et al. 2019; Fagioli et al. 2019; Fernandes and Garcia-Rojas 2020).

18.2.23 Gum Damar

Gum damar is derivative of *Shorea wiesneri* plant, and it belongs to the Dipterocarpaceae family. The main chemical composition of gum damar includes the alcohol-soluble α-resin, β-resin and dammarol acid (Morkhade et al. 2006;

Thombre et al. 2020). Due to its water-resistant capacity and strong binding nature, it is used as coating for tablets and other dental materials (Chick and Cassella 2011). Specifically, it is blended with natural gum copal to develop tablets for controlled release diclofenac sodium as well as ibuprofen and diltiazem hydrochloride (Morkhade et al. 2007).

18.2.24 Grewia Gum

Gum *Grewia* polysaccharide derived from the inner bark of the edible plant *Grewia mollis* belongs to the family Tiliaceae. The polysaccharide gum chemical consists of glucose and rhamnose as main monosaccharide components and galacturonic acid as the main side chain units (Nep EI and Conway BR 2011). Gum Grewia mucilage of leaves and bark was extensively studied for physicochemical properties, surface chemistry, molecular weight, thermal properties and compositional analysis (Nep EI and Conway BR 2011). The binding properties of the gum were evaluated using paracetamol as a model drug. Compressional properties of the formulations were analysed by using Heckel and Kawakita equations. Single polymer matrix tablets of cimetidine were formulated, and gum potentials were evaluated in coating of film using praziquantel as a model drug (Nep EI and Conway BR 2011; Ogaji and Hoag 2011).

18.2.25 **Cassia tora** *Mucilage*

Cassia tora mucilage is obtained from the seeds of *Cassia tora* plant; it belongs to the Caesalpiniaceae family. The chemical constituents of gum are (1→4)-linked-β-D-mannopyranose and (1→6)-α-β-D-galactopyranose along with some sugar side chains (Pawar et al. 2014; Raj Sharma et al. 2002). Due to the potential binding and suspending properties over the other natural gums like acacia, gelatin and tragacanth gum, it has been used for tablet preparation (Sharma et al. 2003; Singh et al. 2018).

18.2.26 Gum Copal

The gum copal is a natural resin-type substance obtained from the plant of *Bursera bipinnata*, which belongs to the family Burseraceae. Copal resin contains agathic acid along with cis-communic acid, trans-communic acid, polycommunic acid, sandaracopimaric acid, agathalic acid, monomethyl ester of agathalic acid, agatholic acid, and acetoxy agatholic acid (Milind et al. 2008; Merali et al. 2018). Copal is enactment against fever, headache, burns and stomach. It exhibits excellent binding, stabilizing and emulsifying properties, which is desirable for paints and production of colours. Copal gum can be evaluated as matrix-forming substance for sustained drug delivery. Also, copal resin is able to form a film with good swelling property. Hence, it can be used in various applications such as coating material for sustained release and colon targeted drug delivery and removal of malachite green from wastewater treatment (Umekar and Yeole 2008; Kaur et al. 2018).

18.2.27 Bhara Gum

Bhara gum is a yellowish nature of gum exudate of *Terminalia bellirica* plant; it belongs to the family Combretaceae. The bhara gum is extracted from the bark of *Terminalia bellirica* in a waste material (Shankar et al. 2008a, b). Its chemical constituents include β-sitosterol, ellagic acid, ethyl gallate, galloyl glucose, chebulagic acid and gallic acid. The bhara gum has been proposed as a new sustained microencapsulated drug delivery system. It has been used to fabricate microcapsules through ionic gelation for controlled release of famotidine drug. Bhara gum microcapsules exhibited slow famotidine release, i.e. over 10 h period (Fizza Rauf et al. 2012; Shankar et al. 2008a, b).

18.2.28 Gellan Gum

The gellan gum is a high molecular weight linear anionic hetero-deacetylated exocellular polysaccharide which belongs to the family *Sphingomonas elodea* with tetra-saccharide repeating unit. Chemical constituents of gellan gum are glucose, rhamnose and glucuronic acid (Wataru Kubo et al. 2003). It is used in commercial products as disintegrating agent, stabilizer and thickening agent, especially in the fields of food industry and pharmaceutics, sustainable drug delivery, ocular drug delivery and colon drug delivery applications (Kuo et al. 1986). In addition, under gelation it is capable to remove metal ions; however, it has more affinity towards divalent cations compared to monovalent cations (Balasubramaniam et al. 2003; Antony and Sanghavi 1997; Agnihotri et al. 2006).

18.2.29 Neem Gum

The neem gum is a purely water-soluble exudate of *Azadirachta indica*, which belongs to the family Meliaceae (Chaudhari et al. 2013). At dried condition it has a clear, brown and white colour. It is widely used in India as well as many other countries for various industries like textile, paper and pharmaceuticals. Also, it has various potential properties such as anti-inflammatory, killing microbes, preventing hypertension and preventing cancer. The chemical constituents of neem gum are glucuronic acid, mannose, arabinose, galactose, fucose, glucose and xylose. However, it is mainly used in Ayurveda in India for medical purpose (Malviya et al. 2019; Velusamy et al. 2015).

18.2.30 Albizia Gum

Albizia gum is extracted from the trunk of *Albizia zygia* tree, which belongs to the family Leguminosae. It is widely spread in various parts of Africa, Australia and India. The chemical composition of *Albizia* gum is β-(1→3)-linked D-galactose unit with some β-(1→6)-linked D-galactose chain units (Odeku. 2005; Swapna and Shailaja 2019). It has been successfully evaluated as suspending agent in sulphadimidine pharmaceutical suspensions; also, the leaves and stem of the tree used in treatment of boils, male sexual and diarrhoea importance. The drug release studies explore that *Albizia* gum is a capable protecting agent of core drug in tablet at the physiological environment of the gastrointestinal tract and intestine (Singh et al. 2018; Femi-Oyewo et al. 2004).

18.2.31 **Abelmoschus** *Gum or Okra Gum*

The *Abelmoschus esculentus* or okra gum is obtained from the fresh fruits of *Abelmoschus esculentus* plant, which belongs to Malvaceae family. It is a bulky crop cultivated annually throughout India; these fruits are green pods of different shapes (Mishra et al. 2008). The chemical composition of its polysaccharide constitutes the different molar ratios of L-galacturonic acid, D-galactose, L-rhamnose, mannose, glucose, xylose and arabinose (Ilango et al. 2010). The gum has been used in various medicinal disorders, i.e., anti-cancer, antimicrobial, anti-ulcer activity. It has been successfully evaluated as suspending agent, emulsion stabilizer, thickener, tablet binder, and disintegrating agent in pharmaceutics and treatment of textile industry waste water.(Kumar et al. 2009; de Carvalho et al. 2011).

18.2.32 **Ocimum basilicum** *Gum*

The genus of *Ocimum* (*Ocimum basilicum*) is collectively called “Basil”; it is one of the largest genera of the family Lamiaceae. Polysaccharide is extracted from basil seeds; and its chemical composition includes glucose, mannose, galactose and glucomannan. The chemical structure constitutes of two main fractions of glucomannan, with a glucose-to-mannose, and (1 → 4)-linked xylan (24.29%) and a minor fraction of glucan. (Keisandokht et al. 2018; Melo and D’souza 2004). Basil is an herbaceous plant which popularly grows and is widely spread in various regions of the world (India, Iran, France, Bulgaria, Egypt and some warm areas of Africa). Basil seeds are most commonly used in culinary and numerous health benefits which include treatment of sore throat, ulcers, insomnia, migraine, depression, skin infections, diarrhoea, kidney disorders and cough. It has potential properties such as emulsion, thickening and mouth feel improvements; in addition, it has various applications in drug delivery, bioabsorption, tableting agent and removal of metal ions (Shamsnejati et al. 2015; Levankumar et al. 2009; Zeynali et al. 2019).

18.2.33 Descurainia sophia *Seed Gum*

Descurainia sophia is known as flixweed with an erect and branched stem and belongs to the family of Brassicaceae. It produces large number of seeds from early to late summer; it is a native of tropical Asia (Nimrouzi M and Zarshenas MM 2016) and also a commonly used herbal medicine in Iranian folk medicine and herbs with prevalent use in the largest traditional pharmacopeia of Persian medicine (Golalikhani M et al. 2014). Gum is broadly utilized in various applications like pharmaceutical and food industry (Soukoulis et al. 2018; Rezan Rezaeian et al. 2016; Kian Aghaabbasi et al. 2014).

18.2.34 Honey Locust Gum

The honey locust gum (HLG) is extracted from *Gleditsia triacanthos* tree, which belongs to the Leguminosae family and deciduous tree native of North America and grown widely around the world. The gum seeds contain mainly fats, proteins, fibres and carbohydrates (Medjekal et al. 2018; Gold and Hanover JW 1993). The chemical composition of HLG is 88% of D-galacto-D-mannoglycan, 6% of protein, 4% pentane and 1% cellulose. The gum galactomannan is composed of (1→4)-linked β-D-mannan backbone with (1→6)-linked α-D-galactose side chain units. Americans used HLG as the supplementary medicine and food source. The honey locust leaves are used for duodenal ulcers, gastric ulcers, colitis and cholecystitis in Bulgaria (Jain et al. 2007). The gum is used to produce tablet matrix at different concentrations by granulation method for theophylline drug delivery (Chuan-Rui Zhang et al. 2015).

18.2.35 Mangifera indica *(Mango)*

Polysaccharide of mango gum is an exudate of *Mangifera indica* plant; it belongs to the Anacardiaceae family. It is a bioactive gum and chemical structure established as 2-C-β-D-glucopyranosyl-1, 3, 6, 7-tetrahydroxyxanthone. Mango gum is used in various applications such as pharmaceutical, food packaging, tablet binder, tablet retardant and disintegrant and also used in sustained release of paracetamol drug (Danalache et al. 2015; Khaliq et al. 2015; Nawab et al. 2016; Rambabu et al. 2019; Danalache et al. 2016; Sharma et al. 2016).

18.2.36 Hakea Gum

Hakea gum is an exudate from the bark of *Hakea acicularis* plant; it belongs to the Proteaceae family. It is widely distributed to western regions of the Cape Province (Eagles et al. 1993; Richardson and Manders 1985). The chemical constituents of hakea gum are D-galactose and L-arabinose linked that are acidic arabino-galactose, and other cases demonstrated linear core in which L-arabinose, L-rhamnose, D-galactose, D-xylose and D-glucuronic acid branched framework are attached. It is used as mucoadhesive polysaccharide for controlled release of drugs in the form of buccal tablets (Hemant et al. 2001; Alur et al. 1999).

18.2.37 Moi Gum

Moi gum is commonly known as "The Indian ash tree"; it is obtained from the plant of *Lannea coromandelica* which belongs to the family Anacardiaceae. It is widely used in India, Bangladesh and some tropical countries (Venkaiah and Shah 1984; Shankar et al. 2008a, b; Nayak et al. 2014). It is hydrophilic and has good swelling property; hence it is used to design controlled drug release dosage forms. Also, the gum is used as a lotion in eruption and obstinate ulcers, skin eruption, heart diseases, mouth sores and dysentery. It is mainly used as drug delivery formulations such as tablets and microspheres for controlled release of losartan potassium and lamivudine, respectively (Baljit Singh et al. 2008; Eswaramma and Murthy 2019).

18.2.38 Balangu Gum

Balangu gum obtained from the *Lallemantia royleana* belongs to the family Lamiaceae. The chemical composition of this gum includes 61 % of carbohydrate, 29 % of crude fibre, 0.87% of protein and 8.33% of ash. It has ability to absorb high water content; hence, it produces sticky mass at short time; also it forms tasteless liquid and turbid. It can be used as a new source of food hydrocolloids in food formulations (Razavi SM et al. 2008).

18.2.39 Bael Gum

The bael gum-based polysaccharide is obtained from the fruits of *Aegle marmelos* which belongs to the family Rutaceae. The bael fruit gum is an indigenous subtropical forest tree growing in India, Sri Lanka, Bangladesh, Thailand and most of the region of Southeast Asia. It comprises anionic polysaccharide chemically consist of -(1→3)-linked-β-D-galactopyranosyl residue, and side chain groups contain rhamnose, galactose, arabinose and galacturonic acid (Sonawane et al. 2020). The gum has greater aqueous solubility and water-holding capacity. The fruits also contain alkaloids, polyphenols, coumarins and carotenoids along with polysaccharides. The gum is mostly used in gelling agent, waterproofing material and adhesives and controlled drug release (Mirza et al. 2018). The gum is used in different applications such as antimicrobial, anticancer and anti-diabetic, triangular nanoplate, bone tissue engineering and films. Recently, ciprofloxacin hydrochloride-loaded bael gum polysaccharide and chitosan nanoparticle gels are developed for sustainable drug delivery (Balasubramanian et al. 2016; Dev et al. 2018; Jindal et al. 2013).

18.2.40 Terminalia Gum

Naturally terminalia gum is obtained from the trunk of *Terminalia randii* tree which belongs to the family Combretaceae. The bark is smooth with beige to grey brown colour, with yellowish or beige slash while the stem is pubescent. Terminalia randii extracted from stem and bark were used for various treatments of diseases such as haemorrhoids, diarrhoea and dysentery and wounds. It has good binding capcaity compared to standard binders such as poly(vinyl pyrrolidone) is a corn starch agent, hence it has used in carvedilol tablet formulations (Bamiro et al. 2010; Meka et al. 2012).

18.2.41 Jackfruit Gum

Jackfruit gum is a hydrophilic polysaccharide extracted from *Artocarpus heterophyllus* Lam tree; it belongs to the Moraceae family. It is a native of India; however, it is widely cultivated in Bangladesh and many parts of Southeast Asia (Anaya-Esparza et al. 2018). The jackfruit gum polysaccharide stores in seeds, the major portion is galactomannan; it consists of β-(1→4) linked mannose residue and single unit side chain attached α-(1→6). (Appukuttan and Basu 1987). It is widely used in the various food products as well as pharmaceutical formulations (Kabir et al. 2018; Baslingappa Swami et al. 2012).

18.2.42 Delonix regia Gum *(Gulmohar Mucilage)*

Delonix regia gum is also known as flamboyant; it is obtained from the seeds of *Delonix regia*, which belongs to the family Fabaceae. The tree mucilage is a water soluble gum and it is yellowish or reddish-brown warty tears. (Igwe and Nwokocha 2014; Dandge et al. 2015). *Delonix regia* gum chemical consists of α-D-mannose-(1→4)-linkage and α-D-galactose-(1→6)-branches. Galactose and mannose properties are similar to guar gum but differ in the OH bond position in the main chain. Gum has different chemical properties and it is a potent material as gelling agent, thickening, emulsifying, binding and suspending agent for different applications such as food industry, textiles, and medicinal; espicially preparation of paracetamol tablet formulations and anti-microbial wound dressings.(Cheirmadurai et al. 2016; Okoye et al. 2014; Nidhi Suhane et al. 2016).

18.2.43 Taro Gum (Colocasia esculenta)

Taro (*Colocasia esculenta*) gum's common name is corns' tubers of several genera which belongs to the family Araceae. It is an extensively refined root crop, available in tropical areas such as Southeast Asia, Africa, Pacific Island, the USA and Mediterranean (Sonia et al. 2019). Taro gum contains high percentage of mucilage and has viscous appearance, so it is used as suspending agent in tablet formulations, drug delivery and mucoadhesive materials. (Andrade et al. 2020; Ghumman et al. 2019).

18.2.44 Gum Acacia

The polysaccharide gum acacia is a plant exudate obtained from *Acacia seyal* or *Acacia senegal* tree which belongs to family Fabaceae. Gum acacia has branched polysaccharide chain, and its chemical constituents are (1→3)-linked β-D-galactopyranosyl units with side chains β-L-rhamnopyranose, α-L-arabinofuranosyl, β-D-di glucuronic acid and β-D-digalactopyranose (Sharma et al. 2020). It is water-soluble, non-toxic, cheaper, renewable, edible and with good physical properties, such as stabilizing agent, emulsifying, thickening and binding properties, and also gum could be incorporated in food formulations. Gum acacia is used in various applications such as pharmaceutical, liquor, water-emulsions, inks, wound dressing, nanoparticles, antimicrobial studies and cosmetic products (lotion and creams) (Jangra and Pothuraju 2020; Mohamed et al. 2020; Patel and Goyal 2015).

18.2.45 Agar Gum

Agar is a hydrocolloid polysaccharide obtained from the species of marine red algae which belongs to the family Rhodophyceae. Its chemical constituents are D-galactose and 3-6-anhydro-L-galastose linked by α-(1→3) and β-(1→6) glycoside bonds with heterogeneous combination of small molecule and repeated linear chain tetra-saccharides (Lee et al. 2019). Agar forms structure cell walls of seaweeds and mixture of agarose (gelling fraction) and agaropectin (non-gelling fraction) units. However, agar is not soluble in water at room temperature but soluble at 80–90 °C; it is dissolved and forms clear solution. Agar is used in various applications such as food packing, antibacterial wound dressing, cartilage regeneration and culture media of microorganism, *salep* films and industrial applications (Akkaya et al. 2020; Mostafavi and Zaeim 2020; Shukla et al. 2012).

18.2.46 Cactus Mucilage

The cactus mucilage can be obtained from *Opuntia ficus-indica* which belongs to the family Cactaceae. Cactus mucilage chemical composition is D-galactose, L-arabinose, D-xylose, D-galacturonic acid and L-rhamnose. In Mexico thousands of people cultivate cactus pear (Lira-Vargas et al. 2014; Del-Valle et al. 2015). Cactus mucilage has dissolving and dispersing ability itself and led viscous or colloid gelation. Cactus mucilage is used in many applications such as cosmetic, food industry, coating films, heavy metal removal and pharmaceuticals (Zegbe et al. 2012; Fox et al. 2012).

18.2.47 Cedrela Gum

Cedrela gum is the most common species obtained from Cedrela odorota belongs to the family Meliaceae. It is commonly called Cuba cedar or Spanish cedar and widely spread in subtropical and dry tropical or tropical life zone forest (Ayorinde and Odeniyi 2018). *Cedrela* gum contains chemical composition of arabinose, galactose, rhamnose and uronic acid residues. These carbohydrate sugars are signified by glucuronic acid and its 4-O-methyl derivative. The *Cedrela* gum is mainly used in insect-repelling resins and honey production and primarily used in household cloth stores. It is mainly used in various applications such as mucosal immune response in chicken vaccinated with infectious bursal diseases vaccine. veterinary vaccine delivery and binder and mucoadhesive component of ibuprofen tablet formulations (Odeniyi et al. 2013; Emikpe et al. 2016).

18.2.48 Galbanum Gum

The *Galbanum* gum is an aromatic gum obtained from the umbelliferous perennial herbs which belong to family Apiaceae (Hamedi et al. 2017). *Galbanum* gum is very bitter and acrid and has peculiar scent followed by intense green, woody, spicy and fragrance. Traditionally it is used for different types of diseases such as anti-flatulent, antiseptic, anti-seizure, pain-killer, antispasm, inflammation reliever and also it is used as tonic for memory improvement by the Egyptians and Iranian. In addition, it is used as acetyl cholinesterase inhibitor applications (Najafabadi et al. 2017; Adhami et al. 2014; Jalali et al. 2013).

18.2.49 Malva Nut Gum

The malva nut gum's common name is *Scaphium*; it belongs to the Malvaceae family and is obtained from the seed mucilaginous substance in the form of hydrocolloid. It is a genus of Scaphium and native from China and Southeast Asian countries such as Myanmar, Thailand, Cambodia, Laos and Malaysia. Malva nut seed medicine is used since a long time in Southeast Asia; Thailand peoples use mucilaginous material for laxative benefits (Srichamroen and Chavasit 2011; Pramualkijja et al. 2016). It is not completely dissolved in water, but swells; however, it is mucilaginously dispersed. The gum polysaccharide contains monosaccharides of galactose, rhamnose and arabinose with molecular weight of 1,62,000 Da, and reports suggest that rhamnose in main chain linked α-(1→3) glycosidic linkage. The seeds have been extensively used for preparation of jellies and sweetened drinks for health benefits and body weight reduction. The malva nut gum is used in reducing high glucose levels by Caxo-2 cell and also used for microporous carbon-based materials for supercapacitors and pasting and textual properties (Srichamroen 2018; Phimolsiripol et al. 2011).

18.2.50 **Mucuna flagellipes** *Gum*

The *Mucuna* genus of around 100 accepted species of climbing shrubs and vines belongs to the family Fabaceae and subfamily Papilionaceae. It is found mainly in tropical rainforest of Nigeria and grown as significant food crop in another part of Asia and Africa (Aviara et al. 2012; Nwokocha and Williams 2009). *Mucuna flagellipes* gum polysaccharide consists of galactomannan, D-galactose and D-mannose. These are highly economic and significant in cosmetics, domestics and pharmaceuticals. In the pharmaceutical industry, seed gum extracted could be used for formulation as tablet binder. This gum exibit the good physico-chemical properties, hence, it is extensively used as emulsifying agent, stabilizer, thickening agent, suspending agent. (Okwu and Okoro 2006; Agba et al. 2019).

18.2.51 Sesbania Gum

Sesbania gum was obtained from the endosperms of seeds of *Sesbania grandiflora* which belongs to the family Leguminosae. The sesbania gum chemically consists of mannose connect β-(1→4)-glycosidic bond and galactose linked –α-(1→6)-glycosidic side chain unit (Patel and Patel 2009; Hongbo et al. 2018). It is widely used in paper making, beneficiation, thickening agent, mining industry, oil drilling food, foods, cosmetic, pharmaceutical applications, painting, petroleum and wastewater treatment. However, the gum is applied in various potential applications such as gelling agent, magnetic carbonaceous nanocomposites and adsorption, warp sizing performance of fine cotton yarns and sustainable drug delivery (Lan et al. 2014; Pal et al. 2018).

18.2.52 Olibanum Gum

Olibanum gum is an oleo-resin exudate obtained from the bark of *Boswellia* which belongs to family Burseraceae, a native of India and Ethiopia. It is available in low cost, eco-friendly, non-toxic, renewable and biodegradable. This gum is used as fumigant and multipurpose in aromatic; it is widely exploited in cosmetic, pharmaceutical, ceramic, food industry, paints and textile industries. The *Olibanum* gum polysaccharide primary structure contains sugars such as arabinose, galactose, xylose and D-glucuronic acid (Assefa et al. 2017). This is traditionally used for several managements of hypolipidaemic reno-protective, inflammatory, arthritic and anticancer disorders and Ayurveda (Kora et al. 2012). *Olibanum* gum is used in various applications such as controlling matrices of controlling release of diclofenac, neural tissue degeneration, degradation of anthropogenic dye pollutants using palladium nanoparticles drug delivery, green synthesis of silver and gold nanoparticles and reductant and stabilizing agent (Kora and Sashidhar 2015; Kora and Rastogi 2016).

18.2.53 Salmalia malabarica Gum

The *Salmalia malabarica* (SM) polysaccharide is an extract of *Bombax ceiba* plant which belongs to the family Malvaceae (Bombacaceae); it is widely available in India. It is easily available, cost-effective, eco-friendly, non-toxic and biodegradable and a renewable gum material (Rani and Khullar 2004; Saleem et al. 2003). SM consists of high molecular weight and negatively charged colloids with various sugar moieties. SM gum has been traditionally used in Ayurveda, especially for the treatment of influenza, pulmonary tuberculosis and menorrhagia. In addition, it has other potential properties such as emetic, expectorant, stimulant, anti-inflammatory, antiangiogenic, astringent, antioxidant, anti-asthma, anticancer and anti-hypertensive. Recent studies demonstrate that SM and its nanocomposites could be successfully utilized for anti-ulcerogenic agent, green synthesis of metal nanoparticles, catalysis and reduction of organic dyes (methylene blue and Congo red dyes) (Jain and Verma 2012; Krishna et al. 2016; Reddy et al. 2015).

18.3 Conclusion

Polysaccharides of natural gums are promising biocompatible and biodegradable polymers. These are potentially functionalized to have biomaterials for drug delivery devices. In addition, they are widely available and cost-effective compared to the synthetic one. However, these are chemically functionalized to have a better material for targeted drug delivery applications. The primary objective of this book chapter is about the classification of gums based on their source, origin, chemical structure and multiple applications. Due to the significant properties of natural gums i.e., water retention, film formation, adhesiveness, binding, suspending, thickening and emulsifying of natural gums, they could successfully utilize for the preparation controlled dug delivery formulations in the form of tablets and capsules. The majorities of the gums are collected in the book chapter and emphasized on their potential drug delivery as well other significant applications.

References

Adedokun MO, Olorunsola EO, Anselem EL (2017) Comparative evaluation of metronidazole suspensions formulated with Raphia africana hydrocolloid and other natural polymers. J Appl Pharm Sci 2:121–126

Aderinola TA, Alashi AM, Nwachukwu ID, Fagbemi TN, Enujiugha VN, Aluko RE (2020) In vitro digestibility, structural and functional properties of Moringa oleifera seed proteins. Food Hydrocoll 101:105574

Adetunji Adeniran G, Ohore OG, Jarikre TA, Olawumi Ola O, Oyebanji V, Emikpe BO (2019) Humoral and mucosal immune responses in challenged chickens vaccinated with infectious bursal disease vaccine using gums from Cedrela odorata and Khaya senegalensis as delivery agents. J Immunoass Immunochem 40:630–641

Adhami HR, Fitz V, Lubich A, Kaehlig H, Zehl M, Krenn L (2014) Acetylcholinesterase inhibitors from galbanum, the oleo gum-resin of Ferula gummosa Boiss. Phytochem Lett 1:10

Adinolfi M, Corsaro MM, Lanzetta R, Parrilli M, Folkard G, Grant W, Sutherland J (1994) Composition of the coagulant polysaccharide fraction from Strychnos potatorum seeds. Carbohydr Res 263:103–110

Agba OA, Eze SC, Atugwu AI, Awere SU, Chukwu C (2019) Influence of combined application of P fertilizer and lime on Mucuna flagellipes nodulation, growth and yield. African J Plant Sci 13:47–58

Aghaabbasi K, Dehghan E, Baghizadeh A, Dashti H (2014) Comparing the effect of ethanol extracts of Descurainia sophia (L.) seed and Althaea officinalis root on Streptococcus pyogenes. Zahedan J Res Med Sci 16:27–32

Agnihotri SA, Jawalkar SS, Aminabhavi TM (2006) Controlled release of cephalexin through gellan gum beads: effect of formulation parameters on entrapment efficiency, size, and drug release. Eur J Pharm Biopharm 63:249–261

Ahlawat KS, Khatkar BS (2011) Processing, food applications and safety of aloe vera products: a review. J Food Sci Technol 48:525–533

Ahmad S, Ahmad M, Manzoor K, Purwar R, Ikram S (2019) A review on latest innovations in natural gums-based hydrogels: preparations & applications. Int J Biol Macromol 136:870–890

Ahuja M, Kumar S, Kumar A (2013) Evaluation of mucoadhesive potential of gum cordia, an anionic polysaccharide. Int J Biol Macromol 55:109–112

Akkaya NE, Ergun C, Saygun A, Yesilcubuk N, Akel-Sadoglu N, Kavakli IH, Turkmen HS, Catalgil-Giz H (2020) New biocompatible antibacterial wound dressing candidates; agar-locust bean gum and agar-salep films. Int J Biol Macromol 155:430–438

Alange VV, Birajdar RP, Kulkarni RV (2017) Functionally modified polyacrylamide-graft-gum karaya pH-sensitive spray dried microspheres for colon targeting of an anti-cancer drug. Int J Biol Macromol 102:829–839

Alle M, Kim TH, Park SH, Lee SH, Kim JC (2020) Doxorubicin-carboxymethyl xanthan gum capped gold nanoparticles: microwave synthesis, characterization, and anti-cancer activity. Carbohydr Polym 229:115511

Alonso-Sande M, Teijeiro-Osorio D, Remuñán-López C, Alonso MJ (2009) Glucomannan, a promising polysaccharide for biopharmaceutical purposes. Eur J Pharm Biopharm 72:453–462

Alur HH, Pather SI, Mitra AK, Johnston TP (1999) Evaluation of the gum from Hakea gibbosa as a sustained-release and mucoadhesive component in buccal tablets. Pharm Dev Technol 4:347–358

Alur HH, Desai RP, Mitra AK, Johnston TP (2001) Inhibition of a model protease—pyroglutamate aminopeptidase by a natural oligosaccharide gum from Hakea gibbosa. Int J Pharm 212:171–176

Anand GT, Renuka D, Ramesh R, Anandaraj L, Sundaram SJ, Ramalingam G, Magdalane CM, Bashir AK, Maaza M, Kaviyarasu K (2019) Green synthesis of ZnO nanoparticle using Prunus dulcis (Almond Gum) for antimicrobial and supercapacitor applications. Surf Interfaces 17:100376

Anaya-Esparza LM, González-Aguilar GA, Domínguez-Ávila JA, Olmos-Cornejo JE, Pérez-Larios A, Montalvo-González E (2018) Effects of minimal processing technologies on jackfruit (Artocarpus heterophyllus Lam.) quality parameters. Food Bioprocess Technol 11:1761–1774

Andrade LA, de Oliveira Silva DA, Nunes CA, Pereira J (2020) Experimental techniques for the extraction of taro mucilage with enhanced emulsifier properties using chemical characterization. Food Chem 19:127095

Antony PJ, Sanghavi NM (1997) A new disintegrant for pharmaceutical dosage forms. Drug Dev Ind Pharm 23:413–415

Anwar F, Latif S, Ashraf M, Gilani AH (2007) Moringa oleifera: a food plant with multiple medicinal uses. Phytother Res 21:17–25

Appukuttan PS, Basu D (1987) A galactomannan-hydrolysing α-galactosidase from jack fruit (Artocarpus integrifolia) seed: affinity chromatographic purification and properties. J Biosci 12:61–69

Assefa AG, Mesfin AA, Akele ML, Alemu AK, Gangapuram BR, Guttena V, Alle M (2017) Microwave-assisted green synthesis of gold nanoparticles using Olibanum gum (Boswellia serrate) and its catalytic reduction of 4-nitrophenol and hexacyanoferrate (III) by sodium borohydride. J Clust Sci 28:917–935

Avachat AM, Dash RR, Shrotriya SN (2011) Recent investigations of plant based natural gums, mucilages and resins in novel drug delivery systems. Ind J Pharm Edu Res 45:86–99
Aviara NA, Onuh OA, Ehiabhi SE (2012) Influence of moisture content and loading orientation on some mechanical properties of Mucuna flagellipes nut. Res Agric Eng 58:66–72
Awang AN, Mohamed AR, Salleh NH, Hoo PY, Kasim NN (2019) Torrefaction of Leucaena Leucocephala under isothermal conditions using the Coats–Redfern method: kinetics and surface morphological analysis. React Kinet Mech Catal 128:663–680
Ayorinde JO, Odeniyi MA (2018) Solid state characterization of two novel gums from Cedrela odorata and Enterolobium cyclocarpum. J Pharm Investig 48:487–496
Balasubramaniam J, Kant S, Pandit JK (2003) In vitro and in vivo evaluation of Gelrite gellan gum-based ocular delivery system for indomethacin. Acta Pharma 53:251–262
Balasubramanian S, Bezawada SR, Raghavachari D (2016) Green, selective, seedless and one-pot synthesis of triangular au nanoplates of controlled size using bael gum and mechanistic study. ACS Sustain Chem Eng 4:3830–3839
Bamiro OA, Sinha VR, Kumar R, ODEKU OA (2010) Characterization and evaluation of Terminalia randii gum as a binder in carvedilol tablet formulation. Acta Pharmaceutica Sciencia 52:3–9
Barak S, Mudgil D (2014) Locust bean gum: processing, properties and food applications—a review. Int J Biol Macromol 66:74–80
Bemiller JN (2008) Gums and related polysaccharides. Glyc 2:1513
Bonnet BR, Wisniewski C, Reissmann CB, Nogueira AC, Andreoli CV, Barbieri SJ (2002) Effects of substrates composed of biosolids on the production of Eucalyptus viminalis, Schinus terebinthifolius and Mimosa scabrella seedlings and on the nutritional status of Schinus terebinthifolius seedlings. Water Sci Technol 46:239–246
Bonsu MA, Ofori-Kwakye K, Kipo SL, Boakye-Gyasi ME, Fosu MA (2016) Development of oral dissolvable films of diclofenac sodium for osteoarthritis using Albizia and Khaya gums as hydrophilic film formers. J Drug Deliv 2:6459280
Bouaziz F, Koubaa M, Kallel F, Chaari F, Driss D, Ghorbel RE, Chaabouni SE (2015) Efficiency of almond gum as a low-cost adsorbent for methylene blue dye removal from aqueous solutions. Ind Crop Prod 74:903–911
Bouaziz F, Koubaa M, Ghorbel RE, Chaabouni SE (2016) Recent advances in Rosaceae gum exudates: from synthesis to food and non-food applications. Int J Biol Macromol 86:535–545
Boudreau MD, Beland FA (2016) An evaluation of the biological and toxicological properties of Aloe barbadensis (miller), Aloe vera. J Environ Sci Health C 24:103–154
Braz L, Grenha A, Corvo MC, Lourenco JP, Ferreira D, Sarmento B, da Costa AM (2018) Synthesis and characterization of Locust Bean Gum derivatives and their application in the production of nanoparticles. Carbohydr Polym 181:974–985
Chamarthy SP, Pinal R (2008) Plasticizer concentration and the performance of a diffusion-controlled polymeric drug delivery system. Colloids Surf A Physicochem Eng Asp 10:25–30
Chaudhari A, Gite V, Rajput S, Mahulikar P, Kulkarni R (2013) Development of eco-friendly polyurethane coatings based on neem oil polyetheramide. Ind Crop Prod 50:550–556
Cheirmadurai K, Thanikaivelan P, Murali R (2016) Highly biocompatible collagen–Delonix regia seed polysaccharide hybrid scaffolds for antimicrobial wound dressing. Carbohydr Polym 137:584–593
Chi M, Liu C, Shen J, Dong Z, Yang Z, Wang L (2018) Antibacterial superabsorbent polymers from Tara gum grafted poly (acrylic acid) embedded silver particles. Polymers 10:945
Chick A, Cassella J (2011) Gum Damar as a substitute for Canada Balsam in mounting media for microscopical specimens. Entomol Mon Mag 147:111–114
da Silva DA, de Paula RC, Feitosa JP (2007) Graft copolymerisation of acrylamide onto cashew gum. Eur Polym J 20:2620–2629
Danalache F, Mata P, Moldao-Martins M, Alves VD (2015) Novel mango bars using gellan gum as gelling agent: rheological and microstructural studies. LWT Food Sci Technol 62:576–583

Danalache F, Carvalho CY, Alves VD, Moldão-Martins M, Mata P (2016) Optimisation of gellan gum edible coating for ready-to-eat mango (Mangifera indica L.) bars. Int J Biol Macromol 84:43–53

Dandge VJ, Joshi UM, Biyani KR (2015) Extraction of a novel seed gum from a locally available plant and its evaluation. Int J Pharm Chem Sci 4:229–303

Dave V, McCarthy SP (1997) Review of konjac glucomannan. J Environ Polym Degrad 5:237

de Carvalho CC, Cruz PA, da Fonseca MM, Xavier-Filho L (2011) Antibacterial properties of the extract of Abelmoschus esculentus. Biotechnol Bioprocess Eng 16:971

Dell B, McComb AJ (1979) Plant resins—their formation, secretion and possible functions. Adv Bot Res 6:277–316

Del-Valle V, Hernández-Muñoz P, Guarda A, Galotto MJ (2015) Development of a cactus-mucilage edible coating (Opuntia ficus indica) and its application to extend strawberry (Fragaria ananassa) shelf-life. Food Chem 91:751–756

Deshmukh AS, Setty CM, Badiger AM, Muralikrishna KS (2012) Gum ghatti: a promising polysaccharide for pharmaceutical applications. Carbohydr Polym 87:980–986

Dev D, Bhardwaj U, Prasad DN (2018) Ciprofloxacin hydrochloride loaded chitosan nanoparticle gel: advanced approach for enhancing permeation and sustainability of drug release. Asian J Pharm Res Dev 6:67–72

Dionísio M, Grenha A (2012) Locust bean gum: exploring its potential for biopharmaceutical applications. J Pharm Bioallied Sci 3:175

Dixit G, Chavhan JI, Kanchan PU, Sushant misra (2015) formulation and characterization of mucoadhesive buccal film of ranitidine Hydrochloride using sterculia foetida gum as polymer. Asian journal of pharmaceutical and clinical research 8(3) 68-71.

Duan M, Ma J, Fang S (2019) Synthesis of hydrazine-grafted guar gum material for the highly effective removal of organic dyes. Carbohydr Polym 4:308–314

Eagles PF, Stephen AM, Churms SC (1993) Molecular structures of gum exudates from Hakea species. Phytochemistry 34:709–713

Eastwood MA, Brydon WG, Anderson DM (1984) The effects of dietary gum tragacanth in man. Toxicol Lett 21:73–81

Emikpe BO, Oyebanji VO, Odeniyi MA, Salaam AM, Oladele OA, Jarikre TA, Akinboade OA (2016) Ex-vivo evaluation of the mucoadhesive properties of Cedrela odorata and Khaya senegalensis gums with possible applications for veterinary vaccine delivery. Springerplus 5:1289

Eswaramma P, Murthy KV (2019) Evaluation of Moi Gum in the formulation of controlled release matrix tablets using losartan potassium. Int J Pharm Sci Res 10:121–129

Fagioli L, Pavoni L, Logrippo S, Pelucchini C, Rampoldi L, Cespi M, Bonacucina G, Casettari L (2019) Linear viscoelastic properties of selected polysaccharide gums as function of concentration, pH, and temperature. J Food Sci 84:65–72

Femi-Oyewo MN, Adedokun MO, Olusoga TO (2004) Evaluation of the suspending properties of Albizia zygia gum on sulphadimidine suspension. Trop J Pharm Res 3:279–284

Fernandes RA, Garcia-Rojas EE (2020) Effect of cosolutes on the rheological and thermal properties of Tara gum aqueous solutions. J Food Sci Technol 58:2773-2782'

Fox DI, Pichler T, Yeh DH, Alcantar NA (2012) Removing heavy metals in water: the interaction of cactus mucilage and arsenate (As (V)). Environ Sci Technol 46:4553–4559

Ganapuram BR, Alle M, Dadigala R, Dasari A, Maragoni V, Guttena V (2015) Catalytic reduction of methylene blue and Congo red dyes using green synthesized gold nanoparticles capped by salmalia malabarica gum. Int Nano Lett 5(4):215–222

Ganesh GN, Sureshkumar R, Jawahar N, Senthil V, Nagasamy Venkatesh D, Shanmukha Srinivas M (2010) Preparation and evaluation of sustained release matrix tablet of diclofenac sodium using natural polymer. J Pharm Sci Res 2:360–368

Ganter JL, Cardoso AT, Kaminski M, Reicher F (1997) Galactomannan from the seeds of Mimosa scabrella: a scale-up process. Int J Biol Macromol 21:137–140

Gbenga BL, Zulikha A (2013) New matrix tablet from okra gum: effects of method of preparation and gum concentration on tablet properties. Pharm Pharm 4:484

Ghori MU, Alba K, Smith AM, Conway BR, Kontogiorgos V (2014) Okra extracts in pharmaceutical and food applications. Food Hydrocoll 42:342–347

Ghumman SA, Bashir S, Noreen S, Khan AM, Malik MZ (2019) Taro-corms mucilage-alginate microspheres for the sustained release of pregabalin: in vitro & in vivo evaluation. Int J Biol Macromol 139:1191–1202

Golalikhani M, Khodaiyan F, Khosravi A (2014) Response surface optimization of mucilage aqueous extraction from flixweed (Descurainia sophia) seeds. Int J Biol Macromol 70:444–449

Gold MA, Hanover JW (1993) Honey locust (Gleditsia triacanthos), a multi-purpose tree for the temperate zone. Int Tree Crops J 7:189–207

Goswami S, Naik S (2014) Natural gums and its pharmaceutical application. J Sci Innov Res 3:112–121

Gowda DC, Neelisiddaiah B, Anjaneyalu YV (1979) Structural studies of polysaccharides from Aloe vera. Carbohydr Res 72:201–205

Hamedi H, Kargozari M, Shotorbani PM, Mogadam NB, Fahimdanesh M (2017) A novel bioactive edible coating based on sodium alginate and galbanum gum incorporated with essential oil of Ziziphora persica: the antioxidant and antimicrobial activity, and application in food model. Food Hydrocoll 72:35–46

Haq MA, Hasnain A, Azam M (2014) Characterization of edible gum cordia film: effects of plasticizers. LWT Food Sci Technol 55:163–169

Hemmati K, Ghaemy M (2016) Synthesis of new thermo/pH sensitive drug delivery systems based on tragacanth gum polysaccharide. Int J Biol Macromol 87:415–425

Hongbo T, Yao Y, Yanping L, Siqing D (2018) Effect of cross-linking and oxidization on structure and properties of sesbania gum. Int J Biol Macromol 114:640–648

Hu Y, Tian J, Zou J, Yuan X, Li J, Liang H, Zhan F, Li B (2019) Partial removal of acetyl groups in konjac glucomannan significantly improved the rheological properties and texture of konjac glucomannan and κ-carrageenan blends. Int J Biol Macromol 123:1165–1171

Hussain A, Qureshi F, Abbas N, Arshad MS, Ali E (2017a) An evaluation of the binding strength of okra gum and the drug release characteristics of tablets prepared from it. Pharmaceutic 9:20

Hussain IS, Gulsonbi M, Jaisankar V (2017b) Synthesis of pH responsive hydrogel–silver nanocomposite for use as biomaterials. International Journal of Research in Pharmacy and Chemistry IJRPC 7:273–282

Igwe OU, Nwokocha LM (2014) Isolation of gum from the seeds of Delonix regia and evaluation of its interactions with cassava and maize starches. Int J Chem Biochem Sci 5:16–21

Ilango K, Mishra M, Devi S, Rajsekaran A, Senthil Kumar M, Subburaju T (2010) In vitro and in vivo evaluation of okra polysaccharide-based colon-targeted drug delivery systems. Int J Pharm Sci Rev Res 5:138–145

Jain V, Verma SK (2012) Pharmacological investigations and toxicity studies. In: Pharmacology of Bombax ceiba Linn. Springer, Berlin, Heidelberg, pp 51–67

Jain A, Gupta Y, Jain SK (2007) Perspectives of biodegradable natural polysaccharides for site-specific drug delivery to the colon. J Pharm Pharm Sci 10:86–128

Jaison D, Chandrasekaran G, Mothilal M (2020) pH-sensitive natural almond gum hydrocolloid based magnetic nanocomposites for theragnostic applications. Int J Biol Macromol 154:256–266

Jalali HT, Petronilho S, Villaverde JJ, Coimbra MA, Domingues MR, Ebrahimian ZJ, Silvestre AJ, Rocha SM (2013) Assessment of the sesquiterpenic profile of Ferula gummosa oleo-gum-resin (galbanum) from Iran. Contributes to its valuation as a potential source of sesquiterpenic compounds. Ind Crop Prod 44:185–191

Jana S, Saha A, Nayak AK, Sen KK, Basu SK (2013) Aceclofenac-loaded chitosan-tamarind seed polysaccharide interpenetrating polymeric network microparticles. Colloids Surf B: Biointerfaces 105:303–309

Jangra S, Pothuraju R (2020) Functional significance of Gum acacia in the Management of Obesity. Curr Pharm Des 26:293–295

Jani GK, Shah DP (2008) Evaluation of mucilage of Hibiscus rosasinensis Linn as rate controlling matrix for sustained release of diclofenac. Drug Dev Ind Pharm 8:807–816

Jani GK, Shah DP, Prajapati VD, Jain VC (2009) Gums and mucilages: versatile excipients for pharmaceutical formulations. Asian J Pharm Sci 4:309–323

Jindal M, Kumar V, Rana V, Tiwary AK (2013) Physico-chemical, mechanical and electrical performance of bael fruit gum–chitosan IPN films. Food Hydrocoll 30:192–199

Kabir SM, Kim SD, Koh J (2018) Application of jackfruit latex gum as an eco-friendly binder to pigment printing. Fibers Polym 19:2365–2371

Kagithoju S, Godishala V, Nanna RS (2015) Eco-friendly and green synthesis of silver nanoparticles using leaf extract of Strychnos potatorum Linn. F. and their bactericidal activities. 3 Biotech 5:709–714

Kaity S, Isaac J, Ghosh A (2013) Interpenetrating polymer network of locust bean gum-poly (vinyl alcohol) for controlled release drug delivery. Carbohydr Polym 94:456–467

Kale SS, Darade V, Thakur HA (2011) Analysis of fixed oil from Sterculia foetida Linn. Int J Pharm Sci Res 2:2908

Kang J, Cui SW, Chen J, Phillips GO, Wu Y, Wang Q (2011) New studies on gum ghatti (Anogeissus latifolia) part I. Fractionation, chemical and physical characterization of the gum. Food Hydrocoll 25:1984–1990

Kaur S, Jindal R, Kaur Bhatia J (2018) Synthesis and RSM-CCD optimization of microwave-induced green interpenetrating network hydrogel adsorbent based on gum copal for selective removal of malachite green from waste water. Polym Eng Sci 58:2293–2303

Kawano A, Sato K, Yamamoto K, Kadokawa JI (2019) Preparation of chitin nanofiber-reinforced xanthan gum hydrogels. J Polym Environ 27:671–677

Keisandokht S, Haddad N, Gariepy Y, Orsat V (2018) Screening the microwave-assisted extraction of hydrocolloids from Ocimum basilicum L. seeds as a novel extraction technique compared with conventional heating-stirring extraction. Food Hydrocoll 74:11–22

Keshani-Dokht S, Emam-Djomeh Z, Yarmand MS, Fathi M (2018) Extraction, chemical composition, rheological behavior, antioxidant activity and functional properties of Cordia myxa mucilage. Int J Biol Macromol 118:485–493

Khaliq G, Mohamed MT, Ali A, Ding P, Ghazali HM (2015) Effect of gum arabic coating combined with calcium chloride on physico-chemical and qualitative properties of mango (Mangifera indica L.) fruit during low temperature storage. Sci Hortic 190:187–194

Kora AJ, Rastogi L (2016) Catalytic degradation of anthropogenic dye pollutants using palladium nanoparticles synthesized by gum olibanum, a glucuronoarabinogalactan biopolymer. Ind Crop Prod 81:1–10

Kora AJ, Sashidhar RB (2010) Arunachalam J. Gum kondagogu (Cochlospermum gossypium): a template for the green synthesis and stabilization of silver nanoparticles with antibacterial application. Carbohydr Polym 4:82670–82679

Kora AJ, Sashidhar RB (2015) Antibacterial activity of biogenic silver nanoparticles synthesized with gum ghatti and gum olibanum: a comparative study. J Antibiot 68:88–97

Kora AJ, Sashidhar RB, Arunachalam J (2012) Aqueous extract of gum olibanum (Boswellia serrata): a reductant and stabilizer for the biosynthesis of antibacterial silver nanoparticles. Process Biochem 47:1516–1520

Krishna IM, Reddy GB, Veerabhadram G, Madhusudhan A (2016) Eco-friendly green synthesis of silver nanoparticles using Salmalia malabarica: synthesis, characterization, antimicrobial, and catalytic activity studies. Appl Nanosci 6(5):681–689

Kristianto H, Rahman H, Prasetyo S, Sugih AK (2019) Removal of Congo red aqueous solution using Leucaena leucocephala seed's extract as natural coagulant. Appl Water Sci 9:88

Kubo W, Miyazaki S, Attwood D (2003) Oral sustained delivery of paracetamol from in situ-gelling gellan and sodium alginate formulations. Int J Pharm 258:55–64

Kuevi DN, Ayertey E, Bartels DA, Owusu FW (2019) Evaluation of the disintegration properties of Khaya senegalensis gum using paracetamol tablets. Asian J Res Med Pharm Sci 15:1–8

Kulkarni Vishakha S, Butte Kishor D, Rathod Sudha S (2012) Natural polymers–a comprehensive review. Int J Res Pharmaceut Biomed Sci 3:1597–1613

Kumar A, Ahuja M (2012) Carboxymethyl gum kondagogu: synthesis, characterization and evaluation as mucoadhesive polymer. Carbohydr Polym 90:637–643

Kumar R, Patil MB, Patil SR, Paschapur MS (2009) Evaluation of Abelmoschus esculentus mucilage as suspending agent in paracetamol suspension. Int J PharmTech Res 1:658–665

Kumar B, Kulanthaivel S, Mondal A, Mishra S, Banerjee B, Bhaumik A, Banerjee I, Giri S (2017) Mesoporous silica nanoparticle-based enzyme responsive system for colon specific drug delivery through guar gum capping. Colloids Surf B: Biointerfaces 150:352–361

Kumar A, Rao KM, Han SS (2018) Application of xanthan gum as polysaccharide in tissue engineering: a review. Carbohydr Polym 15:128–144

Kuo MS, Mort AJ, Dell A (1986) Identification and location of L-glycerate, an unusual acyl substituent in gellan gum. Carbohydr Res 156:173–187

Lan S, Leng Z, Guo N, Wu X, Gan S (2014) Sesbania gum-based magnetic carbonaceous nanocomposites: facile fabrication and adsorption behavior. Colloids Surf A Physicochem Eng Asp 46:163–171

Lee H, Rukmanikrishnan B, Lee J (2019) Rheological, morphological, mechanical, and water-barrier properties of agar/gellan gum/montmorillonite clay composite films. Int J Biol Macromol 141:538–544

Leone G, Consumi M, Lamponi S, Bonechi C, Tamasi G, Donati A, Rossi C, Magnani A (2019) Hybrid PVA-xanthan gum hydrogels as nucleus pulposus substitutes. Int J Polym Mater Polym Biomater 68:681–690

Levankumar L, Muthukumaran V, Gobinath MB (2009) Batch adsorption and kinetics of chromium (VI) removal from aqueous solutions by Ocimum americanum L. seed pods. J Hazard Mater 161:709–713

Lima MR, Paula HC, Abreu FO, da Silva RB, Sombra FM, de Paula RC (2018) Hydrophobization of cashew gum by acetylation mechanism and amphotericin B encapsulation. Int J Biol Macromol 8:523–530

Lira-Vargas AA, Corrales-Garcia JJ, Valle-Guadarrama S, Peña-Valdivia CB, Trejo-Marquez MA (2014) Biopolymeric films based on cactus (Opuntia ficus-indica) mucilage incorporated with gelatin and beeswax. J Prof Assoc Cactus 16:51–70

Madhusudhan A, Reddy GB, Krishana IM (2019) Green synthesis of gold nanoparticles by using natural gums. In: Husen A, Iqbal M (eds) N*anomaterials* and plant potential. Springer, Cham, pp 111–134

Maiti S, Dey P, Banik A, Sa B, Ray S, Kaity S (2010) Tailoring of locust bean gum and development of hydrogel beads for controlled oral delivery of glipizide. Drug Deliv 17:288–300

Makhado E, Pandey S, Nomngongo PN, Ramontja J (2018) Preparation and characterization of xanthan gum-cl-poly (acrylic acid)/o-MWCNTs hydrogel nanocomposite as highly effective re-usable adsorbent for removal of methylene blue from aqueous solutions. J Colloid Interface Sci 1:700–714

Malviya R, Sharma PK, Dubey SK (2019) Microwave-assisted preparation of biodegradable, hemocompatible, and antimicrobial neem gum–grafted poly (acrylamide) hydrogel using (3) 2 factorial design. Emerg Mater 2:95–112

Marvelys L, Maritza M, Lilian S, de Pinto GL, Julio H (2006) Structural elucidation of the polysaccharide from Sterculia apetala gum by a combination of chemical methods and NMR spectroscopy. Food Hydrocoll 20:908–913

Masadome T, Imato T, Itoh S, Asano Y (1997) Flow injection determination of anionic polyelectrolytes using an anionic surfactant-selective plasticized poly (vinyl chloride) membrane electrode detector. Fresenius J Anal Chem 357:901–903

Medjekal S, Bodas R, Bousseboua H, López S (2018) Evaluation of Carob (Ceratonia siliqua) and Honey Locust (Gleditsia triacanthos) pods as a feed for sheep. Iran J Appl Animal Sci 8:247–256

Meka VS, Nali SR, Songa AS, Kolapalli VR (2012) Characterization and in vitro drug release studies of a natural polysaccharide Terminalia catappa gum (Badam gum). AAPS PharmSciTech 13:1451–1464

Melo JS, D'Souza SF (2004) Removal of chromium by mucilaginous seeds of Ocimum basilicum. Bioresour Technol 92:151–155

Merali Z, Cayer C, Kent P, Liu R, Cal V, Harris CS, Arnason JT (2018) Sacred Maya incense, copal (Protium copal-Burseraceae), has antianxiety effects in animal models. J Ethnopharmacol 216:63–70

Mirza S, Zia I, Jolly R, Kazmi S, Owais M, Shakir M (2018) Synergistic combination of natural bioadhesive bael fruit gum and chitosan/nano-hydroxyapatite: a ternary bioactive nanohybrid for bone tissue engineering. Int J Biol Macromol 119:215–224

Mishra A, Malhotra AV (2011) Tamarind xyloglucan: a polysaccharide with versatile application potential. J Mater Chem 19:8528–8536

Mishra A, Yadav A, Pal S, Singh A (2006) Biodegradable graft copolymers of fenugreek mucilage and polyacrylamide: a renewable reservoir to biomaterials. Carbohydr Polym 65:58–63

Mishra A, Clark JH, Pal S (2008) Modification of Okra mucilage with acrylamide: synthesis, characterization and swelling behavior. Carbohydr Polym 72:608–615

Mittal H, Maity A, Ray SS (2015) Effective removal of cationic dyes from aqueous solution using gum ghatti-based biodegradable hydrogel. Int J Biol Macromol 79:8–20

Mittal N, Mattu P, Kaur G (2016) Extraction and derivatization of Leucaena leucocephala (Lam.) galactomannan: optimization and characterization. Int J Biol Macromol 92:831–841

Mohamed AA, Mahmoud GA, ElDin ME, Saad EA (2020) Synthesis and properties of (gum acacia/polyacrylamide/SiO2) magnetic hydrogel nanocomposite prepared by gamma irradiation. Polym Plast Technol Mater 59:357–370

Morkhade DM, Joshi SB (2007) Evaluation of gum Damar as a novel microencapsulating material for ibuprofen and diltiazem hydrochloride. Indian J Pharm Sci 69:263

Morkhade DM, Fulzele SV, Satturwar PM, Joshi SB (2006) Gum copal and gum Damar: novel matrix forming materials for sustained drug delivery. Indian J Pharm Sci 68:53

Mostafavi FS, Zaeim D (2020) Agar-based edible films for food packaging applications-a review. Int J Biol Macromol 159:1165–1176

Moyo B, Masika PJ, Hugo A, Muchenje V (2011) Nutritional characterization of Moringa (Moringa oleifera Lam.) leaves. Afr J Biotechnol 10:12925–12933

Mu RJ, Yuan Y, Wang L, Ni Y, Li M, Chen H, Pang J (2018) Microencapsulation of Lactobacillus acidophilus with konjac glucomannan hydrogel. Food Hydrocoll 76:42–48

Najafabadi AS, Naghavi MR, Farahmand H, Abbasi A (2017) Transcriptome and metabolome analysis of Ferula gummosa Boiss. to reveal major biosynthetic pathways of galbanum compounds. Funct Integr Genomics 17:725–737

Namdeo B, Vidya I, Sushilkumar P (2008) Swelling and erosion modulation of sterculia foetida gum through real time texture probing. Dhaka Univ J Pharm Sci 7:127–132

Nawab A, Alam F, Haq MA, Hasnain A (2016) Effect of guar and xanthan gums on functional properties of mango (Mangifera indica) kernel starch. Int J Biol Macromol 93:630–635

Nayak AK, Pal D, Santra K (2014) Development of calcium pectinate-tamarind seed polysaccharide mucoadhesive beads containing metformin HCl. Carbohydr Polym 101:220–230

Nayak AK, Pal D, Santra K (2015) Screening of polysaccharides from tamarind, fenugreek and jackfruit seeds as pharmaceutical excipients. Int J Biol Macromol 79:756–760

Nayyar T, Sabharwal S, Jain UK, Taneja R (2015) Recent approaches and pharmaceutical applications of natural polysaccharides: a review. Int J Pharm Sci Res (IJPSR) 6:4904–4919

Nep EI, Conway BR (2011a) Grewia gum 2: mucoadhesive properties of compacts and gels. Trop J Pharm Res 10:393–401

Nep EI, Conway BR (2011b) Physicochemical characterization of grewia polysaccharide gum: effect of drying method. Carbohydr Polym 84:446–453

Nep EI, Sims IM, Morris GA, Kontogiorgos V, Smith AM (2016) Evaluation of some important physicochemical properties of starch free grewia gum. Food Hydrocoll 53:134–140

Nimrouzi M, Zarshenas MM (2016) Phytochemical and pharmacological aspects of Descurainia sophia Webb ex Prantl: modern and traditional applications. Avicenna J Phytomed 6:266

Nokhodchi A, Nazemiyeh H, Khodaparast A, Sorkh-Shahan T, Valizadeh H, Ford JL (2008) An in vitro evaluation of fenugreek mucilage as a potential excipient for oral controlled-release matrix tablet. Drug Dev Ind Pharm 34:323–329

Noleto GR, Petkowicz CL, Mercê AL, Noseda MD, Méndez-Sánchez SC, Reicher F, Oliveira MB (2009) Two galactomannan preparations from seeds from Mimosa scabrella (bracatinga): complexation with oxovanadium (IV/V) and cytotoxicity on HeLa cells. J Inorg Biochem 103:749–757

Nwokocha LM, Williams PA (2009) Isolation and rheological characterization of Mucuna flagellipes seed gum. Food Hydrocoll 23:1394–1397

Odeku OA (2005) Assessment of Albizia zygia gum as binding agent in tablet formulations. Acta Pharma 55:263–276

Odeniyi MA, Babalola AO, Ayorinde JO (2013) Evaluation of Cedrela gum as a binder and bioadhesive component in ibuprofen tablet formulations. Braz J Pharm Sci 49:95–105

Ogaji IJ, Hoag SW (2011) Effect of Grewia gum as a suspending agent on ibuprofen pediatric formulation. AAPS PharmSciTech 2:507–513

Okoye EI, Edochie C, Adegbemi JO (2014) Preliminary evaluation of Delonix regia seed gum as a suspending agent in a liquid oral dosage form. Int J Pharm Sci Drug Res 6:114–119

Okwu DE, Okoro E (2006) Phytochemical composition of Brachystegia eurycoma and Mucuna flagellipes seeds. Med Aromat Plant Sci Biotechnol 26:1–4

Padayachee B, Baijnath H (2020) An updated comprehensive review of the medicinal, phytochemical and pharmacological properties of Moringa oleifera. S Afr J Bot 129:304–316

Padil VV, Senan C, Černík M (2015) Dodecenylsuccinic anhydride derivatives of gum karaya (Sterculia urens): preparation, characterization, and their antibacterial properties. J Agric Food Chem 63:3757–3765

Pal P, Pandey JP, Sen G (2018) Sesbania gum-based hydrogel as platform for sustained drug delivery: an 'in vitro' study of 5-Fu release. Int J Biol Macromol 113:1116–1124

Pandey R, Khuller GK (2004) Polymer based drug delivery systems for mycobacterial infections. Curr Drug Deliv 1:195–201

Patel S, Goyal A (2015) Applications of natural polymer gum arabic: a review. Int J Food Prop 18:986–998

Patel GC, Patel MM (2009) Preliminary evaluation of sesbania seed gum mucilage as gelling agent. Int J Pharm Tech Res 1:840–843

Pawar H, Varkhade C, Jadhav P, Mehra K (2014) Development and evaluation of orodispersible tablets using a natural polysaccharide isolated from Cassia tora seeds. Integr Med Res 3:91–98

Phimolsiripol Y, Siripatrawan U, Henry CJ (2011) Pasting behaviour, textural properties and freeze–thaw stability of wheat flour–crude malva nut (Scaphium scaphigerum) gum system. J Food Eng 105:557–562

Prajapati PA, Poddar SS, Patel MM, Patel BK (2010) Ophthalmic Mini-tablet with natural polymer: Sterculia foetida gum. Pharm Lett 2:467–474

Pramualkijja T, Pirak T, Kerdsup P (2016) Effect of salt, rice bran oil and malva nut gum on chemical, physical and physico–chemical properties of beef salt–soluble protein and its application in low fat salami. Food Hydrocoll 53:303–310

Pugh N, Ross SA, ElSohly MA, Pasco DS (2001) Characterization of Aloeride, a new high-molecular-weight polysaccharide from Aloe vera with potent immunostimulatory activity. J Agric Food Chem 49:1030–1034

Raghu K, Naidoo Y, Dewir YH (2019) Secretory structures in the leaves of Hibiscus sabdariffa L. S Afr J Bot 121:16–25

Raj Sharma B, Kumar V, Soni PL (2002) Ceric ammonium nitrate-initiated graft copolymerization of acrylamide onto Cassia tora gum. J Appl Polym Sci 86:3250–3255

Raja W, Bera K, Ray B (2016) Polysaccharides from Moringa oleifera gum: structural elements, interaction with β-lactoglobulin and antioxidative activity. RSC Adv 6:75699–75706

Rambabu K, Bharath G, Banat F, Show PL, Cocoletzi HH (2019) Mango leaf extract incorporated chitosan antioxidant film for active food packaging. Int J Biol Macromol 126:1234–1243
Rana V, Rai P, Tiwary AK, Singh RS, Kennedy JF, Knill CJ (2011) Modified gums: approaches and applications in drug delivery. Carbohydr Polym 83:1031–1047
Rani P, Khullar N (2004) Antimicrobial evaluation of some medicinal plants for their anti-enteric potential against multi-drug resistant Salmonella typhi. Phytother Res 18(8):670–673
Rani P, Sen G, Mishra S, Jha U (2012) Microwave assisted synthesis of polyacrylamide grafted gum ghatti and its application as flocculant. Carbohydr Polym 89:275–281
Rao KM, Kumar A, Rao KS, Haider A, Han SS (2018) Biodegradable tragacanth gum-based silver nanocomposite hydrogels and their antibacterial evaluation. J Polym Environ 26(7):78–88
Rastogi G, Bhalla A, Adhikari A, Bischoff KM, Hughes SR, Christopher LP, Sani RK (2010) Characterization of thermostable cellulases produced by Bacillus and Geobacillus strains. Bioresour Technol 101:8798–8806
Ratcliffe I, Williams PA, Viebke C, Meadows J (2005) Physicochemical characterization of konjac glucomannan. Biomacromolecules 6:1977–1986
Rauf F, Qureshi R, Shaheen H (2012) Folk medicinal uses of indigenous plant species of Barroha, Bhara Kahu and Maanga in Islamabad, Pakistan. J Med Plant Res 6:2061–2070
Razavi SM, Mohammadi Moghaddam T, Mohammad Amini A (2008) Physicomechanic and chemical properties of Balangu seed. Int J Food Eng 4:1–2
Reddy MR, Manjunath K (2013) Pharmaceutical applications of natural gums, mucilages and pectins-a review. Int J Pharm Chem Sci 2:1233–1239
Reddy GB, Madhusudhan A, Ramakrishna D, Ayodhya D, Venkatesham M, Veerabhadram G (2015) Green chemistry approach for the synthesis of gold nanoparticles with gum kondagogu: characterization, catalytic and antibacterial activity. J Nanostruct Chem 5(2):185–193
Rezaeian R, Peyda M, Mohammadian Fazli M (2016) Optimization of extraction method of the natural coagulant from Descurainia Sophia seed, minimization of color generation. J Human Environ Health Promot 1:166–171
Richardson DM, Manders PT (1985) Predicting pathogen-induced mortality in Hakea sericea (Proteaceae), an aggressive alien plant invader in South Africa. Ann Appl Biol 106:243–254
Saif MM, Kumar NS, Prasad MN (2012) Binding of cadmium to Strychnos potatorum seed proteins in aqueous solution: adsorption kinetics and relevance to water purification. Colloids Surf B: Biointerfaces 94:73–79
Saleem R, Ahmad SI, Ahmed M, Faizi Z, Zikr-ur-Rehman S, Ali M, Faizi S (2003) Hypotensive activity and toxicology of constituents from Bombax ceiba stem bark. Biol Pharm Bull 26(1):41–46
Samavati V, Skandari F (2014) Recovery, chemical and rheological characterization of gum from Assyrian pulm. Int J Biol Macromol 67:172–179
Saravanan P, Vinod VT, Sreedhar B, Sashidhar RB (2012) Gum kondagogu modified magnetic nano-adsorbent: an efficient protocol for removal of various toxic metal ions. Mater Sci Eng 32:581–586
Sarojini S, Kunam DS, Manavalan R, Jayanthi B (2010) Effect of natural almond gum as a binder in the formulation of diclofenac sodium tablets. Int J Pharm Sci Res (IJPSR) 1:55–60
Saruchi KV, Rehani V, Kalra GS (2018) Synthesis and characterization of Gum tragacanth, acrylic acid–methacrylic acid based interpenetrating polymer network: its application for controlled release of agrochemicals. J Chin Adv Mater Soc 6:222–233
Sarwar M, Patra JK, Ali A, Maqbool M, Arshad MI (2020) Effect of compost and NPK fertilizer on improving biochemical and antioxidant properties of Moringa oleifera. S Afr J Bot 129:62–66
Seeli DS, Prabaharan M (2016) Guar gum succinate as a carrier for colon-specific drug delivery. Int J Biol Macromol 84:10–15
Seeli DS, Prabaharan M (2017) Guar gum oleate-graft-poly (methacrylic acid) hydrogel as a colon-specific controlled drug delivery carrier. Carbohydr Polym 158:51–57
Semalty M, Semalty A, Kumar G (2008) Formulation and characterization of mucoadhesive buccal films of glipizide. Indian J Pharm Sci 70:43

Sethurajan AK, Ravichandran A, Thangamani SR, Muthukumar K (2011) Adsorptive removal of Congo red dye using ultrasonically pretreated Strychnos potatorum seed powder. Indian Chem Eng 53:1–7
Setia A, Goyal S, Goyal N (2010) Applications of gum karaya in drug delivery systems: a review on recent research. Pharm Lett 2:39–48
Shamsnejati S, Chaibakhsh N, Pendashteh AR, Hayeripour S (2015) Mucilaginous seed of Ocimum basilicum as a natural coagulant for textile wastewater treatment. Ind Crop Prod 69:40–47
Shankar NB, Kumar NU, Balakrishna PK, Kumar RP (2008a) Design and evaluation of controlled release bhara gum microcapsules of famotidine for oral use. Res J Pharm Technol 1:433–436
Shankar NB, Kumar NU, Balakrishna PK, Kumar RP (2008b) Preparation and in vitro evaluation of lamivudine entrapped MOI microspheres for oral administration. Res J Pharm Technol 1:437–440
Sharma BR, Kumar V, Soni PL (2003) Carbamoylethylation of Cassia tora gum. Carbohydr Polym 54:143–147
Sharma K, Kaith BS, Kumar V, Kumar V, Som S, Kalia S, Swart HC (2013) Synthesis and properties of poly (acrylamide-aniline)-grafted gum ghatti based nanospikes. RSC Adv 3:25830–25839
Sharma DR, Sharma A, Kaundal A, Rai PK (2016) Herbal gums and mucilage as excipients for pharmaceutical products. Res J Pharm Phytochem 8:145–152
Sharma S, Virk K, Sharma K, Bose SK, Kumar V, Sharma V, Focarete ML, Kalia S (2020) Preparation of gum acacia-poly (acrylamide-IPN-acrylic acid) based nanocomposite hydrogels via polymerization methods for antimicrobial applications. J Mol Struct 21:128298
Shukla RK, Tiwari A (2012) Carbohydrate polymers: applications and recent advances in delivering drugs to the colon. Carbohydr Polym 88:399–416
Shukla MK, Singh RP, Reddy CR, Jha B (2012) Synthesis and characterization of agar-based silver nanoparticles and nanocomposite film with antibacterial applications. Bioresour Technol 107:295–300
Silva F, Torres L, Silva L, Figueiredo R, Garruti D, Araújo T, Duarte A, Brito D, Ricardo N (2018) Cashew gum and maltodextrin particles for green tea (Camellia sinensis var Assamica) extract encapsulation. Food Chem 30:169–175
Singh B (2007) Psyllium as therapeutic and drug delivery agent. Int J Pharm 334:1–4
Singh P, Agrawal T (2018) Albizia lebbeck (L.) Benth.: a short review. Int J Sci Res Chem Sci 5:8–11
Singh B, Sharma N (2008) Development of novel hydrogels by functionalization of sterculia gum for use in anti-ulcer drug delivery. Carbohydr Polym 74:489–497
Singh B, Varshney L, Francis S (2016) Designing tragacanth gum based sterile hydrogel by radiation method for use in drug delivery and wound dressing applications. Int J Biol Macromol 88:586–602
Singh Y, Sharma A, Tiwari S, Singla A (2018) Optimization of diesel engine performance and emission parameters employing cassia tora methyl esters-response surface methodology approach. Energy 168:909–918
Sonawane A, Pathak SS, Pradhan RC (2020) Physical, thermal, and mechanical properties of bael fruit. J Food Process Eng 20:13393
Sonia S, Julianti E, Ridwansyah R (2019) The characteristic of Taro flour based pasta with addition of modified starch and hydrocolloids. Indones Food Nutr Prog 16:27–35
Soukoulis C, Gaiani C, Hoffmann L (2018) Plant seed mucilage as emerging biopolymer in food industry applications. Curr Opin Food Sci 22:28–42
Srichamroen A (2018) Effect of extracted malva nut gum on reducing high glucose levels by Caco-2 cells. Food Biosci 21:107–116
Srichamroen A, Chavasit V (2011) Rheological properties of extracted malva nut gum (Scaphium scaphigerum) in different conditions of solvent. Food Hydrocoll 25:444–450
Srinivasan R, Mishra A (2008) Okra (Hibiscus esculentus) and fenugreek (Trigonella foenum graceum) mucilage: characterization and application as flocculants for textile effluent treatments. Chin J Polym Sci 26:679–687

Steffy K, Shanthi G, Maroky AS, Selvakumar S (2017) Synthesis and characterization of ZnO phytonanocomposite using Strychnos nux-vomica L.(Loganiaceae) and antimicrobial activity against multidrug-resistant bacterial strains from diabetic foot ulcer. J Adv Res 9:69–77
Suhane N, Shrivastava RR, Singh M (2016) Gulmohar an ornamental plant with medicinal uses. J Pharm Phytochem 5:245–248
Swami SB, Thakor NJ, Haldankar PM, Kalse SB (2012) Jackfruit and its many functional components as related to human health: a review. Compr Rev Food Sci Food Saf 6:565–576
Swapna R, Shailaja P (2019) Formulation in vitro and in vivo evaluation of Cefuroxime Axetil floating tablets using natural gums. Int J Life Sci Pharma Res 9:37–53
Thombare N, Mishra S, Siddiqui MZ, Jha U, Singh D, Mahajan GR (2018) Design and development of guar gum based novel, superabsorbent and moisture retaining hydrogels for agricultural applications (1997). Carbohydr Polym 85:169–178
Thombre N, Aher A, Shimpi P (2020) Formulation development and evaluation of gum Damar based sustained release matrix tablet of metoprolol succinate. Asian J Pharm Res Dev 8:81–86
Ughini F, Andreazza IF, Ganter JL, Bresolin TM (2004) Evaluation of xanthan and highly substituted galactomannan from M. scabrella as a sustained release matrix. Int J Pharm 271:197–205
Umekar MJ, Yeole PG (2008) Characterization and evaluation of natural copal gum-resin as film forming material. Int J Green Pharm (IJGP) 2:37–42
Velusamy P, Das J, Pachaiappan R, Vaseeharan B, Pandian K (2015) Greener approach for synthesis of antibacterial silver nanoparticles using aqueous solution of neem gum (Azadirachta indica L.). Ind Crop Prod 66:103–109
Venkaiah K, Shah JJ (1984) Distribution, development and structure of gum ducts in Lannea coromandelica (Houtt.) Merril. Ann Bot 54:175–186
Venkatesham M, Ayodhya D, Madhusudhan A, Kumari AS, Veerabhadram G, Mangatayaru KG (2014) A novel green synthesis of silver nanoparticles using gum karaya: characterization, antimicrobial and catalytic activity studies. J Clust Sci 25(2):409–422
Vidyasagar G, Jadhav AG, Narkhede SP, Narkhede SB (2010) Isolation and comparative evaluation of Cordia dichotoma Forst. Mucilage as a binding agent with standard binder. J Chem Pharm Res 24:722–726
Vignesh P, Prakash MA, Ponvel MB, Nisha JN, Priyanka RD, Kirubakaran V (2020) Feasibility study of extraction of bio-oil from Leucaena leucocephala seeds for sustainable habitat development. InAIP Conference Proceeding 2225: 040005
Vinod VT, Sashidhar RB, Suresh KI, Rao BR, Saradhi UV, Rao TP (2008) Morphological, physicochemical and structural characterization of gum kondagogu (Cochlospermum gossypium): a tree gum from India. Food Hydrocoll 22:899–915
Vinod VT, Saravanan P, Sreedhar B, Devi DK, Sashidhar RB (2011) A facile synthesis and characterization of Ag, Au and Pt nanoparticles using a natural hydrocolloid gum kondagogu (Cochlospermum gossypium). Colloids Surf B: Biointerfaces 83:291–298
Viyoch J, Sudedmark T, Srema W, Suwongkrua W (2005) Development of hydrogel patch for controlled release of alpha-hydroxy acid contained in tamarind fruit pulp extract. Int J Cosmet Sci 27(8):9–99
Wang B, Han Y, Lin Q, Liu H, Shen C, Nan K, Chen H (2016) In vitro and in vivo evaluation of xanthan gum–succinic anhydride hydrogels for the ionic strength-sensitive release of antibacterial agents. J Mater Chem 4:1853–1861
Xie Y, Yi ZX, Wang JX, Hou TG, Jiang Q (2018) Carboxymethyl konjac glucomannan-crosslinked chitosan sponges for wound dressing. Int J Biol Macromol 112:1225–1233
Yadav KN, Kadam PV, Patel JA, Patil MJ (2014) Strychnos potatorum: phytochemical and pharmacological review. Pharmacogn Rev 8:61
Yadav M, Chiu FC (2019). Cellulose nanocrystals reinforced κ-carrageenan based UV resistant transparent bionanocomposite films for sustainable packaging applications. Carbohydrate polymers, 211, 181-194

Yong H, Wang X, Bai R, Miao Z, Zhang X, Liu J (2019) Development of antioxidant and intelligent pH-sensing packaging films by incorporating purple-fleshed sweet potato extract into chitosan matrix. Food Hydrocoll 90:216–224
Zegbe JA, Mena-Covarrubias J, Dominguez-Canales VS (2012) Cactus mucilage as a coating film to enhance shelf life of unprocessed guavas (Psidium guajava L.). In: VIII international congress on Cactus pear and cochineal, pp 423–427
Zeynali M, Naji-Tabasi S, Farahmandfar R (2019) Investigation of basil (Ocimum basilicum L.) seed gum properties as Cryoprotectant for frozen foods. Food Hydrocoll 90:305–312
Zhang YQ, Xie BJ, Gan X (2005) Advance in the applications of konjac glucomannan and its derivatives. Carbohydr Polym 60:27–31
Zhang CR, Dissanayake AA, Nair MG (2015) Functional food property of honey locust (Gleditsia triacanthos) flowers. J Funct Foods 18:266–274

Chapter 19
Intelligent Packaging Systems: Food Quality and Intelligent Medicine Box Based on Nano-sensors

Ramachandran Chelliah, Imran Khan, Shuai Wei, Inamul Hasan Madar, Ghazala Sultan, Eric Banan-Mwine Daliri, Caroline Swamidoss, and Deog Hwan Oh

19.1 Introduction

To create better materials and products, nanostructure utilizes nanoscale design and device integration of existing materials. In numerous areas such as healthcare, grafts, prostheses, smart fabrics, power generation and maintenance with

R. Chelliah (✉) · E. B.-M. Daliri · D. H. Oh
Department of Food Science and Biotechnology, College of Agriculture and Life Sciences, Kangwon National University, Chuncheon, Gangwon-do, South Korea

I. Khan
Department of Food Science and Technology, The University of Haripur, Haripur, Khyber Pakhtunkh, Pakistan

S. Wei
College of Food Science and Technology, Guangdong Ocean University, Guangdong Provincial Key Laboratory of Aquatic Products Processing and Safety, Guangdong Province Engineering Laboratory for Marine Biological Products, Guangdong Provincial Engineering Technology Research Center of Marine Food, Key Laboratory of Advanced Processing of Aquatic Product of Guangdong Higher Education Institution, Zhanjiang, China

Collaborative Innovation Center of Seafood Deep Processing, Dalian Polytechnic University, Liaoning, China

I. H. Madar
Department of Biomedical Science and Environmental Biology, Kaohsiung Medical University, Kaohsiung, Taiwan

G. Sultan
Department of Computer Science, Faculty of Science, Aligarh Muslim University, Aligarh, Uttar Pradesh, India

C. Swamidoss
Caroline Mercy Andrew Swamidoss, Department of Chemistry, Dr. M.G.R. Educational and Research Institute, Chennai, India

J.-C. Kim et al. (eds.), *Smart Nanomaterials in Biomedical Applications*, Nanotechnology in the Life Sciences,
https://doi.org/10.1007/978-3-030-84262-8_19

power-producing constituents, and extremely well-organized arrays, such as defense, protection, sabotage, and shadowing, nanomaterial applications have ended their presence intensely sensed (Zhu et al. 2015). The study of bio-nanomaterial has developed as an incredible new field, known as a transdisciplinary borderline in the arena of material science and medical field (Siddiqi and Husen 2017; Husen 2019, 2020; Joshi et al. 2019; Husen and Iqbal 2019; Siddiqi et al. 2018; Painuli et al. 2020; Sharma et al. 2020; Bachheti et al. 2019, 2020a, b, c, 2021). In industrial, security, and clinical medicine applications, great developments in nano-biochip materials, nanoscale biomimetic materials, nano-motors, nanocomposite materials, interface biomaterials, nano-biosensors, and nano-drug delivery systems have significant prospects. For example, biomolecular tools applied in genetic engineering and diagnosis have vast pharmaceutical potential for biosensor growth. Biomolecules play a key part in nanoscience. The biosensor consists of a substrate for biosensing and a transducer which can be used to identify biochemical agents. The target analytes are specifically recognized by biosensing materials such as proteins, antibodies, and DNA/RNA probes, while devices can quantitatively track the biochemical processes outlined in Fig. 19.1.

Singh et al. (2017) have documented a disposable biosensor for rapid determination of not only H2O2 but also azide using biotic materials (polyaniline (PANI)) based on bio-electrode. This film does an excellent job of maintaining enzyme activity and preventing leakage. This film is very good at preserving the activity of enzymes and eliminating its leakage from the film (Fig. 19.2).

A new field of interdisciplinary research has become biosensors. In combination with various transmitters, the bio sensing mechanism were applied towards the separation of numerous types of embattled bio-macromolecules, which inherit both advantage and disadvantages (Uludag et al. 2016). Moreover for diverse products in biological analysis and medical diagnostics, the nano-biosensor can be effortlessly

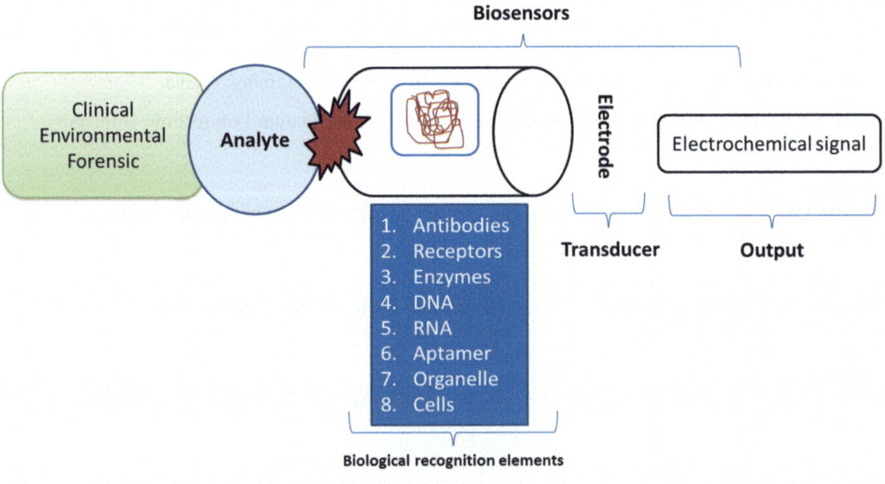

Fig. 19.1 Indicates the fundamental theory of biosensor

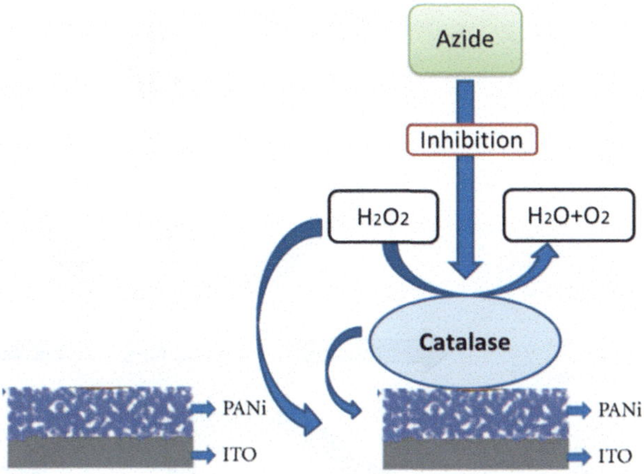

Fig. 19.2 Diagrammatic representation of bio-electrode in the improvement of biosensor for the hydrogen peroxide grit in optimal condition

combined into a single-use polymer-based chip. The recent developments in nano-biosensors and their application in ecology, specifically in healthcare diagnosis, which include the perceptions of synchronized nano-biosensors that incorporate the desired possessions of the distinct constituents: susceptibility and specificity on accuracy polymer machinery and chemistry of peptides or nucleic acid to align the different electron transducing units and nano-electro transducing units. The findings of these systems concentrate on the possible benefits of using nano-biosensors, which would dramatically revolutionize medical diagnosis and managements. Thus in biology and medicine, the creation and application of nano-devices would have tremendous consequences for the good of humanity and healthcare. Biosensors compressed of different field of application such molecular nano-biology, microfluidics, and nanomaterials are used for the rapid, specific, responsive, affordable, in-field, online and/or real-time diagnosis and prevention of chemicals, drugs, microbes, toxins, proteins, bacteria, plants, livestock, foods, in agronomic production, nourishment processing, technical care and the ecology. Ultrasensitive biosensors therefore deliver an outstanding analytical instrument for monitoring systems that make it possible to monitor the biosphere. The structure of bio-electronics, developed based on nano-based molecular technology, which have an outstanding effect on the progress of a novel biosensing stage to solve the forthcoming medical diagnosis and report the tasks related to pollution problems towards human healthcare. There is a current interest in understanding genetic progressions at the solitary fragment level and applying them for potential forthcoming uses in nano-biotechnology. Single molecule-enabled probe studies have opened up exciting research avenues, particularly in nanoscience, for the improvement of versatile biomolecule-based (Morales-Narváez and Merkoçi 2019) technology.

Due to variation on their exterior to capacity ratio, surface characteristics, and ionic conductivity, as well as excellent stability, nanomaterials have enabled the production of ultrasensitive biosensors. They were applied towards targeted delivery and wiring of metabolites to the surface of the electrode, to facilitate biochemical response and to transmit a gesture of proceedings of bio-recognition. Nanomaterials, including nanoparticles, nanowire, nano-needle, nano-sheet, nanotube, nano-rod, and nano-belt, for biosensing have been documented in several

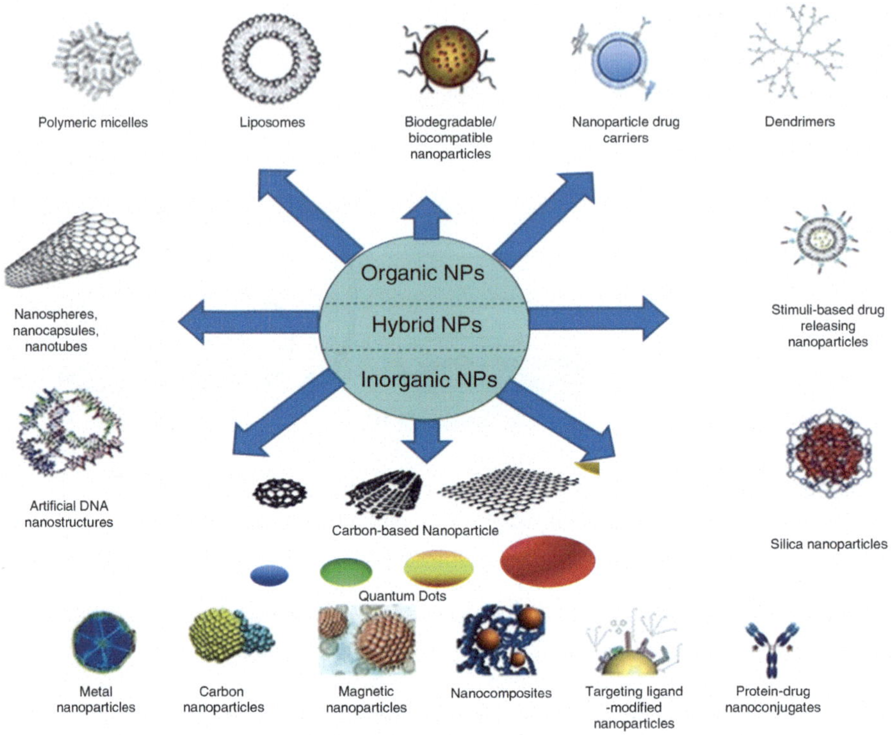

Fig. 19.3 Schematic illustration of different types of nanomaterials used in biomedicine

research papers (Sadik et al. 2009) as shown in representative Fig. 19.3. Application of modified carbon nanotube (CNT) electrodes, gold nanoparticles for ionic, immune-sensors, nanostructured sensors were applied based on Low-potential NADHA (Munge et al. 2011; Jiang et al. 2014), biosensing based on nanoparticles (Loiseau et al. 2019), and nanowire as sensing materials (Ramgir et al. 2010) are just a few significant examples.

19.2 Nanotechnology-Based Sensors for Food Analysis and Monitoring of Food Security

Better health is crucial to human well-being and happiness and makes a significant contribution to economic progress (WHO. 2020). Therefore, the world is focused on the issues of food safety (Wang et al. 2013). Several well-known food detection technologies are already being exploited for detection such as sensor methods, chromatography, chemical detection methods, immune methods, enzyme hydrolysis, and biological methods (Heo et al. 2012; Jadzinsky et al. 2007; Peng et al. 2012). However, these methods need tedious and time-consuming processes for detection and therefore limit their application in foods. With the advent of nanoscience and nanotechnology, an alternative rapid test becomes possible (Mohanty et al. 2009; Viswanath 2004). Nano-sensors allow safety assurance in food processing chain through data curation (Neethirajan and Jayas 2011). Solicitations of nano-sensors in the domain of food sector are discussed below.

19.2.1 Nutrients (Antioxidants and Sugars)

Nano-sensors have been applied in food industry to detect vitamins and antioxidant components in foods as they are easily degraded during food processing and storage. Folic acid (FA) is an important component of the hematopoietic system, and absence of FA gives rise to leucopoenia, gigantocytic anemia, devolution of mentality and psychosis, heart attack, carcinogenesis, etc. (Rao et al. 1978). Multi-walled carbon nanotube (MWNT) and single-walled carbon nanotube (SWCNT)-ionic liquid nanocomposites have been reported for the detection of FA in fruit juices, wheat flour, and milk samples (Wei et al. 2006; Xiao et al. 2008). (Ensafi et al. 2012). Likewise, NiO nanoparticle were towards sensing the ascorbic acid (Vitamin C) (Karimi-Maleh et al. 2014). In another similar study, Taei et al. (2016) designed a MWNT modified with magnetic nanoparticles for the instantaneous grit on ascorbic acid and could be used for pharmaceutical and biological samples. The red wine antioxidant capacity can be attributed to its phenolic contents. Tyrosinase enzyme applied towards immobilized gold-based nanoparticles was developed to identify the quality of phenols (gallic acid, catechol, chlorogenic acid, caffeic acid, and protocatechualdehyde) in white wines (Liu et al. 2003; Sanz et al. 2005). Another important feature of nano-sensors is to determine the fruit maturation and ripening by measuring glucose, sucrose, and ascorbic acid concentrations (Jawaheer et al. 2002).

19.2.2 Toxins

Toxins produced by microorganisms cause a serious health concern around the world. Toxins are generally produced in self-defense by microorganisms (Kumar et al. 2017). Therefore, it is important to develop sensitive and quick methods to detect toxins in foods and related products (Franz et al. 1997; Zhou et al. 2009). Food products contaminated with *Aspergillus parasiticus* and *Aspergillus flavus* produce carcinogenic toxin called aflatoxins. To detect these toxins in food products, gold nanoparticles decorated with anti-aflatoxin antibodies were used. Similarly, to detect aflatoxin M1 in milk sample, Agilent magnetic beads containing anti-aflatoxin M1 antibodies and gold nano-probes have also been used (Dinckaya et al. 2011; Paniel et al. 2010; Radoi et al. 2008; Wang et al. 2011). Seafood generally contaminated with marine toxin is called palytoxin. For the identification of toxin in Pteriomorphia (marine mussels) meat, carbon-based nanotubes-luminous biosensors have been applied (Zamolo et al. 2012).

19.2.3 Adulterants

Food adulteration adversely affects human health and becomes a global health concern. The adulterant present in foods can induce diarrhea and vomiting to paralysis and even death (Kumar et al. 2017). Food adulterant at lower level applied through repetitive recognition system is a competitive work. Melamine is basically compost; however, its uses as contaminant in peptide-enhanced produces, e.g., biscuits, eggs, coffee drinks, and candies, are well documented. Gold-based nanoparticles were designed to sense melamine (at picomolar) in infant formulas and domestic milk products (Vasimalai and John 2013). Sudan I (red dye) was tainted with red pepper powder; it is already recognized as carcinogenic. MWCNT-functionalized biosensors have been used to detect adulteration (Yang et al. 2010).

19.2.4 Residual Veterinary Antibiotics and Pesticides

Chloramphenicol originally isolated from *Streptomyces venezuelae* is cost-effective but a toxic broad-spectrum antibiotic used in bees to cure infection. As a result of continuous use of this antibiotic, honey gets contaminated (Kumar et al. 2017). Nanoparticles produced from polyethylene glycol have been applied for chloramphenicol detection in honey samples (Kara et al. 2013). Xie et al. (2017) designed a hybrid molecularly imprinted polymer with a surface-enhanced Raman spectroscopy (SERS) sensor for the selective detection of chloramphenicol in milk. Nanoparticles-based nano-sensors are already being employed for the detection of pesticides in foods. Among the various pesticides, the most common one is organophosphates (Vamvakaki and Chaniotakis 2007). Nanoparticles fabricated from gold have been used based on fluorometric and colorimetric sensors for the detection of carbamate and organophosphorus pesticides (Lin et al. 2012). Acetamiprid is a pesticide which poses potential health risk to humans. Jokar et al. (2016) designed a biosensor based on aptamer-silver nanoparticles for the detection of acetamiprid. Pesticides residues in agricultural products can also be detected by ligand-free carbon dots sensors (Chang et al. 2017).

19.2.5 Pathogens

The main cause of food spoilage and foodborne diseases is the presence of pathogenic bacteria in foods. The detection of these bacteria can be performed by identifying whole bacterial cell or bacterial deoxyribonucleic acid (DNA) (Ahmed et al. 2014). Nanoparticles-assisted DNA isolation and detection of bacteria has shown promising results as it was more sensitive and less time-consuming as compared to other conventional procedures (Kumar et al. 2017). Banerjee et al. (2016) and

Banerjee et al. (2017) designed multiparametric magneto-fluorescent nano-sensors which can detect *Escherichia coli* O157:H7 contamination with as little as 1 colony-forming unit present in solution within less than 1 h. Magnetic iron oxide nanoparticles were used to isolate DNA from pathogenic *Listeria monocytogenes* bacterium in milk samples and were further quantified by polymerase chain reaction (PCR) (Yang et al. 2007). Varshney et al. (2005) designed magnetic nanoparticle-antibody conjugates and reported that the magnetic nanoparticles showed higher capture efficiency of *E. coli* O157:H7 in ground beef samples. The 16 s ribosomal ribonucleic acid (rRNA) is mainly used as a selective marker for PCR-based microbial detection. However, 16 s rRNA microarray method is expensive and lacks sensitivity, while the nanoparticles-based detection method is easy and sensitive (Call et al. 2003; Joung et al. 2008).

19.2.6 Heavy Metals

Heavy metal ions (Pb^{2+}, Cd^{2+}, Hg^{2+}, Ag^{+}, and Cu^{2+}) present in food resources and water have threatened human safety and environment. Ratiometric sensors for detecting Cu(II) for vegetable and fruit samples have been developed by integrating carbon dots with organic fluorescent dyes, quantum dots, and rare earth metal ions (Rao et al. 2016). Babar et al. (2019) designed a cost-effective gold nano-textured electrode for the selective detection of arsenic in food chain and water. This electrochemical gold nano-textured electrode is also applicable for arsenic detection in a complex system containing Fe^{2+}, Pb^{2+}, Hg^{2+}, Cu^{2+}, Ni^{2+}, and other ions for the selective and sensitive analysis. In a recent study, optical dual-mode nano-sensors based on gold nanoparticles and carbon dots were able to visibly detect arsenic (iii) in water (Li et al. 2020). A turn-on nano-sensor was designed for the detection of Hg^{2+} in tap water and milk samples based on the fluorescence resonance energy transfer between long-strand aptamers-functionalized upconversion nanoparticles and short-strand aptamers-functionalized gold nanoparticles and showed good selectivity and precision (Liu and Corma 2018).

19.3 Application of Nanomaterials in Active and Functional Packaging for Smart Packaging Concepts

With the increasing demand for the quality, safety, and the shelf-life of food from consumers and food industries, effective and innovative food packaging such as smart packaging is gradually taking place with the development of nanotechnology (Biji et al. 2015; Mlalila et al. 2016), which could extend and implement all the principles of packaging and provide novel material to help improve mechanical, barrier, and antimicrobial properties as well as monitor the food during transport and storage (Silvestre et al. 2011). Smart packaging could utilize different kinds of sensors including chemical sensors or biosensors to monitor the quality and safety of the packaged products such as food, pharmaceuticals, or health and household products via testing freshness, microbials, leakages, environmental gas composition, pH, time or temperature (Kuswandi et al. 2011; Schaefer and Cheung 2018), which is divided into active packaging and intelligent packaging (Kerry et al. 2006).

19.3.1 Time-Temperatures Indicators (TTI)

Intelligent packaging monitors the product quality and environment conditions during the distribution, transportation, and storage (Biji et al. 2015). Time-temperature indicators (TTI) are simple, cost-efficient and easy to use for monitoring the effect of temperature on quality and safety from farm to consumers in real time (Giannakourou et al. 2005; Pereira et al. 2015). Various food categories including vegetables (Bobelyn et al. 2006), meat (Kim et al. 2013), milk (Lu et al. 2012), and marine products (Myung-Kee et al. 2018) under chilling or frozen condition were monitored and evaluated by TTI, also named external indicators since they are attached outside the package.

19.3.2 Antimicrobial Active Packaging

Antimicrobials inside the package showed efficient inhibition for the growth of spoilage microorganisms and pathogens, including organic acids, bacteriocins, enzymes, and polysaccharides (Cruz-Romero et al. 2013). The incorporation of antimicrobials and active packaging system provide potential applications of novel packaging materials for shelf life prolonging and quality and safety improvement. Further, nano-sized antimicrobials showed higher antimicrobial properties when compared with normal equivalents (Cruz-Romero et al. 2013). Volatile antimicrobials are also effective with advantage of penetrating most of the food without direct contact with food, which was applied in ground beef package (Nadarajah et al. 2005).

19.3.3 Active Packaging Incorporating Gas Scavengers

Active packaging changes the internal condition of the package for shelf life extension and quality and safety improvement via different strategies such as oxygen scavengers, carbon dioxide absorbers and emitters, and ethylene scavengers (Ahvenainen 2003; Mahieu et al. 2015). The oxygen or carbon dioxide indicators belong to internal indicators since they are inside the package. Oxygen scavengers could help maintain the product quality through absorbing the residual oxygen after packaging, and different chemical principles such as iron powder oxidation, ascorbic acid oxidation, and photosensitive dye oxidation are incorporated in commercial oxygen scavengers, which are easy to see by visual color changes after oxidation (Biji et al. 2015).

It can be applied as single or combined with modified atmosphere packaging. Carbon dioxide may be added inside the package to inhibit the microbial growth and reduce the respiration rate of certain product or removed by carbon dioxide scavengers in fresh roasted coffees (Biji et al. 2015; Lopez-Rubio et al. 2004).

19.3.4 Smart and Intelligent Packaging

As expected, smart packaging will reach $ 26.7 billion in the global market by 2024, and it provides promising opportunities based on digitization and shows potential applications in Industry 4.0 (Schaefer and Cheung 2018). Based on the development of sensors and materials to indicate the product quality, safety, and shelf life, the needs of consumer and industry will be matched by combinations of one or more types of smart packaging.

19.4 Intelligent Medicine Packaging

The most prominent feature of the technologies towards the Internet of Things is the ability to be omnipresent. Patients who are old or with serious diseases cannot follow their doctor's instructions resulting in a non-compliance of medication intake. Doctors often do not detect this problem, and the abuse of the prescription can lead to serious and life-threatening complications. In event of this, an intelligent medicine packaging system is the need of the hour to dispense and remind patients and to track the bestowed medication as well (Hechtman 2018). Hence, the RFID technology is gaining popularity in recent years to cater to all these growing demands (Odważny et al. 2018). In an ideal scenario, it should be able to reach every item in the complete framework (Al-Khafajiy et al. 2018). The system should be designed in a way that it is able to record the medicinal activity of each tablet, track each medicine package, and then also provide the patient with the complete prescription information (Hanina et al. 2018).

In order to achieve this, the smart device should have an intelligent and an interactive packaging. It should be able to execute all the benefits of IoT naturally and be reasonably priced too. Bearing this in mind, the interactive and intelligent medicine packaging (Fig. 19.4a) is a suitable device that caters to all the needs for this purpose (Siu 2018). This medicine packaging integrates features like communication, sensing, display, and RFID, along with other functions of the basic packaging framework (Odważny et al. 2018). Integrating the display and the paper-based actuators results in the packaging being aware of the customers' presence and also informs them about what is going on (Aydindogan et al. 2018). This packaging will result in path breaking transformations in the corporate world.

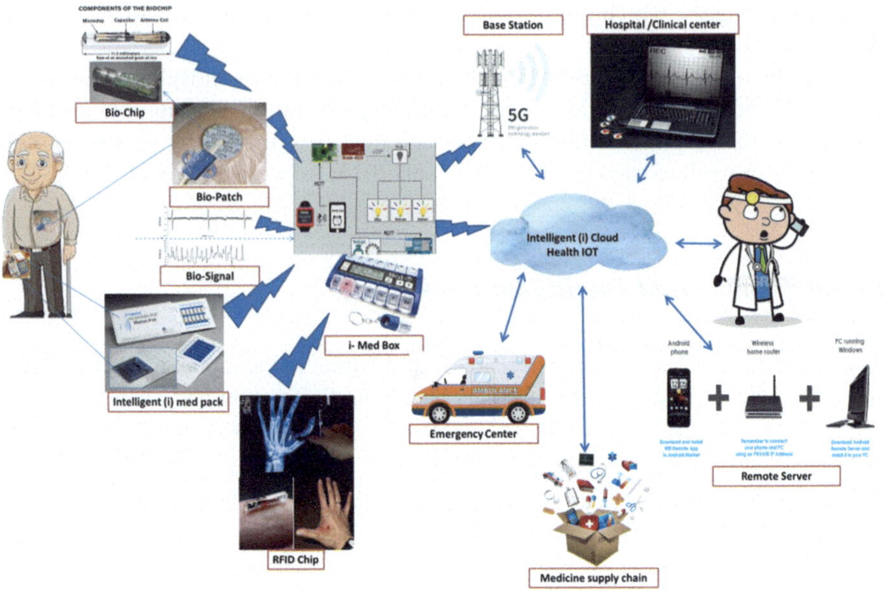

Fig. 19.4a Application scenario for the proposed iHome Health-IoT system

Conventional transformation of information has been static, and that would change into dynamic (Chiang et al. 2018). Traditional flow of information has been unidirectional from the product to the customer, and now it would be bidirectional from both the customer to the product and the product to the customer (Sun et al. 2018). The role of the packaging which was once passive and controlled by the customer would transform to active where it could be either self-controlled by the person or remotely controlled too. Packaging that once had the sole responsibility of containing and protecting the goods can be given more responsibilities and functionalities (Riedel 2018). It assumes the role of a medium of communication between the customer and the manufacturer or supplier. It transforms into a presenter of information, a collector of information, and an on-site seller and also has the ability to execute certain operations.

When we take into account the applications in the health-IoT field, an intelligent and interactive packaging could be accessed electrically through touch sensing. It thus results in offering answers to the management of preventive medication (Siu 2018). The iMedBox sends commands to the respective iMedPack which in turn returns information on the medicine slots opened and intact slots, and consequently patients are given timely reminders to avoid a misuse of the medication (Le et al. 2018).

19.5 . Structural Design of the iMedBox

The system is designed by integrating multiple interfaces together in the iMedBox (Pontes 2018). These include the touch screen, the LCD display, a microphone, speakers, LED lights, vibrators, and a camera (Francois 2018). The following are the distinctive features: It can be connected to the WiFi or 4G/5G communication network for communication, access to the database, and information sharing (Kryszkiewicz et al. 2018) (Fig. 19.4a). It is also connected to the Global Positioning System for support that could be rendered to the exact location. The WBSN feature helps with data collection and star-topology networking (Sodhro et al. 2018). The RFID feature helps with the management of the medicine inventory, its compliance, and CDM controlling.

The architecture of the iMedBox is different from the conventional packages with the inclusion of control circuits and CDM films. The CDM film is made up of a three-layered foil with the top layer made of aluminum, the middle layer made up of an electrochemical epoxy with adhesive properties, and the bottom layer made up of aluminum (Gray et al. 2018) (Fig. 19.4b). When the voltage between the top and bottom layers is greater than the threshold voltage, there is an electrochemical reaction that is instigated in the middle. It is important to consider certain parameters of time and voltage during the manufacture of the material, or else it may result in the epoxy layer getting shattered (Chen and Zhang 2018). The CDM's role is to prevent the function of the intelligent and interactive packaging in case of non-compliance. The adhesive glue on the CDM's middle layer cannot be opened with a human's bare hands and has to only be accessed by sending a command to OPEN to the microchip located in the intelligent package (Karlsson et al. 2018). Thus there is a dedicated part of CDM which is controlled individually for each tablet. This results in an exclusive control of every capsule by the intelligent package (Lee et al. 2018). Consequently, a tablet by tablet compliance is achieved as per the prescription.

The energy required to open the CDM could be provided wirelessly with the near field magnetic resonance, or it could also be self-powered with a battery. Both of these ways are contemporary and will have the desired outcomes (Karlsson et al. 2018). The iTag comprises a WBSN interface, biomedical sensors, batteries, and a low-power MCU (Sodhro et al. 2018). This device has breathtaking advantages in the sensing capabilities and communication distance in comparison with the self-powered and battery-less devices.

This paradigm shift has three aspects, as seen in Fig. 19.4c: (1) Healthcare information systems (HIS) inter-organizational implementation: The information systems (ISs) [87] of all stakeholders participating in the Health-IoT production chain form the backbone of the iHome Health-IoT framework. Cloud computing [87] has provided a viable environment for such inter-organizational convergence against the so-called Health-IoT-in-Cloud. (2) HIS cross-border development: The iHome Health-IoT system's in-home terminal, iMedBox, serves as a connection between in-home healthcare devices and the HIS. The stakeholders' ISs can be efficiently expanded to a patient's home by installing individual applications in the iMedBox.

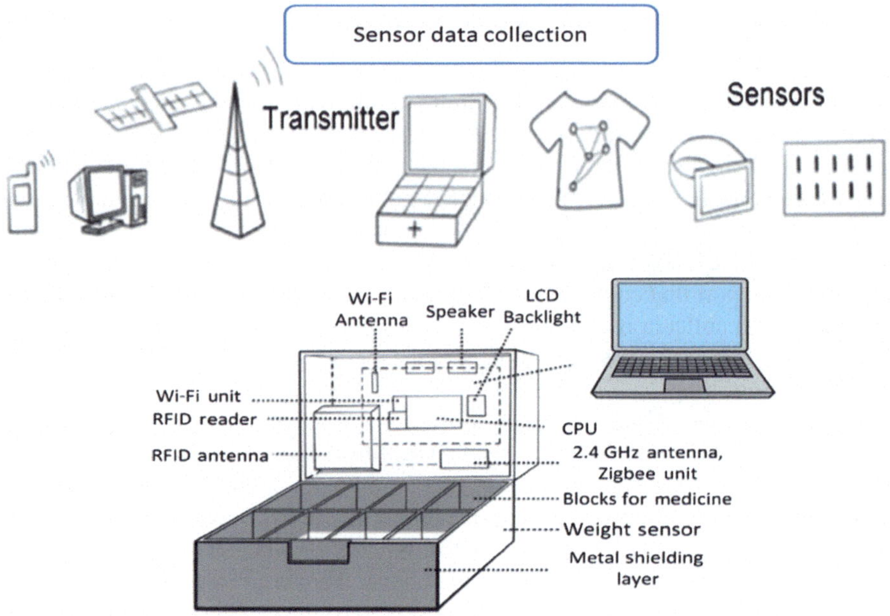

Fig. 19.4b Architectural design of the iMedBox

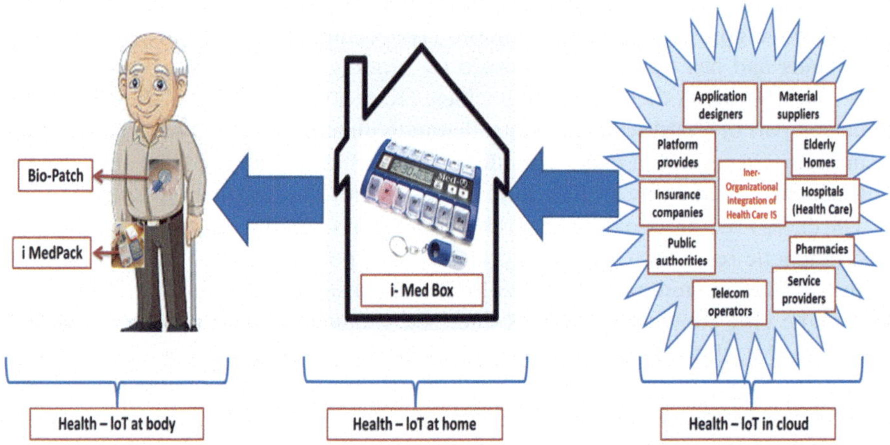

Fig. 19.4c Integrated HIS enabled by the proposed iHome Health-IoT system

As a result, the IHIS exists in the form of both businesses and consumers' homes as the so-called Health-IoT-at-Home. This is in line with the previously stated pattern of medical services shifting from hospital-based to home-based [85]. (3) HIS personalization: Personalized programs will be the predominant method of healthcare in the future. Wearable biomedical applications with ultra-low power and low cost, such as Bio-Patch, allow personalized HIS access to patients' bodies, resulting in the so-called Health-IoT-on-Body. As a result, HIS knowledge and contact can be handled at the level of a single individual's body.

We created the Health-IoT platform with the current and future relevance of IHIS and IoT in the e-health sector in mind. It can be used in patient's homes and nursing homes. The proposed device combines SoC technology, material technology, and advanced printing technology to create a patient-centric, self-assisted, fully automated intelligent in-home healthcare solution. Environmental surveillance, vital sign acquisition, drug control, and healthcare facilities are only a few of the cases where the established functions can be used. The creation of several smart devices, including Bio-Patch, iMedPack, and iMedBox, to realize the vision of iHome Health-IoT is described in this article.

19.6 Unobtrusive Biosensor

The fundamental target of unpretentious detecting is to empower persistent checking of actual exercises and practices, just as physiological and biochemical boundaries during the everyday life of the subject. The most normally estimated imperative signs include ECG, ballistocardiogram (BCG), pulse, circulatory strain (BP), blood oxygen immersion (SpO2), center/surface internal heat level, act, and actual exercises (Zhang and Poon 2013). Unpretentious detecting can be actualized in two different ways: (1) sensors are worn by the subject, e.g., as shoes, eyeglasses, earring, apparel, gloves, and watch, and (2) sensors are installed into the encompassing climate or as savvy objects associating with the subjects, e.g., a seat (Wu et al. 2006), (Baek et al. 2012), vehicle seat (Walter et al. 2011), sleeping pad (Gu et al. 2009), reflect (Poh et al. 2011a, b), controlling wheel (Gomez-Clapers and Casanella 2012), mouse (Lin et al. 2011), latrine seat (Ko Keun et al. 2004), and restroom scale (Inan et al. 2012).

Data can be gathered by a cell phone and communicated remotely to a distant place for capacity and examination. In the accompanying area, we will talk about some inconspicuous detecting techniques for the securing of fundamental signs. Figure 19.5 gives a diagrammatic representation of the distinctive impact classifications and how they identify with organ action just as to the deliberate crucial signs (Leonhardt et al. 2018).

19.6.1 Capacitive Sensing Method

Capacitance-coupled detecting strategy is ordinarily utilized for estimating biopotentials, for example, ECG, electroencephalogram (EEG), and electromyogram (EMG) (Baek et al. 2012; Chi et al. 2009). The skin and the terminal structure, which compresses the two layers of a capacitor without a direct contact on body, which helps in sensing minor issues, such as skin disease. Some run of the mill

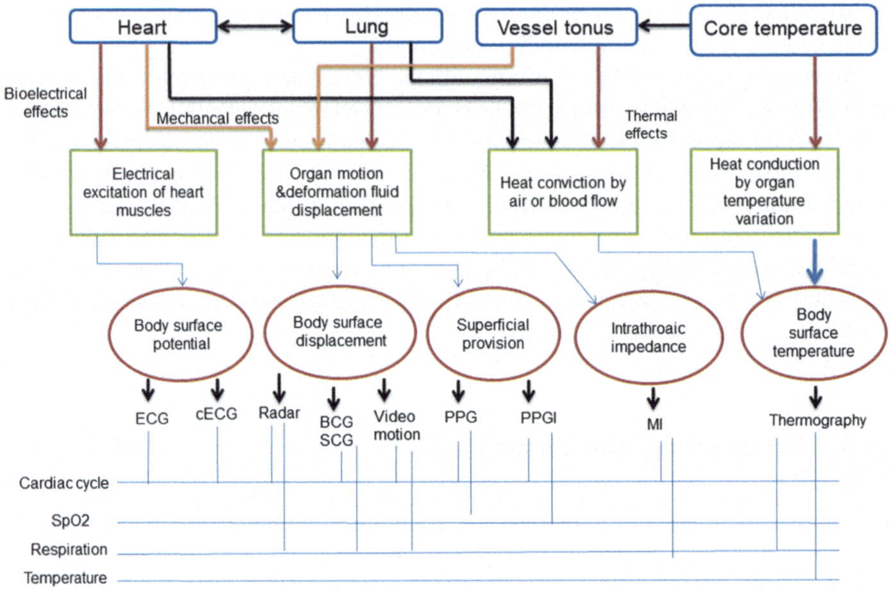

Fig. 19.5 Unobtrusive biosensor impact classification and health monitoring system (point-of-care (POC) diagnostics play an important role in delivering healthcare, particularly for clinical management in different organ and surveillance). Brief sensor platform for health monitoring. A major distinction can be made between non-invasive and invasive applications, including wearable sensors for monitoring biophysical, biochemical, and environment signals and implantable devices for the nervous, cardiovascular, digestive, and locomotors system

Table 19.1 Various usages of capacitive ECG sensors

Systems	Area of ECG electrodes	Measured signals and parameters
BP monitoring chair	Chair pad and arms	ECG,PPG,HR,BP
Non-contact chair-based system	Chair back	ECG,PPG,HR,BP
Sleeping bed	Bed pad	ECG,PPG,HR,BP
Wearable ECG systems	Cloth and bed	ECG
Aachen smart chair	Chair backrest and pad	ECG
Ambulatory ECG monitoring over cloth	Integrated on underwear	ECG
Textile integrated long-term ECG monitor	Incorporated into piece of clothing	ECG
Non-contact ECG/EEG electrodes	Implanted inside texture and attire	EEG,ECG
Remote wearable ECG sensor	Unified into cotton T-shirt	ECG

executions of capacitive ECG detecting are summed up in Table 19.1; capacitive detecting can additionally be utilized for different applications, for example, respiratory estimation, e.g., utilizing a capacitive material power sensor meshed into attire (Hoffmann et al. 2011), or a capacitive electrical field sensor cluster put under a dozing sleeping pad (Wartzek et al. 2011). The significant difficulties in planning these non-contact terminals lie in the high contact impedance because of the roundabout contact and the capacitive befuddle brought about by movement ancient rarities (Chi et al. 2009). It might make low flag clamor proportion and in this manner lead to difficulties in the frontend simple circuit plan. The information impedance of the speaker must be amazingly enormous (>1 Tω) to decrease the shunt impact framed by the capacitor and the information impedance. In addition movement antiquities may turn out to be huge in noncontact detecting. A few strategies have as of late been proposed to overcome these issues to accomplish strong estimation in commonsense circumstances. For example, the gradiometric estimation procedure presented by Ache et al. can significantly decrease movement antiques (GuoChen and Bocko 2013).

19.6.2 Photoplethysmographic Sensing Method

Photoplethysmographic (PPG) detecting, which includes a light source to discharge light into tissue and a photograph locator to gather light reflected from or sent through the tissue, has been generally utilized for the estimation of numerous fundamental signs, for example, SpO2, pulse, pulse rate, and BP. The sign estimated by this strategy speaks to the pulsatile blood volume changes of fringe microvasculature incited by pressure beat inside each heart cycle. Customarily, the detecting unit is in direct contact with skin. Ongoing examination has indicated that sensors can be coordinated into everyday living extras or contraptions like earring, glove, and cap, to accomplish unpretentious estimation. Different kinds of PPG estimating gadgets at various locales of the body are summed up in Table 19.2. As of late, Jae et al. (2009) proposed a backhanded contact sensor for PPG estimation over attire. A control circuit was embraced to adaptively change the light power for different sorts of clothing. Then again, Poh et al. (2011a, b) indicated that pulse and breath rate can be gotten from PPG that was distantly caught from a subject's face utilizing a basic computerized camera. In any case, the fleeting goal of the blood volume distinguished by this technique is limited by the example pace of the camera (up to 30 casings for every second), along these lines influencing its exactness.

19.6.3 Model-Based Cuffless BP Measurement

Sphygmomanometer, which has been utilized longer than a century for BP estimation, works dependent on an inflatable sleeve. Customary strategies, for example, auscultatory, oscillometric, and volume clip are not appropriate for subtle BP measurement. Pulse-wave proliferation strategy is a promising procedure for unpretentious BP estimation. It depends on the connection between beat wave speed (PWV) and blood vessel pressure concurring to Moens-Korteweg condition. Heartbeat

Table 19.2 PPG estimating gadgets at various destinations of the body

Devices	Area of sensor/operation mode	Measured parameters
PPG ring	Finger/reflective	Heart rate variability, SpO2
Pulsear	External ear cartilage/reflective	Heart rate
Forehead mounted sensor	Forehead/reflective	SpO2
e-AR	Posterior and inferior auricular/ reflective	Heart rate
IN-MONIT system	Auditory canal/reflective	Heart activity and heart rate
Glove- and hat-based sensor	Finger and forehead/reflective	Heart rate and pulse wave transit time
Ear-worn monitor	Superior auricular/reflective	Heart rate
Headset	Earlobe/transmissive	Heart rate
Heart-phone	Auditory canal/reflective	Heart rate
Magnetic earing sensor	Earlobe/reflective	Heart rate
Eyeglasses	Nose bridge/reflective	Heart rate and pulse transit time
Ear-worn PPG sensor	Earlobe/reflective	Heart rate
Smartphone	Finger/reflective	Heart rate

travel time (PTT), the complementary of PWV, can be promptly gotten from PPG and ECG in a subtle manner. Different straight and nonlinear models that communicated BP as far as PTT have been created for cuffless BP assessment. The subject-subordinate boundaries in BP-PTT model should be resolved first when utilizing this approach for BP estimation. A very simple way to execute singular alignment is to utilize the hydrostatic weight approach, where the subjects are needed to hoist their hands to explicit statures above/beneath the heart level (Poon et al. 2006).

A hypothetical connection between PTT, BP, and stature can be written in (Zhang and Poon 2013), as appeared at the lower part of the page, where b is the subject ward boundary portraying the conduit properties, L is the separation gone by the beat, Ph is the hydrostatic weight ρgh, and Pi is the inside weight. Utilizing the proposed model, the individualized boundaries in BP-PTT model can be resolved from some straightforward developments. This model-based cuffless BP estimation technique can be actualized in an assortment of stages for unpretentious checking. In spite of the fact that the precision of this technique has been approved in numerous ongoing examinations (Poon and Zhang 2005), there are worries over the utilization of PTT as a substitute estimation of BP. It has been perceived that the major puzzling variables in the current relationship of BP and PTT are vasomotor tone and pre-launch period (Payne et al. 2006). New models ought to be created to remember these frustrating elements for request to investigate the capability of PTT-based strategy for unpretentious BP estimation in future.

19.7 Other Types of Unobtrusive Sensing Methods

Strain sensors are usually used to gauge body movement, for example, breath, heart sound, and BCG. Piezoelectric link sensor, whose detecting component is piezoelectric polymer, has been utilized for breath rate observing (Senior et al. 2010). Adaptable and dainty sensors, for example, piezoresistive texture sensors (Paradiso et al. 2005), and film-type sensors like polyvinylidenefluoride film (PVDF)- and electromechanical film (EMFi)-based sensors (Rajala and Lekkala 2012) have been broadly utilized for cardiopulmonary applications because of the ease of being inserted into dress or everyday objects like seat or bed. Near examination has led to assess the exhibitions of the two distinctive strain sensors in the estimation of body movement (Karki and Lekkala 2009). They have demonstrated that lone little contrasts were found in pulse estimated by PVDF-based and EMFi-based sensors particularly at prostrate stance, which was potentially caused by their various sensitivities to various power parts.

Inductive/impedance plethysmography is another generally utilized strategy for respiratory estimation and has been created in the types of garments and material belt (Karki and Lekkala 2009; Zhang et al. 2011). Two sinusoid wire curls situated at rib confine and the mid-region are driven by a current source that produces high-recurrence sinusoidal current. The development of the chest during breath causes changes of the inductance of the curls and consequently regulates the sufficiency of the sinusoidal current, from which the respiratory sign can be demodulated. An ongoing report thought about the exhibitions of four distinct techniques for wearable breath estimation, including inductive plethysmography, impedance plethysmography, piezoresistive pneumography, and piezoelectric pneumography, and indicated that piezoelectric pneumography gave the best heartiness to movement antiques for respiratory rate estimation (Poon and Zhang et al. 2005).

Optical filaments have likewise been received for unpretentious observing by implanting them into day-by-day articles or garments. As opposed to electronic sensors, they are safe to the electromagnetic obstruction. As of late, fiber Bragg grinding has been proposed as a vibration sensor for BCG estimation, in light of the way that the Bragg frequency is connected with the grinding time frame because of the body vibration brought about by breathing and cardiovascular withdrawals (Lanata et al. 2010). A pneumatic pad dependent on this strategy has been created to screen the physiological states of pilots and drivers (Lanata et al. 2010). D'Angelo built up an optical fiber sensor installed into a shirt for respiratory movement location (Dziuda et al. 2012).

Other far off detecting techniques have likewise been proposed for inconspicuous physiological estimation, for example, recurrence balanced nonstop wave Doppler radar for BCG measurement and radiometric detecting for internal heat level estimation (Postolache et al. 2010).

19.8 Concept of the Flexible Bio-Patch

The extensive use of multimedia healthcare systems and recent advances in nanotechnology have paved the way for more pervasive and personalized healthcare systems in the forms of remote health monitoring and digitalized telemedicine. The integration of nanotechnology and medicine with electronic networks has led to the invention of a vast variety of nano-biosensors for healthcare monitoring through which all the real-time vital physiological signals of the body can be measured with full accuracy and reliability for clinical diagnosis and analysis. As all the bio-signal monitoring devices used in hospitals are bulky with electrodes and wires and limited to one bio-signal per device, various attempts are made to construct new and improved wearable biosensing devices in nanoscale range which can detect all the physiological bio-signals in a single device for continuously monitoring daily life. The flexible, wearable nano-sensors are the part of rapidly growing industry of portable devices for point-of-care (POC) monitoring (Ray and Joseph 2013) (Fig. 19.6).

The characteristics of the new wearable and flexible biosensors are that they should be flexible and less cumbersome and should be able to detect the different biological signals efficiently and should establish a stable sensor network for

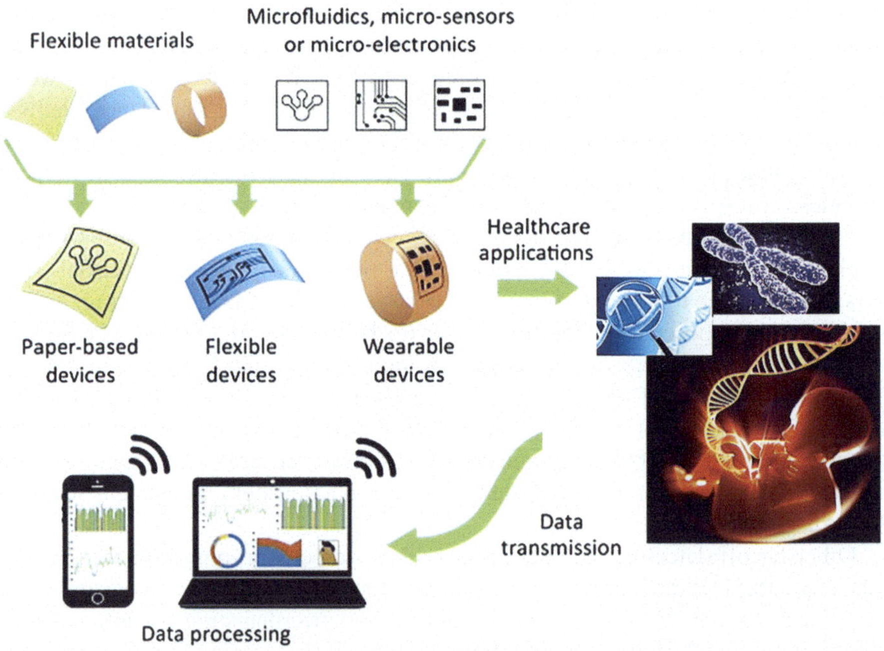

Fig. 19.6 Overview on the working of flexible devices for POC diagnosis. (We discuss recent advancements in the multiplexing capabilities and sensitivity of paper-based POC diagnostics and then discuss the development of flexible polymer-based POC diagnostics for sensing biological targets, as well as their prospective uses as wearable devices)

multichannel measurements. The construction of wearable nano-sensors requires a flexible substrate coated with a conducting polymer and a sensing element (bioactive element) on which the electrode patterns are printed. A transducer is also used to transduce the biological signals from the sensing element. The biosignals reflecting the parameter of interest like temperature, blood glucose level, blood pressure, pulse rate, and O2 and CO2 levels are then converted programmatically into readable outputs. The substrates widely used in the nano-sensors are the Bio-Patches which are used for wide range of medical applications.

The Bio-Patches used as disinfectants to prevent infections after surgery, injections, and skin injury are disc-shaped and cannot be used for constructing the wearable nano-sensors. Hence flexible substrates like polyamide, paper, silk, polyethylene terephthalate (PET), and polydimethylsiloxane (PDMS) should be used (Mata et al. 2005). The Bio-Patches which are to be used as substrate for nano-sensors should be wear-resistant and fracture-resistant and should be bendable and conformable with good stretchability (Wang et al. 2016; Yang et al. 2017; Han et al. 2017). As a result, it would be mandatory to select and synthesize Bio-Patch materials which would be innately flexible and stretchable with good electronic attributes.

Point-of-care (POC): patient care in the emergency room, in primary clinics, at home, or in other nonhospital settings where diagnosis can be made and treatment can be administered. Surface-enhanced Raman scattering (SERS): a spectroscopic approach that can detect chemical and biological species down to a single molecule without labeling.

19.9 Synthetic Substrate Materials

Synthetic polymers are most commonly used as Bio-Patch substrates as they enable the formation of low-cost flexible architecture very efficiently. Polydimethylsiloxane (PDMS) is widely used in biomedical devices and nano-sensors due to its characteristics like non-toxicity, high elasticity, chemical inertness, oxygen permeability, and thermal stability. The electrodes for biosensing can also be easily implanted as it is a good conductive material (Mata et al. 2005). Polyethylene terephthalate (PET), a synthetic polyester fiber, has proved to be a good alternative to silicon Bio-Patches due to its features like inertness, low cost, and mechanical properties (Ahani et al. 2011). It is also conformed into varied structures and shapes by pressing into ultrathin films to create high contact surfaces with optical transparency (MacDonald 2004). Polyethylene naphthalate (PEN), a synthetic polyester, also serves as a good plastic substrate due to its chemical, thermo-oxidative and UV resistance (Murakami et al. 1995). It also has good intrinsic flexibility and bendability though having more rigidity than PET. Both PET and PEN prove as a good layer for attaching functional components like nanoparticles, metal oxides, and conducting elements to form thin structures (Lechat et al. 2006). The screen printing technology has further made it more convenient to assimilate fabricated microelectrodes and other microdevices (Mościcki et al. 2017). With excellent dielectric and structural stability and tensile strength, Polyamides (PI) are also prepossessing substrates with good efficiency, biocompatibility, and efficiency at low cost (Lee et al. 2004).

Paper is an inexpensive substrate which can be used or conformed into paper-based structures to be used as a Bio-Patch for wide variety of sensing techniques like optical, electrochemical, and electrical sensing to detect the bio-target (Parolo and Merkoçi 2013). Due to its thin and porous structure, paper is easily transformed into composites. It also enables easy fabrication of paper-based electrodes using processing modalities like screen printing, inkjet printing, nano-patterning, etc. (Siegel et al. 2010). Since, the tensile strength of paper is weak and is easily receptive to tearing, bioactive paper is produced by incorporating the composites of paper with other biomaterials. For example, cellulose fibers and nanofibers (CNF) are integrated with strengthening polymers like glyoxalated polyacrylamides (GPAM) and/or polyamide epichlorohydrin (PAE) (Pelton 2009).

Textiles and fibers like wool, cotton, and synthetics (nylon, polyester, etc.) serve as attractive novel materials to be used as Bio-Patches in nano-sensors and can also be easily integrated with conductive materials by employing the various fabrication techniques. Also they have appreciable durability, stability, and intrinsic flexibility (Ding et al. 2010). Bio-derived materials like gelatin, silk proteins, and polysaccharides have been proposed as substrates for Bio-Patches for flexible nano-sensors. Silk is a naturally occurring polymer where the silk proteins can be fabricated as gels, fibers, sheets, and films (Altman et al. 2003). It is mechanically strong as textile and provides optical transparency as thin and ultrathin films (Jiang et al. 2007; Lawrence et al. 2008). The films are incorporated with conducting polymer inks to be used as flexible biosensors (Pal et al. 2016). Cellulose and chitin have also been reported to have been used as substrates and components for electrochemical biosensors (Benight et al. 2013).

An enthralling strategy to disseminate flexibility to the Bio-Patches and devices is by the use of specific nanoscale architectures or designs which facilitates bendability and stretchability (Fan et al. 2014; Wang et al. 2015; Ning et al. 2018). The advancements in nanofabrication techniques have enabled the transformation of stiff and highly rigid inorganic and organic compounds like quartz, silicones, and metals to form ultra-thin films and circuits employing ultra-thin (nanoscale) structures. Flexible electronics and nanoscale biosensors are constructed by techniques like cutting, folding, and buckling (Kim et al. 2012). During all these techniques of mechanical deformation (i.e., rolling, twisting, bending), the intrinsic characteristics of the Bio-Patches are retained by the use of highly conductive materials such as gold, copper, and silver (Au, Cu, Ag) (Xu et al. 2019).

Binding of biosensing elements to the Bio-Patch like the other biosensors, even the flexible nano-sensors involve the incorporation of sensing elements like biological components coupled with transducers to the Bio-Patch which is the substrate in the case of small flexible nano-sensors for the detection of specific analytic for the quantitative measurement of biochemical parameters. Enzymes, antibodies, whole cells, aptamers, and polysaccharides are some of the biomolecules used for the detection of the analyte (Kergoat et al. 2012). The flexible Bio-Patches must be integrated with the biomolecules in a way such that the biomolecules do not delaminate and detach from the Bio-Patch during interaction with the analyte and ensure longevity and proper functioning.

Also the Bio-Patch should be compatible with the biological sensing element as it decides the specificity and sensitivity of the nano-sensor. The biomolecule used as sensing element for the analytic detection must also have good stability to remain associated with the Bio-Patch substrate even under mechanical shocks. The biological sensing element can be immobilized onto the Bio-Patch substrate through covalent and non-covalent bonding employing different linkage chemistries (Zhu et al. 2015). At present, owing to their use in electrochemical detection, low cost, high specificity, and strong affinity, enzymes are the most preferred bio-receptors/biomolecules as sensing elements in flexible nano-biosensors (Liao et al. 2015). In flexible nano-biosensors, the biological events between the analytic and the biomolecule are generated as electronic and electrochemical signals to the transducer and to a lesser extent as optical signals due to their high sensitivity, portability, and reliability (Wang et al. 2015). The electrodes are also easily incorporated onto the Bio-Patches by specific techniques in flexible nano-biosensors.

19.10 Flexible Conducting Polymers

The conductivity of the nano-sensors is achieved by the use of conducting polymers with characteristics like redox activity and doping and is capable of mixed electronic and ionic transportation. Doping increases the efficiency and processability of the conducting polymers to several folds in magnitude by modifying the physical and chemical properties of conductive polymers. The conducting polymers entrap the biological sensing elements either by physical methods or by chemically attaching to the functional groups, thus enabling the integration of the biomolecules with the conducting polymers and Bio-Patch substrates for detecting the bio-signals in a nano-biosensor. One of the most extensively studied conductive polymers is PEDOT which is poly (3, 4-ethylenedioxythiophene). It provides great electronic stability and is highly conductive. A conjugate of PEDOT and PSS (Poly (styrene sulfonate)) PEDOT: PSS is employed as a good flexible polymer electrode for sensing of glucose, DNA/RNA, biomarkers, etc. (Kergoat et al. 2012). Metals like Au and Ag are very promising candidates for fabrication of flexible electrodes as they are ductile and can be formed into thin films and can function jointly with the biomolecules used in the sensors and can sustain the mechanical deformations to retain their electrical conductivity by forming different structural configurations (e.g., mesh, serpentine, wavy, etc.) (Le Goff et al. 2011; Reina et al. 2017). Other carbon nanostructures like carbon nanofibers (CNF) and C60 have also found its applications in flexible biosensing technology (Pavinatto et al. 2015).

19.11 Flexible Bio-Patch Nano-sensors

The flexible wearable Bio-Patch nano-sensors bind to the biological surfaces like skin and tissues to measure the physiological signals like breath, blood pressure, temperature, stress, pulse rate, etc. and the biochemical signals like glucose level, oxygen level, uric acid levels, dopamine levels, etc. As discussed previously various synthetic substrates can be used as flexible Bio-Patches on which the nano-sensors are integrated to detect the desired biological or physiological signal. A prototype of the Bio-Patch implemented in consisted of a pair of electrodes printed on a photo paper patch, a SoC (system-on-chip) sensor, and a soft battery. The nano-chips are integrated with the conducting elements like any metal nanostructures (e.g., Ag, Au, etc.) which are most commonly preferred now and are fabricated onto the flexible Bio-Patch. The flexible Bio-Patch nano-sensor which measures the vital physiological parameters is affixed onto the skin surface. The biological signals from the body are detected and transduced into measurable signals like electrochemical, optical, or piezoelectric signals depending on the type of biosensors. These signals are then analyzed and processed programmatically through wireless network (Fig. 19.7). The end results can be accessed digitally by the doctors and the patients. A Bio-Patch can work exclusively or synchronously with other Bio-Patches connected through an active cable printed on a substrate for multichannel bio-signals recording (Yoon et al. 2016).

(a) Image of fabricated sensor. The sensor can be attached to the surface of the meat or be placed onto the inside lining of the package. The status of the meat product can be monitored using smartphone by taking a photo of the sensor.
(b) Schematic representation of self-powered multifunctional electronic skin used for continuous monitoring of lactate, glucose, uric acid, and urea in exercise-induced sweat using piezoelectric-linked enzymatic biosensors. During exercise, this device functions without an additional power supply through piezoelectric-enzymatic reaction coupling.
(c) A wireless glucose sensor incorporated into a contact lens platform with wireless power transfer circuitry and display pixels for a fully integrated and transparent platform that does not hinder vision. This device detects fluctuating tear glucose concentrations through a resistance-based enzymatic mechanism, which was demonstrated in a rabbit model.
(d) Epidermal reverse iontophoretic tattoo-based glucose sensor configuration and operation principle, with picture of device applied to a human subject. Proof-of-concept demonstration of reverse iontophoretic tattoo-based ISF glucose sensor. Indicate working electrode and reference/counter electrode, respectively.
(e) Iontophoretic paper battery and skin-like biosensor for non-invasive blood glucose monitoring, applied to a human subject. Inclusion of hyaluronic acid facilitates enhanced ISF extraction for increased ISF glucose sampling reliability.
(f) Schematic of steps for rapid transfer of silk antennas onto curved substrates: (1) Water vapor is applied to the back of silk films, yielding (2) a film in which the

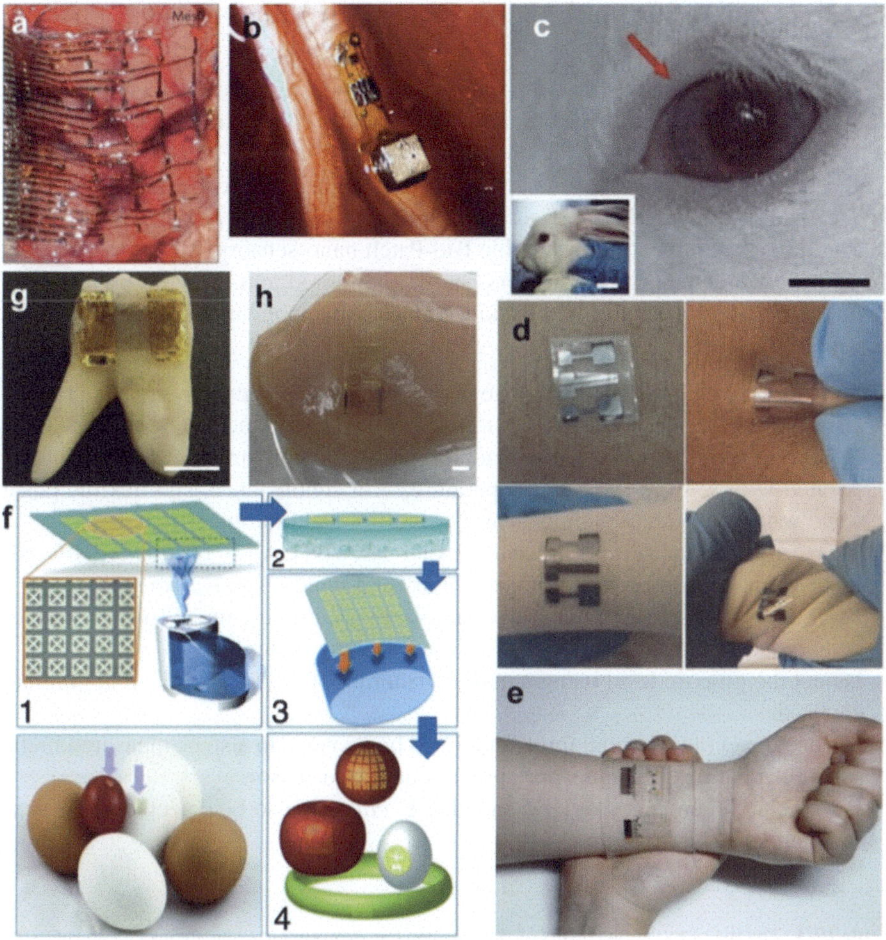

Fig. 19.7 The possible locations where a flexible Bio-Patch nano-sensor can be used

back surface of the film has been partially melted. (3) This melted surface is conformally applied to arbitrary surfaces, yielding (4) applied functional sensors on a variety of surfaces. Photos of THz split ring resonators (SRRs) fabricated on the silk substrate wrapped on an apple.

(g) Mouthguard-based sensor for glucose monitoring in saliva with on-body application and analysis of increasing glucose concentrations. Fully integrated saliva glucose sensor aims towards continuous in-mouth glucose monitoring.

(h) Various demonstrations of biosensors in the food industry. (a) Geometric barcode sensor for monitoring chicken spoilage under different temperature conditions.

19.12 Applications of Flexible Bio-Patch Nano-sensors

Flexible Bio-Patch nano-sensors are manufactured with the sole motive of personalized and pervasive e-healthcare monitoring to obtain round-the-clock information of the patient's physiological and biochemical conditions. The Bio-Patches are also synthesized in such a way for symbiotic functioning of the flexible healthcare device. The skin being the most accessible organ of the body provides a wide surface area for the application of flexible Bio-Patch nano-sensors. All the physiological and most biochemical parameters can be monitored and detected by applying flexible nano-sensor patches on the skin. Some of the applications are listed as follows (Table 19.3):

- To capture the ECG (electrocardiogram) and EMG (electromyogram) signals, also for cardiac health sensor was to monitor bio-impedance along with ECG.
- Human stress monitoring patch made of perforated polyamide membrane incorporated with sensors for skin temperature, skin conductance, and flexible pulse wave sensor to detect the pulse wave by multimodal biological signals.
- Measure the glucose concentration using PVA/BTCA/β-CD/GOx/AuNPs NF hydrogel patch sensor with high sensitivity and rapid response time.
- POC devices, ultra-thin flexible wearable Pt electrodes, and PEDOT: PSS to detect glucose, lactate, and uric acid from the saliva sample.
- Detect the glucose level using a carbon nanotube (CNT)-based flexible graphite electrode.
- Cortisol level (an essential multifunctional hormone) can also be measured in the sweat using a flexible nanospore polyamide-based sensor.
- Pathogenic organisms like HIV and also *E. coli* can also be detected using flexible nano-sensors.

Table 19.3 The list of the applications of flexible Bio Patches other than an anchor/substrate to the flexible nano-sensors

S. No	Type of Bio-Patch used	Application of Bio-Patch sensor
1.	Disposable photo voltaic patches	Provides electric stimulation and promotes regeneration for wound healing
2.	Tenocyte cells-seeded collagen patches	Promotes healing of anterior cruciate ligament (ACL) repair
3.	Si-NN PDMS patch	This microneedle patch is used for intramuscular and intratissue nano-injection of biomolecules
4.	Microneedle patch	Transdermal delivery of dual mineralized peptides for therapy of type 2 diabetes mellitus
5.	Silk fibroin microneedle patches	Sustained transdermal delivery of the contraceptive hormone, Levonogestrel enclosed in microcarriers
6.	3-D printed F-GelMA hydrogel patches	Local delivery of the model nano-medicine PEGylated liposomal doxorubicin (DOX) against cancer
7.	Microneedle patches	Intradermal vaccination of poly (lactic-co-glycolic acid) (PLGA) micro-particles encapsulated antigens

Apart from their applications as flexible and wearable nano-sensors, flexible bio-patches have also found to be very promising candidates for drug and vaccine delivery and healing of wounds. Disposable photovoltaic patches were developed to provide electric stimulation (ES) and promote regenerative activities and contribute to wound healing processes. Thus flexible bio-patches have wide range of application in the field of nano-sensors and medicine.

19.13 Future Perspectives Towards Fusion of Nanotechnology with Biosensors

The food industry has made remarkable progress in recent years in areas such as the improvement of new packaging products with the development of new functional products, transport and controlled release of bioactive substances, and detection of pathogens and pesticides utilizing nano-sensors and indicators. In the world of food packaging, intelligent packaging is still relatively new (still in developmental stage); with further progress being made all the time, it is important to look at what aspects to focus on as underlying objects for effective products. However, it provides a promising future for food safety and also paints a positive picture for years to come with the research into novel approaches such as nanotechnology. The potential applications and benefits of nanotechnology are enormous; however, nanotechnology application in food sector is relatively recent compared with its use in drug delivery and pharmaceuticals. Nanotechnology has a tremendous potential to revolutionize the global food system by generating new food products and better packaging and storage techniques as well as to alter fundamental functionality of food. Intelligent food packaging, using nano-sensors, may also provide consumers with information on the state of the food inside. Food packages are embedded with NPs that alert consumers when a product is no longer safe to eat. In fact, the technological advancement in the nano-industries will probably change the fabrication of the entire packaging industry.

Nevertheless, less effort has been exerted into applications of nanotechnology in food sector, and most of them require a high quality of research and development for their safe application. Ongoing studies are being carried out involving sensors for transmission of required information and stimulating subsequent alterations in packaging materials, environments, or the products for preservation and safety. There are many challenges associated with the future potential success of nanotechnology in food packaging, public approval, economics, and control of food processed with nanomaterials which could build up and cause toxicity. The safety of NPs in food industry is also a challenge for both the government and industry. It cannot be claimed with certainly either that nanotechnology is fully safe for health or that is harmful. The food processing industry must ensure the consumer confidence and acceptance of nano-foods safety. Several evidences regarding food-grade NMs, which supports the toxicity NMs. So, when it comes to the application of nanotechnology at industrial scale, it is important to determine the release of NPs into the environment and to estimate the subsequent levels of exposure to these materials. As the NPs can penetrate into the human organ and organelles, exposure time, exposure concentrations, sites of penetration, immune response, and accumulation and retention of NPs in body and their subsequent effects should be evaluated carefully. Therefore, compulsory testing of nano-modified foods should be carried out before allowing them to be introduced into the market. Moreover, there will also be a great dependence on the actions of governments, regulatory agencies, and manufacturers in relation to the abovementioned challenges. In the long term, nanoscience and nanotechnology in food may be emerged as a new field of research which will be known as 'food nanotechnology'.

References

Ahani A, Saadati-Fard L, Sodagar AM, Boroumad FA (2011) Flexible PET/ITO electrode array for implantable biomedical applications. In: 2011 Annual International Conference of the IEEE Engineering in Medicine and Biology Society. IEEE, pp 2878–2881

Ahmed A, Rushworth JV, Hirst NA, Millner PA (2014) Biosensors for whole-cell bacterial detection. Clin Microbiol Rev 27(3):631–646

Ahvenainen R (2003) Active and intelligent packaging: an introduction. In: Novel food packaging techniques. Wood head Publishing Ltd, Cambridge UK, pp 5–21

Al-Khafajiy M, Webster L, Baker T, Waraich A (2018) Towards fog driven IoT healthcare: challenges and framework of fog computing in healthcare. In: Proceedings of the 2nd international conference on future networks and distributed systems, pp 1–7

Altman GH, Diaz F, Jakuba C, Calabro T, Horan RL, Chen J, Lu H, Richmond J, Kaplan DL (2003) Silk-based biomaterials. Biomaterials 24(3):401–416

Aydindogan E, Guler Celik E, Timur S (2018) Based analytical methods for smartphone sensing with functional nanoparticles: bridges from smart surfaces to global health

Babar NU, Joya KS, Tayyab MA, Ashiq MN, Sohail M (2019) Highly sensitive and selective detection of arsenic using electrogenerated nanotextured gold assemblage. ACS Omega 4(9):13645–13657. https://doi.org/10.1021/acsomega.9b00807

Bachheti RK, Gupta V, Husen A, Joshi A, Konwarh R (2019) Green synthesis of Iron oxide nanoparticles: cutting edge technology and multifaceted applications. In: Husen A, Iqbal M (eds) N*anomaterials* and plant potential. Springer, Cham, pp 239–259. https://doi.org/10.1007/978-3-030-05569-1_9

Bachheti RK, Fikadu A, Bachheti A, Husen A (2020a) Biogenic fabrication of nanomaterials from flower-based chemical compounds, characterization and their various applications: a review. Saudi J Biol Sci 27:2551–2562. https://doi.org/10.1016/j.sjbs.2020.05.012

Bachheti RK, Sharma A, Bachheti A, Husen A, Shanka GM, Pandey DP (2020b) Nanomaterials from various forest tree species and their biomedical applications. In: Husen A, Jawaid M (eds) Nanomaterials for agriculture and forestry applications. Elsevier, Cambridge, MA, pp 81–106. https://doi.org/10.1016/B978-0-12-817852-2.00004-4

Bachheti RK, Godebo Y, Joshi A, Yassin MO, Husen A (2020c) Root-based fabrication of metal and or metal-oxide nanomaterials and their various applications. In: Husen A, Jawaid M (eds) Nanomaterials for agriculture and forestry applications. Elsevier, Cambridge, MA, pp 135–166. https://doi.org/10.1016/B978-0-12-817852-2.00006-8

Bachheti RK, Abate L, Deepti, Bachheti A, Madhusudhan A, Husen H (2021) Algae-, fungi-, and yeast-mediated biological synthesis of nanoparticles and their various biomedical applications. In: Kharisov B, Kharissova O (eds) Handbook of greener synthesis of nanomaterials and compounds, volume 1: fundamental principles and methods. Elsevier Inc, Cambridge, MA, pp 701–734. https://doi.org/10.1016/B978-0-12-821938-6.00022-0

Baek H, Chung G, Kim K, Park K (2012) A smart health monitoring chair for nonintrusive measurement of biological signals. IEEE Trans Inf Technol Biomed 16(1):150–158

Banerjee T, Sulthana S, Shelby T, Heckert B, Jewell J, Woody K, Santra S (2016) Multiparametric magneto-fluorescent nanosensors for the ultrasensitive detection of Escherichia coli O157: H7. ACS Infectious Diseases 2(10):667–673

Banerjee, T., Shelby, T., and Santra, S. (2017). How can nanosensors detect bacterial contamination before it ever reaches the dinner table?

Benight SJ, Wang C, Tok JB, Bao Z (2013) Stretchable and self-healing polymers and devices for electronic skin. Prog Polym Sci 38(12):1961–1977

Biji KB, Ravishankar CN, Mohan CO, Srinivasa Gopal TK (2015) Smart packaging systems for food applications: a review. J Food Sci Technol 52(10):6125–6135. Retrieved from https://doi.org/10.1007/s13197-015-1766-7. https://doi.org/10.1007/s13197-015-1766-7

Bobelyn E, Hertog MLATM, Nicola BM (2006) Applicability of an enzymatic time temperature integrator as a quality indicator for mushrooms in the distribution chain. Postharvest Biol Technol 42(1):104–114

Call DR, Borucki MK, Loge FJ (2003) Detection of bacterial pathogens in environmental samples using DNA microarrays. J Microbiol Methods 53(2):235–243

Chang MMF, Ginjom IR, Ng SM (2017) Single-shot 'turn-off' optical probe for rapid detection of paraoxon-ethyl pesticide on vegetable utilising fluorescence carbon dots. Sensors Actuators B Chem 242:1050–1056

Chen X, Zhang L (2018) Review in manufacturing methods of nanochannels of bio-nanofluidic chips. Sensors Actuators B Chem 254:648–659

Chi YM, Deiss SR, Cauwenberghs G (2009) Non-contact low power EEG/ECG electrode for high density wearable biopotential sensor networks. In: Proc. 6th Int. Workshop Wearable Implantable Body Sensor netw, Berkeley, CA, pp 246–250

Chiang S, Vankov ER, Yeh HJ, Guindani M, Vannucci M, Haneef Z, Stern JM (2018) Temporal and spectral characteristics of dynamic functional connectivity between resting-state networks reveal information beyond static connectivity. PLoS One 13(1):e0190220

Cruz-Romero MC, Murphy T, Morris M, Cummins E, Kerry JP (2013) Antimicrobial activity of chitosan, organic acids and nano-sized solubilisates for potential use in smart antimicrobially-active packaging for potential food applications. Food Control 34(2):393–397

Dinckaya E, Kınık Ö, Sezgintürk MK, Altuğ Ç, Akkoca A (2011) Development of an impedimetric aflatoxin M1 biosensor based on a DNA probe and gold nanoparticles. Biosens Bioelectron 26(9):3806–3811

Ding Y, Invernale MA, Sotzing GA (2010) Conductivity trends of PEDOT-PSS impregnated fabric and the effect of conductivity on electrochromic textile. ACS Appl Mater Interfaces 2(6):1588–1593

Dziuda L, Skibniewski FW, Krej M, Lewandowski J (2012) Monitoring respiration and cardiac activity using fiber bragg grating-based sensor. IEEE Trans Biomed Eng 59(7):1934–1942

Ensafi AA, Karimi-Maleh H, Mallakpour S (2012) Simultaneous determination of ascorbic acid, acetaminophen, and tryptophan by square wave voltammetry using N-(3, 4-Dihydroxyphenethyl)-3, 5-Dinitrobenzamide-modified carbon nanotubes paste electrode. Electroanalysis 24(3):666–675

Fan JA, Yeo W-H, Yewang S, Hattori Y, Lee W, Jung S-Y, Zhang Y (2014) Fractal design concepts for stretchable electronics. Nat Commun 5(1):1–8

Francois BJM (2018) U.S. Patent No. 10,039,198. Washington, DC: U.S. Patent and Trademark Office

Franz DR, Jahrling PB, Friedlander AM, McClain DJ, Hoover DL, Bryne WR et al (1997) Clinical recognition and management of patients exposed to biological warfare agents. JAMA 278(5):399–411

Giannakourou MC, Koutsoumanis K, Nychas GJE, Taoukis PS (2005) Field evaluation of the application of time temperature integrators for monitoring fish quality in the chill chain. Int J Food Microbiol 102(3):323–336

Gomez-Clapers J, Casanella R (2012) A fast and easy-to-use ECG acquisition and heart rate monitoring system using a wireless steering wheel. IEEE Sensors J 12(3):610–616

Gray S, Tsuru T, Cohen Y, Lau WJ (eds) (2018) Advanced materials for membrane fabrication and modification. CRC Press

Gu WB, Poon CCY, Leung HK, Sy MY, Wong MYM, Zhang YT (2009) A novel method for the contactless and continuous measurement of arterial blood pressure on a sleeping bed. In: Proceedings of the Annual International Conference of the IEEE Engineering in Medicine and Biology Society, pp 6084–6086

GuoChen P, Bocko MF (2013) Non-contact ECG sensing employing gradiometer electrodes. IEEE Trans Biomed Eng 60(1):179–183

Han ST, Peng H, Sun Q, Venkatesh S, Chung KS, Lau SC, Roy VAL (2017) An overview of the development of flexible sensors. Adv Mater 29(33):1700375

Hanina A, Kessler G, Guan L (2018) U.S. Patent No. 9,977,870. Washington, DC: U.S. Patent and Trademark Office

Heo DN, Yang DH, Moon H-J, Lee JB, Bae MS, Lee SC et al (2012) Gold nanoparticles surface-functionalized with paclitaxel drug and biotin receptor as theranostic agents for cancer therapy. Biomaterials 33(3):856–866

Herbert R, Kim JH, Kim YS, Lee HM, Yeo WH (2018) Soft material-enabled, flexible hybrid electronics for medicine, healthcare, and human-machine interfaces. Materials 11(2):187

Hoffmann T, Eilebrecht B, Leonhardt S (2011) Respiratory monitoring system on the basis of capacitive textile force sensors. IEEE Sensors J 11(5):1112–1119

Husen A (2019) Medicinal plant-product based fabrication nanoparticles (au and ag) and their anticancer effect. In: Kintzios SE, Barberaki M, Flampouri (eds) Plants that fight cancer – second edition. Taylor & Francis/CRC Press, pp 133–147

Husen A (2020) Introduction and techniques in nanomaterials formulation. In: Husen A, Jawaid M (eds) Nanomaterials for agriculture and forestry applications. Elsevier, Cambridge, MA, pp 1–14. https://doi.org/10.1016/B978-0-12-817852-2.00001-9

Husen A, Iqbal M (2019) N*anomaterials* and plant potential: an overview. In: Husen A, Iqbal M (eds) N*anomaterials* and plant potential. Springer, Cham, pp 3–29. https://doi.org/10.1007/978-3-030-05569-1_1

Inan T, Park D, Giovangrandi L, Kovacs GTA (2012) Noninvasive measurement of physiological signals on a modified home bathroom scale. IEEE Trans Biomed Eng 59(8):2137–2143

Jadzinsky PD, Calero G, Ackerson CJ, Bushnell DA, Kornberg RD (2007) Structure of a thiol monolayer-protected gold nanoparticle at 1.1 Å resolution. Science 318(5849):430–433

Jae BH, Sung CG, Keun KK, Soo KJ, Suk PK (2009) Photoplethysmogram measurement without direct skin-to-sensor contact using an adaptive light source intensity control. IEEE Trans Inf Technol Biomed 13(6):1085–1088

Jawaheer S, White S, Bessant C, Rughooputh S, Cullen D (2002) Determination of fruit status using a disposable multi-analyte biosensor array and principle component analysis. Paper presented at the 7th World Congress on Biosensors. Kyoto, Japan

Jiang C, Wang X, Gunawidjaja R, Lin Y-H, Gupta MK, Kaplan DL, Naik RR, Tsukruk VV (2007) Mechanical properties of robust ultrathin silk fibroin films. Adv Funct Mater 17(13):2229–2237

Jiang J, Fan W, Du X (2014) Nitrite electrochemical biosensing based on coupled graphene and gold nanoparticles. Biosens Bioelectron 51:343–348

Jokar M, Safaralizadeh MH, Hadizadeh F, Rahmani F, Kalani MR (2016) Design and evaluation of an apta-nano-sensor to detect Acetamiprid in vitro and in silico. J Biomol Struct Dyn 34(11):2505–2517

Joshi A, Sharma A, Bachheti RK, Husen A, Mishra VK (2019) Plant-mediated synthesis of copper oxide nanoparticles and their biological applications. In: Husen A, Iqbal M (eds) N*anomaterials* and plant potential. Springer, Cham, pp 221–237. https://doi.org/10.1007/978-3-030-05569-1_8

Joung H-A, Lee N-R, Lee SK, Ahn J, Shin YB, Choi H-S et al (2008) High sensitivity detection of 16s rRNA using peptide nucleic acid probes and a surface plasmon resonance biosensor. Anal Chim Acta 630(2):168–173

Karimi-Maleh H, Moazampour M, Yoosefian M, Sanati AL, Tahernejad-Javazmi F, Mahani M (2014) An electrochemical nanosensor for simultaneous voltammetric determination of ascorbic acid and Sudan I in food samples. Food Anal Methods 7(10):2169–2176

Karki S, Lekkala J (2009) A new method to measure heart rate with EMFi and PVDF materials. J Med Eng Technol 33(7):551–558

Kara E, Fabian AC, Cackett EM, Miniutti G, Uttley P (2013) Revealing the X-ray source in IRAS 13224– 3809 through flux-dependent reverberation lags. Monthly Notices of the Royal Astronomical Society 430(2):1408–1413

Karlsson L, Naess LO, Nightingale A, Thompson J (2018) 'Triple wins' or 'triple faults'? Analysing the equity implications of policy discourses on climate-smart agriculture (CSA). J Peasant Stud 45(1):150–174

Kergoat L, Piro B, Berggren M, Horowitz G, Pham MC (2012) Advances in organic transistor-based biosensors: from organic electrochemical transistors to electrolyte-gated organic field-effect transistors. Anal Bioanal Chem 402(5):1813–1826

Kerry JP, O'Grady MN, Hogan SA (2006) Past, current and potential utilisation of active and intelligent packaging systems for meat and muscle-based products: a review. Meat Sci 74(1):113–130. https://doi.org/10.1016/j.meatsci.2006.04.024

Kim DH, Ghaffari R, Lu N, Rogers JA (2012) Flexible and stretchable electronics for biointegrated devices. Annu Rev Biomed Eng 14:113–128

Kim E, Choi DY, Kim HC, Kim K, Lee SJ (2013) Calibrations between the variables of microbial TTI response and ground pork qualities. Meat Sci 95(2):362–367

Ko Keun K, Yong Kyu L, Kwang-Suk P (2004) The electrically noncontacting ECG measurement on the toilet seat using the capacitively coupled insulated electrodes. In: Proceedings of the 26th Annual International Conference of the IEEE Engineering in Medicine and Biology Society, San Francisco, pp 2375–2378

Kryszkiewicz P, Kliks A, Kułacz Ł, Bogucka H, Koudouridis GP, Dryjański M (2018) Context-based spectrum sharing in 5G wireless networks based on radio environment maps. Wirel Commun Mob Comput 2018

Kumar V, Guleria P, Mehta SK (2017) Nanosensors for food quality and safety assessment. Environ Chem Lett 15(2):165–177

Kuswandi B, Wicaksono Y, Jayus, Abdullah A, Heng LY, Ahmad M (2011) Smart packaging: sensors for monitoring of food quality and safety. Sens & Instrumen Food Qual 5(3):137–146. Retrieved from https://doi.org/10.1007/s11694-011-9120-x, https://doi.org/10.1007/s11694-011-9120-x

Lanata E, Scilingo P, Nardini E, Loriga G, Paradiso R, De-Rossi D (2010) Comparative evaluation of susceptibility to motion artifact in different wearable systems for monitoring respiratory rate. IEEE Trans Inf Technol Biomed 14(2):378–386

Lawrence BD, Cronin-Golomb M, Georgakoudi I, Kaplan DL, Omenetto FG (2008) Bioactive silk protein biomaterial systems for optical devices. Biomacromolecules 9(4):1214–1220

Le Goff A, Holzinger M, Cosnier S (2011) Enzymatic biosensors based on SWCNT-conducting polymer electrodes. Analyst 136(7):1279–1287

Le DN, Van Le C, Tromp JG, Nguyen GN (eds) (2018) Emerging technologies for health and medicine: virtual reality, augmented reality, artificial intelligence, internet of things, robotics, industry 4.0. Wiley

Lechat C, Bunsell AR, Davies P, Piant A (2006) Mechanical behaviour of polyethylene terephthalate and polyethylene naphthalate fibres under cyclic loading. J Mater Sci 41(6):1745–1756

Lee K, Singh A, He J, Massia S, Kim B, Raupp G (2004) Polyimide based neural implants with stiffness improvement. Sensors Actuators B Chem 102(1):67–72

Lee S, Kim S, Kim S, Kim JY, Moon C, Nelson BJ, Choi H (2018) A capsule-type microrobot with pick-and-drop motion for targeted drug and cell delivery. Adv Healthc Mater 7(9):1700985

Leonhardt S, Leicht L, Teichmann D (2018) Unobtrusive vital sign monitoring in automotive environments-A review. Sensors (Basel) 18(9):3080. Published 2018 Sep 13. https://doi.org/10.3390/s18093080

Li J, Yang L, Ruan Y, Chu S, Wang H, Li Z, Zhang Z (2020) Dual-mode optical nanosensor based on gold nanoparticles and carbon dots for visible detection of as (III) in water. ACS Appl Nano Mater 3(8):8224–8231

Liao C, Zhang M, Yao MY, Hua T, Li L, Yan F (2015) Flexible organic electronics in biology: materials and devices. Adv Mater 27(46):7493–7527

Lin J, Shang Y, Ding B, Yang J, Yu J, Al-Deyab SS (2012) Nanoporous polystyrene fibers for oil spill cleanup. Mar Pollut Bull 64(2):347–352

Liu L, Corma A (2018) Metal catalysts for heterogeneous catalysis: from single atoms to nanoclusters and nanoparticles. Chem Rev 118(10):4981–5079

Liu S, Yu J, Ju H (2003) Renewable phenol biosensor based on a tyrosinase-colloidal gold modified carbon paste electrode. J Electroanal Chem 540:61–67

Lin Z, Li X, Kraatz, HB (2011) Impedimetric immobilized DNA-based sensor for simultaneous detection of $Pb2^{+}$, Ag^{+}, and $Hg2^{+}$. Anal Chem 83(17):6896–6901.

Loiseau A, Asila V, Boitel-Aullen G, Lam M, Salmain M, Boujday S (2019) Silver-based plasmonic nanoparticles for and their use in biosensing. Biosensors 9(2):78

Lopez-Rubio A, Almenar E, Hernandez-Munoz P, Lagaron JM, Catala R, Gavara R (2004) Overview of active polymer-based packaging Technologies for Food Applications. Food Rev Intl 20(4):357–387

Lu L, Zheng W, Lv Z, Tang Y (2012) Development and application of time-temperature indicators used on food during the cold chain logistics. Packag Technol Sci

MacDonald WA (2004) Engineered films for display technologies. J Mater Chem 14(1):4–10

Mahieu A, Terrié C, Youssef B (2015) Thermoplastic starch films and thermoplastic starch/polycaprolactone blends with oxygen-scavenging properties: influence of water content. Ind Crop Prod 72(C):192–199

Mata A, Fleischman AJ, Roy S (2005) Characterization of polydimethylsiloxane (PDMS) properties for biomedical micro/nanosystems. Biomed Microdevices 7(4):281–293

Mlalila N, Kadam DM, Swai H, Hilonga A (2016) Transformation of food packaging from passive to innovative via nanotechnology: concepts and critiques. J Food Sci Technol 53(9):3395–3407. https://doi.org/10.1007/s13197-016-2325-6

Mohanty AK, Misra M, Nalwa HS (2009) Packaging nanotechnology. American Scientific Publishers, Los Angeles

Morales-Narváez E, Merkoçi A (2019) Graphene oxide as an optical biosensing platform: a progress report. Adv Mater 31(6):1805043

Mościcki A, Smolarek-Nowak A, Felba J, Kinart A (2017) Ink for ink-jet printing of electrically conductive structures on flexible substrates with low thermal resistance. J Electron Mater 46(7):4100–4108

Munge BS, Coffey AL, Doucette JM, Somba BK, Malhotra R, Patel V et al (2011) Nanostructured immunosensor for attomolar detection of cancer biomarker interleukin-8 using massively labeled superparamagnetic particles. Angew Chem 123(34):8061–8064

Murakami S, Nishikawa Y, Tsuji M, Kawaguchi A, Kohjiya S, Cakmak M (1995) A study on the structural changes during uniaxial drawing and/or heating of poly (ethylene naphthalene-2, 6-dicarboxylate) films. Polymer 36(2):291–297

Myung-Kee J, Chang-Wook H, Jae-Hyuk C, Choi (2018) A time-temperature Indicator for a vision based-detection system for managing the storage temperature of frozen fish products. Korean J Fish Aquat Sci

Nadarajah D, Han JH, Holley RA (2005) Inactivation of Escherichia coli O157:H7 in packaged ground beef by allyl isothiocyanate. Int J Food Microbiol 99(3):269–279

Neethirajan S, Jayas DS (2011) Nanotechnology for the food and bioprocessing industries. Food Bioprocess Technol 4(1):39–47. https://doi.org/10.1007/s11947-010-0328-2

Ning X, Wang X, Zhang Y, Yu X, Choi D, Zheng N, Rogers JA (2018) Assembly of advanced materials into 3D functional structures by methods inspired by origami and kirigami: a review. Adv Mater Interfaces 5(13):1800284

Odważny F, Szymańska O, Cyplik P (2018) Smart Factory: the requirements for implementation of the Industry 4.0 solutions in FMCG environment-case study. LogForum 14(2)

Painuli S, Semwal P, Bachheti A, Bachheti RK, Husen A (2020) Nanomaterials from non-wood forest products and their applications. In: Husen A, Jawaid M (eds) Nanomaterials for agriculture and forestry applications. Elsevier Inc, Cambridge, MA, pp 15–40. https://doi.org/10.1016/B978-0-12-817852-2.00002-0

Pal RK, Farghaly AA, Wang C, Collinson MM, Kundu SC, Yadavalli VK (2016) Conducting polymer-silk biocomposites for flexible and biodegradable electrochemical sensors. Biosens Bioelectron 81:294–302

Paniel N, Radoi A, Marty J-L (2010) Development of an electrochemical biosensor for the detection of aflatoxin M1 in milk. Sensors 10(10):9439–9448

Paradiso R, Loriga G, Taccini N (2005) A wearable health care system based on knitted integrated sensors. IEEE Trans Inf Technol Biomed 9(3):337–344

Parolo C, Merkoçi A (2013) Based nanobiosensors for diagnostics. Chem Soc Rev 42(2):450–457

Pavinatto FJ, Paschoal CW, Arias AC (2015) Printed and flexible biosensor for antioxidants using interdigitated ink-jetted electrodes and gravure-deposited active layer. Biosens Bioelectron 67:553–559

Payne R, Symeonides C, Webb D, Maxwell S (2006) Pulse transit time measured from the ECG: an unreliable marker of beat-to-beat blood pressure. J Appl Phys 100(1):136

Pelton R (2009) Bioactive paper provides a low-cost platform for diagnostics. TrAC Trends Anal Chem 28(8):925–942

Peng C, Zheng L, Chen Q, Shen M, Guo R, Wang H, Shi X (2012) PEGylated dendrimer-entrapped gold nanoparticles for in vivo blood pool and tumor imaging by computed tomography. Biomaterials 33(4):1107–1119

Pereira VA, de Arruda INQ, Stefani R (2015) Active chitosan/PVA films with anthocyanins from Brassica oleraceae (Red Cabbage) as Time–Temperature Indicators for application in intelligent food packaging. Food Hydrocoll 43(Jan):180–188

Poh MZ, McDuff DJ, Picard RW (2011a) Advancements in noncontact, multiparameter physiological measurements using a webcam. IEEE Trans Biomed Eng 58(1):7–11

Poh MZ, McDuff D, Picard R (2011b) A medical mirror for non-contact health monitoring. In: Proc. ACM SIGGRAPH 2011 Emerging Technol.,Vancouver, BC, Canada, p 1–1

Pontes PM (2018) A pattern-based testing framework for IoT ecosystems

Poon CCY, Zhang YT (2005) Cuff-less and noninvasive measurements of arterial blood pressure by pulse transit time. In: Proceedings of 27th Annual International Conference of IEEE Engineering in Medicine and Biology Society, Shanghai, China, pp 5877–5880

Poon CC, Zhang Y-T, Liu Y (2006) Modeling of pulse transit time under the effects of hydrostatic pressure for cuffless blood pressure measurements. In: Proceedings of the 3rd IEEE/EMBS International Summer School and Medical Devices and Biosensors, pp 65–68

Postolache P, Girao S, Madeira RN, Postolache G (2010) Microwave FMCW Doppler radar implementation for in-house pervasive health care system. In: Proc. IEEE Int. Workshop Med. Measurements Appl, Ottawa, pp 47–52

Radoi A, Targa M, Prieto-Simon B, Marty J-L (2008) Enzyme-linked immunosorbent assay (ELISA) based on superparamagnetic nanoparticles for aflatoxin M1 detection. Talanta 77(1):138–143

Rajala S, Lekkala J (2012) Film-type sensor materials PVDF and EMFi in measurement of cardio-respiratory signals: a review. IEEE Sensors J 12(3):439–446

Ramgir NS, Yang Y, Zacharias M (2010) Nanowire-based sensors. Small 6(16):1705–1722

Rao GR, Kanjilal G, Mohan K (1978) Extended application of Folin-Ciocalteu reagent in the determination of drugs. Analyst 103(1230):993–994

Rao H, Liu W, Lu Z, Wang Y, Ge H, Zou P, Wang Y (2016) Silica-coated carbon dots conjugated to CdTe quantum dots: a ratiometric fluorescent probe for copper(II). Microchim Acta 183(2):581–588. https://doi.org/10.1007/s00604-015-1682-6
Ray WJ, Joseph W (2013) Wearable electrochemical sensors and biosensors: a review. Electroanalysis 25(1):29–46
Reina G, González-Domínguez JM, Criado A, Vázquez E, Bianco A, Prato M (2017) Promises, facts and challenges for graphene in biomedical applications. Chem Soc Rev 46(15):4400–4416
Riedel AS (2018) Examining the effects of varying levels of consumer control over advertising intrusiveness on the advertising brand (Doctoral dissertation, Queensland University of Technology)
Sadik OA, Aluoch AO, Zhou A (2009) Status of biomolecular recognition using electrochemical techniques. Biosens Bioelectron 24(9):2749–2765
Sanz VC, Mena ML, González-Cortés A, Yanez-Sedeno P, Pingarrón J (2005) Development of a tyrosinase biosensor based on gold nanoparticles-modified glassy carbon electrodes: application to the measurement of a bioelectrochemical polyphenols index in wines. Anal Chim Acta 528(1):1–8
Schaefer D, Cheung WM (2018) Smart packaging: opportunities and challenges. Procedia CIRP 72:1022–1027
Senior KR (Ed.). (2010) Bone and Muscle: Structure, Force, and Motion. The Rosen Publishing Group, Inc.
Sharma A, Bachheti A, Sharma P, Bachheti RK, Husen A (2020) Phytochemistry, pharmacological activities, nanoparticle fabrication, commercial products and waste utilization of *Carica papaya* L.: a comprehensive review. Curr Res Biotechnol 2:145–160. https://doi.org/10.1016/j.crbiot.2020.11.001
Siddiqi KS, Husen A (2017) Recent advances in plant-mediated engineered gold nanoparticles and their application in biological system. J Trace Elem Med Biol 40:10–23. https://doi.org/10.1016/j.jtemb.2016.11.012
Siddiqi KS, Husen A, Sohrab SS, Yassin MO (2018) Recent status of nanomaterial fabrication and their potential applications in neurological disease management. Nanoscale Res Lett 13:231. https://doi.org/10.1186/s11671-018-2638-7
Siegel AC, Phillips ST, Dickey MD, Lu N, Suo Z, Whitesides GM (2010) Foldable printed circuit boards on paper substrates. Adv Funct Mater 20(1):28–35
Silvestre C, Duraccio D, Cimmino S (2011) Food packaging based on polymer nanomaterials. Prog Polym Sci 36(12):1766–1782
Singh T, Shukla S, Kumar P, Wahla V, Bajpai VK, Rather IA (2017) Application of nanotechnology in food science: perception and overview. Front Microbiol 8:1501
Siu C (ed) (2018) IoT and low-power wireless: circuits, architectures, and techniques. CRC Press
Sodhro AH, Pirbhulal S, Qaraqe M, Lohano S, Sodhro GH, Junejo NUR, Luo Z (2018) Power control algorithms for media transmission in remote healthcare systems. IEEE Access 6:42384–42393
Sun Q, Wang C, Zuo LS, Lu FH (2018) Digital empowerment in a WEEE collection business ecosystem: a comparative study of two typical cases in China. J Clean Prod 184:414–422
Taei M, Salavati H, Hasanpour F, Habibollahi S, Baghlani H (2016) Simultaneous determination of ascorbic acid, acetaminophen and codeine based on multi-walled carbon nanotubes modified with magnetic nanoparticles paste electrode. Mater Sci Eng C 69:1–11
Uludag Y, Narter F, Sağlam E, Köktürk G, Gök MY, Akgün M, Budak S (2016) An integrated lab-on-a-chip-based electrochemical biosensor for rapid and sensitive detection of cancer biomarkers. Anal Bioanal Chem 408(27):7775–7783
Varshney R K, Graner A, Sorrells ME (2005) Genic microsatellite markers in plants: features and applications. Trends Biotechnol 23(1):48–55
Vamvakaki V, Chaniotakis NA (2007) Pesticide detection with a liposome-based nano-biosensor. Biosens Bioelectron 22(12):2848–2853. https://doi.org/10.1016/j.bios.2006.11.024
Vasimalai N, John S (2013) Picomolar melamine enhanced the fluorescence of gold nanoparticles: spectrofluorimetric determination of melamine in milk and infant formulas using functionalized triazole capped gold nanoparticles. Biosens Bioelectron 42:267–272

Viswanath A (2004) Encyclopedia of nanoscience and nanotechnology. United States. American Scientific Publishers, Los Angeles

Walter M, Eilebrecht B, Wartzek T, Leonhardt S (2011) The smart car seat: personalized monitoring of vital signs in automotive applications. Pers Ubiquitous Comput 15(7):707–715

Wang J-J, Liu B-H, Hsu Y-T, Yu F-Y (2011) Sensitive competitive direct enzyme-linked immunosorbent assay and gold nanoparticle immunochromatographic strip for detecting aflatoxin M1 in milk. Food Control 22(6):964–969

Wang J, Wang Z, Liu J, Li H, Li QX, Li J, Xu T (2013) Nanocolloidal gold-based immuno-dip strip assay for rapid detection of Sudan red I in food samples. Food Chem 136(3–4):1478–1483

Wang S, Huang Y, Rogers JA (2015) Mechanical designs for inorganic stretchable circuits in soft electronics. IEEE Trans Compon Packag Manuf Technol 5(9):1201–1218

Wang S, Chinnasamy T, Lifson MA, Inci F, Demirci U (2016) Flexible substrate-based devices for point-of-care diagnostics. Trends Biotechnol 34(11):909–921

Wartzek T, Weyer S, Leonhardt S (2011) A differential capacitive electrical field sensor array for contactless measurement of respiratory rate. Physiol Meas 32(10):1575–1590

Wei S, Zhao F, Xu Z, Zeng B (2006) Voltammetric determination of folic acid with a multi-walled carbon nanotube-modified gold electrode. Microchim Acta 152(3–4):285–290

WHO (2020) World Health Organization. Sustainable development goals. URL: https://www.who.int/health-topics/sustainable-development-goals#tab=tab_1. Accessed on 02.11.2020

Wu KF, Chan CH, Zhang YT (2006) Contactless and cuffless monitoring of blood pressure on a chair using e-textile materials. In: Proceedings of 3rd IEEE/EMBS International Summer School and Symposium on Medical Devices and Biosensors, Cambridge, MA, pp 98–100

Xiao F, Ruan C, Liu L, Yan R, Zhao F, Zeng B (2008) Single-walled carbon nanotube-ionic liquid paste electrode for the sensitive voltammetric determination of folic acid. Sensors Actuators B Chem 134(2):895–901

Xu M, Obodo D, Yadavalli VK (2019) The design, fabrication, and applications of flexible biosensing devices. Biosens Bioelectron 124:96–114

Xie Y, Wang Z, Zhan Y, Zhang P, Wu R, Jiang T, ... Ma X (2017) Controllable growth of monolayer MoS_2 by chemical vapor deposition via close MoO_2 precursor for electrical and optical applications. Nanotechnology 28(8):084001

Yang H, Qu L, Wimbrow AN, Jiang X, Sun Y (2007) Rapid detection of Listeria monocytogenes by nanoparticle-based immunomagnetic separation and real-time PCR. Int J Food Microbiol 118(2):132–138

Yang D, Zhu L, Jiang X (2010) Electrochemical reaction mechanism and determination of Sudan I at a multi wall carbon nanotubes modified glassy carbon electrode. J Electroanal Chem 640(1):17–22. https://doi.org/10.1016/j.jelechem.2009.12.022

Yang Y, Yang X, Tan Y, Yuan Q (2017) Recent progress in flexible and wearable bio-electronics based on nanomaterials. Nano Res 10(5):1560–1583

Yoon S, Sim JK, Cho YH (2016) A flexible and wearable human stress monitoring patch. Sci Rep 6:23468

Zamolo VA, Valenti G, Venturelli E, Chaloin O, Marcaccio M, Boscolo S et al (2012) Highly sensitive electrochemiluminescent nanobiosensor for the detection of palytoxin. ACS Nano 6(9):7989–7997

Zhang YT, Poon CCY (2013) Health informatics: unobtrusive physiological measurement technologies. IEEE J Biomed Health Informat 17(5):893

Zhang ZB, Shen YH, Wang WD, Wang BQ, Zheng JW (2011) Design and implementation of sensing shirt for ambulatory cardiopulmonary monitoring. J Med Biol Eng 31(3):207–215

Zhang J, Wang H, Yan X, Zhang L (2005) Comparison of short-term toxicity between Nano-Se and selenite in mice. Life Sci 76(10):1099–1109

Zhou Y, Pan F-G, Li Y-S, Zhang Y-Y, Zhang J-H, Lu S-Y et al (2009) Colloidal gold probe-based immunochromatographic assay for the rapid detection of brevetoxins in fishery product samples. Biosens Bioelectron 24(8):2744–2747

Zhu C, Yang G, Li H, Du D, Lin Y (2015) Electrochemical sensors and biosensors based on nanomaterials and nanostructures. Anal Chem 87(1):230–249

Index

J.-C. Kim et al. (eds.), *Smart Nanomaterials in Biomedical Applications*, Nanotechnology in the Life Sciences,
https://doi.org/10.1007/978-3-030-84262-8

O

P

R

S

The manufacturer's authorised representative in the EU is Springer Nature Customer Service Centre GmbH, Europaplatz 3, 69115 Heidelberg, Germany. If you have any concerns regarding our products, please contact ProductSafety@springernature.com

Printed and bound by CPI Group (UK) Ltd, Croydon, CR0 4YY
30/09/2026
02225692-0017